T0328934

Course of Theoretical Astrophysics
Volume I: Astrophysical Processes

Graduate students and researchers in astrophysics and cosmology need a solid understanding of a wide range of physical processes. This clear and authoritative textbook has been designed to help them to develop the necessary toolkit of theory. Assuming only an undergraduate background in physics and no detailed knowledge of astronomy, this book guides the reader step by step through a comprehensive collection of fundamental theoretical topics. The book is modular in design, allowing the reader to pick and choose a selection of chapters, if necessary. It can be used alone or in conjunction with the forthcoming accompanying two volumes (covering stars and stellar systems and galaxies and cosmology, respectively).

After the basics of dynamics, electromagnetic theory, and statistical physics are reviewed, a solid understanding of all the key concepts such as radiative processes, spectra, fluid mechanics, plasma physics and magnetohydrodynamics, dynamics of gravitating systems, general relativity, and nuclear physics is developed. Each topic is developed methodically from undergraduate basic physics. Throughout, the reader's understanding is developed and tested with carefully structured problems and helpful hints.

This welcome volume provides graduate students with an indispensable introduction to and reference on all the physical processes they will need to tackle successfully cutting-edge research in astrophysics and cosmology.

THANU PADMANABHAN is a Professor at Inter-University Centre for Astronomy and Astrophysics in Pune, India. His research interests are Gravitation, Cosmology, and Quantum Theory. He has published over hundred technical papers in these areas and has written four books: *Structure Formation in the Universe, Cosomology and Astrophysics Through Problems, After the First Three Minutes*, and, together with J.V. Narlikar, *Gravity, Gauge Theories and Quantum Cosmology*.

He is a member of the Indian Academy of Sciences, National Academy of Sciences, and International Astronomical Union. He has received numerous awards, including the Shanti Swarup Bhatnagar Prize in Physics (1996) and the Millenium Medal (2000) awarded by the Council of Scientific and Industrial Research, India.

Professor Padmanabhan has also written more than 100 popular science articles, a comic strip serial, and several regular columns on astronomy, recreational mathematics, and history of science that have appeared in international journals and papers. He is married, has one daughter, and lives in Pune, India. His hobbies include chess, origami, and recreational mathematics.

COURSE OF THEORETICAL ASTROPHYSICS

Volume I: Astrophysical Processes

T. PADMANABHAN

Inter-University Centre for Astronomy and Astrophysics
Pune, India

CAMBRIDGE
UNIVERSITY PRESS

CAMBRIDGE UNIVERSITY PRESS
Cambridge, New York, Melbourne, Madrid, Cape Town, Singapore, São Paulo

Cambridge University Press
The Edinburgh Building, Cambridge CB2 8RU, UK

Published in the United States of America by Cambridge University Press, New York

www.cambridge.org
Information on this title: www.cambridge.org/9780521562409

First published 2000

A catalogue record for this publication is available from the British Library

Library of Congress Cataloguing in Publication data
Padmanabhan, T. (Thanu), 1957–
Theoretical astrophysics : astrophysical processes / T. Padmanabhan.
p.   cm.
ISBN 0-521-56240-6
1. Astrophysics.   I. Title.
QB461.B33   2000
523.01 – dc21   00-025837

ISBN 978-0-521-56240-9 hardback
ISBN 978-0-521-56632-2 paperback

Transferred to digital printing 2007

Dedicated to the memory of L.D. Landau,
who understood the importance of pedagogy

COURSE OF THEORETICAL ASTROPHYSICS

– in three volumes –

VOLUME I: ASTROPHYSICAL PROCESSES

1: Order-of-magnitude astrophysics; 2: Dynamics; 3: Special Relativity, Electrodynamics, and Optics; 4: Basics of Electromagnetic Radiation; 5: Statistical Mechanics; 6: Radiative Processes; 7: Spectra; 8: Neutral Fluids; 9: Plasma Physics; 10: Gravitational Dynamics; 11: General Theory of Relativity; 12: Basics of Nuclear Physics.

VOLUME II: STARS AND STELLAR SYSTEMS

1: Overview: Stars and Stellar Systems; 2: Stellar Structure; 3: Stellar Evolution; 4: Supernova; 5: White Dwarfs, Neutron Stars, and Black Holes; 6: Pulsars; 7: Binary Stars and Accretion; 8: Sun and Solar System; 9: Interstellar Medium; 10: Globular Clusters.

VOLUME III: GALAXIES AND COSMOLOGY

1: Observational Overview; 2: Galactic Structure; 3: Galactic Dynamics and Interactions; 4: Friedmann Model; 5: Active Galactic Nuclei–Structural Aspects; 6: Thermal History of the Universe; 7: Structure Formation; 8: Cosmic Microwave Background Radiation; 9: Formation of Baryonic Structures; 10: Active Galactic Nuclei–Cosmological Aspects; 11: Intergalactic Medium and Absorption Systems; 12: Cosmological Observations.

Contents

Preface

"...yoyum varo gudham anupravisto,
naanyam thasman Nachiketa vrinithe."
("...Nachiketa does not choose any other boon but
[learning about] that of which Knowledge is hidden.")

Katho Upanishad, Verse 29.

During the past decade or so, theoretical astrophysics has emerged as one of the most active research areas in physics. This advance has also been reflected in the greater interdisciplinary nature of research that is being carried out in this area in the recent years. As a result, those who are learning theoretical astrophysics with the aim of making a research career in this subject need to assimilate considerable amount of concepts and techniques, in different areas of astrophysics, in a short period of time. Every area of theoretical astrophysics, of course, has excellent textbooks that allow the reader to master that *particular* area in a well-defined way. Most of these textbooks, however, are written in a traditional style, focussing on one area of astrophysics (say stellar evolution, galactic dynamics, radiative processes, cosmology etc.) Because different authors have different perspectives regarding their subject matter it is not very easy for a student to understand the key unifying principles behind several different astrophysical phenomena by studying a plethora of separate textbooks, as they do not link up together as a series of core books in theoretical astrophysics covering everything which a student would need. A few books, which *do* cover the whole of astrophysics, deal with the subject at a rather elementary ("first course") level.

What we require is clearly something analogous to the famous Landau–Lifshitz course in theoretical physics, but focussed to the subject of theoretical astrophysics at a fairly advanced level. In such a course, all the key physical concepts (e.g., radiative processes, fluid mechanics, plasma physics, etc.) can be presented from a unified perspective and then applied to different astrophysical situations.

This book is the first of a set of three volumes that are intended to do exactly that. They form one single coherent unit of study through the use of which a student can acquire mastery over all the traditional astrophysical topics. What is more, these volumes emphasise the unity of concepts and techniques in different branches of astrophysics. The interrelationship among different areas and common features in the analysis of different theoretical problems will be stressed throughout. Because many of the basic techniques need to be developed only once, it is possible to achieve a significant economy of presentation and crispness of style in these volumes.

Needless to say, there are some basic "boundary conditions" one has to respect in such an attempt to cover the whole of Theoretical Astrophysics in approximately 3×580 pages. Not much space is available to describe the nuances in greater length or to fill in the details of algebra. For example, I have made conscious choices as to which parts of the algebra can be left to the reader and which need to be worked out explicitly in the text, and I have omitted a detailed discussion of elementary concepts and derivations. However, I do *not* expect the reader to know anything about astrophysics. All astrophysical concepts are developed ab initio in these volumes. The approach used in these three volumes is similar to that used by Gengis Khan, namely, (1) cover as much area as possible, (2) capture the important points, and (3) be utterly ruthless!

To cut out as much repetition as possible, the bulk of the physical principles are presented at one go in Vol. I and are applied in the other two volumes to different situations. This implies that there will be a lot of physics but very little of "concrete" astrophysics in Vol. I; that comes in Vol. II (Stars and Stellar Systems) and Vol. III (Extragalactic Astronomy and Cosmology). The criteria for the selection of material for Vol. I have been the following: (1) Any physical principle that finds application in more than one chapter of Vol. II or Vol. III (for example, bremsstrahlung, Voigt profile, etc.) is discussed in Vol. I. Certain topics that are used in only a specific chapter in Vol. II or Vol. III are discussed in situ rather than in Vol. I. (2) By and large, everything discussed in Vol. I will be utilized directly somewhere in Vols. II and III. On rare occasion, I do cover a topic in Vol. I even if it is not fully utilized in Vol. II or Vol. III because a reader who is going to work in theoretical astrophysics will eventually need an understanding of that particular topic. (3) These three volumes concentrate on *theoretical* aspects. Observation and phenomenology are, of course, discussed in Vols. II and III to the extent necessary to make the motivation clear. However, I do not have the space to discuss how these observations are made, the errors, reliability, etc., of the observations or the astronomical techniques. (Maybe there should be a fourth volume describing observational astrophysics!)

The target audience for this three-volume work will be fairly large and is made up of (1) students in the first year of their Ph.D. Program in theoretical physics, astronomy, astrophysics, and cosmology; (2) research workers in various fields of theoretical astrophysics, cosmology, etc.; and (3) teachers of graduate courses in theoretical astrophysics, cosmology and related subjects. In fact, anyone working in or interested in some area of astronomy or astrophysics will find something useful in these volumes. They are also designed in such a way that parts of the material can be used in modular form to suit the requirements of different people and different courses.

Let me briefly highlight the features which are specific to Vol. I. The reader of Vol. I is assumed to have done basic courses in classical mechanics, nonrelativistic quantum mechanics, and classical electromagnetic theory. Of the 12 chapters in Vol. I, the first one is a broad-brush overview of physical principles in an order-of-magnitude manner and is intended to set the stage. I expect the reader to survey this chapter rapidly but to come back to it periodically at later stages. This chapter is probably the easiest or the most difficult, depending on one's background and aptitude. It is easy in the sense that very little sophisticated mathematics is used; difficult because it takes a high level of maturity to appreciate some of the physical arguments that are presented. Chapters 2 (Dynamics), 3 (Special Relativity, Electrodynamics, and Optics), and 5 (Statistical Mechanics) cover the ground the reader may already be familiar with – but from an advanced perspective. The aim is to introduce powerful techniques in familiar contexts so that the reader can learn and appreciate them. For example, no apologies are made for introducing four-vector notation up front or dealing with distribution functions right from the beginning, so as to get the main results as quickly as possible. The emphasis throughout is on topics relevant in astrophysics, such as the reduced three-body problem, action-angle variables, diffraction and interference, optical systems, propagation in random media, ionisation equilibria, etc. Chapter 4 deals with the basics of radiation theory – both classical and quantum – that is developed from scratch and the reader is *not* assumed to be familiar with quantum field theory.

Chapters 6–12 develop the toolkit for astrophysics in a self-contained manner, virtually ab initio. Chapters 6 (Radiative Processes) and 8 (Fluid Mechanics) are fairly exhaustive and detailed. The short chapter on Spectra (Chap. 7) covers general material that is of astronomical relevance; more specific aspects will be dealt with in Vols. II and III within the appropriate contexts. In Chap. 9 (Plasma Physics) I had to make choices as to which topics are of sufficiently general nature to appear in Vol. I; some specific topics (e.g., instability of axisymmetric systems with magnetic fields, alpha effect, and dynamos) will appear in the relevant chapters of Vols. II and III. Chapter 10 (Gravitational Dynamics) covers the background needed for galactic dynamics, globular cluster evolution, etc. Chapter 11 is a compact introduction to general relativity and *no* previous familiarity with tensor analysis is assumed. Finally, Chap. 12 deals with aspects of nuclear physics that are needed in the study of stellar evolution.

Any one of these topics is fairly vast and often requires a full textbook to do justice to it, whereas I have devoted approximately 60 pages to each of them!

I would like to emphasise that such a crisp, condensed discussion is not only possible but also constitutes a basic matter of policy in these volumes. After all, the idea is to provide the student with the essence of several textbooks in one place. It should be clear to lecturers that these material can be easily regrouped to serve different graduate courses at different levels, especially when complemented by other textbooks.

Because of the highly pedagogical nature of the material covered in Vol. I, I have not given detailed references to original literature except on rare occasions when a particular derivation is not available in standard textbooks. The annotated list of references given at the end of the book cites several other textbooks that I found very useful. Some of these books, of course, contain extensive bibliographies and references to original literature. The selection of core books cited here clearly reflects the personal bias of the author and I apologise to anyone who feels that their work or contribution has been overlooked.

Several people have contributed to the making of these volumes and especially to Vol. I. The idea for these volumes originated over a dinner with J.P. Ostriker in late 1994, while I was visiting Princeton. I was lamenting to Jerry about the lack of a comprehensive set of books covering all of theoretical astrophysics and Jerry said, "Why don't *you* write them?" He was very enthusiastic and supportive of the idea and gave extensive comments and suggestions on the original outline I produced in the next one week. I am grateful to him for the comments and for the moral support that I needed to launch into such a project. I sincerely hope the volumes do not disappoint him.

Adam Black of Cambridge University Press took up the proposal with his characteristic enthusiasm and initiative; this is the third project on which we worked together and I thoroughly enjoyed it. I should also thank him for choosing six excellent (anonymous) referees for this proposal whose support and comments helped to mould the proper framework.

Many of my friends and colleagues carried out the job of reading the earlier drafts and providing comments. Of these, M. Vivekanand has gone through most of the book with meticulous care and has given extensive comments. Many other colleagues, especially Roger Blandford, George Djorgovsky, Peter Goldreich, John Huchra, Donald Lynden-Bell, J.V. Narlikar, R. Nityananda, Sterl Phinney, and Douglas Richstone looked at the whole draft and provided comments and suggestions at different levels of detail. J.S. Bagla, Sai Iyer, Nissim Kanekar, Ben Oppenheimer, K. Subramanian, S. Sankaranarayanan, and K. Srinivasan gave detailed comments on selected chapters; the last two took pains to check most of the derivations and algebraic expressions. I thank all of them for their help.

I have been visiting the Astronomy Department of Caltech during the past several years and the work on this book has benefitted tremendously through my discussions and interactions with the students and staff of the Caltech Astronomy Department. I especially thank Roger Blandford, Peter Goldreich, Shri Kulkarni,

Sterl Phinney, and Tony Readhead for several useful discussions and for sharing with me their insights and experience in physics teaching.

This project would not have been possible but for the dedicated support from Vasanthi Padmanabhan, who not only did the entire TEXing and formatting but also produced most of the figures – often writing the necessary programs for the same. I thank her for the help and look forward to receiving the same for the next two volumes! I also thank Sunu Engineer who was resourceful in solving several computer-related problems that cropped up periodically. It is a pleasure to acknowledge the library and other research facilities available at the Inter-University Centre for Astronomy and Astrophysics, which were useful in this task.

T. Padmanabhan

1

Order-of-Magnitude Astrophysics

1.1 Introduction

The subject of astrophysics involves the application of the laws of physics to large macroscopic systems in order to understand their behaviour and predict new phenomena. This approach is similar in spirit to the application of the laws of physics in the study of, say, condensed-matter phenomena, except for the following three significant differences:

(1) We have far less control over the external conditions and parameters in astrophysics than in, say, condensed-matter physics. It is not possible to study systems under controlled conditions so that certain physical processes dominate the behaviour. Identifying the causes of various observed phenomena in astrophysics will require far greater reliance on statistical arguments than in laboratory physics.

(2) The astrophysical systems of interest span a wide range of parameter space and require inputs from several different branches of physics. Typically, the densities can vary from 10^{-25} gm cm^{-3} (interstellar medium) to 10^{15} gm cm^{-3} (neutron stars); temperatures from 2.7 K (microwave background radiation) to 10^9 K (accreting x-ray sources) or even to 10^{15} K (early universe); radiation from wavelengths of meters (radio waves) to fractions of angstroms (hard gamma rays); typical speeds of particles can go up to $0.99c$ (relativistic jets). Clearly we require inputs from quantum-mechanical and relativistic regimes as well as from more familiar classical physics.

(3) The primary source of information about distant astrophysical sources is the electromagnetic radiation detected from them. Therefore, to obtain a complete picture about any source, it is necessary to examine it in all the wave bands. Because of technological limitations, it is often quite difficult to have uniform coverage across the entire electromagnetic spectrum. Hence the information we have about the sources is often distorted or incomplete.

These considerations suggest that two aspects will be most important in the study of different astrophysical systems. The first is the appreciation of the different states in which bulk matter can exist under different conditions and the dynamics of the matter governed by different equations of state. The second is an understanding of different radiative processes that lead to the emission of photons, which act as prime carriers of information about astronomical objects.

We shall be concerned with these and related topics in several chapters of this book. The purpose of this introductory chapter is twofold: It will first provide – in Sections (1.2) – (1.4) – a rapid overview of several physical processes at an order-of-magnitude level and introduce the necessary concepts. Then we will make an attempt to understand the existence of different astrophysical structures from first principles to the extent possible. Implementing such a plan, of course, has not been possible even in laboratory physics, and it is unlikely to succeed in the case of astrophysics. At present, astrophysics does require a fair amount of observational and phenomenological input, just like any other branch of applied physics. Nevertheless, we will make such an attempt as it is useful in providing the most basic and direct connection between physics and astrophysics.

1.2 Energy Scales of Physical Phenomena

Let us consider a system of N particles ($N \gg 1$), each of mass m, occupying a spherical region of radius R. In dealing with the dynamics of such a large collection of particles, it is useful to introduce the concept of pressure exerted by the system of particles as the momentum transferred per second normal to a (fictitious) surface of unit area. The contribution to the rate of momentum transfer (per unit area) from particles of energy ϵ is $n(\epsilon)\mathbf{p}(\epsilon) \cdot \mathbf{v}(\epsilon)$, where $n(\epsilon)$ denotes the number of particles per unit volume with momentum $\mathbf{p}(\epsilon)$ and velocity $\mathbf{v}(\epsilon)$. We obtain the net pressure by averaging this expression over the angles defined by $\mathbf{p}$ (or $\mathbf{v}$) and summing over all values of the energy. Because the momentum and the velocity are parallel to each other, the vector dot product $\mathbf{p} \cdot \mathbf{v}$ averages to $(1/3)pv$ (in three dimensions), giving

$$P = \frac{1}{3} \int_0^\infty n(\epsilon)p(\epsilon)v(\epsilon)\, d\epsilon, \tag{1.1}$$

where the integration is over all energies. The system is called ideal if the kinetic energy dominates over the interaction energy of the particles. In that case ϵ is essentially the kinetic energy of the particle. With the relations

$$p = \gamma m v, \quad \epsilon = (\gamma - 1)mc^2, \quad \gamma \equiv \left(1 - \frac{v^2}{c^2}\right)^{-1/2}, \tag{1.2}$$

where ϵ is the kinetic energy of the particle, the pressure can be expressed in the form

$$P = \frac{1}{3} \int_0^\infty n\epsilon \left(1 + \frac{2mc^2}{\epsilon}\right)\left(1 + \frac{mc^2}{\epsilon}\right)^{-1} d\epsilon. \tag{1.3}$$

In the nonrelativistic (NR) limit (with $mc^2 \gg \epsilon$), this gives $P_{NR} \approx (2/3) \langle n\epsilon \rangle = (2/3)U_{NR}$, where U_{NR} is the energy density (i.e., energy per unit volume) of the particles. In the relativistic case (with $\epsilon \gg mc^2$ or when the particles are massless), the corresponding expression is $P_{ER} \approx (1/3) \langle n\epsilon \rangle = (1/3)U_{NR}$. Hence, in general, $P \approx U$ up to a factor of order unity.

This result can be converted into a more useful form of equation of state whenever the mean free path of the particles in the system is small compared with the length scales over which the physical parameters of the system change significantly. Then the pressure can be expressed in terms of density and temperature if the energy density can be expressed in terms of these variables. This is possible in several contexts leading to different equations of state. To understand each of these cases it is useful to start by identifying the characteristic energy scales of bulk matter. We now turn to this task.

1.2.1 Rest-Mass Energy

We can associate the rest-mass energy mc^2 with each individual particle of mass m. In normal matter, made up of nucleons and electrons, the lowest value for rest-mass is provided by electrons with $m_e c^2 \approx 0.5$ MeV. For nucleons, the rest-mass energy is $m_p c^2 \approx 1$ GeV. Because the total mass of the system is mostly due to the nucleons, the total rest-mass energy will be $E_{mass} \cong N A m_p c^2 \cong M c^2$, where $A m_p \simeq m$ is the mass of each nucleus and $Nm = M$ is the total mass of the system. Rest-mass energy is extensive – that is, $E_{mass} \propto N$ – in the low-energy phenomena in which masses of individual nuclei do not change.

1.2.2 Atomic Binding Energies

If the particles of the system have internal structure (molecular, atomic, nuclear, etc.) then we get further energy scales that are characteristic of the interactions. The simplest is the atomic binding energy of atoms and molecules, which arises from the electromagnetic coupling between the particles.

The Hamiltonian describing an electron, moving in the Coulomb field of a nucleus of charge Zq, is given by $H_0 = (p^2/2m_e) - (Zq^2/r)$. If this electron is described by a wave function $\psi(\mathbf{x}, L)$, where L denotes the characteristic scale over which ψ varies significantly, then the expectation value for the energy of the electron in this state is $E(L) = \langle \psi | H_0 | \psi \rangle \approx (\hbar^2/2m_e L^2) - (Zq^2/L)$. The first term arises from the fact that $\langle \psi | p^2 | \psi \rangle = -\hbar^2 \langle \psi | \nabla^2 | \psi \rangle \approx (\hbar^2/L^2)$, which is equivalent to the uncertainty principle stated in the form $p \cong \hbar/L$. This expression for $E(L)$ reaches a minimum value of $E_{min} = -Z^2\epsilon_a$ when L is varied, with the minimum occurring at $L_{min} = (a_0/Z)$, where

$$a_0 \equiv \frac{\hbar^2}{m_e q^2} \equiv \frac{\lambdabar_e}{\alpha} \approx 0.52 \times 10^{-8} \text{ cm}, \quad \epsilon_a \equiv \frac{m_e q^4}{2\hbar^2} = \frac{1}{2}\alpha^2 m_e c^2 \approx 13.6 \text{ eV},$$

$$(1.4)$$

with the definitions $\lambdabar_e \equiv (\hbar/m_e c)$ and $\alpha \equiv (q^2/\hbar c)$. a_0 and ϵ_a correspond to the size and the ground-state energy of a hydrogen atom with $Z = 1$. The wavelength λ corresponding to ϵ_a is $\lambda = (hc/\epsilon_a) = 2\alpha^{-2}\lambdabar_e \simeq 10^3$ Å and lies in the UV band. The fine-structure constant $\alpha \approx 7.3 \times 10^{-3}$ plays an important role in the structure of matter and arises as the ratio between several interesting variables:

$$\alpha = (v/Zc) = (2\mu_B/qa_0) = (r_0/\lambdabar_e) = (\lambdabar_e/a_0),$$

where v is the speed of an electron in the atom, $\mu_B \equiv (q\hbar/2m_e c)$ is the Bohr magneton representing the magnetic moment of the electron, and $r_0 \equiv (q^2/m_e c^2)$ is called the classical electron radius.

When atoms of size a_0 are closely packed, the number density of atoms is $n_{solid} \approx (2a_0)^{-3} \approx 10^{24}$ cm^{-3}. The binding energy of such a solid arises essentially because of the residual electromagnetic force between the atoms, and the typical binding energy per particle is $f\epsilon_a$ with $f \approx (0.1-1)$.

1.2.3 Molecular Binding Energy

The simplest molecular structure consists of two atoms bound to each other in the form of a diatomic molecule. The effective potential energy of interaction $U(r)$ between the atoms in such a molecule arises from a residual electrostatic coupling and has a minimum at a separation $r \simeq a_0$, approximately the size of the atom. The depth of the potential well at the minimum is comparable with the electronic-energy level ϵ_a of the atom. In addition to the internal, electronic, binding energies of the atoms comprising the molecule, there are two other contributions to the energy of a diatomic molecule:

(1) The atoms of such a molecule can vibrate at some characteristic frequency ω_{vib} about the mean position along the line connecting them; this will lead to vibrational-energy levels separated by $E_{vib} \approx \hbar\omega_{vib}$. If the displacement is $\sim a_0$ from the minimum, the vibrational energy E_{vib} will be $\sim\epsilon_a$. Writing $\epsilon_a \approx (1/2)\mu\omega_{vib}^2 a_0^2 \cong (\hbar^2/m_e a_0^2)$, where μ is the reduced mass of the two atoms, we get

$$E_{vib} = \hbar\omega_{vib} \approx \frac{\hbar^2}{(\mu m_e)^{1/2} a_0^2} \approx \left(\frac{m_e}{\mu}\right)^{1/2} \epsilon_a \simeq 0.25 \text{ eV} \qquad (1.5)$$

if $\mu \simeq m_p$.

(2) The molecule can also rotate about an axis perpendicular to the line joining them. If the rotational angular momentum is J, this will contribute an energy of approximately

$$E_{\text{rot}} \approx \left(\frac{J^2}{\mu a_0^2} \right) \approx \left(\frac{\hbar^2}{\mu a_0^2} \right) \approx \left(\frac{m_e}{\mu} \right) \epsilon_a \approx 10^{-3} \epsilon_a \simeq 10^{-2} \text{ eV} \qquad (1.6)$$

if $J \simeq \hbar$ and $\mu \simeq m_p$. It follows from these relations that $E_{\text{rot}} : E_{\text{vib}} : E_0 \approx (m_e/\mu) : (m_e/\mu)^{1/2} : 1$ and $E_{\text{vib}} \approx \sqrt{\epsilon_a E_{\text{rot}}}$. Because $(m_e/\mu) \approx 10^{-3}$, the wavelengths of radiation from vibrational transitions are ~ 40 times larger than those of electronic transitions; similarly, the rotational transitions lead to radiation with wavelengths ~ 1000 times larger than those of electronic transitions. These wavelengths are usually in the IR band.

Atomic and molecular energies are also extensive, with the binding energy of a system of N particles scaling as N.

1.2.4 Nuclear-Energy Scales

Atomic nuclei are bound by the strong-interaction force that provides a binding energy per particle of ~ 8 MeV, which is the characteristic scale for nuclear-energy levels. In the astrophysical context, a more relevant energy scale is the one at which nuclear reactions can be triggered in bulk matter, which can be estimated as follows. For two protons to fuse together while undergoing nuclear reaction, it is necessary that they be brought within the range of attractive nuclear force, which is approximately $l \approx (h/m_p c) = (2\pi\hbar/m_p c)$. Because this requires overcoming the Coloumb repulsion, such direct interaction can take place only if the kinetic energy of colliding particles is of the order of the electrostatic potential energy at the separation l. This requires energies of the order of $\epsilon \approx (q^2/l) = (\alpha/2\pi) m_p c^2 \approx 1$ MeV. It is, however, possible for nuclear reactions to occur through quantum-mechanical tunneling when the de Broglie wavelength $\lambda_{\text{deB}} \equiv (h/m_p v) = l(c/v)$ of the two protons overlap. This occurs when the energy of the protons is approximately $\epsilon_{\text{nucl}} \approx (\alpha^2/2\pi^2) m_p c^2 \approx 1$ keV. It is conventional to write this expression as $\epsilon_{\text{nucl}} \approx \eta \alpha^2 m_p c^2$, with $\eta \simeq 0.1$. This quantity ϵ_{nucl} sets the scale for triggering nuclear reactions in astrophysical contexts.

1.2.5 Gravitational Binding Energy

In the nonrelativistic, Newtonian theory for gravity, the gravitational energy of a system of size R and mass M will be $E_{\text{grav}} \approx GM^2/R \approx (Gm_p^2/R)N^2$. This is not extensive with respect to N (for a given R), and the potential energy per particle varies as

$$\epsilon_g \equiv \frac{E_{\text{grav}}}{N} = \left(\frac{Gm_p^2}{R}\right)N = \left(\frac{4\pi}{3}\right)^{1/3} Gm_p^2 N^{2/3} n^{1/3}, \tag{1.7}$$

where $n = (3N/4\pi R^3)$ is the number density of particles. The pressure due to gravitational force near the center of the object will be approximately

$$P_g \approx \frac{(GM^2/R^2)}{(4\pi R^2)} \approx \frac{1}{3}\left(\frac{4\pi}{3}\right)^{1/3} Gm_p^2 N^{2/3} n^{4/3} \cong \frac{1}{3}\left(\frac{E_{\text{grav}}}{V}\right).$$

If the gravitational potential energy is comparable with the rest-mass energy of the system, it is necessary to take general relativistic effects into account. The ratio $\mathcal{R}_{gm} \equiv (E_{\text{grav}}/E_{\text{mass}})$ is $\mathcal{R}_{gm} \simeq 0.7(M/10^{33}\text{ gm})(R/1\text{ km})^{-1}$, which shows that if massive objects (with $M \simeq 10^{33}$ gm) are confined to small regions (with $R \simeq 1$ km), the system will exhibit general relativistic effects. When this ratio is small compared with unity, the system can be treated by Newtonian gravity.

1.2.6 Thermal and Degeneracy Energies of Particles

So far we have not introduced the notion of Temperature or the kinetic energy of the particle. These attributes bring in the next set of energy scales into the problem. For a particle of momentum p and mass m, the kinetic energy is given by

$$\epsilon = \sqrt{p^2c^2 + m^2c^4} - mc^2 = \begin{cases} p^2/2m & (p \ll mc) \\ pc & (p \gg mc) \end{cases}, \tag{1.8}$$

where the two forms are applicable in the non-relativistic (NR) and extreme relativistic (ER) limits. The behaviour of the system depends on the origin of the momentum distribution of the particles.

The familiar situation is the one in which short-range interactions (usually called 'collisions') between the particles effectively exchange the energy so as to randomize the momentum distribution. This will happen if the effective mean free path of the system l is small compared with the length scale L at which physical parameters change. (The explicit form taken by the condition $l \ll L$ can be very different in different cases; this condition is discussed in detail towards the end of this section.) When such a system is in steady state, we can assume that the local thermodynamic equilibrium, characterized by a local temperature T, exists in the system. Then the probability for occupying a state with energy E will scale as $P(E) \propto \exp[-(E/k_BT)]$. The typical momentum of the particle when the temperature is T is given by Eq. (1.8) with $\epsilon \simeq k_BT$, that is,

$$p \cong mc\left[\frac{2k_BT}{mc^2} + \left(\frac{k_BT}{mc^2}\right)^2\right]^{1/2} \cong \begin{cases} (2mk_BT)^{1/2} & (k_BT \ll mc^2;\ \text{NR}) \\ (k_BT/c) & (k_BT \gg mc^2;\ \text{ER}) \end{cases}. \tag{1.9}$$

In this case, the momentum and the kinetic energy of the particles vanish when $T \to 0$.

The situation is actually more complicated for material particles like electrons. The mean energy of a system of electrons will not vanish even at zero temperature because electrons obey the Pauli exclusion principle, which requires that the maximum number of electrons that can occupy any quantum state be two, one with spin up and another with spin down. Because the uncertainty principle requires that $\Delta x \Delta p_x \gtrsim h$, we can associate $(d^3 x d^3 p)/(2\pi\hbar)^3$ microstates with a phase-space volume $d^3 x \, d^3 p$. Therefore the number of quantum states with momentum less than p is $[V(4\pi p^3/3)/(2\pi\hbar)^3]$, where V is the spatial volume available for the system. The lowest energy state will be the one in which the N electrons fill all levels up to some momentum p_F, called Fermi momentum. This requires that

$$n = \left(\frac{N}{V}\right) = 2\frac{\left(4\pi p_F^3/3\right)}{(2\pi\hbar)^3} = \frac{1}{3\pi^2}\left(\frac{p_F}{\hbar}\right)^3, \tag{1.10}$$

giving $p_F = \hbar(3\pi^2 n)^{1/3}$. It is obvious that if $p_F \gtrsim mc$ the system must be treated as relativistic, even in the zero-temperature limit. The energy corresponding to p_F will be

$$\epsilon_F = \sqrt{p_F^2 c^2 + m^2 c^4} - mc^2 = \begin{cases} \dfrac{p_F^2}{2m} = \left(\dfrac{\hbar^2}{2m}\right)(3\pi^2 n)^{2/3} & \text{(NR)} \\[2mm] p_F c = (\hbar c)(3\pi^2 n)^{1/3} & \text{(ER)} \end{cases}. \tag{1.11}$$

The quantity ϵ_F (called the Fermi energy) sets the quantum-mechanical scale of the energy; quantum-mechanical effects will be dominant if $\epsilon_F \gtrsim k_B T$ (degenerate), and the classical theory will be valid for $\epsilon_F \ll k_B T$ (nondegenerate). The relevant ratio $\mathcal{R}_{\mathrm{ft}} \equiv (\epsilon_F/k_B T)$ that determines that the degree of degeneracy is

$$\left(\frac{\epsilon_F}{k_B T}\right) = \frac{mc^2}{k_B T}\left\{\left[\left(\frac{\hbar n^{1/3}}{mc}\right)^2 (3\pi^2)^{2/3} + 1\right]^{1/2} - 1\right\}$$

$$\simeq \begin{cases} \dfrac{1}{2}(3\pi^2)^{2/3}\left(\dfrac{\hbar^2 \, n^{2/3}}{m \, k_B T}\right) \\[3mm] (3\pi^2)^{1/3}\left(\dfrac{\hbar c}{k_B T} n^{1/3}\right) \end{cases}, \tag{1.12}$$

where the two limiting forms are valid for $n \ll (\hbar/mc)^{-3}$ (NR) and $n \gg (\hbar/mc)^{-3}$ (ER), respectively. In the first case $[n \ll (\hbar/m_e c)^{-3} \simeq 10^{31} \text{ cm}^{-3}]$, the system is nonrelativistic; it will also be degenerate if $\mathcal{R}_{\mathrm{ft}} = (\epsilon_F/k_B T) \gg 1$ and classical if $\mathcal{R}_{\mathrm{ft}} \ll 1$. The transition occurs at $\mathcal{R}_{\mathrm{ft}} \approx 1$, which corresponds to $nT^{-3/2} = [(mk_B)^{3/2}/\hbar^3] = 3.6 \times 10^{16}$ in cgs units. In the second case $[n \gg (\hbar/m_e c)^{-3} \simeq 10^{31} \text{ cm}^{-3}; \rho \equiv m_p n \gg 10^7 \text{ gm cm}^{-3}]$, electrons have $p_F \gg m_e c$ and are relativistic irrespective of temperature. The quantum effects will dominate thermal effects if $k_B T \ll (\hbar c) n^{1/3}$, and we will have a relativistic, degenerate gas.

In general, the kinetic energy of the particle will have contributions from the temperature as well as from Fermi energy. If we are interested in only the asymptotic limits, we can take the total kinetic energy per particle to be $\epsilon \approx \epsilon_F(n) + k_B T$. Note that such a system has a minimum energy $N\epsilon_F(n)$ even at $T = 0$.

By using our general result $P \simeq n\epsilon$ [see Eq. (1.3)], we can obtain the equation of state for the different cases discussed above. First, for a quantum-mechanical gas of fermionic particles with $k_B T \ll \epsilon_F$ and $\epsilon \approx \epsilon_F$, it follows from Eq. (1.11) that $P \simeq n\epsilon_F$ varies as the (5/3)rd power of density in the nonrelativistic case and as the (4/3)rd power of density in the relativistic case. Whether the system is relativistic or not is decided by the ratio (p_F/mc) or – equivalently – the ratio (ϵ_F/mc^2). The transition occurs at $n = n_{RQ} \approx (\hbar/mc)^{-3}$. Second, if the system is classical with $k_B T \gg \epsilon_F$ so that $\epsilon \cong k_B T$, then $P \simeq nk_B T$ in both nonrelativistic and extreme relativistic limits.

The energy scale of the individual particles also characterizes the energy involved in the collisions between the particles. If this quantity is larger than the binding energy of the atomic system, the atoms will be ionised and the electrons will be separated from the atoms. The familiar situation in which this happens is at high temperatures with $k_B T \gtrsim \epsilon_a$ when the system will be made of free electrons and positively charged ions, whereas, if $k_B T \ll \epsilon_a$, the system will be neutral. The transition temperature at which nearly half the number of atoms are ionised occurs around $k_B T \approx (\epsilon_a/10)$, which is $\sim 10^4$ K for hydrogen. For $T \gg 10^4$ K, the kinetic energy of the free electrons in the hydrogen plasma will be $\sim k_B T$.

The electrons can be stripped off the atoms in another different context. This occurs if the matter density is so high that the atoms are packed close to each other, with the electrons forming a common pool with $\epsilon_F \gtrsim \epsilon_a$. In this case, the electrons will be quantum mechanical and the relevant energy scale for them will be ϵ_F. The temperature does not enter into the picture if $k_B T \ll \epsilon_F$, and we may call this a zero-temperature plasma. Conventionally, such systems are called degenerate. For normal metals in the laboratory the Fermi energy is comparable with the binding energy within an order of magnitude. If the temperature is below 10^4 K, the properties of the system are essentially governed by Fermi energy.

In the derivation of P in Eq. (1.3) it is assumed that the the gas is ideal, i.e., the mutual interaction energy of the particles is small compared with the kinetic energy. To treat a plasma as ideal, it is necessary that the Coulomb interaction energy of ions and electrons be negligible. The typical Coulomb potential energy between the ions and the electrons in the plasma is given by $\epsilon_{Coul} \approx Zq^2 n^{1/3}$. If the classical high-temperature plasma is to be treated as an ideal gas, this energy should be small compared with the energy scale of the particle $\epsilon \approx k_B T$, which requires the condition $nT^{-3} \ll (k_B/Zq^2)^3 \simeq 2.2 \times 10^8 Z^{-3}$ in cgs units. On the other hand, to treat the high-density quantum gas as ideal, we should require that the Coulomb energy $\epsilon_{\text{Coul}} \approx Zq^2 n^{1/3}$ be small compared with the Fermi energy $\epsilon_F \approx (\hbar^2/2m)n^{2/3}$. The condition now becomes $n \gg 8Z^3 a_0^{-3} \approx Z^3 \times 10^{26}$ cm^{-3}. Note that such a system becomes more ideal at higher densities; this is because the Fermi energy rises faster than the Coulomb energy.

Let us now go back to the tacit assumption we made in the above analysis, viz., that physical interactions between the particles of the system are capable of maintaining the thermal equilibrium. Determining the precise condition that will ensure this is not a simple task; but – naively – we would require that (1) the mean free path for particles, $l = (n\sigma)^{-1}$ based on a relevant scattering process governed by a cross section σ, be small compared with the scale L over which various parameters change significantly, and (2) that the mean time between collisions $\tau = (nv\sigma)^{-1}$ be small compared with the time scale over which physical parameters change.

To apply this condition we need to know the relevant mean free path for the system. For a neutral gas of molecules, this is essentially determined by molecular collisions with $\sigma_0 \approx \pi a_0^2 \approx 8.5 \times 10^{-17}$ cm^2 and $l = (n\sigma_0)^{-1}$. The time scale for the establishment of a Maxwellian distribution of velocities will be approximatelly $\tau_{\text{neu}} \simeq l/v \propto n^{-1}T^{-1/2}$. For an ionized classical gas, the cross section for scattering is decided by Coulomb interaction between charged particles. Because an ionized plasma is made of electrons and ions with vastly different inertia, the interparticle collisions can take different time scales to produce thermal equilibrium between electrons, between ions, and between electrons and ions. Each of these needs to be discussed separately.

The typical impact parameter between two electrons is $b \approx (2Zq^2/m_e v^2)$, where v is the typical velocity of an electron. The corresponding e–e scattering cross section is

$$\sigma_{\text{coul}} \approx \pi b^2 \approx \pi \left(\frac{Zq^2}{m_e}\right)^2 \frac{1}{v^4} \approx 10^{-20} \text{ cm}^2 \, Z^2 \left(\frac{T}{10^5 \text{ K}}\right)^{-2}, \qquad (1.13)$$

and the mean free path varies as $l = (n\sigma_{\text{coul}})^{-1} \propto (T^2/n)$. The mean free time between the electron–electron scattering will be $\tau_{ee} \approx (n\sigma v)^{-1}$, where n is the number density of electrons and $\sigma \approx \pi b^2$. This gives $\tau_{ee} \approx (m_e^2 v^3/2\pi Z^2 q^4 n)$, which is the leading dependence. (A more precise analysis changes the numerical coefficient and introduces an extra logarithmic factor; see Chap. 9.)

Note that $\tau \propto m^2 v^3 \propto T^{3/2} m^{1/2}$ at a given temperature $T \propto (1/2)mv^2$. Therefore the ion–ion collision time scale τ_{pp} will be larger by the factor $(m_p/m_e)^{1/2} \simeq 43$, giving $\tau_{pp} = (m_p/m_e)^{1/2}\tau_{ee} \simeq 43\tau_{ee}$.

The time scale for significant transfer of energy between electrons and ions is still larger because of the following fact. When two particles (of unequal mass) scatter off each other, there is no energy exchange in the centre-of-mass frame. In the case of ions and electrons, the centre-of-mass frame differs from the lab frame only by a velocity $v_{\text{CM}} \simeq (m_e/m_p)^{1/2}v_p \ll v_p$. Because there is no energy exchange in the centre-of-mass frame, the maximum energy transfer in the lab frame (which occurs for a head-on collision) is approximately $\Delta E = (1/2)m_p(2v_{cm})^2 = 2m_p v_{cm}^2 \simeq 2m_e v_p^2$, giving $[\Delta E/(1/2)m_p v_p^2] \simeq (m_e/m_p) \ll 1$. Therefore it takes (m_p/m_e) times more collisions to produce equilibrium between electrons and ions, that is, the time scale for electron–ion collision is $\tau_{pe} = (m_p/m_e)\tau_{ee} \simeq 1836\tau_{ee}$. The plasma will relax to a Maxwellian distribution in this time scale.

Finally, it must be noted that in a high-temperature tenuous plasma, this mean free path can become larger than the size of the system. If that happens, it is necessary to check whether there are any other physical processes that can provide an effective mean free path that is lower. Most astrophysical plasmas host magnetic fields that make the charged particles spiral around the magnetic-field lines. We can estimate the typical radius of a spiraling charged particle in a magnetic field by equating the centrifugal force (mv^2/r) to the magnetic force (qvB/c). This leads to a radius called the Larmor radius, given by

$$r_L = (mcv/qB) = 13 \text{ cm } (T/10^5 \text{ K})^{1/2}(B/1 \text{ G})^{-1}$$

in a thermal plasma. When the Larmor radius is small, it can act as the effective mean free path for the scattering of charged particles. The ratio between the mean free path from Coulomb collisions $[l \propto (T^2/n)]$ and the Larmor radius $[r_L \propto (T^{1/2}/B)]$ varies as $(BT^{3/2}/n)$ and can be large in tenuous high-temperature plasmas with strong magnetic fields. This ratio is unity for a critical magnetic field:

$$B_c = 10^{-19} \text{ G}\left(\frac{T}{10^5 \text{ K}}\right)^{-3/2}\left(\frac{n}{1 \text{ cm}^{-3}}\right). \tag{1.14}$$

The magnetic field in most astrophysical plasmas will be larger than B_c, and hence this effect will be important.

1.3 Classical Radiative Processes

We next turn to the question of gathering information about the cosmic structures from the radiation received from them. To relate the information received through the electromagnetic waves to the properties of the emitting system, it is necessary to understand the process of electromagnetic radiation from different systems and the nature of the spectrum emitted by each of them.

In classical electromagnetic theory, radiation is emitted by any charged particle that is in accelerated motion. A detailed argument given in Chap. 3 shows that the total amount of energy radiated per second in all directions by a particle with charge q moving with acceleration a is given by

$$\frac{d\mathcal{E}}{dt} = \frac{2}{3}\frac{q^2}{c^3}a^2, \tag{1.15}$$

provided the acceleration is measured in the frame in which the particle is instan-teneously at rest. The rate of energy emission, of course, is independent of the frame used to define it. This result is called Larmor's formula and can be used to understand a host of classical electromagnetic phenomena. Because $\mathbf{d} = (q\mathbf{x})$ is the dipole moment related to an isolated charge located at position $\mathbf{x}$, this formula shows that the total power radiated is proportional to the square of $\ddot{\mathbf{d}}$. In bounded motion, if $\mathbf{d}$ varies at frequency ω (so that $\ddot{\mathbf{d}} = -\omega^2\mathbf{d}$), then the energy radiated is given by

$$\frac{d\mathcal{E}}{dt} = \frac{2}{3}\frac{d^2}{c^3}\omega^4. \tag{1.16}$$

Different physical phenomena are essentially characterised by different sources of acceleration in Eq. (1.15) for the charged particle. Let us consider two specific examples.

1.3.1 Thermal Bremsstrahlung

As a first example consider a scattering between an electron of mass m_e and a proton, with an impact parameter b and relative velocity v in a hydrogen plasma. The acceleration of the electron is $a \approx (q^2/m_e b^2)$ and lasts for a time (b/v). Such an encounter will result in the radiation of energy $\mathcal{E} \approx (q^2 a^2/c^3)(b/v) \approx (q^6/c^3 m_e^2 b^3 v) \approx (q^6 n_i/c^3 m_e^2 v)$, as $b \approx n_i^{-1/3}$ on the average. The total energy radiated per unit volume will be $n_e \mathcal{E}$. Because each collision lasts for a time (b/v), there will be very little radiation at frequencies greater than (v/b). For $\omega < (v/b)$, we may take the energy emitted per unit frequency interval to be nearly constant. Further, in the case of plasma in thermal equilibrium, $v \cong (k_B T/m_e)^{1/2}$. Putting all these together, we get

$$j_\omega \equiv \left(\frac{d\mathcal{E}}{d\omega\,dt\,dV}\right) \simeq \left(\frac{q^6}{m_e^2 c^3}\right)\left(\frac{m_e}{k_B T}\right)^{1/2} n_e n_i \propto n^2 T^{-1/2}; \quad \text{(for } \hbar\omega \lesssim k_B T\text{)}. \tag{1.17}$$

This process is called thermal bremsstrahlung. The bremsstrahlung spectrum is flat for $0 < \omega \lesssim (k_B T/\hbar)$ and will fall rapidly for $\omega \gtrsim (k_B T/\hbar)$, where the upper limit comes from the fact that an electron with a typical energy of $(k_B T)$ cannot emit photons with energy higher than $k_B T/\hbar$. The total energy radiated, over all frequencies, from such a plasma can be found by integration of this expression over ω in the range $(0, k_B T/\hbar)$. This gives

$$\left(\frac{d\mathcal{E}}{dt\,dV}\right) = \int_0^{(k_B T/\hbar)} d\omega \left(\frac{d\mathcal{E}}{d\omega\,dt\,dV}\right) \simeq \left(\frac{q^6}{m_e^2 c^3}\right)\left(\frac{m_e k_B T}{\hbar^2}\right)^{1/2} n_e n_i. \tag{1.18}$$

1.3.2 Synchrotron Radiation

Another major source of acceleration for charged particles is the magnetic fields hosted by plasmas. This process, called synchrotron radiation, can be estimated as follows: Consider an electron moving with velocity $\mathbf{v}$ in a magnetic field $\mathbf{B}$. To use Eq. (1.15) we need to estimate the acceleration in the instantaneous rest frame of the particle. In that frame, the magnetic force $(q/c)[\mathbf{v} \times \mathbf{B}]$ is zero, but the magnetic field in the lab frame will lead to an electric field of magnitude $E' \simeq \gamma B$ (see Chap. 3) in the instantaneous rest frame of the charge, inducing an acceleration $a' = (qE'/m_e)$. Accelerated by this field, the charged particle will radiate energy at the rate (which is the same in the rest frame and in the lab frame)

$$\frac{d\mathcal{E}}{dt} = \left(\frac{d\mathcal{E}'}{dt'}\right) = \frac{2}{3}\frac{q^2}{c^3}(a')^2 = \frac{2}{3}\frac{q^2}{c^3}\left(\frac{q^2}{m_e^2}\gamma^2 B^2\right). \tag{1.19}$$

The power radiated by an electron of energy $\epsilon = \gamma m_e c^2$ is $(d\mathcal{E}/dt) \propto \epsilon^2 B^2$. Further, the energy density in the magnetic field is $U_B = (B^2/8\pi)$; hence we can write

$$\left(\frac{d\mathcal{E}}{dt}\right) = \frac{16\pi}{3}\left(\frac{q^2}{m_e c^2}\right)^2 \gamma^2 c U_B \simeq (\sigma_T c U_B)\gamma^2, \tag{1.20}$$

where $\sigma_T \equiv (8\pi/3)(q^2/m_e c^2)^2$ is called the Thomson scattering cross section.

For a nonrelativistic particle spiralling in a magnetic field, the characteristic angular frequency is $\omega = (r_L/v) = (qB/m_e c)$. For relativistic motion with constant v^2, the effective mass is $m_e(1 - v^2/c^2)^{-1/2} = m_e\gamma$, so that the angular frequency becomes $\omega = (qB/mc\gamma) = (qcB/\epsilon)$ for a particle of energy ϵ in a magnetic field of strength B; the synchrotron radiation from an extreme relativistic particle will peak at the frequency

$$\omega_c \approx \omega\gamma^3 \propto B\gamma^2 \propto B\epsilon^2, \tag{1.21}$$

where the extra factor γ^3 arises from special relativistic effects (see Chap. 3). One factor of γ arises from time dilation; the other factor of γ^2 arises from a Doppler factor $[1 - (v/c)]^{-1} \approx 2\gamma^2$ in the $v \to c$ limit.

The total radiation emitted from a bunch of particles will be $j_\omega \propto (B^2\epsilon^2)[n(\epsilon)](d\epsilon/d\omega)$, where the first factor, $B^2\epsilon^2$, is the energy emitted by a single particle, the second factor is the number of particles with energy ϵ, and the last factor is the Jacobian $(d\epsilon/d\omega) \propto \epsilon^{-1} \propto \omega^{-1/2}$ from ϵ to ω. If the spectrum of particles is a power law $n(\epsilon) = C\epsilon^{-p}$, then the radiation spectrum will be

$$j_\nu \approx \frac{e^3}{m_e c^2}\left(\frac{3e}{4\pi m_e^3 c^5}\right)^{(p-1)/2} C B^{(p+1)/2}\nu^{-(p-1)/2}, \tag{1.22}$$

where we have reintroduced all the constants. (A more precise calculation multiplies the expression by a p-dependent factor, which is ~ 0.1.) This leads to a power-law spectrum, $j_\nu \propto \nu^{-\alpha}$ with $\alpha = (p - 1)/2$.

1.4 Radiative Processes in Quantum Theory

The radiation field in quantum theory is described in terms of photons, and the emission or absorption of radiation arises when a physical system makes the transition from one energy level to another. We have already determined the main energy levels of atoms and molecules in Subsection 1.2.2. The transition between these energy levels in an atom will correspond to photon energies upwards of few electron volts, and the corresponding wavelength will be in optical and UV bands in most of the cases.

This estimate of atomic-energy levels was, however, based on the simple form of the Hamiltonian for the electron in an atom. The actual Hamiltonian is a lot more complicated than H_0 used in Subsection 1.2.2. The corrections to H_0 lead to splitting of the original energy levels and allow emission of photons of widely different frequencies by the atomic system. We now consider these corrections.

1.4.1 Fine Structure and Hyperfine Structure

The Hamiltonian for an electron in a hydrogen atom can be expressed as a sum, $H \cong H_0 + H_{\text{rel}} + H_{\text{sp−or}} + H_{\text{sp−sp}}$, where $H_0 = (p^2/2m_e) − (Zq^2/r)$ is the original (zeroth-order) Hamiltonian and the rest are lowest-order corrections. The first correction $H_{\text{rel}} = −(p^4/8m_e^3 c^2)$ is the relativistic correction to the kinetic energy $p^2/2m_e$ that arises as the second term in the Taylor series expansion of $\epsilon(p)$ in Eq. (1.8); the correction $H_{\text{sp−or}}$ arises from the coupling between the spin magnetic moment of the electron $\mu_e \simeq (q\hbar/2m_e c)$ and the magnetic field $B \cong (v/c)E \cong (v/c)(Zq/r^2)$ in the instantaneous rest frame of the electron, obtained by transformation of the Coulomb field. This should have the magnitude

$$\bar{H}_{\text{sp−or}} = \mu B = \frac{Zq^2}{2r^2} \frac{\hbar v}{m_e c^2} = \frac{Zq^2}{m_e^2 c^2 r^3}(\mathbf{l} \cdot \mathbf{s}), \tag{1.23}$$

where $\mathbf{l}$ and $\mathbf{s}$ are the orbital and the spin angular momenta of the electron. The actual result is half of this value where the extra factor arises due to a phenomenon called Thomas precession, to be discussed in Chap. 3, exercise 3.4. The next correction,

$$H_{\text{sp−sp}} = \mu_e \cdot \left(\frac{\mu_N}{r^3} − \frac{3\mathbf{r} \cdot \mu_N}{r^5} \mathbf{r} \right), \tag{1.24}$$

is the coupling between the nuclear magnetic moment and the magnetic moment of the electron. These magnetic moments are given by

$$\mu_e = −\left[2 + \frac{\alpha}{\pi} + \mathcal{O}(\alpha^2) \right] \frac{q\hbar}{2m_e c} \mathbf{s} \equiv −g_e \mu_B \mathbf{s};$$

$$\mu_p \simeq 5.6 \left(\frac{q\hbar}{2m_p c} \right) \mathbf{S} \equiv g_N \mu_N \mathbf{S}, \tag{1.25}$$

where $\mu_B = (q\hbar/2m_e c)$ is called the Bohr magneton, μ_N is the corresponding quantity for the proton, and **s** and **S** are the spin vectors of electron and proton, respectively.

It is now easy to evaluate the order of magnitude of these corrections. The zeroth-order term H_0 is of the order of $(q^2/a_0) \approx 10$ eV, corresponding to $\lambda \simeq 1200$ Å. The first correction gives

$$\frac{p^4}{m_e^3 c^2} \approx \frac{p^2}{m_e}\left(\frac{p}{m_e c}\right)^2 \approx \left(\frac{v}{c}\right)^2 \frac{q^2}{a_0} \approx \alpha^2 E_0 \approx 10^{-3} \text{ eV}, \qquad (1.26)$$

with the corresponding photon wavelength of $\lambda \simeq 1$ mm. The second correction, with $l \simeq s \simeq \hbar$, is

$$\frac{q^2\hbar^2}{m_e^2 c^2 a_0^3} \approx \frac{q^2}{a_0}\left(\frac{\hbar}{m_e c a_0}\right)^2 \approx \alpha^2 E_0, \qquad (1.27)$$

which is of the same order as that of the first correction. These two together are called fine-structure corrections. The third correction is of the order of

$$\frac{\mu_B \mu_N}{a_0^3} \approx \frac{m_e}{m_p}\frac{\mu_B^2}{a_0^3} \approx \frac{m_e}{m_p}\left(\frac{\hbar}{m_e c a_0}\right)^2 \frac{q^2}{a_0} \approx 10^{-3}\alpha^2 E_0 \approx 10^{-6} \text{ eV}, \qquad (1.28)$$

(corresponding to $\lambda \simeq 10^2$ cm), which is smaller and is called the hyperfine correction. A more precise calculation gives the wavelength of radiation emitted in the hyperfine transition of hydrogen to be ~ 21 cm. This radiation, which is in the radio band, is used extensively in astronomy as a diagnostic of atomic (neutral) hydrogen.

1.4.2 Transition Rates and Cross Sections

To complete the quantum-mechanical analysis, we also need to estimate the rate of transition between the various levels. Consider an initial state with an atom in ground state $|G\rangle$ and n photons present. Absorbing a photon, the atom makes the transition to the excited state $|E\rangle$, leaving behind an $(n-1)$ photon state. Let the probability for this process be $\mathcal{P}[|G\rangle|n\rangle \to |E\rangle|n-1\rangle] \propto n \equiv Qn$. The fact that this absorption probability $\mathcal{P}$ is proportional to n seems intuitively acceptable. Consider now the probability $\mathcal{P}'$ for the time-reversed process $[|E\rangle|n-1\rangle \to |G\rangle|n\rangle]$. By principle of microscopic reversibility, we expect $\mathcal{P}' = \mathcal{P}$, giving $\mathcal{P}' \propto n \equiv Qn$. Calling $n-1 = m$, we get $\mathcal{P}'[|E\rangle|m\rangle \to |G\rangle |m+1\rangle] = Qn = Q(m+1)$. Clearly $\mathcal{P}'$ is nonzero even for $m=0$; $\mathcal{P}'[|E\rangle|0\rangle \to |G\rangle|1\rangle] = Q$, which gives the probability for a process conventionally called spontaneous emission. The term Qm gives the corresponding probability for stimulated emission. Thus the fact that absorption probabilities are proportional to n whereas emission probabilities are proportional to $n+1$ originates from the principle of microscopic reversibility.

The basic rate, governed by Q, can be estimated by use of the correspondence with classical theory. If the rate of transition between two energy levels is Q and the energy of the photon emitted during the transition is $\hbar\omega$, then the rate of energy emission is $Q\hbar\omega$. Classically, the same system will emit energy at the rate $(d^2\omega^4/c^3)$, where d is the electric dipole moment of the atom. (This assumes that the radiation is predominantly due to direct coupling between the radiation and the dipole moment of the atom; if not, we have to use the relevant moment – like the electric quadrupole moment or magnetic dipole moment, instead of d – in this equation.) Writing $Q\hbar\omega \approx (d^2\omega^4/c^3)$, we find that

$$Q \cong \frac{\omega^3 d^2}{\hbar c^3} \simeq \frac{\alpha^5}{8}\left(\frac{\lambdabar}{c}\right)^{-1} \simeq \frac{\alpha^5}{8}\left(\frac{m_e c^2}{\hbar}\right) \approx 10^9 \text{ s}^{-1}, \qquad (1.29)$$

where we have used $d = qa_0 \simeq (q\lambdabar_e/\alpha)$ and $\hbar\omega \simeq (1/2)\alpha^2 m_e c^2$. The rate Q can also be expressed in different forms as

$$Q \approx \frac{q^2}{\hbar c}\frac{a_0^2}{c^2}\omega^3 = \frac{q^2}{m_e c^3}\omega^2 = \omega^2\frac{r_0}{c}, \qquad (1.30)$$

where $r_0 = (q^2/m_e c^2)$ is the classical electron radius. A more precise quantum-mechanical calculation corrects this by contributing an extra numerical factor f (called oscillator strength).

The transition rate for other processes, such as the 21-cm radiation, can be estimated in the same manner by using Eq. (1.16). In the relation $Q\hbar\omega_{21} \approx (\omega_{21}^4 d^2/c^3)$, we now have to use the magnetic dipole moment of electron $d \approx (q\hbar/m_e c)$. This gives $Q \approx \alpha(\lambdabar_e/c)^2\omega_{21}^3$. Estimating $\hbar\omega_{21}$ from expression (1.28) derived above for the hyperfine-structure energy level, we can evaluate this transition rate as

$$Q_{21\text{cm}} \simeq \left(\frac{r_0\omega_{21}}{c}\right)\left(\frac{\hbar\omega_{21}}{m_e c^2}\right)\omega_{21} \approx 10^{-15} \text{ s}^{-1}. \qquad (1.31)$$

The ratio between this rate and the transition rate between the primary energy levels of the hydrogen atom computed in relation (1.29) is

$$\frac{Q_{21\text{cm}}}{Q} \simeq \left(\frac{\omega_{21}}{\omega}\right)^3\left(\frac{\lambdabar_e}{a_0}\right)^2 \approx 10^{-24}. \qquad (1.32)$$

Clearly, hyperfine transitions are very slow processes.

This analysis also shows that every excited state has a probability of decaying spontaneously to the ground state with a decay rate Q. Hence the lifetime Δt of the excited state is approximately

$$\Delta t \approx Q^{-1} \approx \left(\frac{q^2}{\hbar c}\right)^{-1}\left(\frac{a_0}{\lambdabar}\right)^{-2}\omega^{-1} \approx \left(\frac{10^8}{\omega}\right) \approx 10^{-9} \text{ s} \qquad (1.33)$$

for optical radiation. (It is correspondingly larger and is approximately $\Delta t = Q_{21\,cm}^{-1} \simeq 10^{15}$ s for 21-cm radiation.) From the uncertainty principle between the energy and time, it follows that the energy level of the excited state will be uncertain by an amount ΔE of the order of $(\hbar/\Delta t) \approx \hbar Q$. In the absence of such an uncertainty, the transition between the two energy levels could lead to infinitely sharp spectral lines at a specific frequency ω. The width of the excited states leads to a corresponding width to the spectral line called the natural width $\Delta \omega$, where $(\Delta \omega/\omega) = (Q/\omega) \approx 10^{-8}$ for the main hydrogen lines. In terms of wavelength, the natural width $\Delta \lambda = 2\pi c (\Delta \omega/\omega^2)$ is of the order of the classical electron radius: $\Delta \lambda \approx 2\pi r_0$.

Because of the finite linewidth, it is convenient to introduce a (frequency-dependent) bound–bound cross section $\sigma_{bb}(\omega)$ for the absorption of radiation by the system, which is sharply peaked at $\omega = \omega_0$ with a width of $\Delta \omega$. If the phase-space density of photons is $dN = n[d^3x\,d^3p/(2\pi\hbar)^3]$, then the number density of photons per unit volume participating in the absorption is $\mathcal{N} = n\,d^3p/(2\pi\hbar)^3 = (n\omega_0^2/\pi c^3)(\Delta\omega/2\pi)$. The corresponding flux is $\mathcal{N}c$, and the rate of absorption is

$$\mathcal{N}\sigma_{bb}c = \frac{n\omega_0^2}{\pi c^2}\sigma_{bb}\frac{\Delta\omega}{2\pi} = Qn = \omega_0^2\left(\frac{r_0}{c}\right)n, \qquad (1.34)$$

giving $\sigma_{bb}(\Delta\omega/2\pi) = \pi r_0 c$. This result can be stated more formally as

$$\int_0^\infty \sigma_{bb}(\omega)\frac{d\omega}{2\pi} = \pi r_0 c = \frac{\pi q^2}{m_e c} = (\pi r_0^2)\left(\frac{c}{r_0}\right); \qquad r_0 \equiv \left(\frac{q^2}{m_e c^2}\right) \qquad (1.35)$$

which suggests expressing the cross section as

$$\sigma_{bb}(\omega) \equiv \frac{\pi q^2}{m_e c}\phi_\omega, \qquad (1.36)$$

where ϕ_ω is called the line-profile function; the integral of ϕ_ω over $(d\omega/2\pi)$ is unity. When the linewidths are ignored, ϕ_ω is proportional to the Dirac delta function; when the finite width of the energy levels is taken into account, ϕ_ω will become a function that is peaked at ω_0 with a narrow width $\Delta\omega$, such that $\phi(\omega_0) \cong (\Delta\omega)^{-1}(2\pi)$. The absorption cross section at the centre of the line $\omega = \omega_0$ is given by $\sigma \simeq 2\pi^2(rc/\Delta\omega) \simeq (\lambda_0^2/2)$, where λ_0 is the wavelength of the photon. A more rigorous quantum-mechanical theory will provide the explicit form for $\phi(\omega)$ (which turns out to be a Lorentzian in this case) and an overall multiplicative factor f called the oscillator strength.

1.4.3 Thermal Radiation

The result regarding the emission and the absorption rates also allows us to determine the the energy distribution of photons in equilibrium with matter at temperature T. In such a situation, photons will be continuously absorbed and emitted by matter. Consider the rate of absorption (or emission) of photons between any two levels, say, a ground state $|G\rangle$ and an excited state $|E\rangle$. We saw above that the absorption rate per atom is given by Qn and the emission rate is $Q(n+1)$, where n is the number of photons present and Q is determined from the quantum theory of radiation. In steady state, the number of upward and downward transitions must match, which requires that product (number of atoms in G) × (rate of upward transitions per atom) equal (number of atoms in E) × (rate of downward transitions per atom), that is, $N_G Qn = N_E Q(n+1)$. Because matter is in thermal equilibrium at temperature T, we must also have $(N_E/N_G) = \exp(-\Delta E/k_B T)$, where ΔE is the energy difference between the two levels. This relation should hold for all forms of matter with arbitrary energy levels; hence we can take the ground-state energy to be zero (i.e, arbitrarily small) and the excited-state energy to be E, leading to

$$n = \frac{1}{(N_G/N_E) - 1} = \frac{1}{\exp(E/k_B T) - 1}. \tag{1.37}$$

This equation gives the number of photons with energy E [or – equivalently – with momentum $p = (E/c)$] in thermal equilibrium with matter. To be more precise, the number of photons with momentum in the interval $[\mathbf{p}, \mathbf{p} + d^3\mathbf{p}]$ is given by $dN = 2n[V d^3 p/(2\pi\hbar^3)]$, where the factor in the square brackets gives the number of quantum states in the phase volume $V d^3 p$ and the factor 2 takes into account the two spin states for each photon. The corresponding energy dE flowing through $d^3 x = dA(cdt)$ will be $dE = h\nu\, dN = 2n(\nu)h\nu\, dA[cdt][d^3 p/h^3]$. Writing $d^3 p = p^2 dp\, d\Omega = (h/c)^3 \nu^2\, d\nu\, d\Omega$, we can determine the intensity (which is the energy per unit area per unit time per solid angle per frequency) of thermal radiation as

$$\frac{dE}{dA\, dt\, d\Omega\, d\nu} \equiv B_\nu = \frac{2h\nu^3}{c^2} n(\nu) = \frac{2h\nu^3}{c^2} \frac{1}{e^{h\nu/k_B T} - 1}. \tag{1.38}$$

The quantity νB_ν reaches a maximum value around $h\nu \approx 4k_B T$, which translates to the fact that a blackbody at 6000 K will have the maximum for νB_ν at 6000 Å. The maximum intensity is $(\nu B_\nu)_{max} \approx (T/100\ \text{K})^4\ \text{W m}^{-2}\ \text{sr}^{-1}$. Such thermal radiation can arise in many different contexts in which a primary source of energy is thermalised because of some physical process, the most important example being stellar radiation.

At low frequencies, the intensity of thermal radiation given by Eq. (1.38) will be $B_\nu \approx (2k_B T/\lambda^2)$. Because of this relation, it is conventional to define a brightness temperature for any source with intensity I_ν as $T_B \equiv (\lambda^2 I_\nu/2k_B)$, which is (in general) a function of frequency.

It is clear from Eq. (1.37) that there are very few photons with momentum greater than $\bar{p} \approx (k_B T/c)$, so that $(N/V) \approx (4\pi/3)(\bar{p}/2\pi\hbar)^3 \approx (k_B T/\hbar c)^3$ and the mean energy is $U_{ER} \approx k_B T(N/V) \approx (k_B T)^4/(\hbar c)^3$.

1.4.4 Photon Opacities in Matter

Let us next determine the conditions under which our original assumption – that the radiation is in thermal equilibrium with matter – holds. If the cross section for the relevant process that scatters or absorbs radiation is given by σ and the number density of scatterers is n, then the mean free path of a photon is given by $l = (n\sigma)^{-1}$. In the case of radiation, it is conventional to define a quantity κ (called opacity) such that

$$\alpha \equiv n\sigma \equiv \rho\kappa, \qquad (1.39)$$

where ρ is the mass density of the scatterers. The optical depth of a system of size R is defined to be $\tau \equiv \alpha R = (R/l)$. From the standard theory of random walk, we know that a photon will traverse a distance R with N_c collisions, where $R = N_c^{1/2} l$; that is, the number of collisions is given by $N_c = (R/l)^2 = \tau^2$, provided that $\tau \gg 1$.

The opacity for the photons is provided mainly by three different processes: (1) scattering by free electrons, (2) the free–free absorption of photons, and (3) the bound–free transitions induced in matter by the photons that are passing through it.

(1) The simplest case is the one in which the charged particle is accelerated by an electromagnetic wave that is incident upon it. Consider a charge q placed on an electromagnetic wave of amplitude E. The wave will induce an acceleration $a \simeq (qE/m)$, causing the charge to radiate. The power radiated will be $P = (2q^2a^2/3c^3) = (2q^4/3m^2c^3)E^2$. Because the incident power in the electromagnetic wave is $S = (cE^2/4\pi)$, the scattering cross section (for electrons with $m = m_e$) is

$$\sigma_T \equiv \frac{P}{S} = \frac{8\pi}{3}\left(\frac{q^2}{m_ec^2}\right)^2 \approx 6.7 \times 10^{-25}\ \text{cm}^2, \qquad (1.40)$$

which is the Thomson scattering cross section. This cross section governs the basic scattering phenomena between charged particles and radiation. The corresponding mean free path for photons through a plasma is $l_T = (n_e\,\sigma_T)^{-1}$, and the Thomson scattering opacity, defined to be $\kappa_T \equiv (n_e\sigma_T/\rho)$, is

$$\kappa_T = \left(\frac{n_e}{n_p}\right)\left(\frac{\sigma_T}{m_p}\right) = 0.4\ \text{cm}^2\ \text{gm}^{-1} \qquad (1.41)$$

for ionised hydrogen with $n_e = n_p$.

The opacity for the process in (2) and (3) can be determined by the principle of detailed balance which allows us to relate the rate for certain processes to the rate for the corresponding 'inverse' process.

(2) The time-reversed process corresponding to bremsstrahlung is the one in which a photon is absorbed by an electron while in the proximity of an ion.

In equilibrium, the rate for this free–free absorption should match that of thermal bremsstrahlung. We write the free–free absorption rate as $n \sigma_{ff} B(\nu)$, where σ_{ff} is the free–free absorption cross section and $B(\nu) \propto \nu^3 (e^{\hbar\nu/k_B T} - 1)^{-1} \propto \nu^2 T$ (when $\hbar\nu \ll k_B T$) is the intensity of the thermal radiation. Equating this to the bremsstrahlung emissivity found in Eq. (1.17), $j_\nu \propto n^2 T^{-1/2}$, we get $\sigma_{ff} \propto (n/\nu^2 T^{3/2})$. Taking the typical frequency of the photon to be proportional to T, we find that $\sigma_{ff} \propto n T^{-3.5}$. The corresponding opacity can be written in the form

$$\kappa_{ff} \propto \rho T^{-3.5}. \tag{1.42}$$

(3) To obtain the bound–free opacity, which arises when a photon ionizes an atom, we begin by relating the photoionization rate to the recombination rate in which an ion and an electron get bound together, releasing the excess energy as radiation. The recombination rate per unit volume of a plasma will be proportional to (1) the number density of electrons n_e, (2) the number density of ions n_i, (3) the relative velocity of encounter v, and (4) the cross section for the process, which will be $\sim \pi \lambda^2$, where λ gives the effective range of interaction between electron and proton. For Coulomb interaction, this range is $\lambda_1 \cong (2Zq^2/m_e v^2)$; on the other hand, an electron of speed v has a de Broglie wavelength of $\lambda_2 \cong (\hbar/m_e v)$. We should choose λ to be the larger of λ_1 or λ_2, depending on the context. Let us first consider the case with $\lambda \simeq \lambda_2$, which corresponds to $(v/c) \gtrsim (q^2/\hbar c)$. Each recombination will release an energy of approximately $(1/2)m_e v^2$. Hence,

$$\left(\frac{dE_{\text{rec}}}{dV dt}\right) \propto \left(\frac{m_e v^2}{2}\right)(n_e n_i)\left(\frac{h}{m_e v}\right)^2 v \propto n_e n_i T^{1/2} \propto \rho^2 T^{1/2}, \tag{1.43}$$

where we have assumed that $m_e v^2 \approx k_B T$, $n_e \propto \rho$, and $n_i \propto \rho$. In equilibrium, the photoionisation rate (which removes energy from the radiation field) should match the recombination rate. The amount of energy removed by photoionisation is proportional to $dE_{\text{ion}} \propto n_{\text{atom}} \sigma_{bf} E_{\text{rad}}$, where σ_{bf} is the photoionisation cross section. Equating dE_{ion} to dE_{rec} and using $E_{\text{rad}} \propto T^4$, we get

$$n_{\text{atom}} \sigma_{bf} T^4 \propto n_e n_i T^{1/2}. \tag{1.44}$$

Introducing the bound–free opacity κ_{bf} by the definition $\kappa_{bf} = (n_{\text{atom}} \sigma_{bf}/\rho)$ and taking $n_e \propto \rho$ and $n_i \propto \rho$, we find that

$$\kappa_{bf} \propto \rho T^{-3.5} \tag{1.45}$$

which scales just like relation (1.42).

In a radiation bath with temperature T, the typical energy of photons is $h\nu \simeq k_B T$. The result of relation (1.45) suggests that the frequency dependence of $\sigma_{bf}(\nu) \propto \kappa_{bf} \propto T^{-3.5}$ will be of the form $\sigma(\nu) = \sigma_{bf}(\nu/\nu_I)^{-s}$ for $\nu > \nu_I$ (and zero otherwise). Here $s \approx 3$–3.5 and $\nu_I = (\epsilon_a/h)$ is the frequency corresponding to the ionisation energy ϵ_a of the atom. Because the cross section for photoionisation $\sigma(\nu)$ satisfies constraint (1.35),

$$\int_0^\infty \sigma(\nu)\,d\nu = \pi r_0 c f = \left(\frac{\pi e^2}{m_e c}\right) f, \qquad (1.46)$$

where the term within the parentheses comes from classical theory and the oscillator strength f is supplied by the quantum theory, we get

$$\sigma_{bf} = \pi(s-1)f r_0 \lambda_I \simeq 2\pi f r_0 \lambda_I, \qquad (1.47)$$

where $\lambda_I = (c/\nu_I)$ and $s \simeq 3$. This shows that the photoionisation cross section is essentially the product of the classical electron radius and the wavelength at ionisation threshold. Using $\lambda_I = 912$ Å, we get $\sigma_{bf} \approx 10^{-17}$ cm^2 for hydrogen. For heavier elements σ_{bf} will scale as

$$\sigma(\nu) = \sigma_{bf}\left(\frac{\nu_I}{\nu}\right)^3 \propto f r_0 \left(\frac{c}{\nu_I}\right)\left(\frac{\nu_I}{\nu}\right)^3 \propto \left(\frac{\nu_I^2}{\nu^3}\right) \propto Z^4 \nu^{-3} \qquad (1.48)$$

if we take $s = 3$.

The cross section for recombination, $\sigma_{\rm rec}$, is related to σ_{bf} by $4\sigma_{\rm rec}d^3 p_e = 8\sigma_{bf}d^3 p_\gamma$ where $4 = 2 \times 2$ is due to electron and proton spins, $8 = 2 \times 2 \times 2$ is due to electron, proton, and photon spins, and p_e and p_γ refer to the momenta of the electron and the photon, respectively. Because $dp_e \simeq dp_\gamma$, this gives $\sigma_{\rm rec} = 2\sigma_{bf}(p_\gamma/p_e)^2$; if the plasma is at some temperature T, then

$$\sigma_{\rm rec} = \frac{2(\hbar\omega)^2}{c^2(m_e v)^2}\sigma_{bf} \simeq 2\left(\frac{\epsilon_a}{m_e c^2}\right)\left(\frac{\epsilon_a}{k_B T}\right)\sigma_{bf} \simeq 10^{-22}\left(\frac{k_B T}{10\ \text{eV}}\right)^{-1} \text{cm}^2$$

$$(1.49)$$

at the threshold with $\hbar\omega \simeq \epsilon_a \simeq 10$ eV. The rate of recombination per unit volume per second is $n^2 \sigma_{\rm rec} v \equiv n^2 \alpha_R$, where $\alpha_R \simeq 2 \times 10^{-14}(k_B T/10\ \text{eV})^{-1/2}$ cm^3 s^{-1}.

In the analysis so far we have assumed that the recombination proceeds directly to the ground state. Such a process, however, will release a photon with energy $\hbar\omega \gtrsim \epsilon_a$ that will immediately ionise another atom. The net recombination actually proceeds through electron and proton, forming an excited state that decays to ground state later on. All the photons emitted in the process will have $\hbar\omega < \epsilon_a$ and cannot ionise the neutral atoms in the ground state. In this case, we can write a relation similar to relation (1.43) but with $\lambda = \lambda_1 = (2Zq^2/m_e v^2)$. Then the energy loss that is due to recombination is

$$\left(\frac{dE_{\rm rec}}{dV\,dt}\right) \propto \left(\frac{1}{2}m_e v^2\right)(n_e n_i)\left[\frac{Zq^2}{(1/2)m_e v^2}\right]^2 v \propto n_e n_i v^{-1} \propto n^2 T^{-1/2}.$$

This can be one source of cooling for a plasma at $T \gtrsim 10^4$ K in addition to the bremsstrahlung cooling $(dE_{\rm bre}/dV\,dt) \propto n^2 T^{1/2}$ obtained in Eq. (1.18). The net rate of cooling for a plasma can be expressed in the form $(dE/dV\,dt) = n^2 \Lambda(T)$, where $\Lambda(T) = aT^{1/2} + bT^{-1/2}$; the first term which arises due to bremsstrahlung dominates for $T \gtrsim 10^6$ K.

When both photoionisation and recombinations occur in a region, the equilibrium is described by the relation $n_e n_i \alpha_R \simeq \sigma_{bf} n_H F$, where F is the flux of ionising photons with $\nu > \nu_I$. Taking $n_e = n_i = xn_0$, $n_H = (1 - x)n_0$, we can write this equation in the form

$$\frac{x^2}{(1 - x)} \simeq \left(\frac{\sigma_{bf}c}{\alpha_R}\right)\left(\frac{F}{n_0 c}\right) \simeq 5 \times 10^4 \left(\frac{T}{10^4 \text{ K}}\right)^{1/2}\left(\frac{F}{n_0 c}\right), \qquad (1.50)$$

which determines the ionisation fraction x in many astrophysical contexts. If a source of photons emitting $\dot{N}_\gamma$ ionising photons per second (with $\nu > \nu_I$) ionises a region of volume V around it, then the same argument gives $n_e^2 \alpha_R V \simeq \dot{N}_\gamma$. Taking $n_e = xn_0 \simeq n_0$, we get $V = (\dot{N}_\gamma / \alpha_R n_0^2)$.

It is also possible to imagine a situation in which the mean free path for collisions between atoms is small (so that the matter is in thermal equilibrium) but the mean free path for collisions between photons and matter is large (so that radiation is *not* in equilibrium with matter). In that case, the photons emitted by the system will escape from it with negligible scattering and there could arise radiation from atomic or molecular transitions, characteristic of composition. Then the amount of energy emitted per second by unit volume of gas, per unit frequency range, and solid angle can be written as

$$j_\nu = \left(\frac{dE}{d\Omega\,d\nu\,dt\,dV}\right) = \left[\frac{g_j}{Z}\exp\left(\frac{-h\nu_{ij}}{kT_s}\right)\right]\left(\frac{h\nu_{ij}}{4\pi}\right)Q_{ij}n\phi(\nu). \qquad (1.51)$$

The first factor in the brackets gives the fraction of atoms in the excited state, with Z (the partition function) providing the normalisation. Because all the factors are known in this expression, it can be used to relate the source properties to the observed intensity. An important example of this is the 21-cm radiation from neutral hydrogen atoms. The intensity of this line and its width arising due to various effects, make it a useful probe of several systems.

1.5 Varieties of Astrophysical Structures

We now turn to the question of trying to determine broad features of astronomical systems from the description of the physical principles given above. It is clear, right at the outset, that we are interested in systems that are massive enough so that gravity plays a significant role in their dynamics. In the most extreme limit, we can think of the entire universe as a physical system and try to determine its structure. At sufficiently large scales, we can ignore the graininess in the distribution of matter and think of the universe as reasonably homogeneous and isotropic. Further, because no location in such a universe can be considered as special, any possible motion of matter on a large scale should also maintain the same characteristics with respect to any observer. It immediately follows that the most general motion consistent with these requirements must have the form $\mathbf{v}(t) = f(t)\mathbf{r}$, where $\mathbf{r}$ and $\mathbf{v}$ denote the position and the velocity, respectively, of any material body in the universe and $f(t)$ is an arbitrary function of time. This is the only kind of motion that is consistent with the requirement that from *any* point in the universe an observer will see matter moving in an identical manner. Using the fact that $\mathbf{v} = \dot{\mathbf{r}}$, this equation can be integrated to give $\mathbf{r} = a(t)\mathbf{x}$, where $a(t)$ is another arbitrary function related to $f(t)$ by $f(t) = (\dot{a}/a)$ and $\mathbf{x}$ is a constant for any given material body in the universe. It is conventional to call $\mathbf{x}$ and $\mathbf{r}$ the comoving and proper coordinates of the body and $a(t)$ the expansion factor (even though, if $\dot{a} < 0$, it acts as a contraction factor).

The dynamics of the universe is entirely determined by the function $a(t)$. The simplest choice will be $a(t) = $ constant, in which case there will be no motion in the universe and all matter will be distributed uniformly in a static configuration. It is, however, clear that such a configuration will be violently unstable when the mutual gravitational forces of the bodies are taken into account. Any such instability will eventually lead to random motion of particles in localised regions, thereby destroying the initial homogeneity. Observations, however, indicate that this is not true and that the relation $\mathbf{v} = (\dot{a}/a)\mathbf{r}$ does hold in the observed universe. In that case, the dynamics of $a(t)$ can be qualitatively understood along the following lines. Consider a particle of unit mass at the location $\mathbf{r}$, with respect to some coordinate system. Equating the sum of its kinetic energy $v^2/2$ and gravitational potential energy that is due to the attraction of matter within a sphere of radius r to a constant, we find that $a(t)$ should satisfy the condition

$$\frac{1}{2}\dot{a}^2 - \frac{4\pi G\rho(t)}{3}a^2 = \text{constant}, \tag{1.52}$$

where ρ is the mean density of the universe, that is,

$$\frac{\dot{a}^2}{a^2} + \frac{k}{a^2} = \frac{8\pi G}{3}\rho(t), \tag{1.53}$$

where k is a constant. Although the argument given above to determine this equation is fallacious, Eq. (1.53) happens to be exact and arises from the proper application of Einstein's theory of relativity to a homogeneous and isotropic distribution of matter. Observations suggest that our universe today (at $t = t_0$) is governed by this equation with $\rho(t_0) \approx 10^{-30}$ gm cm^{-3} and $(\dot{a}/a)_0 \equiv H_0 = 0.3 \times 10^{-17}h$ s^{-1}, where $h \approx 0.5$–1. This is equivalent to $H_0 = 100h$ km s^{-1} Mpc^{-1} where 1 Mpc $\approx 3 \times 10^{24}$ cm is a convenient unit for cosmological distances. (We will also use the units 1 kpc $= 10^{-3}$ Mpc and 1 pc $= 10^{-6}$ Mpc in our discussion.) From H_0 we can form the time scale $t_{\text{univ}} \equiv H_0^{-1} \approx 10^{10}h^{-1}$ yr and the length scale $cH_0^{-1} \approx 3000h^{-1}$ Mpc; t_{univ} characterises the evolutionary time scale of the universe and H_0^{-1} gives the largest length scales currently accessible in the universe.

The light emitted at an earlier epoch by an object will reach us today with the wavelengths stretched because of the expansion. If the light was emitted at $a = a_e$ and received today (when $a = a_0$), the wavelength will change by the factor $(1 + z_e) = (a_0/a_e)$, where z_e is called the redshift of the emitting object. The observed luminosity L of a source will decrease as $(1 + z)^{-4}$, as L is proportional to $(p_\gamma c) d^3 p_\gamma \propto v^3 dv \propto (1 + z)^{-4}$, where $p_\gamma = (\epsilon/c) = (h\nu/c)$ is the photon momentum.

If neither particles nor photons are created or destroyed during the expansion, then the number density of particles or photons will decrease as a^{-3} as a increases. In the case of photons, the wavelength will also get stretched during expansion with $\lambda \propto a$; because the energy density of material particles is nmc^2 whereas that of photons of frequency ν will be $nh\nu = (nhc/\lambda)$, it follows that the energy densities of matter and radiation vary as $\rho_{rad} \propto a^{-4}$ and $\rho_{matter} \propto a^{-3}$. Combining this result $\rho_{rad} \propto a^{-4}$ with the result $\rho_{rad} \propto T^4$ for thermal radiation, it follows that any thermal spectrum of photons in the universe will have its temperature varying as $T \propto a^{-1}$. In the past, when the universe was smaller, it would also have been (1) denser (2) hotter, and – at sufficiently early epochs – (3) dominated by radiation energy density. When the temperature of the universe was higher than the temperatures corresponding to the ionisation energy, the matter content in the universe was a high-temperature plasma.

Starting from such a hot initial plasma stage, the universe cools as it expands and eventually a variety of physical structures form in it. In the hot early phase, the radiation is in thermal equilibrium with matter; as the universe cools below $k_B T \simeq (\epsilon_a/10)$ the electrons and ions combine to form neutral atoms and radiation will decouple from matter. This occurs at $T_{rec} \simeq 3 \times 10^3$ K, and the temperature of this radiation will continue to fall as $T \propto a^{-1}$. Observations show that the present universe is indeed bathed in such a thermal radiation field, with the current temperature being about ~ 2.7 K. The energy density of this radiation today is $\rho_\gamma \simeq (k_B T)^4/(\hbar c)^3 \simeq 5.7 \times 10^{-13}$ erg cm^{-3}, which corresponds to a mass density of $(\rho_\gamma/c^2) = 5.7 \times 10^{-34}$ gm cm^{-3}. Taking the matter density today as $\rho_0 = 10^{-30}$ gm cm^{-3}, we find that $\rho_\gamma \simeq 5.7 \times 10^{-4} \rho_0$; radiation would have dominated over matter when the redshift was more than $z_{eq} = 1.7 \times 10^3$. Also note that photons decoupled from matter when the universe was $(T_{rec}/T_0) \simeq 10^3$ times smaller, i.e., at $z_{rec} \simeq 10^3$.

These considerations were independent of the explicit form of $a(t)$. We now turn to the solutions of Eq. (1.53) that determine $a(t)$. The simplest solution to Eq. (1.53) will occur for $k = 0$ if we take the matter density in the universe as decreasing as a^{-3} with expansion. Then we get $a(t) = (t/t_0)^{2/3}$, where $t_0^{-2} = (6\pi G\rho_0)$ and $a(t)$ is normalised to $a = 1$ at the present epoch $t = t_0$. Such a totally uniform universe, of course, will never lead to any of the inhomogeneous structures seen today. However, if the universe had even the slightest inhomogeneity, then gravitational instability could amplify the density perturbations. To see how this comes about in the simplest context, consider Eq. (1.53) written in the equivalent form as

$$\ddot{a} = -\frac{4\pi G\rho_0}{3a^2} = -\left(\frac{2}{9t_0^2}\right)\frac{1}{a^2}, \tag{1.54}$$

where we have put $\rho = (\rho_0 a_0^3/a^3)$. If we perturb $a(t)$ slightly to $a(t) + \delta a(t)$ such that the corresponding fractional density perturbation is $\delta \equiv (\delta\rho/\rho) = -3(\delta a/a)$, we find that δa satisfies the equation

$$\frac{d^2}{dt^2}\delta a = \left(\frac{4}{9t_0^2}\right)\frac{\delta a}{a^3} = \frac{4}{9}\frac{\delta a}{t^2}. \tag{1.55}$$

This equation has the growing solution $\delta a \propto t^{4/3} \propto a^2$. Hence the density perturbation $\delta = -3(\delta a/a)$ grows as $\delta \propto a$. When the perturbations have grown sufficiently, their self-gravity will start dominating and the matter can collapse to form a gravitationally bound system. This can happen, for example, if the plasma in some local region can cool sufficiently fast. We now estimate the conditions for this.

1.5.1 $t_{cool} \approx t_{grav}$: Existence of Galaxies

The cooling of a plasma occurs mainly through two processes. The first is the radiation emitted during recombination of electrons and ions which, if it escapes the plasma, can be a source of recombination cooling. From the discussion in Subsection 1.4.4, we know that the recombination rate varies as $n^2T^{-1/2}$. The second is the bremsstrahlung cooling with an energy-loss rate proportional to $n^2T^{1/2}$ [see Eq. (1.18) with $n_i = n_e = n$]. For systems with temperature $k_BT \simeq (GMm_p/R)$, which is much higher than the ionisation potential, $\alpha^2 m_e c^2$, the dominant cooling mechanism is thermal bremsstrahlung. The cooling time for this process is

$$t_{cool} \simeq \frac{nk_BT}{(d\mathcal{E}/dt\,dV)} = \left(\frac{\hbar}{m_ec^2}\right)\left(\frac{1}{n\,\lambda_e^3}\right)\left(\frac{k_BT}{m_ec^2}\right)^{1/2}\frac{1}{\alpha^3}, \tag{1.56}$$

and the time scale for gravitational collapse is

$$t_{grav} \simeq \left(\frac{GM}{R^3}\right)^{-1/2}. \tag{1.57}$$

The condition for efficient cooling $t_{cool} < t_{grav}$, coupled with $k_BT \simeq GMm_p/R$, leads to the constraint $R < R_g$, where

$$R_g \simeq \alpha^3 \alpha_G^{-1} \lambda_e \left(\frac{m_p}{m_e}\right)^{1/2} \simeq 74 \text{ kpc}, \tag{1.58}$$

where $\alpha_G \equiv (Gm_p^2/\hbar c) \approx 6 \times 10^{-39}$ is the gravitational (equivalent) of the fine-structure constant. In the above analysis we assume that $k_BT > \alpha^2 m_e c^2$; for

$R \simeq R_g$ this constraint is equivalent to the condition $M > M_g$, where

$$M_g \simeq \alpha_G^{-2} \alpha^5 m_p \left(\frac{m_p}{m_e} \right)^{1/2} \simeq 3 \times 10^{44} \text{ gm.} \qquad (1.59)$$

This result suggests that systems having a mass of $\sim 3 \times 10^{44}$ gm and a radius of ~ 70 kpc could rapidly cool, fragment, and form gravitationally bound structures. Most galaxies have masses around this region. This is one possible scenario for forming galaxies. Note that the mass and the length scales in relations (1.59) and (1.58) arise entirely from the fundamental physical constants. We now consider some more properties of these structures.

The original (maximum) radius of the cooling plasma estimated above is ~ 70 kpc. After the matter has cooled and contracted, the final radius is more like 10–20 kpc, which is the typical radii of large galaxies. For $M_g \simeq 3 \times 10^{44}$ gm and $R_g \simeq 20$ kpc, the density is $\rho_{gal} \simeq 10^{-25}$ gm cm^{-3}, which is $\sim 10^5$ times larger than the current mean density, $\rho_0 \simeq 10^{-30}$ gm cm^{-3}, of the universe. If we assume that high-density regions with $\bar{\rho} \gtrsim 100 \bar{\rho}_{\text{univ}}$ collapsed to form the galaxies, then the galaxy formation must have taken place when the density of the universe was ~ 1000 times larger; the value of $a(t)$ would have been 10 times smaller and the redshift of the galaxy formation should have been $z_{gal} \lesssim 9$. If the protogalactic plasma condensations were almost touching each other at the time of formation, these centres (which would have been at a separation of ~ 150 kpc) would have now moved apart to a distance of $150(1 + z_{gal})$ kpc ≈ 1500 kpc $= 1.5$ Mpc. This is indeed the mean separation between the large galaxies today. The nearest galaxy with a radius of ~ 10 kpc, at a distance of 1 Mpc, would subtend an angle of $\theta_{gal} \approx 10^{-2}$ rad $\approx 30'$. A galaxy at a distance of ~ 4000 Mpc will subtend 0.5''.

Observations have indicated that many different kinds of structures exist at redshifts of $z \lesssim 5$. Because the process of gravitational instability, which leads to the condensation of galaxy like objects, cannot be 100% efficient, it would leave some amount of matter uniformly distributed in between the galaxies. The light from distant galaxies will have to pass through this matter and will contain signature of the state of such an intergalactic medium (IGM). The photons (with $\nu > \nu_I$) produced in the first-generation objects could cause a significant amount of ionization of the IGM, especially the low-density regions. When a flux of photons (with $\nu > \nu_I$) impinge on a gas of neutral hydrogen with number density n_H, it will have an ionisation optical depth of $\tau = n_H \sigma_{bf} R$. Setting $\tau = 1$ gives a critical column density for ionisation to be $N_c \equiv n_H R = \sigma_{bf}^{-1} \simeq 10^{17}$ cm^{-2}. Regions with a hydrogen column density $N_c \simeq nR$ greater than 10^{17} cm^{-2} will appear as patches of neutral regions in the ionized plasma of the IGM. Such regions can be studied by absorption of light from more distant sources, (especially through Lyman alpha absorption corresponding to the transition between $n = 1$ and $n = 2$ levels) and are called Lyman alpha clouds.

Still larger structures than galaxies, called galaxy clusters, with masses of $\sim 10^{47}$ gm, a radius of ~ 3 Mpc, and a mean density of 10^{-27} gm cm^{-3}, exist in the universe as gravitationally bound systems. Our argument given above shows that the gas in these structures could not have yet cooled and will have a virial temperature of $T \approx (GMm_p/Rk_B) \approx 4 \times 10^7$ K.

Let us next investigate the nature of smaller-scale structures that can form inside a galaxy. Here, the existence of the different energy scales and equations of state allow for the possible existence of several – widely different – astrophysical systems, and most of these systems can be understood by systematic comparison of the energy scales (ϵ, ϵ_a, $\epsilon_{\text{nucl}} \cdots$) with ϵ_g.

1.5.2 $\epsilon_{\text{grav}} \approx \epsilon_a$: Existence of Giant Planets

The laboratory systems have negligible gravitational potential energy; in a plot of (ϵ_F/k_BT) against (GMm_p/Rk_BT) they exist (almost) along the y axis [see Fig. 1.1 (top)]. We now see how gravity affects the structures as they get bigger. The atomic binding energy (per particle) of a system is approximately $\epsilon_a \approx \alpha^2 m_e c^2 \approx (q^2/a_0) \approx q^2 n^{1/3}$ if the atoms are closely packed (with $na_0^3 \simeq 1$), and the gravitational energy per particle is $\epsilon_g = (4\pi/3)^{1/3} Gm_p^2 N^{2/3} n^{1/3}$. Their ratio is given by

$$\mathcal{R}_{ga} \equiv \frac{\epsilon_a}{\epsilon_g} \approx \left(\frac{\alpha}{\alpha_G}\right)\left(\frac{1}{N^{2/3}}\right) = \left(\frac{N_G}{N}\right)^{2/3} \approx \left(\frac{10^{54}}{N}\right)^{2/3}. \tag{1.60}$$

Clearly, the number $N_G \equiv \alpha^{3/2} \alpha_G^{-3/2} \approx 10^{54}$, arising out of fundamental constants, sets the smallest scale in astrophysics, in which the gravitational binding energy becomes as important as the electromagnetic binding energy of matter. The corresponding mass and length scales are $M_{\text{planet}} \simeq N_G m_p \simeq 10^{30}$ gm and $R_{\text{planet}} \simeq N_G^{1/3} a_0 \simeq 10^{10}$ cm and correspond to those of a large planet; for larger masses, gravitational interaction changes the structure significantly whereas for smaller masses gravity is ignorable and matter is homogeneous with constant density so that $M \propto R^3$.

Because we used Newtonian gravity to arrive at this conclusion, it is necessary to verify that the parameter $\mathcal{R}_{gm} \equiv (E_g/Mc^2)$ is small for this scale; this ratio for $M \simeq M_{\text{planet}}$, $R \simeq R_{\text{planet}}$ is $\mathcal{R}_{gm} = \alpha^2(m_e/m_p) \ll 1$. Note that the smallness of $\mathcal{R}_{gm}$ follows purely from the values of fundamental constants.

Most of the astrophysically interesting systems have larger mass and require the gravitational force to be balanced by forces other than normal solid-state forces. In general, such systems can be classified into two categories. The first set has the gravitational force balanced by the kinetic energy of classical motion, whereas the second one has the gravitational force balanced by degeneracy pressure. For a system with $na_0^3 \approx 1$, the non-relativistic Fermi energy of electrons is comparable with the atomic binding energy and we can compare ϵ_a or ϵ_F with ϵ_g. This is meaningful as long as the temperature of the system is low

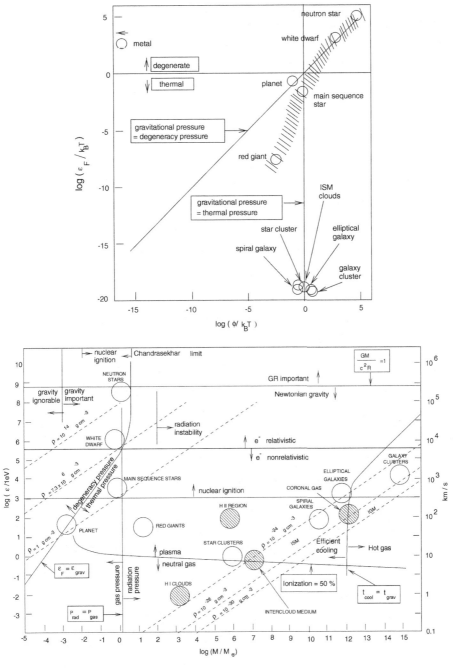

Fig. 1.1.   Aspects governing the dynamics of various cosmic structures are summarised. See text for a detailed description.

and $k_B T \ll \epsilon_F \simeq \epsilon_a \approx 10$ eV. If the temperature is significantly higher, then it is the thermal energy $k_B T$ that should be compared with ϵ_g; that is, the gravitational pressure will be balanced by the Fermi pressure at low temperatures and by thermal pressure at high temperatures. We now examine such structures.

1.5.3 $\epsilon_{grav} \simeq \epsilon_{nucl}$: Existence of Stars

When the mass of the system is increased further, the gravitational pressure increases and – to balance it – both the Fermi pressure and the thermal pressure will increase. The dynamics of the system will then depend on the relative significance of these two quantities. To take into account both thermal and quantum degeneracy contributions, we take the matter pressure to be $P \approx nk_B T + n\epsilon_F$, which is a simple interpolation between the two limits. This pressure can balance the gravitational pressure if $(k_B T + \epsilon_F) \simeq Gm_p^2 N^{2/3} n^{1/3}$. Using expression (1.11) for ϵ_F for the non-relativistic electrons, we get

$$k_B T \simeq Gm_p^2 N^{2/3} n^{1/3} - \frac{(3\pi^2)^{2/3}}{2} \frac{\hbar^2}{m_e} n^{2/3}. \tag{1.61}$$

For a classical system, the first term on the right-hand side dominates, and we see that the gravitational potential energy and kinetic energy corresponding to the temperature T are comparable; this is merely a restatement of the virial theorem. As the radius R of the system is reduced, the second term on the right-hand side $(\propto n^{2/3})$ grows faster than the first $(\propto n^{1/3})$ and the temperature of the system will increase, reach a maximum, and decrease again; equilibrium is possible for any of these values with gravity balanced by thermal and degeneracy pressure. The maximum temperature T_{max} is reached when $n = n_c$, with

$$n_c^{1/3} \simeq \frac{\alpha_G}{(3\pi^2)^{2/3}} \left(\frac{N^{2/3}}{\lambdabar_e} \right); \quad k_B T_{max} \simeq \frac{\alpha_G^2}{2(3\pi^2)^{2/3}} (N^{4/3} m_e c^2), \tag{1.62}$$

where $\lambdabar_e \equiv (\hbar/m_e c)$ and $\alpha_G \equiv Gm_p^2/\hbar c$.

An interesting phenomenon arises if the maximum temperature T_{max} is sufficiently high to trigger nuclear fusion in the system; then we obtain a gravitationally bound, self-sustained nuclear reactor. The condition for triggering nuclear reaction has to come from detailed study of the atomic nucleus and occurs typically at energy scales higher than $\epsilon_{nucl} \approx \eta\alpha^2 m_p c^2$, with $\eta \simeq 0.1$. The energy corresponding to the maximum temperature $k_B T_{max}$ will be larger than ϵ_{nucl} when

$$N > (2\eta)^{3/4} (3\pi^2)^{1/2} \left(\frac{m_p}{m_e} \right)^{3/4} \left(\frac{\alpha}{\alpha_G} \right)^{3/2} \approx 4 \times 10^{56} \tag{1.63}$$

for $\eta \simeq 0.1$. The corresponding condition on mass is $M > M_*$, where

$$M_* \approx (2\eta)^{3/4} (3\pi^2)^{1/2} \left(\frac{m_p}{m_e} \right)^{3/4} \left(\frac{\alpha}{\alpha_G} \right)^{3/2} m_p \approx 4 \times 10^{32} \text{ gm}, \tag{1.64}$$

which is comparable with the mass of the smallest stars observed in our universe. The mass of the Sun, for example, is $M_\odot = 2 \times 10^{33}$ gm.

Comparison of relations (1.59) and (1.64) shows that the number of stars $N_* \simeq (M_g/M_*)$ in a typical galaxy will be given by the combination of fundamental constants $N_* = \alpha^{7/2}\alpha_G^{-1/2}(m_e/m_p)^{1/4} \simeq 10^{12}$. Typical galaxies indeed have approximately $10^{11} - 10^{12}$ stars, although there is a fair amount of spread in this number. The mean distance between stars in a galaxy will be $d_{star} \approx (R_{gal}/N_*^{1/3}) \approx$ 1 pc. A star like the Sun, with a radius of 10^{11} cm, located at a distance of 10 pc, will subtend an angle of about ~ 1 milliarcsecond; it is clear that most stars will look like point objects.

1.5.4 Existence of H–R Diagram for Stars

Once the nuclear reactions occur at the hot central region of the gas cloud, its structure changes significantly. If the transport of this energy to the outer regions is through photon diffusion, then the opacity of matter will play a vital role in determining the stellar structure. In particular, the opacities determine the relation between the luminosity and the mass of the star.

A photon with mean free path $l = (n\sigma)^{-1}$, random walking through the plasma, will have $N_{coll} \simeq (R/l)^2$ collisions in traversing the radius R. This will take the time $t_{esc} \simeq (lN_{coll}/c) \simeq (R/c)(R/l)$ for the photon to escape. The luminosity of a star L is the ratio between the radiant energy content of the star E_γ and t_{esc}. Because $E_\gamma \simeq (aT^4)R^3$, where $a = (\pi^2 k_B^4/15\hbar^3 c^3) = (\pi^2/15) \simeq 1$ (in units with $k_B = \hbar = c = 1$), we find that $L = (aR^3T^4l/R^2) \simeq RT^4l$. For a wide class of stars, we may assume that the central temperature $T \simeq (GMm_p/R)$ is reasonably constant because nuclear reactions – which depend strongly on T – act as a thermostat. If Thomson scattering dominates, then $\sigma = \sigma_T$, and we get

$$L \simeq \frac{RT^4}{\sigma_T n} \simeq \frac{T^4 R^4}{\sigma_T N} \simeq \frac{G^4}{\alpha^2}m_p^5 m_e^2 M^3 \simeq 10^{34} \text{ erg s}^{-1}\left(\frac{M}{M_*}\right)^3. \quad (1.65)$$

If, on the other hand, the plasma is only partially ionised, we should use the opacities in expression (1.45), with $l \propto T^{7/2}n^{-2} \propto T^{7/2}R^6M^{-2}$, and we have

$$L \propto RT^4l \propto R^7 T^{15/2}M^{-2} \propto M^{11/2}R^{-1/2}.$$

Taking $R \propto M$ gives $L \propto M^5$. It is convenient to define the surface temperature T_s of the star by the relation $L \propto R^2 T_s^4$, so that $T_s \propto L^{1/4}R^{-1/2} \propto L^{1/4}M^{-1/2}$. Combining this result with the relation $M \propto L^{1/5}$, when the interior is only partially ionised, we get $T_s \propto L^{1/4}L^{-1/10} \propto L^{3/20}$. On the other hand, if Thomson scattering dominates with $L \propto M^3$, we get $T_s \propto L^{1/12}$. When the stars are plotted in a log T_s– log L plane, (called the *H–R diagram*) we expect them to lie within the lines with slopes $(3/20) = 0.15$ and $(1/12) \simeq 0.08$. The observed slope is ~ 0.13.

The temperature inside a star varies from approximately 10^8 K at the core to approximately a few thousand degrees Kelvin at the surface. The physical conditions, equation of state, and the opacity of matter vary significantly inside a star even when it is in a steady state. Further, when the stars evolve because of different nuclear processes, the response of matter to changing physical conditions can be dramatically different. We shall encounter many of these features in Volume II of this series.

The condensation of stars from galactic matter cannot also be a totally efficient phenomenon, and we do expect a fair amount of matter to be distributed in the galaxy in different forms. This constitutes the interstellar medium (ISM) in which structures of very different densities and temperatures exist in pressure equilibrium. In our galaxy, the ISM contributes a mass of $\sim 10^9 M_\odot$ and has a pressure of approximately $P = nk_B T \simeq 10^{-12}$ dyn cm^{-2}. There exist a hot diffuse component ($T \simeq 10^6$ K, $n \simeq 10^{-3}$ cm^{-3}), a warm ionized component ($T \simeq 8000$ K, $n \simeq 10^{-1}$ cm^{-3}), a warm neutral component ($T \simeq 5000$ K, $n \simeq 10^{-1}$ cm^{-3}), a cold neutral component ($T \simeq 80$ K, $n \simeq 10$–100 cm^{-3}), and giant molecular clouds ($T \simeq 10$ K, $n \simeq 10^2$–10^5 cm^{-3}) in pressure equilibrium in the ISM. There are also processes that strongly couple the stars and the ISM. Consider, for example, the region around a hot star with $L = 3.5 \times 10^{36}$ erg s^{-1}, and $T_s \simeq 3 \times 10^4$ K. The number of ionizing photons $\dot{N}_\gamma$ (with $\nu > \nu_I$) emitted by such a star can be estimated from the Planck spectrum and will be approximately 3×10^{48} s^{-1}. In Subsection 1.4.4 we saw [see the discussion following relation (1.50)] that the volume of the region ionised by such a flux is given by $V = (\dot{N}_\gamma / \alpha n_0^2)$. Using $n_0 \simeq 10$ cm^{-3}, we find that matter will be fully ionised for a region of radius $R = (3V/4\pi)^{1/3} \simeq 10$ pc. At a distance of 5 pc from the star, relation (1.50) gives $(1 - x) \simeq 10^{-3}$, indicating nearly total ionisation. Such a local island of plasma in the ISM is called a HII region. The radiation from the plasma in HII region is one of the probes of the conditions of the ISM.

1.5.5 $\epsilon_{\mathrm{grav}} \simeq \epsilon_F$: Existence of Stellar Remnants

We saw above that stars are gravitationally bound systems in which self-sustaining nuclear reactions are taking place in the centre. For such systems, the kinetic energy and the potential energy are comparable and $Nk_B T \approx (GM^2/R)$. The process of combining four protons into a helium nuclei releases $\sim 0.03 m_p c^2$ of energy, which is $\sim 0.7\%$ of the original energy, $4m_p c^2$. Assuming that a fraction $\epsilon \approx 0.01$ of the rest-mass energy can be made available for nuclear reactions, we find that the lifetime of the nuclear burning phase of the star will be $t_{\mathrm{star}} = \epsilon M/L \approx 3 \times 10^9$ yr $(\epsilon/0.01)(M/M_*)^{-2}$ if the opacity is due to Thomson scattering. This defines the characteristic time scale in stellar evolution.

When the nuclear fuel in the star is exhausted, the gravitational force will start contracting the matter again and the density will increase. Eventually, the density will be sufficiently high so that the quantum degeneracy pressure will dominate over thermal pressure. The equilibrium condition for such a system will require the degeneracy pressure of matter to be large enough to balance gravitational pressure. Equivalently, the Fermi energy $\epsilon_F(n)$ must be larger than the gravitational potential energy $\epsilon_g \cong Gm_p^2 N^{2/3} n^{1/3}$. When the particles are nonrelativistic, $\epsilon_F(n) = (\hbar^2/2m_e)(3\pi^2)^{2/3} n^{2/3}$ and the condition $\epsilon_F \geq \epsilon_g$ can be satisfied (at equality) if

$$n^{1/3} = \frac{2}{(3\pi^2)^{2/3}} \left(\frac{Gm_p^2 m_e}{\hbar^2} \right) N^{2/3}. \tag{1.66}$$

With $n = (3N/4\pi R^3)$ and $N = (M/m_p)$, this reduces to the following mass–radius relation:

$$RM^{1/3} \simeq \alpha_G^{-1} \lambdabar_e m_p^{1/3} \simeq 8.7 \times 10^{-3} R_\odot M_\odot^{1/3}. \tag{1.67}$$

Such structures are called white dwarfs. A white dwarf with $M \simeq M_\odot$ will have $R \simeq 10^{-2} R_\odot$ and density $\rho \simeq 10^6 \rho_\odot$.

As the density increases, electrons combine with protons through inverse beta decay to form neutrons, which can provide the degeneracy pressure. Equation (1.66) is still applicable with m_e replaced with m_n; correspondingly the right-hand side of relation (1.67) is reduced by $(\lambdabar_n/\lambdabar_e) = (m_e/m_n) \simeq 10^{-3}$. Such objects – called neutron stars – will have a radius of $R \simeq 10^{-5} R_\odot$ and a density of $\rho \simeq 10^{15} \rho_\odot$ if $M \simeq M_\odot$. For such values $\mathcal{R}_{gm} \simeq 1$ and general relativistic effects are beginning to be important.

When the density is still higher, the Fermi energy has to be supplied by relativistic particles, and ϵ_F now becomes $\epsilon_F \simeq \hbar c n^{1/3}$, which scales as $\epsilon_F \propto n^{1/3}$, just like ϵ_g. Therefore the condition $\epsilon_F \geq \epsilon_g$ can be satisfied only if $\hbar c \geq Gm_p^2 N^{2/3}$ or $N \leq \alpha_G^{-3/2} \simeq N_G \alpha^{-3/2}$. The corresponding mass bound (called Chandrasekhar limit) is $M \lesssim m_p \alpha_G^{-3/2} \simeq 1 M_\odot$. (A more precise calculation gives a slightly higher value.)

If the mass of the stellar remnant is higher than $\alpha_G^{-3/2} m_p$, no physical process can provide support against the gravitational collapse. In such a case, the star will form a black hole and is likely to exert a very strong gravitational influence on its surroundings. More complicated processes can lead to the formation of black holes with masses significantly higher than stellar masses in the centres of galaxies. Whenever such a localised centre of a high gravitational field is formed in the form of neutron stars or black holes, a wide variety of new physical phenomena can take place in that vicinity, essentially involving accretion of matter. In an accretion process, the gravitational potential energy is converted into the kinetic energy of matter and dissipated as thermal radiation. Some of the very high-energy sources of radiation – both galactic and extra galactic – are generally believed to be powered by such an accretion process. On the galactic scale, accretion discs around stars can be a source of thermalised x-ray emission;

in the extragalactic domain there are objects called active galactic nuclei and quasars that have a luminosity of $\sim 10^{44}$ erg s^{-1}, which are thought to be powered by accretion discs around very massive black holes.

The dynamical aspects of different structures in the universe discussed so far are summarised in Fig. 1.1. In the top part of the figure, various cosmic structures are classified by two ratios: (ϵ_F/k_BT) and (GMm_p/Rk_BT). Objects with negligible self-gravity lie along the y axis (with $x \simeq 0$); a simple example of this is small-scale matter in the laboratory, like a piece of a metallic solid with the size of a few cubic meters, say. (Such a metal actually goes out of scale in the figure and is shown on the top left with an arrow indicating the fact that it is out of scale.) The line $x = 1$ corresponds to $k_BT = (GMm_p/R)$, implying that in these structures gravity is supported by thermal pressure. Among such objects, those with negligible Fermi energy will lie in the lower part of the figure. It can be seen that star clusters, clouds in the ISM, spiral and elliptical galaxies, clusters of galaxies, etc., all lie around the line $x = 1$ in the lower half of the diagram. For these systems, we have interpreted the mean kinetic energy as providing an equivalent temperature, when necessary. The line $y = 1$ corresponds to $\epsilon_F = k_BT$ so that the upper part of the diagram corresponds to systems dominated by degeneracy effects ($\epsilon_F > k_BT$) and the lower part is dominated by thermal effects ($\epsilon_F < k_BT$). We note that objects such as neutron stars and white dwarfs are dominated by degeneracy effects whereas the rest of astrophysical structures can be interpreted in classical terms. Finally, the line $x = y$ corresponds to $\epsilon_F = (GMm_p/R)$ and represents structures in which gravity is supported by degeneracy pressure. They lie along in the upper right half of the figure.

The bottom part of the figure describes these structures from a different and more detailed perspective. The y axis is the the dominant internal energy per particle, and the x axis denotes the mass of different structures. For unfilled circles (planets, main-sequence star, red giants, white dwarfs, neutron stars, star clusters, spiral and elliptical galaxies, and clusters of galaxies) the internal energy is gravitational potential energy per particle. The filled circles denote the components in the ISM (molecular clouds, intercloud medium and hot ionised gas, HII regions) and for these objects the internal energy is taken to be the thermal energy. In the latter case the density n and the temperature T are used to obtain an effective mass and radius by means of the relations $M = (4\pi/3)m_p n R^3$, $GMm_p/R = k_BT$. The equivalent velocity corresponding to the internal energy is shown on the y axis along the right-hand side; it is computed by the relation $\epsilon = (1/2)m_p v^2$.

The nearly horizontal line around 1 eV separates ionised hydrogen gas from neutral with the line corresponding to 50% ionization. The other three horizontal lines around 1 keV, 1 MeV, and 1 GeV demarcate energy scales corresponding to nuclear ignition, pair production of e^+e^-, and the black-hole limit. Assuming that the internal energy ϵ is comparable with gravitational self-energy

GMm_p/R, we can determine the scalings for the radius $R \propto M/\epsilon$ and density $\rho \propto M/R^3 \propto \epsilon^3/M^2$; the lines of constant ρ are given by $\epsilon \propto M^{2/3}$, which are marked for different values of density by dashed lines.

The unbroken curve on the left is obtained when Fermi energy is equated to gravitational energy and corresponds to the $x = y$ line in the upper part of the figure. We see that neutron stars, white dwarfs, and even large planets lie along this line and also that neutron stars are close to the $GM/c^2 R = 1$ line, which decides whether general relativistic effects are important. For the planet, the Fermi energy is comparable with the atomic binding energy and gravity is just marginally important; this fact is indicated by a small vertical line near the left top. The main-sequence stars lie away from the degeneracy line and are essentially thermally supported; they are also, of course, above the line for nuclear ignition.

To the right-hand side of the figure we have a bending, unbroken curve that we obtain by equating the cooling time scale for plasma to the gravitational collapse time scale. Material to the left of this curve cool efficiently and we find that spiral and elliptical galaxies lie on this side; the galaxy clusters, on the other hand, lie to the right and will contain hot plasma.

In between the two extremes, some of the components of the ISM and the IGM have been marked. The molecular clouds, intercloud medium, and the coronal gas are actually in pressure equilibrium and lie along a line corresponding to $p \propto nT =$ constant. (Because, for these objects, $M \propto nR^3$ and $\epsilon \propto T \propto M/R$, it follows that $p \propto nT \propto \epsilon^4/M^2$. Lines of constant p correspond to $\epsilon \propto M^{1/2}$.)

Much of the discussion in the preceding sections is summarized in this Fig. 1.1 and Table 1.1, which also shows the large dynamic range spanned by the astrophysical systems, with, for example, the density varying from 10^{-30} g cm^{-3} in the IGM to 10^{15} g cm^{-3} in the neutron star.

1.6 Detecting the Photons

The universe has been studied in a wide variety of wave bands from very long waves ($\simeq 10$ m) to ultrahigh-frequency γ-ray bands. We now summarize the main sources in various wave bands.

1.6.1 Role of Earth's Atmosphere

To begin with, it must be noted that the energy levels of atoms and molecules have an important implication for observational astronomy. Ground-based observations can detect only radiation that can penetrate through the Earth's atmosphere. The atoms of most elements have energy levels of the order of $E_0 \approx (1/2)\alpha^2 m_e c^2 \approx 10$ eV. Using the relation between photon energy and wavelength, $(E/1 \text{ eV}) \approx (\lambda/12345)^{-1}$ Å, we conclude that photons with $\lambda \lesssim 10^3$ Å will be absorbed by the atmosphere, leading to ionisation of the upper layers.

Table 1.1. *Structures in the universe*

Object	Mass (gm)	Radius (cm)	$\sigma = (GM/R)^{1/2}$ (km s^{-1})	$\rho = 3M/4\pi R^3$ (gm cm^{-3})
Jupiter	2×10^{30}	6×10^9	47	2.3
Sun	2×10^{33}	7×10^{10}	470	1.4
Red giant	$(2\text{–}6) \times 10^{34}$	10^{14}	37–63	$(4.8\text{–}14.3) \times 10^{-9}$
White dwarf	2×10^{33}	10^8	3×10^4	5×10^8
Neutron star	3×10^{33}	10^6	1.4×10^5	7×10^{14}
Global cluster	1.2×10^{39}	1.5×10^{20}	8.4	8.5×10^{-23}
Open cluster	5×10^{35}	3×10^{19}	0.3	4.4×10^{-24}
Spiral	$2 \times (10^{44}\text{–}10^{45})$	$(6\text{–}15) \times 10^{22}$	150–300	$(14\text{–}22) \times 10^{-26}$
Elliptical	$2 \times (10^{43}\text{–}10^{45})$	$(1.5\text{–}3) \times 10^{23}$	30–210	$(0.14\text{–}1.8) \times 10^{-26}$
Group	4×10^{46}	3×10^{24}	300	3.5×10^{-28}
Cluster	2×10^{48}	1.2×10^{25}	10^3	2.7×10^{-28}
Universe	$7.5 \times 10^{55}\Omega h^{-1}$	$10^{28}h^{-1}$	$2.2 \times 10^5 \Omega^{1/2}$	$1.8 \times 10^{-29}\Omega h^2$

Further, the rotational- and the vibrational-energy levels of molecules such as H_2O and CO_2 (which exist in the atmosphere) fall within the IR band; this causes the IR radiation also to be absorbed by the atmosphere, although to somewhat lesser degree than the higher energy radiation. Because of these effects, the ground-based observations are essentially limited to visible ($\lambda \approx 3000\text{–}6000$ Å, $\nu \approx 10^{15}\text{–}5 \times 10^{14}$ Hz) and radio ($\lambda > 1$ cm, $\nu < 3 \times 10^{10}$ Hz) waves.

There is, however, another limitation arising from the fact that very long wavelength radiation ($\lambda \gtrsim 100$ m) cannot propagate through the plasma in the ionosphere and is reflected back. Consider an electromagnetic wave that moves the electrons in a plasma (relative to ions) by a small distance δx along the x axis. This deposits a charge $Q \simeq e(nA\delta x)$ on a fictitious surface of area A perpendicular to the x axis. This charge density, in turn, will lead to an electric field $E_x \simeq 4\pi(Q/A) \simeq 4\pi en\delta x$ that acts on the electrons in this small volume, pulling them back. Such a restoring force, proportional to displacement, gives electrons a characteristic frequency of oscillation (called plasma frequency):

$$\omega_p = \left(\frac{4\pi e^2 n}{m}\right)^{1/2} = 5.64 \times 10^4 \, \text{Hz}\left(\frac{n}{1 \, \text{cm}^{-3}}\right)^{1/2}. \tag{1.68}$$

Waves with frequencies lower than the plasma frequency cannot propagate through a plasma as the electrons can redistribute themselves sufficiently quickly to cancel the field of such an electromagnetic wave. We can estimate the number density n of electrons in the ionosphere by equating the ionisation rate that is due to solar radiation with the recombination rate, as done above for HII regions. This gives an electron density of approximately 4×10^5 cm^{-3}; the corresponding plasma frequency is approximately $\nu_p = (\omega_p/2\pi) = 6$ MHz. Thus we cannot observe radio waves with frequencies lower than $\sim$6 MHz, corresponding to wavelengths larger than about $\sim$50 m. To obtain information about all other wavelength regimes, it is necessary to make observations at high altitudes: from balloons, aircrafts, spacecrafts, satellites, etc. We now consider each band separately.

1.6.2 Radio

$(\lambda = 3$ cm–10 m, $\quad \nu \simeq 3 \times 10^7$–$10^{10}$ Hz, $\quad T \simeq 10^{-3}$–0.5 K$)$.

Several discrete sources (supernova remnants, radio galaxies, quasars, etc.) emit radio waves essentially because of synchrotron process. As a specific example, consider the diffuse radio background in our galaxy that is due to synchrotron radiation from electrons in the ISM. Observations indicate that the spectrum of the electrons is given by $(dN/dE) \simeq 3 \times 10^{-11} E^{-3.3}$ particles cm^{-3}, if E is measured in giga–electron-volts. If the magnetic field is approximately 6×10^{-6} G, the relation (1.22) will predict a volume emissivity of

$$j_\nu \simeq 3 \times 10^{-38} \left(\frac{\nu}{10 \text{ MHz}} \right)^{-1.1} \text{erg s}^{-1} \text{ cm}^{-3} \text{ Hz}^{-1}. \qquad (1.69)$$

This is close to the observed background emission. Over a line of sight of 1 kpc, this will give a flux of 10^{-20} W m^{-2} rad^{-2} Hz^{-1}. Similar power-law radio spectra are seen in the case of radio galaxies, quasars, etc., and are thought to be due to synchrotron radiation.

Along the spiral arms of galaxies, there exist clouds of HII regions that emit thermal bremsstrahlung radiation with a relatively flat spectrum. For example, the HII regions in Orion have $T \simeq 10^4$ K, $n_i \simeq 2 \times 10^3$ cm^{-3}, and an effective line-of-sight thickness of 6×10^{-4} pc. From Eq. (1.17) we can estimate the flux to be approximately 3×10^{-21} erg m^{-2} s^{-1} Hz^{-1} rad^{-2} = 300 Jansky, where 1 Jansky $\equiv$ 1 Jy $= 10^{-26}$ W m^{-2} sr^{-1} Hz^{-1}. This is typical of emission from HII regions. The total diffuse radio emission from the galaxy is in the range 10^{29}–10^{33} W.

Radio galaxies are another main class of radiators in this band with a complicated pattern of emission. The radiation usually arises from two blobs on either side of the central galaxy with a separation ranging from 3 kpc to 1 Mpc. The power from these radio sources is quite high: 10^{33}–10^{39} W; special models are needed to explain this emission. The flux that is due to such a source at a distance of 3000 Mpc will be approximately 10^{-16} W m^{-2} sr^{-1}.

Radio observations can also detect the presence of neutral hydrogen in the universe through the 21-cm radiation discussed in Subsections 1.4.1 and 1.4.2. Because $h\nu \ll k_B T$ in most cases involving 21-cm radiation, g/Z in Eq. (1.51) is essentially the ratio of spin states $g_1/(g_1 + g_0)$, which is 3/4 for the hydrogen hyperfine transition. The total intensity from an optically thin column of length L will be (for $h\nu \ll k_B T$)

$$I_\nu \cong j_\nu L \cong \frac{3}{16\pi}\left(\frac{hc}{\lambda}\right)Q_{21cm}(nL)\phi_\nu. \tag{1.70}$$

The corresponding brightness temperature $T_B(\nu) \equiv (\lambda^2 I_\nu/2k_B)$ is given by

$$T_B(\nu) = \frac{3}{32\pi}\left(\frac{hc\lambda}{k_B}\right)Q_{21}N\phi(\nu), \tag{1.71}$$

where N is the column density of the hydrogen along the line of sight. The observed form of $T_B(\nu)$ will contain information about N as well as the features of the source (such as its motion) that lead to the broadening of the line.

Neutral hydrogen at a redshift of z can be detected by observations at the wavelength of $21(1 + z)$ cm, which is in the radio band for sources with the redshift in the range of, say, $z \simeq 0$–10.

The faintest detectable flux in this band is ~ 1 mJy; there are $\sim 10^6$ discrete sources in the sky up to this level. The diffuse background in the radio band arises from our galactic disk, halo, and unresolved extragalactic radio sources. The background flux varies from approximately 3×10^4 Jy at 100 cm to 6×10^5 Jy at 10^4 cm.

1.6.3 Microwave and Submillimeter

$$(\lambda = 0.02\text{–}3 \text{ cm}, \quad \nu \simeq 10^{10}\text{–}3 \times 10^{12} \text{ Hz}, \quad T = 0.5\text{–}300 \text{ K})$$

The discrete sources in this band are usually dust clouds, hydrogen gas, and quasars. The background radiation in this band, however, is of tremendous theoretical importance and has been studied extensively. (In fact, this band has the maximum intensity of background radiation.) It turns out that the major component of the background radiation in the microwave band can be fitted very accurately by a thermal spectrum at a temperature of about ~ 2.7 K. It seems reasonable to interpret this radiation as a relic arising from the early, hot phase of the evolving universe. The νB_ν for this radiation peaks at a wavelength of 1 mm and has a maximum intensity of 5.3×10^{-7} W m^{-2} rad^{-2} over the entire sky. The intensity per square arcsecond of the sky is approximately 1.33×10^{-17} W m^{-2} arc sec^{-2}.

Microwave radiation is also a sensitive probe of the structure-formation scenario along the following lines: The gravitational potential that is due to a density perturbation $\delta\rho = \bar{\rho}\delta$ in a region of size R will be $\phi \propto \bar{\rho}\delta R^2$. In an expanding universe $\bar{\rho} \propto a^{-3}$ and $R \propto a$ and the perturbation δ grows as $\delta \propto a$ [see discussion following Eq. (1.55)], making ϕ constant in time. Because photons climbing out of a potential well of size ϕ will lose energy and undergo a redshift

$(\Delta\nu/\nu) \approx (\phi/c^2)$, we would expect to see a temperature anisotropy in the microwave radiation of the order of $(\Delta T/T) \approx (\Delta\nu/\nu) \approx (\phi/c^2)$. The largest potential wells would have left their imprint on the cosmic background radiation at the time of decoupling of radiation and matter. The galaxy clusters constitute the deepest gravitational potential wells in the universe from which the escape velocities are $v_{clus} \approx (GM/R)^{1/2} \approx 10^3$ km s^{-1}. This will lead to a temperature anisotropy of $\Delta T/T \approx (v_{clus}/c)^2 = 10^{-5}$. Such a temperature perturbation has indeed been observed in microwave background radiation, vindicating the ideas for structure formation.

1.6.4 Infrared

$$(\lambda \simeq 8000 \text{ Å–}0.01 \text{ cm}, \quad \nu \simeq 3 \times 10^{12}\text{–}10^{14} \text{ Hz}, \quad T = 300\text{–}4000 \text{ K})$$

Several interesting astrophysical processes contribute in this band, the most prominent being dust, which, when irradiated by some luminous source and heated to the temperature range of 400–4000 K emits IR. However, this is one of the most difficult ranges of frequencies to study owing to the enormous opacity of Earth's atmosphere as well as contamination that is due to emission from interstellar and interplanetary dust. In addition, hydrogen and dust clouds in our galaxy and outside galaxies as well as star-forming regions in our galaxy contribute to this band. The near-IR band of $(1\text{–}10) \times 10^{-4}$ cm will also receive contributions from the redshifted light associated with the initial epoch of galaxy formation. There have been several attempts to subtract out the galactic contamination and obtain the extragalactic *IR* flux. Although firm upper bounds are available, there is still substantial uncertainty about the actual shape of the background *IR* spectrum. The faintest detectable flux is $\sim$1.0 Jy and there are about $\sim$10^4 discrete sources up to this limit.

1.6.5 Optical and Ultraviolet

$$(\lambda \simeq 100\text{–}8000 \text{ Å}, \quad \nu \simeq 8 \times 10^{14}\text{–}3 \times 10^{16} \text{ Hz}, \quad T = (4000\text{–}3 \times 10^4 \text{ K})$$

This range of wavelengths includes visible, *UV*, far *UV*, and even what may be called soft x-ray. Most of the astronomical observations are still carried out in the optical band, in which we can reach up to 10^{-6} Jy. Stars, galaxies, and quasars contribute dominantly in this band; there are $\sim$10^{10} discrete sources in the sky. This spectral region also gets a large amount of line radiation from atomic gaseous systems.

Given the luminosity L and the radius R of a star, its effective surface temperature is determined by the equation $L = (4\pi R^2)(\sigma T_{eff}^4)$. For the Sun, this gives $T_{eff} \approx 5500$ K; the Planckian radiation corresponding to this temperature will peak around $\lambda \approx 5500$ Å (in the visible band). The flux of radiation from such a star located at a distance of 10 pc will be approximately $F = 3 \times 10^{-10}$ W m^{-2}. It is conventional in astronomy to use a unit called the bolometric magnitude, m, to indicate the flux where F and m are related approximately by $\log F \cong -8 - 0.4(m-1)$ when F is measured in Watts per square meter. When the flux changes by 1 order of magnitude, the magnitude changes by 2.5. A sunlike star at 10 pc will have a magnitude of ~ 4.6.

Starlight also contributes to the sky background in the optical band. A solid angle $d\Omega$ will intercept a volume $(1/3)R^3 d\Omega$ of our galaxy if R is the radius of the galaxy. If the number density of bright stars with luminosity $L_\odot$ is $n(\simeq 0.1$ pc$^{-3})$, then the flux per steradian is $[(1/3)n R^3 L_\odot/4\pi(R/2)^2] \simeq (1/3\pi)n L_\odot R \simeq 2.4 \times 10^{-5}$ W m^{-2} rad^{-2} if $R = 10$ kpc. Using 1 rad$^2 \simeq 4 \times 10^{10}$ arc sec^2, we get a sky brightness that is due to integrated starlight of approximately 6×10^{-16} W m^2 arc sec^{-2}. In other words, the sky background that is due to integrated starlight provides ~ 21 mag arc sec^{-2}.

We have seen above that a galaxy consists of $\sim 10^{11}$ stars and could be located at distances ranging from 1 to 4000 Mpc. At 10 Mpc, such a galaxy will subtend an angle of approximately $\theta \simeq (2R/d) \simeq 200''$ and will have a flux of approximately 3×10^{-11} W m^{-2} if the size of the galaxy $R_{\text{gal}} \simeq 10$ kpc. The surface brightness of the galaxy will be $\sim 10^{-16}$ W m^{-2} arc sec^{-2} or, equivalently, ~ 21 mag arc sec^{-2}. These galaxies also contribute to the background light in the optical band. Repeating the above analysis we did for stars with $L = 10^{11} L_\odot$, $R \simeq 4000$ Mpc, and $n \simeq 1$ Mpc^{-3}, we get a background of 2×10^{-17} W m^{-2} arc sec^{-2}, equivalent to 24.5 mag arc sec^{-2}.

We still do not have conclusive evidence that suggests the existence of a smooth background in this band. As there is a large amount of contamination from zodiacal light, backscattering of radiation from interstellar gas, and hot stars in the field of view, these observations are quite difficult.

1.6.6 X Ray and γ Ray

$$(\lambda = 3 \times 10^{-3}\text{--}100 \text{ Å}, \quad \epsilon \simeq 0.12\text{--}400 \text{ keV}, \quad T = 3 \times 10^4\text{--}10^9 \text{ K})$$

Because x rays and γ rays are strongly absorbed in the Earth's atmosphere, observations in this wave band need to be carried out from outside the atmosphere from x-ray satellites. Equivalent temperatures of $\sim 10^8$ K are required for producing hard x-rays. Besides in the central cores of stars, such high energies can be usually found only in binary stars and in supernova remnants. The accretion of matter from one star to a compact companion can lead to the production of x-rays; so can the explosion of a supernova.

The process of accretion works along the following lines. When a mass m falls from infinite distance to a radius R, in the gravitational field of a massive object with mass M, it gains the kinetic energy $E \simeq (GMm/R)$. If this kinetic energy is converted into radiation with efficiency ϵ, then the luminosity of the accreting system will be $L = \epsilon(dE/dt) = \epsilon(GM/R)(dm/dt)$. The photons that are emitted by this process will be continuously interacting with the in-falling particles and will be exerting a force on the ionised gas. When this force is comparable with the gravitational force attracting the gas towards the central object, the accretion will effectively stop. The number density $n(r)$ of photons crossing a sphere of radius r, centred at the accreting object of luminosity L, is $(L/4\pi r^2)(\hbar\omega)^{-1}$, where ω is some average frequency. The rate of collisions between photons and the electrons in the ionised matter will be $[n(r)\sigma_T]$ and each collision will transfer a momentum $(\hbar\omega/c)$. Because electrons and ions are strongly coupled in a plasma, this force will be transferred to the protons. Hence the outward force on an in-falling proton at a distance r will be

$$f_{\text{rad}} \simeq (n\sigma_T)\left(\frac{\hbar\omega}{c}\right) = \left(\frac{L}{4\pi r^2}\right)\left(\frac{1}{\hbar\omega}\right)\sigma_T\left(\frac{\hbar\omega}{c}\right) = \left(\frac{L\sigma_T}{4\pi c r^2}\right). \quad (1.72)$$

This force will exceed the gravitational force attracting the proton, $f_g = (GMm_p/r^2)$, if $L > L_E$, where

$$L_E = \frac{4\pi Gm_p c}{\sigma_T}M \simeq 1.3 \times 10^{46}\left(\frac{M}{10^8 M_\odot}\right) \text{ erg s}^{-1} \quad (1.73)$$

is called the Eddington luminosity. The temperature of a system of size R radiating at L_E will be determined by $(4\pi R^2)\sigma T^4 = L_E$, that is,

$$T \simeq 1.8 \times 10^8 \text{ K}\left(\frac{M}{M_\odot}\right)^{1/4}\left(\frac{R}{1 \text{ km}}\right)^{-1/2}. \quad (1.74)$$

For a solar-mass compact ($R \simeq 1$ km) star this radiation will peak in the x-ray band.

The main extragalactic sources of x-rays are quasars and hot ionised gas in clusters of galaxies. Taking the cluster gas to be fully ionized hydrogen with a mass of approximately $M_{\text{gas}} = 10^{48}$ gm spread over a sphere of radius $R \simeq 3$ Mpc, we can estimate the number density of ions and electrons as $n_e \approx (M_{\text{gas}}/m_p V) \approx 10^{-4}$ cm^{-3}, where V is the volume of the cluster. Such a gas will be a source of thermal bremsstrahlung radiation at the rate of $L = 1.42 \times 10^{-27} n_e^2 T^{1/2}$ erg s^{-1} cm^{-3} [see Eq. (1.18)] where all quantities are in cgs units. Using the estimated values of n_e and T, we get the total luminosity from the whole cluster to be $\mathcal{L} = LV \approx 5 \times 10^{44}$ erg s^{-1}. This radiation will be in the wave band corresponding to the temperature of 10^8 K, which is in x-rays. Several such x-ray–emitting clusters have been observed.

In addition, there exists a well-defined diffuse x-ray background in the range of 1 keV to 100 MeV. Part of this background could be due to unresolved pointlike sources and another part may be due to hot ($T \simeq 10^9$ K) diffuse, intergalactic plasma. In the range 3–50 keV it can be fitted by an optically thin thermal bremsstrahlung at the temperature 40 ± 5 keV.

No object in the universe is hot enough to produce high-energy γ rays by thermal radiation. The γ rays are produced by accretion of matter on compact objects and by the collision of high-energy particles in the cosmic rays with the nuclei of atoms in our galaxy.

Figure 1.2 summarizes the electromagnetic spectra of the universe. In the top part of the figure the y axis is the flux per logarithmic band, $F \equiv \nu F_\nu$ in units of watts per square meter. The corresponding bolometric magnitude m, related to flux by $\log F = -8-0.4(m-1)$, is shown along the y axis on the right. The

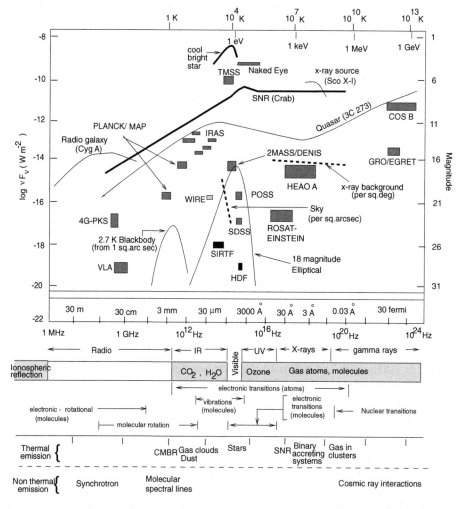

Fig. 1.2.   This figure summarises the variety of radiative phenomena in astrophysics. The top part of the figure gives the spectra of a class of objects and the effective sensitivities reached in the surveys in different wave bands. The bottom part gives (1) the atmospheric absorption, (2) the key processes contributing to line radiation, (3) major thermal sources, and (4) main sources of nonthermal emission in different spectral bands.

main x axis is marked in both frequency and wavelength at the bottom and in equivalent temperature, obtained by $k_B T = h\nu$, along the top. The bottom part of the figure gives several general features discussed above. The first panel below the x axis divides the frequency range into different wave bands and indicates the major components of the atmosphere that absorb the radiation in each band. The next panels indicates the processes which can lead to line emission in each of the wave bands. Finally, the main sources of thermal and nonthermal radiation in different wavebands are summarized in the last two panels.

In the main figure, the typical spectra from different sources as well as the detection limit of different instruments in various bands are shown. A bright star, close to $m = 1$ and peaked in optical, has the maximum flux. Other sources are (1) a typical supernova remnant marked SNR (Crab), (2) a radio galaxy Cygnus-A, (3) a quasar (3C273) with emission in several wave bands, (4) an 18th-magnitude elliptical galaxy, and (5) a bright–x-ray source Sco X-1. In addition to these sources, also marked are (1) the flux from the cosmic microwave background radiation from one square arc second of the sky, (2) x-ray background per square degree that is due to unresolved sources, and (3) the sky background per square arc second in the optical band. The filled rectangular boxes are the detection limits of different probes operating in various bands. Many of these, like IRAS, HEAO, ROSAT-Einstein, GRO-EGRET, COS-B, and of course, the Hubble space telescope, are satellite-based instruments. Some of the points marked in the optical band like the Hubble deep field (HDF), come from integrated–pencil-beam kind of surveys, whereas many other limits are applicable to wider-band surveys. This figure summarizes our current technological capabilities as well as the expected flux in different astrophysical contexts.

Exercise 1.1

Why the rest of the book? Arguments similar to the ones given in this chapter are appealing as they *seem* to capture the essence of the physics behind each of the phenomena. Investigate carefully whether this is indeed true. In particular, ask (1) Are the scaling relations obtained by such arguments trustworthy? (2) Are the numerical estimates trustworthy – especially because, for example, $(2\pi)^3 \approx 10^2$? (3) Do these arguments have predictive power? [Answers: (1) The scaling relations are usually trustworthy. (2) No. In several places, the numerical estimates can be wrong by more than an order of magnitude. (3) No. None of these effects were ever derived by such arguments. However, once the correct result is known from painstaking, rigorous, mathematical analysis – which we shall encounter in rest of the book – such arguments can be provided to "explain" the "physical origin". The true value of these arguments lies in acting as useful mnemonics.]

2

Dynamics

2.1 Introduction

This chapter develops several basic ideas of dynamics, emphasizing general principles that are useful in classifying the behaviour of dynamical systems. The reader is assumed to be familiar with elementary concepts of classical mechanics. Concepts developed here will be needed in the study of special relativity (Chap. 3), statistical mechanics (Chap. 5), general relativity (Chap. 11), Sun and solar system (Vol. II), binary stars (Vol. II), and galactic dynamics (Vol. III).

2.2 Time Evolution of Dynamical Systems

Many systems encountered in nature can be described by a finite set of N real variables $[q_1(t), q_2(t), \ldots, q_i(t), \ldots, q_N(t)]$ that evolve in time. For example, in the study of two stars, moving under the influence of their mutual gravitational force, we are interested in the positions of the stars as functions of time. The position of each star can be described by three coordinates (in three-dimensional space) so that the full system can be described by a total of six functions of time. The quantities $q_i(t)$ (with $i = 1, 2, \ldots, N$) are called dynamical variables; obviously, we are free to choose any other set of N independent, single-valued functions of q_i as dynamical variables to describe the system, with the particular choice often dictated by mathematical convenience. The central problem of dynamics is related to determining the time dependence of $q_i(t)$ and studying the general characteristics of motion.

It turns out that the evolution of any physical system can be described in terms of differential equations that are (at most) second order in time in some suitably chosen variables. This implies that we could determine the functions $q_i(t)$ for all $t > t_0$ if we know the $2N$ quantities $[q_1(t_0), q_2(t_0), \ldots, q_N(t_0), \dot{q}_1(t_0), \dot{q}_2(t_0), \ldots, \dot{q}_N(t_0)]$ at $t = t_0$. In general, the initial conditions could also be specified in terms of any other set of $2N$ functions that are independent and invertible (in a unique manner) to provide the above set. Because t_0 is arbitrary, it is clear that the state of the system at any given moment can be specified uniquely by giving $2N$ independent quantities; that is, the state of the system may be described by a point in a $2N$-dimensional space, called phase space, at any given instant. The evolution of the system is described by a curve in this $2N$-dimensional space.

Although this curve is unique, the coordinates used to describe it do not have any unique significance and are essentially decided by mathematical convenience. It is conventional to introduce in this space a general set of $2N$ coordinates denoted by $[q_1(t), q_2(t), \ldots, q_i(t), \ldots, q_N(t); p_1(t), p_2(t), \ldots, p_i(t), \ldots, p_N(t)]$ that provide the same amount of information as the dynamical variables and their first-time derivatives. Because the N-dynamical variables satisfy second-order differential equations, there must exist a set of $2N$ first-order differential equations governing the evolution of the $2N$ variables $[q_i, p_j]$. The form of these equations depends on the nature of the system and on the choice of coordinates in the phase space. We are concerned with systems called Hamiltonian systems that satisfy the following criterion: For these systems, it is possible to choose (in many different ways) the coordinates in phase space such that the differential equations can be cast in the form

$$\dot{q_i} = \frac{\partial H}{\partial p_i}, \quad \dot{p_i} = -\frac{\partial H}{\partial q_i}, \tag{2.1}$$

where $H(p, q)$ is a given function called the Hamiltonian. (We will omit the indices i, etc., on q_i and merely write q for q_i, etc., when no confusion is likely to arise. Also note that we use the symbols q_i and q^i, etc., interchangeably in this chapter.) As we shall see in Section 2.4, this does *not* uniquely specify the coordinates in phase space and a wide class of them are still possible. The N quantities q_i are the dynamical variables and the second set of N quantities, p_j, are called canonical momenta.

The Hamiltonian is assumed to be an additive quantity in the following sense: The total Hamiltonian H for a system made of two noninteracting subcomponents A and B, with variables $[q^{(A)}, p^{(A)}]$ and $[q^{(B)}, p^{(B)}]$, will be taken to be the sum of the individual Hamiltonians, i.e., $H[p^{(A)}, p^{(B)}; q^{(A)}, q^{(B)}] = H^{(A)}[p^{(A)}, q^{(A)}] + H^{(B)}[p^{(B)}, q^{(B)}]$.

The time evolution of any other function, $f(p, q, t)$, defined on the phase space (and possibly depending explicitly on time) can be computed using Eq. (2.1). We have

$$\frac{df}{dt} = \frac{\partial f}{\partial t} + \sum_{i=1}^{N} \left(\frac{\partial f}{\partial q_i} \dot{q_i} + \frac{\partial f}{\partial p_i} \dot{p_i} \right) \equiv \frac{\partial f}{\partial t} + \frac{\partial f}{\partial q_i} \dot{q_i} + \frac{\partial f}{\partial p_i} \dot{p_i}, \tag{2.2}$$

where the second equality is based on the summation convention, which states that any index that is repeated in a term should be summed over. By using

Eqs. (2.1) we can write this as

$$\frac{df}{dt} = \frac{\partial f}{\partial t} + [H, f], \tag{2.3}$$

where the quantity $[H, f]$ – called the Poisson bracket between H and f – is defined as

$$[H, f] \equiv \left(\frac{\partial H}{\partial p_i} \frac{\partial}{\partial q_i} - \frac{\partial H}{\partial q_i} \frac{\partial}{\partial p_i} \right) f. \tag{2.4}$$

This is a linear first-order differential operator acting on f. In general, we can define the Poisson bracket between any two functions g and f of phase-space coordinates by the relation

$$[g, f] = \left(\frac{\partial g}{\partial p_i} \frac{\partial}{\partial q_i} - \frac{\partial g}{\partial q_i} \frac{\partial}{\partial p_i} \right) f. \tag{2.5}$$

It can be easily verified that the Poisson brackets have the following properties:

$$[f, g] = -[g, f], \tag{2.6}$$

$$[f_1 + f_2, g] = [f_1, g] + [f_2, g], \tag{2.7}$$

$$[g, f_1 f_2] = [g, f_1] f_2 + [g, f_2] f_1, \tag{2.8}$$

$$[f, [g, h]] + [g, [h, f]] + [h, [f, g]] = 0. \tag{2.9}$$

The usefulness of the Poisson brackets is closely related to the fact that the dynamical equations governing the system can be derived from a Hamiltonian. It is obvious from Eq. (2.3) that $(dH/dt) = (\partial H/\partial t) = 0$ if the Hamiltonian does not explicitly depend on time; thus the Hamiltonian is a conserved quantity for a closed system.

Exercise 2.1
Filling in the details. Prove Eqs. (2.6)–(2.9). A more transparent notation for a Poisson bracket is $D_g f \equiv [g, f]$, which shows that $[g, f]$ is the action of a differential operator (dependent on g) on f. Rewrite Eqs. (2.6)–(2.9) in this notation and interpret the results.

The dynamical equations (2.1) can also be obtained from a variational principle, which turns out to be very useful. To obtain this variational principle, let us consider the value of the quantity (called action)

$$\mathcal{A} = \int_{q_1, t_1}^{q_2, t_2} dt [p_i \dot{q}^i - H(p, q)] = \int_{q_1, t_1}^{q_2, t_2} [p_i dq^i - H(p, q) dt] \tag{2.10}$$

for all possible functions $p_i(t), q^i(t)$ that satisfy the condition $q^i(t_1) = q_1^i$, $q^i(t_2) = q_2^i$. The evolution of the system is postulated to be such that it makes the value of $\mathcal{A}$ a local extremum. Setting the variation of $\mathcal{A}$ to zero, we get

$$0 = \delta\mathcal{A} = \int_{\mathcal{P}_1}^{\mathcal{P}_2} dt \left(\delta p_i \dot{q}^i + p_i \delta \dot{q}^i - \frac{\partial H}{\partial q_i} \delta q^i - \frac{\partial H}{\partial p_i} \delta p^i \right)$$

$$= p_i \delta q^i \Big|_{\mathcal{P}_1}^{\mathcal{P}_2} + \int_{\mathcal{P}_1}^{\mathcal{P}_2} dt \left[\left(\dot{q}_i - \frac{\partial H}{\partial p_i} \right) \delta p^i - \left(\dot{p}_i + \frac{\partial H}{\partial q_i} \right) \delta q^i \right], \quad (2.11)$$

where $\mathcal{P}_1 = (q_1, t_1)$, etc. The first term vanishes because δq vanishes at the end points. The second term can vanish for arbitrary δp^i and δq^i if and only if Eqs. (2.1) are satisfied, thereby proving the equivalence of the two descriptions. Note that the variational principle only demands that the end values of the dynamical variables q_i be fixed; the momenta p_j are not constrained in any manner.

For a wide class of Hamiltonians, it is possible to invert the relation $\dot{q}_i = (\partial H / \partial p_i)$ and express p_i in terms of $\dot{q}_j$. In that case the quantity $L = p\dot{q} - H$, called the Lagrangian, can be expressed as a function of q and $\dot{q}$; i.e., $L = L(q, \dot{q})$. Then the action principle involves extremizing the quantity

$$\mathcal{A} = \int_{\mathcal{P}_1}^{\mathcal{P}_2} L(q, \dot{q}) dt \quad (2.12)$$

with respect to all functions $q(t)$ that satisfy the conditions $q(t_1) = q_1$, $q(t_2) = q_2$. The variation now gives

$$\delta\mathcal{A} = \int_{\mathcal{P}_1}^{\mathcal{P}_2} dt \left(\frac{\partial L}{\partial q_i} \delta q^i + \frac{\partial L}{\partial \dot{q}_i} \delta \dot{q}^i \right)$$

$$= \int_{\mathcal{P}_1}^{\mathcal{P}_2} dt \left[\frac{\partial L}{\partial q_i} - \frac{d}{dt} \left(\frac{\partial L}{\partial \dot{q}_i} \right) \right] \delta q^i + \left(\frac{\partial L}{\partial \dot{q}_i} \right) \delta q^i \Big|_{\mathcal{P}_1}^{\mathcal{P}_2}. \quad (2.13)$$

The second term vanishes because $\delta q = 0$ at the end points, giving

$$\frac{\partial L}{\partial q_i} = \frac{d}{dt} \left(\frac{\partial L}{\partial \dot{q}_i} \right). \quad (2.14)$$

Thus, in the case of systems for which p_i can be expressed in terms of $\dot{q}_j$, we can study the dynamics either by using the Hamiltonian $H(p, q)$ or by using the Lagrangian $L(q, \dot{q})$. Given the Lagrangian $L(q, \dot{q})$, we can obtain the Hamiltonian by first computing the quantities $p = (\partial L / \partial \dot{q})$ and then computing $H = p\dot{q} - L$ and finally expressing the $\dot{q}$'s in terms of p and q. It follows from the definition that the Lagrangian is also additive in the same sense as the Hamiltonian.

Equation (2.13) also allows us to draw another important conclusion. Suppose we evaluate the value of the action $\mathcal{A}[q_2^i, t_2; q_1^i, t_1]$ along the actual trajectory of the particle and ask how this function varies when the end point q_2^i

is varied. In this case, the first term in Eq. (2.13) vanishes as the integrand is identically zero along the actual trajectory; from the second term, we find that $(\partial \mathcal{A}/\partial q_2^i) = (\partial L/\partial \dot{q}^i) = p_2^i$. Denoting the upper limit of integration by just (q, t) and treating $\mathcal{A}[q, t]$ as a function of the end point, we have the relation

$$p_i = \frac{\partial \mathcal{A}}{\partial q_i} = \frac{\partial L}{\partial \dot{q}_i}. \tag{2.15}$$

Thus we can determine both the equations of motion and the canonical momenta from the action principle based on the Lagrangian. This result will be of use in later chapters.

The Lagrangian is not unique in the sense that it is possible to come up with very different Lagrangians that lead to the same set of differential equations (see exercises 2.3 and 2.4). One simple example of this nonuniqueness is provided by two Lagrangians that differ by a total time derivative $[dF(q, t)/dt]$ of a function $F(q, t)$ of the dynamical variables and (possibly) time. Such a total time derivative contributes to the action a quantity that is independent of the path and hence does not affect the equations of motion. But this does change the definition of canonical momenta.

Exercise 2.2
Symmetries and conservation laws: (1) Show that the quantity $E(q, \dot{q}) \equiv \dot{q}(\partial L/\partial \dot{q}) - L$ is conserved if L does not explicitly depend on time. This quantity E, called the energy of the system, is numerically the same as the Hamiltonian; note, however, that the Hamiltonian is a function of q and p whereas E is a function of q and $\dot{q}$.

(2) A Lagrangian for a free particle cannot depend on the position of the particle $\mathbf{x}$ or the direction of the velocity; so $L = L(v^2)$. Further, let us assume that the equations of motion are invariant under a Galelian transformation, defined by $\mathbf{x}_a \rightarrow \mathbf{x}_a + \mathbf{V}t$. This requires $L(v^2)$ to change (at most) by a total time derivative under Galelian transformations. Show that this is possible only if $L \propto v^2$. It is conventional to write $L = (1/2)mv^2$, where m is called the mass of the particle. Also show that for a system of interacting particles, we can maintain Galelian invariance if the Lagrangian is taken to be of the form

$$L = \sum \frac{1}{2}m_a v_a^2 - V(\mathbf{x}_1 - \mathbf{x}_2, \dots, \mathbf{x}_a - \mathbf{x}_b, \dots,), \tag{2.16}$$

where the function V depends on only the pairwise separation of the particle coordinates. (This result shows how symmetry considerations allow us to determine the form of the Lagrangian. We shall see several more examples of this feature in Chaps. 3 and 11.)

(3) A Lagrangian L, describing the interaction of N particles, depends on the coordinates $\mathbf{x}_1, \mathbf{x}_2, \dots, \mathbf{x}_a, \dots, \mathbf{x}_N$ and the corresponding time derivatives $\dot{\mathbf{x}}_a, a = 1, 2, \dots, N$ as in part (2) above. If the origin of the coordinate system is shifted by an amount $\mathbf{l}$ so that $\mathbf{x}_a \rightarrow \mathbf{x}_a + \mathbf{l}$, then the Lagrangian should remain invariant for a closed system. Show that this implies the conservation of the quantity $\mathbf{P} = \sum_a m_a \mathbf{v}_a$, called the total momentum of the system.

(4) Consider next an infinitesimal rotation of the coordinate system by an angle $\delta\Omega$ about some axis with unit vector $\mathbf{n}$. This changes the coordinate by $\delta\mathbf{x}_a = \delta\Omega \, \mathbf{n} \times \mathbf{x}_a \equiv$

$\delta\Omega \times x_a$. For a closed system, this change should not have any effect on the dynamics. Show that this implies the conservation of the quantity $J \equiv \sum m_a x_a \times v_a$, called the total angular momentum of the system. E, P, and J are all additive quantities in the same sense as the Lagrangian.

Exercise 2.3
Equivalent Lagrangians: Consider a Lagrangian $L(q, \dot{q})$ given by

$$L = \dot{q} \int^{\dot{q}} \frac{d\alpha}{\alpha^2} F(\epsilon); \quad \epsilon \equiv \left[\frac{1}{2}\alpha^2 + V(q)\right], \tag{2.17}$$

where $F(\epsilon)$ is an arbitrary nonconstant function of its argument. Show that this Lagrangian leads to the equation of motion $\ddot{q} = -(dV/dq)$, independently of the choice for F. This result illustrates that there can be several different Lagrangians that lead to the same equation of motion.

Exercise 2.4
More on equivalent Lagrangians: Given a Lagrangian $L = L(q, \dot{q})$, we can construct a modified Lagrangian L' defined by the relation

$$L' = L - \frac{d}{dt}\left(q \frac{\partial L}{\partial \dot{q}}\right). \tag{2.18}$$

Show that varying the path $q(t)$ in L', keeping the momentum $(\partial L/\partial \dot{q})$ fixed at the end points, will lead to the same equations of motion as varying the path $q(t)$ in L keeping the coordinates fixed at end points. What is the crucial difference between L and L'?

Exercise 2.5
Maupertuis' principle: For a particle moving with constant energy $E = H(p, q)$, the relevant part of the action that should be varied, is the first term in Eq. (2.10). Show that this variation can lead to the differential equation for the spatial trajectory of the particle moving with energy E.

2.3 Examples of Dynamical Systems

As examples of simple dynamical systems, we consider some important cases that find application in different areas of astrophysics. All these Lagrangians have the form of Eq. (2.16).

2.3.1 Motion under a Central Force

As the first example, let us study a particle moving in a two-dimensional plane under the action of a central force derivable from a potential $V(r)$. The dynamical variable describing the system can be taken to be the two (polar) coordinates $q_i = (r, \theta)$. The phase space is four dimensional and has the coordinates $(r, \theta; p_r, p_\theta)$ and the Hamiltonian is given by the function

$$H(p, q) = \frac{1}{2m}\left(p_r^2 + \frac{p_\theta^2}{r^2}\right) + V(r) \tag{2.19}$$

in suitable units. The equations of motion, corresponding to Eqs. (2.1), are

$$\dot{r} = \frac{p_r}{m}, \quad \dot{\theta} = \frac{p_\theta}{mr^2}, \quad \dot{p}_r = \frac{p_\theta^2}{mr^3} - V'(r), \quad \dot{p}_\theta = 0. \qquad (2.20)$$

Given the initial position in phase space, these equations can be integrated to determine the future time evolution of the system.

For this particular case, we can invert the relations $\dot{q}_i = (\partial H / \partial p_i)$ and obtain the Lagrangian:

$$L(q, \dot{q}) = \frac{1}{2}m\,(\dot{r}^2 + r^2\dot{\theta}^2) - V(r). \qquad (2.21)$$

The second-order differential equations, corresponding to Eq. (2.14), are

$$\frac{d}{dt}(m\dot{r}) = mr\dot{\theta}^2 - V'(r); \quad \frac{d}{dt}(mr^2\dot{\theta}) = 0. \qquad (2.22)$$

These equations can be integrated if we know the coordinates and their first derivatives at any given instant of time.

2.3.2 Motion in a Rotating Frame

As a second example, let us consider the motion of a particle acted on by a potential $V(\mathbf{x})$ but viewed from a frame of reference that is rotating with a constant angular velocity $\mathbf{\Omega}$. To obtain the Lagrangian for such a system, we begin with the Lagrangian in the inertial frame given by

$$L(\mathbf{x}, \dot{\mathbf{x}}) = \frac{1}{2}m\dot{\mathbf{x}}^2 - V(\mathbf{x}) \qquad (2.23)$$

and note that the velocities transform under rotation by the law

$$\mathbf{v}_{\text{inertial}} = \mathbf{v}_{\text{rot}} + \mathbf{\Omega} \times \mathbf{x}, \qquad (2.24)$$

where the subscripts inertial and rot denote the velocities in the inertial and the rotating frames, respectively. Substituting Eq. (2.24) into Eq. (2.23), we find the Lagrangian in the rotating frame to be

$$L = \frac{1}{2}m\mathbf{v}^2 + m\mathbf{v} \cdot (\mathbf{\Omega} \times \mathbf{x}) + \frac{1}{2}m(\mathbf{\Omega} \times \mathbf{x})^2 - V(\mathbf{x}), \qquad (2.25)$$

where all quantities are measured in the rotating frame and $\mathbf{v} \equiv \mathbf{v}_{\text{rot}}$. By varying this Lagrangian, we get the equations of motion:

$$m\frac{d\mathbf{v}}{dt} = -\frac{\partial V}{\partial \mathbf{x}} + 2m\mathbf{v} \times \mathbf{\Omega} + m\mathbf{\Omega} \times (\mathbf{x} \times \mathbf{\Omega}). \qquad (2.26)$$

The second term on the right-hand side is called Coriolis force, and the third term is the centrifugal force. (This analysis also shows that the Lagrangian or Hamiltonian description is not tied to inertial frames of reference and can be easily generalised for any coordinate system.) The momentum and energy are now given by

$$\mathbf{p} = \frac{\partial L}{\partial \mathbf{v}} = m\mathbf{v} + m\boldsymbol{\Omega} \times \mathbf{x}, \quad E = \frac{1}{2}mv^2 - \frac{1}{2}m(\boldsymbol{\Omega} \times \mathbf{x})^2 + V. \quad (2.27)$$

Note that rotation changes the form of the potential energy by adding to it a term $[-(1/2)m(\boldsymbol{\Omega} \times \mathbf{x})^2]$ that is quadratic in the coordinates.

For a system of particles, constituting a rigid body, the Lagrangian in Eq. (2.25) has to be summed over all the particles in the system (with the same $\boldsymbol{\Omega}$ for all particles). Then $\mathbf{v}_{\text{rot}} = \mathbf{V}$ gives the velocity of the centre of mass of the rigid body and $\boldsymbol{\Omega}$ denotes the angular velocity of rotation of the body. In that case, the kinetic energy will be

$$T = \sum \frac{1}{2}m(\mathbf{V} + \boldsymbol{\Omega} \times \mathbf{x})^2 = \sum \frac{1}{2}mV^2 + \sum m\mathbf{V} \cdot \boldsymbol{\Omega} \times \mathbf{x} + \sum \frac{1}{2}m(\boldsymbol{\Omega} \times \mathbf{x})^2,$$
$$(2.28)$$

where the summation is over all particles in the rigid body. The first term is the kinetic energy of the centre of mass motion, and the second term, which can also be expressed as $(\mathbf{V} \times \boldsymbol{\Omega}) \cdot \sum m\mathbf{x}$, vanishes when $\mathbf{x}$ is measured from the centre of mass. Therefore the rotation of the system essentially contributes only the third term to the kinetic energy. This term can be written in the form

$$T_{\text{rot}} = \frac{1}{2}I_{ik}\Omega_i\Omega_k, \quad I_{ik} = \sum m(x^2\delta_{ik} - x_ix_k). \quad (2.29)$$

The symmetric tensor I_{ik} is called moment of inertia tensor and plays a role in rotation that is similar to the one played by the mass in the case of translational motion. In the expression for T_{rot}, summation over $i, k = 1, 2, 3$ is assumed as per summation convention. The sum in the expression for I_{ik} is over all particles in the system. In the continuum limit, I_{ik} becomes

$$I_{ik} = \int d^3\mathbf{x}\, \rho(\mathbf{x})\, (x^2\delta_{ik} - x_ix_k). \quad (2.30)$$

The angular momentum of the system, measured with respect to the centre of mass of the body, is given by

$$(M)_i = \left[\sum m\mathbf{x} \times (\boldsymbol{\Omega} \times \mathbf{x})\right]_i = \sum m[x^2\boldsymbol{\Omega} - \mathbf{x}(\mathbf{x} \cdot \boldsymbol{\Omega})]_i = I_{ik}\Omega_k. \quad (2.31)$$

Any symmetric second-rank tensor can be reduced to the diagonal form by a suitable choice of axis. These directions for a given rigid body are called the principal axes of inertia and the corresponding values $I_1, I_2,$ and I_3 are called the principal moments of inertia. The moment I_1, for example, will be

$$I_1 \equiv I_{11} = \sum m(x^2 - x_1^2) = \sum m(x_1^2 + x_3^2) = \sum mx_\perp^2,$$

where $\mathbf{x}_\perp$ is the perpendicular distance to mass m from the x_1 axis. This is the usual formula for computing moments of inertia of bodies with significant degree of symmetry. Clearly T_{rot} becomes the sum of three diagonal terms in this case, and we also have $M_1 = I_1 \Omega_1$, etc., in this limit.

From the variational principle applied to the basic Lagrangian, it is trivial to verify that the angular momentum of the body changes in accordance with the law $(d\mathbf{M}/dt) = \mathbf{K}$, where $\mathbf{K} = -\sum \mathbf{x} \times \nabla V$ is called the torque, corresponding to the force $\mathbf{F} = -\nabla V$ derived from the potential $V(\mathbf{x})$. Since the time derivative of any vector $\mathbf{A}$ in the inertial and the rotating frames differs by the term $\mathbf{\Omega} \times \mathbf{A}$, the equation of motion for the angular momentum in the rotating frame is given by

$$\frac{d'\mathbf{M}}{dt} + \mathbf{\Omega} \times \mathbf{M} = \mathbf{K}, \tag{2.32}$$

where the prime on (d'/dt) is a reminder of the fact that the derivative is evaluated in the moving system of coordinates. Taking the components of this equation along the principle axis of the system, we get

$$I_1 \frac{d\Omega_1}{dt} + (I_3 - I_2)\Omega_2\Omega_3 = K_1,$$

$$I_2 \frac{d\Omega_2}{dt} + (I_1 - I_3)\Omega_3\Omega_1 = K_2, \tag{2.33}$$

$$I_3 \frac{d\Omega_3}{dt} + (I_2 - I_1)\Omega_1\Omega_2 = K_3.$$

These are called Euler's equations and govern the rotational motion of a rigid body.

As an example of the use of Eqs. (2.33), let us consider two simple cases: (1) To begin with, if the body is totally symmetric with $I_1 = I_2 = I_3$, then its free motion (corresponding to $\mathbf{K} = 0$) is trivial and is given by a uniform rotation $\mathbf{\Omega} = \text{constant}$ about some axis. (2) A more interesting case is the free motion of a system with axial symmetry, for which $I_1 = I_2 \neq I_3$. Then $\Omega_3 = \text{constant}$ and Ω_1 and Ω_2 satisfy the equation $\dot{\Omega}_1 = -\omega\Omega_2$, $\dot{\Omega}_2 = \omega\Omega_1$, where $\omega = \Omega_3(I_3 - I_1)/I_1$ is a constant. Multiplying one of the equations by i and adding to the second equation, we can easily solve the set and obtain $\Omega_1 = A \cos \omega t$; $\Omega_2 = A \sin \omega t$. The motion therefore consists of the vector $\mathbf{\Omega}$ precessing with an angular velocity ω around the symmetry axis of the body while remaining constant in magnitude. In the case in which all the principal moments of inertia are unequal, the motion can be extremely complex.

Exercise 2.6

Effects of Coriolis force: (1) Find the deflection of a freely falling body from the vertical, produced because of the rotation of the Earth, to the lowest order of approximation. (2) Determine the effect of earth's rotation on the plane of oscillation of a pendulum to the lowest order of approximation. [Answers: (1) Taking the gravitational potential to be $U = -m\mathbf{g} \cdot \mathbf{r}$, we need to solve the equation of motion $\dot{\mathbf{v}} = 2\mathbf{v} \times \mathbf{\Omega} + \mathbf{g}$, where we have ignored the quadratic terms in $\mathbf{\Omega}$. Ignoring the term that is due to rotation, solving for $\mathbf{v}$, substituting into this equation, and solving again, we get the solution as

$$\mathbf{r} = \mathbf{h} + \mathbf{v}_0 t + \frac{1}{2}\mathbf{g}t^2 + \frac{1}{3}t^3\mathbf{g} \times \mathbf{\Omega} + t^2\mathbf{v}_0 \times \mathbf{\Omega}. \tag{2.34}$$

If the z axis points vertically upwards and the x axis towards the pole and λ is a northern latitude, the net deflection is given by $y = -(1/3)(2h/g)^{3/2}g\Omega \cos \lambda$, with the negative sign indicating eastward deflection. (2) Ignoring the vertical displacement of the pendulum and terms quadratic in Ω, the equations of motions are $\ddot{x} + \omega^2 x = 2\Omega_z \dot{y}$, $\ddot{y} + \omega^2 y = -2\Omega_z \dot{x}$, where Ω is the original frequency of the pendulum and $\Omega_z = \Omega \sin \lambda$. The solution to this equation is concisely given by $x(t) + iy(t) = [x_0(t) + iy_0(t)] \exp(-i\Omega_z t)$, where $x_0(t)$, $y_0(t)$ is the trajectory of the pendulum in the absence of Earth's rotation. It is clear that the net effect of rotation is to cause a shift in the plane of rotation at the rate $\Omega_z = \Omega \sin \lambda$.]

Exercise 2.7
Feynman's plate: A thin plate is thrown into the air spinning with a slight wobble. Show that it will wobble at a rate that is approximately twice as fast as it rotates. (In Feynman's memoirs, he says that "the medallion rotates twice as fast as the wobble rate"; he has mistakenly interchanged the rotation rate and the wobble rate.)

2.3.3 The Reduced Three-Body Problem

To see the effect of rotation in a physical context, we consider the motion of three-point particles under their mutual gravitational attraction subject to the following approximations: (1) We assume that, of the three bodies, the first two, with masses m_1 and m_2, are heavy and the third one, with mass m_3, is a test particle, i.e., $m_3 \ll m_1, m_2$. (2) We take m_1 and m_2 to describe circular orbits around their common centre of mass; this orbit is assumed to be unaffected by m_3. (3) The motion of m_3 takes place in the orbital plane of m_1 and m_2. This problem (called the reduced planar three-body problem) is of interest in astrophysics because it is tractable and approximates more realistic situations. For example, the motion of an asteroid (m_3) in the mutual gravitational field of the Sun (m_1) and Jupiter (m_2) can be treated along these lines.

To simplify the notation, we make a choice of units as follows. First, we take $(m_1 + m_2)$ as the unit of mass; the two masses m_1 and m_2 can now be called $(1 - \mu)$ and μ. Next, we take the unit of length to be the constant distance (diameter of the circular orbit) between m_1 and m_2. It follows that the radii of the orbits are μ and $(1 - \mu)$ for m_1 and m_2, respectively. Finally, we choose the unit of time such that $G = 1$. From these choices, it is easy to see that the angular velocity of m_1 and m_2 is also unity.

It is convenient to use a system of axes (x, y) that corotates with m_1 and m_2 with the x axis pointing towards m_2, say. In this frame m_1 and m_2 are stationary with fixed coordinates $(-\mu, 0)$ and $(1 - \mu, 0)$ respectively. If the coordinates of m_3 are (x, y) then the equations of motion obtained from Eq. (2.26) can be easily shown to be

$$\ddot{x} - 2\dot{y} = \frac{\partial \Phi}{\partial x}, \quad \ddot{y} + 2\dot{x} = \frac{\partial \Phi}{\partial y}, \tag{2.35}$$

where

$$\Phi(x, y) = \frac{1}{2}(x^2 + y^2) + U + \frac{1}{2}\mu(1 - \mu), \tag{2.36}$$

with

$$U = \frac{1 - \mu}{\rho_1} + \frac{\mu}{\rho_2}, \quad \rho_1 = [(x + \mu)^2 + y^2]^{1/2}, \quad \rho_2 = [(x - 1 + \mu)^2 + y^2]^{1/2}. \tag{2.37}$$

By multiplying the two equations of Eqs. (2.35) by $\dot{x}$ and $\dot{y}$, respectively, and adding them, we see that

$$2\Phi - \dot{x}^2 - \dot{y}^2 = C = \text{constant}. \tag{2.38}$$

This constant C, called the Jacobi constant, is essentially the energy defined by the second equation of Eqs. (2.27). No other integral for this system is known, and hence the solutions have to be found by numerical integration. We shall discuss some characteristics of the numerical solution in Section 2.8. Right now we concentrate on the equilibrium solutions that can be obtained directly from the properties of the Lagrangian.

Given a value for the Jacobi constant C, we must satisfy the inequality $\dot{x}^2 + \dot{y}^2 = 2\Phi - C > 0$. This shows that the motion is restricted to a region bounded by the curve $\Phi = C/2$, called the Hill curve. If we rewrite Eq. (2.36) for Φ, expressing x and y in terms of ρ_1 and ρ_2, we get

$$\Phi = \frac{3}{2} + (1 - \mu)\left(\frac{1}{2} + \frac{1}{\rho_1}\right)(\rho_1 - 1)^2 + \mu\left(\frac{1}{2} + \frac{1}{\rho_2}\right)(\rho_2 - 1)^2. \tag{2.39}$$

Clearly, Φ is always positive and has a minimum value $\Phi_{\min} = 3/2$ when $\rho_1 = \rho_2 = 1$. These points (at which $\rho_1 = \rho_2 = 1$) form the vertices of an equilateral triangle with the positions of masses m_1 and m_2. It follows that if the third body is kept at either of these two points with zero velocity in the rotating frame, the configuration is stationary with the particle located at the minimum of a potential. We thus find that there exists an equilibrium solution to the restricted planar three-body problem with the three particles located at the vertices of an equilateral triangle. The configuration rotates rigidly with a constant angular velocity when viewed from the inertial frame. This case corresponds to $C = 3$ for which the Hill curve degenerates into two isolated points that are usually denoted by the symbols L_4 and L_5 (see Fig. 2.1).

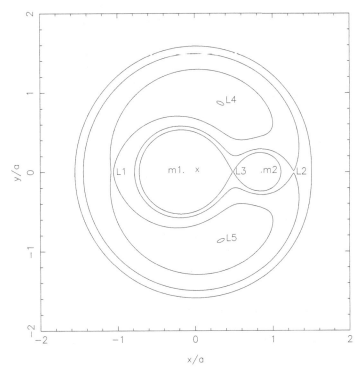

Fig. 2.1. The Lagrange points for the reduced three-body problem in a rotating frame.

More generally, the equilibrium solutions will correspond to points at which the condition $\dot{x}=0$, $\dot{y}=0$, $(\partial\Phi/\partial x)=0$, and $(\partial\Phi/\partial y)=0$ are satisfied simultaneously. These equations can be solved in a straightforward manner, and we will find that, in addition to L_4 and L_5, there are also three other points L_1, L_2, and L_3 along the line joining m_1 and m_2 at which these conditions are satisfied. It is conventional to take L_3 to be between the two masses. All these points correspond to equilibrium situations and are called Lagrange points.

The Hill curve can be thought of as an equipotential line, and its characteristics depend critically on the value of C. Figure 2.1 shows the variation of Hill curves as the value of C is increased from $C=3$. The point marked X is the centre of mass of the system with masses m_1 and m_2 where $(m_1/m_2)=5$. Close to the masses, the gravitational potential is dominated by the individual masses and has circular contours. The same result holds far away from the two masses. At intermediate scales, the contours for potentials have the shapes shown in the figure.

Exercise 2.8
Filling in the details. Prove Eqs. (2.35)–(2.39).

It is obvious that the Lagrange points L_1, L_2, and L_3 correspond to unstable equilibria and are local maxima of the potentials. For future reference, we give a fitting function for the distance to the inner Lagrangian point L_3 from the two masses:

$$l_1 = a\left[0.500 - 0.227 \log_{10}\left(\frac{m_2}{m_1}\right)\right], \quad l_2 = a\left[0.500 + 0.227 \log_{10}\left(\frac{m_2}{m_1}\right)\right],$$

$$(2.40)$$

where a is the distance between the two masses m_1 and m_2. The stability of the equilibrium solution corresponding to the Lagrange points L_4 or L_5 is more difficult to determine. Linear stability analysis can be performed by examination of the eigenvalues of the matrix of second derivatives $(\partial^2\Phi/\partial x_i \partial x_j)$. This analysis is straightforward, although tedious, and leads to the conclusion that the solutions are not linearly unstable for $\mu(1-\mu) < (1/27)$. This corresponds to the condition

$$\mu < \left(\frac{1}{2} - \sqrt{\frac{23}{108}}\right) \approx 0.0385, \qquad (2.41)$$

that is, the upper limit on the ratio of masses is $\sim 1/25$. A more complex analysis shows that the motion in the equilateral solution is actually stable if this condition is satisfied except for two specific values of the ratio μ. Thus, for all practical purposes, we can take the above criteria as the one for stability of the equilateral solution.

Exercise 2.9
Stabilising influence of rotation: A system of masses produces a gravitational potential $\phi(\mathbf{x}, t)$. Prove that such a gravitational potential cannot have a local maxima or minima at any $\mathbf{x}$ unoccupied by the mass points. Explain how the rotation, combined with gravity, can lead to a Lagrange point that is stable against small perturbations.

2.4 Canonical Transformations

It is clear from Eqs. (2.1) that the problem of Hamiltonian dynamics boils down to solving $2N$-coupled first-order ordinary differential equations. The structure of these equations will depend very much on the choice of the dynamical variables and conjugate momenta. As it may be easier to solve these equations with one particular choice rather than with another, it is important to ask whether any general formalism can be developed to facilitate solving these equations. We now discuss some of these general techniques.

It was mentioned in Section (2.2) that the form of the Eqs. (2.1) presupposes a judicious choice of coordinates in phase space. An arbitrary transformation of coordinates from the set (q, p) to some other set (q', p') will not maintain the form of Eqs. (2.1). We begin by asking the following question: What is the most general class of transformation from a set (q, p, H) to another set (Q, P, H'), that will preserve the form of equations of motion (2.1)?

Because the same dynamical equations are obtained by varying the two different actions, made from $(p\dot{q} - H)$ and $(P\dot{Q} - H')$, the two Lagrangians can differ by a total time derivative (dF/dt) of some function F, so that

$$p\dot{q} - H - (P\dot{Q} - H') = \frac{dF}{dt} \tag{2.42}$$

or

$$dF = pdq - PdQ + (H' - H)dt. \tag{2.43}$$

If $F = F(q, Q, t)$, it follows that

$$p_i = \frac{\partial F}{\partial q_i}, \quad P_i = -\frac{\partial F}{\partial Q_i}, \quad H' = H + \frac{\partial F}{\partial t}. \tag{2.44}$$

The function F is called the generating function for the canonical transformation from (q, p) to (Q, P). Given any function $F(q, Q, t)$, we can generate a transformation from one set of coordinates in phase space to another such that the structure of Eqs. (2.1) remains the same. In particular, if F has no explicit dependence on time, then the Hamiltonian does not change in its numerical value.

In the above case, we have taken the generating function to depend on old and new dynamical variables. It is also possible to have a generating function that depends on the old coordinates q and new canonical momenta P. To obtain the formulas for this case, we only have to note that Eq. (2.43) can be rewritten as

$$d(F + P_i Q_i) = p_i dq_i + Q_i dP_i + (H' - H)dt. \tag{2.45}$$

The argument of the differential on the left-hand side, expressed in terms of the variables q and P, is a new generating function $\Phi(q, P, t)$, say. Then,

$$p_i = \frac{\partial \Phi}{\partial q_i}, \quad Q_i = \frac{\partial \Phi}{\partial P_i}, \quad H' = H + \frac{\partial \Phi}{\partial t}. \tag{2.46}$$

We can also obtain similar formulas with other combinations of variables.

The existence of canonical transformations, under which the form of the dynamical equations remains invariant, denies any special status to the dynamical variables and conjugate momenta. For example, by using a generating function $F = qQ$, we can produce the transformation $Q = p$, $P = -q$, thereby interchanging the roles of coordinates and momenta.

The Poisson brackets remain invariant under the canonical transformation. Let $[f, g]_{p,q}$ be the Poisson bracket for two quantities f and g, in which the differentiation is with respect to the variables p and q, and let $[f, g]_{P,Q}$ be the bracket in which the differentiation is with respect to P and Q. Then

$$[f, g]_{p,q} = [f, g]_{P,Q}. \tag{2.47}$$

To prove this, we may argue as follows: Because time appears only as a parameter in the canonical transformation, it is sufficient to prove the invariance for quantities that do not depend explicitly on time. Let us now regard g as a

Hamiltonian for some fictitious system. Then we know that $df/dt = -[f, g]_{p,q}$. But the derivative df/dt can depend on only the properties of motion of the system and not on the choice of coordinates in phase space. Therefore the Poisson bracket $[f, g]$ is invariant under canonical transformations.

Exercise 2.10
Invariant volumes in phase space: The canonical transformation also leaves volumes of certain regions in phase space invariant. Let $dV = dq_1 \cdots dq_N dp_1 \cdots dp_N$ denote an element of volume in phase space. The integral of dV over some region of phase space represents the volume of that region. If we now replace the variables p, q with P, Q through a canonical transformation, show that the volumes of corresponding regions of space remain equal:

$$\int \cdots \int dq_1 \cdots dq_N \, dp_1 \cdots dp_N = \int \cdots dQ_1 \cdots dQ_N \, dP_1 \cdots dP_N. \quad (2.48)$$

Next prove the invariance of a larger class of phase-space integrals of the form

$$\iint \sum_i dq_i \, dp_i, \quad \iiiint \sum_{i \neq k} dq_i \, dp_i \, dq_k \, dp_k, \ldots, \quad (2.49)$$

in which the integration is over manifolds of two, four, etc., dimensions in phase space. (These are called Poincaré invariants. Because the dynamical variables and conjugate momenta get mapped to the corresponding quantities during the time evolution, it follows that the evolution of the system itself can be thought of as a canonical transformation from the initial set of variables to the final set of variables. Because the phase volume remains invariant under any canonical transformation, it follows that the volume of the phase space remains invariant during the evolution.)

Having determined the most general set of transformation that leaves the dynamical equations invariant, we now turn to the question of using this information to solve for the dynamical evolution of the system. The simplest possible Hamiltonian we can think of is one that vanishes identically in the new set of variables, that is, $H' = 0$. In this case P and Q will be constant and the evolution will be frozen in the new coordinate system. The vanishing of the new Hamiltonian can be achieved by use of a generating function S that satisfies

$$\frac{\partial S}{\partial t} + H\left(\frac{\partial S}{\partial q}, q\right) = 0. \quad (2.50)$$

We obtained this by setting $H' = 0$ in the third equation of Eqs. (2.46) and substituting for p in $H(p, q)$ by $p = (\partial \Phi/\partial q)$; it is conventional to use the symbol S rather than Φ for the generating function that makes $H' = 0$. This equation (called the Hamilton–Jacobi equation) determines the generating function; by using this function we can go to a new set of coordinates and vanishing Hamiltonian. Suppose that this equation can be solved to obtain a particular solution $S = S(t, q^i, \alpha_i)$, where α_i (with $i = 1, 2, \ldots, N$) are constants of integration. This generating function, of course, makes $H' = 0$ by Eqs. (2.44) and (2.50). We now treat α_i as the new momenta P_i (which are constant in time because $H' = 0$). From Eqs. (2.46) we see that the new coordinates are $Q_i = (\partial S / \partial \alpha_i)$. But because $H' = 0$, these coordinates should also be constant; $Q^i = \text{constant} \equiv \beta^i$, say. We thus get the equations $\beta_i = [\partial S(t, q_i, \alpha_i) / \partial \alpha_i]$, which implicitly determine the trajectories $q_i(t)$ in terms of the $2N$ constants (α_i, β_i). Thus we have reduced the problem of dynamics to one of solving the partial differential equation (2.50).

The generating function S has a simple physical meaning; it actually corresponds to the action $\mathcal{A}$ treated as a function of the coordinates and time at the end point, that is, the functional dependence of S on q and t will be the same as that of the quantity

$$A(q, t) = \int_{q_1, t_1}^{q, t} L \, dt, \tag{2.51}$$

where the integral is taken along the extremal path that is a solution to the equation of motion. To prove this, we only have to show that

$$p = \left(\frac{\partial \mathcal{A}}{\partial q}\right), \quad H = -\left(\frac{\partial \mathcal{A}}{\partial t}\right). \tag{2.52}$$

The first relation was proved earlier; see Eq. (2.15) in Section 2.2. To prove the second relation, we note that $d\mathcal{A}/dt = L$ by definition, and we can also write

$$\frac{d\mathcal{A}}{dt} = \frac{\partial \mathcal{A}}{\partial t} + \frac{\partial \mathcal{A}}{\partial q_i} \dot{q}_i = \frac{\partial \mathcal{A}}{\partial t} + p_i \dot{q}_i. \tag{2.53}$$

Comparison gives $(\partial \mathcal{A}/\partial t) = L - p_i \dot{q}_i$ or

$$\frac{\partial \mathcal{A}}{\partial t} = -H. \tag{2.54}$$

This shows that the solution to Eq. (2.50) is actually the action treated as a function of the end points.

As an example of the use of Eq. (2.50), let us consider the motion of a particle in one dimension under the action of a potential $V(x)$. The Hamiltonian for the system is

$$H(p, x) = \frac{p^2}{2m} + V(x), \tag{2.55}$$

and the Hamilton–Jacobi equation is

$$\frac{\partial S}{\partial t} + \frac{1}{2m} \left(\frac{\partial S}{\partial x}\right)^2 + V(x) = 0. \tag{2.56}$$

This partial differential equation can be solved by the method of separation of variables with the ansatz $S = -Et + A(x)$, where E is a constant. By substituting this ansatz into Eq. (2.56) and rearranging terms, we can express $A(x)$ as an integral, giving

$$S = -Et + \int dx \sqrt{2m[E - V(x)]}. \tag{2.57}$$

This gives the solution to the Hamilton–Jacobi equation in terms of the coordinate x, time t, and a constant E. We think of this function as a generating function, with the new canonical momentum identified with E. We also know that, by construction, the new Hamiltonian vanishes. From our canonical transformations in Eqs. (2.46), the new coordinates are given by $(\partial S/\partial E)$. However, because the new Hamiltonian vanishes, the new coordinates are also constant in time. Equating $(\partial S/\partial E)$ to some constant t_0, we get

$$t + t_0 = \int dx \left\{ \frac{m}{2[E - V(x)]} \right\}^{1/2}. \tag{2.58}$$

This equation provides the complete solution to our problem. Given the form of $V(x)$, we can now express x as a function of t and the two constants t_0 and E. The constants can be determined if x and $\dot{x}$ are given at some initial instant of time. We thus note that the problem of a particle, moving in one dimension under the action of an arbitrary potential, can be reduced to a quadrature.

Exercise 2.11
Potentials for which period is independent of amplitude: (1) Consider a particle of mass m moving in a potential $V(x) \propto |x|^n$, where $n > 0$. The energy of the particle is such that it oscillates between $x = a$ and $x = -a$. How does the period of oscillation scale with the amplitude a? (2) In particular, note that the period is independent of amplitude for such a power-law potential only if $n = 2$. There exists, however, very many different potentials $V(x)$ (which are not pure power laws) such that a particle can oscillate about a minimum of the potential with a period independent of amplitude. Give a general characterisation of such a potential. (3) One particular example of such a potential is $V(x) \propto (x^{-2} + x^2)$. Argue why the period of oscillation should be independent of amplitude in such a potential without doing any calculation. (Hint: Think of a two-dimensional harmonic oscillator.)

As a second example, consider the motion of a particle in a plane under the action of a central force with potential $V(r)$. By use of the Hamiltonian in Eq. (2.19), the Hamilton–Jacobi equation becomes

$$\frac{\partial S}{\partial t} + \frac{1}{2m}\left[\left(\frac{\partial S}{\partial r}\right)^2 + \frac{1}{r^2}\left(\frac{\partial S}{\partial \theta}\right)^2 \right] + V(r) = 0. \tag{2.59}$$

We now substitute the ansatz $S = -Et + J\theta + A(r)$ into the Hamilton–Jacobi equation and obtain

$$A = \int dr \left[2m\left(E - V - \frac{J^2}{2mr^2} \right) \right]^{1/2}, \tag{2.60}$$

so that

$$S = -Et + J\theta + \int dr \left[2m \left(E - V - \frac{J^2}{2mr^2} \right) \right]^{1/2}. \qquad (2.61)$$

Once again $S(t; r, \theta; E, J)$ can be thought of as a generating function for a canonical transformation depending on the old coordinates r, θ and new momenta E, J. Because the new Hamiltonian vanishes, the new coordinates, given by the derivatives of S with respect to E and J, must be constants. With these constants denoted by t_0 and θ_0, the equations $t_0 = (\partial S/\partial E)$ and $\theta_0 = (\partial S/\partial J)$ are reduced to

$$t + t_0 = \int dr \left(\frac{m}{E - V - J^2/2mr^2} \right)^{1/2}, \qquad (2.62)$$

$$\theta - \theta_0 = \int dr \frac{J/r^2}{[2m(E - V - J^2/2mr^2)]^{1/2}}. \qquad (2.63)$$

These equations provide the complete solution to the problem. Equation (2.62) relates r to t, and Eq. (2.63) relates θ and r. There are also four arbitrary constants, J, E, t_0, and θ_0, that need to be determined in terms of r, θ, $\dot{r}$, $\dot{\theta}$ at an initial instant. The problem has again been reduced to quadrature.

The two examples above show how the method of the Hamilton–Jacobi equation works in practice. It must, however, be remembered that the success was due to the fact that we could separate the variables in the partial differential equation. It is this feature that allowed us to reduce the partial differential equation to an ordinary differential equation (for the function A) and to solve the problem. In the second example, the choice of coordinates also played a key role. If we had used the Cartesian coordinates x, y rather than the polar coordinates r, θ, the problem would have been less tractable. There has been extensive study of the separability of the Hamilton–Jacobi equation in different coordinate systems. It can be shown that there are 11 real coordinate systems that satisfy the criteria for separability, but only a few of them are useful in studying interesting physical phenomena. If the equation is not separable, then we must resort to numerical techniques, and it is better to start with Eqs. (2.1) rather than with the Hamilton–Jacobi equation.

Exercise 2.12
Separability of Hamilton–Jacobi equation in an unfamiliar coordinate system: Consider the motion of a particle under the action of a potential $U = (\alpha/r) - Fz$, which is a combination of a Coulomb field and a uniform field. (1) To study the motion of such a particle, it is convenient to use *parabolic* coordinates (ξ, ϕ, η) that are related to the standard cylindrical coordinates (ρ, ϕ, z) by the equations $z = (\xi - \eta)/2$ and $\rho = \sqrt{\xi\eta}$. Show that the Hamiltonian for a particle can be written in these coordinates as

$$H = \frac{2}{m} \frac{\xi p_\xi^2 + \eta p_\eta^2}{\xi + \eta} + \frac{p_\phi^2}{2m\xi\eta} + U(\xi, \eta, \phi). \qquad (2.64)$$

(2) Show that the Hamilton–Jacobi equation is separable in these coordinates for any potential of the form

$$U = \frac{a(\xi) + b(\eta)}{\xi + \eta} = \frac{a(r+z) + b(r-z)}{2r}. \tag{2.65}$$

The potential $U = (\alpha/r) - Fz$ has this form, with $a(\xi) = \alpha - (F\xi^2/2), b(\eta) = \alpha + (F\eta^2/2)$. Hence obtain the complete solution to the problem.

The fact that there are only a limited number of cases in which a complete solution of the problem is possible gives rise to two important questions that are central to the investigations in modern mechanics. Under what circumstances can we reduce a given dynamical problem to quadrature? Can we say anything about the general characteristics of motion for systems that can be reduced to quadrature? We now turn to the discussion of these questions.

2.5 Integrable Systems

To understand the nature of systems for which the problem of motion can be reduced to quadrature, we need to introduce the concept of an integral of motion. Note that, if the equations of motion can be integrated, then we can obtain the functions $q_i = q_i(t; q_i^0, p_i^0)$ and $p_i = p_i(t; q_i^0, p_i^0)$ in terms of the initial conditions (q_i^0, p_i^0). In principle we can invert these relations and express the $2N$ quantities q_i^0 and p_i^0 in terms of q_j, p_j and t. By their very construction, these $2N$ functions of q_j, p_j and t are constants, that is, their values remain the same as the system evolves. Out of these $2N$ constants, one constant is trivial as the origin of time is immaterial for a closed system. We can always change t to $(t + t_0)$ in a parameterised curve $\mathcal{C}(t)$, keeping the physics invariant. Thus any dynamical system has $(2N - 1)$ nontrivial constants of motion that restrict the system to a $[2N - (2N - 1)] = 1$ dimensional surface in phase space. This one-dimensional surface is the trajectory $\mathcal{C}(t)$.

Most of the $(2N - 1)$ constants of motion described above are, however, useless in the sense that these constants can be determined only after the equations of motion are integrated (and hence are of no help in studying the dynamics). For certain physical systems there might exist a subset of k constants of motion $f_A(q, p), A = 1, 2, \ldots, k$ that are independent of t and that can be detected from general symmetry considerations. Such constants of motion, which are independent of time, are called integrals of motion. Even among the integrals of motion, all are not of equal importance. Certain integrals of motion isolate the time evolution of the system to a restricted region in phase space, thereby simplifying the physics. Such integrals are called isolating integrals. The integrals of motion that do not isolate the motion to a well-defined region in phase space are called nonisolating.

To illustrate these concepts, let us consider the example of a two-dimensional harmonic oscillator with the Lagrangian

$$L = \frac{1}{2}(\dot{x}^2 + \dot{y}^2) - \frac{1}{2}(\omega_x^2 x^2 + \omega_y^2 y^2). \tag{2.66}$$

In this particular case, we can easily integrate the equations of motion and obtain the solutions

$$x(t) = A \cos(\omega_x t + \epsilon_x), \quad y(t) = B \cos(\omega_y t + \epsilon_y), \tag{2.67}$$

where $(A, B, \epsilon_x, \epsilon_y)$ are the four constants that are to be determined in terms of $[x(0), y(0), p_x(0), p_y(0)]$. From the above solution we get

$$\begin{aligned} x &= A \cos(\omega_x t + \epsilon_x), & p_x &= -A\omega_x \sin(\omega_x t + \epsilon_x), \\ y &= B \cos(\omega_y t + \epsilon_y), & p_y &= -B\omega_y \sin(\omega_y t + \epsilon_y). \end{aligned} \tag{2.68}$$

We can now express $(A, B, \epsilon_x, \epsilon_y)$ in terms of (x, p_x, y, p_y, t):

$$A^2 = x^2 + \frac{p_x^2}{\omega_x^2}, \quad B^2 = y^2 + \frac{p_y^2}{\omega_y^2}, \tag{2.69}$$

$$\epsilon_x = -\omega_x t - \tan^{-1}\left(\frac{p_x}{x\omega_x}\right), \quad \epsilon_y = -\omega_y t - \tan^{-1}\left(\frac{p_y}{y\omega_y}\right). \tag{2.70}$$

These are all constants of motion, among which ϵ_x and ϵ_y depend on t explicitly; by using the arbitrariness in the origin of t, we can eliminate one of them.

Because the Lagrangian depends on both x and y we expect only the energy to be a conserved quantity. This will define a three-dimensional region of four-dimensional phase space in which the motion takes place. The above analysis shows, however, that it is possible to write two independent integrals A and B for the system because the x and y motions are uncoupled. We have the integrals (A, B) or, equivalently, (E_x, E_y), where

$$E_x \equiv \frac{1}{2}\omega_x^2 A^2 = \frac{p_x^2}{2} + \frac{1}{2}\omega_x^2 x^2, \quad E_y \equiv \frac{1}{2}\omega_y^2 B^2 = \frac{p_y^2}{2} + \frac{1}{2}\omega_y^2 y^2. \tag{2.71}$$

[The total energy E is just $(E_x + E_y)$ and hence is not an independent integral.] These integrals clearly isolate the motion in the $x - p_x$ and $y - p_y$ planes. Because there are two integrals, the motion is confined to a $4 - 2 = 2$ dimensional surface in the phase space. The equations of this surface, given by Eqs. (2.71), can be written parametrically as

$$p_x = \sqrt{2E_x} \sin\alpha, \quad x = \sqrt{\frac{2E_x}{\omega_x^2}} \cos\alpha \quad (0 \le \alpha < 2\pi), \tag{2.72}$$

$$p_y = \sqrt{2E_y} \sin\beta, \quad y = \sqrt{\frac{2E_y}{\omega_y^2}} \cos\beta \quad (0 \le \beta < 2\pi). \tag{2.73}$$

This is the two-dimensional surface of a torus on which the motion takes place.

We can ask whether the existence of two more integrals ϵ_x and ϵ_y puts any further restriction on the nature of the motion. From solutions (2.67), it is clear that

$$\frac{\epsilon_x}{\omega_x} - \frac{1}{\omega_x}\cos^{-1}\left(\frac{x}{A}\right) = \frac{\epsilon_y}{\omega_y} - \frac{1}{\omega_y}\cos^{-1}\left(\frac{y}{B}\right) \qquad (2.74)$$

or

$$\cos^{-1}\left(\frac{x}{A}\right) - \frac{\omega_x}{\omega_y}\cos^{-1}\left(\frac{y}{B}\right) = \text{constant} \equiv c. \qquad (2.75)$$

This quantity c is clearly another integral of motion. But – in general – this does not isolate the region where the motion takes place any further, because $\cos^{-1} z$ is a multiple-valued function. To see this more clearly, let us write Eq. (2.75) as

$$x = A\cos\left\{c + \frac{\omega_x}{\omega_y}\left[\cos^{-1}\left(\frac{y}{B}\right) + 2\pi n\right]\right\}, \qquad (2.76)$$

where $\text{Cos}^{-1} z$ (with an uppercase C) denotes the principle value. For a given value of y we will get an infinite number of x's as we take $n = 0, \pm 1, \pm 2 \ldots$. Thus, in general, the curve in Eq. (2.76) will fill a region in the (x, y) plane.

A special situation arises if (ω_x/ω_y) is a rational number. In that case, the curve closes on itself after a finite number of cycles. Then c is also an isolating integral and we have three isolating integrals: (E_x, E_y, c). The motion is confined to a closed (one-dimensional) curve on the surface of the torus.

We began with a system with two dynamical variables x and y so that the phase space is four dimensional. In general, such a system will have only energy as a conserved quantity; thus, in general, we would have expected the motion to take place in a three-dimensional subspace of the full four-dimensional phase space. However, in this particular case, we had two integrals of motion, E_x and E_y. This restricts the motion to a $(4 - 2) = 2$ dimensional subspace of the phase space, which happens to be a torus. For generic values of ω_x and ω_y, the motion fills the surface of the torus. But if the ratio of the frequencies is rational, then the motion is restricted to a space of one lower dimension: in this case, to a $(2 - 1) = 1$ dimensional curve.

All these features happen to be quite generic for a large class of systems called integrable Hamiltonian systems. A system with N-dynamical variables is called integrable if there exist N independent, isolating integrals of motion with mutual Poisson brackets vanishing. (In the above example, $N = 2$ and E_x and E_y are the two isolating independent integrals of motion.) This immediately implies that the motion is confined to a $(2N - N) = N$ dimensional subspace of the phase space. More importantly, this subspace always happens to be an N-dimensional torus. Given the N integrals of motion, it is also possible to reduce the problem to quadrature by a suitable canonical transformation. We now explore these ideas in detail.

Let us consider a system with a $2N$-dimensional phase space and N independent integrals of motion, $F_m(q, p)$, where $m = 1, 2, \ldots, N$. We assume that the Poisson brackets $[F_m, F_n]$ vanish for all values of n and m. The existence of these integrals suggests the following approach to the dynamics of such a system. Given the N integrals of motion, $F_i(q, p)$ with $i = 1, 2, \ldots, N$, it is convenient to make a canonical transformation to a new set of variables (Q, P), with $P_j = F_j(q, p)$. Because F's are constants of motion, we must have

$$\dot{P}_j = -\frac{\partial \mathcal{H}}{\partial Q_j} = \dot{F}_j = 0. \tag{2.77}$$

Therefore the new Hamiltonian $\mathcal{H}$ cannot depend on Q_i and is purely a function of the P's. Let $(\partial \mathcal{H}/\partial P_j) = \Omega_j(P)$. Because the P's are independent of t, the Ω's are also independent of t, allowing the equations of motion $\dot{q}_j = (\partial \mathcal{H}/\partial P_j) = \Omega_j$ to be integrated to give

$$Q_j = \Omega_j t + \delta_j. \tag{2.78}$$

To solve the problem completely, we only have to express the old coordinates q in terms of the new coordinates Q. We do this by demanding that our transformation should be canonical with a suitable generating function that can be obtained as follows. The constancy of F_j can be expressed by the set of equations $F_j(q, p) = f_j$, where the f_j's are N constants. Inverting this equation, we can obtain p_i in terms of q's and f's as $p_i = p_i(q_j, f_k)$. The generating function can be taken to be

$$S(q, P) = \int^q p_k(q, f) \, dq^k = S(q, f), \tag{2.79}$$

with a summation over k assumed. The new coordinates Q_i are given by

$$Q_i = \frac{\partial S(q, f)}{\partial f_i}, \tag{2.80}$$

which relates the new and the old coordinates and contains N arbitrary constants. Equation (2.78) gives the new coordinates as a function of time and other N constants. Combining the two, we can find the trajectories $q_i(t)$ in terms of $2N$ arbitrary constants of motion, which completely solves the problem.

Thus, for integrable systems, the procedure for solving the dynamical problem can be summarised as follows: (1) Invert the relation $F_j(q, p) = f_j$ to obtain $p_i = p_i(q_j, f_k)$; here F_j's are the N given integrals of motion. (2) Integrate $p_i(q_j, f_k)$ over dq_i to determine the function $S(q, f)$ [see Eq. (2.79)]. (3) Use Eq. (2.80) to relate the new coordinates Q_j to the old coordinates q_i and N constants f_i. (4) The new coordinates are linear functions of time [see Eq. (2.78)] that allow us to determine the time evolution of old coordinates $q_i(t)$.

The existence of N integrals restricts the motion to an N-dimensional manifold $\mathcal{M}$. To understand the topology of this manifold $\mathcal{M}$, let us construct the following

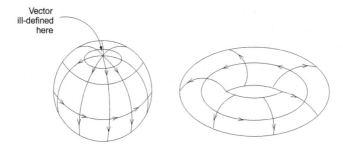

Fig. 2.2. Two parallelisable vector fields can exist on a two torus but not on a two sphere.

N vector fields V_a:

$$V_a \equiv \left(\frac{\partial F_a}{\partial p^i}, -\frac{\partial F_a}{\partial q^j} \right). \tag{2.81}$$

Note that a runs from 1 to N, specifying which vector field; for each value of a, the vector field has $2N$ components given by Eq. (2.81) so that we now have N vectors (each having $2N$ components) defined at every location in the phase space. In the submanifold $\mathcal{M}$ defined by the equation $F_i(q, p) = f_i$, the vector fields are smooth and independent; moreover, the $V_a's$ are tangential to $\mathcal{M}$ in the sense that V_a is perpendicular to the normal of $\mathcal{M}$:

$$V_a \cdot n^b = \left(\frac{\partial F_a}{\partial p^i}, -\frac{\partial F_a}{\partial q^j} \right) \cdot \left(\frac{\partial F_b}{\partial q^i}, \frac{\partial F_b}{\partial p^j} \right) = [F_a, F_b] = 0 \tag{2.82}$$

for all a, b. If we restrict ourselves to bounded motion in which the region of phase space accessible to the particle is finite, the manifold $\mathcal{M}$ will be compact. There is a theorem in topology that states that a compact manifold that is parallelisable with N smooth independent vector fields must be an N torus. Figure 2.2 illustrates the situation for $N = 2$ and shows why the manifold could not, for example, be a sphere.

These tori are called invariant tori because an orbit starting out in one remains in it forever. A possible coordinate system on the phase space is now given by the set (Q, P), with $P = f$ labelling the torus and Q being the coordinates on the torus. This choice, however, is not yet unique because we could have started with any other function of the original integrals of the motion. There is, however, a very natural coordinatisation of the torus $\mathcal{M}$ called the action-angle variables (I, Θ). The definition of these coordinates is based on the observation that the generating function in Eq. (2.79) is generally multivalued because the momenta can be multivalued. In other words, the generating function can depend on the path chosen to define the integral in Eq. (2.79). But if two different paths in $\mathcal{M}$ are continuously deformable into one another, then the generating function must be the same for these two because it satisfies the (local) Hamilton–Jacobi equation. We may say that S is locally single valued. It immediately follows that for all closed paths on $\mathcal{M}$ that can be shrunk to a point, $S = 0$. On an N torus, there exist N independent irreducible closed paths C_a that cannot be deformed to a point. The action variables I_a are defined by

$$I_a(f) = \frac{1}{2\pi} \int_{C_a} p_i(q, f) \, dq^i, \tag{2.83}$$

where the integral is taken over these irreducible closed curves. This defines the N coordinates I_a in terms of the f's and vice versa. Because the Hamiltonian is one of the f's, we can invert this relation and express the Hamiltonian as a function of the I_a's: $H = H(I)$.

Once we have the actions I_a defining the torus, the coordinates on the torus are defined to be the angles $\Theta_a = [\partial S(q, I)/\partial I_a]$. [Note that we are now expressing the action S in terms of (q, I) by expressing the f's in terms of I.] It can be easily verified that as we traverse the closed path C_a, the angle Θ_a changes by 2π while all other angles Θ_b with $b \neq a$ remain the same. Because of this feature, the canonical transformation from q, p to Θ, I expresses the original coordinates as periodic functions of Θ with period 2π:

$$q = \sum_{\mathbf{m}} q_{\mathbf{m}}(I) \exp(i\mathbf{m} \cdot \Theta), \quad p = \sum_{\mathbf{m}} p_{\mathbf{m}}(I) \exp(i\mathbf{m} \cdot \Theta). \tag{2.84}$$

To illustrate the above concepts, let us work out two examples by using action-angle variables. We start with the case of the two-dimensional harmonic oscillator with the Hamiltonian

$$H(q, p) = \frac{p_1^2}{2} + \frac{p_2^2}{2} + \frac{\omega_1^2 q_1^2}{2} + \frac{\omega_2^2 q_2^2}{2}. \tag{2.85}$$

The two constants of motion are given by

$$F_1 = \frac{p_1^2}{2} + \frac{\omega_1^2 q_1^2}{2}, \quad F_2 = \frac{p_2^2}{2} + \frac{\omega_2^2 q_2^2}{2}. \tag{2.86}$$

To effect a canonical transformation from (q, p) to (Θ, I), we have to find the actions defined by Eq. (2.83). They are

$$I_a = \frac{1}{2\pi} \oint p_i dq^i = \frac{1}{2\pi} \oint \sqrt{2\left(F_a - \frac{1}{2}\omega_a^2 q_a^2\right)} \, dq_a = \frac{F_a}{\omega_a} \quad (a = 1, 2). \tag{2.87}$$

Because the original Hamiltonian is just $H = F_1 + F_2$, we find that the Hamiltonian can be expressed in terms of the actions as

$$H(I) = F_1 + F_2 = I_1\omega_1 + I_2\omega_2. \tag{2.88}$$

From the general theory of action angle-variables, we know that

$$\dot{\Theta}_a = \frac{\partial H}{\partial I_a} = \omega_a = \text{constant}, \tag{2.89}$$

so that the trajectories are given by

$$\Theta_a = \omega_a t + \epsilon_a. \tag{2.90}$$

All that remains is to express the old coordinates q_a in terms of Θ_a by use of relations (2.79) and (2.80). In this case, we have

$$S(q_1, Q_2; f_1, f_2) = \int \sqrt{2\left(f_1 - \frac{1}{2}\omega_1^2 q_1^2\right)} \, dq_1 + \int \sqrt{2\left(f_2 - \frac{1}{2}\omega_2^2 q_2^2\right)} \, dq_2. \tag{2.91}$$

Because $\Theta_a = (\partial S/\partial I_a) = \omega_a(\partial S/\partial f_a)$, we have

$$\Theta_a = \omega_a \int \frac{dq_a}{\sqrt{2(f_a - \frac{1}{2}\omega_a^2 q_a^2)}} = \sin^{-1}\left(\frac{\omega_a q_a}{\sqrt{2f_a}}\right) \quad (a = 1, 2). \tag{2.92}$$

By using Eq. (2.90) we find that

$$q_a = \sqrt{\frac{2f_a}{\omega_a^2}} \sin(\omega_a t + \epsilon_a), \tag{2.93}$$

which solves the problem completely.

As a second example, let us consider the motion of a particle in two-dimensions under the action of a central potential. For the Hamiltonian

$$H = \frac{p_r^2}{2m} + \frac{p_\theta^2}{2mr^2} + V(r), \tag{2.94}$$

the two integrals of motion are

$$F_1 = H, \quad F_2 = p_\theta \ (= mr^2\dot{\theta}), \tag{2.95}$$

leading to the actions

$$I_2 = \frac{1}{2\pi} \oint p_\theta \, d\theta = \frac{F_2}{2\pi} \int_0^{2\pi} d\theta = F_2, \tag{2.96}$$

$$I_1 = \frac{1}{2\pi} \oint p_r \, dr = \frac{1}{\pi} \int \left\{ 2m\left[F_1 - \frac{I_2^2}{2mr^2} - V(r) \right] \right\}^{1/2} dr, \tag{2.97}$$

where the second integral is to be taken between the turning points in r. Given any $V(r)$, we can perform this integral and find I_1 in terms of I_2 and F_1. Inverting this relation, we can express the Hamiltonian (which is the same as F_1) in terms of I_1 and I_2. The rest of the analysis will proceed as described above. As a particular case, let us consider the motion under the potential

$$V(r) = -\frac{\alpha}{r} - \frac{\beta}{r^2}. \tag{2.98}$$

The two actions are

$$I_2 = F_2, \tag{2.99}$$

$$I_1 = \frac{1}{2\pi} \oint p_r \, dr = \frac{1}{2\pi} \oint \left(F_1 - \frac{F_2^2}{2mr^2} + \frac{\alpha}{r} + \frac{\beta}{r^2} \right)^{1/2} \sqrt{2m} \, dr$$

$$= -\sqrt{-2m\beta + F_2^2} + \alpha \left(\frac{m}{2|F_1|} \right)^{1/2}. \tag{2.100}$$

Solving for $H = F_1$ in terms of I_1, I_2, we get

$$H(I_1, I_2) = \frac{1}{2} \frac{m\alpha^2}{\left(I_1 + \sqrt{I_2^2 - 2m\beta} \right)^2}. \tag{2.101}$$

From the general theory, we know that the corresponding angular coordinates are linear functions of time: $\Theta_a = \omega_a t + \epsilon_a$, where the two frequencies are given by

$$\omega_1 = \frac{\partial H}{\partial I_1} = -\frac{m\alpha^2}{\left(I_1 + \sqrt{I_2^2 - 2m\beta} \right)^3}, \tag{2.102}$$

$$\omega_2 = \frac{\partial H}{\partial I_2} = -\frac{m\alpha^2}{\left(I_1 + \sqrt{I_2^2 - 2m\beta} \right)^3} \frac{I_2}{\sqrt{I_2^2 - 2m\beta}}. \tag{2.103}$$

We can easily express the old coordinates r, θ in terms of Θ_a as in the case of a two-dimensional oscillator and thus solve the problem completely.

Expressions (2.102) and (2.103) have some interesting features. We know that, in general, the motion will fill the surface of a two torus. An exception arises when the two frequencies are commensurable, i.e, their ratio is a rational number. This ratio is given by

$$\frac{\omega_1}{\omega_2} = \frac{\sqrt{I_2^2 - 2m\beta}}{I_2} = \sqrt{1 - \frac{2m\beta}{I_2^2}}. \tag{2.104}$$

If $\beta \neq 0$, then, this ratio can be rational only if $(1 - 2m\beta/I_2^2)^{1/2}$ is rational. If that happens, then the motion will be confined to a closed curve on the surface of the two torus, making the motion one dimensional rather than two dimensional. This reduction in the dimensionality of motion implies the existence of another isolating integral. To find this integral explicitly, we can proceed as in the case of the two-dimensional oscillator. By using Eq. (2.63) we can find the orbit of the particle to be

$$\frac{1}{r} = \frac{\alpha m}{J^2}\left(1 - \frac{2\beta m}{J^2}\right)^{-1} + Q\cos\left[\left(1 - \frac{2\beta m}{J^2}\right)^{1/2}(\theta - \theta_0)\right], \quad (2.105)$$

where $Q \equiv [1 - (2m\beta/J^2)]^{-1}\{\alpha^2m^2/J^4 + (2mE/J^2)[1 - (2m\beta/J^2)]\}^{1/2}$ and θ_0 is another constant. Solving for θ_0, we get

$$\theta_0 = \theta + \left(1 - \frac{2\beta m}{J^2}\right)^{-1/2}\cos^{-1}\left[\frac{1}{Q}\left\{\frac{1}{r} - \frac{\alpha m}{J^2}\left(1 - \frac{2\beta m}{J^2}\right)^{-1}\right\}\right]. \quad (2.106)$$

This quantity θ_0 is clearly an integral of the motion. If it is single valued, there will exist another isolating integral of motion, reducing the motion from two- to one-dimensional space. In general, $\cos^{-1} z$ will introduce a $2\pi n$ factor, giving

$$\theta = \theta_0 - \left(1 - \frac{2\beta m}{J^2}\right)^{-1/2}\left[\text{Cos}^{-1}\frac{1}{Q}\left\{\frac{1}{r} - \frac{\alpha m}{J^2}\left(1 - \frac{2\beta m}{J^2}\right)^{-1}\right\} \pm 2\pi n\right]$$

$$= \theta_0 - \left(1 - \frac{2\beta m}{J^2}\right)^{-1/2}\text{Cos}^{-1}\frac{1}{Q}\left\{\frac{1}{r} - \frac{\alpha m}{J^2}\left(1 - \frac{2\beta m}{J^2}\right)^{-1}\right\}$$

$$\pm 2\pi\left(1 - \frac{2\beta m}{J^2}\right)^{-1/2}n, \quad (2.107)$$

where $\text{Cos}^{-1}z$ (with an uppercase C) denotes the principal value. If $\beta \neq 0$, then θ can be single valued only if $(1 - 2\beta m/J^2)^{1/2}$ is a rational number. In this case we have an extra isolating integral.

The nature of the orbit in real space reflects the dimensionality of the motion in phase space. In general, the projection of phase-space motion (which fills a two-dimensional torus) will fill a two-dimensional region in the r, θ plane. But when the frequencies are commensurable, the trajectory spans a one-dimensional region in phase space and its projection will be a closed orbit in the r, θ plane.

The situation is more interesting in the case of $\beta = 0$, which corresponds to the Kepler problem. In this case, the two frequencies are identical and the motion always spans a one-dimensional region of the phase space; correspondingly, the orbit is a closed curve in real space. Once again, there must exist an additional integral of motion, which can be found along the following lines.

Consider a particle moving under a central force $\mathbf{f}(r) = -V'(r)\hat{\mathbf{r}}$. By direct differentiation, it is easy to see that

$$\frac{d}{dt}(\mathbf{p} \times \mathbf{J}) = -mf(r)r^2\frac{d\hat{\mathbf{r}}}{dt}, \quad (2.108)$$

where $\mathbf{p}$, $\mathbf{J}$, and $\mathbf{r}$ are the momentum, the angular momentum, and the position of the particle. For the special case of inverse-square-law force, $\alpha \equiv f(r)r^2$ is a constant and Eq. (2.108) implies that

$$\mathbf{A} \equiv \mathbf{p} \times \mathbf{J} - \frac{\alpha m}{r}\mathbf{r} \tag{2.109}$$

is conserved. This conserved vector, called the Runge–Lenz vector, has three components, but it satisfies the constraints

$$A^2 = 2mJ^2E + \alpha^2 m^2, \quad \mathbf{A} \cdot \mathbf{J} = 0. \tag{2.110}$$

The first one shows that the magnitude of $\mathbf{A}$ can be expressed in terms of other constants of motion; and the second one shows that $\mathbf{A}$ lies in the orbital plane. These two constraints reduce the number of independent components in $\mathbf{A}$ from three to one. This extra, independent constraint lowers the dimensionality of motion from two to one.

In this particular case, we can easily find the orbit of the particle by taking the dot product of $\mathbf{A}$ with $\mathbf{r}$ and by using the identity $\mathbf{r} \cdot (\mathbf{p} \times \mathbf{J}) = \mathbf{J} \cdot (\mathbf{r} \times \mathbf{p}) = J^2$. This gives

$$\mathbf{A} \cdot \mathbf{r} = Ar \cos\theta = J^2 - \alpha mr, \tag{2.111}$$

or, in more familiar form,

$$\frac{1}{r} = \frac{\alpha m}{J^2}\left(1 + \frac{A}{\alpha m}\cos\theta\right). \tag{2.112}$$

We see that $\mathbf{A}$ is in the direction of the major axis of the ellipse.

For future reference, the explicit solution for the Kepler problem in different cases is given here. By using the first equation of Eqs. (2.110), we can write the trajectory of Eq. (2.112) as

$$\frac{1}{r} = \frac{1}{a|(\epsilon^2 - 1)|}(1 + \epsilon\cos\theta); \quad (\epsilon^2 \neq 1), \tag{2.113}$$

where the semimajor axis a and the eccentricity ϵ of the ellipse are given by

$$a = \frac{\alpha}{2|E|}, \quad \epsilon^2 = 1 + \frac{2EJ^2}{m\alpha^2}. \tag{2.114}$$

The relation between the coordinate r and time t can be expressed in parametric form, depending on whether E is positive or negative:

$$r = a(\epsilon\cosh\xi - 1), \quad t = \left(\frac{ma^3}{\alpha}\right)^{1/2}(\epsilon\sinh\xi - \xi) \quad (E > 0),$$

$$r = a(1 - \epsilon\cos\xi), \quad t = \left(\frac{ma^3}{\alpha}\right)^{1/2}(\xi - \epsilon\sin\xi) \quad (E < 0). \tag{2.115}$$

The above examples clearly illustrate that, if the system is integrable, then the motion takes place on an N-dimensional torus and – in general – fills the surface. The exceptional situation arises when the ratio between any two of the frequencies is a rational number. In such a case, the motion is confined to a space with a lower dimension than N. When such a situation arises, there will exist another isolating integral of motion. In the case of the two-dimensional harmonic oscillator, this is given by Eq. (2.75); for the Kepler problem, it is the Runge–Lenz vector.

Exercise 2.13
Hodograph for Kepler problem: The curve traced by the tip of the velocity vector of a particle moving under some force law is called a hodograph. Show that the hodograph of a particle moving in the Kepler problem is a circle in the case of bounded motion and part of a circle in the case of unbounded motion.

Exercise 2.14
Kepler motion of a binary system: Consider two stars of masses m_1 and m_2 orbiting around the common centre of mass (in an elliptical path) under their mutual gravitational attraction. The plane of the orbit is inclined at an angle i to the line of sight connecting the observer to the focus. From the Doppler shift of the radiation emitted by star 1, it is possible to determine the velocity of that star along the line of sight. Show how this information will help us to determine the combination

$$C = \frac{m_2^3 \sin^3 i}{(m_1 + m_2)^2}. \tag{2.116}$$

This procedure is of importance in determining the mass of an invisible companion in a binary stellar system.

2.6 Adiabatic Invariance

The Hamiltonian of any physical system depends on some of the parameters that characterise the system and are treated as constants during the time evolution. For example, the frequencies ω_a of a two-dimensional oscillator or the mass of the Sun in planetary motion belong to this set. There are physical situations in which such parameters may change slowly with time rather than remain strictly constant (for example, the mass of the Sun may slowly decrease because of solar wind). We would like to understand how the slow change in the parameters of the problem affects our solution. It turns out that the formalism developed above – in terms of action-angle variables – comes in handy in this case.

To illustrate this idea let us consider a one-dimensional system executing a finite motion with period T. Let λ be a parameter in the Hamiltonian that varies slowly with time in the sense that $T(d\lambda/dt) \ll 1$. If λ were strictly constant, the energy E of the system would be conserved; when λ varies slowly, E would not be conserved but we expect the rate of change of energy with time $\dot{E}$ to be small. If we now average this rate over a period T, the rapid variations will be averaged out and the resulting value of $\langle \dot{E} \rangle$ will give the secular slow change in E that is due to the variation $\dot{\lambda}$. Because the variation of λ leads to an average variation of E, we can express the averaged E as some function of λ; equivalently, we can express this relation as the constancy of some function of E and λ. Such functions, which remain constant when the parameters of the system are varied slowly, are called adiabatic invariants. We now show that the actions I_a are adiabatic invariants and remain unchanged when the parameters of the system are varied slowly.

Let $H[q, p, \lambda(t)]$ be the Hamiltonian of the system that depends on the slowly varying parameter λ. The rate of change of energy of the system is

$$\frac{dE}{dt} = \frac{\partial H}{\partial t} = \frac{\partial H}{\partial \lambda}\frac{d\lambda}{dt} = \frac{\partial H}{\partial \lambda}\dot{\lambda}. \tag{2.117}$$

On the right-hand side, $\dot{\lambda}$ is a slowly varying quantity but $(\partial H/\partial\lambda)$ depends on the rapidly varying quantities q and p. To determine the steady, secular change in E we average this expression over the period of motion. Because $\dot{\lambda}$ varies slowly, we can take it outside the averaging and write

$$\left\langle\frac{dE}{dt}\right\rangle = \frac{d\lambda}{dt}\left\langle\frac{\partial H}{\partial\lambda}\right\rangle, \tag{2.118}$$

where we treat λ as a constant in the function $(\partial H/\partial\lambda)$ that is being averaged. Expressing the time average explicitly as

$$\left\langle\frac{\partial H}{\partial\lambda}\right\rangle = \frac{1}{T}\int_0^T \frac{\partial H}{\partial\lambda}\,dt \tag{2.119}$$

and transforming from dt to dq by $dt = dq/[(\partial H/\partial p)]$, we find that

$$\left\langle\frac{dE}{dt}\right\rangle = \left(\frac{d\lambda}{dt}\right)\frac{\oint(\partial H/\partial\lambda)\,dq/(\partial H/\partial p)}{\oint dq/(\partial H/\partial p)}. \tag{2.120}$$

The circle on the integral sign denotes that the integration is taken over the complete range of variation of the coordinates during one period. Because these integrals are taken along paths in which the Hamiltonian has a constant value E, we can express the momentum as a definite function of the variable coordinate q and the two constant parameters E and λ. Setting $p = p(q; E, \lambda)$ and differentiating the equation $H(q, p, \lambda) = E$ with respect to λ, we get $(\partial H/\partial\lambda) = -[(\partial H/\partial p)][(\partial p/\partial\lambda)]$ or

$$\frac{(\partial H/\partial\lambda)}{(\partial H/\partial p)} = -\frac{\partial p}{\partial\lambda}. \tag{2.121}$$

By using this in Eq. (2.120) and writing the integrand of the denominator as

$(\partial p/\partial E)$, we get

$$\left\langle \frac{dE}{dt} \right\rangle = -\left(\frac{d\lambda}{dt} \right) \frac{\oint (\partial p/\partial \lambda)\, dq}{\oint (\partial p/\partial E)\, dq}, \tag{2.122}$$

that is,

$$\oint \left(\frac{\partial p}{\partial E} \left\langle \frac{dE}{dt} \right\rangle + \frac{\partial p}{\partial \lambda} \frac{d\lambda}{dt} \right) dq = 0. \tag{2.123}$$

This is equivalent to the condition $\langle dI/dt \rangle = 0$, where

$$I(E, \lambda) \equiv \oint p(q; E, \lambda) \frac{dq}{2\pi}, \tag{2.124}$$

and the integral is taken over a path with given E and λ. This result can be directly generalised to systems with more degrees of freedom and shows that the actions introduced by Eq. (2.83) are adiabatic invariants.

Note that adiabatic invariance of actions requires that the change in the parameter be small during a complete cycle of motion, that is, we must have $T\dot{\lambda} \ll 1$, where T is the period. We can imagine situations in which this condition is violated even though the external parameter changes slowly in some other pre-specified manner. Adiabatic invariance will not be a valid approximation under such circumstances.

As a first example of adiabatic invariance, consider a harmonic oscillator whose frequency is varying slowly. Because the action is E/ω, we can conclude that the energy changes in proportion to the frequency.

A more interesting example is given by the Kepler problem. In this case, the orbit is an ellipse with the equation

$$\frac{l}{r} = 1 + \epsilon \cos \theta, \quad l = \frac{J^2}{m\alpha}, \quad \epsilon = \sqrt{1 + \left(\frac{2EJ^2}{m\alpha^2} \right)}, \tag{2.125}$$

where $2l \equiv 2a(\epsilon^2 - 1)$ is the latus rectum, ϵ is the eccentricity [see Eq. (2.113)], and J and E are the conserved angular momentum and energy, respectively. The actions in this case are

$$I_2 = J, \quad I_1 = -J + \alpha \left(\frac{m}{2|E|} \right)^{1/2} \tag{2.126}$$

[see Eqs. (2.99) and (2.100) with $\beta = 0$], it is easy to see that

$$l = \frac{I_2^2}{m\alpha}, \quad \epsilon^2 = 1 - \left(\frac{I_2}{I_2 + I_1} \right)^2. \tag{2.127}$$

Suppose now that the parameter m or α changes slowly with time. Because the actions I_1 and I_2 remain invariant, the eccentricity of the orbit does not change. On the other hand, the scale of the orbit determined by l changes inversely as $m\alpha$. These conclusions, which are straightforward to obtain from action-angle variables and adiabatic invariance, are not so easy to derive from more direct approaches.

2.7 Perturbation Theory for Nonintegrable Systems

For a closed system with $2N$-dimensional phase space, the only obvious constant of motion is the Hamiltonian. This restricts the motion of the system to a $(2N - 1)$ submanifold. On the other hand, for a fully integrable system, the motion is confined to an N-dimensional submanifold, which is of lower dimensionality compared with $(2N - 1)$ except in the trivial case of $N = 1$. If the system is made of several particles, each moving in three-dimensional physical space, then homogeneity of time and space and isotropy of space will lead to the existence of a total of seven integrals of motion corresponding to the conservation of energy, three components of linear momentum, and three components of angular momentum. For a system with k particles, the phase space is $6k$ dimensional and the existence of the above conserved quantities confines the motion to a $(6k - 7)$ dimensional space. If the system is integrable, then the motion will be confined to $3k$-dimensional space, which is of lower dimensionality for all $k > 2$. It is clear that integrability puts a severe restriction on the region of phase space that the dynamical system can explore.

This prompts the following question: Is integrability a rule or an exception? In other words, do all Hamiltonian dynamical systems in $2N$-dimensional phase space possess N integrals of motion? The question is nontrivial because it is possible to construct systems for which fairly complicated integrals of motion can exist. For example, in the case of the Kepler problem, it is impossible to decide by inspection the conservation of the Runge–Lenz vector. So we could take the point of view that all systems are integrable and our inability to find N integrals of motion is only due to lack of mathematical expertise. On the other hand, we could also take the point of view that the class of integrable Hamiltonian is a very small subset of the class of all possible Hamiltonians. In that case, the smallest perturbation to an integrable system will destroy the structure of the tori, and the system will – in due course of time – fill a region of phase space of higher dimensionality than $\mathcal{M}$.

These issues are of importance in the study of several astrophysical systems, especially in the study of the long-term evolution of planetary systems. A planet moving under the inverse-square-law force of the Sun constitutes an integrable system, and its trajectory in phase space is confined to a torus. Even if the force law were changed by the addition of an inverse cube term, say, the system would still be integrable. (In the first case, the trajectory is closed on the torus, whereas, in the second case, it will fill the torus.) However, from the point of view of celestial mechanics, it is important to know what happens to the trajectory when the perturbations from the other planets are taken into account. The system can no longer be described by a central potential and hence we cannot, by inspection, obtain N integrals of motion. We would like to know, in this case, whether the perturbation destroys the tori completely or whether they survive.

Detailed investigations have shown that most of the tori persist under perturbations, although in a distorted form. Some of them are destroyed (and these do not form a set of measure zero), with the measure of destruction growing with the perturbation. The destroyed tori are distributed among those that are preserved in a very pathological manner. These assertions were rigorously proved in what is now called the Kolmogorov–Arnold–Moser (KAM) theorem. Some features of this analysis are now described.

Let us consider the effect of a perturbation on an integrable system by assuming that the total Hamiltonian is given by

$$H(I, \Theta) = H_0(I) + \epsilon H_1(I, \Theta). \tag{2.128}$$

Here $H_0(I)$ is the Hamiltonian corresponding to an integrable system and I, Θ are the action-angle variables for this system. [Both I and Θ denote N tuplets (I_a, Θ_a) with $a = 1, 2, \ldots, N$. We suppress the subscripts when no confusion is likely to arise.] The second term $\epsilon H_1(I, \Theta)$ is a perturbation characterized by the strength ϵ. Because the total Hamiltonian depends on Θ (through the perturbation), I and Θ are no longer action-angle variables for the whole system, although they are perfectly good canonical coordinates. If tori exist for the perturbed system, then there must exist a new set of action-angle variables I', Θ' such that

$$H(I, \Theta) = H'(I'), \tag{2.129}$$

and the new variables must be related to the old by a canonical transformation produced by some generating function $S(\Theta, I')$, with

$$I_a = \frac{\partial S}{\partial \Theta_a}, \quad \Theta'_a = \frac{\partial S}{\partial I'_a}. \tag{2.130}$$

Substituting these relations into Eq. (2.129), we get the condition that S must satisfy:

$$H[\nabla_\Theta S(\Theta, I'), \Theta] = H'(I'). \tag{2.131}$$

Whether the tori exist or not is thus reduced to the problem of finding acceptable solutions to this equation.

Let us examine whether a solution can be found as a power series in the perturbation parameter ϵ. The zeroth-order term must be $S_0 = \Theta_a I'_a$, which generates the identity transformation. Therefore, we write

$$S = \Theta_a I'_a + \epsilon S_1(\Theta, I') + \cdots. \tag{2.132}$$

Substituting this expression in Eq. (2.131), we get

$$H_0(I' + \epsilon \nabla_\Theta S_1 + \cdots) + \epsilon H_1(I' + \cdots, \Theta) = H'(I'), \tag{2.133}$$

with the notation $\nabla_\Theta S = (\partial S/\partial \Theta_a)$, etc. To first order in ϵ,

$$H_0(I') + \epsilon[\nabla_{I'}H_0(I') \cdot \nabla_\Theta S_1 + H_1(I', \Theta)] = H'(I'). \qquad (2.134)$$

We now note that $\omega_{0a}(I') = [\partial H_0(I')/\partial I_a']$ is the frequency of the unperturbed motion and that H_1 and S_1 can be expanded as periodic functions in Θ:

$$H_1(I', \Theta) = \sum_{\mathbf{m}} H_{1\mathbf{m}}(I') \exp(i\mathbf{m} \cdot \Theta),$$

$$S_1(\Theta, I') = \sum_{\mathbf{m} \neq 0} S_{1\mathbf{m}}(I') \exp(i\mathbf{m} \cdot \Theta). \qquad (2.135)$$

(Note that H_1 and S_1 are functions of q and p, which in turn are periodic functions of Θ.) Substituting Eq. (2.135) into Eq. (2.134) and equating the Fourier coefficients, we get

$$H'(I') = H_0(I') + \epsilon H_{10}(I') + \cdots \quad (\mathbf{m} = 0), \qquad (2.136)$$

$$S_{1\mathbf{m}}(I') = +\frac{iH_{1\mathbf{m}}(I')}{m_a\omega_{0a}(I')} + \cdots \quad (\mathbf{m} \neq 0). \qquad (2.137)$$

Thus, to the lowest order, the generator of the new tori is given by

$$S(\Theta, I') = \Theta \cdot I' + i\epsilon \sum_{\mathbf{m} \neq 0} \frac{H_{1\mathbf{m}}(I')}{m_a\omega_{0a}(I')} \exp(i\mathbf{m} \cdot \Theta) + \cdots. \qquad (2.138)$$

It might appear now that we have achieved what we originally wanted; namely, to the lowest order in perturbation theory, we have a generating function that will lead to a new set of action-angle variables for the full system and hence to a new set of tori. By repeating this process order by order in ϵ, we will find the full generating function and the tori, thereby showing that perturbations leave an integrable system still integrable.

These conclusions, however, are wrong. The key difficulty arises from the denominators in Eq. (2.137) that have the terms $m_a\omega_{0a}$. Whenever these quantities vanish, the perturbation theory breaks down. This situation arises when the frequencies of motion on the unperturbed torus are commensurable – that is, if the original orbits are closed and the fundamental frequencies are in resonance. In fact, even when the frequencies are incommensurable, it is always possible to find a set of m_a such that $m_a\omega_{0a}$ is arbitrarily small. This raises the doubt as to whether the series ever converges. Note that we have to worry about both the convergence of the sum over individual terms as well as the convergence of the series in powers of ϵ.

In the study of planetary systems etc., the unperturbed motion usually does not have commensurable frequencies (although there are some very important and interesting exceptions in the solar system, as we shall see in Vol. II). In that case, the trouble arises for only large values of m_a. For such large values of m_a the corresponding Fourier coefficients $H_{1\mathbf{m}}$ are very small. If we truncate the sums before these large values of m_a come into play and work with a few low-order terms, we can predict the motion for a long time. The success of these predictions is important from the practical point of view of celestial mechanics, but does not help in answering the key theoretical question of *arbitrarily* late-time behaviour of the system.

Significant progress in this problem can be made if the nature of the perturbative analysis is changed. Instead of expanding in a power series about the original Hamiltonian, we can do a perturbation series with the nth term defined as a perturbation around the system defined by all terms up to the $(n - 1)$th term. From such an analysis, we can prove the following result: The process of generating (perturbed) tori does converge almost always for a small but finite ϵ. So most of the trajectories stay on an N-dimensional manifold for all times and do not explore the $2N - 1$ dimensional energy surface. More precisely, all initial tori, whose frequency ratios are sufficiently irrational, are preserved. The condition for sufficient irrationality may be stated (in the case of $N = 2$, for example) as the following condition on the ratio of the two frequencies ω_{01} and ω_{02}:

$$\left| \frac{\omega_{01}}{\omega_{02}} - \frac{r}{s} \right| > \frac{K(\epsilon)}{s^{2.5}} \tag{2.139}$$

for all integers r and s, where K is a number that is independent of r and s and tends to zero as ϵ tends to zero. Note that this condition says the frequency ratio should be sufficiently far away from any rational number r/s. The tori that are excluded by this condition are the ones that are likely to be seriously affected by the perturbation.

Let us estimate the measure of these tori. Taking $0 \leq (\omega_{01}/\omega_{02}) \leq 1$ (without any loss of generality), we remove from the interval $[0, 1]$ all patches that violate condition (2.139), i.e., all patches that satisfy

$$\left| \frac{\omega_{01}}{\omega_{02}} - \frac{r}{s} \right| < \frac{K(\epsilon)}{s^{2.5}}. \tag{2.140}$$

Thus we take off a region of length $(K/s^{2.5})$ around each rational r/s in the interval $[0, 1]$. This amounts to taking away a total length that is less than

$$\sum_{s=1}^{\infty} \frac{K}{s^{2.5}} s = K \sum_{s=1}^{\infty} \frac{1}{s^{1.5}} = K\zeta(1.5) \cong 2.6K, \tag{2.141}$$

which tends to zero as the perturbation gets smaller. [The factor s in Eq. (2.141) denotes the number of r values within r/s on 0 to 1; actually, the result in Eq. (2.141) is an overestimate because we have double counted neighbourhoods which overlap.] Clearly, we are left with a set of nonzero measure, even after we remove the tori that are close to rationals.

What happens to these tori, which originally had commensurable [or nearly commensurable, in the sense of condition (2.140)] frequencies? Numerical investigations suggest that these tori are destroyed and the orbits in them are redistributed in phase space. The low-order resonances in the original system are almost always destroyed by the perturbation (and can lead to interesting physical effects, which we will study in Vol. II), although the fate of higher-order resonances is difficult to study.

2.8 Surface of Section

It is clear from the discussion in Section 2.7 that the nature of motion in the phase space for nonintegrable systems can be extremely complex. In principle, we can attempt to study such a system by numerically integrating the equations of motion starting from some initial condition. Although this approach is of great utility (and is often resorted to) in understanding the behaviour of specific systems, it does not, by itself, help us to understand the general characteristics of motion. To do so, it is convenient to represent the key features of the motion by use of a technique called surface of section.

To illustrate this technique, let us consider the motion of a particle in the $x-y$ plane under the action of a potential $\phi(x, y)$. The phase space for this system is four dimensional. Conservation of energy will restrict the motion to a three-dimensional surface in phase space. The path of the system $C(t)$ on this three-dimensional surface will cut the $y = 0$ plane in a two-dimensional surface, which can be characterised by the two coordinates x, and p_x. We can obtain valuable insight into the nature of the motion by plotting the intersection of the curve $C(t)$ with the $y = 0$ surface. In general, the motion will be spread over a bounded region in the x, p_x plane, as we expect the motion to span a two-dimensional surface. If the system has any other hidden integrals of motion, then the motion can become one dimensional and the points of intersection of $C(t)$ with the $y = 0$ surface will lie on a curve.

For a system with $N = 2$, the phase space is four dimensional with the coordinates x, y, p_x, and p_y. Conservation of energy restricts the motion to a three-dimensional surface defined by

$$\frac{1}{2}(p_x^2 + p_y^2) + \phi(x, y) = E = \text{constant.} \tag{2.142}$$

(We shall take $m = 1$.) The intersection of this three-dimensional surface with the surface $y = 0$ defines the surface of section for the dynamical problem, on which any point has the coordinates x, and p_x. Given (x, p_x) and the condition $y = 0$, we can determine p_y as

$$p_y = +\{2[E - \phi(x, 0)] - p_x^2\}^{1/2}, \tag{2.143}$$

where we have chosen the positive square root by convention. Given the energy E and suitable initial conditions, the equations of motion can be integrated to give the orbit of the particle. This orbit will repeatedly cut the surface of section at different values of x and p_x. If there are no other integrals of motion, we expect these points to be scattered in a two-dimensional region bounded by the curve

$$\frac{1}{2} p_x^2 + \phi(x, 0) \le E. \tag{2.144}$$

In case the system has some other hidden integrals of motion, the orbit will be confined to a lower-dimensional surface in phase space and hence will form a smooth curve in the xp_x plane. In general, very different kinds of surface of section are generated by different Hamiltonians. Even for a given Hamiltonian, the nature of the points (induced by the motion) on the surface of section can be widely different for different initial conditions.

To illustrate these concepts, we consider the motion of three-point particles under their mutual gravitational attraction subject to the approximations discussed in Section 2.3. As in that section, it is convenient to use a system of axes (x, y) that corotates with m_1 and m_2, with the x axis pointing toward m_2, say. In this frame m_1 and m_2 are stationary with fixed coordinates $(-\mu, 0)$ and $(1 - \mu, 0)$, respectively, where $\mu = m_2/(m_1 + m_2)$. If the coordinates of m_3 are (x, y) then the equations of motion are given by Eqs. (2.35). This system can be brought into a Hamiltonian form with 2 degrees of freedom by choosing

$$q_1 = x, \quad q_2 = y, \quad p_1 = \dot{x} - y, \quad p_2 = \dot{y} + x, \tag{2.145}$$

$$H = \frac{1}{2}(p_1^2 + p_2^2) + p_1 q_2 - p_2 q_1 - \frac{1-\mu}{\rho_1} - \frac{\mu}{\rho_2}, \tag{2.146}$$

with ρ_1 and ρ_2 defined as in Eq. (2.37). It is obvious that the Hamiltonian is an integral of motion; it is conventional to define it in terms of a quantity called the Jacobi integral, which is given by $C = -2H$. In terms of the original variables, C is given by

$$C = x^2 + y^2 + \frac{2(1-\mu)}{\rho_1} + \frac{2\mu}{\rho_2} - \dot{x}^2 - \dot{y}^2. \tag{2.147}$$

Here C differs from that in Eq. (2.37) by only a constant $\mu(1 - \mu)$.

No other integral for this system is known, and hence the trajectories have to be computed by numerical integration. Let us consider the surface of section defined by $y = 0$, $\dot{y} > 0$ for the case of equal masses ($\mu = 1/2$). Figure 2.3 shows the surface of section for $C = 4.5$. It appears as though the system is nearly integrable and all accessible regions are covered by well-connected regular curves. The orbit itself shows great regularity, with m_3 staying comparatively close to m_1.

When the value of C is lowered slightly to $C = 4$, the surface of section changes to that shown in Fig. 2.4. Now the curves do not fill the accessible region in a regular fashion. All the isolated points correspond to a single trajectory that has a chaotic character. For a still lower value of $C = 3.5$ (see Fig. 2.5) the chaotic region increases in extent. The orbit has a very disordered character. In a typical

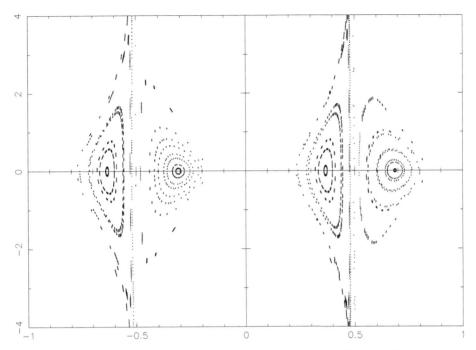

Fig. 2.3.   Surface of section for restricted three-body problem for $C = 4.5$.

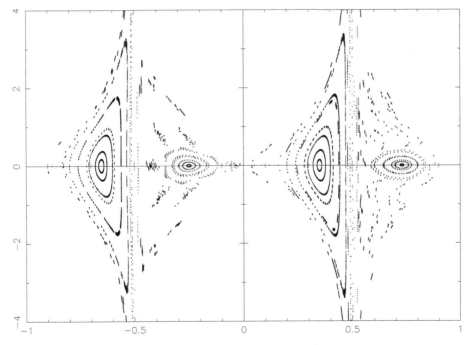

Fig. 2.4.   Surface of section for restricted three-body problem for $C = 4$.

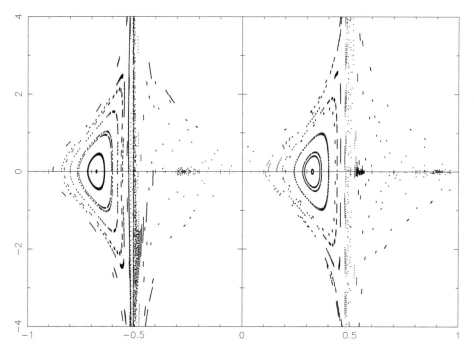

Fig. 2.5.   Surface of section for restricted three-body problem for $C = 3.5$.

motion, m_3 will describe a few loops around m_1, then jump to m_2, then back to m_1, etc.

This example shows clearly that the dynamical evolution of even simple look-ing systems can be highly irregular. What is more, the behaviour of the Poincaré section changes drastically when some parameters in the problem change slightly. The motion of the particle as depicted in the Poincaré section changes from a fairly regular form to an aperiodic, chaotic form when the parameters of the problem change.

Some aspects of this transition to chaos can be understood along the following lines. The points in the surface of section are generated when the equations of motion are integrated for a particle. Mathematically, we can model such a system by giving a map of the form $\mathbf{x}_{n+1} = f(\mathbf{x}_n)$, where $\mathbf{x}_n$ gives the coordinates of the nth point in the surface of section and the function f is a rule that allows us to determine the $(n + 1)$th point from the nth point. This suggests that we should study the evolution of the coordinates under iterative maps in order to understand the dynamics of complex systems. Simplifying the model still further, let us consider one-dimensional maps of the type $x_{n+1} = f(x_n, \mu)$, where μ is some parameter in the problem. Given a particular value of μ, the form of the function $f(x, \mu)$, and a starting value x_0, this rule will generate a succession of points $x_0, x_1, x_2 \dots$. We are interested in studying the general features of this sequence.

To begin with, it is clear that the sequence will repeat itself with a period k if $x_{n+k-1} = f(x_n, \mu)$. Given the functional form of f, we could examine whether such nontrivial solutions exist for specific values of x and μ. Further, if $k=1$, that is, if $x_n = f(x_n, \mu)$, such a value of x_n is called a fixed point of the map. The structure of mappings can usually be understood in terms of the fixed points and the periods in the sequences.

As an example, consider the mapping generated by the function $f(z, \mu) = 4\mu z$ $(1 - z)$, where μ is a parameter in the range $0 \le \mu \le 1$. This function maps the interval $(0, 1)$ to itself. The fixed points of this map are the roots of the equation $z = 4\mu z(1 - z)$ and are given by (1) $z = 0$ and (2) $z = [1 - (4\mu)^{-1}]$. The first solution is trivial and exists for all values of μ, whereas the second solution exists in the relevant interval $(0,1)$ only if $\mu > (1/4)$.

These fixed points are the solutions to the equation $z = f(z, \mu)$. Let us consider the effect of varying the value of x slightly around the fixed point. We do this by writing $x_n = (z + \xi_n)$ and expanding the map in a Taylor series in ξ. This leads to the relation $\xi_{n+1} = f'(z, \mu)\xi_n$, where the prime denotes the first derivative with respect to z. This equation admits solutions of the form $\xi_n = c[f'(z, \mu)]^n$, where c is a constant. The perturbation ξ_n will decay with successive iterations if and only if $|f'(z, \mu)| < 1$. Such a fixed point is called stable. For the map discussed above, the nontrivial fixed point in (2) is stable if $\mu < (3/4)$. Thus in the range $(1/4) < \mu < (3/4)$ there exists a nontrivial, stable fixed point for our mapping at $z = [1 - (4\mu)^{-1}]$. Figure 2.6(a) shows a geometrical way of identifying the successive values of the coordinate for a particular value of μ in this range. The thick curve is described by the equation for our map $y = 4\mu x(1 - x)$. We start with a particular value x_0 in the x axis and identify the point to which x_0 is mapped by drawing a vertical line at $x = x_0$ and finding the point of intersection with the thick curve. The diagonal dashed line corresponds to $y = x$; hence a line parallel to x axis from the point of intersection will cut this diagonal line at an x coordinate that corresponds to $x_1 = 4\mu x_0(1 - x_0)$. The procedure can be now repeated from x_1 to find x_2, etc. It is clear that the intersection of the dashed line and the thick curve will correspond to a fixed point. The evolution proceeds towards the fixed point x_{11} and winds around it.

When the value of μ is beyond $3/4$, a new feature emerges in the evolution [see Fig. 2.6(b)]. A new solution with period 2 takes over from the original stable fixed point. The sequence now oscillates between two values, x_{2-} and x_{2+}. These values are determined by the simultaneous solutions of the equations $x_{2+} = 4\mu x_{2-}(1 - x_{2-})$ and $x_{2-} = 4\mu x_{2+}(1 - x_{2+})$. The solution is given by

$$x_{2\pm} = \frac{1}{8\mu}[1 + 4\mu \pm (16\mu^2 - 8\mu - 3)^{1/2}]. \tag{2.148}$$

This behaviour persists as long as $\mu < (1 + \sqrt{6})/4 \cong 0.86$. When μ is increased above this value, another new solution appears that has a period of 4. This pattern of evolution goes on ad infinitum; there is a series of critical values $\mu = 0.75, 0.86, \ldots$, such that as μ crosses these values the period of the oscillation doubles.

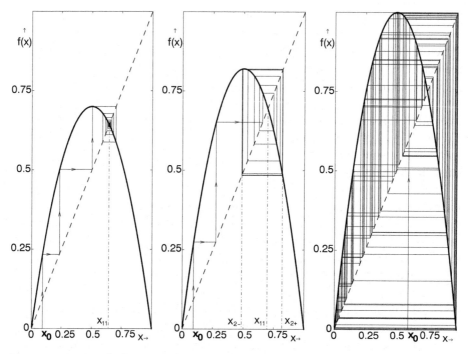

Fig. 2.6. Geometrical construction to determine successive values of the logistic map: (a) $\mu = 0.7$ (b) $\mu = 0.775$, (c) $\mu = 1$.

Finally, let us consider the maximum allowed value: $\mu = 1$. In this case, the sequence from a generic starting value will lead to a complex, aperiodic behaviour, shown in Fig. 2.6(c). The actual terms in the sequence depend sensitively on the starting value, but the overall nature of the evolution remains the same.

This example illustrates that very complex evolution can be exhibited by a seemingly simple system. It also shows that period doubling at critical values of the parameter can lead to chaotic behaviour eventually. Because the dynamical system considered in Eq. (2.146) (the reduced three-body problem) is at least as complicated as this example, it is understandable that the sequence of points in the Poincaré section shows complex behaviour and is sensitive to the value of the input parameter.

Exercise 2.15

Lorenz equations: One of the early discoveries of chaos was in a model for atmospheric convection that was governed by three variables x, y, and z evolving according to the equations $\dot{x} = 10(y - x)$, $\dot{y} = 28x - xz - y$, and $\dot{z} = xy - (8z/3)$. Integrate these equations numerically for different initial conditions and show that x, y, and z follow nonrepeating orbits in the phase space that can be chaotic.

3

Special Relativity, Electrodynamics, and Optics

3.1 Introduction

This chapter begins with a rapid overview of concepts from special relativity and develops the four-vector notation. Several aspects of electrodynamics are then introduced using special relativity and four-vector notation. Finally, this formalism is used to discuss some aspects of principles of optics that are relevant to astronomy. The concept of distribution functions, developed in Section 3.6, will be used extensively in several later chapters. This chapter will also be needed in the development of radiative processes (Chaps. 4 and 6), general relativity (Chap. 11), and in the study of cosmology in Vol. III.

3.2 The Principles of Special Relativity

The description of physical processes requires the specification of spatial and temporal coordinates of events that may be combined into a single entity characterised by the four numbers $x^i = (t, \mathbf{x})$. Throughout this chapter, the Latin indices $a, b, \ldots, i, j$, etc., run over 0, 1, 2, and 3, with the 0 index denoting the 'fourth' dimension and 1, 2, and 3 denoting the standard space dimensions. The actual values of x^i, attributed to any given event $\mathcal{P}$, will depend on the specific coordinate system which is used. We pay special attention to a subset of all possible coordinate systems called inertial coordinate systems. Such coordinate systems are defined by the property that a material particle, far removed from all external influences, will move with uniform velocity in such systems. This definition is convenient and useful but is inherently flawed, as we can never operationally verify this criterion. In fact, there is no fundamental reason why any one class of coordinate systems should be preferred over others except for mathematical convenience. In the development of general relativity in Chap. 11 we shall drop this restrictive assumption and develop the physical principles, treating all coordinate systems as equivalent. For the purpose of this chapter, however, we shall postulate the existence of inertial coordinate systems that enjoy a special status. It is obvious from the definition that any coordinate frame moving with uniform velocity with respect to an inertial frame will also constitute an inertial frame.

Experiments then show that the following two statements are true: (1) All laws of nature remain identical in all inertial frames of reference, that is, the equations expressing the laws of nature are invariant in form with respect to the coordinate transformation connecting any two inertial frames. (2) Interaction between material particles does not take place instantaneously and there exists a maximum possible speed of propagation for interactions. We denote this speed by the letter c and shall show later in Section (3.12) that ordinary light waves, described by Maxwell's equations, propagate at this speed. Anticipating this result we may talk of light rays propagating in straight lines with the speed c.

From statement (1), it follows that the maximum velocity of propagation c should be the same in all inertial frames. This fact has profound consequences. To begin with, this result rules out any absolute nature for simultaneity; two events that appear to occur at the same time in one inertial frame will not, in general, appear to occur at the same time in another inertial frame. Consider two inertial frames K and K', with K' moving relative to K along the x axis. Let A, B, and C be three points along the common x axis, with $AB = AC$ in the primed frame K'. Two light signals start from point A and go towards B and C and will reach B and C at the same instant of time as measured in K'. But the two events, namely the arrival of signals at B and C, cannot be simultaneous for an observer in K. This is because, in frame K, point B moves towards the signal whereas C moves away from the signal, but the speed of the signal is postulated to be the same in both the frames.

The second consequence of the constancy of speed of light is the following: Consider two events $\mathcal{P}$ and $\mathcal{Q}$ with coordinates x^i and $x^i + dx^i$. We define a quantity ds – called the space–time interval – between these two events by the relation

$$ds^2 = c^2 dt^2 - dx^2 - dy^2 - dz^2. \tag{3.1}$$

If $ds = 0$ in one frame, it follows that these two infinitesimally separated events $\mathcal{P}$ and $\mathcal{Q}$ can be connected by a light signal. Because light travels with the same speed c in all inertial frames, $ds' = 0$ in any other inertial frame. Treating $(ds)^2$ as a function of $(ds')^2$, we can expand ds^2 in a Taylor series in ds'^2, as $ds^2 = \alpha + a ds'^2 + \cdots$. The fact that $ds = 0$ when $ds' = 0$ implies that $\alpha = 0$; the coefficient a can be a function of the relative velocity $\mathbf{V}$ only between the frames. Further, homogeneity and isotropy of space requires that only the magnitude $|\mathbf{V}|$ enter into this function. Thus we conclude that $ds^2 = a(V) ds'^2$, where the coefficient $a(V)$ can depend on the absolute value of the relative velocity between the inertial frames. Now consider three inertial frames K, K_1, K_2, where K_1 and K_2 have relative velocities $\mathbf{V}_1$ and $\mathbf{V}_2$, respectively, with respect to K. It is easy to see that $a(V_2)/a(V_1) = a(V_{12})$, where $\mathbf{V}_{12}$ is the relative velocity of K_1 with respect to K_2. This is, however, impossible because the relative velocity V_{12} depends not only on the magnitudes of $\mathbf{V}_1$ and $\mathbf{V}_2$ but also on the angle between the velocity vectors. Hence the function $a(V)$ must be a constant; further, this constant should be equal to unity to satisfy the above relation.

It follows that the quantity ds has the same value in all inertial frames; $ds^2 = ds'^2$, i.e., the space–time interval is an invariant quantity. Events for which ds^2 is greater than, equal to, or less than zero are said to be separated by timelike, null, or spacelike intervals, respectively.

To understand the physical meaning of ds, consider a clock that is moving relative to an inertial frame K in an arbitrary trajectory. During a time interval dt, the clock moves through a distance $|d\mathbf{x}|$ as measured in K. In another inertial coordinate system, K', which is moving with respect to K with the same velocity as the clock at that instant of time t, the clock is momentarily at rest, giving $d\mathbf{x}' = 0$. If the clock indicates a lapse of time dt' when the time interval measured in K is dt, the invariance of space–time intervals shows that

$$ds^2 = c^2\, dt^2 - dx^2 - dy^2 - dz^2 = ds'^2 = c^2\, dt'^2 \qquad (3.2)$$

or,

$$dt' = \frac{ds}{c} = dt\sqrt{1 - \frac{v^2}{c^2}}. \qquad (3.3)$$

Hence ds/c may be thought of (for timelike intervals) as the lapse of time in a moving clock; this is usually called the proper time along the trajectory of the clock.

In defining ds^2 we have used a positive sign for $c^2 dt^2$ and a negative sign for the spatial terms dx^2, etc. The sequence of signs is called signature, and it is usual to say that the signature of space–time is $(+ - - -)$. We can, equivalently, use the signature $(- + + +)$, which will require a change of sign in several expressions. We should keep this point in mind while comparing formulas in different textbooks.

3.3 Transformation of Coordinates and Velocities

Because the concept of simultaneity has no invariant significance, the coordinates of any two inertial frames will be related by a transformation in which space and time coordinates will, in general, be different. With future applications in mind, we study the general question of determining the coordinates appropriate for an arbitrary observer moving along the x axis.

Let (t, x, y, z) be an inertial coordinate system. Consider an observer travelling along the x axis in a trajectory $x = f(\tau)$, $t = h(\tau)$, where f and h are specified functions and τ is the proper time in the clock carried by the observer. We would like to assign a suitable coordinate system to this observer. Let $\mathcal{P}$ be some event with inertial coordinates (t, x). The observer sends a light signal from the event $\mathcal{A}$ (at $\tau = \tau_A$) to the event $\mathcal{P}$. The signal is reflected at $\mathcal{P}$ and reaches back the observer at event $\mathcal{B}$ (at $\tau = \tau_B$). Because the light has travelled for a time interval $(\tau_B - \tau_A)$, it is reasonable to attribute the coordinates

$$t' = \frac{1}{2}(\tau_B + \tau_A), \quad x' = \frac{1}{2}(\tau_B - \tau_A)c \qquad (3.4)$$

to the event $\mathcal{P}$.

To relate (t', x') to (t, x) we proceed as follows: Because the events $\mathcal{P}(t, x)$, $\mathcal{A}(t_A, x_A)$ and $\mathcal{B}(t_B, x_B)$ are connected by light signals travelling in forward and backward directions, it follows that

$$x - x_A = c(t - t_A), \quad x - x_B = -c(t - t_B) \qquad (3.5)$$

or

$$x - ct = x_A - ct_A = f(\tau_A) - ch(\tau_A) = f\left(t' - \frac{x'}{c}\right) - ch\left(t' - \frac{x'}{c}\right), \quad (3.6)$$

$$x + ct = x_B + ct_B = f(\tau_B) + ch(\tau_B) = f\left(t' + \frac{x'}{c}\right) + ch\left(t' + \frac{x'}{c}\right). \quad (3.7)$$

Given f and h, these equations can be solved to find (x, t) in terms of (x', t'). This procedure is applicable to any observer.

For an observer moving with uniform velocity V, the trajectory is $x = Vt$, with the proper time given by $\tau = t[1 - (V^2/c^2)]^{1/2}$ [see Eq. (3.3)]. So the parameterised trajectory is

$$f(\tau) = \frac{V}{\sqrt{1 - (V^2/c^2)}}\, \tau, \quad h(\tau) = \frac{1}{\sqrt{1 - (V^2/c^2)}}\, \tau \qquad (3.8)$$

giving

$$x \pm ct = f\left(t' \pm \frac{x'}{c}\right) \pm ch\left(t' \pm \frac{x'}{c}\right)$$

$$= \frac{1}{\sqrt{1 - (V^2/c^2)}}\left[\left(Vt' \pm \frac{V}{c}x'\right) \pm \left(ct' \pm \frac{x'}{c}\right)\right] = \sqrt{\frac{1 \pm (V/c)}{1 \mp (V/c)}}\,(x' \pm ct').$$

$$(3.9)$$

Solving these equations, we get the transformation between the inertial frames, called the Lorentz transformation (LT) as

$$x = \gamma(x' + Vt'), \quad y = y', \quad z = z',$$

$$t = \gamma\left(t' + \frac{V}{c^2}x'\right), \quad \gamma \equiv [1 - (V^2/c^2)]^{-1/2}. \qquad (3.10)$$

This transformation leaves the quantity $s^2 \equiv c^2 t^2 - |\mathbf{x}|^2$ invariant, as this is the space–time interval between the origin and any event $(t, \mathbf{x})$. [This fact can also be directly verified from Eqs. (3.10).] A quadratic expression of this form is similar to the length of a vector in three dimensions that – as is well known – is invariant under rotation of the coordinate axes. This suggests that the transformation between inertial frames can be thought of as a rotation in four-dimensional space. For two inertial frames K and K' with a relative velocity $\mathbf{V}$, we can always align the coordinates in such a way that the relative velocity vector is along the common (x, x') axis. Then, from symmetry, it follows that the transverse directions are not affected and $y' = y$ and $z' = z$. The rotation must be in the tx plane characterised by a parameter, say ψ. The LT in Eqs. (3.10) can be written as

$$x = x' \cosh\psi + ct' \sinh\psi, \qquad ct = x' \sinh\psi + ct' \cosh\psi, \qquad (3.11)$$

where $\tanh\psi = V/c$, which determines the parameter ψ in terms of the relative velocity between the two frames.

In general, the LT mixes up space and time coordinates and reduces to the standard Galelean transformations of nonrelativistic mechanics ($x' = x - Vt$, $t' = t$, $y' = y$, $z' = z$) in the limit of $c \to \infty$.

Exercise 3.1
Lorentz transformations in arbitrary directions: Consider a LT along a direction indicated by the unit vector $\hat{\mathbf{n}}$. (1) Show that the transformation of coordinates can be written in the matrix form $x^{i'} = \Lambda^i_j x^j$, where the components of the matrix Λ are given by

$$\Lambda^0_0 = \gamma = (1 - \beta^2)^{-1/2}, \qquad \Lambda^0_\alpha = \Lambda^\alpha_0 = -\gamma\beta n^\alpha, \qquad \beta = V/c, \qquad (3.12)$$

$$\Lambda^\alpha_\beta = \Lambda^\beta_\alpha = (\gamma - 1) n^\alpha n^\beta + \delta^{\alpha\beta}. \qquad (3.13)$$

Verify that the Jacobian of this transformation is unity. (2) Prove that the inverse of the matrix Λ is obtained when β is changed to $-\beta$.

Exercise 3.2
Four vectors and matrices: To any event $x^i = (x^0, \mathbf{x})$ we can associate a 2×2 matrix $P \equiv x^i \sigma_i$, where σ_0 is the identity matrix and σ_i are the Pauli matrices. Under a LT along the direction $\hat{\mathbf{n}}$ with speed V, the event x^i goes to $x^{i'}$ and P goes to P'. (By convention σ_i's do not change.) Show that $P' = LPL^*$, where $L = \cosh(\alpha/2) + (\mathbf{n} \cdot \sigma) \sinh(\alpha/2)$ and $\tanh\alpha = (V/c)$.

Given the LT, we can compute the transformation law for any other physical quantity that depends on the coordinates. As an example, consider the transformation of the velocities between two inertial frames: Taking the differential form of the LT,

$$dx = \gamma(dx' + V dt'), \quad dy = dy', \quad dz = dz', \quad dt = \gamma\left(dt' + \frac{V}{c^2}dx'\right),$$

(3.14)

and forming the ratios $\mathbf{v} = d\mathbf{x}/dt$, $\mathbf{v}' = d\mathbf{x}'/dt'$, we find the transformation law for the velocities to be

$$v_x = \frac{v_x' + V}{1 + (v_x' V/c^2)}, \quad v_y = \gamma^{-1}\frac{v_y'}{1 + (v_x' V/c^2)}, \quad v_z = \gamma^{-1}\frac{v_z'}{1 + (v_x' V/c^2)}.$$

(3.15)

From this result, it is easy to verify the following facts:

(1) The transformation law reduces to the familiar addition of velocities in the limit of $c \to \infty$. But in the relativistic case, none of the velocity components can exceed c.

(2) The addition law for the x component of the velocity is equivalent to the additivity in $\psi = \tanh^{-1}(V/c)$, viz., $\psi_{12} = \psi_1 + \psi_2$ for two successive rotations in the tx plane. Even the transverse velocities are, however, affected by motion along the x axis; this is quite unlike the situation in nonrelativistic physics.

(3) The velocities $\mathbf{v}'$ and $\mathbf{V}$ combine in an asymmetrical way except in the trivial case in which they are both along the same direction. This is a reflection of the fact that the net effect of two Lorentz transformations along two (noncollinear) axes depends on the order in which the transformations are performed.

(4) The transformation of velocities shows that the direction of motion of a particle will appear to be different in different inertial frames. If $v_x = v \cos\theta$ and $v_y = v \sin\theta$ are the components in the coordinate frame K (with primes denoting corresponding quantities in frame K'), then it is easy to see from Eqs. (3.15) that

$$\tan\theta = \gamma^{-1}\frac{v' \sin\theta'}{v' \cos\theta' + V}.$$

(3.16)

For a particle moving with relativistic velocities ($v' \approx c$) and for a ray of light, this formula reduces to

$$\tan\theta = \gamma^{-1}\frac{\sin\theta'}{(V/c) + \cos\theta'}.$$

(3.17)

We shall see later (in Section 3.12.1) that this formula has important applications in radiative processes.

Exercise 3.3

Superluminal motion: Consider a blob of plasma moving with at a speed v along a direction that makes an angle ψ with respect to the line of sight. (1) Show that the apparent transverse speed of the source will be related to the actual speed by

$$v_{app} = \frac{v \sin \psi}{1 - (v/c) \cos \psi}.$$ (3.18)

From this, conclude that the apparent speed can exceed the speed of light. How does v_{app} vary with ψ for a constant value of v? (2) If the blob of plasma is ejected from a source with equal probability in all directions in the rest frame of the observer, show that the probability that the observer will see the blob move with an apparent velocity greater than v_{app} is given by

$$P(>\beta_a) = \frac{1}{1 + \beta_a^2}\left(1 - \frac{\beta_a^2}{\gamma^2 - 1}\right), \quad \gamma = \frac{1}{\sqrt{1 - (v^2/c^2)}}, \quad \beta_a \equiv \frac{v_{app}}{c}.$$ (3.19)

Exercise 3.4
Thomas precession: (1) Let $\Lambda(\mathbf{v})$ denote the matrix corresponding to a Lorentz boost along the direction $\hat{\mathbf{v}}$ with speed v. Show that $\Lambda(\mathbf{v}_1)\Lambda(\mathbf{v}_2) = R(\theta\hat{\mathbf{n}})\Lambda(\mathbf{v}_3)$, where $R(\theta\hat{\mathbf{n}})$ is the spatial rotation matrix for rotation by an angle θ about the direction $\hat{\mathbf{n}}$. This shows that the effect of two LTs is (in general) the same as that of an LT and a rotation. Find θ, $\hat{\mathbf{n}}$, and $\mathbf{v}_3$ in terms of $\mathbf{v}_1$ and $\mathbf{v}_2$. (2) Consider a particle undergoing acceleration $\mathbf{a}_{comoving}$ in the comoving, instantaneous rest frame. Because of the acceleration, the comoving frames for the particle at two different instants of time t and $(t + dt)$ will be different. In the lab frame, the combination of two boosts with velocities $\mathbf{v}$ and $(\mathbf{v} + \delta\mathbf{v})$ will not be a pure boost but will be a combination of pure boost and rotation by an angle $\delta\theta = \Omega dt$, thereby leading to a precession (called Thomas precession) by an amount Ω. Use the results of (1) to show that

$$\mathbf{a}_{comoving} = \hat{\mathbf{n}}_\alpha \frac{d\alpha}{d\tau} + (\sinh\alpha)\frac{d\hat{\mathbf{n}}_\alpha}{d\tau}, \quad \Omega = \left(2\sinh^2\frac{\alpha}{2}\right)\left(\frac{d\hat{\mathbf{n}}_\alpha}{dt} \times \hat{\mathbf{n}}_\alpha\right),$$ (3.20)

where $\tanh\alpha = (v/c)$. In the nonrelativistic limit, this gives a precession angular velocity $\Omega \cong (1/2c^2)(\mathbf{a} \times \mathbf{v})$. (Hint: Use the result of Exercise 3.2.)

Exercise 3.5
Light echoes: A large, planar sheet of dust exists between an observer and a source of radiation with the plane inclined at an angle i to the line of sight. Radiation from the source is scattered by dust in this plane towards the observer. Suppose the source flares up in intensity at $t = 0$ for some reason. Show that the observer will see, at time $t > 0$, a circle of scattered light with the radius

$$\rho = ct\left(\csc^2 i + \frac{2d \csc i}{ct}\right)^{1/2},$$ (3.21)

where d is the perpendicular distance from the source to the dust plane.

3.4 Four Vectors

Equations like $\mathbf{F} = m\mathbf{a}$, which are written in vector notation, remain valid in any three-dimensional coordinate system without change of form. For example, consider two Cartesian coordinate systems with the same origin and the axes rotated with respect to each other. The components of the vectors **F** and **a** will be different in these two coordinate systems but the equality between the two sides of the equation will continue to hold.

If the laws of physics are to be expressed in a form that remains invariant under LT, we should similarly construct vectorial quantities with four components and treat LTs as rotations in a four-dimensional space. Such vectors are called four vectors and will have one time component and three spatial components. The spatial components, of course, will form ordinary three vectors and transform as such under spatial rotations with the time component remaining unchanged.

Let us denote a generic four vector as A^i with components $(A^0, \mathbf{A})$ in some inertial frame K. The simplest example of such a four vector is the space–time coordinates of an event $x^i = (ct, \mathbf{x})$. (We have used ct rather than t in order to give the same dimensions to all four components.) Knowing the transformation properties of the four vector x^i, we define the transformation law for a general four vector to be

$$A^0 = \gamma\left(A'^0 + \frac{V}{c}A'^1\right), \quad A^1 = \gamma\left(A'^1 + \frac{V}{c}A'^0\right), \quad A^2 = A'^2, \quad A^3 = A'^3.$$

$$(3.22)$$

It is obvious from our construction that, under these transformations, the square of the length of the vector defined by $(A^0)^2 - |\mathbf{A}|^2$ remains invariant.

It is convenient to introduce at this stage two different types of components of four vectors denoted by A^i and A_i, with $A^i \equiv (A^0, \mathbf{A})$ and $A_i \equiv (A^0, -\mathbf{A})$. In other words, lowering of the index changes the sign of the spatial components. Given this definition, we can write the squared length of the vector as $A^i A_i$ with the understanding that any index that is repeated in an expression is summed over, that is, $A^i A_i$ stands for the expression

$$A^i A_i \equiv \sum_{i=0}^{3} A^i A_i = A^0 A_0 + A^1 A_1 + A^2 A_2 + A^3 A_3. \quad (3.23)$$

This quantity need not be positive definite. A four vector is called timelike, null, or spacelike depending on whether this quantity is positive, zero, or negative, respectively.

More generally, given two four vectors $A^i = (A^0, \mathbf{A})$ and $B^i = (B^0, \mathbf{B})$, we can define a dot product between them by a rule similar to that for $A^i B_i$, with

$$A^i B_i = A^0 B_0 + A^1 B_1 + A^2 B_2 + A^3 B_3 = A^0 B^0 - \mathbf{A} \cdot \mathbf{B}. \quad (3.24)$$

The dot product is invariant under LTs, as can be easily verified. The squared length of the vector, of course, is just the dot product of the vector with itself.

To illustrate the above formalism, consider two examples of four vectors. Let $x^i(s)$ denote the space–time coordinates of a particle when the proper time carried by a clock moving with the particle is s. The four velocity of the particle can be defined to be $u^i = dx^i/ds$. Because $ds^2 = dx^i dx_i$ is a scalar, it follows that u^i transforms as a four vector. The components of this vector are

$$u^i = \left(\gamma, \gamma \frac{\mathbf{v}}{c}\right). \tag{3.25}$$

From this definition it follows that $u^i u_i = dx^i dx_i / ds^2 = 1$ so that the four velocity has only three independent components.

In a similar manner we can define the four acceleration to be $a^i = d^2 x^i / ds^2 = du^i / ds$. Differentiating the relation $u^i u_i = 1$ with respect to s, we easily see that $a^i u_i = 0$.

Just as the concept of the three vector has been generalised to that of a four vector, we can also generalise the gradient operator to four dimensions. The ordinary three-dimensional gradient, $\nabla = [(\partial/\partial x), (\partial/\partial y), (\partial/\partial z)]$, transforms as a vector under spatial rotations. To define the four-dimensional gradient, we have to only note that the differential of a scalar quantity $d\phi = (\partial\phi/\partial x^i) dx^i$ is also a scalar. Because this expression is a dot product between dx^i and $(\partial\phi/\partial x^i)$, it follows that the latter quantity transforms like a four vector under LT. Explicitly, the components of the four gradient of a scalar are given by

$$\frac{\partial\phi}{\partial x^i} = \left(\frac{1}{c}\frac{\partial\phi}{\partial t}, \nabla\phi\right). \tag{3.26}$$

As an example, consider the notion of a normal to the surface. A three-dimensional surface in four-dimensional space is given by an equation of the form $f(x^i) = 0$. The normal vector $n_i(x^a)$ at any event x^a on this surface is given by $n_i = (\partial f/\partial x^i)$, which is an example of a four gradient. It is conventional to call a surface spacelike, null, or timelike at x^a, depending on whether n_i is timelike, null, or spacelike at x^a. In a similar manner, the four divergence of a four vector is defined to be $(\partial A^i/\partial x^i)$ with summation over index i.

Gauss's theorem in three dimensions is generalised to four dimensions as

$$\int_{\mathcal{V}} d^4 x \frac{\partial A^i}{\partial x^i} = \int_{\partial\mathcal{V}} d^3\Sigma \, (n^a A_a), \tag{3.27}$$

where $\mathcal{V}$ is a region of four-dimensional space bounded by a three surface $\partial\mathcal{V}$ and $d^3\Sigma$ is an element of a three surface with a normal vector n^a. The left-hand side is a volume integral and the right-hand side is a surface integral. The boundaries of a four-dimensional region $\mathcal{V}$ can be conveniently taken to be made of the following components: (1) Two three-dimensional surfaces at $t = t_1$ and $t = t_2$, both of which are spacelike; the coordinates on these surfaces are the regular spatial coordinates (r, θ, ϕ). (2) One timelike surface at a large spatial distance $(r = R \to \infty)$ at any time t satisfying $t_1 < t < t_2$; the coordinates on this three-dimensional surface could be (t, θ, ϕ). On the right-hand side of Eq. (3.27) the integral has to be taken over the surfaces in components (1) and (2). If the vector field A_a vanishes at large spatial distances, then the integral over the surface in component (2) vanishes for $R \to \infty$. For the integral over the surfaces in component (1), the normal vector n^a has components $(1, 0, 0, 0)$. It follows that

$$\int_V d^4x \frac{\partial A^i}{\partial x^i} = \int_{t=t_1} d^3\mathbf{x}\, A_0 - \int_{t=t_2} d^3\mathbf{x}\, A_0, \qquad (3.28)$$

with the minus sign arising from the fact that the normal must always be treated as outwardly directed. It follows that if $(\partial A^i/\partial x^i) = 0$, then the integral of $A_0(= A^0)$ over all space is conserved in time. We can obtain the same result, of course, by writing the equation $(\partial A^i/\partial x^i) = 0$ in the form $[\partial A^0/\partial(ct)] + \nabla \cdot \mathbf{A} = 0$ and integrating the terms over all space.

At the next level of complication, we can define four tensors as quantities that transform like the product of four vectors. For example, consider an entity C_{ik} defined to be $C_{ik} = A_i B_k$, where A_i and B_k are four vectors. Knowing the transformation law for the four vectors, we can easily find out how the components of C_{ik} get mixed under a LT. A second-rank tensor T_{ik} is defined to be a set of $4 \times 4 = 16$ quantities that transform like the product $A_i B_k$ of two four vectors under a LT. Because we have defined two types of components for four vectors, we can have second-rank tensors like T^{ik}, T_k^i, or T_{ik} differing in the position of indices. Whenever an index corresponding to a spatial coordinate is raised or lowered, the sign of the component changes. Higher-rank tensors, with more indices, can be defined along similar lines.

A tensor is symmetric (antisymmetric) on two indices (a, b) if interchanging the indices leaves the value of the component same (changes the sign of the component). Any tensor can be written as the sum of a symmetric and an antisymmetric part with respect to any pair of indices. For example, $T^{ik} = A^{ik} + S^{ik}$, where $A^{ik} \equiv (T^{ik} - T^{ki})/2$ is antisymmetric and $S^{ik} = (T^{ik} + T^{ki})/2$ is symmetric in (i, k).

We can make the raising and the lowering of the indices more formal by introducing a special tensor with components $\eta_{ik} \equiv \eta^{ik} = \text{dia}(1, -1, -1, -1)$ and writing the relation connecting the two types of components as $A_i = \eta_{ik}A^k$ and $A^i = \eta^{ik}A_k$. (We stress the convention that indices which are repeated are summed over in such expressions.) It can be trivially verified that (1) this relation reproduces our original definition of the components, and (2) the components η^{ik} and η_{ik} have the same numerical value in all inertial frames (see Exercise 3.6). The raising and the lowering of tensor indices follow the obvious generalisation of the above rule, e.g., $T_k^i = \eta_{ak}T^{ia}$, etc. The dot product between two vectors can now be expressed as $\eta_{ik}A^i B^k$ and the norm of a vector by $\eta_{ik}A^i A^k$; in particular, $ds^2 = \eta_{ik}dx^i dx^k$. These concepts and notation will be of tremendous use when we develop general relativity in Chap. 11.

Exercise 3.6

Index gymnastics: (1) Show that η_{ik} and η^{ik} have the same numerical value for the components in all inertial frames. (2) The four-index tensor ϵ_{ijkl} is defined to be completely antisymmetric in any pair of indices with $\epsilon_{0123} = 1$. Show that this tensor also has the same components in all coordinate frames. We can also define a three-index object $\epsilon_{\alpha\beta\gamma}$ (where the Greek subscripts take the values 1,2,3) along similar lines. It is antisymmetric in any pair of indices and $\epsilon_{123} = 1$. Show that, in three-dimensional space, $\epsilon_{\alpha\beta\gamma} A^\beta B^\gamma$ gives the αth component of $\mathbf{A} \times \mathbf{B}$. (3) Write down the transformation law for a two-index object A^{ik} under LTs, i.e., express the components A'^{ik} explicitly in terms of A^{ik}. How does it simplify if $A^{ik} = -A^{ki}$? (4) Show that $A^{iklm} \cdots S_{ikab} \cdots = 0$ if $A^{ikl} \cdots$ is antisymmetric in the pair (i, k) and $S_{ika} \cdots$ is symmetric in the pair (i, k). This result will be used extensively in future discussions.

3.5 Particle Dynamics

To determine the laws governing the motion of a free particle, we need an expression for the action that can be varied. This action should be constructed from the trajectory $x^i(s)$ of the particle and should be invariant under LTs. The only possibility is some quantity proportional to the integral of ds, so the action must be

$$\mathcal{A} = -\alpha \int_a^b ds = -\int_{t_1}^{t_2} \alpha c \sqrt{1 - \frac{v^2}{c^2}} \, dt, \tag{3.29}$$

where α is a constant. In arriving at the second equality, we have expressed ds in terms of dt by using Eq. (3.3), which shows that the Lagrangian is given by $L \equiv d\mathcal{A}/dt = -\alpha c\sqrt{1 - v^2/c^2}$. When $c \to \infty$, this Lagrangian reduces to $L = \alpha v^2/2c + $ constant. Comparing this with the Lagrangian $(1/2)mv^2$ for a free particle in nonrelativistic mechanics, we find that $\alpha = mc$, where m is the mass of the particle. Substituting this result back in Eq. (3.29), we find that the action for a relativistic particle becomes

$$\mathcal{A} = -mc \int ds = -\int_{t_1}^{t_2} mc^2 \sqrt{1 - \frac{v^2}{c^2}} \, dt, \tag{3.30}$$

where the second equation identifies the Lagrangian to be $L = -mc^2\sqrt{1 - v^2/c^2}$. This action in relativistic mechanics has a clear geometric meaning, unlike its nonrelativistic counterpart.

To determine the dynamics, we vary the action with respect to the trajectory $x^i(s)$ and get

$$\delta\mathcal{A} = -mc \int_a^b \delta(ds) = -mc \int_a^b \delta(\sqrt{dx_i dx^i}) = -mc \int_a^b \frac{dx_i \delta dx^i}{ds}$$

$$= -mc \int_a^b u_i d\delta x^i = -mc u_i \delta x^i \Big|_a^b + mc \int_a^b \delta x^i \frac{du_i}{ds} \, ds. \tag{3.31}$$

If we now assume that δx^i vanishes at the end points, we obtain the equation of motion $du^i/ds = 0$, which is a generalisation of force-free motion to relativistic mechanics. Further, by treating the action as a function of the end points of the trajectory that satisfies the equation of motion, we obtain $(\partial \mathcal{A}/\partial x^i) = -mcu_i$. Because the derivatives of the action with respect to the coordinates define the momenta, the four-momentum vector is given by

$$p_i = -\frac{\partial \mathcal{A}}{\partial x^i} = mcu_i = (\gamma mc, -\gamma m\mathbf{v}) \equiv \left(\frac{E}{c}, -\mathbf{p}\right), \tag{3.32}$$

$$p^i = mcu^i = (\gamma mc, \gamma m\mathbf{v}) = \left(\frac{E}{c}, \mathbf{p}\right), \quad \gamma = \left(1 - \frac{v^2}{c^2}\right)^{-1/2}. \tag{3.33}$$

To obtain the physical significance of the time component, $E = \gamma mc^2$, we note that, in the nonrelativistic limit, $E \approx mc^2 + mv^2/2$. This suggests that E corresponds to the relativistic energy of the particle. Such an identification is further vindicated by the fact that, for a Lagrangian $L = -mc^2/\gamma$, the Hamiltonian $\mathcal{H} = \mathbf{p} \cdot \mathbf{v} - L$ is numerically the same as E. We thus conclude that the three momentum $\mathbf{p} = \gamma m\mathbf{v}$ and energy (divided by c) form the components of a four vector.

Because $u^i u_i = 1$, it follows that $p^i p_i = m^2 c^2$, giving the following relations connecting momentum, energy, and velocity:

$$E = \sqrt{p^2 c^2 + m^2 c^4}, \quad \mathbf{p} = E\left(\frac{\mathbf{v}}{c^2}\right). \tag{3.34}$$

The first relation allows the existence of massless particles with $m \to 0$, $E = pc$, and $v = (pc^2/E) = c$.

Finally, the Hamilton–Jacobi equation for the relativistic particle can be obtained from the definition $p_i = -(\partial \mathcal{A}/\partial x_i)$ and the condition $p^i p_i = m^2 c^2$; we get

$$\frac{1}{c^2}\left(\frac{\partial \mathcal{A}}{\partial t}\right)^2 - \left(\frac{\partial \mathcal{A}}{\partial x}\right)^2 - \left(\frac{\partial \mathcal{A}}{\partial y}\right)^2 - \left(\frac{\partial \mathcal{A}}{\partial z}\right)^2 = m^2 c^2. \tag{3.35}$$

To obtain the nonrelativistic limit of this equation, we substitute $\mathcal{A} = -mc^2 t + S(x^i)$ into Eq. (3.35). Simplification then gives

$$\frac{1}{2m}(\nabla S)^2 + \frac{\partial S}{\partial t} = \frac{1}{c^2}\left(\frac{\partial S}{\partial t}\right)^2 \cong 0, \tag{3.36}$$

where the last equality arises in the limit of $c \to \infty$. This is exactly the Hamilton–Jacobi equation for the free particle in the nonrelativistic theory.

Exercise 3.7

Practice with four vectors: An observer with four velocity u^a measures the properties of a particle with four momentum p^b. Express (1) the energy, (2) the magnitude of the three momentum, and (3) the magnitude of the three velocity attributed to the particle by the observer in terms of u^a and p^b. [Answers: (1) Consider the dot product $p^a u_a$ that is Lorentz invariant and has the same numerical value in all frames. We evaluate it in the rest frame of the observer in which $u_a = (1, 0, 0, 0)$. In this frame $p^a u_a = p^0 = E/c$, where E is the energy of the particle. Because $p^a u_a$ is Lorentz invariant, it follows that the energy is $E = c p^a u_a$, evaluated in any frame. The rest follows easily from the relation $pc = \sqrt{E^2 - m^2 c^4}$ and $\mathbf{v} = (\mathbf{p} c^2 / E)]$.

3.6 Distribution Functions and Moments

So far, we have discussed the dynamics of a single particle. Often in astrophysics, we need to deal with a large collection of particles undergoing nearly identical physical processes. In nonrelativistic mechanics, we deal with this situation by using a distribution function. It is necessary to generalise this concept in a Lorentz invariant manner to take into account a system of relativistic particles.

To do that, we first obtain several relativistically invariant quantities that will serve as basic building blocks. To begin with, an element of four-dimensional volume, $d^4 x = c \, dt \, dx \, dy \, dz$ is invariant under LTs. Although this is obvious from the definition of LTs as rotations in space–time, it can also be explicitly proved by checking that the Jacobian of a LT is unity (see Exercise 3.1).

Let us next consider a set of N particles, each of mass m, described by a distribution function $f(p^i)$ at any given location in space. The total number of particles can be written in terms of the distribution function as

$$N = \int d^4 p \; \theta(p^0) \delta_D(p^a p_a - m^2 c^2) f(p^i), \tag{3.37}$$

where $d^4 p = dp^0 d^3 \mathbf{p}$; the Dirac delta function $\delta_D(p^a p_a - m^2 c^2)$ ensures that all the particles have mass m, and the theta function $\theta(p^0)$ (which is unity for $p^0 > 0$ and vanishes for $p^0 < 0$) ensures that $p^0 > 0$. The quantities $N, d^4 p, \theta$, and $\delta_D(p^a p_a - m^2 c^2)$ are all individually Lorentz invariant, implying that f is Lorentz invariant. Writing the Dirac delta function as

$$\delta_D(p_i p^i - m^2 c^2) \equiv \delta_D\left(p_0^2 - \frac{E_{\mathbf{p}}^2}{c^2}\right) = \frac{c}{2 E_{\mathbf{p}}} \left[\delta_D\left(p^0 - \frac{E_{\mathbf{p}}}{c}\right) + \delta_D\left(p^0 + \frac{E_{\mathbf{p}}}{c}\right)\right],$$

$$\tag{3.38}$$

where $E_{\mathbf{p}}^2 \equiv m^2 c^4 + p^2 c^2$ and noting that integration over dp^0 will merely replace p^0 with $(E_{\mathbf{p}}/c)$ because of the condition $p^0 > 0$, we get

$$N = \int d^3 \mathbf{p} \, dp^0 \theta(p^0) \frac{c}{2 E_{\mathbf{p}}} \left[\delta_D\left(p^0 - \frac{E_{\mathbf{p}}}{c}\right) + \delta_D\left(p^0 + \frac{E_{\mathbf{p}}}{c}\right)\right] f(p^0, \mathbf{p})$$

$$= \frac{c}{2} \int \frac{d^3 \mathbf{p}}{E_{\mathbf{p}}} \, f(p^0 = E_{\mathbf{p}}/c, \mathbf{p}). \tag{3.39}$$

Because N and f are invariant, the combination $(d^3 p / E_{\mathbf{p}})$ must be invariant under Lorentz transformations.

Finally, from the relations $E_p = \gamma mc^2$, $(ds/cdt) = \gamma^{-1}$, we find that $E_p(ds/dt)$ is Lorentz invariant. Multiplying by $d^3\mathbf{x}$ in the numerator and denominator, we get

$$E_p \frac{ds}{dt} \frac{d^3\mathbf{x}}{d^3\mathbf{x}} = (E_p d^3\mathbf{x})\left(\frac{ds}{d^4x}\right), \tag{3.40}$$

showing that the combination $E_p d^3\mathbf{x}$ is also invariant. Combined with the earlier result that $d^3\mathbf{p}/E_p$ is Lorentz invariant, we conclude that the product $(E_p d^3\mathbf{x})$ $(d^3\mathbf{p}/E_p) = d^3\mathbf{x}d^3\mathbf{p}$ is Lorentz invariant. In other words, an element of phase volume is relativistically invariant, even though neither the spatial volume nor the volume in momentum space is individually invariant.

This result allows us to introduce distribution functions in relativistic theory in exact analogy with nonrelativistic mechanics. We define the distribution function f such that

$$dN = f(x^i, \mathbf{p})d^3\mathbf{x}d^3\mathbf{p}, \tag{3.41}$$

represents the number of particles in a small phase volume $d^3\mathbf{x}d^3\mathbf{p}$. The x^i here stands for the four-vector components $(t, \mathbf{x})$, and $\mathbf{p}$ is the three-momentum vector; the fourth component of the momentum vector (E_p/c) does not appear as it is completely determined by $\mathbf{p}$ and mass m of the particle. Each of the quantities dN, f, and $d^3\mathbf{x}d^3\mathbf{p}$ are individually Lorentz invariant.

Given the Lorentz invariant distribution function f, we can construct several other invariant quantities by taking moments of this function. Of particular importance are the moments constructed by integration of the distribution function over various powers of the four momentum. We now construct a few examples.

The simplest Lorentz invariant quantity that we can obtain from the distribution function by integrating out the momentum is the harmonic mean $\bar{E}_{\text{har}}$ of the energy of the particles at an event x^i. This is defined by the relation

$$\frac{1}{\bar{E}_{\text{har}}(x^i)} \equiv \int \frac{d^3\mathbf{p}}{E_p} f(x^i, \mathbf{p}), \quad E_p \equiv (p^2c^2 + m^2c^4)^{1/2}, \tag{3.42}$$

which is clearly Lorentz invariant because of our preceding result. Unfortunately, this quantity does not seem to play any important role in physics.

Taking the first power of the four momentum, we can define the four vector

$$J^a(x^i) \equiv c \int \frac{d^3\mathbf{p}}{E_p} p^a f(x^i, \mathbf{p}). \tag{3.43}$$

The components of this vector are $(J^0, \mathbf{J})$, where

$$J^0(x^i) = \int d^3\mathbf{p} f(x^i, \mathbf{p}) \equiv n(x^i),$$

$$\mathbf{J}(x^i) = \frac{1}{c} \int d^3\mathbf{p} f(x^i, \mathbf{p})\mathbf{v} \equiv c^{-1}n(x^i)\langle\mathbf{v}\rangle, \tag{3.44}$$

where we have used the relation $(p^\alpha/E) = (v^\alpha/c^2)$. The time component of this vector, J^0, gives the particle number density n in a given frame; the spatial components give the flux of the particles in cach direction. The factor c was introduced in definition (3.43) to facilitate this interpretation.

Taking quadratic moments allows us to define the quantity

$$T^{ab}(x^i) \equiv c^2 \int \frac{d^3\mathbf{p}}{E_\mathbf{p}} p^a p^b f(x^i, \mathbf{p}), \qquad (3.45)$$

called the energy-momentum tensor or stress tensor of the system. This tensor is clearly symmetric. When one of the indices is zero, we get

$$T^{b0}(x^i) = T^{0b}(x^i) = c \int \frac{d^3\mathbf{p}}{E_\mathbf{p}} (E_\mathbf{p} p^b) f(x^i, \mathbf{p}) = c \int d^3\mathbf{p}\, p^b f(x^i, \mathbf{p}), \quad (3.46)$$

which is (c times) the total four momentum of the particles per unit volume. The time–time component $T^{00}(x^i)$ gives the energy density and the time–space component $T^{0\alpha}(x^i)$ gives the density of the α component of the three momentum.

The space–space components of this tensor represent the stresses within the medium. The component $T^{\alpha\beta}$ gives the α component of the momentum that crosses unit area orthogonally to the β direction per unit time.

Given a distribution function, we can construct the four vector $J^a(x^i)$ at any given event through definition (3.43). It is also possible, always, to choose a Lorentz frame such that the spatial components of this vector vanish at that event (i.e., $\langle \mathbf{v} \rangle = 0$) so that an observer with that Lorentz frame does not see any mean flux of particles around a given event. If the gradient of the mean velocity $\langle \mathbf{v} \rangle$ is sufficiently small, then such a Lorentz frame can be defined even globally for the whole system. (Such a definition is approximate and valid only when physical processes that depend on the gradient of mean velocity, mean kinetic energy, etc., are ignored. We shall discuss in Chap. 8, Section 8.4, the contributions that arise from the gradients.) Let us suppose that we are working in such a Lorentz frame and also that the distribution function is isotropic in momentum in this frame; that is, it depends only on the magnitude p of the momentum $\mathbf{p}$. In such a frame,

$$J^0 = \int d^3\mathbf{p}\, f(x^i, \mathbf{p}) = 4\pi \int_0^\infty p^2\, dp\, f(x^i, p), \quad J^\alpha = 0, \qquad (3.47)$$

$$T^{00} = \int d^3\mathbf{p}\, E_\mathbf{p} f(x^i, \mathbf{p}) = 4\pi \int_0^\infty p^2 E(p)\, f(x^i, p)\, dp, \quad T^{0\alpha} = 0. \quad (3.48)$$

As regards the space–space part of the energy-momentum tensor, it has to be an isotropic, symmetric, three-dimensional tensor. Hence T^α_β must have the form $T^\alpha_\beta = P(x^i)\delta^\alpha_\beta$, as δ^α_β is the only tensor available that satisfies these conditions.

To find an expression for $P(x^i)$, note that

$$T^\alpha_\alpha = P(x^i)\delta^\alpha_\alpha = 3P(x^i) = c^2 \int \frac{d^3\mathbf{p}}{E_\mathbf{p}} p^2 f(x^i, p) = 4\pi c^2 \int_0^\infty dp \frac{p^4}{E(p)} f(x^i, p).$$

$$(3.49)$$

Hence,

$$P(x^i) = \frac{4\pi c^2}{3} \int_0^\infty dp \frac{p^4}{E(p)} f(x^i, p).$$

$$(3.50)$$

This quantity represents the pressure of the fluid and has simple limits in two ex-treme cases. In the nonrelativistic limit, the energy of the particle is $E(p) \cong mc^2 + (p^2/2m)$. Substituting this form for $E(p)$ in the expression for T^{00}, we find that the energy density can be written as $T^{00} \equiv mc^2 n + \epsilon_{nr}$, where the nonrelativistic contribution ϵ_{nr} to the kinetic energy is

$$\epsilon_{nr} \equiv 4\pi \int_0^\infty p^2 \frac{p^2}{2m} f(p) dp = \frac{2\pi}{m} \int_0^\infty p^4 f(p) dp.$$

$$(3.51)$$

In the same limit, expression (3.50) for pressure reduces to

$$P_{nr} \cong \frac{4\pi c^2}{3} \int_0^\infty dp \frac{p^4}{mc^2} f(p) = \frac{4\pi}{3m} \int_0^\infty p^4 f(p) dp.$$

$$(3.52)$$

Comparing expressions (3.51) and (3.52), we see that $\epsilon_{nr} = (3/2)P_{nr}$, which is the relation between energy density and pressure in nonrelativistic theory. Note that pressure has nothing to do with interparticle collisions a priori, but arises from momentum transfer across a surface. In the other extreme limit of highly relativistic particles, $E(p) \cong pc$. In this case,

$$\rho \equiv T^{00}_{rel} = 4\pi c \int_0^\infty p^3 f(x^i, p) dp, \qquad P = \frac{4\pi c}{3} \int_0^\infty p^3 f(x^i, p) dp, \quad (3.53)$$

which shows that, for extreme relativistic particles, the pressure and the energy density are related by

$$P = \frac{1}{3}\rho.$$

$$(3.54)$$

In particular, this equation is exact for particles with zero mass, for which $E(p) = pc$ is an exact relation.

Given the components of the stress tensor in the special frame in which bulk flow vanishes, it is easy to obtain the results in any other frame in which the observer has a four velocity u^a. The result, obtained by a LT, is

$$T^a_b = (P + \rho)u^a u_b - P\delta^a_b, \qquad J^a = n_{prop} u^a.$$

$$(3.55)$$

Here n_{prop} is the proper number density – i.e, the number density in the frame comoving with the particles – and is a scalar; it is related to n in Eqs. (3.44) by $n = \gamma n_{\text{prop}}$.

To avoid misunderstanding, the following fact is stressed. The above form of T^{ab} is valid only when physical processes that arise because of gradients in u^a, etc., are ignored so that a global Lorentz frame could be defined. The spatial dependence of quantities like $P(x^i)$, etc., should be interpreted with this caveat in mind. When the gradients are not ignorable, there will be additional terms to T^{ab} that we will discuss in Chap. 8, Section 8.4.

Exercise 3.8
Filling in blanks: Prove the result of Eqs. (3.55).

3.7 External Fields of Force

In nonrelativistic mechanics, we can introduce the effect of an external force field by adding to the Lagrangian the term $-V(t, \mathbf{x})$, thereby adding to the action the integral of $-V dt$. Such a modification is, however, not Lorentz invariant and hence cannot be used in the relativistic theory. To find the kind of interactions that are permitted by Lorentz invariance, we may proceed as follows.

The action for the free particle is the integral of ds, which is Lorentz invariant. We can modify this expression to the form

$$A = -mc \int \mathcal{L}(x^a, u^a)\, ds, \tag{3.56}$$

where $\mathcal{L}(x^a, u^a)$ is a Lorentz invariant scalar made of the position and the velocity of the particle, and still maintain Lorentz invariance. We obtain a possible choice for $\mathcal{L}(x^a, u^a)$ by taking the polynomial in u^a as

$$\mathcal{L} = 1 + c_0\phi(x) + c_1 A_i(x)u^i + c_2 g_{ab}(x)u^a u^b + \cdots, \tag{3.57}$$

where $\phi(x)$ is a scalar, $A_i(x)$ is a four vector, $g_{ab}(x)$ is a second-rank tensor, etc., and $c_0, c_2, \ldots$, etc., are constants introduced for later convenience. [Quantities like ϕ depend on the four vector x^i but for convenience of notation we write $\phi(x)$ instead of $\phi(x^i)$.] In this expansion, ϕ, A_i, g_{ab}, etc., are externally specified fields that influence the trajectory of the particle.

It turns out that, in nature, we come across only a vector field A_i describing electromagnetism and a second-rank tensor field $g_{ab}(x)$ describing gravity; in other words, the Taylor series expansion of $\mathcal{L}$ in the variable u^a terminates after the quadratic term and no higher-degree terms arise. Further, the scalar field ϕ can be included in the term with g_{ab} by the addition of a part $\phi(x)\eta_{ab}$. Thus we need to deal with only $A_i(x)$ and $g_{ab}(x)$ as externally imposed fields of force that influence the trajectory of a particle. The study of $g_{ab}(x)$ (which describes the gravitational field) is postponed to Chap. 11, and we concentrate on $A_i(x)$

here. The action for a particle influenced by such a field is given by

$$A = \int_a^b \left(-mc\,ds - \frac{q}{c} A_i dx^i \right), \tag{3.58}$$

where we have written the constant c_1 as $-q/c$ and used the fact that $u^i = dx^i/ds$. The quantity q is called the electric charge of the particle, which is a Lorentz invariant number (like m) that characterises the particle. The corresponding Lagrangian is

$$L = -mc^2\sqrt{1 - (v^2/c^2)} - q\phi + \frac{q}{c}\mathbf{A}\cdot\mathbf{v} \tag{3.59}$$

if $A^i = (\phi, \mathbf{A})$ and $A_i = (\phi, -\mathbf{A})$.

To find the equations of motion in this particular case, we vary the action of Eq. (3.58) with respect to the trajectory $x^i(s)$ and get

$$\delta A = -\int_a^b \left(mc\frac{dx_i d\delta x^i}{ds} + \frac{q}{c}A_i d\delta x^i + \frac{q}{c}\delta A_i dx^i \right) = 0. \tag{3.60}$$

By integrating the first two terms by parts and using the relations

$$\delta A_i = \frac{\partial A_i}{\partial x^k}\delta x^k, \quad dA_i = \frac{\partial A_i}{\partial x^k}dx^k, \tag{3.61}$$

we find that

$$\int_a^b \left[mc\frac{du^i}{ds}\delta x^i + \frac{q}{c}(\partial_k A_i)u^k\delta x^i - \frac{q}{c}(\partial_k A_i)u^i\delta x^k \right]ds$$
$$- \left(mcu^i + \frac{q}{c}A_i \right)\delta x^i \Big|_a^b = 0, \tag{3.62}$$

where $\partial_k A_i \equiv (\partial A_i/\partial x^k)$, etc. In the third term, we interchange indices i and k (which changes nothing, as they are summed over) to obtain

$$\int_a^b \left(mc\frac{du^i}{ds} - \frac{q}{c}F_{ik}u^k \right)\delta x^i ds - \left(mcu^i + \frac{q}{c}A_i \right)\delta x^i \Big|_a^b = 0, \tag{3.63}$$

with the definition

$$F_{ik} = \frac{\partial A_k}{\partial x^i} - \frac{\partial A_i}{\partial x^k}. \tag{3.64}$$

The variations for which δx^i vanishes at the end points lead to the equations of motion:

$$mc\frac{du^i}{ds} = \frac{q}{c}F^{ik}u_k. \tag{3.65}$$

The variation of the action as a function of the end points, with the trajectories satisfying the equations of motion, allows us to identify the canonical momenta as

$$P_i = -\frac{\partial \mathcal{A}}{\partial x^i} = mcu_i + \frac{q}{c}A_i = p_i + \frac{q}{c}A_i. \tag{3.66}$$

To understand these equations in more familiar terms, we take the components of the four vector A^i to be $(\phi, \mathbf{A})$. From the definition in Eq. (3.64) it is clear that F_{ik} is antisymmetric and hence has only six independent components. It is easy to verify from the definition in Eq. (3.64) that the time–space components are expressible in terms of the three vector $\mathbf{E}$ (electric field) as $F^{\alpha 0} = E^\alpha$, with

$$\mathbf{E} = -\frac{1}{c}\frac{\partial \mathbf{A}}{\partial t} - \text{grad } \phi. \tag{3.67}$$

Similarly, the spatial components $F_{\alpha\beta}$ are given by the components of the magnetic-field vector,

$$\mathbf{B} = \text{curl } \mathbf{A}, \tag{3.68}$$

with $F_{\mu\nu} = -\epsilon_{\mu\nu\alpha}B^\alpha$, where $\epsilon_{\mu\nu\alpha}$ is the completely antisymmetric tensor in three dimensions, introduced in Exercise 3.6. In terms of the components of $\mathbf{E}$ and $\mathbf{B}$, the matrix structure of F_{ik} is

$$F_{ik} = \begin{bmatrix} 0 & E_x & E_y & E_z \\ -E_x & 0 & -B_z & B_y \\ -E_y & B_z & 0 & -B_x \\ -E_z & -B_y & B_x & 0 \end{bmatrix}, \quad F^{ik} = \begin{bmatrix} 0 & -E_x & -E_y & -E_z \\ E_x & 0 & -B_z & B_y \\ E_y & B_z & 0 & -B_x \\ E_z & -B_y & B_x & 0 \end{bmatrix}. \tag{3.69}$$

The definition of $\mathbf{E}$ and $\mathbf{B}$ in terms of ϕ and $\mathbf{A}$ implies that

$$\nabla \times \mathbf{E} = -\frac{1}{c}\frac{\partial \mathbf{B}}{\partial t}, \quad \text{div } \mathbf{B} = 0. \tag{3.70}$$

Finally, expressing the components of F^{ik} in terms of $\mathbf{E}$ and $\mathbf{B}$ in the equations of motion, we can write the spatial part of Eq. (3.65) in three-dimensional form as

$$\frac{d\mathbf{p}}{dt} = q\mathbf{E} + \frac{q}{c}\mathbf{v} \times \mathbf{B}. \tag{3.71}$$

In the nonrelativistic limit, $\mathbf{p} \approx m\mathbf{v}$, and this reduces to the familiar Lorentz force equation for a charged particle in an electromagnetic field.

Because the energy $\mathcal{E}$ of the particle is related to the momentum by $\mathcal{E}^2 = p^2c^2 + m^2c^4$, it follows that

$$\frac{d\mathcal{E}}{dt} = \mathbf{v} \cdot \frac{d\mathbf{p}}{dt}, \tag{3.72}$$

where we have used Eqs. (3.34). Hence the time component of Eq. (3.65), which gives

$$\frac{d\mathcal{E}}{dt} = q\mathbf{E} \cdot \mathbf{v},\tag{3.73}$$

represents only energy conservation and does not give any additional information. Incidentally, this shows that $\mathcal{E}$ is a constant for a particle moving in a purely magnetic field.

Because a second-rank tensor like F^{ik} transforms like the product of two four vectors $V^i U^k$ we can easily find how the electric and the magnetic fields change under LT. A simple calculation shows that

$$\mathbf{E}'_\perp = \gamma \left(\mathbf{E} + \frac{\mathbf{V}}{c} \times \mathbf{B} \right)_\perp, \quad \mathbf{E}'_\| = \mathbf{E}_\|, \quad \mathbf{B}'_\perp = \gamma \left(\mathbf{B} - \frac{\mathbf{V}}{c} \times \mathbf{E} \right)_\perp, \quad \mathbf{B}'_\| = \mathbf{B}_\|,\tag{3.74}$$

where the subscripts $\|$ and $\perp$ represent the components of the vectors along and transverse to the direction of the velocity vector $\mathbf{V}$, respectively. Clearly, the electric and the magnetic fields are not Lorentz invariant; the vanishing of the electric or the magnetic field in one frame does not necessarily imply its vanishing in other inertial frames.

There are, however, two combinations made out of electric and magnetic fields that remain invariant under LTs. These are $E^2 - B^2$ and $\mathbf{E} \cdot \mathbf{B}$. The first of these two invariants is proportional to $F_{ik}F^{ik} = 2(\mathbf{B}^2 - \mathbf{E}^2)$, and the second one is proportional to $\epsilon_{iklm} F^{ik} F^{lm} = -8(\mathbf{E} \cdot \mathbf{B})$, where ϵ_{iklm} is the completely antisymmetric fourth-rank tensor introduced in Exercise 3.6.

It is also easy to write the three-dimensional expression for any other physical quantity. For example, the spatial component $\mathbf{P}$ of the canonical momentum, defined by Eq. (3.66) is given as

$$\mathbf{P} = \gamma m\mathbf{v} + \frac{q}{c}\mathbf{A} = \mathbf{p} + \frac{q}{c}\mathbf{A}.\tag{3.75}$$

The corresponding Hamiltonian $\mathcal{H}$, expressed in terms of the canonical momenta, will be

$$\mathcal{H} = \sqrt{m^2 c^4 + c^2 \left(\mathbf{P} - \frac{q}{c}\mathbf{A} \right)^2} + q\phi.\tag{3.76}$$

We obtain the nonrelativistic limit of $\mathcal{H}$ by taking the $c \to \infty$ limit and subtracting the rest energy mc^2. This gives

$$\mathcal{H}_{\text{nr}} = \frac{1}{2m} \left(\mathbf{P} - \frac{q}{c}\mathbf{A} \right)^2 + q\phi,\tag{3.77}$$

which governs the interaction of nonrelativistic particles with the electromagnetic field. Finally, the Hamilton–Jacobi equation for a particle in an electromagnetic field will be

$$\left(\nabla \mathcal{A} - \frac{q}{c}\mathbf{A}\right)^2 - \frac{1}{c^2}\left(\frac{\partial \mathcal{A}}{\partial t} + q\phi\right)^2 + m^2c^2 = 0. \tag{3.78}$$

So far we have considered a single charge in an external electromagnetic field. If there are several charges, we have to add up the terms for each of the particles, which will give, in place of Eq. (3.58),

$$\mathcal{A} = -\sum_a \int m_a c\, ds_a - \sum_a \int \frac{q_a}{c} A_k u^k ds_a, \tag{3.79}$$

where m_a, q_a, and s_a correspond to the mass, charge, and proper time of the ath particle, respectively. When a large number of charges are present, it is convenient to introduce a charge density ρ such that $dq = \rho\, dV$ gives the amount of charge in an infinitesimal region of volume dV. Multiplying both sides of this relation by dx^i, we can write

$$dq\, dx^i = \rho\, dV\, dx^i = \rho\, dV\, dt\frac{dx^i}{dt}. \tag{3.80}$$

The left-hand side is a four vector (as dq is Lorentz invariant), and on the right-hand side $d^4x/c = dt\, dV$ is a scalar; so the combination

$$j^i = \rho\frac{dx^i}{dt} \tag{3.81}$$

must be a four vector and is called the current vector. In the action, the summation over the charges q_a involving $q_a u^k ds_a$ can be replaced with an integration over $u^k ds_a \rho\, dV = dx^k(\rho\, dV\, dt)/dt = c^{-1} j^k d^4x$. Hence we can write the action of Eq. (3.79) as

$$\mathcal{A} = -\sum_a \int m_a ds_a - \int \frac{1}{c^2} A_i j^i d^4x. \tag{3.82}$$

This form is useful for further generalisations. The Lagrangian and the Hamiltonian corresponding to the second term are

$$L_{\text{int}} = -\frac{1}{c}\int d^3\mathbf{x}(\rho\phi - \mathbf{A}\cdot\mathbf{j}), \qquad H_{\text{int}} = \frac{1}{c}\int d^3\mathbf{x}(\rho\phi - \mathbf{A}\cdot\mathbf{j}).$$

Clearly, L_{int} is a generalisation of last two terms of Eq. (3.59) for a collection of charges.

Exercise 3.9

Filling in blanks: (1) Verify Eqs. (3.69), (3.71), and (3.74). (2) Evaluate $F_{ik}F^{ik}$ and $\epsilon_{iklm}F^{ik}F^{lm}$ in terms of $\mathbf{E}$ and $\mathbf{B}$.

Exercise 3.10

Pure electric or magnetic fields: Let the electric and magnetic fields at a given event be
E and **B**. We attempt to make an LT to a different frame such that, in the neighborhood of
this event, the electromagnetic field is either purely electric or purely magnetic. (1) When
is this impossible? (2) When it is possible, obtain a condition on **E** and **B** that decides
whether the field will be purely electric or purely magnetic in the new frame. (3) Express
the velocity of the new Lorentz frame (in which the field is purely electric or magnetic)
in terms of **E** and **B**.

Exercise 3.11

Current produced by a moving charge: Consider a charge q moving along a trajectory
$z^a(s)$, where s is the proper time. Show that the current that is due to this charge can be
written in a manifestly Lorentz invariant notation

$$j^i(x^a) = q \int_{-\infty}^{\infty} ds \, \delta_D[x^a - z^a(s)] \left(\frac{dz^i}{ds} \right). \tag{3.83}$$

{Hint: The Dirac delta function is the product of three spatial ones and $\delta_D[x^0 - z^0(s)]$.
Integrate over s, eliminating the delta function on time coordinate.}

3.8 Motion of Charged Particles in External Fields

The formalism developed above allows the study of motion of charged particles
in any externally specified electromagnetic field. The basic equation governing
the motion can be taken to be either the Hamilton–Jacobi equation (3.78) or the
Lorentz force equation (3.71). In this section several features of the motion of
charged particles are illustrated within different contexts. As to be expected, these
equations allow simple, analytic solutions in only very special cases; in general,
we must resort to numerical techniques or approximate solutions. The broad
qualitative ideas developed below will, however, be of use in more complicated
cases as well.

3.8.1 Motion in a Coulomb Field

As a first application, let us consider the motion of a charged particle (with charge
q and mass m) in an external field given by $\phi = e/r$, $\mathbf{A} = 0$, called the Coulomb
field. In the nonrelativistic limit, this would correspond to the Kepler problem
and the trajectory would be a conic section. In the exact, relativistic case, the
Hamilton–Jacobi equation (3.78) is

$$-\frac{1}{c^2}\left(\frac{\partial \mathcal{A}}{\partial t} + \frac{\alpha}{r}\right)^2 + \left(\frac{\partial \mathcal{A}}{\partial r}\right)^2 + \frac{1}{r^2}\left(\frac{\partial \mathcal{A}}{\partial \theta}\right)^2 + m^2 c^2 = 0, \tag{3.84}$$

where (r, θ) are the polar coordinates on the plane of the motion and $\alpha = qe$.

Writing $A = -\mathcal{E}t + J\theta + f(r)$ and solving for $f(r)$, we find that

$$A = -\mathcal{E}t + J\theta + \int dr \sqrt{\frac{1}{c^2}\left(\mathcal{E} - \frac{\alpha}{r}\right)^2 - \frac{J^2}{r^2} - m^2c^2}. \tag{3.85}$$

The behaviour is now quite different from that of the nonrelativistic Kepler problem. Expansion of $c^{-2}[\mathcal{E} - \alpha/r]^2$ will lead to a $(1/r)$ term and $(1/r^2)$ term; the latter will combine with the (J^2/r^2) term, and the net sign of this term will depend on whether Jc is greater than $|\alpha|$ or less. In either case, we have a situation with both a $(1/r)$ term and a $(1/r^2)$ term, and the motion will not – in general – be closed on the torus in phase space (see Chap. 2, Section 2.5).

We can obtain the trajectories for all the cases by the usual procedure of setting $(\partial A/\partial J) = \theta_0 = $ constant. This gives, for the three cases $Jc > |\alpha|$, $Jc < |\alpha|$, and $Jc = |\alpha|$,

$$(c^2 J^2 - \alpha^2)\frac{1}{r} = c\sqrt{(J\mathcal{E})^2 - m^2c^2(J^2c^2 - \alpha^2)}\cos\left(\theta\sqrt{1 - \frac{\alpha^2}{c^2 J^2}}\right) - \mathcal{E}\alpha,$$

$$\tag{3.86}$$

$$(\alpha^2 - c^2 J^2)\frac{1}{r} = \pm c\sqrt{(J\mathcal{E})^2 + m^2c^2(\alpha^2 - J^2c^2)}\cosh\left(\theta\sqrt{\frac{\alpha^2}{c^2 J^2} - 1}\right) + \mathcal{E}\alpha,$$

$$\tag{3.87}$$

$$\frac{2\mathcal{E}\alpha}{r} = \mathcal{E}^2 - m^2c^4 - \theta^2\left(\frac{\mathcal{E}\alpha}{cJ}\right)^2, \tag{3.88}$$

respectively. We have also chosen the initial conditions such that $\theta_0 = 0$ for simplicity.

For $Jc < |\alpha|$, we take the positive root for $\alpha < 0$ and the negative root for $\alpha > 0$. In the former case, the trajectory spirals in to the origin, which is quite unlike the nonrelativistic case and arises because the sign of the $(1/r^2)$ term is always positive in the nonrelativistic case but can be negative in the relativistic case. This possible change of sign makes the angular momentum inadequate – in general – for preventing the collapse to the origin.

Consider next the case with $Jc > |\alpha|$; in this case, if $\alpha < 0$, the field is attractive and we have bound motion for $\mathcal{E} < mc^2$. The motion, of course, is not closed and the orbit precesses. The trajectory can be expressed in the form

$$\frac{1}{r} = \frac{1}{R}\cos\left[\left(1 - \frac{\alpha^2}{c^2 J^2}\right)^{1/2}\theta\right] - \frac{\mathcal{E}\alpha}{c^2 J^2}\left(1 - \frac{\alpha^2}{c^2 J^2}\right)^{-1}, \tag{3.89}$$

where

$$R \equiv \frac{J}{mc}\left(1 - \frac{\alpha^2}{c^2 J^2}\right)\left[\left(\frac{\mathcal{E}}{mc^2}\right)^2 - 1 + \frac{\alpha^2}{c^2 J^2}\right]^{-1/2}. \tag{3.90}$$

If we further assume that $\alpha^2 \ll c^2 J^2$, then this expression simplifies to

$$\frac{1}{r} \cong \frac{1}{R} \cos\theta - \frac{\alpha\mathcal{E}}{c^2 J^2}, \qquad R \cong \frac{J}{p_\infty}, \tag{3.91}$$

where p_∞ is the momentum of the particle that corresponds to energy $\mathcal{E}$ when it is far away from the origin. The quantity R is the impact parameter for the collision of the two charged particles. Because, in the absence of the field ($\alpha = 0$), the trajectory is a straight line with $r\cos\theta = x = R$, the second term ($\alpha\mathcal{E}/c^2 J^2$) on the right-hand side of relation (3.91) introduces a small correction to this straight-line trajectory when $\alpha^2 \ll c^2 J^2$. In this limit, the above trajectory describes the (small) deflection of a charged particle in the external field. We can obtain the asymptotic directions $\pm\theta_c$ of the charged particle by setting $r^{-1} = 0$ in the trajectory, which gives $\cos\theta_c \approx (\alpha R\mathcal{E}/c^2 J^2)$. If we write $\theta_c = (\pi/2) - \psi$, so that $\cos\theta_c = \sin\psi \cong \psi = (\alpha R\epsilon/c^2 J^2)$, the total deflection χ is

$$\chi = 2\psi \cong \frac{2\alpha}{Rc^2}\frac{\mathcal{E}}{p_\infty^2} = \frac{2\alpha}{mRv_\infty^2}\left(1 - \frac{v_\infty^2}{c^2}\right)^{1/2}, \tag{3.92}$$

where v_∞ denotes the speed corresponding to the momentum p_∞. Therefore the deflection is $\chi = [(\alpha/R)/(mv^2/2)]$ in the nonrelativistic case (which is the ratio between electrostatic potential energy at the impact parameter and the kinetic energy at infinity) and goes to zero for an ultrarelativistic particle as $v_\infty \to c$. Also note that significant scattering ($\chi \simeq 1$) occurs in Coulomb interaction only for impact parameters $R_c \simeq (\alpha/mv^2)$. This fact will find application in many contexts.

This expression can be used to determine the scattering cross section in the Coulomb field for small deflections. If a deflection $\chi = 2\psi$ is produced in a scattering with impact parameter R, then the angular-scattering cross section is defined by

$$d\sigma \equiv 2\pi R\,dR = 2\pi R\left|\frac{dR}{d\chi}\right|d\chi = \frac{R}{\sin\chi}\left|\left(\frac{dR}{d\chi}\right)\right|d\Omega \cong \frac{R}{\chi}\left|\left(\frac{dR}{d\chi}\right)\right|d\Omega, \tag{3.93}$$

where $d\Omega = 2\pi\sin\chi\,d\chi$ is an element of solid angle and the last equality is valid for small deflections. In our case, $\chi(R) = (2\alpha\mathcal{E}/p_\infty^2 c^2)R^{-1} = (2\alpha/p_\infty v_\infty R)$, giving

$$\frac{d\sigma}{d\Omega} = \frac{1}{\chi^4}\left(\frac{2\alpha}{p_\infty v_\infty}\right)^2 \cong \frac{1}{\chi^4}\left(\frac{2\alpha}{mv_\infty^2}\right)^2, \tag{3.94}$$

where the second equality is valid for nonrelativistic motion. In the case of scattering between two moving charges, v_∞ should be interpreted as the magnitude of the relative velocity.

3.8.2 Motion in a Constant, Uniform, Electric Field

Taking the direction of the electric field along the x axis and the plane of motion as the x–y plane, we find that force equation (3.71) becomes $\dot{p}_x = q\,E$, $\dot{p}_y = 0$, which may be integrated to give $p_x = q\,Et$, $p_y = p_0$ with a suitable choice of integration constants. The energy of the particle is given by

$$\mathcal{E} = \sqrt{m^2 c^4 + c^2 p_0^2 + (cq\,Et)^2} \equiv \sqrt{\mathcal{E}_0^2 + (cq\,Et)^2}. \tag{3.95}$$

Using Eqs. (3.34), we find that v_x is given by

$$v_x = \frac{dx}{dt} = \frac{p_x c^2}{\mathcal{E}} = \frac{c^2 q\,Et}{\sqrt{\mathcal{E}_0^2 + (cq\,Et)^2}}. \tag{3.96}$$

Integrating, we obtain the trajectory that is a hyperbola,

$$x^2 - c^2 t^2 = \left(\frac{\mathcal{E}_0}{q\,E}\right)^2, \tag{3.97}$$

with a suitable choice for the origin. The motion along the y axis can also be determined by the same method to give

$$y = \frac{p_0 c}{q\,E} \sinh^{-1}\left(\frac{cq\,Et}{\mathcal{E}_0}\right). \tag{3.98}$$

Eliminating t between Eqs. (3.97) and (3.98), we find that the trajectory in the x–y plane is given by $x = (\mathcal{E}_0/q\,E)\cosh(q\,Ey/p_0 c)$, which is a catenary. This is a simple example of motion with constant acceleration in special relativity. When $c \to \infty$, the trajectory will reduce to a parabola, which is a well-known nonrelativistic result.

3.8.3 Motion in a Constant, Uniform, Magnetic Field

In this case, which is of considerable astrophysical importance, the equation of motion is $\dot{\mathbf{p}} = (q/c)\mathbf{v} \times \mathbf{B}$. Using the relation $\mathbf{p} = (\mathcal{E}/c^2)\mathbf{v}$ and noting that the energy $\mathcal{E}$ is a constant for motion under the action of a magnetic field [see Eq. (3.73)], we can write the equation of motion as

$$\frac{d\mathbf{v}}{dt} = -\frac{qc}{\mathcal{E}}(\mathbf{B} \times \mathbf{v}) \equiv -(\boldsymbol{\Omega} \times v), \quad \boldsymbol{\Omega} \equiv \frac{qc}{\mathcal{E}}\mathbf{B}. \tag{3.99}$$

This is identical to the equation of motion for a particle that is rotating with an angular velocity $\boldsymbol{\Omega}$. Taking the magnetic field to be along the z axis, we find that the solution is given by

$$v_x = v_\perp \cos(\Omega t + \alpha), \quad v_y = -v_\perp \sin(\Omega t + \alpha), \quad v_z = v_\|, \tag{3.100}$$

$$x = x_0 + \frac{v_\perp}{\Omega} \sin(\Omega t + \alpha), \quad y = y_0 + \frac{v_\perp}{\Omega} \cos(\Omega t + \alpha), \quad z = z_0 + v_\| t, \tag{3.101}$$

where $v_\parallel$, $v_\perp$, and α are determined by the initial conditions and are constants. The motion along the z axis has uniform velocity; the motion in the transverse x–y plane is a circle around the point (x_0, y_0), where the radius of the circle is

$$r_L = \frac{v_\perp}{\Omega} = \frac{v_\perp \mathcal{E}}{q c B} = \frac{c p_\perp}{q B}. \tag{3.102}$$

This r_L is called the Larmor radius and is determined by the ratio of the transverse component of the momentum to the magnetic field. In the nonrelativistic limit, $\mathcal{E} \approx mc^2$, and we get

$$\Omega \cong \frac{q}{mc} \mathbf{B}, \quad r_L \cong \left(\frac{mc}{q} \right) \left(\frac{v_\perp}{B} \right). \tag{3.103}$$

We now study variants of these results in more complicated situations.

3.8.4 Motion in a Slowly Varying Magnetic Field

In many astrophysical scenarios, the magnetic field will vary both in magnitude and direction in space but can be taken to be independent of time. The motion of charged particles in such fields can be described by using the concepts developed in Subsection (3.8.3) above. By and large, such a motion involves slow changes in the Larmor radius or a slow drift superposed over the basic motion. We now consider such examples one by one.

Let us first consider a case in which the magnetic field remains uniform in space but changes slowly in time. We can then use the concept of adiabatic invariants developed in Chap. 2, Section 2.6. Because the motion in the plane perpendicular to the magnetic field is quasiperiodic, the adiabatic invariant is given by the integral

$$I = \frac{1}{2\pi} \oint \mathbf{P}_\perp \cdot d\mathbf{r} = \frac{1}{2\pi} \oint \mathbf{p}_\perp \cdot d\mathbf{r} + \frac{q}{2\pi c} \oint \mathbf{A} \cdot d\mathbf{r}. \tag{3.104}$$

Because $\mathbf{p}_\perp$ is constant in magnitude and is directed along $d\mathbf{r}$, the first term gives the contribution $p_\perp r_L$. In the second term, we apply Stokes theorem and use $\mathbf{B} = \nabla \times \mathbf{A}$ to obtain

$$I = r_L p_\perp + \frac{q}{2c} B r_L^2 = \frac{3 c p_\perp^2}{2 q B}. \tag{3.105}$$

The constancy of I shows that $p_\perp^2 \propto B$. Because $p_\perp = q B r_L / c$, it follows that $B r_L^2$ remains constant as the magnetic field varies. Thus, for adiabatic changes, the flux of the magnetic field going through the orbit of the particle does not change.

The above result can be used to obtain the nature of motion in a different case that occurs frequently in astrophysics. Consider a situation in which the magnetic field is constant in time but the magnetic-field lines converge towards a particular

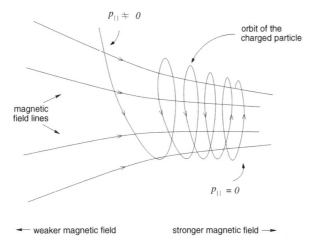

$p_{\parallel} \neq 0$

orbit of the
charged particle

magnetic
field lines

$p_{\parallel} = 0$

← weaker magnetic field   stronger magnetic field →

Fig. 3.1.   Motion of a charged particle towards a region of increasing magnetic field.
The particle is reflected and starts moving back to left from the location where $p_{\parallel} = 0$.

direction (see Fig. 3.1). We assume that the particle is initially moving with a
velocity $v_{\parallel}$ directed towards the region of the stronger magnetic field. In a frame
that moves with speed $v_{\parallel}$ towards the region of stronger field, the particle will
experience a magnetic field that is becoming stronger (at a slow rate). Using the
result obtained above in this frame, we expect the transverse momentum to scale
as $p_{\perp}^2 = kB$, where k is some constant. Further, for motion in such a magnetic
field, the energy of the particle is a constant, implying that $p^2 = p_{\parallel}^2 + p_{\perp}^2$ remains
constant. So the longitudinal momentum in the direction of the field is given by

$$p_{\parallel}^2 = p^2 - p_{\perp}^2 = p^2 - kB(\mathbf{x}). \tag{3.106}$$

Because $p_{\parallel}^2 \geq 0$, the particle cannot penetrate into regions of high field strength
where $B > p^2/k$. As the particle moves towards the direction of the increasing
field, the radius of the trajectory decreases in proportion to $p_{\perp}/B \propto B^{-1/2}$. On
reaching the boundary where $p_{\parallel}$ vanishes, the particle is reflected: It will continue
to rotate in the same direction but begin to move longitudinally towards the region
of low field density.

   If the magnetic field is such that it is stronger at two different points with
a comparatively weaker regime in between, then the charged particles can be
trapped between the two strong field regimes and will move back and forth.
Such field configurations are called magnetic traps or magnetic bottles.

3.8.5 Drifts in Magnetic Fields

The spatial variation of the magnetic field can also lead to several other phenomena, which we now discuss. In this section, we assume that the motion of the
particle is nonrelativistic.

Consider a case in which a charged particle is acted on by a magnetic field **B**, an electric field **E** that is in a direction perpendicular to the magnetic field, and some other force that is again in a direction perpendicular to the magnetic field, producing an acceleration **g**. (This extra force could be the gravitational force in some astrophysical contexts.) The equation of motion for the particle is then given by

$$\dot{\mathbf{v}} = \frac{q}{m}\left(\mathbf{E} + \frac{\mathbf{v}}{c} \times \mathbf{B}\right) + \mathbf{g}$$

$$= \frac{q}{mc}\left(\mathbf{v} - c\frac{\mathbf{E} \times \mathbf{B}}{B^2} - \frac{mc}{q}\frac{\mathbf{g} \times \mathbf{B}}{B^2}\right) \times \mathbf{B}, \tag{3.107}$$

where the validity of the second line follows from the vector identity for triple products and the fact that **E** and **g** are transverse to **B**. When a new velocity $\mathbf{v}'$ is defined such that

$$\mathbf{v} = \mathbf{v}' + c\frac{\mathbf{E} \times \mathbf{B}}{B^2} + \frac{mc}{q}\frac{\mathbf{g} \times \mathbf{B}}{B^2}, \tag{3.108}$$

the equation of motion reduces to

$$\dot{\mathbf{v}}' = -\frac{q}{mc}(\mathbf{B} \times \mathbf{v}'). \tag{3.109}$$

This equation has the same form as that of Eq. (3.99); hence the motion described by $\mathbf{v}'$ is essentially a rotation around the magnetic field.

The full motion described by **v** has two more additional terms. The second term in Eq. (3.108) represents a drift in the direction of $\mathbf{E} \times \mathbf{B}$, that is, the charged particle moves in a circle whose centre drifts with a velocity $(c/B^2)(\mathbf{E} \times \mathbf{B})$. This drift is independent of the charge and the mass of the particle. All charged particles move with the same drift velocity.

The third term in Eq. (3.108) gives a similar drift in the direction of $\mathbf{g} \times \mathbf{B}$. This drift, however, depends on the charge and the mass of the particle and has different signs for positively and negatively charged particles.

It is clear from the above discussion that whenever there is an external force field **F** that is transverse to the magnetic field, it will lead to a drift with velocity

$$\mathbf{v}_D = \frac{c}{q}\frac{\mathbf{F} \times \mathbf{B}}{B^2}. \tag{3.110}$$

This fact can be used to study several effects arising because of the inhomogeneities in the magnetic field.

As a first example, consider a magnetic-field line that has a slight curvature, that is, the field bends in the direction of unit vector **n** with a radius of curvature R [see Fig. 3.2(a)]. We assume that the Larmor radius $r_L \ll R$. Then the charged particle will move around the field line in a circle with radius r_L while the centre of the circle will move along the curved field line. Such a motion of the centre results in a centripetal acceleration given by $\mathbf{g} = (v_{\parallel}^2/R)\mathbf{n}$, where $v_{\parallel}$ is the speed

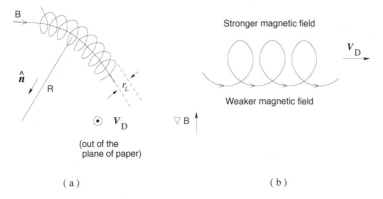

Fig. 3.2.   (a) Curvature and (b) gradient drifts of a charged particle in a magnetic field.

of motion along the field lines. From the above discussion [see Eq. (3.110)], it follows that this will cause a drift (called curvature drift) with velocity

$$\mathbf{v}_{\text{curv}} = \frac{c}{q} \frac{m\mathbf{g} \times \mathbf{B}}{B^2} = \frac{mc}{q} \frac{v_\parallel^2}{B^2} \left(\frac{\mathbf{n}}{R} \times \mathbf{B} \right). \tag{3.111}$$

By using the geometrical relation $(\mathbf{n}/R) = B^{-2}(\mathbf{B} \cdot \nabla)\mathbf{B}$, we can express the net drift as

$$\mathbf{v}_{\text{curv}} = -\frac{mv_\parallel^2 c}{q} \frac{\mathbf{B} \times (\mathbf{B} \cdot \nabla)\mathbf{B}}{B^4}. \tag{3.112}$$

The magnitude of this curvature drift is easily seen to be $v_{\text{curve}} = (r_L/R)(v_\parallel/v_\perp)v_\parallel$, where $v_\perp$ is the speed of the motion perpendicular to the field lines.

A second kind of drift arises because of the transverse spatial gradient in the magnitude of $\mathbf{B}$ [see Fig. 3.2(b)]. Such a gradient causes the circular motion of the charged particle to be tighter in the region of larger $\mathbf{B}$ than in the region of smaller $\mathbf{B}$, thereby producing a drift. There are several ways of deriving this effect, and we use a method that illustrates a useful technique. This technique involves averaging the Lagrangian over relatively fast motion, thereby obtaining an effective Lagrangian that describes the drift.

Let $A^\alpha(x^\mu)$ be a component of a vector potential describing a slowly varying magnetic field. (All Greek indices range over 1, 2, and 3, and we do not distinguish between the superscripts and the subscripts.) Let us separate the instantaneous position x^μ of the particle as $x^\mu = R^\mu + l^\mu$, where l^μ represents the rapid rotation at Larmor frequency around the field lines and R^μ is the slower drift motion. In that case, the components of the velocity are given by

$$v^\alpha = \dot{R}^\alpha + \dot{l}^\alpha = \dot{R}^\alpha + \Omega\epsilon^{\alpha\mu\nu}l_\mu n_\nu, \tag{3.113}$$

where we have substituted for the Larmor rotation by using Eqs. (3.100), with $n^\alpha = B^\alpha / B$ as the unit vector in the direction of the field. Clearly the charged particle samples the field in a region of size l around R. Because the field is supposed to vary slowly over the Larmor radius, we can expand the vector potential in a Taylor series as

$$A^\alpha \cong A^\alpha(R^\mu) + l^\beta \partial_\beta A^\alpha(R^\mu), \qquad (3.114)$$

retaining up to the first order. The Lagrangian describing the coupling between the charged particle and the field, in the nonrelativistic limit, is $L = (q/c)\mathbf{A} \cdot \mathbf{v}$ [see Eq. (3.59)]. This Lagrangian governs both the dynamics of the vector $\mathbf{R}$ as well as that of vector $\mathbf{l}$. To find an effective Lagrangian governing the slow drift, we can average L over the rapid motion described by the Larmor rotation. To do this, we have to compute the average value $\langle A^\alpha v_\alpha \rangle$. By using relations (3.113) and (3.114), we find that the product $A^\alpha v_\alpha$ has four terms. Of these, the zeroth-order term is irrelevant for the study of the drift; the terms linear in l^μ vanish on averaging because $\langle l^\mu \rangle = 0$; in the term quadratic in l, we use the result $\langle l^\mu l^\nu \rangle = (1/2)\delta^{\mu\nu}\langle l^2 \rangle = (1/2)\delta^{\mu\nu} r_L^2$ to get

$$\langle A^\alpha v_\alpha \rangle = \frac{1}{2} r_L^2 \Omega n^\mu \epsilon_{\alpha\beta\mu}(\partial^\beta A^\alpha) = \frac{1}{4}\Omega r_L^2 n^\mu \epsilon_{\alpha\beta\mu} F^{\beta\alpha}, \qquad (3.115)$$

where the second equality follows from the antisymmetry of $\epsilon_{\alpha\beta\mu}$ and the definition of $F^{\beta\alpha}$. We can simplify this expression further by using the relations

$$F^{\beta\alpha} = -\epsilon^{\beta\alpha\sigma} B_\sigma, \qquad \epsilon_{\alpha\beta\mu}\epsilon^{\beta\alpha\sigma} = -2\delta_\mu^\sigma \qquad (3.116)$$

to provide the effective Lagrangian

$$L_{\text{eff}} \equiv \langle L \rangle = \frac{q}{c}\langle A^\alpha v_\alpha \rangle = \frac{q}{c}\left(\frac{1}{2}\Omega r_L^2 B\right) \equiv \mu B, \qquad (3.117)$$

with the definition

$$\mu \equiv \frac{q}{2c}\Omega r_L^2 = \frac{1}{2}\frac{m v_\perp^2}{B}. \qquad (3.118)$$

We have seen above in Subsection (3.8.4) that $v_\perp^2 \propto B$ for small variations. Therefore, in varying the effective Lagrangian, we can treat μ as a constant to the same order of accuracy. Then the extra force contributed by this effective Lagrangian will be

$$\mathbf{f} = -\nabla U = \nabla L_{\text{eff}} = \mu \nabla B = \frac{1}{2}\frac{m v_\perp^2}{B}\nabla B, \qquad (3.119)$$

which arises because of the gradient in the magnitude of the field. From the general result of Eq. (3.110), we conclude that this will cause a gradient drift with velocity

$$\mathbf{v}_{\text{grad}} = -\frac{mv_{\perp}^2}{2q}\left(\frac{c}{B^3}\right)(\mathbf{B} \times \nabla B). \tag{3.120}$$

In general, both the curvature drift and the gradient drift will be present when the magnetic field varies slowly in space. If any other force field (such as the electric field) exists transverse to the magnetic field, there will be an additional drift such as the $\mathbf{E} \times \mathbf{B}$ drift.

3.9 Maxwell's Equations

The action principle developed in Section 3.7 couples the charged particle to the electromagnetic field but treats the field as an externally specified entity. Such an external field can act on the charged particle and change its energy, momentum, angular momentum, etc. However, because the conservation of these quantities is ensured for a closed system from general symmetry considerations, it is clear that the electromagnetic field must possess energy, momentum, and angular momentum, which should also become changed during the interaction with the charged particle. In other words, the field must be a dynamic entity and change in response to the interaction, obeying certain equations of motion.

The action in Eq. (3.82) does not allow us to determine the evolution of the field. To achieve this, we should treat the field as a dynamical variable and add a term to the action that will produce the equations of motion determining the field when the action is varied with respect to the field variables. Let us consider the form of this extra term.

The action representing the field will be expressible as an integral over the four volume d^4x of some scalar Lagrangian (density) $L = L(A^i, F^{ik})$, which could be a function of the potential A^i and the electromagnetic field F^{ik}. We note that the equations of motion for the charged particle (3.65) contain only the fields and not the potentials. Therefore, by measuring the trajectories of the charges, we will be able to determine only the electromagnetic fields and not the potentials. In fact, several different potentials can lead to the same electromagnetic field: If we add to a potential A_i a gradient $\partial_i f$ of a scalar f, then the resulting potential $A_i' = A_i + \partial_i f$ will have the same field tensor, that is, $F_{ik}' = F_{ik}$, as can be trivially verified. Such a change is called a gauge transformation, and the fields are invariant under gauge transformation. This shows that the potential is not directly observable; hence, it need not appear in L and we should be able to construct L out of F_{ik} alone.

Experiments show that electromagnetic fields obey the principle of superposition, viz., the field that is due to two independently specified charge distributions is the sum of the fields produced by each of them in the absence of the other. For this to be true the differential equations in electrodynamics have to be linear in the field; alternatively, the Lagrangian can be at most quadratic in the field variable.

We saw in Section 3.8 that there are only two independent, scalar quantities that can be made out of F_{ik} that are quadratic, viz., $Q_1 \equiv F_{ik}F^{ik} \propto \mathbf{E}^2 - \mathbf{B}^2$ and $Q_2 = \epsilon_{iklm}F^{ik}F^{lm} \propto \mathbf{E} \cdot \mathbf{B}$. Therefore, in general the Lagrangian can be a linear combination of these two quantities. Of these, Q_2 can be expressed as a four divergence:

$$\epsilon^{iklm}F_{ik}F_{lm} = 4\frac{\partial}{\partial x^i}\left(\epsilon^{iklm}A_k\frac{\partial}{\partial x^l}A_m\right). \tag{3.121}$$

When the Lagrangian is integrated over the four volume to obtain the action, any four divergence will give contributions only at the surface; because the fields are treated as fixed on the surface, this term Q_2 will not contribute to the variation in the action. Therefore, only the scalar $F_{ik}F^{ik}$ survives as a possible choice for the Lagrangian of the electromagnetic field, and hence the action must be proportional to the integral of this term over d^4x. It is conventional to write this part of the action as

$$\mathcal{A}_f = -\frac{1}{16\pi c}\int F_{ik}F^{ik}d^4x = \frac{1}{8\pi c}\int (\mathbf{E}^2 - \mathbf{B}^2)\,d^4x. \tag{3.122}$$

The magnitude of the constant in front is arbitrary and merely decides the units used for measuring the electromagnetic field. The sign is chosen so that the term $(\partial \mathbf{A}/\partial t)^2$ has a positive coefficient. If not, we can make the action arbitrarily small by varying $\mathbf{A}$ rapidly in time so that there is no minimum for $\mathcal{A}_f$. We can thus choose the sign by invoking the principle that the action should have a minimum rather than just an extremum. In this action, the dynamical variables are the potentials A^i; the space–time coordinates $x^i = (ct, \mathbf{x})$ are just parameters. The second equality in Eq. (3.122) allows us to identify the Lagrangian for the field as the integral over $d^3\mathbf{x}$ of the quantity $\mathcal{L} \equiv (8\pi)^{-1}(\mathbf{E}^2 - \mathbf{B}^2)$.

The full action for a system of charged particles interacting with electromagnetic fields now is

$$A = -\sum \int mc\,ds - \frac{1}{c^2}\int A_i j^i d^4x - \frac{1}{16\pi c}\int F_{ik}F^{ik}\,d^4x, \tag{3.123}$$

where the first two terms are same as those in Eq. (3.82). To find the equations of motion for the field, we have to vary the potentials A_i in this action. The first term is independent of A_i and does not contribute; the next two terms give

$$\delta A = -\int \frac{1}{c}\left(\frac{1}{c}j^i\delta A_i + \frac{1}{8\pi}F^{ik}\frac{\partial}{\partial x^i}\delta A_k - \frac{1}{8\pi}F^{ik}\frac{\partial}{\partial x^k}\delta A_i\right)d^4x, \tag{3.124}$$

where we have used the definition of F_{ik} and the relation $F^{ik}\delta F_{ik} = F_{ik}\delta F^{ik}$. In the second term we interchange i and k and replace F_{ki} with $-F_{ik}$ to obtain

$$\delta A = -\int \frac{1}{c}\left(\frac{1}{c}j^i\delta A_i - \frac{1}{4\pi}F^{ik}\frac{\partial}{\partial x^k}\delta A_i\right)d^4x. \tag{3.125}$$

Finally, we integrate the second term by parts and convert it to a surface integral:

$$\delta A = -\frac{1}{c}\int\left(\frac{1}{c}j^i + \frac{1}{4\pi}\frac{\partial F^{ik}}{\partial x^k}\right)\delta A_i\, d^4x - \frac{1}{4\pi c}\int F^{ik}\delta A_i\, dS_k\bigg|. \tag{3.126}$$

Let us assume that the surface term vanishes for a suitable class of variations. In that case, we can set to zero the integrand of the first term, thereby obtaining the equations of motion for the field:

$$\frac{\partial F^{ik}}{\partial x^k} = -\frac{4\pi}{c}j^i. \tag{3.127}$$

Taking the components, with $j^i = (\rho c, \mathbf{j})$, we can write Eq. (3.127) in three-dimensional notation as

$$\nabla \times \mathbf{B} = \frac{1}{c}\frac{\partial \mathbf{E}}{\partial t} + \frac{4\pi}{c}\mathbf{j}, \quad \text{div } \mathbf{E} = 4\pi\rho. \tag{3.128}$$

These are the equations connecting the field to its source. These equations, along with Eqs. (3.70), determine the evolution of electromagnetic field and are called Maxwell's equations.

Note that if we differentiate Eq. (3.127) with respect to x^i, the left-hand side $\partial_i\partial_k F^{ik}$ vanishes identically because of the antisymmetry F^{ik}. Hence we get the constraint $\partial_i j^i = 0$ on the current vector. In three-dimensional form, this becomes

$$\frac{\partial \rho}{\partial t} + \text{div } \mathbf{j} = 0. \tag{3.129}$$

Integrating this equation over a large three volume d^3x and using Gauss theorem, we can see that the total charge does not change with time. Thus Maxwell's equations imply the conservation of electric charge.

Using the relation between F_{ik} and A_j, we can write Eq. (3.127) as

$$\frac{\partial}{\partial x^k}\left(\frac{\partial A^k}{\partial x_i} - \frac{\partial A^i}{\partial x_k}\right) = \frac{\partial}{\partial x_i}\left(\frac{\partial A^k}{\partial x^k}\right) - \frac{\partial^2}{\partial x^k\partial x_k}A^i = -\frac{4\pi}{c}j^i. \tag{3.130}$$

The freedom in the choice of gauge can be now used to set $(\partial A^k/\partial x^k) = 0$ (which is called the Lorentz gauge), thereby reducing Maxwell's equations to the form

$$\frac{\partial^2}{\partial x^k\partial x_k}A^i \equiv \left(\frac{1}{c^2}\frac{\partial^2}{\partial t^2} - \nabla^2\right)A^i \equiv \Box A^i = \frac{4\pi}{c}j^i. \tag{3.131}$$

In Chap. 2, we saw that the canonical momentum for a system can be expressed as the derivative of the action with respect to the variation of the coordinate at one of the end points. This concept can be extended even for a field. In the case of electromagnetic fields, we can easily show [by using the form of the second term in Eq. (3.126)] that the canonical four-momentum density is given by $P^i = (F^{i0}/4\pi)$. Thus the canonical momentum corresponding to **A** is proportional to the electric field **E**, while the canonical momentum corresponding to A^0 vanishes identically. The latter result is a consequence of the fact that the scalar potential A^0 can always be made to vanish by a suitable gauge and does not represent a true degree of freedom of the system.

Exercise 3.12

General solution to Maxwell's equations: Because Maxwell's equations are linear in the electromagnetic field, it is possible to provide the most general solution to these equations in Fourier space. (1) To do this, it is convenient to deal with the variables $\phi_{\mathbf{k}}(t)$, $\mathbf{A}_{\mathbf{k}}(t)$, $\mathbf{E}_{\mathbf{k}}(t)$, $\mathbf{B}_{\mathbf{k}}(t)$, $\rho_{\mathbf{k}}(t)$, and $\mathbf{j}_{\mathbf{k}}(t)$, which are the spatial Fourier transforms of the scalar potential, vector potential, electric field, magnetic field, charge density, and current density, respectively. Write the gauge condition $\nabla \cdot \mathbf{A} = 0$, law of conservation of charge, and the Maxwell's equations in Fourier space. (2) Manipulate these to show that (i) The scalar potential is related to the charge by

$$\phi_{\mathbf{k}}(t) = \frac{4\pi}{k^2}\rho_{\mathbf{k}}(t). \tag{3.132}$$

How do you reconcile Eq. (3.132) with the fact that signals cannot travel with infinite speed? (ii) The vector potential is related to the transverse part of the current by

$$\ddot{\mathbf{A}}_{\mathbf{k}}^{\perp} + c^2 k^2 \mathbf{A}_{\mathbf{k}}^{\perp} = 4\pi c \mathbf{j}_{\mathbf{k}}^{\perp}, \quad \mathbf{A}_{\mathbf{k}}^{\parallel} = 0, \tag{3.133}$$

where the symbol $\perp$ indicates the component perpendicular to **k**, etc. This analysis shows that the true dynamical degree of freedom of a electromagnetic field is the transverse component $\mathbf{A}_{\mathbf{k}}^{\perp}$ of the vector potential. [The scalar potential $\phi_{\mathbf{k}}$ is completely determined by the charge density instantaneously through Eq. (3.132) and the longitudinal component $\mathbf{A}_{\mathbf{k}}^{\parallel}$ of the vector potential can be made to vanish by a gauge condition.] Thus the electromagnetic field has only 2 independent degrees of freedom per space point. Because the magnetic field has a one-to-one correspondence with the transverse part of vector potential, the true dynamical degrees of freedom of the electromagnetic field are contained in the (divergence-free) magnetic field.

Exercise 3.13

Coordinates and momentum: Study the second term of Eq. (3.126) carefully and show that magnetic field is the quantity that is kept fixed at the end points of the variation. Thus **B** is the coordinate and **E** is the momentum for the electromagnetic field.

3.10 Energy and Momentum of the Electromagnetic Field

By including the electromagnetic fields in the action, we are treating them as dynamical entities with degrees of freedom of their own. When a charge moves under the influence of an electromagnetic field, its momentum and energy change.

Because the total momentum (or energy) of a closed system is a constant, it follows that the change of energy and momentum of the charged particle must be compensated for by a change of the energy and the momentum of the field. To show this explicitly, we need to obtain the expressions for the energy and the momentum of the electromagnetic field.

Two of Maxwell's equations give the time derivative of the electric and the magnetic fields:

$$\frac{1}{c}\frac{\partial \mathbf{B}}{\partial t} = -(\nabla \times \mathbf{E}), \qquad \frac{1}{c}\frac{\partial \mathbf{E}}{\partial t} = \nabla \times \mathbf{B} - \frac{4\pi}{c}\mathbf{j}.$$

By taking the dot product of these equations with $\mathbf{B}$ and $\mathbf{E}$ and adding, we get

$$\frac{1}{c}\mathbf{E}\cdot\frac{\partial \mathbf{E}}{\partial t} + \frac{1}{c}\mathbf{B}\cdot\frac{\partial \mathbf{B}}{\partial t} = -\frac{4\pi}{c}\mathbf{j}\cdot\mathbf{E} - (\mathbf{B}\cdot\nabla\times\mathbf{E} - \mathbf{E}\cdot\nabla\times\mathbf{B}), \quad (3.134)$$

which can be rewritten as

$$\frac{\partial}{\partial t}\left(\frac{E^2 + B^2}{8\pi}\right) = -\mathbf{j}\cdot\mathbf{E} - \text{div } \mathbf{S}, \quad (3.135)$$

where

$$\mathbf{S} = \frac{c}{4\pi}\mathbf{E}\times\mathbf{B}, \quad (3.136)$$

is called the Poynting vector. To understand the meaning of Eq. (3.135), let us integrate it over a three-dimensional volume and apply the Gauss theorem to the second term on the right, thereby obtaining

$$\frac{\partial}{\partial t}\int\frac{E^2 + B^2}{8\pi}dV = -\int\mathbf{j}\cdot\mathbf{E}\,dV - \oint\mathbf{S}\cdot d\mathbf{f}. \quad (3.137)$$

The first term on the right-hand side can be written as $\sum q_a\mathbf{v}_a\cdot\mathbf{E}$, where the sum is over all charges. Because this represents the amount of work done on the charged particles by the electromagnetic field, it is equal to the rate of change of the kinetic energy $\mathcal{E}$ of the charges. Therefore Eq. (3.137) can be written as

$$\frac{\partial}{\partial t}\left(\int\frac{E^2 + B^2}{8\pi}dV + \mathcal{E}\right) = -\oint\mathbf{S}\cdot d\mathbf{f}. \quad (3.138)$$

The left-hand side can be interpreted as the rate of change of the total energy of the system contained in a volume, and the right-hand side gives the flux of this energy through the surface bounding the given volume. This suggests that the quantity

$$W = \frac{E^2 + B^2}{8\pi} \quad (3.139)$$

is the energy density of the electromagnetic field; similarly, $\mathbf{S}$ is the energy flux of the electromagnetic field. The integral of W over d^3x represents the

Hamiltonian for the electromagnetic field, corresponding to the Lagrangian $L = (8\pi)^{-1}(\mathbf{E}^2 - \mathbf{B}^2)$ in Eq. (3.122). We will use this fact in Chap. 4, Section 4.5.

The procedure given above does not uniquely determine either of the quantities; it also does not make clear the transformation properties of these quantities under LTs. To correct these deficiencies, we need to introduce the concept of an energy-momentum tensor T^{ik} for the electromagnetic field just as we did it for a set of particles in Section 3.6. The T^{00} component of this tensor will give the energy density, $T^{\alpha 0} = T^{0\alpha}$ will give the momentum density, and the spatial component $T^{\alpha\beta}$ will give the flux of the α component of the momentum passing through a surface perpendicular to the x^β axis. Given the Lagrangian of a system, there is a well-defined procedure for obtaining the energy-momentum tensor, which we shall discuss in Chap. 11. Postponing the derivation to Chap. 11, we now give the final result for the energy-momentum tensor of the electromagnetic field:

$$T^{ik} = \frac{1}{4\pi}\left(-F^{il}F^k_l + \frac{1}{4}\eta^{ik}F_{lm}F^{lm}\right). \tag{3.140}$$

This tensor is obviously symmetric and is traceless; i.e., $T^i_i = 0$. Further, $T^{00} = W$, $T^{0\alpha} = S^\alpha/c$, and

$$T_{\alpha\beta} = \frac{1}{4\pi}\left[-E_\alpha E_\beta - B_\alpha B_\beta + \frac{1}{2}\delta_{\alpha\beta}(E^2 + B^2)\right]. \tag{3.141}$$

It is also easy to verify, by use of Eq. (3.127), that

$$\frac{\partial T^k_i}{\partial x_k} = -\frac{1}{c}F_{ik}j^k. \tag{3.142}$$

This equation relates the change in electromagnetic energy momentum to the work done by the field on the charged particles. We shall use this result in Chap. 9 to study the effects of a magnetic field on matter.

Exercise 3.14
Filling in the blanks: (1) Verify that the components of T^{ik} are the ones stated above. (2) Prove Eq. (3.142). (3) Compute $T^i_k T^k_j$.

3.11 Time-Independent Electromagnetic Fields

The full action in Eq. (3.123) represents the complicated interaction between a system of charges and the electromagnetic field. The motion of charges generates electromagnetic fields, which in turn act on the charged particles, modifying their trajectories. To understand various features of this system, it is convenient to divide the study into three parts: (1) the motion of charged particles in a given electromagnetic field, (2) the generation of electromagnetic fields by the motion of charged particles, and (3) the dynamics of the electromagnetic fields in the absence of the charged particles. Of these, part (1) was discussed in Section 3.8; some aspects of part (2), dealing with time-independent electromagnetic fields, are discussed in this section. The radiation and the propagation of electromagnetic waves will be discussed in Chaps. 4 and 6; aspects related to part (3) are discussed in Sections 3.12–3.16.

The simplest solutions to Maxwell's equations (3.128) and (3.70) arise when electric and magnetic fields are independent of time. In this case, because $\nabla \times \mathbf{E} = 0$, we can express $\mathbf{E}$ as $\mathbf{E} = -\nabla\phi$. Similarly, for the magnetic field we can write $\mathbf{B} = \nabla \times \mathbf{A}$ and impose the gauge condition $\nabla \cdot \mathbf{A} = 0$. We therefore need to solve the equations

$$\nabla^2\phi = -4\pi\rho, \quad \nabla^2\mathbf{A} = -\frac{4\pi}{c}\mathbf{J}. \tag{3.143}$$

In this case, the electric and the magnetic fields have decoupled from each other and can be discussed separately. Some of the solutions to the above equations, which are needed in future sections, are summarised below.

3.11.1 Coulomb Field of a Charged Particle

For a charged particle at rest at the origin of an inertial frame $\rho = q\delta(\mathbf{x})$ and $\mathbf{J} = 0$. Then $\phi = (q/|\mathbf{x}|) = (q/R)$, and the corresponding electric field is given by $\mathbf{E} = q\mathbf{R}/R^3$. The vector potential and the magnetic field vanish in this case.

Let us next consider the field produced by the same charge moving with a velocity $\mathbf{V}$ in the laboratory frame K. Taking the x axis to be the direction of the velocity, we introduce another frame K' in which the charge is at rest. In K', the potentials are $\phi' = q/R'$ and $\mathbf{A}' = 0$. Transforming the four vector $A^{i'} = (\phi', \mathbf{A}')$ from K' to K and expressing the coordinates of K' in terms of that in K by a LT, we can easily compute the potentials and the electromagnetic fields in K:

$$\mathbf{E} = \frac{q\mathbf{R}}{R^3}\frac{(1 - V^2/c^2)}{[1 - (V^2/c^2)\sin^2\theta]^{3/2}}, \quad \mathbf{B} = \frac{1}{c}\mathbf{V} \times \mathbf{E}, \tag{3.144}$$

where θ is the angle between the direction of motion and the radius vector $\mathbf{R}$. The vector $\mathbf{R}$ has the components $(x - Vt, y, z)$. The electric field is radially directed from the instantaneous position of the charged particle.

Exercise 3.15
Electric field of a moving charge: Carry out the derivation of Eqs. (3.144) and plot the electric-field lines for $(V/c) = 0.9$.

3.11.2 Dipole and Multipole Moments

Let us next consider the electric field produced by a system of N stationary charged particles having coordinates $\mathbf{r}_a$, with $a = 1, 2, 3, \ldots, N$, at a distance that is large compared with the dimensions of the system. The potential at a point $\mathbf{R}_0$ is given by

$$\phi = \sum_a \frac{q_a}{|\mathbf{R}_0 - \mathbf{r}_a|} \approx \frac{1}{R_0} \sum q_a - \left(\nabla \frac{1}{R_0}\right) \cdot \sum_a q_a \mathbf{r}_a + \phi^{(2)} + \cdots. \qquad (3.145)$$

In arriving at the second equality, the denominator has been expanded in powers of $1/R_0$. The first term is the Coulomb potential that is due to the total charge $Q = \sum q_a$ and falls as $1/R_0$; the second term is called the dipole potential and is given by

$$\phi^{(1)} = -\mathbf{d} \cdot \nabla \frac{1}{R_0} = \frac{\mathbf{d} \cdot \mathbf{R}_0}{R_0^3}, \qquad (3.146)$$

where $\mathbf{d} = \sum q_a \mathbf{r}_a$ is the dipole moment of the system. This term decreases as $1/R_0^2$. The next term in this series is the quadrapole potential and is given by

$$\phi^{(2)} = \frac{1}{2} \sum q x_\alpha x_\beta \frac{\partial^2}{\partial X_\alpha \partial X_\beta}\left(\frac{1}{R_0}\right), \qquad (3.147)$$

where X^α denotes the components of $\mathbf{R}_0$. The sum is over all charges, and we have dropped the subscript a on q_a, etc. Because the function $1/R_0$ satisfies the Laplace equation, this equation can be rewritten as

$$\phi^{(2)} = \frac{D_{\alpha\beta}}{6} \frac{\partial^2}{\partial X_\alpha \partial X_\beta}\left(\frac{1}{R_0}\right), \qquad (3.148)$$

where the tensor

$$D_{\alpha\beta} = \sum q(3x_\alpha x_\beta - r^2 \delta_{\alpha\beta}) \qquad (3.149)$$

is called the quadrapole moment of the system. Performing the differentiation in Eq. (3.148) and using the fact that $D_{\alpha\alpha} = 0$, we can write this potential as

$$\phi^{(2)} = \frac{D_{\alpha\beta} n_\alpha n_\beta}{2 R_0^3}, \qquad (3.150)$$

where $\mathbf{n}$ is a unit vector along $\mathbf{R}_0$. The quadrapole potential decreases as $1/R_0^3$.

The procedure indicated above can be continued for higher multipole fields in a straightforward manner and gives the potential that is due to a system of charges at large distances as a series in powers of $1/R_0$.

3.11.3 Magnetic Field of a Steady Current

Let us next consider time-independent magnetic fields. Because currents are produced by motion of charged particles, we can expect the magnetic field to be independent of time only when the motion is stationary in character. This is equivalent to assuming that the actual velocities of the charged particles, when averaged over the time scale of interest, remain stationary. In that case, the second equation in Eqs. (3.143) has the solution

$$\bar{\mathbf{A}} = \frac{1}{c} \int \frac{\bar{\mathbf{j}}}{R} \, dV = \frac{1}{c} \sum \frac{\overline{q_a \mathbf{v}_a}}{R_a}, \tag{3.151}$$

where the overbar denotes the averaging over time. We can expand this expression in inverse powers of the distance, as in Section 3.10. For example, the lowest-order terms will be

$$\bar{\mathbf{A}} = \frac{1}{cR_0} \sum q\bar{\mathbf{v}} - \frac{1}{c} \sum \overline{q\mathbf{v}\left(\mathbf{r} \cdot \nabla \frac{1}{R_0}\right)}. \tag{3.152}$$

But the first term is the total time derivative of the dipole moment and will give zero on time averaging. We can simplify the remaining term further by noting that $\mathbf{v} = \dot{\mathbf{r}}$ and $\mathbf{R}_0$ is a constant. Writing

$$\sum q(\mathbf{R}_0 \cdot \mathbf{r})\mathbf{v} = \frac{1}{2}\frac{d}{dt} \sum q\mathbf{r}(\mathbf{r} \cdot \mathbf{R}_0) + \frac{1}{2} \sum q[\mathbf{v}(\mathbf{r} \cdot \mathbf{R}_0) - \mathbf{r}(\mathbf{v} \cdot \mathbf{R}_0)], \tag{3.153}$$

and substituting into Eq. (3.152), we find that the time derivative again vanishes on averaging. Thus

$$\bar{\mathbf{A}} = \frac{1}{2cR_0^3} \sum \overline{q[\mathbf{v}(\mathbf{r} \cdot \mathbf{R}_0) - \mathbf{r}(\mathbf{v} \cdot \mathbf{R}_0)]}. \tag{3.154}$$

This result can be expressed as

$$\bar{\mathbf{A}} = \frac{\bar{\mathbf{m}} \times \mathbf{R}_0}{R_0^3} = \left(\nabla \frac{1}{R_0}\right) \times \bar{\mathbf{m}}, \tag{3.155}$$

where

$$\mathbf{m} = \frac{1}{2c} \sum q\mathbf{r} \times \mathbf{v} \tag{3.156}$$

is called magnetic moment of the system. This quantity is analogous to the dipole moment for the electric field. If all charges in the system have the same value for (q/m), the magnetic moment becomes

$$\mathbf{m} = \frac{1}{2c} \sum q\mathbf{r} \times \mathbf{v} = \frac{q}{2mc} \sum m\mathbf{r} \times \mathbf{v} = \frac{q}{2mc} \sum \mathbf{r} \times \mathbf{p} = \frac{q}{2mc}\mathbf{J}, \tag{3.157}$$

where **J** is the angular momentum of the system and we have assumed that the motion of the particles is nonrelativistic. In this case, the ratio between the magnetic moment and the angular momentum of the system is $q/2mc$.

3.11.4 Maxwell's Equations in a Polarisable Medium

When a material medium containing large number of charged particles is present, it is convenient to recast Maxwell's equations in a different form. In the simplest context, the effect of an external field **E** is to induce a polarisation **P** in the medium through redistribution of bound charges of the material medium, thereby leading to a polarisation charge ρ_{pol} and polarisation current $\mathbf{j}_{pol}$. When **P** is defined through the relation $\nabla \cdot \mathbf{P} \equiv -\rho_{pol}$, the second of Maxwell's equations (3.128) becomes

$$\nabla \cdot \mathbf{E} = 4\pi(\rho_{external} + \rho_{pol}) = 4\pi\rho_{ex} - \nabla \cdot (4\pi\mathbf{P}) \qquad (3.158)$$

or

$$\nabla \cdot \mathbf{D} = 4\pi\rho_{ext}, \quad \mathbf{D} \equiv \mathbf{E} + 4\pi\mathbf{P}. \qquad (3.159)$$

From the conservation of charges in the material medium, $\dot{\rho}_{pol} + \nabla \cdot \mathbf{j}_{pol} = 0$, and the definition $\nabla \cdot \mathbf{P} = -\rho_{pol}$, it follows that $\mathbf{j}_{pol} = (\partial\mathbf{P}/\partial t)$. This allows the first equation in Eqs. (3.128) to be written in the form

$$\nabla \times \mathbf{B} = \frac{4\pi}{c}(\mathbf{j}_{ext} + \mathbf{j}_{pol}) + \frac{1}{c}\frac{\partial\mathbf{E}}{\partial t} = \frac{4\pi}{c}\mathbf{j}_{ext} + \frac{1}{c}\frac{\partial\mathbf{D}}{\partial t}. \qquad (3.160)$$

In a polarisable medium Eqs. (3.160) and (3.159) replace Eqs. (3.128); Eqs. (3.70) remain unchanged. To solve these equations, we need an extra relation between **D** and **E** or, equivalently, between **P** and **E**. This equation depends on the nature of the medium and has to be obtained from the study of its properties.

Exercise 3.16

Model for a polarisable medium: We can construct a simple but useful model for a dielectric medium by assuming that the electrons (bound to atoms) in a medium respond to the external electric field like classical harmonic oscillators with a natural frequency ω_0 and a damping coefficient γ. When an electric field varying at a frequency ω acts on such a system, it will induce a dipole moment $\mathbf{d} \equiv (\alpha/4\pi)\mathbf{E}$, where α is called the atomic polarisability. (1) Show that in the harmonic-oscillator model described above, $\alpha = (4\pi q^2/m)(\omega_0^2 - \omega^2 + i\gamma\omega)^{-1}$. (2) If an electric field $\mathbf{E}_{applied}$ is imposed externally on such a medium, containing n atoms per unit volume, then the net polarisation will be $\mathbf{P} = (n\alpha/4\pi)\mathbf{E}_{tot}$, where $\mathbf{E}_{tot}$ is the local electric field acting on the atom, which includes the external electric field as well as the field that is due to the dipoles induced in the vicinity. Show that $\mathbf{E}_{tot} = \mathbf{E}_{applied} + (4\pi/3)\mathbf{P}$. Hence show that $\mathbf{P} = \epsilon\mathbf{E}_{applied}$, where $\epsilon = (n\alpha/4\pi)[1 - (n\alpha/3)]^{-1}$.

3.12 Electromagnetic Waves

We next discuss the solutions to Maxwell's equations in the absence of charges. We can take the point of view that *all* electromagnetic fields are generated by charged particles and hence there is no such thing as free electromagnetic fields. Strictly speaking, this is correct. However, the radiation field behaves, for all practical purposes, as a dynamical entity unconnected with the source. It is convenient therefore to understand its behaviour as an independent entity.

In the absence of charged particles, $j^i = 0$ and F^{ik} satisfies the equation $\partial_k F^{ik} = 0$. Expressing F^{ik} in terms of the potential A^i and imposing Lorentz gauge, we get

$$\Box A^i = \frac{\partial^2}{\partial x^k \partial x_k} A^i = \left(\frac{1}{c^2} \frac{\partial^2}{\partial t^2} - \nabla^2 \right) A^i = 0. \tag{3.161}$$

Thus each component A^i satisfies a wave equation with the speed of propagation equal to c.

To solve this equation, we use the ansatz $A^j = a^j \exp i(k_b x^b)$, where k_b and a^j are constant four vectors. Substituting this ansatz into the wave equation, we find that $k^b k_b = 0$. The Lorentz gauge condition $\partial_j A^j = 0$ implies that $k_b a^b = 0$. Thus the solution is parameterised by two four vectors subject to two constraints. When the components of k^b are denoted by $(\omega/c, \mathbf{k})$, the second constraint becomes $|\mathbf{k}|^2 = (\omega/c)^2$; i.e., $\omega = \pm\omega_{\mathbf{k}} \equiv \pm kc$. For the sake of simplicity, let us choose the amplitude to be $a^b = (0, \mathbf{a})$, that is, we take this vector to be purely spatial in some Lorentz frame. In that case, the first constraint becomes $\mathbf{k} \cdot \mathbf{a} = 0$ and the solution to the wave equation becomes

$$\mathbf{A}_{\mathbf{k}}(t, \mathbf{x}) = \mathbf{a}_{\mathbf{k}} \exp i(\pm\omega_{\mathbf{k}} t - \mathbf{k} \cdot \mathbf{x}), \quad \omega_{\mathbf{k}} = |\mathbf{k}|c, \tag{3.162}$$

where $\mathbf{a}$ is confined to a plane perpendicular to the direction of $\mathbf{k}$; thus the vector potential has only two independent components.

To find the general solution to Eq. (3.161), we note that it is linear in A^i and hence solutions can be superposed. Because each solution is parameterised by a wave vector $\mathbf{k}$, such a superposition can be written as

$$\mathbf{A}(t, \mathbf{x}) = \int \frac{d^3\mathbf{k}}{(2\pi)^3} \{ \mathbf{f}(\mathbf{k}) \exp i[\omega_{\mathbf{k}} t - \mathbf{k} \cdot \mathbf{x}] + \mathbf{g}(\mathbf{k}) \exp - i[\omega_{\mathbf{k}} t + \mathbf{k} \cdot \mathbf{x}] \},$$

$$\tag{3.163}$$

where $\mathbf{f}(\mathbf{k})$ and $\mathbf{g}(\mathbf{k})$ are arbitrary functions that can be determined in terms of the initial conditions. This provides the complete solution to the problem of propagation of free electromagnetic waves. We now study various special cases and properties. For our discussion, it is convenient to choose the initial conditions such that $\mathbf{g} = 0$, giving

$$\mathbf{A}(t, \mathbf{x}) = \int \frac{d^3\mathbf{k}}{(2\pi)^3} \mathbf{f}(\mathbf{k}) \exp i[\omega_{\mathbf{k}} t - \mathbf{k} \cdot \mathbf{x}]. \tag{3.164}$$

Setting $t = 0$, we see that $\mathbf{f}(\mathbf{k})$ is the Fourier transform of $\mathbf{A}(0, \mathbf{x})$.

Exercise 3.17

Waves in the material medium: (1) Show that, when $\rho_{\text{ext}} = 0$ and $\mathbf{j}_{\text{ext}} = 0$, Maxwell's equations in a material medium, discussed in Subsection 3.11.4, imply that

$$\nabla^2 \mathbf{E} - \frac{1}{c^2} \frac{\partial^2 \mathbf{E}}{\partial t^2} = -4\pi \nabla(\nabla \cdot \mathbf{P}) + \frac{4\pi}{c^2} \frac{\partial^2 \mathbf{P}}{\partial t^2}. \tag{3.165}$$

(2) The simplest type of material medium has $\mathbf{P} = \epsilon \mathbf{E}$, where ϵ is a constant. In this case, show that Maxwell's equations admit plane-wave solutions of the form $\exp(i\mathbf{k} \cdot \mathbf{x} - i\omega t)$ for $\mathbf{E}$ (say) with the dispersion relation $c^2 k^2 = \omega^2(1 + 4\pi\epsilon)$. The phase velocity of these waves is $v^2 \equiv (\omega^2/k^2) = c^2(1 + 4\pi\epsilon)^{-1} \equiv (c^2/n^2)$, where the refractive index n is defined to be $n = (1 + 4\pi\epsilon)^{1/2}$. (3) Use the model developed in Exercise 3.16 for ϵ and α to determine the refractive index n as a function of the frequency ω. How can we interpret the imaginary part of the refractive index? The above model is also a reasonable approximation for metals if we take $\omega_0 = 0$. Explain why metals in general have high reflectivity. (4) For a more challenging task, work out the wave solutions and phase velocity for an anisotropic medium with $P_\alpha = \epsilon_{\alpha\beta} E_\beta$.

3.12.1 Monochromatic Plane Waves

Many astrophysical applications are concerned with (nearly) monochromatic waves that have a definite frequency ω. Further, if the wave is propagating along a definite direction with unit vector $\mathbf{n}$ [so that $\mathbf{k} = (\omega/c)\mathbf{n}$], it is called a monochromatic plane wave. The central quantity that characterises such a wave is the wave (four) vector k^a. Because the components of this vector determine both the frequency and the direction of propagation of the wave, it follows that different observers will see the wave as having different frequencies and directions of propagation.

As a specific example, consider two Lorentz frames S and S', with S' moving along the positive x axis of S with velocity v. We call S' the rest frame and S the lab frame and denote quantities in these frames by subscripts R and L, respectively. To bring out the relativistic effects clearly, we assume that v is close to c so that

$$\left(1 - \frac{v}{c}\right) = \frac{(1 - v^2/c^2)}{(1 + v/c)} \simeq \frac{1}{2\gamma^2}, \tag{3.166}$$

where $\gamma = (1 - v^2/c^2)^{-1/2} \gg 1$. Consider a plane wave with frequency ω_L travelling along the direction (θ_L, ϕ_L) in the lab frame. Let the four vector describing the wave be k^i with components $(\omega_L/c, \mathbf{k}_L)$ in the lab frame and $(\omega_R/c, \mathbf{k}_R)$ in the rest frame. For motion along the x axis, only the angle made by $\mathbf{k}$ with the x axis will change; the azimuthal angle is clearly invariant because of the symmetry. Writing $ck_L^x = \omega_L \cos\theta_L$, $ck_R^x = \omega_R \cos\theta_R$ and using the LTs for the four vector k^i, we find that

$$\omega_R = \gamma \omega_L [1 - (v/c) \cos \theta_L], \quad (3.167)$$

$$\omega_R \cos \theta_R = \gamma \omega_L [\cos \theta_L - (v/c)]. \quad (3.168)$$

These equations relate the directions of propagation and frequencies in the two frames.

Using Eqs. (3.167) and (3.168), we can express $\mu_R \equiv \cos \theta_R$ in terms of $\mu_L \equiv \cos \theta_L$:

$$\mu_R = \frac{\mu_L - (v/c)}{1 - (v\mu_L/c)}. \quad (3.169)$$

[This equation is equivalent to Eq. (3.17).] Figure 3.3 shows a plot of μ_R against μ_L when $v \lesssim c$. When $\mu_L = 1$ we have $\mu_R = 1$ and when $\mu_L = -1$ we have $\mu_R = -1$, that is, waves travelling along the x axis appear to do so in both the frames. The frequencies, however, are different in the two cases. When $\mu_L = 1$, $\omega_R = \gamma \omega_L [1 - (v/c)]$, indicating that $\omega_R < \omega_L$; the radiation is blue shifted, as seen in the lab frame. For $\mu_L = -1$, $\omega_R = \gamma \omega_L [1 + (v/c)]$ and the radiation is red shifted. When $\mu_L = (v/c)$, we have $\mu_R = 0$. Because $v \lesssim c$, we

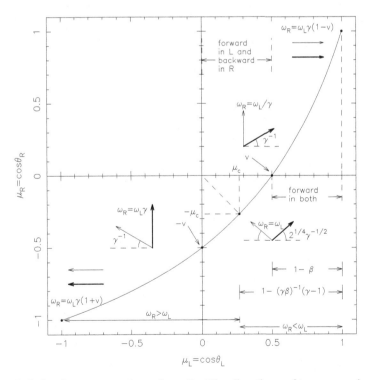

Fig. 3.3. Relation between $\cos \theta_R$ and $\cos \theta_L$. The directions of two waves in the rest frame and the lab frame are indicated by thin and thick arrows, respectively.

expect $\mu_L = \cos\theta_L \simeq (1 - \theta_L^2/2)$, with $\theta_L \ll 1$. Comparing with the relation $1 - (v/c) \simeq (2\gamma^2)^{-1}$ we find that $\theta_L \simeq \gamma^{-1} \ll 1$. Hence this direction of propagation is almost along the positive x axis in the lab frame. However, because $\mu_R = 0$, it is clear that $\theta_R = \pi/2$ and the wave is propagating perpendicularly to the x axis in the rest frame. Thus a wave emitted in the direction perpendicular to the direction of motion will be turned around to (almost) forward direction by the relativistic motion of the source. Similarly, when $\mu_L = 0$, $\mu_R = -(v/c)$. In this case, a wave that has been travelling almost along the negative x axis has been turned around to travel orthogonally to the x axis. It is clear from the diagram that the wave will appear to propagate forward in both frames only if $\theta_L < \gamma^{-1}$.

In the two cases discussed above, the frequency of the wave changes. When $\mu_L = (v/c)$, $\omega_R = \omega_L \gamma^{-1}$ and the wave is still blue shifted. When $\mu_L = 0$, $\omega_R = \gamma \omega_L$ and the wave is red shifted. To find the angle of propagation at which there is no frequency change, we have to set $\omega_L = \omega_R$ and solve for μ. This gives $\theta_c \simeq \gamma^{-1/2}$. For $\theta < \theta_c$ the wave is blue shifted. A wave propagating along $\theta = \theta_c$ appears to make the same angle with respect to the x axis in both the frames.

The discussion above shows that the motion of a source drags the wave forward. (A corollary to this result is that a charged particle, moving relativistically, will beam most of the radiation it emits in the forward direction.) Correspondingly, a charged particle moving relativistically through an isotropic bath of radiation will see most of the radiation as hitting it in the front.

Finally, by using Eq. (3.140), we can find the energy-momentum tensor for the plane wave that is given by

$$T^{ab} = \frac{Wc^2}{\omega^2} k^a k^b, \tag{3.170}$$

where $W = (E^2 + B^2)/8\pi = E^2/4\pi$. Because the left-hand side of Eq. (3.170) and $k^a k^b$ are tensors, it follows that the combination (E/ω) is Lorentz invariant. (This is an interesting result, especially as it is valid for any LT – not merely for the ones along the direction of propagation.) The momentum flux of a plane wave is given by

$$\mathbf{S} = \frac{c}{4\pi}\mathbf{E} \times \mathbf{B} = \frac{c}{4\pi}\mathbf{E} \times (\mathbf{n} \times \mathbf{E}) = \frac{c}{4\pi}E^2\hat{\mathbf{n}} = \frac{c}{4\pi}B^2\hat{\mathbf{n}}, \tag{3.171}$$

or $\mathbf{S} = cW\mathbf{n}$. The relation between momentum density and energy density $S/c^2 = W/c$ is the same as that for a particle of mass zero moving with the speed of light.

3.12.2 Polarisation of Light

The vector potential for the plane wave has the form given in Eq. (3.162). Since $a^0 = 0$ it follows that $\phi = 0$. (For convenience we omit the subscript $\mathbf{k}$ in the notation). The corresponding electric and magnetic fields are given by

$\mathbf{E} = ik\mathbf{A}$, $\mathbf{B} = i\mathbf{k} \times \mathbf{A}$. More explicitly, the electric field is

$$\mathbf{E} = \mathbf{E}_0 \exp[i(\mathbf{k} \cdot \mathbf{x} - \omega t)]. \tag{3.172}$$

Such a field has definite polarisation, which we now discuss.

Because $\mathbf{E}_0$ is a complex vector, so also is its square, which we write as $\mathbf{E}_0^2 = |\mathbf{E}_0|^2 \exp(-2i\alpha)$. Defining a complex vector $\mathbf{b}$ by $\mathbf{E}_0 \equiv \mathbf{b}e^{-i\alpha}$, we see that $\mathbf{b}^2 = |\mathbf{E}_0|^2$ is real; if $\mathbf{b} = \mathbf{b}_1 + i\mathbf{b}_2$, where $\mathbf{b}_{1,2}$ are real, it follows that $\mathbf{b}_1 \cdot \mathbf{b}_2 = 0$ so that the two vectors are perpendicular. Further, because the electric field is orthogonal to the direction of propagation of the wave (which is, say, the x axis), we can take $\mathbf{b}_1$ along the y axis and $\mathbf{b}_2$ along the z axis. Finally, noting that the physical electric-field components are the real parts of the complex exponents, we can write the electric field as

$$\begin{aligned} E_y &= b_1 \cos(\omega t - \mathbf{k} \cdot \mathbf{x} + \alpha), \\ E_z &= \pm b_2 \sin(\omega t - \mathbf{k} \cdot \mathbf{x} + \alpha). \end{aligned} \tag{3.173}$$

This gives the relation

$$\frac{E_y^2}{b_1^2} + \frac{E_z^2}{b_2^2} = 1 \tag{3.174}$$

between the components of the wave, showing that the tip of the vector rotates on an ellipse in the x–y plane as t varies. Such a wave is called an elliptically polarised wave. If $b_1 = b_2$, the ellipse becomes a circle, and we have a circularly polarised wave. If b_1 or b_2 vanishes, the field is along one of the axes and is called plane polarised.

The monochromatic wave is necessarily polarised in a manner discussed above. However, we often come across in nature sources of radiation that are not strictly monochromatic but contain frequencies in a narrow band $\Delta\omega$ around the mean frequency. In that case, the time variation of the electric field will be of the form $\mathbf{E} = \mathbf{E}_0(t) \exp(-i\omega t)$, where $\mathbf{E}_0(t)$ is a slowly varying function of time [compared to $\exp(-i\omega t)$]. To determine the degree of polarisation of such a wave, we should consider expressions that are quadratic in the electric field; however, it is now necessary to average these expressions over the rapidly varying part to determine the mean state of polarisation. The quadratic expressions are made of $E_\alpha E_\beta$, $E_\alpha E_\beta^*$, or their complex conjugates. Of these, the first one and its conjugate vary with a frequency 2ω and hence will average to zero. Therefore the polarisation properties are decided by the average of the product $E_\alpha E_\beta^* = E_{0\alpha} E_{0\beta}^*$.

We define a quantity $J_{\alpha\beta} = \langle E_{0\alpha} E_{0\beta}^* \rangle$ that has four independent components (because the indices take the values 1,2 in the plane perpendicular to the direction of propagation). The trace of this quantity, $J \equiv \sum J_{\alpha\alpha} = \langle \mathbf{E}_0 \cdot \mathbf{E}_0 \rangle$ measures the energy density of the field and is not directly related to the polarisation. Hence it is convenient to divide $J_{\alpha\beta}$ by its trace and define a polarisation tensor as

$\rho_{\alpha\beta} = (J_{\alpha\beta}/J)$. From this definition it follows that $\rho_{\alpha\beta} = \rho_{\beta\alpha}^*$, that is, the matrix is Hermitian with unit trace. Any such matrix can be written in the form

$$\rho_{\alpha\beta} = \frac{1}{2}\begin{bmatrix} 1+\xi_3 & \xi_1 - i\xi_2 \\ \xi_1 + i\xi_2 & 1-\xi_3 \end{bmatrix}, \tag{3.175}$$

where the quantities ξ_μ are called the Stokes parameters. From the definition, it is clear that the determinant of $\rho_{\alpha\beta}$, which is given by

$$\det \rho = \frac{1}{4}(1 - \xi_1^2 - \xi_2^2 - \xi_3^2) \equiv \frac{1}{4}(1 - P^2), \tag{3.176}$$

is positive (see Exercise 3.18). So each of the Stokes parameters vary in the range $(-1, 1)$.

Some general properties of the polarisation tensor can be easily ascertained. To begin with, if the wave is completely polarised, $\mathbf{E}_0$ is independent of t and the averaging has no effect on the definition of the polarisation tensor; in this case, the polarisation tensor is expressible as the direct product of two vectors. The necessary and sufficient condition for this is that $\det|\rho|$ vanishes, implying that $P = 1$. On the other hand, a completely unpolarised wave will have – by symmetry – the polarisation tensor $\rho_{\alpha\beta} = (1/2)\delta_{\alpha\beta}$, so that $P = 0$. Because of this feature P is called the degree of polarisation.

In general, the quantity ξ_2 represents the degree of circular polarisation; ξ_3 gives the degree of linear polarisation along the y or the z axis, with $\xi_3 = 1$ representing linear polarisation along the y axis and $\xi_3 = -1$ giving linear polarisation along the z axis. The parameter ξ_1 quantifies the linear polarisation along directions that make $45°$ with the y axis; a value of $\xi_1 = 1$ corresponds to complete polarisation along $\phi = (\pi/4)$ and $\xi_1 = -1$ corresponds to polarisation along $\phi = -(\pi/4)$.

Exercise 3.18
Properties of the polarisation tensor: Prove that the polarisation tensor must have positive determinant. Verify the various results stated above explicitly.

Exercise 3.19
Stokes parameters: Another way of characterising the polarisation of a wave is by using certain variables defined along the following lines. Consider an elliptically polarised light for which the major axis of the ellipse is inclined at an angle ψ with the arbitrarily chosen x axis. Let the amplitude of the electric field along the major and the minor axes of the ellipse be E_a and E_b with $\chi \equiv \tan^{-1}(E_a/E_b)$. We now define the quantities S_i $(i = 0, \ldots, 3)$ and three variables (U, V, W) by the relations

$$\begin{aligned} S_0 = I = E_a^2 + E_b^2, \quad & S_1 = Q = S_0 \cos 2\chi \cos 2\psi, \\ S_2 = U = S_0 \cos 2\chi \sin 2\psi, \quad & S_3 = V = S_0 \sin 2\chi. \end{aligned} \tag{3.177}$$

(1) Show that the parameters S_i are not all independent. (2) Consider a sphere of radius S_0 in the space with Cartesian axes (S_1, S_2, S_3). Every point on the sphere (called a Poincaré

sphere) defines a state of polarisation. Where do the following lie on the Poincare sphere: right circularly polarised light, left circularly polarised light, linearly polarised light? How does this description in terms of S_i relate to the one given in the text?

Exercise 3.20
Angular momentum of the wave: A circularly polarised electromagnetic wave of frequency ω impinges on a charged particle. Average the motion of the charged particle (which is assumed to be nonrelativistic) over a time T that is large compared with the period of the wave and show that the wave transfers an amount of energy $\mathcal{E}$ and angular momentum J to the charged particle, where $J = (\mathcal{E}/\omega)$.

3.13 Diffraction

The propagation of free electromagnetic waves is completely described by Eq. (3.164). By reformulating this equation in a more convenient form, we can understand a host of optical phenomena in which the wave nature of the light plays a vital role. We begin with the first of these phenomena, which is usually called diffraction.

Consider a monochromatic wave of frequency ω. Because the vector nature of the electromagnetic field is not very important in our discussion, we deal with just one component of the vector potential. In accordance with Eq. (3.164), any one component of the vector potential can be represented as

$$A(t, \mathbf{x}) = \int F_1(\mathbf{k}) e^{i\mathbf{k}\cdot\mathbf{x}} e^{-i\omega t} \frac{d^3 k}{(2\pi)^3}. \tag{3.178}$$

In the study of optical phenomena we are often concerned with waves that are propagating, by and large, in some given direction, which can be taken as the positive z axis. Mathematically, this means that the function $F_1(\mathbf{k})$ is significantly nonzero only for wave vectors with $k_z > 0$ and $(k_x, k_y) \ll k_z$. Further, because the wave has a definite frequency ω, the magnitude of the wave vector is fixed at the value ω/c. It follows that one of the components of the wave vector, say k_z, can be expressed in terms of the other three. Therefore the function F_1 has the structure

$$F_1(k_z, \mathbf{k}_\perp) = 2\pi f(\mathbf{k}_\perp) \delta_D \left(k_z - \sqrt{\omega^2/c^2 - k_\perp^2} \right), \tag{3.179}$$

where the subscript $\perp$ denotes the components of the vector in the transverse x–y plane. Substituting this expression into Eq. (3.178) we find that

$$A(t; z, \mathbf{x}_\perp) \equiv a(z, \mathbf{x}_\perp) e^{-i\omega t}$$

$$= e^{-i\omega t} \int \frac{d^2 \mathbf{k}_\perp}{(2\pi)^2} f(\mathbf{k}_\perp) e^{i\mathbf{k}_\perp \cdot \mathbf{x}_\perp} \exp \left(\frac{iz}{c} \sqrt{\omega^2 - c^2 k_\perp^2} \right). \tag{3.180}$$

Because the time variation of a monochromatic wave is always $\exp(-i\omega t)$, we ignore this factor and concentrate on the spatial dependence of the amplitude,

$a(z, \mathbf{x}_\perp)$. For a wave travelling, by and large, along the z direction, the transverse components of $\mathbf{k}$ are small compared with its magnitude that is, $c^2 k_\perp^2 \ll \omega^2$. Using the Taylor series

$$\sqrt{\omega^2 - c^2 k_\perp^2} \cong \omega\left(1 - \frac{1}{2}\frac{c^2 k_\perp^2}{\omega^2}\right) = \omega - \frac{1}{2}\frac{c^2 k_\perp^2}{\omega} \tag{3.181}$$

in Eq. (3.180), we get

$$a(z, \mathbf{x}_\perp) \cong e^{i\omega z/c} \int \frac{d^2 \mathbf{k}_\perp}{(2\pi)^2} f(\mathbf{k}_\perp) \exp\{i[\mathbf{k}_\perp \cdot \mathbf{x}_\perp - (c/2\omega)k_\perp^2 z]\}. \tag{3.182}$$

This equation describes the propagation of a wave along the positive z axis with a small spread in the transverse direction. The function $f(\mathbf{k}_\perp)$ can be determined by a simple Fourier transform if the amplitude $a(z', \mathbf{x}'_\perp)$ is given at some location z'. Doing this, we can relate the amplitudes of the wave at two planes with coordinates z and z' by

$$a(z, \mathbf{x}_\perp) = e^{i\omega(z-z')/c} \int d^2\mathbf{x}'_\perp a(z', \mathbf{x}'_\perp) \, G(z - z'; \mathbf{x}_\perp - \mathbf{x}'_\perp), \tag{3.183}$$

where

$$G(z - z'; \mathbf{x}_\perp - \mathbf{x}'_\perp) = \int \frac{d^2 \mathbf{k}_\perp}{(2\pi)^2} e^{i\mathbf{k}_\perp \cdot (\mathbf{x}_\perp - \mathbf{x}'_\perp)} e^{-(ic/2\omega)k_\perp^2(z-z')}$$

$$= \left(\frac{\omega}{2\pi i c}\right) \frac{1}{|z - z'|} \exp\left[\frac{i\omega}{2c}\frac{(\mathbf{x}_\perp - \mathbf{x}'_\perp)^2}{(z - z')}\right]. \tag{3.184}$$

The function G may be thought of as a propagator that propagates the amplitude from the location $(z', \mathbf{x}'_\perp)$ to the location $(z, \mathbf{x}_\perp)$. The factor $e^{i\omega(z-z')/c}$ in Eq. (3.183) does not contribute to the intensity, and we will drop it when it is not needed.

In general, the amplitude of the wave satisfies a second-order differential equation, and we cannot determine its evolution by just knowing the amplitude (i.e., one single function) at a given location. This could be done in Eq. (3.183) only because of the assumption that the wave is travelling forward in the z direction. The actual form of the propagator depends on the assumption that the transverse components of the wave vector are small compared with k_z. The study of wave propagation under these approximations is called paraxial optics.

The propagator G introduces a factor $|z - z'|^{-1}$ to the amplitude and contributes an amount

$$\phi = \frac{\omega}{2c}\frac{(\mathbf{x}_\perp - \mathbf{x}'_\perp)^2}{(z - z')} \tag{3.185}$$

to the phase. The change in the amplitude merely reflects the r^{-2} falloff of the intensity (which is proportional to the square of the amplitude) of the wave.

To understand the change in phase, note that a path difference Δs between two points in space will introduce a phase difference of $k\Delta s$. In this case, it is clear that the phase difference is

$$k\Delta s = \frac{\omega}{c}\left[\sqrt{(\mathbf{x}_\perp - \mathbf{x}'_\perp)^2 + (z - z')^2} - (z - z')\right] \cong \frac{\omega}{c}\left[\frac{1}{2}\frac{(\mathbf{x}_\perp - \mathbf{x}'_\perp)^2}{(z - z')}\right],$$

(3.186)

provided the transverse displacements are small compared with the longitudinal distance – an assumption that is central to paraxial optics.

Equation (3.183) can be used to study the amplitude distribution $a(z, \mathbf{x}_\perp)$ at some location z given different kinds of wave fields at an initial location $a(z', \mathbf{x}'_\perp)$. In general, such phenomena can be classified into two categories, depending on the relative values of the parameters of the problem. To understand this classification, let us rewrite the propagation equation (3.183) as

$$a(z, \mathbf{x}_\perp) = \left(\frac{\omega}{2\pi i c}\right)\frac{1}{r}e^{(ikx_\perp^2/2r)}\int d^2\mathbf{x}'_\perp a(z', \mathbf{x}'_\perp)\exp\left(\frac{ikx_\perp'^2}{2r}\right)\exp\left(\frac{ik}{r}\mathbf{x}_\perp \cdot \mathbf{x}'_\perp\right),$$

(3.187)

where $r = z - z'$. Let us suppose that the initial amplitude $a(z', \mathbf{x}'_\perp)$ is nonzero only in a region of size D in the transverse direction. This will be the situation, for example, if a plane wave was transmitted through a hole of diameter D in an opaque screen. The first exponential factor inside the integrand contributes a maximum phase of $kD^2/r = 2\pi D^2/\lambda r$. The second exponential contributes a maximum phase $(kDx_\perp/r) = (2\pi D^2/\lambda r)(x_\perp/D)$. If (1) $D^2 \ll \lambda r$, then the phase change that is due to the first exponential can be ignored; this happens for a small aperture (small D) or at large distances (large r). On the other hand, if (2) $D^2 \gtrsim \lambda r$, then we cannot ignore either of the exponentials. Phenomena corresponding to case (1) are called Fraunhofer diffraction and phenomena corresponding to case (2) are called Fresnel diffraction. The quantity $r_F \equiv \sqrt{r\lambda}$ is called the Fresnel length for a given separation $r = z - z'$. Fresnel diffraction occurs for $D \gtrsim r_F$.

The distinction between these two cases has a simple physical meaning. Suppose a plane wave, travelling along the z axis, is blocked by an opaque screen with a hole of size D. From elementary theory of Fourier transforms, we know that confining the wave in transverse direction within size D will introduce uncertainty in the transverse wave-vector components of the order of $\Delta k_x \approx 1/D$. When the wave propagates a farther distance r, its vertical spread will increase by the amount $(\Delta k_x/k)r = (r/kD) \approx (\lambda r/D)$. Adding this spread to the original spread D, we get the total spread as

$$D + \frac{\lambda r}{D} \approx \begin{cases} D & (\text{for } r \ll D^2/\lambda) \\ \lambda r/D & (\text{for } r \gg D^2/\lambda) \end{cases}.$$

(3.188)

Thus, in the Fresnel diffraction, the original size of the aperture is dominant, whereas in the Fraunhofer diffraction, the spreading that is due to the uncertainty in the transverse component makes the dominant contribution. The transition from Fresnel diffraction to Fraunhofer diffraction takes place at a distance $r_c \approx (D^2/\lambda)$. It is obvious from the above discussion that the Fraunhofer limit does not exist for initial amplitude distributions that are *not* confined to a finite region (i.e., if $D \to \infty$). We now study different aspects of these diffraction phenomena in detail.

3.13.1 Fraunhofer Diffraction ($r \gg D^2/\lambda$)

In this case, the first exponential factor inside the integrand in Eq. (3.187) can be replaced with unity. Defining $\mathbf{q} \equiv (k/r)\mathbf{x}_\perp$, we note that the intensity pattern is essentially determined by the quantity

$$|a(\mathbf{q})|^2 \cong \frac{N}{r^2} \left| \int d^2\mathbf{x}'_\perp a(z', \mathbf{x}'_\perp) e^{i\mathbf{q}\cdot\mathbf{x}'_\perp} \right|^2, \tag{3.189}$$

where N is an unimportant normalisation constant. Thus the intensity is determined by the Fourier transform of the original amplitude and all the information is contained in the two-dimensional Fourier transform

$$u(\mathbf{q}) \equiv \int d^2\mathbf{x}'_\perp a(\mathbf{x}'_\perp) e^{i\mathbf{q}\cdot\mathbf{x}'_\perp} \tag{3.190}$$

taken at the initial plane z'. Hereafter we shall deal with a fixed z and z' and suppress them in the formulas. Most of the practical applications will be concerned with waves transmitted through apertures of different kinds, so that $a(\mathbf{x}'_\perp)$ will be zero in the opaque parts in the screen and nonzero (with a constant value u_0) in the transparent parts of the aperture. In that case, Eq. (3.190) gives the two-dimensional Fourier transform of the shape of the aperture. The relative intensity diffracted into a solid angle $d\Omega = d\theta_y d\theta_x$ (where $\theta_x = q_x/k$, $\theta_y = q_y/k$) is given by

$$\frac{|u_{\mathbf{q}}|^2}{u_0^2} \frac{dq_y dq_x}{(2\pi)^2} = \frac{|u_{\mathbf{q}}|^2}{u_0^2} \frac{d\theta_x d\theta_y}{(2\pi k)^2} = \left(\frac{\omega}{2\pi c}\right)^2 \left|\frac{u_{\mathbf{q}}}{u_0}\right|^2 d\Omega. \tag{3.191}$$

Eq. (3.188) shows that the lateral spread of the beam, in traversing a distance r after passing through an aperture of size D, is $x_\perp \simeq (\lambda r/D)$. The corresponding angular spread is $\theta_{\text{diff}} \equiv (x_\perp/r) \simeq (\lambda/D)$. We now consider some simple examples of these phenomena.

As a first example, we calculate the diffraction pattern that arises when a plane wave is normally incident upon an infinite slit of width $2a$. We choose the plane of the slit as the transverse x–y plane with the x axis along the slit. Because the slit is infinitely long, the diffraction occurs in only the y–z plane, with the Fourier transform vanishing for $q_x \neq 0$. The Fourier transform on the y

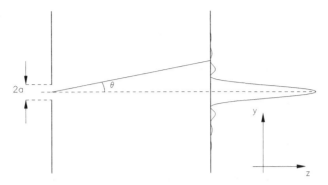

Fig. 3.4. Geometry of diffraction by a slit.

coordinate gives

$$u_q = u_0 \int_{-a}^{a} e^{-iqy} dy = \frac{2u_0}{q} \sin qa. \tag{3.192}$$

The intensity of the diffracted light in an angular range $d\theta$ is given by

$$dI \propto \left| \frac{u_q}{u_0^2} \right|^2 \frac{dq}{2\pi} = \frac{I_0}{2a} \left| \frac{u_q}{u_0} \right|^2 \frac{dq}{2\pi} = \frac{I_0}{\pi ak} \frac{\sin^2 ka\theta}{\theta^2} d\theta, \tag{3.193}$$

where the normalisation is chosen such that I_0 is the total intensity. The geometry of this phenomena is shown in Fig. 3.4. As $|\theta|$ increases, the intensity goes through a series of maxima with rapidly decreasing height. The successive maxima are separated by minima at the points $\theta = (n\pi/ka)$, where n is an integer. The width of the central peak is $2\theta(n=1) = (2\pi/ka) = (\lambda/a)$, as expected.

The above result can be used to calculate the intensity pattern arising from a diffraction grating, which can be modelled as a plane screen in which a series of identical parallel slits is cut. Let the width of each of these slits be $2a$, the width of the opaque screen between the neighboring slits be $2b$, and the total number of slits be N. In this case, the total amplitude is the sum

$$u_q = u_q' \sum_{n=0}^{N-1} e^{-2inqd} = u_q' \frac{1 - e^{-2iNqd}}{1 - e^{-2iqd}}, \qquad d = a + b, \tag{3.194}$$

where u_q' is the intensity pattern from a single slit. By using expression (3.193), we obtain the intensity distribution as

$$dI = \frac{I_0 a}{N\pi} \left(\frac{\sin Nqd}{\sin qd} \right)^2 \left(\frac{\sin qa}{qa} \right)^2 dq = \frac{I_0}{N\pi ak} \left(\frac{\sin Nk\theta d}{\sin k\theta d} \right)^2 \frac{\sin^2 ka\theta}{\theta^2} d\theta, \tag{3.195}$$

where I_0 is the total intensity passing through all the slits.

As a second example, consider the diffraction of light that is incident normal to the plane of a circular aperture of radius a. In this case, it is convenient to use polar coordinates, with the z axis passing through the centre of aperture and perpendicular to its plane and with (r, ϕ) denoting the polar coordinates in the transverse plane. Because of axial symmetry, the vector $\mathbf{q}$ has only a radial component, $q_r = q = k\theta$. The Fourier transform now gives

$$u_q = u_0 \int_0^a \int_0^{2\pi} e^{-iqr\cos\phi} r\,d\phi\,dr = 2\pi u_0 \int_0^a J_0(qr) r\,dr \tag{3.196}$$

where $J_n(z)$ is the Bessel function of the order of n. Using the identity

$$\int_0^a J_0(qr)r\,dr = \frac{a}{q} J_1(aq), \tag{3.197}$$

we get

$$u_q = 2\pi \frac{u_0 a}{q} J_1(aq). \tag{3.198}$$

The intensity of light diffracted into a solid angle $d\Omega$ is

$$dI = I_0 \frac{J_1^2(ak\theta)}{\pi\theta^2} d\Omega, \tag{3.199}$$

where I_0 is the total intensity falling on the aperture. This intensity goes to zero for the first time at $\theta \approx 1.22(\lambda/2a)$. This formula is often used to determine the resolving power of telescopes with a given diameter.

3.13.2 Fresnel Diffraction ($r \ll D^2/\lambda$)

When $r\lambda \ll D^2$, we are interested in the behaviour of a wave field (comparatively) close to an obstruction. In this case, it is better to use Eq. (3.183) with Eq. (3.184) as neither of the exponential factors in Eq. (3.187) can be ignored. As an example of this situation, let us consider a plane wave that is incident upon a screen that covers the semi-infinite plane $x \leq 0$, $-\infty < y < \infty$, $z = 0$. The direction of propagation is the positive z axis, and we are interested in the intensity distribution in the plane $z = D > 0$.

We can calculate this from Eq. (3.183) by setting $a(z', \mathbf{x}'_\perp) = 1$, $z' = 0$, $z = D$, and limiting the two-dimensional integration to the range $0 < x' < \infty$ and $-\infty < y' < \infty$. The integral reduces to

$$a(D, \mathbf{x}_\perp) = \left(\frac{k}{2\pi i D}\right) \int_{-\infty}^{\infty} dy' \int_0^{\infty} dx' \exp\frac{ik}{2D}[(x-x')^2 + (y-y')^2]$$

$$= \left(\frac{k}{2\pi i D}\right)\left(\frac{2i\pi D}{k}\right)^{1/2} \int_0^{\infty} dx' \exp\frac{ik}{2D}(x-x')^2. \tag{3.200}$$

where $k = (\omega/c)$. Except for unimportant constant factors, this expression can be transformed to give

$$a \propto A(x) = \int_{-w}^{\infty} e^{i\eta^2} d\eta, \quad w = x\sqrt{\frac{k}{2D}}. \tag{3.201}$$

To study the properties of the intensity variation with x, it is necessary to understand the behaviour of this function. This is conventionally done through a graphical construction called Cornu's spiral; however, it is easier to understand the behaviour analytically along the following lines.

To begin with, let us consider the intensity at negative values of w that corresponds to the region of geometrical shadow. (If we approximate the wave field by a ray of light, we would expect zero intensity in this region.) One integration by parts gives

$$\int_{|w|}^{\infty} e^{i\eta^2} d\eta = -\frac{1}{2i|w|} e^{iw^2} + \frac{1}{2i} \int_{|w|}^{\infty} e^{i\eta^2} \frac{d\eta}{\eta^2}. \tag{3.202}$$

Repeating the process for the integral on the right-hand side, we get an asymptotic expansion for the integral in the form

$$\int_{|w|}^{\infty} e^{i\eta^2} d\eta = e^{iw^2} \left(-\frac{1}{2i|w|} + \frac{1}{4|w|^3} - \cdots \right). \tag{3.203}$$

The first term in this series already gives a good approximation to the integral in Eq. (3.201). Taking the absolute value, we find that the intensity for large negative values of w is given by

$$I = \frac{I_0}{4\pi w^2}, \tag{3.204}$$

where I_0 is a constant. Thus the intensity of a wave falls as a square of the distance from the edge of the geometrical shadow.

To find the variation of intensity at positive values of w, we write

$$\int_{-w}^{\infty} e^{i\eta^2} d\eta = \int_{-\infty}^{\infty} e^{i\eta^2} d\eta - \int_{-\infty}^{-w} e^{i\eta^2} d\eta = (1+i)\sqrt{\frac{\pi}{2}} - \int_{w}^{\infty} e^{i\eta^2} d\eta \tag{3.205}$$

and use the asymptotic expansion obtained in Eq. (3.203) for the second integral. This gives

$$\int_{-w}^{\infty} e^{i\eta^2} d\eta \cong \sqrt{\pi}\, e^{i\pi/4} + \frac{1}{2iw} e^{iw^2}, \tag{3.206}$$

so that the intensity distribution becomes

$$I = I_0 \left(1 + \sqrt{\frac{1}{\pi}} \frac{\sin\left(w^2 - \frac{\pi}{4}\right)}{w} \right). \tag{3.207}$$

Thus, even in the illuminated region, the wave nature of light produces an infinite sequence of maxima and minima, with the amplitude decreasing inversely as the distance of the maxima and minima. The quantity I_0 represents the value of intensity far away from the screen.

The exact expression for the intensity can be expressed in the form

$$I = \frac{I_0}{2}\left|\sqrt{\frac{2}{\pi}}\int_{-w}^{\infty} e^{i\eta^2}d\eta\right|^2 = \frac{I_0}{2}\left\{\left[C(w^2)+\frac{1}{2}\right]^2 + \left[S(w^2)+\frac{1}{2}\right]^2\right\}, \quad (3.208)$$

where

$$C(z) = \sqrt{\frac{2}{\pi}}\int_0^{\sqrt{z}} \cos\eta^2 d\eta, \quad S(z) = \sqrt{\frac{2}{\pi}}\int_0^{\sqrt{z}} \sin\eta^2 d\eta \quad (3.209)$$

are called the Fresnel integrals. In general, these functions need to be evaluated numerically.

Exercise 3.21

Lunar occultation: The motion of the Moon past a distant source of electromagnetic radiation, called occultation, is used as a helpful technique in the astronomical positioning of cosmic sources. To first approximation, this reduces to the problem of Fresnel diffraction by a screen. Suppose that the observing wavelength λ is $\sim$20 cm. Estimate the Fresnel length and the time it takes for the diffraction pattern to pass across the telescope. [Answer: Because the Moon is $z \approx 4 \times 10^{10}$ cm away, the Fresnel length $r_F \equiv (\lambda z)^{1/2} \approx 10$ km. Taking the orbital speed of Moon to be $v \approx 200$ m/s, the entire diffraction pattern would have taken $(2r_F/v) \approx 2$ min to cross the telescope.]

3.14 Interference and Coherence

Consider a wave of unit amplitude incident upon an opaque screen with two small holes separated by a distance l. We are interested in the amplitude at large distances along a direction that makes a small angle θ with the original direction of propagation. This is a problem in Fraunhofer diffraction, in which the amplitude in the source plane is approximated as being due to two Dirac delta functions separated by distance l. Because the Fraunhofer diffraction pattern is essentially the Fourier transform of the initial amplitude, the resultant wave amplitude is given by

$$a(\theta) = e^{-ikl\theta/2} + e^{ikl\theta/2} \propto \cos\left(\frac{1}{2}kl\theta\right), \quad (3.210)$$

except for unimportant constants. The intensity variation, given by the square of the amplitude, has successive maxima and minima, with the intensity dropping to zero at the minimum. These are conventionally called interference fringes and the phenomena is called interference. The way we have approached the problem is only a special case of Fraunhofer diffraction. (In the above discussion, we have ignored the finite size of the hole and thus arrived at infinite number of fringes. When the finite size w of the holes is taken into account, we also must deal with diffraction that is due to the holes and the number of fringes will be of the order of l/w when $l \gg w$; see Exercise 3.22.)

In Eq. (3.210), the initial wave front was taken to be incident normal to the screen. If it is incident at a small angle α to the plane of the slits, then the resultant intensity distribution will be proportional to

$$I(\theta) \propto \left| e^{-ikl(\theta-\alpha)/2} + e^{ikl(\theta-\alpha)/2} \right|^2 \propto \cos^2\left(\frac{kl(\theta - \alpha)}{2} \right). \qquad (3.211)$$

When the direction of the incident beam changes, the location θ of the maxima of the fringes changes; thus the location of the interference fringes contains information about the direction of the source.

The discussion so far has dealt with strictly monochromatic and coherent waves incident upon the screen. In realistic astrophysical situations, the sources are seldom coherent. An incoherent wave field can be represented by an amplitude that varies with time as

$$a(t) \propto \exp i \left[\omega t + \phi(t) \right], \qquad (3.212)$$

where the phase $\phi(t)$ is expected to fluctuate randomly at a time scale t_c, which is long compared with the period $2\pi/\omega$; the time scale t_c is called the coherence time of the wave. (Strictly speaking, such a wave is not monochromatic either, and we could have described it by a superposition of waves of different frequencies in a narrow band around ω.) In the case of waves emitted by a realistic, extended, astrophysical source, the waves arriving from different locations of the source will be completely uncorrelated in phase. Therefore, when we average the total intensity of the wave arriving at a given location from two different directions, say, α_1 and α_2, we can add the intensities of the individual waves. Mathematically

$$\langle |a(\alpha_1, t) + a(\alpha_2, t)|^2 \rangle \cong \langle |a(\alpha_1, t)|^2 \rangle + \langle |a(\alpha_2, t)|^2 \rangle, \qquad (3.213)$$

where the averages are taken over time scales that are long compared with the coherence time t_c.

Suppose such an incoherent wave from an extended astrophysical source is sent through the interference slits. We assume that the coherence time of the wave is also long compared with the difference in the light travel time to the screen from the two slits. In that case, the waves arriving from each point on the source [characterised by an angular intensity distribution function $S(\alpha)$] will form interference fringes as described by Eq. (3.211). The contributions from different directions now add incoherently and thus the net intensity on the screen is given by

$$I(\theta) \propto \int d\alpha \, S(\alpha) \cos^2 \left[\frac{kl(\theta - \alpha)}{2} \right] \propto \int d\alpha \, S(\alpha)\{1 + \cos[kl(\theta - \alpha)]\}.$$

(3.214)

This is conventionally written in the form

$$I(\theta) \propto F_S\{1 + \mathrm{Re}[\gamma_\perp(kl)e^{-ikl\theta}]\}, \quad F_S = \int d\alpha \, S(\alpha),$$ (3.215)

where F_S is the total intensity from the source and

$$\gamma_\perp(kl) = \frac{\int d\alpha \, S(\alpha) e^{ikl\alpha}}{F_S}$$ (3.216)

is called the degree of spatial coherence. This quantity is nonzero even though the source is assumed to have a complete lack of angular coherence. Lack of angular coherence, in general, reduces the degree of spatial coherence but does not make it vanish.

Eq. (3.216) also shows that the degree of spatial coherence is the Fourier transform of the angular-intensity pattern of the source. Alternatively, if we know the degree of spatial coherence as a function of kl, then we can reconstruct the angular intensity of the source by Fourier inversion:

$$S(\alpha) = F_S \int \frac{d(kl)}{2\pi} \gamma_\perp(kl) e^{-ikl\alpha}.$$ (3.217)

The above analysis can be easily extended to two dimensions to give the corresponding formulas:

$$\gamma_\perp(k\mathbf{l}) = \frac{\int d\Omega_\alpha \, S(\alpha) e^{ik\mathbf{l}\cdot\alpha}}{F_S}, \quad S(\alpha) = F_S \int \frac{d^2(k\mathbf{l})}{(2\pi)^2} \gamma_\perp(k\mathbf{l}) e^{-ik\mathbf{l}\cdot\alpha}. \quad (3.218)$$

These formulas are used in astronomical observations to reconstruct the source's angular intensity pattern from the observed degree of spatial coherence. The modulus of $\gamma_\perp$ is called the visibility and is related to the maximum and the minimum values of the observed intensity by

$$|\gamma_\perp| \equiv V = \frac{I_{\max} - I_{\min}}{I_{\max} + I_{\min}}.$$ (3.219)

The visibility can lie in the range from zero (which corresponds to the absence of any contrast with fringes being completely undetectable) to unity (which corresponds to monochromatic plane wave with the contrast as large as possible).

To get complete information about $\gamma_\perp$ we also need to know its phase ϕ. When $\phi = 0$, there will be a bright fringe right on the optic axis, which will happen, for example, if the source is symmetric about the optic axis. If we tilt the plane of the slits such that the symmetry point of the source is moved off the optic axis by an angle δ, the fringe pattern will shift by the same angle δ and the phase ϕ will change by $kl\delta$. Thus we can, in principle, measure both the modulus and the phase of the complex number $\gamma_\perp$ and – by using Eq. (3.217) – we can reconstruct the angular-intensity pattern of the source.

The above discussion applies equally well to the situation in which the source has a degree of coherence in time rather than in space. Consider two points separated by a distance s along the direction of propagation of the wave. If we calculate the two-point correlation function of the wave at these two locations at the same instant of time, we are essentially measuring the correlation of the wave field at times differing by the time of propagation $\tau = s/c$. We can now define the degree of temporal coherence by

$$\gamma_\parallel(\tau) = \frac{\langle \psi(t)\psi^*(t+\tau)\rangle}{|\psi|^2}, \tag{3.220}$$

where the average is taken over sufficiently long time scales. If the incident wave is not strictly monochromatic but has significant power in a small bandwidth $\delta\omega$ around the mean frequency ω_0, then its coherence time will be $\tau_c \approx (1/\delta\omega)$. If we sample the wave at two points separated by a distance greater than $s_c = c\tau_c \approx (c/\delta\omega)$, then the wave will be essentially incoherent and $\gamma_\parallel$ will be significantly small. On the other hand, when the wave field is sampled at distances small compared with s_c, $\gamma_\parallel$ will be typically of the order of unity. It is easy to see that equation corresponding to Eq. (3.216), in the case of temporal coherence, is

$$\gamma_\parallel(\tau) = \frac{\int d\omega F(\omega)e^{i\omega\tau}}{F_S}, \tag{3.221}$$

where $F(\omega)$ represents the power spectrum (in the frequency space) of the incident wave field. The inverse relation is given by

$$F(\omega) = F_S \int \frac{d\tau}{2\pi}\gamma_\parallel(\tau)e^{-i\omega\tau}. \tag{3.222}$$

In a simple interferometer, for example, we arrange an incident wave field to go through two paths differing by some length s and then make them interfere. By studying the interference fringes as a function of $s = c\tau$, we can obtain information about $\gamma_\parallel(\tau)$. The inverse Fourier transform, Eq. (3.222), will then give information about the spectral density of the source and, in particular, about the width $\delta\omega$ of the spectral line.

Exercise 3.22

Effect of hole size: In the two-slit interference discussed in the text, the finite size of the slit was ignored, thereby leading to an infinite number of fringes. If the slits have a finite size w, then the diffraction that is due to the slit has to be taken into account. Show that the number of distinct fringes will be of the order of (l/w), where l is the separation between the slits and $l \gg w$.

3.15 Linear Optical Systems

Equation (3.183) allows us to compute the wave amplitude at any location on the plane $z = z_2$ if the amplitude on a plane $z = z_1 < z_2$ is given. The following situation arises very often in astrophysics: A wave front propagates freely up to a plane $z = z_1$, where it passes through an optical system (say a lens, mirror, atmosphere, etc.) that modifies the wave in a particular fashion. The optical system extends from $z = z_1$ to $z = z_2$ and the wave propagates freely for $z > z_2$. We are interested in the amplitude at $z > z_2$, given the amplitude at $z < z_1$.

It is clear that Eq. (3.183) can be used to propagate the amplitude from some initial plane $z = z_O < z_1$ to $z = z_1$ and from $z = z_2$ to some final plane $z = z_I > z_2$. (The subscripts O and I stand for object and image, respectively, based on the idea that the optical system is a lens.) The propagation of a wave from z_1 to z_2 depends entirely on the optical system and, in fact, defines the particular optical system. An optical system is called linear if the output is linear in input. In such a case, the amplitude at the exit point of the optical system is related to the amplitude at the entrance point by a relation of the kind

$$a(z_2, \mathbf{x}_2) = \int d^2\mathbf{x}_1 \, P(z_2, z_1; \mathbf{x}_2, \mathbf{x}_1) a(z_1, \mathbf{x}_1), \qquad (3.223)$$

where the functional form of P defines the kind of optical system. (Here and in what follows, we omit the subscript $\perp$ with the understanding that the vector $\mathbf{x}$ is in the transverse plane and is two dimensional.) In this case, the amplitude at the image plane can be expressed in terms of the amplitude at the object plane by the relation

$$a(z_I, \mathbf{x}_I) = \int d^2\mathbf{x}_O \, \mathcal{G}(z_I, z_O; \mathbf{x}_I, \mathbf{x}_O) a(z_O, \mathbf{x}_O), \qquad (3.224)$$

where

$$\mathcal{G}(z_I, z_O; \mathbf{x}_I, \mathbf{x}_O) = \int d^2\mathbf{x}_2 d^2\mathbf{x}_1 \, G(z_I - z_2, \mathbf{x}_I - \mathbf{x}_2) \, P(z_2, z_1; \mathbf{x}_2, \mathbf{x}_1)$$
$$\times \, G(z_1 - z_O, \mathbf{x}_1 - \mathbf{x}_O). \qquad (3.225)$$

Given the properties of any linear optical system, we can compute the quantity P and thus evaluate $\mathcal{G}$ and determine the properties of wave propagation.

As a simple example, let us find out the form of the function P for a convex lens. If the lens is sufficiently thin, P will be nonzero only at the plane of the lens $z_2 = z_1 = z_L$. Because the lens does not absorb radiation, it cannot change the amplitude $|a(z_L, \mathbf{x}_L)|$ of the incident wave and can modify only the phase.

Therefore P must have the form $P = \exp[i\theta(\mathbf{x}_L)]$. Then the amplitude at the image plane is given by

$$a(z_I, \mathbf{x}_I) = \int d^2\mathbf{x}_L a(z_L, \mathbf{x}_L) P(z_L, \mathbf{x}_L) G(z_I - z_L, \mathbf{x}_I - \mathbf{x}_L)$$

$$= a \int d^2\mathbf{x}_L e^{i\theta(z_L, \mathbf{x}_L)} G(z_I - z_L, \mathbf{x}_I - \mathbf{x}_L), \qquad (3.226)$$

where we have used the fact that the amplitude $a(z_L, \mathbf{x}_L)$ on the lens plane is constant for a plane wave incident from a large distance. To determine the form of $\theta(\mathbf{x}_L)$, we use the basic defining property of lens of focal length f: If a plane wave front of constant intensity is incident upon the lens plane $z = z_L$, the rays will be focused at a point $z_I = z_L + f$ if the wave nature of the light is ignored. In the limit of zero wavelength for the wave, most of the contributions to the integral come from points at which the phase of the integrand is stationary. Because the phase of G is $(k/2)[(\Delta\mathbf{x})^2/\Delta z]$ the principle of stationary phase gives

$$\frac{\partial\theta}{\partial\mathbf{x}_L} = \frac{k}{f}(\mathbf{x}_I - \mathbf{x}_L), \qquad (3.227)$$

where $f = z_I - z_L$. For the image to be formed along the z axis, this equation should be satisfied for $\mathbf{x}_I = 0$. Setting $\mathbf{x}_I = 0$ and integrating this equation, we find that $\theta = (-kx_L^2/2f)$ and

$$P(\mathbf{x}_L) = \exp\left(-\frac{ik}{2f}x_L^2\right). \qquad (3.228)$$

Thus the effect of a lens is to introduce a phase variation that is quadratic in the transverse coordinates. Such a lens will focus the light to a point on the z axis in the limit of zero wavelength.

A geometrical interpretation of this result is given in Fig. 3.5. The constant phase surfaces are planes to the left of the lens and they are arcs of circles (centred on the focus F) on the right-hand side of the lens. Changing the plane to a circle (of radius f) at $z = z_L$ introduces a path difference of $\Delta l = [f - (f^2 - x_L^2)^{1/2}] \simeq (x_L^2/2f)$ at a transverse distance x_L. This corresponds to a phase difference $k\Delta l = (kx_L^2/2f) = \theta$ introduced by the lens.

Let us next consider the effect of this lens on a point source of radiation along the z axis at $z = z_O$. [That is, we take the initial amplitude to be $a(z_O, \mathbf{x}_O) \propto \delta_D(\mathbf{x}_O)$.] We can obtain this by first propagating the field from z_O to z_L, modifying the phase that is due to the lens at $z = z_L$, and propagating it farther to some point z with the transverse coordinate set to zero. The net result is given by

$$a(z, 0) = -\frac{k^2}{4\pi^2 uv} \int d^2\mathbf{x}_L \exp\left(-\frac{ik}{2f}x_L^2\right) \exp\left(\frac{ikx_L^2}{2u} + \frac{ikx_L^2}{2v}\right), \qquad (3.229)$$

where $u = z_L - z_O$ and $v = z - z_L$. In the limit of zero wavelength (called ray

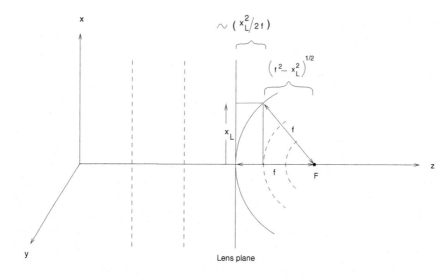

Fig. 3.5.   The focusing action of a convex lens in terms of the phase change of wave fronts.

optics), we can again obtain the maximum contribution to this integral by setting the variation of the phase to zero. This gives

$$-\frac{k}{f}\mathbf{x}_L + \frac{k}{u}\mathbf{x}_L + \frac{k}{v}\mathbf{x}_L = 0 \tag{3.230}$$

or

$$\frac{1}{u} + \frac{1}{v} = \frac{1}{f}, \tag{3.231}$$

which is a familiar formula in the theory of lenses.

The above result was obtained in the limit of ray optics. To study the wave propagation through the lens from a source plane at a distance u to an image plane at a distance v with $v = fu/(f - u)$, we first propagate the wave from the source plane S (at a distance u from the lens) to the lens and then to the focal plane (at a distance f behind the lens) and finally through a distance $v - f$ from the focal plane to the image plane. The net propagation amplitude from the source to the focal plane is then

$$P_{FS} = \int d^2 x_L \, \frac{k}{2\pi i f} \exp\left[\frac{ik(\mathbf{x}_F - \mathbf{x}_L)^2}{2f}\right] \exp\left(-\frac{ik|\mathbf{x}_L|^2}{2f}\right)$$

$$\times \frac{k}{2\pi i u} \exp\left[\frac{ik(\mathbf{x}_S - \mathbf{x}_L)^2}{2u}\right]$$

$$= -\frac{ik}{2\pi f} \exp\left[-\frac{ikx_F^2}{2(v - f)}\right] \exp\left(-\frac{ik\mathbf{x}_f \cdot \mathbf{x}_S}{f}\right). \tag{3.232}$$

The amplitude on the focal plane is

$$a_F(\mathbf{x}_F) = \int d^2x_S \, P_{FS} a_S(\mathbf{x}_S) = -\frac{ik}{2\pi f} \exp\left[-\frac{ikx_F^2}{2(v-f)}\right] \tilde{a}_S(\mathbf{x}_F/f),$$

(3.233)

where

$$\tilde{a}_S(\mathbf{q}) \equiv \int d^2x_S a_S(\mathbf{x}_S) e^{-i\mathbf{q}\cdot\mathbf{x}_S}.$$

(3.234)

We thus arrive at the important result that the intensity distribution at the focal plane is determined by the Fourier transform of the incident intensity distribution. This is similar to the result in the case of Fraunhofer diffraction. In fact, the use of a lens in many optical devices is essentially to bring points at infinity to a finite distance, which is achieved by compensating for the quadratic part of the propagation amplitude by adding an appropriate phase-distortion.

This is a general result and arises as follows: The action of a lens on the phase of an initial intensity distribution is governed by the integral

$$a(z, \mathbf{x}) \propto \int d^2x_L a(z_L, \mathbf{x}_L) \exp\left(-\frac{ik}{2f}x_L^2\right) \exp\frac{ik}{2(z-z_L)}(\mathbf{x}-\mathbf{x}_L)^2,$$

(3.235)

where $a(z_L, \mathbf{x}_L)$ is the incident amplitude upon the lens, the first exponential gives the distortion in phase produced by the lens, and the second exponential gives the propagation amplitude z_L to z. On the focal plane, which is a plane located at a distance f from the lens at $z = z_L + f$, the second exponential characterising the propagation becomes

$$\exp\frac{ik(\mathbf{x}-\mathbf{x}_L)^2}{2(z-z_L)} = \exp\frac{ik}{2f}(x^2 + x_L^2 - 2\mathbf{x}\cdot\mathbf{x}_L).$$

(3.236)

The quadratic term $(ikx_L^2/2f)$ in the propagation amplitude is now precisely cancelled by the phase distortion introduced by the lens, so that the resultant amplitude can be written as

$$a(z_L + f, \mathbf{x}) \propto \exp\left(\frac{ik}{2f}x^2\right) \int d^2x_L a(z_L, \mathbf{x}_L) \exp\left(\frac{ik}{f}\mathbf{x}\cdot\mathbf{x}_L\right).$$

(3.237)

The intensity at the focal plane is given by $|a(z_L+f, \mathbf{x})|^2$ in which the phase factor $\exp(ikx^2/2f)$ does not contribute. This is clearly determined by the Fourier transform of the incident amplitude.

Further propagation of the amplitude from the focal plane to the image plane leads to

$$a_I = \int d^2x_F G(z_I - z_F, \mathbf{x}_I - \mathbf{x}_F) a_F$$

$$= -\left(\frac{u}{v}\right) \exp\left[\frac{ikx_I^2}{2(v-f)}\right] a_S(\mathbf{x}_S = -\mathbf{x}_I u/v).$$

(3.238)

Ignoring the unimportant phase factor, we see that the wave amplitude in the image plane is just a magnified version of the wave in the source plane. The lens action thus involves taking two Fourier transforms, one to the focal plane and the other from focal plane to the image plane. Because of this reason, we can often use the focal plane to alter the characteristics of the wave by introducing, for example, spatial filtering devices at the focal plane.

If the lens has an aperture D, then we may take $a(z_L, \mathbf{x}_L)$ to be unity within a circle of diameter D and zero outside. In such a case, we are obtaining a Fourier transform of the aperture shape and the results obtained in Section 3.13 in the case of Fraunhofer diffraction will be applicable. We saw in Section 3.13 that a circular aperture of diameter D will produce a Fraunhofer diffraction pattern, with the intensity falling to zero at an angular scale $\theta = 1.22(\lambda/D)$. The same phenomena will take place for a lens of diameter D as well. On the focal plane, at a distance f from the lens, this will lead to a spot of diameter $f\theta \approx f\lambda/D$.

Exercise 3.23

Geometrical and wave optics: Consider a wave $\psi(t, \mathbf{x})$ that satisfies the differential equation

$$-\frac{\partial}{\partial t}\left(\frac{n^2}{c^2}\frac{\partial \psi}{\partial t}\right) + \nabla^2 \psi = 0, \tag{3.239}$$

where (c/n) is the local speed of propagation of the wave. (In general, n can be a function of space and time.) We can obtain the geometrical optics limit of such a wave, formally, by writing the solution in the form

$$\psi = (A + \epsilon B + \cdots)\, e^{i\phi/\epsilon}, \tag{3.240}$$

where ϵ is a small parameter. (1) Assuming that a suitable small parameter exists in the problem, show that, to the lowest order, we must have

$$\omega = \Omega(\mathbf{k}, \mathbf{x}, t) \equiv \frac{c}{n(\mathbf{x}, t)}k, \quad \omega \equiv -\frac{\partial \phi}{\partial t}, \quad \mathbf{k} \equiv \nabla \phi. \tag{3.241}$$

Interpret this result. (2) What does the next higher-order term in the wave equation give?

3.16 Wave Propagation through a Random Medium

Because the angular resolution of, say, a telescope of diameter D will be of the order of λ/D, where λ is the operating wavelength, we may think that, by increasing D sufficiently, we can obtain high angular resolution. Unfortunately, this is not strictly true for ground-based observations in which the light is transmitted through the atmosphere. Because of turbulence in atmosphere, the refractive index of air undergoes fluctuations in a random fashion. This, in turn, degrades the image obtained by any telescope and could even be the limiting factor in resolution. In this section we discuss some of these effects.

The propagation of the wave amplitude in the absence of any intervening optical system is given by

$$a(z, \mathbf{x}) = \int d^2\mathbf{x}_A \, a(z', \mathbf{x}_A) \, G(z - z'; \mathbf{x} - \mathbf{x}_A). \qquad (3.242)$$

While propagating through an atmosphere, the wave will be continuously scattered by the inhomogeneities from some point $z = H$ to $z = 0$, where H is the height of the atmosphere and $z = 0$ represents the ground level. As a first approximation, we may assume that the entire effect of this scattering is to introduce a random phase to the amplitude at some effective height h. In that case the equation for propagation will be modified to

$$a(0, \mathbf{x}) = \int d^2\mathbf{x}_A \, a(h, \mathbf{x}_A) \, e^{i\theta(\mathbf{x}_A)} G(h; \mathbf{x} - \mathbf{x}_A) = \int d^2\mathbf{x}_A \, e^{i\theta(\mathbf{x}_A)} G(h; \mathbf{x} - \mathbf{x}_A),$$

$$(3.243)$$

where $\theta(\mathbf{x}_A)$ is a random phase. In arriving at the second equality, we have assumed that the incident amplitude is constant and is normalised to unity. The statistical properties of such a random phase can be characterised by the different orders of the correlation functions. In particular, for a stationary random process that is translationally invariant, we can write

$$\left\langle e^{i\theta(\mathbf{x}_A)} \right\rangle = 0, \quad \left\langle e^{i[\theta(\mathbf{x}_A) - \theta(\mathbf{x}_B)]} \right\rangle = F(\mathbf{x}_A - \mathbf{x}_B), \qquad (3.244)$$

where $F(\mathbf{x}_A - \mathbf{x}_B)$ is some function related to the atmospheric turbulence. We are interested in determining the statistical property of the final amplitude $a(0, \mathbf{x})$ in terms of the function F, which is treated as known.

In this particular case, we first show that the scale h is irrelevant in determining the statistical properties of the final amplitude. To see this, we consider the two-point correlation function $\langle a(0, \mathbf{x})a^*(0, \mathbf{y}) \rangle$ of the wave amplitude at the receiving plane. This is given by

$$\langle a(0, \mathbf{x})a^*(0, \mathbf{y}) \rangle = \int d^2\mathbf{x}_A \, d^2\mathbf{x}_B \, F(\mathbf{x}_A - \mathbf{x}_B) \, G(h; \mathbf{x} - \mathbf{x}_A)G^*(h; \mathbf{y} - \mathbf{x}_B).$$

$$(3.245)$$

In this expression the phase contributed by the product of the propagators is

$$\frac{ik}{2h}[(\mathbf{x} - \mathbf{x}_A)^2 - (\mathbf{y} - \mathbf{x}_B)^2] = \frac{ik}{2h}[(\mathbf{x} - \mathbf{y} - \mathbf{u}) \cdot (\mathbf{x} + \mathbf{y} - \mathbf{v})], \quad (3.246)$$

where $\mathbf{u} = \mathbf{x}_A - \mathbf{x}_B$ and $\mathbf{v} = \mathbf{x}_A + \mathbf{x}_B$. The integral over $\mathbf{x}_A$ and $\mathbf{x}_B$ can be transformed into an integral over $\mathbf{u}$ and $\mathbf{v}$ to give

$$\langle a(0, \mathbf{x})a^*(0, \mathbf{y}) \rangle \propto \int d\mathbf{u} \, F(\mathbf{u}) \exp\left[\frac{ik}{2h}(\mathbf{x} + \mathbf{y}) \cdot (\mathbf{x} - \mathbf{y} - \mathbf{u})\right]$$

$$\times \int d\mathbf{v} \exp\left[\frac{ik}{2h}\mathbf{v} \cdot (\mathbf{x} - \mathbf{y} - \mathbf{u})\right]. \qquad (3.247)$$

The integration over $\mathbf{v}$ gives a Dirac delta function that can be used to simplify the integration over **u**, leading to

$$\langle a(0, \mathbf{x})a^*(0, \mathbf{y})\rangle \propto F(\mathbf{x} - \mathbf{y}) \propto \langle e^{i[\theta(\mathbf{x}_A) - \theta(\mathbf{x}_B)]}\rangle, \tag{3.248}$$

which shows that the correlation function of the wave amplitudes propagates without change as z varies. Because of this result, we can simplify this analysis by assuming that the effect of atmosphere is to introduce a random phase in the wave amplitude just before it is incident upon a optical system.

One of the questions that is of importance in optical astronomy is related to the limitations on the resolving power of a telescope that are due to atmospheric fluctuations. To understand these limitations let us consider the propagation of a wave amplitude through a lens of focal length f and aperture D. We have seen in Section 3.15 that the amplitude at the focal plane is related by a Fourier transform [see expression (3.237)] to the incident amplitude. In the presence of a turbulent atmosphere, the incident wave amplitude at the lens is multiplied by a random-phase factor, and this equation becomes

$$a(z_L + f, \mathbf{x}) = \int_{\mathcal{A}} d^2\mathbf{x}_L \, a(z_L, \mathbf{x}_L) \exp\left[i\theta(\mathbf{x}_L) + \frac{ik}{f}\mathbf{x} \cdot \mathbf{x}_L\right], \tag{3.249}$$

where $\theta(\mathbf{x}_L)$ is the random phase that is due to the turbulent atmosphere and the integration is over the aperture $\mathcal{A}$ of the lens. To obtain an analytic solution in terms of simple functions, we replace the circular aperture with a Gaussian aperture of width D, that is, we multiply the integrand in Eq. (3.249) by a factor $\exp(-x_L^2/2D^2)$ and extend the integration over the entire range of $\mathbf{x}_L$. Taking the initial amplitude $a(z_L, \mathbf{x}_L)$ to be unity, we find that the integral becomes

$$a(z_L + f, \mathbf{x}) = \int d^2\mathbf{x}_L \exp\left[-\frac{x_L^2}{2D^2} + \frac{ik}{f}\mathbf{x} \cdot \mathbf{x}_L + i\theta(\mathbf{x}_L)\right]. \tag{3.250}$$

This equation represents the effect of atmospheric turbulence on a telescopic system with focal length f and (Gaussian) aperture D. We can find the intensity pattern produced by such a system by averaging the intensity $I = |a(z_L + f, \mathbf{x})|^2$ over the random phases. This gives

$$\langle I(z_L + f, \mathbf{x})\rangle = \int d^2\mathbf{x}_L \, d^2\mathbf{x}_L' \, F(\mathbf{x}_L - \mathbf{x}_L')$$

$$\times \exp\left(-\frac{x_L^2 + x_L'^2}{2D^2}\right) \exp\left[\frac{ik}{f}\mathbf{x} \cdot (\mathbf{x}_L - \mathbf{x}_L')\right]. \tag{3.251}$$

Introducing $\mathbf{u} = \mathbf{x}_L - \mathbf{x}_L'$ and $\mathbf{v} = \mathbf{x}_L + \mathbf{x}_L'$, we find that this integral becomes

$$\langle I\rangle \propto \int d\mathbf{u} \, F(\mathbf{u}) \exp\left(-\frac{u^2}{4D^2} + \frac{ik}{f}\mathbf{x} \cdot \mathbf{u}\right) \int d\mathbf{v} \, e^{-(v^2/4D^2)}$$

$$\propto \int d\mathbf{u} \, F(\mathbf{u}) \exp\left(-\frac{u^2}{4D^2} + \frac{ik}{f}\mathbf{x} \cdot \mathbf{u}\right). \tag{3.252}$$

Given a model for atmospheric turbulence, the correlation function $F(\mathbf{u})$ can be computed, and by using expression (3.252) we can determine the effect of turbulence on the intensity pattern.

We can easily understand the essential features of the turbulence by evaluating the integral in a saddle-point limit. This consists of approximating F by the Gaussian

$$F(\mathbf{u}) \propto \exp\left(-\frac{u^2}{2l^2}\right), \tag{3.253}$$

where l represents the length scale over which atmospheric turbulence is correlated. The integrals are now easy to do and the final result, including the constant of proportionality, is given by

$$I(\mathbf{x}) = \frac{k^2 l^2 D^2}{2 f^2 [1 + (l^2/D^2)]} \exp\left[-\frac{k^2 l^2}{2 f^2}\left(1 + \frac{l^2}{2D^2}\right)^{-1} x^2\right]. \tag{3.254}$$

This result has interesting implications. It shows that the intensity pattern is a Gaussian with a width given by

$$R = \frac{f}{kl}\left(1 + \frac{l^2}{2D^2}\right)^{1/2} = \frac{f\lambda}{2\pi}\left(\frac{1}{l^2} + \frac{1}{2D^2}\right)^{1/2}. \tag{3.255}$$

The corresponding angular width, which determines the angle of resolution of the telescope, is given by $\theta = R/f$. The resolving angle has two limiting forms, depending on whether the aperture of the telescope D is smaller or larger than the scale of atmospheric turbulence l:

$$\theta \simeq \begin{cases} \dfrac{1}{2\pi}\left(\dfrac{\lambda}{\sqrt{2}D}\right) & (D \ll l) \\[2ex] \dfrac{1}{2\pi}\left(\dfrac{\lambda}{l}\right) & (D \gg l) \end{cases}. \tag{3.256}$$

In the first case $(D \ll l)$ the resolution is essentially limited by the size of the telescope aperture. This is the case we came across earlier in Section 3.13 and corresponds to the standard diffraction pattern. [The numerical constants that appear in front of λ/D have no specific significance as they depend on the detailed assumptions regarding the shape of aperture and the definition of D.] As long as $D \ll l$, the resolution can be improved by an increase in D. It is, however, clear that there is no significant advantage in increasing the aperture when D becomes comparable with l. When $D \gg l$, the resolution is limited not by the aperture but by the atmospheric effects and is of the order of λ/l. For Earth's atmosphere in optical wavelengths, $l \simeq 10$ cm, $\lambda \simeq 4000$ Å, and $(\lambda/l) \simeq 1''$. This feature describes an intrinsic limitation in resolution for all ground-based optical telescopes.

For more realistic modelling of the atmospheric effects, we need to obtain the form of F in expression (3.248) from the structure of the atmosphere. In many cases, we can assume that $\theta(\mathbf{x})$ is a Gaussian random variable for which

$$\langle e^{i[\theta(\mathbf{x})-\theta(\mathbf{y})]} \rangle = \exp\left\{-\frac{1}{2}\langle[\theta(\mathbf{x})-\theta(\mathbf{y})]^2\rangle\right\} \equiv \exp\left[-\frac{1}{2}D(\mathbf{x}-\mathbf{y})\right], \quad (3.257)$$

where $D(\mathbf{x}-\mathbf{y})$ is called the structure function. If the turbulence is described by a Kolmogorov spectrum, to be studied in Chap. 8, Section 8.15, then $D(\mathbf{u}) \propto u^{5/3} = (u/u_0)^{5/3}$, where u_0 is the length over which the atmospheric fluctuations are strongly correlated (see Exercise 3.24). The integral in expression (3.252) cannot be evaluated in terms of elementary functions now, but the key result in relation (3.256) still holds, with $l \simeq u_0$.

The result that atmospheric fluctuations introduces an effective aperture of size u_0 can be interpreted in a different manner that provides some physical insight. The actual aperture D of a telescope can be thought of as divided into N smaller patches, each of radius u_0 with $N \approx (u_0/D)^2$. At any given instant of time, the wave will be coherent over each of the patches of size u_0 and will produce N individual images of a point source in the image plane. Each of these images will, of course, be spread to a size λ/D because of the telescopic diffraction. These N images (called speckles) will together be spread over a region of size λ/u_0 on the image plane. Atmospheric fluctuations will cause these speckles to fluctuate randomly, thereby producing an image smeared over the size λ/u_0 on the image plane, which is the result obtained above. It is, however, now clear that at any given instant in time the speckles contain information at a finer resolution than is available if the fluctuation of the speckles is allowed to smear the image. In fact, by taking a very short exposure image (that is, an image taken over a time scale shorter than the fluctuation scale of the speckles), we can retrieve this information at higher resolution. This idea forms the basis of a technique called speckle interferometry.

Exercise 3.24

Fluctuations of atmospheric refractive index: In the text it was assumed that the phase change occurs at a given plane with a fixed value of z. More generally, we can think of the phase change $\theta(\mathbf{x})$ as arising because of an integrated effect along the z axis. If the phase change is due to a random variation of the refractive index n in the atmosphere, then the integrated phase change will be

$$\theta(\mathbf{x}) = k \int_0^l n(\mathbf{x}, z)\, dz, \quad (3.258)$$

where l is the distance travelled along the z axis in the atmosphere and $n(\mathbf{x}, z)$ is the varying refractive index, with $\mathbf{x}$ denoting the two transverse coordinates. The structure function for the fluctuations in the refractive index can be modelled by the law $D_n(\mathbf{x}, z) = C_n^2(x^2 + z^2)^{1/3}$. Show that this corresponds to an intensity fluctuation pattern given by

$$\exp\left[-\frac{1}{2}D(\mathbf{x} - \mathbf{y})\right] = \exp\left[-\left(\frac{|\mathbf{x} - \mathbf{y}|}{u_0}\right)^{5/3}\right], \qquad (3.259)$$

where $u_0 \approx (1.45\, k^2 C_n^2 l)^{-3/5}$. (For atmosphere, $C_n^2 l \approx 10^{-13}$ cm$^{1/3}$ at $\lambda = 5000$ Å, giving $u_0 \approx 9.6$ cm. Note that u_0 is strongly chromatic and varies as $\lambda^{6/5}$.)

4

Basics of Electromagnetic Radiation

4.1 Introduction

This chapter and Chap. 6 describe several radiative processes that are of importance in astrophysics. This chapter covers the basics, and Chap. 6 discusses more detailed applications of the principles to bulk matter. This chapter will be needed extensively in the study of several astrophysical systems in Vols. II and III.

4.2 Radiation from an Accelerated Charge

We saw in Chap. 3 that the electric field of a stationary point charge, as well as that of a charge moving with uniform velocity, falls as r^{-2}. The situation, however, changes drastically when a charged particle moves with an acceleration. The field of an accelerated charge picks up a part which falls only as $1/r$, usually called the radiation field. A field with $E \propto r^{-1}$ has an energy flux $S \propto E^2 \propto r^{-2}$; because the surface area of a sphere increases as r^2, same amount of energy will flow through spheres of different radii. This fact allows the accelerating charge to transfer energy to large distances and thus provides the radiation field with an independent dynamical existence.

Because this phenomenon is vital in the study of radiative processes, two separate derivations of the field of an arbitrarily moving charged particle are provided. The first derivation is quite simple and illustrates the physical origin of the radiation field; the second one is a concise and mathematically elegant derivation that is fully four dimensional.

4.2.1 Why Does an Accelerated Charge Radiate?

The fact that the electric field of an accelerated charge should have an r^{-1} term can be seen along the following lines: From Eq. (3.131) it is clear that $\mathbf{A}$ scales as the velocity $\mathbf{v}$ of the charge; that is, if we multiply the velocity $\mathbf{v}$ by a factor μ, the field $\mathbf{A}$ will also change by the same factor μ. Because $\dot{\mathbf{A}}$ contributes to $\mathbf{E}$, there will be one term in **E** that is linear in the acceleration **a**. (This is, of course, in addition to the usual Coulomb term that is independent of **a** and falls as r^{-2}.) Because the electric field is linear in the charge q as well, this term should also be linear in q. Let us consider this electric field in the instantaneous rest frame of the charge. It has to be constructed from q, a, c, and r and hence *must* have the general form

$$E = C(\theta)\frac{qa}{c^n r^m} = C(\theta)\left(\frac{q}{r^2}\right)\left(\frac{a}{c^n r^{m-2}}\right), \tag{4.1}$$

where C is a dimensionless factor, depending on only the angle θ between **r** and **a**, and n and m need to be determined. (Because $\mathbf{v} = 0$ in the instantaneous rest frame, the field cannot depend on the velocity.) From dimensional analysis, it immediately follows that $n = 2$ and $m = 1$. Hence

$$E = C(\theta)\frac{qa}{c^2 r}. \tag{4.2}$$

Thus dimensional analysis and the fact that **E** must be linear in q and a imply the r^{-1} dependence for the radiation term.

We can determine the factor $C(\theta)$ by studying a special case: Consider a charge that was at rest, at the origin, from $t = -\infty$ to $t = 0$ and undergoes constant acceleration **a** along the x axis for a short time Δt. For $t > \Delta t$, it moves with constant velocity $v = a\Delta t$ along the x axis. Let us study the electric field produced by this charge at some time $t \gg \Delta t$. Because Δt is arbitrarily small, we have $a\Delta t \ll c$ and we can use the nonrelativistic approximation throughout.

The "news" that the charge was accelerated at $t = 0$ could have travelled only to a distance $r = ct$ in time t. Thus, at $r > ct$, the electric field should be that due to a charge located at the origin :

$$\mathbf{E} = \frac{q}{r^2}\hat{\mathbf{r}} \quad (\text{for } r > ct). \tag{4.3}$$

At $r \lesssim ct$, the field is that due to a charge moving with velocity v along the x axis. This will be a Coulomb field radially directed from the *instantaneous* position of the charge (see Chap. 3, Subsection 3.11.1):

$$\mathbf{E} = \frac{q}{r'^2}\hat{\mathbf{r}}' \quad (\text{for } r < ct). \tag{4.4}$$

Around $r = ct$ there exists a small shell of thickness $c\Delta t$ in which neither result holds good. From Fig. 4.1(a) it is clear that the electric field in the transition region should interpolate between the two Coulomb fields. Let $E_{\parallel}$ and $E_{\perp}$ be the magnitudes of the electric fields parallel and perpendicular to the direction $\hat{\mathbf{r}}$, respectively. From the geometry, we have

$$\frac{E_{\perp}}{E_{\parallel}} = \frac{v_{\perp}t}{c\Delta t}. \tag{4.5}$$

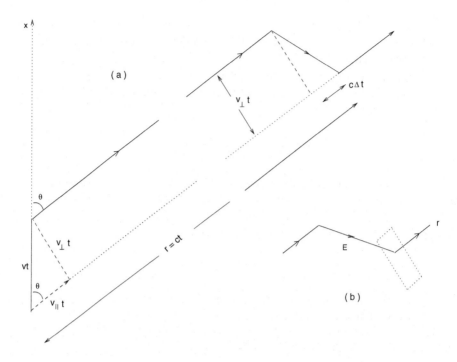

Fig. 4.1. (a) The electric field produced by a charged particle that was accelerated for a small time interval Δt. For $t > \Delta t$, the particle is moving with a uniform nonrelativistic velocity v along the x axis. At $r > ct$, te field is that of a charge at rest in the origin. At $r < c(t - \Delta t)$ the field is directed towards the instantaneous position of the particle. The radiation field connects these two Coulomb fields in a small region of thickness $c\Delta t$. (b) Pillbox construction to relate the normal component of the electric field around the radiation zone.

But $v_\perp = a_\perp \Delta t$ and $t = (r/c)$, giving

$$\frac{E_\perp}{E_\|} = \frac{(a_\perp \Delta t)(r/c)}{c\Delta t} = a_\perp \left(\frac{r}{c^2}\right). \tag{4.6}$$

We can determine the value of $E_\|$ by using Gauss theorem to a small pillbox, as shown in Fig. 4.1(b). This gives $E_\| = E_r = (q/r^2)$; thus we find that

$$E_\perp = a_\perp \left(\frac{r}{c^2}\right)\frac{q}{r^2} = \frac{q}{c^2}\left(\frac{a_\perp}{r}\right). \tag{4.7}$$

This is the radiation field located in a shell at $r = ct$, which is propagating with a velocity c. The above argument clearly shows that the origin of the r^{-1} dependence lies in the necessity of interpolating between two Coulomb fields. We can express this result in the vector notation as

$$\mathbf{E}_{\text{rad}}(t, \mathbf{r}) = \frac{q}{c^2}\left[\frac{1}{r}\hat{\mathbf{n}} \times (\hat{\mathbf{n}} \times \mathbf{a})\right]_{\text{ret}}, \tag{4.8}$$

where $\mathbf{n} = (\mathbf{r}/r)$ and the subscript ret imply that the expression in brackets should be evaluated at $t' = t - r/c$. Comparison with Eq. (4.2) shows that $C(\theta) = \sin\theta$.

The full electric field *in the frame in which the charge is instantaneously at rest* is $\mathbf{E} = \mathbf{E}_{\text{coul}} + \mathbf{E}_{\text{rad}}$. We can obtain the velocity dependence of this field by making a Lorentz transformation.

4.2.2 *Relativistically Invariant Derivation*

We now obtain the electric and the magnetic fields produced by a charged particle moving in an arbitrary trajectory by directly integrating Maxwell's equations in a four-dimensional notation.

Consider a charge q moving along a trajectory $z^a(\tau)$, where τ is the proper time. It contributes a current (see Problem 3.11)

$$J^a(x) = q \int_{-\infty}^{\infty} d\tau\, \delta_D[x - z(\tau)] u^a(\tau), \tag{4.9}$$

where $u^a(\tau)$ is the four velocity of the particle. (In this subsection we use units with $c = 1$ and denote four vectors x^a, z^a, etc., as x, z, etc., without the superscript when no confusion can arise.) To find the field that is due to this charged particle, we have to solve the equation $\Box A^a = 4\pi J^a$. To solve an equation of the type

$$\Box Q = P(x), \tag{4.10}$$

we can use the method of the Green's function. We define a retarded Green's function D_R to be the solution to the equation $\Box D_R = \delta(x)$. The subscript R implies that we choose the boundary conditions so as to ensure that $D_R(t, \mathbf{x}) = 0$ for $t < 0$. Given $D_R(x)$, we can relate Q to P by

$$Q(x) = \int d^4y\, D_R(x - y) P(y). \tag{4.11}$$

Thus we need to find only the retarded Green's function $D_R(x)$.

The conventional way of solving the equation $\Box D_R = \delta(x)$ is by Fourier transforming both sides. There is, however, a simpler way of getting the result. Assume for a moment we are working in four-dimensional *Euclidean* space (rather than *Minkowskian* space) so that the distance from the origin to x^i is $s^2 = (\tau^2 + |\mathbf{x}|^2)$ rather than $(-t^2 + \mathbf{x}^2)$. For the sake of convenience, we temporarily use the signature $(-, +, +, +)$ and set $\tau = it$. In such Euclidean space, it is trivial to verify that the spherically symmetric solution to $\Box D_R = 0$ is proportional to s^{-2} except at the origin. Consider now the four-volume integral of $\Box D_R$ over a region bounded by a sphere of radius R. We have

$$\int d^4x\, \Box D_R = \int d^3x\, \hat{\mathbf{n}} \cdot \nabla D_R = (2\pi^2 R^3)\left(-\frac{2}{R^3}\right) = -4\pi^2. \tag{4.12}$$

In arriving at the last result, we have used the fact that the surface area of a three sphere of radius R is $2\pi^2 R^3$ and $\nabla D_R = (-2/R^3)\hat{n}$. It follows that

$$\Box\left(\frac{-1}{4\pi^2 s^2}\right) = \delta(x), \tag{4.13}$$

giving $D_R = (-4\pi^2 s^2)^{-1}$. By using Eq. (4.11), we can write the solution to the equation of the form $\Box Q = P$ as

$$Q(x) = -\int \frac{d^4 y}{4\pi^2} \frac{P(y)}{(x-y)^2}, \tag{4.14}$$

where x stands for x^i, etc. This is in exact analogy with the solution to the Poisson equation in three dimensions and should be intuitively obvious. If we now continue analytically from the Euclidean to the Minkowski space by using $(d^4 y)_E = i(dt d^3 \mathbf{y})_M$ we get

$$Q(x) = -i\int \frac{d^4 y}{4\pi^2} \frac{P(y)}{(x-y)^2}. \tag{4.15}$$

In the case of Maxwell's equations this relation becomes

$$A^a(x) = -\frac{i}{\pi}\int d^4 y \frac{J^a(y)}{(x-y)^2} = -\frac{iq}{\pi}\int_{-\infty}^{\infty} d\tau \frac{u^a(\tau)}{s^2}, \tag{4.16}$$

where we have used Eq. (4.9), integrated over y, eliminating the delta function, and defined as

$$s^2 = [x - z(\tau)]^2 \equiv R^a R_a, \qquad R^a \equiv x^a - z^a(\tau). \tag{4.17}$$

The integrand in Eq. (4.16) has a pole along the real axis that corresponds to the points at which $s^2 = (x - z)^2 = 0$. Given any point x, this equation will determine two unique proper times τ_A and τ_R such that light signals connect $z(\tau_R)$ to x and x to $z(\tau_A)$. Our boundary condition requires that only the contribution at the retarded time τ_R is chosen, that is, the integral in Eq. (4.16) should be interpreted as providing the complex residue at the first-order pole, which satisfies the retarded condition $s^2 = 0$ and $z^0 < x^0$.

We now convert the integration over τ to integration over s^2 by using the Jacobian $(ds^2/d\tau) = 2R^a(-u_a) \equiv -2l$. This gives

$$A^a(x) = -\frac{iq}{\pi}\int ds^2 \frac{u^a(\tau)}{|ds^2/d\tau| s^2} = -\frac{iq}{2\pi}\int ds^2 \frac{(u^a/l)}{s^2}. \tag{4.18}$$

Taking the contribution from the first-order pole at $s^2 = 0$ as $2\pi i$ times the residue at the retarded time, we find that the answer is

$$A^a(x) = \left(\frac{qu^a}{l}\right)_{\text{ret}} = \left(\frac{qu^a}{R^b u_b}\right)_{\text{ret}}, \qquad s^2[x, z(\tau)] = 0, \qquad x^0 > z^0. \tag{4.19}$$

This four potential, of an arbitrarily moving charged particle, is called the Lienard–Wiechert potential. Its three-dimensional form is

$$\phi = \frac{q}{[R - (\mathbf{v} \cdot \mathbf{R}/c)]}, \quad \mathbf{A} = \frac{q\mathbf{v}}{c[R - (\mathbf{v} \cdot \mathbf{R}/c)]}, \tag{4.20}$$

where all the quantities on the right-hand side are evaluated at the retarded time.

To find the electromagnetic field, $F^{ba} = \partial^b A^a - \partial^a A^b$, we have to evaluate $\partial^b A^a$. From Eq. (4.16), this is given by

$$\partial^b A^a = \frac{iq}{\pi} \int_{-\infty}^{\infty} d\tau \frac{u^a(\tau)}{s^4} \frac{\partial s^2}{\partial x_b} = \frac{2iq}{\pi} \int_{-\infty}^{\infty} d\tau \frac{R^b u^a}{s^4}. \tag{4.21}$$

We now use the result

$$\frac{d}{d\tau}\left(\frac{1}{s^2}\right) = -\frac{1}{s^4}\frac{ds^2}{d\tau} = \frac{2l}{s^4} \tag{4.22}$$

to substitute for $1/s^4$ in the integrand of Eq. (4.21) and obtain

$$\partial^b A^a = \frac{2iq}{\pi} \int_{-\infty}^{\infty} d\tau \left(\frac{R^b u^a}{2l}\right) \frac{d}{d\tau}\left(\frac{1}{s^2}\right) = -\frac{iq}{\pi} \int_{-\infty}^{\infty} \frac{d\tau}{s^2} \frac{d}{d\tau}\left(\frac{R^b u^a}{l}\right). \tag{4.23}$$

The integral is exactly in the same form as that in Eq. (4.16), with u^a replaced with $d(R^b u^a/l)/d\tau$. Therefore, in analogy with the result obtained in Eqs. (4.19), we get

$$\partial^b A^a = \left[\frac{q}{l}\frac{d}{d\tau}\left(\frac{R^b u^a}{l}\right)\right]_{\text{ret}}. \tag{4.24}$$

Antisymmeterising the right-hand side, we find that the electromagnetic fields of an arbitrarily moving charged particle is given by the manifestly four-dimensional expression

$$F^{ba} = \left[\frac{q}{l}\frac{d}{d\tau}\left(\frac{R^b u^a - R^a u^b}{l}\right)\right]_{\text{ret}}. \tag{4.25}$$

This completely solves the problem. To obtain three-dimensional expressions, we have to carry out the differentiation in Eq. (4.25), which we can do by using the result

$$\frac{d}{d\tau}\left(\frac{R^{[k}u^{i]}}{l}\right) = \frac{R^{[k}a^{i]}}{l} - \frac{R^{[k}u^{i]}}{l^2}\frac{dl}{d\tau} = \frac{R^{[k}a^{i]}}{l} - \frac{R^{[k}u^{i]}}{l^2}(R_b a^b - 1), \tag{4.26}$$

where the brackets denotes antisymmeterisation. To convert from proper time to coordinate time, we use the result of differentiating the relation $x^0 = z^0 + |\mathbf{R}|$;

we have

$$l_{\text{ret}} = R^0 u^0 - \mathbf{R} \cdot \mathbf{u} = |\mathbf{R}| u^0 (1 - v_R), \quad dt = dx^0 = dz^0 (1 - v_R), \quad (4.27)$$

where $v_R = (\mathbf{v} \cdot \mathbf{R}/Rc)$. Combining these, we get

$$l_{\text{ret}} \, d\tau = dz^0 (1 - v_R) = R \, dt, \quad (4.28)$$

which is useful in carrying out the differentiation. Doing this and changing over to three-dimensional notation, we get the final result:

$$\mathbf{E} = \left[\frac{q\mathbf{n}}{R^2} + q \frac{R}{c} \frac{d}{dt} \left(\frac{\mathbf{n}}{R^2} \right) + \frac{q}{c^2} \frac{d^2\mathbf{n}}{dt^2} \right]_{\text{ret}}. \quad (4.29)$$

This formula for the electric field has a curious interpretation. The first term represents the Coulomb field of the particle evaluated at the retarded time. The second term can be thought of as a first-order correction to the retarded Coulomb field: We obtain this term by multiplying the time taken for the light signal to travel the distance R and the time rate of change of the Coulomb term $(\mathbf{n}/R^2)$. These two terms fall as R^{-2}. The third term depends on the acceleration of the charge and represents the radiation field of the charge.

The final result can also be written in the form

$$\mathbf{E} = \frac{q(1 - v^2/c^2)}{R^2(1 - v_R)^3} \left(\mathbf{n} - \frac{\mathbf{v}}{c} \right) + \frac{q}{R(1 - v_R)^3} \mathbf{n} \times \left[\left(\mathbf{n} - \frac{\mathbf{v}}{c} \right) \times \frac{d\mathbf{v}}{c^2 dt} \right], \quad \mathbf{B} = \mathbf{n} \times \mathbf{E},$$

$$(4.30)$$

where all quantities on the right-hand side refer to the retarded time. The first term in the electric field is independent of the acceleration and depends on only the velocity. This is the field of a charged particle moving with uniform velocity obtained in Eqs. (3.144) and falls as R^{-2} at large distances. The second term is linear in acceleration and falls as $1/R$ for large R. In the instantaneously retarded rest frame $\mathbf{v} = 0$ and the first term reduces to the Coulomb field; the second term reduces to the acceleration field obtained in Subsection (4.21) above by more physical considerations. (Note that the magnetic field of a *single* charged particle, in *any* state of motion, is always perpendicular to the electric field.) In the final formulas, we have reintroduced the c factor.

Exercise 4.1
Filling in the details: Prove Eqs. (4.20), (4.26), (4.29), and (4.30).

Exercise 4.2
Čerenkov radiation. In the presence of material medium, it is possible for radiation to be emitted even by a charged particle moving with uniform velocity. One such case occurs for a particle with charge q moving with a speed v in a medium with a refractive index n, provided that $v > (c/n)$. Derive this result along the following lines: (1) Take the current $\mathbf{J}(t, \mathbf{x})$ to be that of a particle with speed v moving along the x axis. Relate the Fourier

transform of **J** to the Fourier transform of the vector potential and hence to the electric field. (2) Show that the total energy radiated through a sphere at a large distance can be expressed in the form

$$U = \int_0^\infty \int_\Omega \frac{n \sin^2 \theta}{2\pi c^3} \left| \int [\dot{\mathbf{J}}(\omega)] \, d^3 r' \right|^2 d\Omega \, d\omega, \tag{4.31}$$

where $\dot{\mathbf{J}}(\omega)$ is the Fourier transform of $\dot{\mathbf{J}}(t) \equiv (d\mathbf{J}/dt)$. (3) Show that the integral over the current controlling the radiation can be expressed as

$$\left| \int [\dot{\mathbf{J}}(\omega)] \, d^3 r' \right|^2 = \left| \frac{\omega q}{(2\pi)^{1/2}} \int \exp \left[i \left(\mathbf{k} \cdot \mathbf{x} + \frac{\omega x}{v} \right) \right] dx \right|^2. \tag{4.32}$$

Hence show that the radiation exists only if $v > (c/n)$ and is beamed along a cone of angle $\theta = \cos^{-1}(c/nv)$. Perform the relevant integrals and show that the total energy emitted at frequency ω per unit length of the path of the charged particle is given by

$$\frac{dU(\omega)}{dx} = \frac{\omega q^2}{c^3} \left(1 - \frac{c^2}{n^2 v^2} \right). \tag{4.33}$$

4.3 General Properties of the Radiation Field

The expressions obtained in Section (4.2) for the radiation field take simple forms in several important special cases. We now discuss these special cases and obtain several formulas that will be needed in later chapters.

4.3.1 Radiation in the Nonrelativistic Case

Let a system of charges, confined to a finite region of size L, execute nonrelativistic motion with $v \ll c$. We consider the radiation field of this system at a distance R with $R \gg L$ and $R \gg \lambda$, where λ is the typical wavelength of radiation. The field at any event $(t, \mathbf{r})$ depends on the behaviour of the charge at a time t', where $ct' = ct - |\mathbf{r} - \mathbf{x}(t')|$ and $\mathbf{x}(t)$ is the trajectory of the charge. At large distances from the system of charges, R can be written as

$$R \equiv |\mathbf{r} - \mathbf{x}(t')| \cong r - \mathbf{x} \cdot \mathbf{n} \tag{4.34}$$

where $\mathbf{n}$ is the unit vector $(\mathbf{r}/r)$. Also, in calculating the field at large distances, we can replace the R^{-1} factor in Eqs. (4.20) with r^{-1} and ignore $(\mathbf{v} \cdot \mathbf{n}/c)$ in the denominator to obtain

$$\mathbf{A}(t, \mathbf{r}) = \frac{1}{cr} \sum_i q_i \mathbf{v}_i(t'), \quad t' = t - \frac{r}{c} + \frac{\mathbf{x} \cdot \mathbf{n}}{c}. \tag{4.35}$$

In the expansion for t', $\mathbf{x} \cdot \mathbf{n}$ can be ignored compared with r if the charge distribution does not change too fast. If T is the characteristic time in which the charge distribution changes, then the typical wavelength of radiation emitted will be $\lambda \simeq (cT)$. The quantity $(\mathbf{x} \cdot \mathbf{n}/c)$ is of the order of L/c, where L is the size of the system. This term can be ignored if $(L/c) \ll T$ or when $L \ll \lambda$. This condition is clearly satisfied if $(L/T) \ll c$, that is, if $v \ll c$, where v is the typical velocity of the charges. Hence this approximation can be used whenever the motion is nonrelativistic. In that case, we get the simple formula

$$\mathbf{A}(t, \mathbf{r}) = \frac{1}{cr} \sum_i q_i \mathbf{v}_i [t - (r/c)] = \frac{\dot{\mathbf{d}}[t - (r/c)]}{cr}, \qquad (4.36)$$

where $\mathbf{d} = \sum q_i \mathbf{x}_i$ is the dipole moment of the system and the sum is over all the charges in the system. (The overdot denotes the derivative with respect to t.) Such a radiation field is called the dipole radiation field.

The angular distribution of dipole radiation can be obtained as follows: At large distances from the system of charges, the electromagnetic wave may be treated as a plane wave. Then, noting that the vector potential depends on only $[t - (r/c)]$, we can write $\mathbf{B} = \nabla \times \mathbf{A} \cong (\dot{\mathbf{A}} \times \mathbf{n})/c$ and $\mathbf{E} = \mathbf{B} \times \mathbf{n}$. Therefore, for dipole radiation,

$$\mathbf{B} = \frac{1}{c^2 r}(\ddot{\mathbf{d}} \times \mathbf{n}), \quad \mathbf{E} = \frac{1}{c^2 r}(\ddot{\mathbf{d}} \times \mathbf{n}) \times \mathbf{n} \qquad (4.37)$$

and the energy flux is $\mathbf{S} = c(\mathbf{E} \times \mathbf{B})/4\pi = c(B^2/4\pi)\,\mathbf{n}$. The amount of energy propagating into a solid angle $d\Omega$ in unit time is $|\mathbf{S}|r^2 d\Omega$, giving

$$\frac{d\mathcal{E}}{dt\,d\Omega} = |\mathbf{S}|r^2 = \frac{cB^2 r^2}{4\pi} = \frac{1}{4\pi c^3}(\ddot{\mathbf{d}} \times \mathbf{n})^2 = \frac{|\ddot{\mathbf{d}}|^2}{4\pi c^3}\sin^2\theta. \qquad (4.38)$$

In equations (4.37) and (4.38), the right-hand side should be evaluated at the retarded time. The total energy radiated, found by integration over the solid angle, will be

$$\frac{d\mathcal{E}}{dt} = \frac{|\ddot{\mathbf{d}}|^2}{4\pi c^3}\int_0^\pi 2\pi(\sin^2\theta)(\sin\theta\,d\theta) = \frac{2}{3c^3}|\ddot{\mathbf{d}}|^2. \qquad (4.39)$$

For a single charge, $\mathbf{d} = q\mathbf{x}$ and

$$\frac{d\mathcal{E}}{dt} = \frac{2q^2}{3c^3}|\mathbf{a}|^2 \quad \text{(single charge)}, \qquad (4.40)$$

where $\mathbf{a}$ is the acceleration. This is called the Larmor formula.

The spectral composition of radiation, that is, the amount of energy that is radiated by the system between frequencies ω and $\omega + d\omega$, can be obtained as follows: The Fourier transform of $\mathbf{B}(t)$ in Eqs. (4.37) is $\mathbf{B}(\omega) = (c^2 r)^{-1}[\ddot{\mathbf{d}}(\omega) \times \mathbf{n}]$, where $\ddot{\mathbf{d}}(\omega) = -\omega^2 \mathbf{d}(\omega)$. Using the relation

$$\int_{-\infty}^{+\infty} B^2(t)\,dt = \int_{-\infty}^{+\infty} |\mathbf{B}(\omega)|^2 \frac{d\omega}{2\pi} = 2\int_0^\infty |\mathbf{B}(\omega)|^2 \frac{d\omega}{2\pi}, \qquad (4.41)$$

we can write Eq. (4.38)

$$\frac{d\mathcal{E}}{d\Omega} = \frac{cr^2}{4\pi} \int_{-\infty}^{+\infty} B^2 dt = \frac{cr^2}{2\pi} \int_0^\infty |\mathbf{B}(\omega)|^2 \frac{d\omega}{2\pi}, \tag{4.42}$$

which gives

$$\frac{d\mathcal{E}}{d\omega d\Omega} = \frac{cr^2 |\mathbf{B}(\omega)|^2}{4\pi^2} = \frac{\omega^4 |\mathbf{d}(\omega)|^2}{4\pi^2 c^3} \sin^2 \theta. \tag{4.43}$$

This expression gives the amount of energy radiated into a solid angle $d\Omega$ and frequency range $d\omega$. To find the spectral energy density, we integrate this expression over the solid angle $d\Omega$ and find [by using $\langle \sin^2 \theta \rangle = (2/3)$]

$$\frac{d\mathcal{E}}{d\omega} = \frac{\omega^4 |\mathbf{d}(\omega)|^2}{4\pi^2 c^3} \int d\Omega \sin^2 \theta = \frac{\omega^4 |\mathbf{d}(\omega)|^2}{4\pi^2 c^3} 4\pi \langle \sin^2 \theta \rangle = \frac{2}{3\pi} \frac{\omega^4}{c^3} |\mathbf{d}(\omega)|^2. \tag{4.44}$$

In some special cases, it is convenient to obtain the spectral distribution of energy directly from the vector potential rather than through the formulas obtained above. To do this, we start with Eq. (4.35) for a single charge, multiply both sides by $\exp(i\omega t)$, and integrate over t to get

$$\mathbf{A}(\omega) \equiv \int_{-\infty}^{\infty} dt\, \mathbf{A}(t, \mathbf{r}) e^{i\omega t} = \frac{q}{cr} \int_{-\infty}^{\infty} dt\, \mathbf{v}[t'] e^{i\omega t}$$

$$= \frac{q}{cr} \int_{-\infty}^{\infty} dt\, \mathbf{v}[t] \exp i\omega \left[t + \frac{r}{c} - \frac{\mathbf{n} \cdot \mathbf{x}(t)}{c} \right]$$

$$= \frac{q}{rc} e^{ikr} \int_{-\infty}^{\infty} dt\, \mathbf{v}(t) e^{i[\omega t - \mathbf{k} \cdot \mathbf{x}(t)]}, \tag{4.45}$$

and by writing $\mathbf{v} dt = d\mathbf{x}$, we can express this relation as a line integral over the trajectory of the charge,

$$\mathbf{A}(\omega) = q \frac{e^{ikr}}{cr} \int e^{i(\omega t - \mathbf{k} \cdot \mathbf{x})} \, d\mathbf{x}. \tag{4.46}$$

When the particle executes periodic motion, we should use a Fourier *series* rather than a Fourier *integral*. Then we will obtain nonzero values for $\mathbf{A}(\omega)$ only when $\omega = n\omega_0$, where ω_0 is the fundamental frequency of periodic motion. The formulas then become

$$\mathbf{A}_n = \frac{q e^{ikr}}{cr} \oint \exp[i(\omega_0 n t - \mathbf{k} \cdot \mathbf{x})] \, d\mathbf{x}, \tag{4.47}$$

$$\frac{dI_n}{d\Omega} = \frac{c}{2\pi} |\mathbf{B}_n|^2 r^2 = \frac{c}{2\pi} |\mathbf{k} \times \mathbf{A}_n|^2 r^2, \tag{4.48}$$

where $dI_n/d\Omega$ is the amount of energy radiated per unit time per unit solid angle at the frequency $n\omega_0$.

4.3.2 Radiation in the Relativistic Case

Let us next consider the case of a charged particle moving with a relativistic velocity. (In this case, it is convenient to work with units in which $c = 1$.) At sufficiently large distances from the charged particle, we have to retain only the term in the electromagnetic field that falls as $1/R$. This field is given by second term in Eqs. (4.30),

$$\mathbf{E} = \frac{q}{r}\mu^3 \mathbf{n} \times [(\mathbf{n} - \mathbf{v}) \times \mathbf{a}], \quad \mathbf{B} = \mathbf{n} \times \mathbf{E}, \quad \mu = (1 - \mathbf{v} \cdot \mathbf{n})^{-1}. \quad (4.49)$$

To find the power emitted into a solid angle $d\Omega$ we must first compute E^2, which requires the result

$$\mu^6\{\mathbf{n} \times [(\mathbf{n} - \mathbf{v}) \times \mathbf{a}]\}^2 = 2(\mathbf{n} \cdot \mathbf{a})(\mathbf{v} \cdot \mathbf{a})\mu^5 + a^2\mu^4 - (1 - v^2)(\mathbf{n} \cdot \mathbf{a})^2\mu^6, \quad (4.50)$$

where we have used the relations

$$\mathbf{n} \times (\mathbf{n} \times \mathbf{a}) = \mathbf{n}(\mathbf{a} \cdot \mathbf{n}) - \mathbf{a}, \quad \mathbf{n} \times (\mathbf{v} \times \mathbf{a}) = \mathbf{v}(\mathbf{a} \cdot \mathbf{n}) - \mathbf{a}(\mathbf{v} \cdot \mathbf{n}). \quad (4.51)$$

Then Eq. (4.38) gives

$$\frac{d\mathcal{E}}{dt d\Omega} = \frac{q^2}{4\pi}[2\mu^5(\mathbf{n} \cdot \mathbf{a})(\mathbf{v} \cdot \mathbf{a}) + \mu^4 a^2 - \mu^6\gamma^{-2}(\mathbf{n} \cdot \mathbf{a})^2]. \quad (4.52)$$

The radiation intensity is largest along the directions for which $\mu \gg 1$, that is, where $(1 - \mathbf{v} \cdot \mathbf{n}) \ll 1$. If θ is the angle between $\mathbf{v}$ and $\mathbf{n}$, then by using $\beta = (1 - \gamma^{-2})^{1/2} \simeq 1 - (1/2\gamma^2)$ we find that

$$(1 - \beta\cos\theta)^{-1} \simeq \left[1 - \left(1 - \frac{1}{2\gamma^2}\right)\left(1 - \frac{1}{2}\theta^2\right)\right]^{-1} = \frac{2\gamma^2}{1 + \gamma^2\theta^2} \quad (4.53)$$

for $\theta \ll 1$ and $\beta \simeq 1$. For $\gamma \gg 1$, this expression is sharply peaked around $\theta = 0$ and has a width $\Delta\theta \simeq \gamma^{-1} = (1 - v^2)^{1/2}$. We also note that the θ dependence is through *only* the combination $\gamma\theta$; we will need this result in Chap. 6.

The above expressions simplify further if $\mathbf{v}$ and $\mathbf{a}$ are parallel or perpendicular. When $\mathbf{v}$ is perpendicular to $\mathbf{a}$, Eq. (4.52) becomes

$$\frac{d\mathcal{E}}{dt d\Omega} = \frac{q^2 a^2}{4\pi}[\mu^4 - \mu^6\gamma^{-2}\sin^2\theta\cos^2\Phi], \quad (4.54)$$

where θ is the angle between $\mathbf{n}$ and $\mathbf{v}$ and Φ is the azimuthal angle of $\mathbf{n}$ relative to the plane containing **a** and **v**. Similarly, when **a** and **v** are parallel, we get

$$\frac{d\mathcal{E}}{dt d\Omega} = \frac{q^2 a^2 \mu^6}{4\pi} \left[\frac{2}{\mu} v \cos\theta + \frac{1}{\mu^2} - \gamma^{-2} \cos^2\theta \right]. \tag{4.55}$$

The expression in brackets is actually

$$2v \cos\theta (1 - v \cos\theta) + (1 - v \cos\theta)^2 - (1 - v^2) \cos^2\theta = \sin^2\theta, \tag{4.56}$$

leading to

$$\frac{d\mathcal{E}}{dt d\Omega} = \frac{q^2 a^2 \mu^6}{4\pi} \sin^2\theta. \tag{4.57}$$

The formulas obtained above deal with the case of a single charged particle moving with relativistic velocities. Very often in astrophysics, we have to deal with a system of charged particles (say a blob of plasma) that has a bulk velocity that is relativistic in the lab frame. In the rest frame of such a plasma, the charged particles will be emitting radiation with some angular distribution given by $(d\mathcal{E}'/dt' d\Omega') \equiv f(\theta' \phi')$. We are interested in the corresponding radiation pattern in the laboratory frame, in which the blob is moving with a velocity v, which may be found as follows.

Consider two Lorentz frames S and S', with S' moving along the z axis with velocity v. Any given direction is represented by angles (θ, ϕ) and (θ', ϕ') in the two frames, with θ being the direction with respect to the z axis. The relation between (θ, ϕ) and (θ', ϕ') was found in Chap. 3, Section 3.12:

$$\cos\theta' = \frac{\cos\theta - v}{1 - v \cos\theta}, \quad \phi' = \phi. \tag{4.58}$$

To relate the intensities $d\mathcal{E}/dt d\Omega$ in the two frames we use the transformations

$$d\Omega' = d(\cos\theta') d\phi' = \frac{1 - v^2}{(1 - v \cos\theta)^2} d(\cos\theta) d\phi = \frac{1}{\gamma^2} \frac{d\Omega}{(1 - v \cos\theta)^2},$$

$$d\mathcal{E}' = \gamma(d\mathcal{E} - \mathbf{v} \cdot d\mathbf{P}) = \gamma d\mathcal{E}(1 - v \cos\theta), \tag{4.59}$$

$$dt' = dt\sqrt{1 - v^2} = \gamma^{-1} dt$$

to obtain

$$\frac{d\mathcal{E}'}{dt' d\Omega'} = \frac{\gamma(1 - v \cos\theta) d\mathcal{E}}{(\gamma^{-1} dt) \gamma^{-2} (1 - v \cos\theta)^{-2} d\Omega} = \gamma^4 (1 - v \cos\theta)^3 \frac{d\mathcal{E}}{dt d\Omega} \tag{4.60}$$

or

$$\left(\frac{d\mathcal{E}}{dt d\Omega} \right)_{\text{lab}} = \frac{(1 - v^2)^2}{(1 - v \cos\theta)^3} \left(\frac{d\mathcal{E}'}{dt' d\Omega'} \right)_{\text{rest}}, \tag{4.61}$$

which is the required formula. As a check, consider the case in which $d\mathcal{E}'/dt' d\Omega'$

is independent of (θ', ϕ'), i.e., the emission is isotropic in the rest frame. Then

$$\left(\frac{d\mathcal{E}}{dt}\right)_{lab} = \int d\Omega \left(\frac{\mathcal{E}}{dt\,d\Omega}\right)_{lab} = \int d\Omega \frac{(1 - v^2)^2}{(1 - v\cos\theta)^3} \left(\frac{d\mathcal{E}'}{dt'\,d\Omega'}\right)_{rest}$$

$$= \left(\frac{d\mathcal{E}'}{dt'\,d\Omega'}\right)_{rest} \int_{-1}^{+1} 2\pi\,d\mu \frac{(1 - v^2)^2}{(1 - v\mu)^3}. \tag{4.62}$$

The integral is elementary and gives (4π). Hence

$$\left(\frac{d\mathcal{E}}{dt}\right)_{lab} = 4\pi \left(\frac{d\mathcal{E}'}{dt'\,d\Omega'}\right)_{rest} = \left(\frac{d\mathcal{E}'}{dt'}\right)_{rest}, \tag{4.63}$$

as expected.

4.3.3 Radiation during an Impulsive Motion

Consider a situation in which the velocity of a charged particle changes from $\mathbf{v}_1$ to $\mathbf{v}_2$ in a short time τ. In that case, very little radiation will be emitted at frequencies higher than $\omega_0 \approx (1/\tau)$ and some general results can be obtained for the low-frequency ($\omega \ll \omega_0$) regime.

The Fourier transform of the magnetic field can now be approximated as

$$\mathbf{B}(\omega) \equiv \int_{-\infty}^{+\infty} \mathbf{B}(t)e^{-i\omega t}\,dt \simeq \int_{-\infty}^{+\infty} \mathbf{B}(t)\,dt \tag{4.64}$$

for $\omega \ll \tau^{-1}$, where τ is the characteristic time scale of the variation of $\mathbf{B}(t)$. When $\mathbf{B} = (\dot{\mathbf{A}} \times \mathbf{n})/c$ is used, this becomes

$$c\mathbf{B}(\omega) = -\mathbf{n} \times \int_{-\infty}^{+\infty} \dot{\mathbf{A}}\,dt = -\mathbf{n} \times [\mathbf{A}_2 - \mathbf{A}_1], \tag{4.65}$$

where $\mathbf{A}_1$ and $\mathbf{A}_2$ are the initial and the final values, respectively, of the vector potential. The radiated energy, given by Eq.(4.43), is

$$\frac{d\mathcal{E}}{d\omega\,d\Omega} = \frac{cr^2|\mathbf{B}(\omega)|^2}{4\pi^2} = \frac{r^2}{4\pi^2 c}[(\mathbf{A}_2 - \mathbf{A}_1) \times \mathbf{n}]^2. \tag{4.66}$$

The vector potentials can be related to the velocities $\mathbf{v}_1$ and $\mathbf{v}_2$ of the charge by the Lienard–Wiechert formula (4.20):

$$\mathbf{A} = \frac{q\mathbf{v}}{cr[1 - (\mathbf{v} \cdot \mathbf{n})/c]} \simeq \frac{q\mathbf{v}}{cr}, \tag{4.67}$$

where the second relation is valid for nonrelativistic motion. Substituting for $\mathbf{A}$

from Eq. (4.67), we get

$$\frac{d\mathcal{E}}{d\omega d\Omega} = \frac{q^2}{4\pi^2 c^3} \left[\frac{\mathbf{v}_2 \times \mathbf{n}}{1 - \mathbf{n} \cdot \mathbf{v}_2/c} - \frac{\mathbf{v}_1 \times \mathbf{n}}{1 - \mathbf{n} \cdot \mathbf{v}_1/c} \right]^2 \quad \text{(exact)}$$

(4.68)

$$\simeq \frac{q^2}{4\pi^2 c^3} [(\mathbf{v}_2 - \mathbf{v}_1) \times \mathbf{n}]^2 = \frac{q^2}{4\pi^2 c^3} (\Delta v)^2 \sin^2 \theta \quad \text{(nonrelativisitic)}.$$

The total energy emitted in all directions can be found by integration over $d\Omega$, which is equivalent to replacing $\sin^2 \theta$ with $4\pi \langle \sin^2 \theta \rangle = (4\pi)(2/3) = (8\pi/3)$, so that

$$\frac{d\mathcal{E}}{d\omega} = \frac{2}{3\pi} \frac{q^2}{c^3} (\Delta v)^2.$$

(4.69)

The energy emitted per unit frequency interval is independent of ω for $\omega \ll \tau^{-1}$; for $\omega \gtrsim \tau^{-1}$, there is very little radiation.

4.3.4 Relativistic Formula for Radiated Four Momentum

Finally, we obtain the expression for the four momentum radiated by a charged particle, which is a generalisation of the result in Subsection 4.3.1 for the radiated energy. The Larmor formula [Eq. (4.40)] for radiation can be written in the form

$$d\mathcal{E} = \frac{2}{3} \frac{q^2}{c^3} a^2(t') \, dt,$$

(4.70)

where $t' = (t - r/c)$ is the retarded time. Because $d\mathcal{E}$ and dt are fourth components of four vectors, this form suggests a simple relativistic generalisation along the following lines.

Let us choose an instantaneous rest frame for the charge in which this non-relativistic formula is valid at $t = t'$. Because of symmetry, the net momentum radiated $d\mathbf{P}$ will vanish in this instantaneous rest frame. Clearly this result should be valid even for relativistic motion if it is expressed in a Lorentz invariant manner. If a^i is the four acceleration so that $-a^2/c^4 = a^i a_i$ in the instantaneous rest frame of the charge, then we can express Eq. (4.70), as well as the condition $d\mathbf{P} = 0$, by the relation

$$dP^k = -\frac{2}{3} \frac{q^2}{c} (a^i a_i) \, dx^k = -\frac{2}{3} \frac{q^2}{c} (a^i a_i) u^k \, ds,$$

(4.71)

where dP^k is the four momentum radiated by the particle during the proper time interval ds. Being relativistically invariant, this result is true for arbitrary velocities.

This formula does not specify the source of acceleration for a charged particle. In the case in which the source is an electromagnetic field, then we can obtain expressions for radiated energy and momentum in terms of the external field. Consider a frame S in which the particle has velocity $\mathbf{v}$ and acceleration $\mathbf{a}$.

We now make a Lorentz transformation to a frame S' in which the charge is instantaneously at rest. In this frame $\mathbf{E}'_\parallel = \mathbf{E}_\parallel$, $\mathbf{E}'_\perp = \gamma(\mathbf{E}_\perp + \mathbf{v} \times \mathbf{B})$ [see Chap. 3, Eqs. (3.74)], and the acceleration is $\mathbf{a}' = (q/m)\mathbf{E}'$. (We have temporarily set $c = 1$ to simplify the expressions.) Hence the instantaneous power radiated is

$$\frac{2}{3}q^2 a^2 = \frac{2}{3}\frac{q^4}{m^2}[\mathbf{E}_\parallel^2 + \gamma^2(\mathbf{E}_\perp + \mathbf{v} \times \mathbf{B})^2] = \frac{2}{3}\frac{q^4}{m^2}[\mathbf{E}_\parallel^2 + \gamma^2(\mathbf{E} + \mathbf{v} \times \mathbf{B} - \mathbf{E}_\parallel)^2]$$

$$= \frac{2}{3}\frac{q^4}{m^2}[\gamma^2(\mathbf{E} + \mathbf{v} \times \mathbf{B})^2 - \gamma^2 E_\parallel^2 v^2]. \tag{4.72}$$

In arriving at the last equation we have used the relations $\mathbf{E} \cdot \mathbf{E}_\parallel = E_\parallel^2$ and $\mathbf{E}_\parallel \cdot (\mathbf{v} \times \mathbf{B}) = 0$. Writing $E_\parallel^2 v^2 = (\mathbf{E} \cdot \mathbf{v})^2$, we get

$$\Delta\mathcal{E} = \frac{2}{3}\left(\frac{q^4}{m^2}\right)\gamma^2[(\mathbf{E} + \mathbf{v} \times \mathbf{B})^2 - (\mathbf{E} \cdot \mathbf{v})^2]\Delta t. \tag{4.73}$$

We can obtain the relativistic form of this result by using $a^i = (q/m)F^i{}_k u^k$ in Eq. (4.71), which leads to

$$\Delta P^k = -\frac{2}{3}\left(\frac{q^4}{m^2}\right)(F^{ia}F_{ib})u_a u^b u^k \Delta s. \tag{4.74}$$

4.4 Radiation Reaction

The radiation of electromagnetic waves transports energy from the charged particle to large distances in the course of time. This energy has to be eventually supplied by the agency that is accelerating the charged particle. Hence there has to be some extra drag force $\mathbf{f}$ acting on the charged particle because it is radiating energy. The rate of work done against this drag, $\mathbf{f} \cdot \mathbf{v}$, should account for the energy radiated.

The determination of the form of the drag force, called the radiation reaction force or radiation damping force, is riddled with several conceptual problems. Because we are interested in astrophysical applications, we ignore these deep (and interesting) conceptual issues and obtain the radiation reaction force by a fairly simple approach. In particular, the formulas derived below can be used only when the motion of the particle is bounded, that is, the particle should be confined to a finite region of space at all times. We first derive the radiation damping force for nonrelativistic motion and then generalise the result for the relativistic case.

If the damping force is $\mathbf{f}$, then the work done by the damping force is expected to be equal to the mean power radiated, i.e.,

$$\langle \mathbf{f} \cdot \mathbf{v} \rangle = \left\langle \left(\frac{\Delta\mathcal{E}}{\Delta t}\right) \right\rangle = -\left\langle \frac{2q^2 a^2}{3c^3} \right\rangle \tag{4.75}$$

when averaged over a period of time. Averaging a^2 over a time interval T, we get

$$\langle a^2 \rangle = \frac{1}{T} \int_0^T dt\, a^2 = \frac{1}{T} \int_0^T dt(\dot{\mathbf{v}} \cdot \dot{\mathbf{v}})$$

$$= \frac{1}{T} \int_0^T dt \left[\frac{d}{dt}(\mathbf{v} \cdot \dot{\mathbf{v}}) - \mathbf{v} \cdot \ddot{\mathbf{v}} \right] = \frac{1}{T}[\mathbf{v} \cdot \dot{\mathbf{v}}]_0^T - \langle \mathbf{v} \cdot \ddot{\mathbf{v}} \rangle. \qquad (4.76)$$

The first term vanishes as $T \to \infty$ for any bounded motion, giving $\langle a^2 \rangle = -\langle \mathbf{v} \cdot \ddot{\mathbf{v}} \rangle$. It follows that, if the damping force is

$$\mathbf{f}_{\text{damp}} = \frac{2}{3} \frac{q^2 \ddot{\mathbf{v}}}{c^3} = \frac{2}{3} \frac{q^2}{c^3} \dddot{\mathbf{x}}, \qquad (4.77)$$

then the average work done by the damping force exactly accounts for the mean energy radiated.

Note that this expression was obtained after the energy radiated over a period of time was averaged. In other words, this expression is capable of maintaining equality between the total energy radiated during a finite interval of time and the total amount of work done during the same period of time. Careless use of this formula can easily lead to wrong results. For example, consider a charged particle that is moving with uniform acceleration $\mathbf{a}$. Because the radiation reaction force depends on the second derivative of $\mathbf{v}$, it vanishes in the case of uniform acceleration. But the Larmor formula implies that a charge undergoing uniform acceleration will be radiating energy at a steady rate. This contradiction arises because a charge moving with uniform acceleration at all times will not be bounded and we cannot apply the results obtained above. If the uniform acceleration occurs for only a finite duration of time, then a more careful consideration of the conditions at the beginning and the ending of the acceleration will reveal that there is no inherent conceptual difficulty (see Exercise 4.4).

Exercise 4.3
Radiation damping in electromagnetic fields: Expression (4.77) for the damping force is valid for any source of acceleration. When the particle is accelerated by an external electromagnetic field, $\dddot{\mathbf{x}}$ can be expressed in terms of the time derivatives of the electric and the magnetic fields. Obtain an expression for the damping force to the lowest order in v/c in the instantaneous rest frame of the particle and show that

$$\mathbf{f}_{\text{damp}} = \frac{2}{3} \frac{q^3}{mc^3} \dot{\mathbf{E}} + \frac{2}{3} \frac{q^4}{m^2 c^4} (\mathbf{E} \times \mathbf{B}). \qquad (4.78)$$

The above result is valid provided the damping force is small compared with the original source of acceleration. Show that this requires (1) $(q^2 \omega / mc^3) \ll 1$ or, equivalently, $\lambda \gg (q^2 / mc^2)$, where λ is the wavelength of the external electric field and ω is the frequency of variation of the external electric field and (2) $B \ll (m^2 c^4 / q^3)$. The right-hand side represents a critical field strength expressed entirely in terms of fundamental constants. For fields higher than this, classical electrodynamics ceases to be valid. It must be noted that all these conditions are to be invoked in the Lorentz frame in which the charge is *instantaneously* at rest.

Let us next consider the case of radiation reaction for a particle moving with relativistic velocities. In this case, the radiation reaction force **f** can be expressed in terms of a four force g^i with components $(\gamma \mathbf{f} \cdot \mathbf{v}, \gamma \mathbf{f})$ where $\gamma = (1 - v^2)^{-1/2}$. (We have temporarily set $c = 1$.) Therefore we need to find a four vector g^i that reduces to $[0, (2/3)q^2 \dot{\mathbf{v}}]$ in the rest frame of the charge. This condition is satisfied by any vector of the form $g^i = (2q^2/3)[(d^2 u^i/ds^2) - Au^i]$, where A is to be determined. To find A, we use the condition that $g^i u_i = 0$, which is valid for any particle trajectory. (See Chap. 3, Section 3.4). This gives $A = u^k(d^2 u_k/ds^2)$, leading to

$$g^i = \left(\frac{2q^2}{3}\right) \left[\frac{d^2 u^i}{ds^2} - u^i u^k \frac{d^2 u_k}{ds^2} \right]. \tag{4.79}$$

We can rewrite the second term by using the result

$$u^k \frac{d^2 u_k}{ds^2} = u^k \frac{da_k}{ds} = \frac{d}{ds}(u^k a_k) - a^k a_k = -a^k a_k, \tag{4.80}$$

as $u^k a_k = 0$. This gives another equivalent form for g^i:

$$g^i = \frac{2}{3} q^2 \left[\frac{d^2 u^i}{ds^2} + u^i (a^k a_k) \right]. \tag{4.81}$$

If the acceleration is caused by an external electromagnetic field, described by a field tensor F^{ik} that varies slowly in space and time, then the radiation reaction force can be expressed in a more transparent form. In this case we have

$$a^i = \left(\frac{q}{m}\right) F^i{}_k u^k, \qquad \frac{da^i}{ds} \simeq \left(\frac{q}{m}\right)^2 F^i{}_k F^k{}_j u^j, \tag{4.82}$$

where the second equation treats F^i_k as approximately constant, the precise condition being $u^a \partial_a F^i_k \ll (q/m)(F^i_b F^b_k)$. Substituting these expressions into Eq. (4.79) and rearranging the terms gives

$$g^i = \frac{2}{3} \left(\frac{q^2}{m}\right)^2 [(F^{ka} F_{kj}) u_a u^j u^i - F^{ki} F_{kj} u^j]. \tag{4.83}$$

We next use the definition of electromagnetic stress tensor T^{ab} [see Eq. (3.140), Chap. 3] to write

$$F^{il} F_{kl} = F^{li} F_{lk} = -(4\pi) T^i_k + \frac{1}{4} \delta^i_k (F_{ab} F^{ab}), \tag{4.84}$$

which allows us to express g^i in terms of T^{ab} alone. In the combination (with

$F^2 \equiv F_{ab}F^{ab})$

$$(F^{ka}F_{kj})u_a u^j u^i - F^{ki}F_{kj}u^i = u_a u^j u^i \left[-4\pi T_j^a + \frac{1}{4}\delta_j^a F^2 \right]$$

$$- u^j \left[-4\pi T_j^i + \frac{1}{4}\delta_j^i F^2 \right] = -4\pi (T^{aj}u_a u_j)u^i + 4\pi T^{ij}u_j, \quad (4.85)$$

the term involving $F^2 = F_{ab}F^{ab}$ cancels out. Therefore

$$g^i = \frac{8\pi}{3}\left(\frac{q^2}{m}\right)^2 [T^{ij}u_j - (T^{ab}u_a u_b)u^i] = \left(\frac{\sigma_T}{c}\right)[T^{ij}u_j - (T^{ab}u_a u_b)u^i]$$

$$(4.86)$$

where $\sigma_T \equiv (8\pi/3)(q^2/mc^2)^2$ and we have reintroduced the c factor in the last line. This quantity, σ_T, called the Thomson scattering cross section, will figure prominently in our discussions in Chap. 6.

Exercise 4.4
Radiation reaction: Consider a charged particle that is at rest until $t = 0$, moves with uniform acceleration g during the time $0 \le t \le T$, and moves with uniform velocity for $t > T$. (1) Compute the total amount of energy radiated by the particle to infinity. (2) When does the particle experience radiation reaction force? (3) Compute the work done by the radiation reaction force on the particle. Is this equal to the energy radiated by the particle? (4) Sketch roughly the behaviour of electromagnetic fields at all distances from the particle at different stages of motion. How exactly is the energy transferred from finite distance to infinity?

4.5 Quantum Theory of Radiation

The quantum-mechanical description of a radiative process has three separate ingredients. First, we must provide a fully quantum-mechanical treatment of the electromagnetic field. We will see that the electromagnetic field can be thought of as a collection of harmonic oscillators, with each oscillator labelled by a wave vector **k** and polarisation state. Each of these harmonic oscillators can be quantised using the standard rules of quantum mechanics. The stationary states of any given oscillator will then be described by an integer. The collection of these integers, for all the oscillators put together, will then determine the stationary-state basis for any quantum state of the electromagnetic field. We shall see that the concept of photons that carry specific energy and momentum arises naturally in this approach. The integers that specify the quantum state can be interpreted as the number of photons with given energy, momentum, and polarisation.

Second, matter that is interacting with the electromagnetic field should also be described in quantum-mechanical terms. If matter is relativistic then this requires a description in terms of (what is known as) the fermionic field. The study of fermionic fields interacting with the electromagnetic field is called quantum electrodynamics. However, when matter is nonrelativistic – but quantum mechanical – we can use the simpler description in terms of the ordinary Schrödinger equation. In that case, a suitable basis for the quantum state of the matter will be described by the eigenvalues for the Hamiltonian and other operators. In what follows, we shall assume that matter can be described by nonrelativistic quantum mechanics.

Finally, the interaction between an electromagnetic field and matter will be described by a Hamiltonian that depends on both the variables pertaining to both. The full interacting system, which comprises the electromagnetic field and quantized matter, does not allow an exact tractable solution. To make progress, we have to treat the coupling between matter and the electromagnetic field as a small perturbation. In that case, the effect of the coupling is to induce transitions between the quantum states of matter and the quantum states of the electromagnetic field. We shall see that, in the lowest order of approximation, the transitions in the (basis) quantum states of the electromagnetic fields are between consecutive levels of harmonic oscillators, that is, transitions are between states in which some oscillator changes from $|n\rangle$ to $|n \pm 1\rangle$. This will be interpreted as either an emission or an absorption of a photon (with a specific energy, momentum, and polarisation) by the material system. In the process, the quantum state of the matter will also change; if a photon is emitted, for example, the matter state will change from one of higher energy to one of lower energy. Quantum-mechanical formalism allows us to compute the probability for such transitions to occur. Given the description of the material systems, these probabilities can be converted to describe more useful, macroscopic, quantities characterising the emission and the absorption of radiation.

We begin by developing the quantum theory of the electromagnetic field and then work out the probabilities for emission and absorption of photons. It is convenient to work with units $c = \hbar = 1$, which we shall adopt.

4.5.1 Quantisation of an Electromagnetic Field

Consider a free electromagnetic field represented by a vector potential $A^i = (0, \mathbf{A})$, with $\nabla \cdot \mathbf{A} = 0$. (This choice is called the transverse gauge). It is convenient to assume that the field is confined to a cubic box of volume L^3 and vanishes on the boundaries of the box. The size of this box L will be much larger than all other length scales of the problem and none of the final answers will, of course, depend on L. In that case, $\mathbf{A}(t, \mathbf{x})$ can be expanded as

$$\mathbf{A}(t, \mathbf{x}) = \sum_{\mathbf{k}} \mathbf{q_k}(t) \exp i\mathbf{k} \cdot \mathbf{x} = \sum_{\mathbf{k}} [\mathbf{a_k}(t)e^{i\mathbf{k} \cdot \mathbf{x}} + \mathbf{a_k^*}(t)e^{-i\mathbf{k} \cdot \mathbf{x}}], \quad (4.87)$$

with $\mathbf{k} = (2\pi/L)\mathbf{n}$, where $\mathbf{n}$ is a vector with integer components. The first equality is merely the Fourier series expansion of the vector potential; the second relation follows from the fact that the vector potential is real. In terms of $\mathbf{q_k}$, the reality condition is $\mathbf{q}_{-\mathbf{k}} = \mathbf{q}_{\mathbf{k}}^*$. By using this condition, we can easily relate the two variables:

$$\mathbf{q}_{\mathbf{k}} = \mathbf{a}_{\mathbf{k}} + \mathbf{a}_{-\mathbf{k}}^*. \tag{4.88}$$

The condition $\nabla \cdot \mathbf{A} = 0$ translates into

$$\mathbf{k} \cdot \mathbf{q}_{\mathbf{k}} = \mathbf{k} \cdot \mathbf{a}_{\mathbf{k}} = \mathbf{k} \cdot \mathbf{a}_{\mathbf{k}}^* = 0, \tag{4.89}$$

i.e., for every value of $\mathbf{k}$, the vector $\mathbf{a}_{\mathbf{k}}$ is perpendicular to $\mathbf{k}$. This allows the vector $\mathbf{a}_{\mathbf{k}}$ to have two components $a_{\mathbf{k}d}(d = 1, 2)$ in the plane perpendicular to $\mathbf{k}$ with $\{\mathbf{a}_{\mathbf{k}1}, \mathbf{a}_{\mathbf{k}2}, \mathbf{k}\}$ forming an orthogonal system. The electric and the magnetic fields corresponding to this vector potential $A^i = (0, \mathbf{A})$ are

$$\mathbf{E} = -\dot{\mathbf{A}} = -\sum_{\mathbf{k}} \dot{\mathbf{q}}_{\mathbf{k}} e^{i\mathbf{k} \cdot \mathbf{x}}, \quad \mathbf{B} = \nabla \times \mathbf{A} = i \sum_{\mathbf{k}} (\mathbf{k} \times \mathbf{q}_{\mathbf{k}}) e^{i\mathbf{k} \cdot \mathbf{x}}. \tag{4.90}$$

To calculate the Hamiltonian for the system, we have to integrate $(E^2 + B^2)/8\pi$ over the volume of the box. These integrals will reduce to expressions of the form

$$I = \int_0^L dx \, \exp[i(k_x - k_x')x] = \int_0^L dx \, \exp\left[\frac{2\pi i}{L}(n_x - n_x')x\right], \tag{4.91}$$

which vanishes when $n_x \neq n_x'$ and gives $I = L$ for $n_x = n_x'$. By using this result and the identity

$$(\mathbf{k} \times \mathbf{q}_{\mathbf{k}}) \cdot (\mathbf{k} \times \mathbf{q}_{\mathbf{k}}^*) = \mathbf{q}_{\mathbf{k}}^* \cdot [(\mathbf{k} \times \mathbf{q}_{\mathbf{k}}) \times \mathbf{k}] = (\mathbf{q}_{\mathbf{k}}^* \cdot \mathbf{q}_{\mathbf{k}})k^2, \tag{4.92}$$

we find that

$$H = \frac{1}{8\pi} \int_V d^3x (E^2 + B^2) = \frac{V}{8\pi} \sum_{\mathbf{k}} (|\dot{\mathbf{q}}_{\mathbf{k}}|^2 + k^2 |\mathbf{q}_{\mathbf{k}}|^2). \tag{4.93}$$

It is obvious from Eq. (4.93) that the Hamiltonian for the electromagnetic field can be expressed as a sum over the Hamiltonians for harmonic oscillators, with each oscillator labelled by the wave vector $\mathbf{k}$ and a polarisation that will depend on the direction of the amplitude. In fact, Eq. (4.93) provides this description except for the fact that $\mathbf{q}_{\mathbf{k}}$'s are complex. The dynamical variables in a Hamiltonian need to be real, and hence we have to choose a different linear combination of $\mathbf{q}_{\mathbf{k}}$'s that will ensure this. This is best done in two steps. First, we express the Hamiltonian in terms of $\mathbf{a}_{\mathbf{k}}$ and $\mathbf{a}_{\mathbf{k}}^*$ and then introduce two new real variables in terms of these.

The wave equation satisfied by $\mathbf{A}(t, \mathbf{x})$ implies that $\mathbf{q}_{\mathbf{k}}(t)$ satisfies the harmonic oscillator equation

$$\ddot{\mathbf{q}}_{\mathbf{k}} + k^2 \mathbf{q}_{\mathbf{k}} = 0, \quad \mathbf{q}_{\mathbf{k}} \propto e^{-ikt}. \tag{4.94}$$

Therefore, $\mathbf{a}_{\mathbf{k}}$ also satisfies the harmonic oscillator equation. It is conventional to take $\mathbf{a}_{\mathbf{k}} \propto \exp(-ikt)$ [so that $\mathbf{a}_{\mathbf{k}}^* \propto \exp(ikt)$], giving

$$\dot{\mathbf{q}}_{\mathbf{k}} = -ik(\mathbf{a}_{\mathbf{k}} - \mathbf{a}_{-\mathbf{k}}^*). \tag{4.95}$$

When Eqs. (4.88) and (4.95) are substituted into Eq. (4.93), terms like $\mathbf{a_k} \cdot \mathbf{a_{-k}}$ cancel out whereas terms like $\mathbf{a_k} \cdot \mathbf{a_k^*}$ and $\mathbf{a_{-k}} \cdot \mathbf{a_{-k}^*}$ differ only in relabelling. So the Hamiltonian becomes

$$H = \sum_{\mathbf{k}} \frac{k^2 V}{2\pi} \mathbf{a_k} \cdot \mathbf{a_k^*}. \tag{4.96}$$

Let us now define two new *real* variables $\mathbf{Q_k}$ and $\mathbf{P_k}$ by the relations

$$\mathbf{Q_k} = \left(\frac{V}{4\pi}\right)^{1/2} (\mathbf{a_k} + \mathbf{a_k^*}), \quad \mathbf{P_k} = -ik\left(\frac{V}{4\pi}\right)^{1/2} (\mathbf{a_k} - \mathbf{a_k^*}) = \dot{\mathbf{Q}}_\mathbf{k}. \tag{4.97}$$

Then H becomes

$$H = \sum_{\mathbf{k}} \frac{1}{2} \left(\mathbf{P_k^2} + k^2 \mathbf{Q_k^2}\right). \tag{4.98}$$

Because $\mathbf{P_k}$ and $\mathbf{Q_k}$ are also orthogonal to $\mathbf{k}$ they also have only two independent components in the plane perpendicular to $\mathbf{k}$. Denoting these components by $\{Q_{\mathbf{k}\alpha}, P_{\mathbf{k}\alpha}\}$, with $\alpha = 1, 2$, we get

$$H = \sum_{\alpha=1}^{2} \sum_{\mathbf{k}} \frac{1}{2} \left(P_{\mathbf{k}\alpha}^2 + k^2 Q_{\mathbf{k}\alpha}^2\right) = \sum_{\alpha} \sum_{\mathbf{k}} H_{\mathbf{k}\alpha}, \tag{4.99}$$

which is the final result we were seeking. This equation shows that the Hamiltonian governing the free electromagnetic field can be expressed as a sum of Hamiltonians for harmonic oscillators, with each oscillator labelled by a wave vector $\mathbf{k}$ and polarisation index α. The frequency of the oscillator labelled by wave vector $\mathbf{k}$ is $\omega_\mathbf{k} = k = kc$ (in conventional units). It must be noted that we need an infinite number of harmonic oscillators as the wave vector $\mathbf{k} = (2\pi/L)\mathbf{n}$ takes infinite number values.

So far, the entire discussion has been classical. To provide the quantum version of this theory, we have to treat the dynamical variables of each of the oscillators as operators and impose commutation relations between the coordinates and momenta. Because the total Hamiltonian is a sum of independent oscillators, the operators corresponding to any one oscillator commute with the operators of any other oscillator. Therefore we can treat $P_{\mathbf{k}\alpha}$ and $Q_{\mathbf{k}\alpha}$ as quantum-mechanical operators that obey the commutation rules

$$[Q_{\mathbf{j}\beta}, P_{\mathbf{k}\alpha}] = i\delta_{\mathbf{k}\mathbf{j}}\delta_{\alpha\beta}. \tag{4.100}$$

By using the relation between $\mathbf{a}_{\mathbf{k}\alpha}$ and $Q_{\mathbf{k}\alpha}$, we can find the corresponding commutation rules for $\mathbf{a}_{\mathbf{k}\alpha}$ as

$$[\mathbf{a}_{\mathbf{k}\alpha}, \mathbf{a}_{\mathbf{p}\beta}^\dagger] = \left(\frac{2\pi}{V\omega_\mathbf{k}}\right)\delta_{\mathbf{k}\mathbf{p}}\delta_{\alpha\beta}. \tag{4.101}$$

with $\omega_\mathbf{k} = |\mathbf{k}|$. These are identical to the commutation relation between the creation and annihilation operators of a harmonic oscillator except for a constant factor on the right-hand side. We can remove this constant by rescaling the a's and defining a set of quantities $c_{\mathbf{k}\alpha}$. Separating out the time dependence of $\mathbf{a}_\mathbf{k}(t)$ as well and writing

$$\mathbf{a}_{\mathbf{k}\alpha} = c_{\mathbf{k}\alpha}\hat{\mathbf{e}}_\alpha \left(\frac{2\pi}{V\omega_\mathbf{k}}\right)^{1/2} \exp(-i\omega_\mathbf{k}t), \tag{4.102}$$

where $\hat{\mathbf{e}}_1$ and $\hat{\mathbf{e}}_2$ are two unit vectors chosen such that $[\hat{\mathbf{e}}_1, \hat{\mathbf{e}}_2, \mathbf{k}/k]$ form an orthogonal system and $c_{\mathbf{k}\alpha} = (v\omega_\mathbf{k}/2\pi)^{1/2}\mathbf{a}_{\mathbf{k}\alpha}$, we find it follows that

$$[c_{\mathbf{k}\alpha}, c_{\mathbf{p}\beta}^\dagger] = \delta_{\mathbf{k}\mathbf{p}}\delta_{\alpha\beta}. \tag{4.103}$$

This is the standard commutation rule for the creation and the annihilation operators of a harmonic oscillator. The $c_{\mathbf{k}\alpha}$ and $c_{\mathbf{k}\alpha}^\dagger$ can be treated as annihilation and creation operators, respectively, for photons with wave vectors $\mathbf{k}$ and polarization state α. The expansion of the vector potential in Eq. (4.87) now becomes, on use of Eq. (4.102),

$$\mathbf{A}(t, \mathbf{x}) = \sum_{\mathbf{k}\alpha}(c_{\mathbf{k}\alpha}\mathbf{A}_{\mathbf{k}\alpha} + c_{\mathbf{k}\alpha}^\dagger\mathbf{A}_{\mathbf{k}\alpha}^*), \quad \mathbf{A}_{\mathbf{k}\alpha} = \left(\frac{2\pi}{V\omega_\mathbf{k}}\right)^{1/2}\hat{\mathbf{e}}_\alpha e^{i(\omega t - \mathbf{k}\cdot\mathbf{x})}. \tag{4.104}$$

From nonrelativistic quantum mechanics, it is known that the quantum states of a harmonic oscillator can be labelled by an integer n, where $n = 0$ corresponds to the ground state and the energy of the nth state is $\hbar\omega[n + (1/2)]$. For a bunch of oscillators, we can specify the quantum state by giving a set of integers $(n_1, n_2, \ldots,)$, where n_j denotes the eigenstate of the jth oscillator. In our case, a generic oscillator is labelled by a wave vector $\mathbf{k}$ and polarisation state α. Let $n_{\mathbf{k}\alpha}$ be the integer denoting the quantum state of this oscillator. To specify the full quantum state of the electromagnetic field, we have to provide one integer for each value of $\mathbf{k}\alpha$. We write such a state symbolically as $|\{n_{\mathbf{k}\alpha}\}\rangle$. Because the energy of the oscillator labelled by $\mathbf{k}\alpha$ is given by

$$E_{\mathbf{k}\alpha} = \hbar\omega_\mathbf{k}\left(n_{\mathbf{k}\alpha} + \frac{1}{2}\right), \tag{4.105}$$

the total energy of the system is

$$E = \sum_{\mathbf{k}\alpha}E_{\mathbf{k}\alpha} = \sum_{\mathbf{k}\alpha}\hbar\omega_\mathbf{k}\left(n_{\mathbf{k}\alpha} + \frac{1}{2}\right) = \sum_{\mathbf{k}\alpha}n_{\mathbf{k}\alpha}\hbar\omega_\mathbf{k} + E_0, \tag{4.106}$$

where E_0 is the energy of the system when all the oscillators are in the ground state. Because there are infinite number of oscillators, the ground-state energy E_0 is divergent. This is the first sign of difficulty arising out of an attempt to quantise a field that has infinite number of degrees of freedom. No *conceptually* satisfactory solution to this problem exists at present. A satisfactory *operational* solution, however, is available, which involves dropping the ground-state energy E_0. The results obtained from such a formalism agree remarkably well with observations, and hence we shall follow that procedure here. Henceforth we drop E_0 and take the energy of the electromagnetic field in a quantum state $|\{n_{\mathbf{k}\alpha}\}\rangle$ to be

$$E_{\text{ren}} = \sum_{\mathbf{k}\alpha} n_{\mathbf{k}\alpha}\, \hbar\omega_{\mathbf{k}}, \tag{4.107}$$

where the subscript ren stands for renormalised.

This expression has a remarkable interpretation. Suppose the electromagnetic field is made of a bunch of quanta (called photons); we associate each photon with a wave vector $\mathbf{k}$, polarisation α, and energy $\hbar\omega_{\mathbf{k}}$. If there are $n_{\mathbf{k}\alpha}$ photons with wave vector $\mathbf{k}$ and polarisation α, each contributing an energy $\hbar\omega_{\mathbf{k}}$, then the total energy is given precisely by E_{ren}. We can therefore think of the integers $n_{\mathbf{k}\alpha}$ – which were introduced as integers characterising the energy state of the oscillator labelled by $\mathbf{k}\alpha$ – as giving the number of photons with a particular energy and polarisation. By calculating the momentum of the electromagnetic field, we can also show that each quantum carries the momentum $\hbar\mathbf{k}$. Because $\mathbf{k}^2 c^2 = \omega_{\mathbf{k}}^2$, it follows that the energy ϵ and the momentum $\mathbf{p}$ of the individual photons obey the relationship $\epsilon = pc$; in other words, photons are particles with zero rest mass travelling at the speed of light.

Exercise 4.5

Compton scattering: Consider the scattering between a photon of frequency ω and a relativistic electron with velocity $\mathbf{v}$ leading to a photon of frequency ω' and electron with velocity $\mathbf{v}'$. Let α be the angle between the incident and the scattered photon and θ and θ' be the angles between the direction of propagation of photon and the velocity vector of the electron before and after the collision. Using the conservation of energy and momentum in the scattering, show that

$$\frac{\omega'}{\omega} = \frac{1 - (v/c)\cos\theta}{[1 - (v/c)\cos\theta' + (\hbar\omega/\gamma m_e c^2)(1 - \cos\alpha)]}. \tag{4.108}$$

When $\hbar\omega \ll \gamma m_e c^2$, show that the frequency shift of the photon can be written as

$$\frac{\omega' - \omega}{\omega} = \frac{\Delta\omega}{\omega} = \frac{v}{c}\frac{(\cos\theta - \cos\theta')}{[1 - (v/c)\cos\theta']}. \tag{4.109}$$

This expression shows that, if θ and θ' are randomly distributed, then there is no net increase in photon energy to first order in v/c. Use the exact expression and compute the net energy change of the photon to the order of v^2/c^2 when θ and θ' are randomly distributed.

Exercise 4.6

Photon–photon collision: A purely relativistic process corresponds to the production of electron–positron pairs in a collision of two high-energy gamma-ray photons. If the energies of the photons are ϵ_1 and ϵ_2 and the relative angle between them is θ, then show (by using the conservation of momentum and energy) that the process can occur only if

$$\epsilon_1 \epsilon_2 > \frac{2m_e c^4}{1 - \cos\theta}. \tag{4.110}$$

Exercise 4.7

Poynting–Robertson effect: A dust particle is orbiting a star in a circular path with velocity v. It absorbs photons from the star, heats up, and emits the excess energy isotropically in its rest frame. Show that the angular momenta of the dust particle before and after the emission of a single photon (denoted by l_0 and l, respectively) are related by

$$l = l_0 \left(1 + \frac{2h\gamma v}{mc^2} \right)^{-1/2}. \tag{4.111}$$

This effect is purely relativistic even to the lowest order when $v \ll c$. Provide a physical explanation for this result.

Exercise 4.8

Filling in the details: Obtain Eqs. (4.93), (4.96), (4.98) and (4,101).

4.5.2 *Interaction of Matter and Radiation*

Having quantised the free electromagnetic field and developed the concept of the photon, we next turn our attention to the coupling between the electromagnetic field and charged particles. A charged particle, in quantum mechanics, will have its own dynamical variable $\mathbf{x}$ and momentum $\mathbf{p}$ that obey standard commutation rules. Depending on the nature of the system, such a charged particle can exist in different (basis) quantum states, each of which will be labelled by the eigenvalues of a complete set of commuting variables. (For example, the quantum state $|nlm\rangle$ of an electron in a hydrogen atom is usually labelled by three quantum numbers n, l, and m.) For the sake of simplicity, let us assume that the quantum states of the charged particle are labelled by the energy eigenvalues $|E\rangle$; the formalism can be easily generalised when more labels are needed to specify the quantum state. The coupling between the electromagnetic field and the atomic system is described by the Hamiltonian (with $c = 1$)

$$H_{\text{int}} = -\int d^3x \mathbf{J} \cdot \mathbf{A}, \tag{4.112}$$

where, in the lowest order, in the nonrelativistic limit, $\mathbf{J} = q\mathbf{v} \cong (q/m)\mathbf{p} = (q/m)(-i\nabla)$. [The canonical momentum in the presence of $\mathbf{A}$ will have an extra piece

$(q\mathbf{A}/c)$ that will lead to a term proportional to $q^2\mathbf{A}^2$ in $\mathbf{J}\cdot\mathbf{A}$. This quadratic term in $q\mathbf{A}$ is ignored in the lowest order.] We are interested in the transitions caused in the quantum states of the electromagnetic field and matter that is due to this interaction.

Let the initial state of the system be $|E_i, \{n_{\mathbf{k}\alpha}\}\rangle$, where E_i represents the initial energy of the matter state and the set of integers $\{n_{\mathbf{k}\alpha}\}$ denote the quantum state of the electromagnetic field. We now turn on the interaction Hamiltonian H_{int}. Because of the coupling, the system can make a transition to a final state $|E_f, \{n'_{\mathbf{k}\alpha}\}\rangle$, where E_f represents the final energy of the matter state and a new set of integers $\{n'_{\mathbf{k}\alpha}\}$ denotes the final quantum state of the electromagnetic field. To the lowest order in the perturbation theory, the probability amplitude for this process is governed by the matrix element

$$\langle E_f, \{n'_{\mathbf{k}\alpha}\}|H_{\text{int}}|E_i, \{n_{\mathbf{k}\alpha}\}\rangle. \tag{4.113}$$

To see the nature of this matrix element, let us substitute the expansion of the vector potential of Eqs. (4.104) into the interaction Hamiltonian; then H_{int} becomes the sum of two terms:

$$H_{\text{int}} = -\int d^3\mathbf{x}\,\mathbf{J}\cdot\mathbf{A} = -\int d^3\mathbf{x}\,\mathbf{J}\cdot\sum_{\mathbf{k}\alpha}(c_{\mathbf{k}\alpha}\mathbf{A}_{\mathbf{k}\alpha} + c^\dagger_{\mathbf{k}\alpha}\mathbf{A}^*_{\mathbf{k}\alpha}) \equiv H_{ab} + H_{em}.$$
$$\tag{4.114}$$

Because the creation and annihilation operators can change the energy eigenstate of the oscillator by only one step, it is clear that the probability amplitude in matrix element (4.113) will be nonzero only if the set of integers characterising the initial and the final states differ by unity for some oscillator labelled $\mathbf{k}\alpha$. In other words, the lowest-order transition amplitudes describe either the emission or the absorption of a single photon with a definite momentum and polarisation. Because the creation operator $c^\dagger_{\mathbf{k}\alpha}$ changes the integer $n_{\mathbf{k}\alpha}$ to $(n_{\mathbf{k}\alpha}+1)$, the term proportional to $c^\dagger_{\mathbf{k}\alpha}$ governs the emission; similarly, the term proportional to $c_{\mathbf{k}\alpha}$ governs the absorption of the photon. Let us work out the amplitude for emission in detail.

The emission process, in which the quantum system makes the transition from $|E_i\rangle$ to $|E_f\rangle$ and the electromagnetic field goes from a state $|n_{\mathbf{k}\alpha}\rangle$ to $|n_{\mathbf{k}\alpha}+1\rangle$, during the time interval $(0, T)$ is governed by the amplitude

$$\mathcal{A} = \int_0^T dt\langle E_f|\langle n_{\mathbf{k}\alpha}+1|H_{em}|n_{\mathbf{k}\alpha}\rangle|E_i\rangle$$

$$= -\int_0^T dt\int d^3\mathbf{x}\langle E_f|\mathbf{J}\cdot\mathbf{A}^*_{\mathbf{k}\alpha}|E_i\rangle(n_{\mathbf{k}\alpha}+1)^{1/2}, \tag{4.115}$$

where we have used the fact that $\langle n_{\mathbf{k}\alpha}+1|A|n_{\mathbf{k}\alpha}\rangle = A^*_{\mathbf{k}\alpha}(n_{\mathbf{k}\alpha}+1)^{1/2}$. Using the expansion for $\mathbf{A}_{\mathbf{k}\alpha}$ in Eqs. (4.104) and the fact that the energy eigenstates have the time dependence $\exp(-iEt)$, we can write the amplitude $\mathcal{A}$ as

$$\mathcal{A} = -\left(\frac{2\pi}{V\omega_k}\right)^{1/2} (n_{\mathbf{k}\alpha} + 1)^{1/2} \int_0^T dt$$

$$\times \int d^3\mathbf{x}\, \phi_f^*(\mathbf{x})\big(\mathbf{J}\cdot\hat{\mathbf{e}}_\alpha e^{-i\mathbf{k}\cdot\mathbf{x}}\big)\phi_i(\mathbf{x})e^{-i(E_i - E_f - \omega)t}, \qquad (4.116)$$

where $\phi_i(\mathbf{x})$ and $\phi_f(\mathbf{x})$ denote the wave functions of the two states. With the matrix element

$$\int d^3\mathbf{x}\, \phi_f^*(\mathbf{x})(\mathbf{J}e^{-i\mathbf{k}\cdot\mathbf{x}})\phi_i(\mathbf{x}) = \frac{q}{m}\int d^3\mathbf{x}\, \phi_f^*(\mathbf{x})\,\mathbf{p}\,e^{-i\mathbf{k}\cdot\mathbf{x}}\phi_i(\mathbf{x}) \qquad (4.117)$$

denoted by the symbol $\mathbf{M}_{fi}$, the probability of transition $|\mathcal{A}|^2$ becomes

$$|\mathcal{A}|^2 = P(T) = \left(\frac{2\pi}{V\omega_k}\right)|\mathbf{e}_{\mathbf{k}\alpha}\cdot\mathbf{M}_{fi}|^2(n_{\mathbf{k}\alpha} + 1)|F(T)|^2, \qquad (4.118)$$

where

$$|F(T)|^2 = \left|\int_0^T dt\, e^{-i(E_i - E_f - \omega)t}\right|^2 = \left[\frac{\sin(QT/2)}{Q/2}\right]^2, \qquad (4.119)$$

with $Q \equiv (E_i - E_f - \omega)$. In the limit of $T \to \infty$, for any smooth function $S(\omega)$, we have the result

$$\int_0^\infty d\omega\, S(\omega)\frac{\sin^2[(\omega - \nu)T/2]}{[(\omega - \nu)/2]^2} \simeq 2T\, S(\omega_{\mathrm{fi}})\int_{-\infty}^\infty \frac{\sin^2 \eta}{\eta^2}\, d\eta = 2\pi\, T\, S(\omega_{\mathrm{fi}}), \qquad (4.120)$$

which shows that

$$\lim_{T\to\infty} \frac{\sin^2[(\omega - \nu)T/2]}{[(\omega - \nu)/2]^2} \to 2\pi\, T\delta_D(\omega - \nu) \qquad (4.121)$$

in a formal sense. Hence Eq. (4.118) becomes, as $T \to \infty$,

$$P(T) \cong T\left(\frac{2\pi}{V\omega_k}\right)|\mathbf{e}_{\mathbf{k}\alpha}\cdot\mathbf{M}_{fi}|^2(n_{\mathbf{k}\alpha} + 1)\delta_D(E_i - E_f - \omega). \qquad (4.122)$$

The corresponding rate of transitions is $R = P(T)/T$, which gives a finite rate for the emission of photons:

$$R \equiv \frac{dP}{dt} = \left(\frac{2\pi}{V\omega_k}\right)(n_{\mathbf{k}\alpha} + 1)|\hat{\mathbf{e}}_{\mathbf{k}\alpha}\cdot\mathbf{M}_{fi}|^2 2\pi\,\delta_D(E_i - E_f - \omega). \qquad (4.123)$$

[The integration over an infinite range of t implies, in practice, an integration over a range $(0, T)$ with $\omega T \gg 1$. If the energy levels have a characteristic width $\Delta\omega \ll \omega$, then the above analysis is valid for $\omega^{-1} \ll T \ll (\Delta\omega)^{-1}$. The effect of a finite width of energy levels, as well as a more rigorous derivation of the above result, will be provided in Chap. 7, Section 7.2.]. Equation (4.123) gives the rate for emission of a photon with a specific wave vector $\mathbf{k}$ and polarisation α. The delta function, $2\pi\delta_D(E_i - E_f - \omega)$, expresses conservation of energy and shows that the probability is nonzero only if the energy difference between the states $E_i - E_f$ is equal to the energy of the emitted photon $\hbar\omega$.

Usually we will be interested in the probability for emission of a photon in a frequency range ω, $\omega + d\omega$ and in a direction defined by the solid angle element $d\Omega$. To obtain this quantity, we have to multiply the rate of transition by the density of states available for the photon in this range. The density of states is given by

$$\frac{dN}{d\omega\,d\Omega} = \frac{V\,d^3k}{(2\pi)^3}\frac{1}{d\omega\,d\Omega} = \frac{V}{(2\pi)^3}\frac{k^2\,dk\,d\Omega}{d\omega\,d\Omega} = \frac{Vk^2}{(2\pi)^3} = \frac{V\omega^2}{(2\pi)^3}. \quad (4.124)$$

Hence

$$\left[\frac{dP}{dt\,d\omega\,d\Omega}\right]_{emi} = \frac{dP}{dt}\frac{dN}{d\omega\,d\Omega} = \frac{dP}{dt}\frac{V\omega^2}{(2\pi)^3}$$

$$= \left(\frac{2\pi}{V\omega}\right)(n_{\mathbf{k}\alpha} + 1)|\hat{\mathbf{e}}_{\mathbf{k}\alpha}\cdot\mathbf{M}_{fi}|^2 2\pi\delta_D(\omega - \omega_{fi})\frac{V\omega^2}{(2\pi)^3}$$

$$= \left(\frac{\omega}{2\pi\,\hbar c^3}\right)(n_{\mathbf{k}\alpha} + 1)|\hat{\mathbf{e}}_{\mathbf{k}\alpha}\cdot\mathbf{M}_{fi}|^2\delta_D(\omega - \omega_{fi}), \quad (4.125)$$

with $\hbar\omega_{fi} \equiv E_i - E_f$. In the last line we have introduced the $\hbar$ and c factors appropriately. [The correctness of this reinsertion can be verified as follows: The left-hand side is dimensionless. On the right-hand side, $\omega\delta_D(\omega - \omega_{fi})$ is dimensionless; the matrix element has dimensions of q^2v^2. Because $(q^2v^2/\hbar c^3) = (q^2/\hbar c)(v/c)^2$ is dimensionless, the $\hbar c^3$ factor in the last line is indeed correct.]

The analysis for the absorption rate of photons is identical except that only the annihilation operator $c_{\mathbf{k}\alpha}$ contributes. Because $\langle n_{\mathbf{k}\alpha} - 1|c_{\mathbf{k}\alpha}|n_{\mathbf{k}\alpha}\rangle = n_{\mathbf{k}\alpha}^{1/2}$, we get $n_{\mathbf{k}\alpha}$ rather than $(n_{\mathbf{k}\alpha} + 1)$ in the final result:

$$\left[\frac{dP}{d\Omega\,dt\,d\omega}\right]_{abs} = \left(\frac{\omega}{2\pi\,\hbar c^3}\right)n_{\mathbf{k}\alpha}|\hat{\mathbf{e}}_{\mathbf{k}\alpha}\cdot\mathbf{M}_{fi}|^2\delta_D(\omega - \omega_{fi}). \quad (4.126)$$

There are several features about these results that are worth noting. To begin with, the probabilities for absorption and emission differ in only the dependence on $n_{\mathbf{k}\alpha}$. The probability for absorption scales in proportion with $n_{\mathbf{k}\alpha}$. Clearly, if $n_{\mathbf{k}\alpha} = 0$, this probability vanishes; this is obvious as no photons can be absorbed if there were none in the initial state to begin with. However, the probability for emission is proportional to $n_{\mathbf{k}\alpha} + 1$ and does not vanish even when $n_{\mathbf{k}\alpha} = 0$. Hence there is a nonzero probability for a system at an excited state to emit a photon and come down to a lower state spontaneously. If the initial state of the electromagnetic field has a certain number of photons already present, then the probability for emission is further enhanced. The emission of a photon by an excited system when no photons were originally present is called spontaneous emission, and the emission of a photon in the presence of initial photons is called stimulated emission. Both these processes exist and contribute to electromagnetic transitions. The above analysis shows that this is an elementary consequence of the properties of the harmonic-oscillator matrix element. (A more physical interpretation, based on microscopic reversibility, was given in Chap. 1, Subsection 1.4.2.)

We will make some general comments about quantisation of fields at this juncture, although a detailed proof of these assertions is not relevant to us. The existence of a factor n for absorption and $1 + n$ for emission is a very general feature and arises in quantising a very large class of fields. Quanta of such fields carry a spin angular momentum that is an integral multiple of $\hbar$ and are called bosons. In contrast, there also exist quanta of fields that carry half-integral multiples of $\hbar$ as their spin, called fermions. The simplest example of such a quantum is the electron. Classically, we are accustomed to thinking of electrons as particles and photons as waves. However, in the rigorous quantum-mechanical formulation, both photons and electrons arise as quanta when certain fields are quantised. We saw that photons arise when we quantise the electromagnetic field. Electrons arise when we quantise a field called the Dirac field.

There is, however, one vital difference between the processes involving bosons and fermions. It turns out that although any number of bosons can exist in a given quantum state, not more than one fermion can be present at a time in a specific quantum state. For example, in the case of photons, the number $n_{k\alpha}$, which denotes the number of photons in a state labelled by $\mathbf{k}$ and α, can take any integer value. In the case of electrons, however, the corresponding number $n_e(\mathbf{k}, \alpha)$, which denotes the number of electrons in a state with wave vector $\mathbf{k}$ and spin α, can only be 0 or 1. The absorption (or annihilation) of fermions will produce a term proportional to the number of fermions in a given quantum state, just as in the case of bosons; but the emission (or creation) of fermions is proportional to a term $1 - n$, where n is the number of fermions originally present in the quantum state to which emission is taking place. If a fermion is already present in a particular state (that is, if $n = 1$), no emission can take place to that state. This result is a generalised version of a principle called the Pauli exclusion principle.

4.5.3 Quantum Dipole Radiation

To proceed further, we have to evaluate the matrix element $\mathbf{M}_{fi}$ that appears in the probabilities. This can always be done, in principle, given the wave functions of the initial and final states. Under certain circumstances, however, we can simplify these results still further and obtain useful formulas. These formulas also show the correspondence between the quantum-mechanical treatment of the radiation and the classical dipole radiation obtained earlier in Section (4.3.1). We now discuss this case.

Very often we are interested in situations in which the wavelength λ of the radiation is much larger than the typical atomic size a. When $\lambda \gg a$, we can use the approximation $\exp(\pm i\mathbf{k} \cdot \mathbf{x}) \simeq 1$ in matrix element (4.117) and express $\mathbf{M}_{fi}$ in terms of the matrix element $\mathbf{p}_{fi}$ of the momentum operator. But because $\mathbf{p} = (im/\hbar)[H_0, \mathbf{x}]$ it follows that $\mathbf{p}_{fi} = (im/\hbar)[H_0, \mathbf{x}]_{fi} = im\omega_{fi}\mathbf{x}_{fi}$. Therefore the rate of spontaneous emission [corresponding to $n_{\mathbf{k}\alpha} = 0$ in Eq. (4.125)] can be written as

$$\frac{dP}{d\Omega\, dt} = \left(\frac{\omega}{2\pi\,\hbar c^3}\right)\frac{q^2}{m^2}|\hat{\mathbf{e}}_{\mathbf{k}\alpha} \cdot \mathbf{p}_{fi}|^2 [\delta_D(\omega - \omega_{fi})\, d\omega]$$

$$= \left(\frac{\omega^3}{2\pi\,\hbar c^3}\right)|\hat{\mathbf{e}}_{\mathbf{k}\alpha} \cdot q\mathbf{x}_{fi}|^2 [\delta_D(\omega - \omega_{fi})\, d\omega]$$

$$= \frac{\omega^3}{2\pi\,\hbar c^3}|\hat{\mathbf{e}}_{\mathbf{k}\alpha} \cdot \mathbf{d}_{fi}|^2 [\delta_D(\omega - \omega_{fi})\, d\omega], \tag{4.127}$$

where $\mathbf{d}_{fi} = \langle f|q\mathbf{x}|i\rangle$ is the matrix element of the dipole moment operator. Because the Dirac delta function ensures only energy conservation, we can integrate this expression over ω to eliminate the Dirac delta function; this integration will replace ω^3 with ω_{fi}^3. The result still depends explicitly on the polarisation vector of the photon that is emitted. In practical situations we are often not interested in this information and hence can average over all possible orientations of the polarisation vector. By using the relation

$$\langle(\mathbf{a} \cdot \mathbf{e})(\mathbf{b} \cdot \mathbf{e})\rangle = \frac{1}{2}[\mathbf{a} \times \mathbf{n}] \cdot [\mathbf{b} \times \mathbf{n}], \tag{4.128}$$

where $\mathbf{n} = (\mathbf{k}/k)$, we get

$$\left\langle\frac{dP}{dt\, d\Omega}\right\rangle = \frac{\omega_{fi}^3}{4\pi\,\hbar c^3}|\mathbf{n} \times \mathbf{d}_{fi}|^2 = \frac{\omega_{fi}^3}{4\pi\,\hbar c^3}|\mathbf{d}_{fi}|^2 \sin^2\theta. \tag{4.129}$$

We find the average energy emitted per polarisation state by multiplying by $\hbar\omega_{fi}$:

$$\left\langle\frac{dE}{dt\, d\Omega}\right\rangle = \hbar\omega_{fi}\left\langle\frac{dP}{dt\, d\Omega}\right\rangle = \frac{\omega_{fi}^4}{4\pi c^3}|\mathbf{d}_{fi}|^2 \sin^2\theta. \tag{4.130}$$

This expression matches with the classical result in Eq. (4.38), provided that $|\ddot{\mathbf{d}}|^2$ is identified with $\omega_{fi}^4|\mathbf{d}_{fi}|^2$. Such an identification will arise automatically if we assume that the charged particle behaves as a harmonic oscillator with the equation of motion $\ddot{\mathbf{x}} = -\omega_{fi}^2\mathbf{x}$. This approach turns out to be remarkably fertile in providing insights into quantum-mechanical processes along classical lines. In

other words, a quantum-mechanical system undergoing transitions between two
levels i and f behaves similarly to a charged particle moving under the action
of a harmonic restoring force with frequency ω_{fi}.

Let us next consider the corresponding absorption probability between states
i and f when $n_{\mathbf{k}\alpha}$ photons are originally present. We will assume that the initial
radiation field is isotropic and unpolarised so that $n_{\mathbf{k}\alpha} = [n(\omega)/2]$. We can obtain the absorption rate for unpolarised, isotropic radiation from Eq. (4.129) by
multiplying by $n_{\mathbf{k}\alpha} = [n(\omega)/2]$. To take into account two polarisation states, we
multiply by 2 and finally integrate over the solid angle by replacing $\sin^2 \theta$ with
$4\pi \langle \sin^2 \theta \rangle = (8\pi/3)$. This gives

$$\left(\frac{dP}{dt}\right)_{abs} = \frac{\omega_{fi}^3}{4\pi \hbar c^3}|\mathbf{d}_{fi}|^2 \left(\frac{n}{2}\right) 2\frac{8\pi}{3} = \frac{2}{3}\frac{q^2}{\hbar c^3}n\omega_{fi}^3|\mathbf{x}_{fi}|^2. \tag{4.131}$$

This result is conventionally expressed with the concept of an absorption
cross section σ_{bb} (where bb stands for bound–bound) and a quantity called
oscillator strength. The bound–bound absorption cross section $\sigma_{bb}(\omega)$ for the
atom is defined by the relation

$$\binom{\text{number of photons}}{\text{absorbed per second}} = \int_0^\infty \sigma_{bb}(\omega)(\text{flux of photons})(\text{density of states})\, d\omega.$$

$$\tag{4.132}$$

Because the flux of photons is given by $cn(\omega)$ and the density of states per unit
frequency range is

$$\frac{d^3k}{(2\pi)^3 d\omega} = \frac{4\pi k^2 dk}{(2\pi)^3 d\omega} = \frac{4\pi (\omega/c)^2 d(\omega/c)}{(2\pi)^3 d\omega}, \tag{4.133}$$

we get

$$\frac{dP_{if}}{dt} = \int_0^\infty \sigma_{bb}(\omega)[cn(\omega)]\frac{4\pi (\omega/c)^2 d(\omega/c)}{(2\pi)^3}. \tag{4.134}$$

It is conventional to write σ_{bb} as

$$\sigma_{bb}(\omega) = \frac{\pi q^2}{mc} f_{fi}[2\pi \delta_D(\omega - \omega_{fi})] \equiv \frac{\pi q^2}{mc} f_{fi}\phi_\omega, \tag{4.135}$$

where f_{fi} is called the oscillator strength for the given initial and final states
and ϕ_ω is called the line-profile function. [In the present case, ϕ_ω is proportional
to the Dirac delta function. When the finite width of the energy levels is taken
into account (see Chap. 7), ϕ_ω will become a function that peaks at ω_{if} with a
narrow width; the definition here anticipates that result.] Because q^2/mc has the
dimension of $\text{cm}^2\ \text{s}^{-1}$ and ϕ_ω has the dimension of ω^{-1}, viz., second, it follows
that σ_{bb} has the dimension of area, as to be expected of a cross section. It should

also be noted that

$$\int_0^\infty \sigma_{bb}(\omega) \frac{d\omega}{2\pi} = \frac{\pi q^2}{mc} f_{fi} = f_{fi}(\pi r_0^2)\left(\frac{c}{r_0}\right), \quad r_0 \equiv \left(\frac{q^2}{mc^2}\right), \quad (4.136)$$

as $\phi_\omega/2\pi$ is normalised to unity on integration over ω. This shows that the to-tal, integrated cross section is essentially the product of the area corresponding to classical electron radius r_0 and the frequency c/r_0 corresponding to this ra-dius. When definition (4.135) is substituted into Eq. (4.134), the absorption rate becomes

$$\left(\frac{dP_{abs}}{dt}\right) = \left(\frac{q^2}{mc^3}\right) n\omega^2 f_{fi} = \left(\frac{2q^2}{mc^3}\right) n_\alpha(\omega)\omega^2 f_{fi}. \quad (4.137)$$

Similarly, the emission rate becomes

$$\left(\frac{dP_{em}}{dt}\right) = \left(\frac{2q^2}{mc^3}\right)[1 + n_\alpha(\omega)]\omega^2 f_{fi}, \quad (4.138)$$

where $n_\alpha(\omega) = [n(\omega)/2]$ is the number of photons per polarisation state. All the quantum-mechanical information regarding the absorption (and emission) is now contained in the oscillator strength f_{fi}. Comparing Eq. (4.137) with Eq. (4.131), we find that

$$f_{fi} = \frac{2}{3}\frac{m\omega_{fi}}{\hbar}|\mathbf{x}_{fi}|^2 = \frac{2}{3}\left[\frac{m\omega_{fi}^2|\mathbf{x}_{fi}|^2}{\hbar\omega_{fi}}\right]. \quad (4.139)$$

If the electron is treated as a classical oscillator with frequency ω, then it will have an energy $m\omega^2 a^2$, where a is the amplitude of the oscillation. The energy of radiation $\hbar\omega$, emitted or absorbed, will be approximately the same order. Therefore, whenever $a \approx |\mathbf{x}_{fi}|$, it is reasonable to expect f_{fi} to be of the order of unity. When the argument is inverted, $|\mathbf{x}_{fi}|^2$ can be expressed in terms of f_{fi} as

$$|\mathbf{x}_{fi}|^2 = \frac{3}{2}\left(\frac{\hbar}{m\omega}\right) f_{fi}. \quad (4.140)$$

These are the basic results describing quantum radiative processes in the simplest possible situation. When the matrix element of the dipole moment between two states i and f does not vanish, we have to calculate only the oscillator strength f_{fi} between the two states by using Eq. (4.139). The absorption and the emission rates are then provided by Eqs. (4.137) and (4.138).

Exercise 4.9

Sum rule: Show that the sum of f_{fi} over all f is unity for a given i. The sum goes over states with $E_f > E_i$ (absorption) and $E_f < E_i$ (emission). Treat f as positive for emission and negative for absorption.

We emphasize that the probability for absorption of a photon is proportional to the number of photons n in that state and the probability for emission of photons is proportional to $n + 1$. There exists a certain probability for emission into a particular state even when no photons are present in that state. But when some photons are already present, the probability of emission to that state is enhanced.

Exercise 4.10

Example of oscillator strength: As a specific example, let us compute the value of f for the hydrogen atom when the initial state is the $n = 1$ ground state and the final state is any one of the three states with $n = 2, l = 1, m = -1, 0, 1$ by evaluating the matrix element x_{ij} between these two atomic states. The initial state has the wave function corresponding to $n = 1, l = 0$, and $m = 0$,

$$\Psi_{1,0,0} = \frac{1}{\sqrt{\pi a^3}} \exp\left(-\frac{r}{a}\right), \quad a = \left(\frac{\hbar^2}{me^2}\right), \tag{4.141}$$

and the final states could be any of the following:

$$\Psi_{2,1,-1} = \frac{R_{21}(r)}{r} Y_{1,-1}(\theta, \phi), \quad \Psi_{2,1,1} = \frac{R_{21}(r)}{r} Y_{1,1}(\theta, \phi), \quad \Psi_{2,1,0} = \frac{R_{21}(r)}{r} Y_{1,0}(\theta, \phi). \tag{4.142}$$

Because the final state is degenerate, we also have to sum over all possible final states with the same energy. The matrix elements x, y, and z between these states can be now computed by direct integration. However, we can simplify the analysis by noting that

$$x \pm iy = r(\sin\theta)e^{\pm i\phi} = r\left(\frac{8\pi}{3}\right)^{1/2} Y_{1,\pm 1},$$

$$z = r(\cos\theta) = r\left(\frac{4\pi}{3}\right)^{1/2} Y_{1,0}, \tag{4.143}$$

$$|x_{fi}|^2 = |z_{fi}|^2 + \frac{1}{2}|(x + iy)_{if}|^2 + \frac{1}{2}|(x - iy)_{if}|^2. \tag{4.144}$$

Use these features to compute the oscillator strength and show that

$$f_{fi} = \frac{2^{14}}{3^9} \simeq 0.83, \tag{4.145}$$

which is of the order of unity.

The matrix element of the dipole operator can vanish under certain circumstances. To begin with, it will vanish if the emission of a photon cannot satisfy standard conservation laws. For example, the photon carries an angular momentum of one unit of $\hbar$; hence, if the initial and the final states have the same amount of angular momentum, then the transition process cannot take place by the emission of a single photon. The emission of radiation can still occur through higher-order terms in the perturbation theory; for example, through two-photon emission. Or it can proceed through electric quadrapole or magnetic dipole type of radiation. Some of these details will be discussed in Chap. 7.

The radiative processes described so far did not involve direct conversion of matter into energy. It is also possible to produce radiation by direct annihilation of a charged particle and its antiparticle, such as, for example, an electron and a positron. This will be discussed later in Chap. 6, Section 6.4.

5

Statistical Mechanics

5.1 Introduction

This chapter discusses physical systems involving large number of particles, conventionally called statistical mechanics. Because these concepts are used in several later chapters, a complete and pedagogical discussion is presented here. We begin with the equilibrium statistical mechanics of classical systems and derive macroscopic thermodynamics from statistical mechanics. In the second half of the chapter we deal with quantum statistical mechanics, including the physics of Fermi gas. This chapter depends on the concepts developed in the previous three chapters and will be needed for Chaps. 6 (radiative processes), 8 (neutral fluids), 9 (plasma physics) as well as in the study of stellar evolution and stellar remnants (Vol. II) and the thermal history of the universe (Vol. III).

5.2 Operational Basis of Statistical Mechanics

The dynamical evolution of any system can be studied most conveniently by use of the concept of phase space developed in Chap. 2, Section 2.2. For a system of N-point particles the phase space will be $6N$ dimensional. Given the initial state of the system as a point in the phase space, the dynamical evolution of the system traces out a one-dimensional curve in the $6N$-dimensional phase space, starting from the given point. If the equations of motion for the system are solved exactly (with the given initial conditions), then this curve in the phase space can be determined exactly.

For any realistic N-particle system, this task is impossible even with the best computers available today. Further, there are three important reasons why such an approach is inappropriate: To begin with, finding the exact trajectory requires knowledge of the exact initial condition for the system. For example, to predict the behaviour of a gas of 10^{23} molecules used in a particular experiment, we will need to know the exact positions and velocities of 10^{23} molecules at the start of the experiment. This is impossible to obtain operationally, as the initial microscopic configuration of most macroscopic bodies will not be available. Second, even if we know the initial phase point and possessed computing facilities to integrate the equations of motion, we still cannot make progress unless we know all the forces that are influencing the system. For example, the gravitational influence of an electron – located at the edge of the visible universe – on a gas in laboratory is significant enough to kill predictability in ~60 collision times (see Exercise 5.1)! Thus it is not really possible to compute the exact phase trajectory, even in principle. Finally, solving for the exact trajectory will give us the $6N$ pieces of information (q_i, p_i) of the system. In most practical situations, we will not be interested in so many bits of information and will have to compute averages over large quantities of information to obtain macroscopic properties, like mean density, mean energy, etc. Therefore it will be a waste of effort if we obtain the microscopic parameters while we are really interested in only a small number of macroscopic quantities.

It is therefore convenient to develop a theoretical formalism that is capable of answering macroscopic questions without requiring the information about the initial conditions *or even the precise form of the total Hamiltonian*. Let us consider the nature of such a theoretical formalism and the conditions under which it will be useful.

Consider an experiment in which some macroscopic parameter of a body say, the pressure of a gas, is measured. The pressure, at any instant, depends on the momentum transferred to the wall by all the molecules that are colliding with the wall at that instant; because the number of molecules hitting the wall and their velocity will vary instant by instant the *precise* value of the pressure will fluctuate from instant to instant. We cannot ascertain the precise value of *any* macroscopic parameter $f(t)$ at an instant t without knowing the trajectory in phase space. The best we can hope for is the following: Let us suppose that the precise values of the microscopic variables fluctuate in time around a mean value in such a manner that the fluctuations are small compared with the mean value. To make this notion more rigorous, we define the time average of an observable $f(t)$ as

$$\langle f(t) \rangle = \frac{1}{T} \int_{t-T/2}^{t+T/2} f(t') \, dt', \tag{5.1}$$

where T is large compared with the time scale of the fluctuations. The fluctuations are small if

$$\left\{ \frac{\langle [f(t) - \langle f(t) \rangle]^2 \rangle}{\langle f(t) \rangle^2} \right\}^{1/2} \ll 1 \tag{5.2}$$

for some reasonable value of T. We have tacitly assumed that the value of the observable f fluctuates around the mean value $\langle f \rangle$ at some time scale τ such that $\tau \ll T$, where T represents a time scale chosen such that we are not interested in following the evolution of the system for $t \lesssim T$. We can certainly follow slower variations (i.e., variations that take place at time scales that are large compared with T) in the mean value of the observable $\langle f(t) \rangle$.

We are often interested in studying closed systems in which the external macroscopic parameters are not changing with time. In that case, it turns out that the mean values of the observables attain values that are independent of time at sufficiently late times. More precisely, physical systems are characterised by a time scale t_R (called relaxation time) such that for $t \gg t_R$ the following statements are true to a high degree of accuracy: (1) The mean values $\langle f(t) \rangle$ of macroscopic observables become independent of t for $t \gg t_R$. The macroscopic observables are, by definition, built out of several microscopic parameters. The approximate constancy of $\langle f(t) \rangle$ also depends on the fact that a large number of particles are involved in the definition of the observable $f(t)$. (2) The fluctuations, around the mean value of the macroscopic observables, turn out to be small for $t \gg t_R$. In general the fluctuations are lower if a large number of particles are involved in the definition of the observables. Given the Hamiltonian governing the system, we should be able to estimate t_R and verify whether the above features are true. In practise, however, this is an extremely hard (and, in general, unsolved) problem. We shall merely assume that physical systems we are interested in exhibit the behaviour outlined above.

The two features listed above suggest that the late-time behaviour of closed systems is independent of the initial conditions, that is, the system "forgets" the initial condition in a time scale $t \simeq t_R$ and the macroscopic observables reach mean values, which are independent of the initial conditions used to prepare the systems, for $t \gtrsim t_R$. This feature, however, is true only for those macroscopic variables that are not constrained in any manner. For example, if two identical systems were started off with different values for the total energy, then – even at late times – they will continue to have different values for the total energy. Therefore the values of conserved quantities, (like energy) certainly depend on the initial condition.

For a closed system with an arbitrary Hamiltonian, there are seven such constants: the energy E, three components of total angular momentum $\mathbf{J}$, and three components of total linear momentum $\mathbf{P}$. (Special kinds of Hamiltonians may allow the existence of other constants of motion. However, we should not use any feature that is specific to a Hamiltonian in developing our formalism as we are uncertain about the exact form of the total Hamiltonian.) Often, however, we will confine the particles by a rigid box of volume V and arbitrary shape. Such a box is mathematically characterised by an infinite potential outside some compact region in three space and zero potential inside. The existence of such an external potential invalidates translational and rotational invariance of the Hamiltonian and hence $\mathbf{J}$ and $\mathbf{P}$ (of the system excluding the rigid box) will no longer be conserved. Thus, for a system confined by a rigid box of arbitrary shape, the only robust constant of motion is the energy $E(p, q)$.

The motion of the system in phase space will therefore be confined to the constant-energy surface defined by $H(p, q) = E$, with the coordinates q of the particles being confined to the inside of the box. If we start off the system at $t = 0$ with different initial conditions, i.e., from different points on the constant-energy surface in phase space, it will be found at different locations at any later time t. Let us choose a large number M of initial points $P_1, P_2, \ldots, P_M$ and study the behaviour of the phase trajectory at late times $t \gg t_R$. Any given macroscopic observable f will attain a constant mean value $\langle f \rangle$ at late times; for any one of the initial conditions, we can calculate this mean value by using Eq. (5.1). On the other hand, we can also calculate the mean value of a physical variable at a given time t by averaging it over different initial conditions. If $n(f)$ of the initial conditions (among M) leads to the value f for the observable, then the mean value could also be defined as

$$\bar{f} = \frac{1}{M} \sum n(f) f. \tag{5.3}$$

If the memory of initial conditions is wiped out for $t \gg t_R$ and if the Hamiltonian has no special features, then it seems reasonable to assume that the two kinds of averages should give the same result, that is, we expect the time average $\langle f \rangle$ to be equal to the ensemble average, $\bar{f}$ obtained by, say, averaging over initial conditions. This equivalence is called the ergodic hypothesis. We shall assume that this hypothesis is true in the form stated above.

Given the above assumptions regarding the behaviour of systems at late times and the ergodic hypothesis, we can develop the formalism of equilibrium statistical mechanics. This structure can be used to deduce, from first principles, the laws of thermodynamics and to calculate fluctuations around the equilibrium. We first develop the formalism in the classical context and then introduce the modifications from quantum theory.

Exercise 5.1

Classical unpredictability: Consider a large number N of molecules kept inside a cubical box. At $t = 0$, all the particles are arranged to move parallel to the x axis. We approximate the molecules as hard spheres of radius r, mean separation l, and mean velocity v. The collisions are elastic and occur head-on when two particles, which are moving along the same line in the x direction, meet. (That is, the particles are distributed in such a way that there are no glancing collisions.) The collisions with the walls are also elastic. It may seem that, in this situation, the initial conditions can be preserved for arbitrarily long periods of time and no relaxation will ensue. Prove that the above conclusion is erroneous. Consider the gravitational force on this system that is due to an electron located at the edge of the universe at a distance $D \simeq 6000$ Mpc. Show that this perturbation is enough to destroy the orderly state within ~ 60 collisions if the system is made of molecules with $r \simeq 10^{-8}$ cm and $m \simeq 10^{-23}$ gm at room temperature and pressure.

5.3 The Density of States and Microcanonical Distribution

Consider a system at $t \gg t_R$, when mean values of physical observables have reached steady values. We are interested in computing the mean value of some macroscopic observable $f(q, p)$. Because the mean value – defined in terms of time averages – is the same as that defined by a variety of initial conditions, we could use the latter to estimate the mean values. Then the mean value of an observable $f(q, p)$, evaluated at a time t, is

$$\langle f \rangle = \int dp \, dq \, f(q, p) \, \rho(q, p, t) \tag{5.4}$$

where $\rho(q, p, t) \, dp \, dq$ is the fraction of cases in which the system will be found with dynamical variables and momenta having values in the range $(q, q + dq; p, p + dp)$ at time t, if we consider all possible initial conditions. Let us determine the function $\rho(q, p, t)$.

We begin by noting that ρ obeys a conservation law as a direct consequence of the fact that the total number M of prototypical initial conditions (using which we define the average) is fixed. At any given time, there must exist these M points in phase space; if the number of points inside any compact region of phase space changes, it could be only because these points have moved out (or in) through the surface around the compact region. Because the phase space has the coordinates (q^i, p^i) the conservation law for these points becomes

$$\frac{\partial \rho}{\partial t} + \frac{\partial}{\partial q^i}(\rho \dot{q}^i) + \frac{\partial}{\partial p^i}(\rho \dot{p}^i) = 0. \tag{5.5}$$

Expanding the derivatives and using Hamilton's equations for $\dot{q}^i$ and $\dot{p}^i$ we find that

$$\frac{\partial \rho}{\partial t} + \dot{q}^i \frac{\partial \rho}{\partial q^i} + \dot{p}^i \frac{\partial \rho}{\partial p^i} = \frac{d\rho}{dt} = 0. \tag{5.6}$$

This relation, called Liouville's theorem, shows that ρ must be a function of the constants of motion of the system. Because the mean value in Eq. (5.4) should be independent of time in equilibrium, it is also necessary that $\rho(q, p, t) = \rho(q, p)$; that is, ρ cannot have any explicit time dependence. Therefore $\rho = \rho[I_1(q, p), I_2(q, p), \ldots,]$, where all the I_j's are time-independent single-valued integrals of motion. For a general Hamiltonian describing particles confined by a rigid box of arbitrary shape, energy is the only quantity that satisfies this requirement. Therefore ρ must be a function of the Hamiltonian alone: $\rho = \rho[H(q, p)]$.

The functional form can be determined quite easily. Because Eq. (5.4) is applicable to any observable f, it is also applicable to the Hamiltonian and various powers of the Hamiltonian such as H^2, H^3, ... However, because the system has constant energy E, the energy cannot show fluctuations, requiring that $\langle H^n \rangle = \langle H \rangle^n$. This requires ρ to be proportional to a Dirac delta function in energy, that is,

$$\rho(q, p) = C(E)\delta_D[E - H(p, q)].\tag{5.7}$$

The constant $C(E)$ can be determined by the normalisation condition for ρ, and we get

$$C^{-1} \equiv g(E) = \int dp\, dq\, \delta_D[E - H(p, q)].\tag{5.8}$$

The mean value of any observable can be then determined from Eq. (5.4) as

$$\langle f \rangle = \frac{1}{g(E)} \int dp\, dq\, \delta_D[E - H(p, q)]f(q, p).\tag{5.9}$$

This expression has a simple physical interpretation. The probability in Eq. (5.7) shows that (1) the system is confined to the constant-energy surface, and (2) it has equal probability of being found in any infinitesimal volume $dp\, dq$ of this constant-energy surface. Because $g(E)$ is the volume of the constant-energy surface, the probability that the system will be found in some volume ΔV on the constant-energy surface is just $\Delta V/g(E)$. Any small volume on the constant-energy surface may be thought of as a microstate that is consistent with the macroscopic parameters of the system. Equation (5.7) shows that all microstates are equally probable. The probability for any given macroscopic configuration will be proportional to the number of microstates that is consistent with that configuration. The probability distribution in Eq. (5.7) is called micro-canonical. [To make this interpretation rigorous, we need a way of counting the actual number of microstates on the energy surface or – more generally – a way of determining the measure of integration in Eq. (5.9). This cannot be done in classical theory; as we shall see later in Section 5.7, quantum mechanics makes this notion precise.]

The quantity $g(E)$ also has an interpretation as density of states in phase space. We can explain this nomenclature by noting that if any function $A(q, p)$ depends on q and p only through $H(p, q)$, i.e., if $A(q, p) = A[H(p, q)]$, then

$$I = \int dp\, dq\, A[H(q, p)] = \int dE\, g(E)A(E).\tag{5.10}$$

To evaluate $g(E)$ we often use the following procedure. We write

$$g(E) = \frac{d}{dE} \int \theta[E - H(p, q)]\, dp\, dq = \frac{d\Gamma}{dE},\tag{5.11}$$

where the function $\theta(z)$, called the Heaviside theta function, is unity if $z \geq 0$ and vanishes for $z < 0$ and

$$\Gamma(E) \equiv \int \theta[E - H(p, q)]\, dp\, dq\tag{5.12}$$

is the volume of phase space enclosed by the energy surface. Very often it is more convenient to calculate $\Gamma(E)$ and obtain $g(E)$ by differentiation rather than work with $g(E)$ directly.

As an example, let us consider a system of nonrelativistic particles with the Hamiltonian $H = \sum_{i=1}^{3N}(p_i^2/2m)$. Then

$$\Gamma(E) = \int d^{3N}p \, d^{3N}q \, \theta\left(E - \sum_{i=1}^{3N} \frac{p_i^2}{2m}\right) = V^N \int d^{3N}p \, \theta\left(E - \sum_{i=1}^{3N} \frac{p_i^2}{2m}\right).$$

(5.13)

The integral gives the volume of a $3N$-dimensional sphere of radius $(2mE)^{1/2}$. Because the volume $\mathcal{V}_M(R)$ of an M-dimensional sphere of radius R is (see Exercise 5.2),

$$\mathcal{V}_M(R) \equiv C_M R^M = \frac{\pi^{M/2}}{(M/2)\Gamma(M/2)} R^M,$$

(5.14)

where $\Gamma(M/2)$ is the standard gamma function [and not to be confused with the phase volume $\Gamma(E)$], we find that

$$\Gamma(E) = C_{3N}(2mE)^{3N/2}V^N, \quad g(E) = \frac{d\Gamma}{dE} \simeq C_{3N}\left(\frac{3N}{2}\right)(2mE)^{3N/2}V^N,$$

(5.15)

where we have set $[(3N/2) - 1] \approx (3N/2)$ in the second equation. This expression gives $g(E)$ for nonrelativistic ideal gas.

The E and V dependence of Γ can be obtained by the following scaling argument: by changing the integration variable in Eq. (5.13) from p to $k = (p/\sqrt{2mE})$, we find that $\Gamma(E) \propto E^{3N/2}V^N$, which fixes the dependence on E and V. If, instead of nonrelativistic particles, we had ultrarelativistic particles with energies $E \cong c|\mathbf{p}|$, then the corresponding scaling to a variable $k = (|\mathbf{p}|c/E)$ shows that $\Gamma(E)_{\text{rel}} \propto E^{3N}V^N$. Because $N \gg 1$, $g(E)$ has the same scaling as $\Gamma(E)$.

Exercise 5.2
Volume of M-dimensional sphere: Prove the assertion (5.14) regarding the volume of an M-dimensional sphere. [Hint: An easy way will be to evaluate the integral

$$I = \int d^M x \, \exp\left(-\sum_{i=1}^{M} x_i^2\right)$$

(5.16)

in two different ways: first, as a product of M one-dimensional Gaussian integral and next, in polar coordinates in the M-dimensional space. Equating the two results will give the required formula.]

As a second example we take up the following important question: When the system has reached steady state, what is the probability $\mathcal{P}(\epsilon)$ that a specified subsystem of the total system has the energy ϵ? To answer this question in a general context, let us assume that the total Hamiltonian can be split up as the sum of two parts:

$$H(p, q) = H_1(p_1, q_1) + H_2(p_2, q_2), \qquad (5.17)$$

where (p_1, q_1) denotes the coordinates of particles belonging to the subsystem 1, etc. If this statement is strictly true, then systems 1 and 2 are uncoupled and there is no way they could exchange energy. In reality, the full Hamiltonian will also have another small part $H_I(p_1, q_1; p_2, q_2)$ that couples the two systems. However, we will assume that the interaction is weak and does not contribute to the total energy, that is, $H_I \ll H_1, H_2$. [Note that (a nearly) ideal gas satisfies this criterion quite well. The interaction energy between the molecules is negligible compared with the kinetic energy in the gaseous phase. All the same, it is the interaction during close collisions that redistributes the energy in an ideal gas and leads to the establishment of equilibrium.] The density of state for the full system can now be expressed as

$$g(E) = \int dp \, dq \delta_D[E - H(p, q)] = \int dp_1 \, dq_1 \, dp_2 \, dq_2 \delta_D[E - H_1 - H_2].$$

$$(5.18)$$

We can rewrite this expression by introducing two more Dirac delta functions in the form

$$
\begin{aligned}
g(E) &= \int dp_1 \, dq_1 \, dp_2 \, dq_2 \int d\epsilon \int d\epsilon' \delta_D[E - \epsilon - \epsilon'] \delta_D[\epsilon - H_1] \delta_D[\epsilon' - H_2] \\
&= \int d\epsilon \, d\epsilon' \delta_D[E - \epsilon - \epsilon'] \int dp_1 \, dq_1 \delta_D[\epsilon - H_1] \int dp_2 \, dq_2 \delta_D[\epsilon' - H_2] \\
&= \int d\epsilon \, d\epsilon' \delta_D[E - \epsilon - \epsilon'] g_1(\epsilon) g_2(\epsilon') = \int d\epsilon g_1(\epsilon) g_2(E - \epsilon). \qquad (5.19)
\end{aligned}
$$

The final relation

$$g(E) = \int d\epsilon g_1(\epsilon) \, g_2(E - \epsilon) \qquad (5.20)$$

shows that the density of state for the full system can be expressed as the sum over different configurations, each of which is parameterised by the energy ϵ of the subsystem. The volume of the constant-energy surface occupied by a configuration in which the subsystem has energy in the range $(\epsilon, \epsilon + d\epsilon)$ is proportional to $dV \propto g_1(\epsilon) g_2(E - \epsilon) \, d\epsilon$. From our preceding analysis, we know that the probability of the subsystem's having energy ϵ is proportional to dV, so

$$\mathcal{P}(\epsilon) \propto g_1(\epsilon) g_2(E - \epsilon), \qquad (5.21)$$

which is the required answer.

As an application of this result, let us consider the following case: An ideal gas of N particles is located in a volume V and has total energy E. We want to compute the probability that there exist k particles with a total energy ϵ, among this original collection of N particles. This is given by

$$P = g(\epsilon)g(E - \epsilon; N - k) \propto (E - \epsilon)^{(3/2)(N-k)}g(\epsilon), \qquad (5.22)$$

where we have used Eqs. (5.15) and retained only the terms that depend on ϵ. This expression is exact. However, most often, we will be interested in the case of a *small* subsystem for which $k \ll N$ and $\epsilon \ll E$. Using the first condition, we get

$$P(\epsilon)\, d\epsilon \propto \left(1 - \frac{\epsilon}{E}\right)^{(3/2)N} g(\epsilon)\, d\epsilon. \qquad (5.23)$$

Taking the second limit $\epsilon \ll E$ requires some care. Normally an expression like $(1 - \mu)^N$ can be expanded in a Taylor series as

$$Q = (1 - \mu)^N = 1 - N\mu + \frac{1}{2}N(N - 1)\mu^2 + \cdots + \qquad (5.24)$$

when $\mu \ll 1$. However, if $N \gg 1$ and $\mu \ll 1$, we could have a situation in which $N\mu \gtrsim 1$, thereby making the above expansion invalid or inaccurate. In such a case, it is best to take logarithms and write

$$\ln Q = N \ln(1 - \mu) \approx N\left(-\mu + \frac{1}{2}\mu^2 \cdots\right), \qquad (5.25)$$

which uses *only* the condition $\mu \ll 1$ and is valid for arbitrarily large $N\mu$. Then

$$Q = \exp\left[N\left(-\mu + \frac{1}{2}\mu^2 \cdots\right)\right] = \exp\left[-N\mu\left(1 - \frac{1}{2}\mu\right) \cdots\right] \approx \exp(-N\mu). \qquad (5.26)$$

By using this procedure in expression (5.23) we can write the probability as

$$P(\epsilon) \propto g(\epsilon)\exp[-(3N/2E)\epsilon]. \qquad (5.27)$$

The factor $3N/2E$ in the exponent depends on the properties of only the larger system of N particles. We write this quantity as

$$\frac{3N}{2E} \equiv \beta \equiv \frac{\partial}{\partial E}\ln g_N(E) \qquad (5.28)$$

so that β^{-1} is proportional to the mean energy of the system.

All these results carry over to the general case in which $\epsilon \ll E$. Because g's are expected to be steeply rising functions of their arguments [like E^M with $M \gg 1$; see Eqs. (5.15)] it is preferable to use the expansion in $\ln g$. Defining a quantity $S(E) \equiv \ln g(E)$, we can write the probability in expression (5.21) as

$$P(\epsilon) \propto \exp[S_1(\epsilon)]\exp[S_2(E - \epsilon)]. \qquad (5.29)$$

Now with the Taylor series,

$$S_2(E - \epsilon) \simeq S_2(E) - \epsilon \frac{\partial S_2}{\partial E} = S_2(E) - \beta\epsilon, \qquad (5.30)$$

with β defined by the relation

$$\beta \equiv \frac{\partial S_2}{\partial E} = \frac{\partial}{\partial E}[\ln g_2(E)], \qquad (5.31)$$

the probability becomes

$$\mathcal{P}(\epsilon) \propto g_1(\epsilon) \exp(-\beta\epsilon). \qquad (5.32)$$

We have reabsorbed the ϵ-independent factor, $\exp(S_2)$, into the proportionality constant. We can determine the net constant of proportionality in expression (5.32) by normalising the probability and we get

$$\mathcal{P}(\epsilon) \, d\epsilon = \frac{1}{Z} g(\epsilon) \, d\epsilon \, \exp(-\beta\epsilon) = \frac{dp \, dq}{Z} e^{-\beta H(p,q)}, \qquad (5.33)$$

where

$$Z(\beta) = \int_0^\infty g(\epsilon) \exp(-\beta\epsilon) \, d\epsilon = \int dp \, dq \, e^{-\beta H(p,q)} \qquad (5.34)$$

is called the partition function. Note that $Z(\beta)$ is the Laplace transform of the density of states.

Thus we find – quite generally – that the probability of a subsystem (of a larger system) having an energy ϵ is proportional to $\exp(-\beta\epsilon)$ per microstate; because the number of microstates in the energy interval $d\epsilon$ is $g(\epsilon) \, d\epsilon$, the net probability is given by expression (5.32). The only parameter of the full system that enters this probability is the quantity β. Hereafter, while discussing the subsystem, we omit the subscript 1 in g_1, etc. The following points should be stressed regarding the above derivation:

(1) The two systems are described by two *different* Hamiltonians H_1 and H_2 and hence can be quite different. All we have assumed is that there exists some mechanism for energy exchange between the systems.
(2) The most probable configuration for the two systems will be the one in which the integrand in expression (5.21) is maximised. Taking logarithms and setting the derivative with respect to ϵ zero, we get

$$0 = \frac{\partial S_1(\epsilon)}{\partial \epsilon} + \frac{\partial S_2(E - \epsilon)}{\partial \epsilon} = \frac{\partial S_1(\epsilon)}{\partial \epsilon} - \frac{\partial S_2(E - \epsilon)}{\partial (E - \epsilon)} = \beta_1 - \beta_2. \quad (5.35)$$

Thus the most probable configuration will have the β's of the two systems equal.
(3) We have assumed that the total energy of the system can be taken to be the sum of the energies of the components. This does not hold for self-gravitating

systems and hence our derivation of Eq. (5.33) fails for systems dominated by gravity. The statistical mechanics of gravitating systems is discussed in Chap. 10.

The above discussion can be easily generalised to a situation in which a system (with total energy E_t and total number N_t) contains a subsystem that is capable of exchanging not only the energy but also particles with the full system. From our general principles we know that the probability of the subsystem's having energy E and number of particles N is given by

$$P(N, E) \propto g_{E,N} g_{\text{tot}}(E_t - E, N_t - N) \propto g_{E,N} \exp\left[\beta(-E) + \frac{\partial S}{\partial N}(-N)\right]$$

$$\propto g_{E,N} \exp(-\beta E + \beta\mu N), \tag{5.36}$$

where we have used the relation $g = \exp S$ and defined μ by the equation $\beta\mu \equiv -(\partial S/\partial N)$. The normalisation can be fixed by the usual procedure

$$\sum_N \int dp\, dq\, P[N, E(p, q)] = 1, \tag{5.37}$$

giving

$$P(N, E) = \mathcal{G}^{-1} g_{E,N} \exp -\beta(E - \mu N) \tag{5.38}$$

with

$$\mathcal{G}(\beta, \mu) = \sum_N e^{\beta\mu N} \int dp\, dq\, e^{-\beta E} = \sum_{N=0}^{\infty} e^{\beta\mu N} Z(N; \beta) \tag{5.39}$$

where $Z(N; \beta)$ is the partition function for the system with N particles. $\mathcal{G}$ is called the grand partition function of the system.

The probability distributions in Eqs. (5.7), (5.33) and (5.38) are called microcanonical distribution, canonical distribution, and grand canonical distribution, respectively. It is obvious from the derivations that they correspond to systems with fixed total energy, a subsystem that can exchange energy with the full system, and a subsystem that can exchange energy and particles with the full system, respectively.

Exercise 5.3

Maxwellian velocity distribution: One simple application of canonical ensemble is to study the velocity or the momentum distribution for molecules of a gas. When a single particle is treated as an extreme limit of a subsystem that can exchange energy with the rest of the system, it follows that the number of molecules having momenta in the interval $(\mathbf{p}, \mathbf{p} + d^3\mathbf{p})$ will be proportional to $\exp[-\beta\epsilon(p)]d^3p$, where the proportionality constant can be determined by the normalisation condition. (1) Obtain the velocity distribution for a system of nonrelativistic particles (called the Maxwell distribution) with $\epsilon = (p^2/2m)$ and $v = (p/m)$. What happens if the particles are relativistic? (2) Consider a system of nonrelativistic particles that obey the above distribution. Show that the distribution of *relative* velocities of the molecules also obey the Maxwellian distribution.

Exercise 5.4
A paradox: Consider a large, closed system with a fixed value of energy E. Divide this into K small subsystems. Each of these systems individually can be made to satisfy the necessary condition for the validity of canonical distribution. Therefore the probability that the ith system has energy E_i is given by $P_i \propto \exp(-\beta E_i)$. The probability of the subsystems' having the energies $E_1, E_2, \ldots, E_i, \ldots, E_K$ simultaneously will be

$$ P \propto \prod P_i \propto \exp\left(-\beta \sum E_i \right) \propto \exp(-\beta E), \tag{5.40} $$

suggesting that the full system can also show fluctuations in energy! What is the fallacy in the above argument?

5.4 Mean Values in Canonical Distribution

We have seen that the probability of a system $\mathcal{S}$, interacting with much larger system B, having an energy ϵ is given by Eqs. (5.33) and (5.34). Let us next compute the mean values of various physical parameters relevant to $\mathcal{S}$.

We begin with the mean value of energy. Because $g(\epsilon)$ is a rapidly growing function and $e^{-\beta\epsilon}$ is a rapidly decreasing function, the product will have a sharp maximum at some $\epsilon = \bar{\epsilon}$. Therefore, $\mathcal{P}(\epsilon)$ in Eq. (5.33) can be approximated as

$$ \mathcal{P}(\epsilon) = \frac{1}{Z} \exp(\ln g - \beta\epsilon) \equiv \frac{1}{Z} \exp f(\epsilon) $$

$$ = \frac{1}{Z} \exp\left[f(\bar{\epsilon}) + \frac{1}{2}(\epsilon - \bar{\epsilon})^2 f''(\bar{\epsilon}) + \cdots \right], \tag{5.41} $$

where $\bar{\epsilon}$ is determined by the condition

$$ \frac{\partial}{\partial\epsilon} \ln g(\epsilon)|_{\epsilon=\bar{\epsilon}} = \beta. \tag{5.42} $$

This condition is just the equality of β's for the subsystem and the full system. Further, $f''(\bar{\epsilon})$ can be written as

$$ f'' = \frac{d}{d\epsilon}\left(\frac{d\ln g}{d\epsilon} \right)\bigg|_{\epsilon=\bar{\epsilon}} = \left(\frac{d\beta}{d\epsilon} \right)\bigg|_{\bar{\epsilon}} = \left(\frac{d\epsilon}{d\beta} \right)^{-1}\bigg|_{\bar{\epsilon}} \equiv -\frac{\beta^2}{C_V}, \tag{5.43} $$

where C_V is defined as $C_V \equiv -\beta^2(d\bar{\epsilon}/d\beta)$. [The notation will become meaningful later when we identify $T = \beta^{-1}$ with temperature, making $C_V = (d\bar{\epsilon}/dT)$ the specific heat.] Thus

$$ \mathcal{P}(\epsilon) = \mathcal{P}(\bar{\epsilon}) \exp\left[-\frac{\beta^2}{2C_V}(\epsilon - \bar{\epsilon})^2 \right], \tag{5.44} $$

with $\mathcal{P}(\bar{\epsilon}) = Z^{-1} \exp[\ln g(\bar{\epsilon}) - \beta\bar{\epsilon}]$, showing that the energy in canonical distribution is sharply peaked at a value $\bar{\epsilon}$ with the fluctuation $(\Delta\epsilon)^2 = C_V\beta^{-2}$.

For normal systems $\bar{\epsilon} \propto N$, $C_V \propto N$, and β will be independent of N. Hence the relative fluctuations in energy $(\Delta\epsilon/\bar{\epsilon}) \propto (\sqrt{N}/N) \propto 1/(\sqrt{N})$. So, for sufficiently large systems, there is very little fluctuation in the energy if the system is in canonical distribution. On the other hand, in the microcanonical distribution, the energy fluctuation is strictly zero. This suggests that for systems with sufficiently large N, canonical and microcanonical distributions will give approximately the same results as far as calculations of mean values are concerned.

The above calculation shows that our formalism not only determines the mean values of observables but also provides information about fluctuations around the mean values. This is an important feature of statistical mechanics to which we shall return later in the next section.

The quantity $\bar{\epsilon}$ described above is the *most probable* value, which we obtain by maximising the probability $\mathcal{P}(\epsilon)$. The *mean* value is defined by the relation

$$\langle\epsilon\rangle = \frac{1}{Z}\int_0^\infty d\epsilon\,\epsilon\,g(\epsilon)\,e^{-\beta\epsilon} = -\frac{\partial\,\ln Z}{\partial\beta}. \tag{5.45}$$

Because the fluctuations around $\bar{\epsilon}$ are small (for large N), the most probable energy and mean energy will be approximately equal. To prove this, we first evaluate Z. Because $ge^{-\beta\epsilon}$ is a sharply peaked function, the integrand for Z can be approximated as a Gaussian to give

$$Z = \int_0^\infty g(\epsilon)e^{-\beta\epsilon}\,d\epsilon \simeq g(\bar{\epsilon})e^{-\beta\bar{\epsilon}}(\beta^{-2}C_V)^{1/2}. \tag{5.46}$$

Therefore

$$\ln Z \cong \ln g(\bar{\epsilon}) - \beta\bar{\epsilon} + \frac{1}{2}\ln(\beta^{-2}C_V). \tag{5.47}$$

For normal systems $\bar{\epsilon} \propto N$, $C_V \propto N$, $\ln g \propto N$, and the last term scales as $\ln N$; hence

$$\ln Z \cong \ln g(\bar{\epsilon}) - \beta\bar{\epsilon}, \tag{5.48}$$

accurate to $\mathcal{O}(\ln N/N)$. To the same accuracy, we get

$$\langle\epsilon\rangle \equiv \frac{\partial\,\ln Z}{\partial(-\beta)} = -\frac{\partial\,\ln g}{\partial\bar{\epsilon}}\frac{\partial\bar{\epsilon}}{\partial\beta} + \beta\frac{\partial\bar{\epsilon}}{\partial\beta} + \bar{\epsilon} = -\beta\frac{\partial\bar{\epsilon}}{\partial\beta} + \beta\frac{\partial\bar{\epsilon}}{\partial\beta} + \bar{\epsilon} = \bar{\epsilon}, \tag{5.49}$$

where we have used Eqs. (5.42). Hereafter we use $\bar{\epsilon}$ and $\langle\epsilon\rangle$ interchangeably.

Finally, let us consider the mean values of a special class of variables, which can be expressed in the form

$$F_i = \frac{\partial H(p, q; \lambda_i)}{\partial\lambda_i}, \tag{5.50}$$

where λ_i are some parameters that appear in the Hamiltonian. The mean value of any such quantity is

$$\langle F_i \rangle = \int dp\, dq \left(\frac{\partial H}{\partial \lambda_i} \right) \frac{e^{-\beta H}}{Z} = \frac{1}{\beta} \frac{\partial \ln Z}{\partial (-\lambda_i)}. \tag{5.51}$$

For example, the pressure exerted by a gas can be expressed as $P = -(\partial H/\partial V)$. To prove this, note that the force exerted on the walls of a container parameterised by coordinates $\mathbf{r}$ is

$$\mathbf{F} = -\frac{\partial H}{\partial \mathbf{r}} = -\frac{\partial H}{\partial V} \frac{\partial V}{d\mathbf{r}} = -\frac{\partial H}{\partial V} \cdot d\mathbf{s} = P\,d\mathbf{s}, \tag{5.52}$$

as $dV = d\mathbf{s} \cdot d\mathbf{r}$ and $\mathbf{F} = P\,d\mathbf{s}$. Therefore the mean pressure exerted by the system will be

$$\bar{P} = \frac{1}{\beta} \frac{\partial \ln Z}{\partial V}. \tag{5.53}$$

With these preliminaries, we are now in a position to establish the connection between statistical mechanics and classical thermodynamics.

Exercise 5.5

Fluctuations around the equilibrium: The probability $w(x_i)$ for a set of macroscopic quantities x_i, where $i = 1, \ldots, n$, to have values in the interval $(x_1, x_1 + dx_1, \ldots, x_n + dx_n)$ is proportional to $\exp[S(x_i)]$, where $S = \ln g$ is treated as a function of the variables $(x_1, \ldots, x_n)$. This principle allows us to evaluate the fluctuations around the equilibrium, from the knowledge of equilibrium configuration. For small deviations from the equilibrium, show that

$$w = \frac{\sqrt{\beta}}{(2\pi)^{n/2}} \exp\left[-\frac{1}{2} \sum_{i,k=1}^{n} \beta_{ik}(x_i - \bar{x}_i)(x_k - \bar{x}_k) \right]; \quad \beta_{ik} \equiv \frac{\partial^2 S}{\partial x^i \partial x^k}, \tag{5.54}$$

where $\bar{x}_i$ is the mean value of the variable x_i at equilibrium and $\beta = \det \beta_{ik}$. Use this to compute the root-mean-square fluctuation in the volume of a subsystem in thermal equilibrium.

5.5 Derivation of Classical Thermodynamics

Using the expressions for the mean energy $\langle \epsilon \rangle \equiv \bar{E} = -(\partial \ln Z/\partial \beta)$ and mean pressure $\bar{P} = (1/\beta)(\partial \ln Z/\partial V)$, we can express the differential $d \ln Z$ as

$$d(\ln Z) = -\bar{E}d\beta + \beta\bar{P}dV = -d(\bar{E}\beta) + \beta d\bar{E} + \beta\bar{P}dV. \tag{5.55}$$

With relation (5.48), this becomes

$$d(\ln Z + \bar{E}\beta) \cong d[\ln g(\bar{E})] = \beta(d\bar{E} + \bar{P}dV) \tag{5.56}$$

or

$$\beta^{-1} d \ln g = d\bar{E} + \bar{P} dV, \tag{5.57}$$

which is accurate to $\mathcal{O}(\ln N/N)$. This relation, *obtained purely from microscopic considerations*, is quite significant. In a process that changes the volume of the system by dV, the system does a mechanical work of amount $\bar{P}dV$ and its energy changes by an amount $d\bar{E}$. Naively, we might have expected the sum of these two terms to vanish. Equation (5.57) shows that – for macroscopic systems with large number of particles – this is *not* true. The sum cannot even be expressed as an exact differential of a physical quantity but is of the form $\beta^{-1} d(\ln g)$.

Historically this fact had been noted even before the development of statistical mechanics and the quantity $\beta^{-1} d \ln g$ used to be written as $\mathcal{T} d\mathcal{S}$, where $\mathcal{T}$ was called the temperature and $\mathcal{S}$ was called the entropy. The combination $\mathcal{T} d\mathcal{S}$ was called the change in the heat content of the system. The above derivation allows us to write $\beta^{-1} d(\ln g) = T d(\ln g) = T dS$ and make the identifications

$$\mathcal{T} = T = \left(\frac{\partial \ln g}{\partial E} \right)_V^{-1}, \quad d\mathcal{S} = dS = d(\ln g) = \frac{dg}{g}. \tag{5.58}$$

The second relation gives the entropy in terms of the density of states:

$$S = \ln g(E) + q(N), \tag{5.59}$$

where $q(N)$ is some function independent of E and V and possibly dependent on the number of particles N. By definition, $g(E)$ is not dimensionless; hence the logarithm should be interpreted as $\ln[g(E)/g_0]$, where g_0 is a constant with the dimensions of g. (We shall see later in Section 5.7 that quantum theory provides a value for g_0 and allows us to define the entropy as a logarithm of the number of microstates available to the system). Also note that, for systems such as ideal gases with $N \gg 1$, we have $\ln g \approx \ln \Gamma$ within an accuracy of $\mathcal{O}(\ln N/N)$. Hence entropy may be defined as $\ln \Gamma$ within the same degree of accuracy.

The identifications of entropy and temperature, of course, allow a trivial change of units between (say) T and $\mathcal{T}$. We could have put $T = k_B \mathcal{T}$ and set $S = k_B \ln g$ for some constant k_B. This scale change is of no fundamental importance and we will often set $k_B = 1$, measuring the thermodynamic temperature in energy units. In terms of the historical unit for temperature, viz., Kelvin, the constant k_B has the units of ergs per Kelvin and is given by $k_B \approx 1.4 \times 10^{-16}$ ergs $\mathrm{K}^{-1} \approx 1.4 \times 10^{-23}$ J K^{-1}. With this understanding, we identify the thermodynamic entropy $(\mathcal{S})$ and temperature $(\mathcal{T})$ with the corresponding quantities S and T defined by means of statistical mechanics and use the latter symbols.

The state that maximises the probability (viz., the one that maximises $S = \ln g$) is the one in which the pressure and the temperature of the subsystem are the same as those of the full system. In general, if S depends on any quantity λ that is *additive* for the subsystems (such as energy, volume, etc.), then the most probable state will have the same value for $(\partial S/\partial \lambda)$ for the subsystem and the full system. The proof is obvious from the corresponding result for $(\partial S/\partial E)$ given in Eq. (5.35)

Let us examine what the identifications in Eqs. (5.58) imply for an ideal gas for which the density of states is given by [see Eqs. (5.15)]

$$g(E) = \frac{3N}{2} \frac{\pi^{\frac{3N}{2}}}{\left(\frac{3N}{2}\right)\left(\frac{3N}{2}\right)!}(2m)V^N(2mE)^{\frac{3N}{2}}. \qquad (5.60)$$

Then

$$S(E, V, N) = \ln g = N \ln V + \frac{3N}{2} \ln \left(\frac{2E}{3N}\right) + N \ln C + q(N), \quad (5.61)$$

where C is a constant independent of E, V, and N, and

$$\bar{P} = \frac{1}{\beta}\left(\frac{\partial S}{\partial V}\right)_{E,N} = \frac{N}{\beta V} = \frac{NT}{V}, \quad \frac{1}{T} = \left(\frac{\partial S}{\partial E}\right)_{V,N} = \frac{3N}{2E}, \quad (5.62)$$

which are the usual ideal-gas equations: $\bar{E} = (3/2)NT$ and $\bar{P}V = NT$, giving $\bar{P} = (2/3)(E/V)$. In the relativistic case, with $g \propto V^N E^{3N}$, we get $\bar{P} = (1/3)$ (E/V). (This result was obtained earlier in Chap. 3, Section 3.6.) Thus the identification of thermodynamic variables makes sense, at least in this simple case.

To determine the form of $q(N)$ in Eq. (5.61), we can proceed as follows. We first note that when we divide the system into subsystems, the $g's$ multiply or – equivalently – the entropies add. (Quantities that have this property are called extensive.) The extensivity of entropy implies that we must have the relation

$$S(\lambda E, \lambda V, \lambda N) = \lambda S(E, V, N). \qquad (5.63)$$

Using this fact in Eq. (5.61), we see that q has to satisfy the condition

$$\frac{q(\lambda N)}{\lambda} - q(N) = -N \ln \lambda. \qquad (5.64)$$

To solve this equation, we differentiate this relation with respect to N and denote the derivative of $q(N)$ with respect to its argument by the function $F(N)$; that is, $F(N) = q'(N)$. Then F satisfies the equation $F(\lambda N) - F(N) = -\ln \lambda$. Differentiating this relation again with respect to λ and denoting the derivative of F with respect to its arguments by F', we get $NF'(\lambda N) = -\lambda^{-1}$ or $F'(x) = -x^{-1}$. Integrating this relation, we get $F(x) = -\ln x + k$, where k is some constant. Integrating once again, we can determine the function q to be

$$q(x) = -x \ln x + x + kx \simeq -\ln x! + kx \qquad (5.65)$$

within the accuracy of $(\ln N/N)$ at which we are working. We could have added another integration constant to q, which, however, would not have changed any of our results. This shows that the $q(N)$, which appears in the expression for entropy, has the form

$$q(N) = -\ln N! + kN. \tag{5.66}$$

The term kN can be combined with the term $N \ln C$ in Eq. (5.61) and hence is not relevant; the only extra bit is the term $(-\ln N!)$.

Adding this factor to S is equivalent to changing our original definition of g by $g_{old} \to g_{new} = (g_{old}/N!)$. In other words, configurations that differ by the interchange of only two particles in the system are treated as indistinguishable in g_{new} whereas they are considered distinct in g_{old}. We hereafter assume that the expression for g, etc., are modified by the $N!$ factor.

The contribution proportional to N cannot be fixed by the description developed so far. The definition of g gives it the dimension A^{3N}, where A is some constant with the dimension of action. This will contribute to S a term $3N \ln A$ that cannot be determined until we know A. This issue will be settled later in Section 5.7 when we produce a dimensionless $g(E)$ from quantum theory.

Finally, we mention an expression for the entropy of a system in terms of the probability distribution $\mathcal{P}(E)$ in the canonical distribution. The relation is

$$S = \int_0^\infty dE \, \mathcal{P}(E) \ln \mathcal{P}(E). \tag{5.67}$$

We easily verify this by substituting $\mathcal{P}(E) = Z^{-1} \exp(-\beta E)$ into Eq. (5.67) and by using the definition $S = \ln g$ and relation (5.48).

Exercise 5.6
Virial and equipartition theorems for interacting systems: It is possible to obtain two general results in equilibrium thermodynamics by using the formalism we have developed so far. Consider a system described by a Hamiltonian $H(q, p)$, with q, p, running over $3N$ variables each. We denote the phase-space coordinates collectively as $q^A = (q^i, p^j)$, with $A = 1, 2, \ldots, 6N$. (1) Show that the mean value of the observable $q^A(\partial H/\partial q^B)$ is given by $\delta_{AB}(k_B T)$, where q^A may refer to a coordinate or momentum. (2) Consider a Hamiltonian of the form

$$H(q, p) = \sum_j \left(A_j p_j^2 + B_j q_j^2 \right).$$

Use the above result to show that $\langle H \rangle = f(k_B T/2)$, where f is the number of degrees of freedom. This result (equipartition theorem) shows that each degree of freedom contributes $(k_B T/2)$ to the mean energy. (3) When q^A refers to the coordinates of particles of an ideal gas, show that

$$\mathcal{V} \equiv \sum_i q_i \dot{p}_i = -3Nk_B T,$$

where $\mathcal{V}$ is called virial. Show that for an ideal gas $\mathcal{V} = -3PV$. (4) Consider a system of particles interacting by means of a two-body potential $u(\mathbf{r}_j - \mathbf{r}_i)$. Let $g(r)$ denote the probability of finding a pair of particles separated by a distance r. Show that, in this case, the above results are generalised to

$$PV = Nk_BT\left[1 - \frac{2\pi N}{3Vk_BT}\int_0^\infty \frac{\partial u(r)}{\partial r} g(r)r^3\,dr\right],\qquad(5.68)$$

$$E = \frac{3}{2}Nk_BT + \frac{1}{2}N^2\int\int u(r)g(r)\frac{d\mathbf{r}_1\,d\mathbf{r}_2}{V^2}$$

$$= \frac{3}{2}Nk_BT\left[1 + \frac{4\pi N}{3Vk_BT}\int_0^\infty u(r)\,g(r)r^2\,dr\right].\qquad(5.69)$$

5.6 Description of Macroscopic Thermodynamics

Given the microscopic Hamiltonian of the system, statistical mechanics allows us to construct the function $S(E, V, N)$ from which all the macroscopic variables of the system can be determined. We now summarise several useful, macroscopic, thermodynamic quantities and explore their interrelationships.

The function $S(E, V, N)$ can be inverted to give the energy in terms of other variables: $E = E(N, V, S) = Nf_1(V/N, S/N)$, where the second relation follows from the extensivity of energy, volume, and entropy. Hence the energy density $\rho \equiv (E/V)$ can be expressed purely as a function of the number density $n \equiv (N/V)$ and entropy per particle $s \equiv (S/N)$. Such a functional form $\rho = \rho(n, s)$ clearly shows that there are really only two independent variables in the thermodynamic description. To see what this implies, let us take the differential of $E(N, V, S)$ to get

$$dE = \frac{\partial E}{\partial S}dS + \frac{\partial E}{\partial V}dV + \frac{\partial E}{\partial N}dN \equiv TdS - PdV + \mu dN,\qquad(5.70)$$

where we have reintroduced the quantity $\mu \equiv (\partial E/\partial N)_{S,V}$. It is clear from Eq. (5.70) that $(\partial S/\partial N)_{E,V} = -\beta\mu$; this is the definition used in Section 5.3 in expression (5.36) (μ is called the chemical potential). Because there are only two truly independent variables, all three partial derivatives cannot be independent. We can find a relation among them by writing the differential of ρ with respect to n and s and substituting for ρ, n, and s in terms of E, N, V, S. This gives

$$dE = \left[\frac{E}{V} - \frac{N}{V}\left(\frac{\partial\rho}{\partial n}\right)_s\right]dV + \frac{V}{N}\left(\frac{\partial\rho}{\partial s}\right)_n dS + \left[\left(\frac{\partial\rho}{\partial n}\right)_s - \left(\frac{\partial\rho}{\partial s}\right)_n\frac{VS}{N^2}\right]dN.$$
$$(5.71)$$

Comparing Eq. (5.71) with Eq. (5.70), we identify the variables as

$$T = \frac{V}{N}\left(\frac{\partial\rho}{\partial s}\right)_n,\quad P = \frac{N}{V}\left(\frac{\partial\rho}{\partial n}\right)_s - \frac{E}{V},\quad \mu = \left(\frac{\partial\rho}{\partial n}\right)_s - \left(\frac{\partial\rho}{\partial s}\right)_n\frac{VS}{N^2},$$
$$(5.72)$$

which allows the chemical potential to be expressed in terms of other variables as

$$\mu = \frac{P + \rho}{n} - Ts = \frac{1}{N}(E + PV - TS). \tag{5.73}$$

Thus the chemical potential need not be treated as an independent thermodynamic entity. Note that this is a direct consequence of the extensivity of energy.

It is possible, of course, to use other variables in place of N, V, and S to describe the thermodynamics of the system. To begin with, let us retain the N dependence but change the variables from (V, S) to (V, T), (P, S), or (P, T). These transformations can be implemented through standard rules of differentiation. The change of the independent variables from (S, V) to (T, V), for example, can be effected by writing

$$dE = d(TS) - S\, dT - P\, dV, \quad d(E - TS) \equiv dF = -S\, dT - P\, dV + \mu\, dN, \tag{5.74}$$

where we have defined the free energy $F = E - TS = F(N, V, T)$. From the differentials, we find that $P = -(\partial F/\partial V)$ and $S = -(\partial F/\partial T)$. Because of extensivity, we can express the density of free energy $a \equiv (F/V)$ in terms of n and T as $a = a(n, T)$. Similarly, we can define two other thermodynamic potentials called enthalpy, $H(N, P, S) \equiv E + PV$, and Gibbs free energy, $G(N, P, T) \equiv E + PV - TS$, when the independent variables are (P, S) or (P, T) respectively. These thermodynamic potentials and their properties are given in Table 5.1, The various properties listed there are briefly summarised in the following paragraphs.

The first column of the table gives the relevant thermodynamic potential, the independent variables, and the corresponding energy density. If the independent variables are N, V, S, we use energy E; for N, V, T we use free energy F; for N, P, S, we use enthalpy H; for N, P, T we use G. We obtain the corresponding energy densities by dividing the potential by V; these are denoted by $\rho(n, s)$, $a(n, T)$, $w(P, s)$, and $g(P, T)$. Throughout the table, partial derivatives are evaluated with the other independent variables kept constant.

The second column gives the definition of various potentials and energy densities in terms of the fundamental variables. The first three entries (for E, F, and G) are straightforward. The definition of G shows that it is actually equal to μN because of Eq. (5.73). The w and g give enthalpy and Gibbs free energy per particle, respectively. The enthalpy and Gibbs free energy per unit volume are obtained by multiplying by n.

The third column shows how other thermodynamic variables can be obtained by differentiating the potential with respect to the independent variables. All the potentials E, F, G, H, when differentiated with respect to N, give μ; so this relation is indicated in only the first row. Also note that μ can be expressed in terms of other variables by use of Eq. (5.73). This relation – and

Table 5.1. Summary of thermodynamics

Thermodynamic Potential and Variables	Definition of the Potential	Secondary Thermodynamic Variables	Thermodynamic Identities
Energy (E) $E = E(N, V, S)$ $\rho = E/V, n = N/V$ $s \equiv S/V$	$E(N, V, S)$ $\rho = \rho(n, s)$	$P = -(\partial E/\partial V)$, $T = (\partial E/\partial S)$ $\mu = (E + PV - TS)/N = (\partial E/\partial N)$ $P(n, s) = n(\partial \rho/\partial n) - \rho$ $T(n, s) = n^{-1}(\partial \rho/\partial s)$ $\mu = n^{-1}(P + \rho) - Ts$	$\left(\dfrac{\partial T}{\partial V}\right)_s = -\left(\dfrac{\partial P}{\partial S}\right)_v$
Free energy (F) $F = F(N, V, T)$ $a \equiv (F/V)$	$F \equiv E - TS = F(N, V, T)$ $a(n, T) \equiv (\rho/n) - Ts$	$P = -(\partial F/\partial V)$, $S = (\partial F/\partial T)$ $P(n, T) = n^2(\partial a/\partial n)$ $s(n, T) = -(\partial a/\partial T)$ $\rho(n, T) = -nT^2[(\partial(a/T)/\partial T)]$	$\left(\dfrac{\partial S}{\partial V}\right)_T = \left(\dfrac{\partial P}{\partial T}\right)_v$
Enthalpy (H) (Heat function) $H = H(N, P, S)$ $w \equiv (H/V)$	$H \equiv E + PV = H(N, P, S)$ $w(P, s) \equiv n^{-1}(P + p)$	$V = (\partial H/\partial P)$, $T = (\partial H/\partial S)$ $n^{-1}(P, s) = (\partial w/\partial P)$; $T(P, s) = (\partial w/\partial s)$ $\rho(P, s) = [w/(\partial w/\partial P)] - P$	$\left(\dfrac{\partial V}{\partial S}\right)_P = \left(\dfrac{\partial T}{\partial P}\right)_s$
Gibbs free energy (G) $G = G(N, P, T)$ $g \equiv (G/V)$	$G \equiv E + PV - TS$ $= G(N, P, T) = \mu N$ $g(P, T) \equiv n^{-1}(P + \rho) - Ts$	$V = -(\partial G/\partial P)$, $S = (\partial G/\partial T)$ $n^{-1}(P, T) = (\partial g/\partial P)$ $s(P, T) = -(\partial g/\partial T)$ $\rho(P, T) = [(\partial g/\partial P)]^{-1}$ $[g - T(\partial g/\partial T)] - P$	$\left(\dfrac{\partial V}{\partial T}\right)_P = -\left(\dfrac{\partial S}{\partial P}\right)_T$

its equivalent in terms of energy densities – is also given in the first row and not repeated in other rows. All these relations can be derived directly from the definitions.

The last column gives some useful differential identities between the thermodynamic variables. The first one, for example, follows from using the relations $P = -(\partial E/\partial V)$ and $T = (\partial E/\partial S)$ and the identity $(\partial^2 E/\partial S \partial V) = (\partial^2 E/\partial V \partial S)$. All other relations can be obtained in a similar manner.

Among the potentials defined so far, the free energy F is closely related to the partition function by

$$\ln Z = S - \beta E = -\beta(E - TS) = -\beta F, \tag{5.75}$$

giving

$$F = -T \ln Z = -T \ln \left(\int dp\, dq\, e^{-\beta H} \right). \tag{5.76}$$

This relation allows us to write the basic probability distribution as $P(\epsilon) = Z^{-1} e^{-\beta\epsilon} = e^{\beta(F-E)}$.

In all the potentials defined in Table 5.1, N has been retained as a independent variable. It is possible to trade off the N dependence for μ dependence and define another quantity, called the thermodynamic potential, $\Phi(\mu, V, T) \equiv E - TS - \mu N$. From Eq. (5.73), it follows that $\Phi = -PV$. We can obtain all other thermodynamic variables from Φ by taking suitable derivatives:

$$P = -(\partial\Phi/\partial V), \quad S = -(\partial\Phi/\partial T), \quad N = -(\partial\Phi/\partial\mu). \tag{5.77}$$

Just as the free energy was related to the partition function by Eq. (5.76), this potential Φ can be related to the grand partition function in Eq. (5.39). To show this connection, we carry out an approximate evaluation of $\mathcal{G}$ in the same manner as we did for Z in the thermodynamic limit [see relation (5.48)] to obtain

$$\ln \mathcal{G} \cong \beta\mu\bar{N} + \ln Z(\bar{N}; \beta) = -\beta(F - \mu\bar{N}) \equiv -\beta\Phi. \tag{5.78}$$

The original probability can now be written as

$$\mathcal{P}(N, E) = \mathcal{G}^{-1} \exp \beta(\mu N - E) = \exp \beta(\Phi + \mu N - E). \tag{5.79}$$

Two more macroscopic variables that are heavily used in thermodynamics are the specific heats C_V and C_P at constant volume and pressure:

$$C_V \equiv \left(\frac{\partial E}{\partial T} \right)_V = T \left(\frac{\partial S}{\partial T} \right)_V = C_V(T, V),$$

$$C_P \equiv \left(\frac{\partial E}{\partial T} \right)_P = T \left(\frac{\partial S}{\partial T} \right)_P = C_P(T, P). \tag{5.80}$$

These specific heats satisfy a simple relation that can be derived from the identities in Table 5.1 (see Exercise 5.7):

$$C_V = C_P - VT\left(\frac{\alpha^2}{k}\right), \quad \alpha = \frac{1}{V}\left(\frac{\partial V}{\partial T}\right)_P, \quad k = -\frac{1}{V}\left(\frac{\partial V}{\partial P}\right)_T. \quad (5.81)$$

We conclude this section with a discussion of several dimensionless exponents extensively used in the thermodynamics of stellar interiors. The first two of these exponents, χ_ρ and χ_T, are defined through the equation of state written in the form

$$P = P_0\rho^{\chi_\rho}T^{\chi_T}, \quad (5.82)$$

so that

$$\chi_T = \left(\frac{\partial \ln P}{\partial \ln T}\right)_\rho, \quad \chi_\rho = \left(\frac{\partial \ln P}{\partial \ln \rho}\right)_T. \quad (5.83)$$

[Often, it is convenient to define the derivatives at constant ρ rather than at constant V; when N is fixed, $\rho \propto (1/V)$.] In this case, it is easy to show that

$$C_P - C_V = \frac{NP}{\rho T}\frac{\chi_T^2}{\chi_\rho}, \quad \frac{C_P}{C_V} = 1 + \frac{NP}{\rho T C_V}\frac{\chi_T^2}{\chi_\rho} \equiv \gamma. \quad (5.84)$$

The first relation is same as that of Eqs. (5.81), and the second relation also defines the symbol γ as the ratio of specific heats.

To define the next set of exponents we need to introduce the concept of an adiabatic process. In discussing the correspondence between classical thermodynamics and statistical mechanics, we saw that the change in the internal energy dE is contributed to by both the external work PdV and the internal change in the density of states TdS. Processes in which $dS = 0$ are called adiabatic processes. Because no internal change in the phase volume occurs, these processes are fully reversible. Using the conventional terminology, we may say that these processes do not involve any heat exchange with the surroundings. To study adiabatic processes, it is convenient to define logarithmic derivatives of physical variables evaluated at constant S. These adiabatic exponents Γ_1, Γ_2, and Γ_3 are defined by the relations

$$\Gamma_1 = \left(\frac{\partial \ln P}{\partial \ln \rho}\right)_S, \quad \frac{\Gamma_2}{\Gamma_2 - 1} = \left(\frac{\partial \ln P}{\partial \ln T}\right)_S \equiv \frac{1}{\nabla_{\text{ad}}}, \quad \Gamma_3 - 1 = \left(\frac{\partial \ln T}{\partial \ln \rho}\right)_S, \quad (5.85)$$

where we have also defined a quantity ∇_{ad} in terms of Γ_2. The Γ's are not independent; it is easily verified that

$$\frac{\Gamma_3 - 1}{\Gamma_1} = \frac{\Gamma_2 - 1}{\Gamma_2} = \nabla_{\text{ad}}, \quad (5.86)$$

so that knowledge of Γ_1 and Γ_2 allows us to determine Γ_3. Each of these Γ's contain useful information about the thermodynamics of the system, especially when it is not an ideal gas. Γ_3 describes the response of the heat content of the gas to compression; Γ_1 has relevance to the dynamical features of the gas such as the speed of sound (see Chap. 8). Finally, Γ_2 plays a crucial role in deciding whether convection takes place or not (see Chap. 8).

Given the equation of state of the form in Eq. (5.82), we can compute all the adiabatic exponents by evaluating the derivatives at constant S. This calculation is straightforward (see Exercise 5.7) and leads to the following results. We find that

$$\Gamma_3 - 1 = \frac{NP}{\rho T} \frac{\chi_T}{C_V} = \frac{N}{\rho}\left(\frac{\partial P}{\partial E}\right)_\rho. \tag{5.87}$$

This can be integrated to give the pressure as $P = (\Gamma_3 - 1)\rho E$. The other adiabatic exponents are

$$\frac{\Gamma_2}{\Gamma_2 - 1} = C_P\left(\frac{\rho T}{NP}\right)\frac{\chi_\rho}{\chi_T} = \frac{\chi_\rho}{\Gamma_3 - 1} + \frac{\chi_T}{N}, \tag{5.88}$$

$$\Gamma_1 = \frac{\chi_T}{N}(\Gamma_3 - 1) + \chi_\rho = \frac{\chi_\rho}{1 - (\chi_T \nabla_{\text{ad}}/N)}. \tag{5.89}$$

Finally, the ratio of specific heats is related to the Γ's by

$$\gamma = \frac{\Gamma_1}{\chi_\rho} = 1 + \frac{\chi_T}{N\chi_\rho}(\Gamma_3 - 1). \tag{5.90}$$

From the definition of Γ_1, it follows that $P \propto \rho^{\Gamma_1}$; when $\chi_\rho = 1$ this relation could also be written as $P \propto \rho^\gamma$.

Many of these expressions simplify for an ideal gas, which is summarised for future reference. In astrophysical contexts, we write the equation for ideal gas in several equivalent forms:

$$P = \frac{N}{V}k_B T = \left(\frac{Nm_H}{V\rho}\right)\left(\frac{k_B}{m_H}\right)\rho T = \frac{\mathcal{R}}{\mu}(\rho T),$$

$$\mu \equiv \frac{V\rho}{Nm_H}, \quad \mathcal{R} \equiv \frac{k_B}{m_H} \equiv N_A k_B, \tag{5.91}$$

where we have defined the gas constant $\mathcal{R} \equiv (k_B/m_H) \equiv N_A k_B$, Avagadro's number $N_A = m_H^{-1}$, and the mean molecular weight μ (not to be confused with the chemical potential). The quantity k_B is the Boltzmann constant, which allows the temperature to be measured in Kelvin. The inverse of the molecular weight gives the effective number of particles per hydrogen atom of the gas. Clearly the value of μ depends on the number of particles contributed by each atom, which, in turn, will depend on whether the atom is ionised or not. (Note that both free electrons and ions contribute to pressure whereas the mass is contributed mainly by the ions.) Therefore

$$\frac{1}{\mu m_H} = \frac{\text{total number of particles}}{\text{total mass of gas}}$$

$$= \sum_j \left(\frac{\text{number of particles of species } j}{\text{mass of particles of species } j} \times \frac{\text{mass of particles of species } j}{\text{total mass}} \right)$$

$$\equiv \sum_j \left(\frac{N_j(1 + Z_j)}{N_j A_j m_H} \times X_j \right) = \frac{1}{m_H} \sum_j \left(\frac{1 + Z_j}{A_j} \right) X_j, \qquad (5.92)$$

where A_j is the atomic weight of the jth species of the particles, X_j is the fraction of the total mass contributed by the jth species, and Z_j is the effective number of electrons supplied by each atom of the jth species. If the particular species is fully ionised, Z_j will be the atomic number of that species, whereas if the atom is neutral, Z_j will be zero for that species. Separating out the contribution from ions and electrons, we can formally write the molecular weight as

$$\frac{1}{\mu} = \sum_j \left(\frac{X_j}{A_J} \right) + \sum_j \left(\frac{Z_j}{A_j} \right) X_j = \frac{1}{\mu_{\text{ions}}} + \frac{1}{\mu_{\text{electrons}}}. \qquad (5.93)$$

We will have occasions to use this in Vol. II. Note that if we are interested in the pressure contributed by only the electrons (or ions), we can use Eqs. (5.91) with μ replaced with μ_{ele} (or μ_{ion}).

In several astrophysical contexts, the material may be thought of as a mixture of hydrogen, helium, and other heavier elements with mass fractions of approximately $0.75, 0.24$, and 0.01. In this case, $(Z_j/A_j) \simeq (1/2)$ for helium and heavier elements and $(Z_j/A_j) = 1$ for hydrogen. If the gas is fully ionised, then the electronic contribution to μ is given by

$$\frac{1}{\mu_{\text{ele}}} = X_H + \frac{1}{2}(X_{\text{He}} + X_{\text{others}}) = X_H + \frac{1}{2}(1 - X_H) = \frac{1}{2}(1 + X_H). \qquad (5.94)$$

The number density of electrons is

$$n_e = \left(\frac{\rho}{m_H} \right) \mu_e^{-1} \simeq \left(\frac{\rho}{2m_H} \right)(1 + X_H), \qquad (5.95)$$

which is a useful relation in estimating processes related to electrons.

The internal energy of the gas depends on whether the gas is monoatomic, diatomic, etc. When $P \propto \rho T$, we have $\chi_\rho = \chi_T = 1$. Substituting these values into Eqs. (5.84) we get $C_P - C_V = (k_B/\mu m_H)$ and $C_V = [k_B/\mu m_H(\gamma - 1)] = (N_A k_B/\mu)(\gamma - 1)$. Integrating the equation $C_V = (\partial E/\partial T)_V$ gives the total energy as $E = [k_B T/\mu m_H(\gamma - 1)]$. Using this result in $T(\partial S/\partial E)_V = 1$ and integrating, we obtain the entropy as $S = C_V \ln T + f(V)$. Finally, using the relation $T(\partial S/\partial V)_E = P$ and the equation of state, we find that $f(V) = C_V(\gamma - 1) \ln V$, giving the entropy of the ideal gas:

$$S = C_V \ln\left[TV^{(\gamma-1)}\right] \propto \ln\left[T^{1/(\gamma-1)}V\right] \propto \ln[PV^\gamma] \propto \ln\left[PT^{\gamma/(\gamma-1)}\right]. \tag{5.96}$$

It follows that for adiabatic processes, PV^γ and other combinations inside the brackets are constant. The relation $P \propto \rho^\gamma$ for adiabatic processes will be used extensively in later chapters.

Given the equation of state and the form of C_V, all other thermodynamic variables can be computed. The enthalpy, for example, is given by

$$H = E + PV = \frac{k_B T}{\mu m_H (\gamma - 1)} + \frac{k_B T}{\mu m_H} = \frac{\gamma}{\gamma - 1} \frac{P}{\rho}. \tag{5.97}$$

We also note that, for an ideal gas, $\gamma = \Gamma_1 = (1 - \nabla_{ad})^{-1}$, giving $\nabla_{ad} = (\gamma - 1)/\gamma$. For an ideal monotonic gas, it is also easy to verify that

$$\Gamma_1 = \Gamma_2 = \Gamma_3 = \gamma = \frac{5}{3}. \tag{5.98}$$

Exercise 5.7

For long winter evenings I: (1) Verify the entries in Table 5.1. (2) Prove Eqs. (5.80)–(5.84). (3) Prove Eqs. (5.87)–(5.90).

5.7 Quantum Statistical Mechanics

A classical particle, confined inside a cubic box of volume $V = L^3$, can have any value for the (nonrelativistic) kinetic energy $\epsilon = p^2/2m$, as p^2 can take any positive value. This is not the case in quantum mechanics. The energy levels of such a particle are labelled by three integers, n_x, n_y, and n_z, and are given by

$$\epsilon = \frac{\hbar^2}{2m} \frac{\pi^2}{L^2} (n_x^2 + n_y^2 + n_z^2), \tag{5.99}$$

that is, single-particle energy levels are quantised. Let us denote the allowed energy levels for a single particle as $\{\epsilon_1, \epsilon_2 \cdots \epsilon_j \cdots\}$ where the label j specifies energy level. In general, a particular energy level will correspond to different quantum states; such an energy level is called degenerate. In the above case, to specify a given quantum state, we need to give the triplet of integers; merely specifying the energy ϵ will not be enough.

A more profound effect of quantum mechanics arises from the following fact: To define the trajectory of a particle, we need to specify the position and the momentum simultaneously – which is impossible in quantum mechanics because of the uncertainty principle. As the concept of trajectory does not exist in quantum mechanics, it is not possible to track the evolution of any single particle in time and the particles have to be treated as strictly indistinguishable. Consider, for example, the wave function $\psi(x_1, x_2, x_3, \ldots, x_N)$ representing N particles. The state obtained by interchanging, say, the first two particles $\psi(x_2, x_1, \ldots, x_N)$, must lead to the same probability distribution as the original state, as the particles are indistinguishable. Therefore $|\psi(x_1, x_2, \ldots, x_N)|^2 = |\psi(x_2, x_1, \ldots, x_N)|^2$, requiring that $\psi(x_2, x_1, \ldots, x_N) = \psi(x_1, x_2, \ldots, x_N) \exp(i\delta)$. Interchanging the particles once again, we get $\exp(2i\delta) = 1$ or $\exp(i\delta) = \pm 1$. Thus we must have

$$\psi(x_1, x_2, \ldots, x_N) = \pm\psi(x_2, x_1, \ldots, x_N). \tag{5.100}$$

The wave function must be totally symmetric or antisymmetric under the exchange of *any* pair of particles, as the choice of 1 and 2 is arbitrary.

The particles that require an antisymmetric wave function are called fermions and particles that require a symmetric wave function are called bosons. Because $\psi(x_1, x_2, \ldots, x_N) = -\psi(x_2, x_1, \ldots, x_N)$ for fermions, it is clear that two fermions cannot be at the same location: $\psi(x, x, \ldots, x_N) = 0$. Similarly, two fermions cannot be in the same quantum state.

These differences are easily illustrated by a simple example. With two nondegenerate energy levels ϵ_1 and ϵ_2 and two particles, a state with energy $E = \epsilon_1 + \epsilon_2$ can be produced in two different ways in classical statistical mechanics; by putting the first particle in ϵ_1 and second in ϵ_2 or vice versa. In quantum mechanics it can be produced in only one way: By putting *a* particle in ϵ_1 and *a* particle in ϵ_2. Further, we are allowed to put both the particles in the first level or both in the second level (to obtain energies $2\epsilon_1$ or $2\epsilon_2$) in classical statistical mechanics. This is true in quantum statistics as well, provided the particles are bosons; but no two fermions can be put in the same quantum state. In evaluating sums for partition function describing a quantum system, we must take these features into account explicitly. We now define the density of states in quantum theory by incorporating the above elements.

Most of the discussion in earlier sections continues to be valid even if the energy of a subsystem interacting with a bigger system takes discrete values. In this case, the probability that the system can be found in any given microstate with energy E_r is given by

$$P_r = Z^{-1} e^{-\beta E_r}, \quad Z = \sum_r e^{-\beta E_r} = \sum_n g_n e^{-\beta E_n}. \tag{5.101}$$

In the sum over r, we are summing over all quantum states whereas in the sum over n we are summing over all energy levels. They differ by the degeneracy factor g_n, which gives the number of distinct quantum states with energy E_n. For macroscopic systems the energy levels will be highly degenerate. For example, consider the ideal gas of N particles, with energy levels of each particle given by Eq. (5.99). A given quantum state is specified by the set of $3N$ integers $(n_{1x}, n_{1y}, n_{1z}, \ldots, n_{Nx}, n_{Ny}, n_{Nz})$. If we specify only the energy E, then we only know that the sum of the squares of these $3N$ integers has a fixed value $(2mE/\hbar^2)(L^2/\pi^2)$. For macroscopic values of N and E, there are enormous numbers of ways of choosing the $3N$ integers such that this condition is satisfied; in other words the degeneracy of the energy level E is a very large integer.

The degeneracy factor for an energy level is very closely related to the density of states $g(\epsilon)$, used in the earlier formalism. To establish the precise connection, we equate the two definitions for the partition functions:

$$Z = \sum_{\text{energy levels}} g_n e^{-\beta E_n} \equiv \int_0^\infty g(\epsilon) e^{-\beta \epsilon} d\epsilon, \qquad (5.102)$$

which gives

$$g(\epsilon) = \sum_{\text{microstates}} \delta_D(\epsilon - E_r) = \sum_{\text{energy levels}} g_n \delta_D(\epsilon - E_n). \qquad (5.103)$$

This expression, however, is highly singular; it is zero whenever $\epsilon \neq E_n$ and is g_n times a delta function at $\epsilon = E_n$. It is more convenient to work with a smoother function $\Gamma(E)$ defined as

$$\Gamma(E) = \sum_{n=0}^\infty g_n \theta(E - E_n) = \frac{\text{total number of distinct quantum}}{\text{states with energy less than } E}. \qquad (5.104)$$

This object is also often easier to compute than g_n or g, and we can always obtain a smooth approximation to g by using a smooth approximation for Γ or θ and evaluating $d\Gamma/dE$.

The Γ, g_n, and Z in quantum theory are dimensionless numbers, and the corresponding classical quantities have the dimension related to $(\text{action})^{3N}$. We saw in Section 5.5 that the N dependence of the Z or g could not be uniquely fixed in the classical theory. This problem does not arise in the quantum-mechanical definition of Γ, etc. To see how this comes about, consider, for example, $\Gamma(E)$ for a system of N nonrelativistic particles confined in a cubic box of volume $V = L^3$. The quantum state of each particle is specified by three integers $n_x, n_y, n_z = 1, 2, \ldots$, with single-particle energy given by Eq. (5.99). Therefore a general energy level is specified by a set of $3N$ integers with

$$E = \frac{\hbar^2 \pi^2}{2mL^2} \sum_{k=1}^{3N} n_k^2, \quad \sum_{k=1}^{3N} n_k^2 = \frac{(2mE)L^2}{\pi^2 \hbar^2} = \mathcal{N}. \qquad (5.105)$$

To compute $\Gamma(E)$ we need to know the number of ways of partitioning integers up to $\mathcal{N}$ in $3N$ groups of perfect squares. It is clear that for macroscopic values of N and E, $\mathcal{N}$ is a large positive integer. If $\mathcal{N}$ is sufficiently large, then we expect this number to be well approximated by the volume of the N-dimensional octant of radius $\mathcal{N}^{1/2}$. Therefore, by using Eq. (5.14), we obtain

$$\Gamma(E) = \left(\frac{1}{2}\right)^{3N} V_{3N}\left(\sqrt{2mE}\frac{L}{\pi\hbar}\right) = \left(\frac{1}{2}\right)^{3N} \frac{\pi^{\frac{3N}{2}}}{\left(\frac{3N}{2}\right)\Gamma\left(\frac{3N}{2}\right)}(2mE)^{\frac{3N}{2}}\frac{(L^3)^N}{(\pi\hbar)^{3N}}$$

$$= C_{3N}(2mE)^{\frac{3N}{2}}\frac{V^N}{(2\pi\hbar)^{3N}} = \frac{\Gamma_{\text{class}}(E)}{(2\pi\hbar)^{3N}}. \tag{5.106}$$

In this computation each configuration of $\{n_k\}$ was counted as separate. However, because the configurations in which two particles are merely interchanged are not to be treated as distinct in quantum theory, the answer must be divided by $N!$ to obtain the final expression:

$$\Gamma(E) = \frac{\Gamma_{\text{class}}(E)}{(2\pi\hbar)^{3N}N!}. \tag{5.107}$$

Comparing this with the analysis in Section 5.5, we note that the constant with the dimension of action, which was missing in classical theory, was $2\pi\hbar$. Quantum theory automatically provides this factor and fixes the N dependence of the entropy, etc., uniquely.

This result can be interpreted in a more useful manner. Consider a single particle with wave vectors $\mathbf{k} = (\pi/L)\mathbf{n}$ and corresponding momenta $\mathbf{p} = (\pi\hbar/L)\mathbf{n}$. The number of microstates between the momenta $\mathbf{p}$ and $\mathbf{p} + d^3\mathbf{p}$ will be $(\Delta n)^3 \equiv (\Delta n_x \Delta n_y \Delta n_z)$, where

$$d^3\mathbf{p} = \left(\frac{\pi\hbar}{L}\right)^3 \Delta n_x \Delta n_y \Delta n_z \equiv \frac{(\pi\hbar)^3}{V}(\Delta n)^3, \tag{5.108}$$

giving

$$(\Delta n)^3 = \frac{V}{(\pi\hbar)^3}d^3p = \frac{V}{\pi^3\hbar^3}\frac{1}{8}4\pi p^2\,dp = \frac{V}{(2\pi\hbar)^3}(4\pi p^2\,dp) \tag{5.109}$$

where the factor $(1/8)$ arises from the fact that only the octant in which Δn_x, Δn_y, and Δn_z are all positive contributes to the volume element. This suggests the correspondence

$$\frac{d^3x\,d^3p \text{ in}}{\text{phase volume}} \rightarrow \frac{d^3x\,d^3p}{(2\pi\hbar)^3} \text{ microstates}, \tag{5.110}$$

which we shall use repeatedly. This expression allows the counting of the actual number of microstates inside a compact region of phase space bounded by the energy surface, with each cell of volume $(2\pi\hbar)^3$ counted as a distinct microstate.

The above analysis has led to *smooth* functions for Γ, g, etc., only because we have approximated the *jumpy* $\Gamma(E)$ by its smooth profile. In this process Γ remains dimensionless but $g(E) = d\Gamma/dE$ will acquire the dimensions of E^{-1}. To avoid this, it is better to redefine $g(E)$ as follows. Let the typical spacing between the levels be ΔE. We choose some energy interval δE such that $\Delta E \ll \delta E \lesssim E$ (which can always be done for macroscopic systems; see Exercise 5.8) and define

$$\frac{\text{number of microstates}}{\text{between } E \text{ and } E + \delta E} = g(E)\delta E. \qquad (5.111)$$

In fact, very often we will choose $\delta E \approx E$; note that if $\Gamma(E) \propto E^{3N/2}$ then $g(E) \propto E^{(3N/2)-1}$; but for $N \gg 1$, we usually approximate $g(E) \propto E^{3N/2}$. This approximation changes the dimensions of g and is equivalent to taking $\delta E \simeq E$. The crucial point is that $\ln \Gamma$, $\ln g$, and $\ln(gE)$ are all equal within the accuracy of $\mathcal{O}(\ln N/N)$ and can be treated as dimensionless.

We can estimate the accuracy of the continuum approximation used in changing the summation over energy levels to integration over q, p for an ideal gas by computing the partition function in the continuous and discrete cases separately. In the continuum approximation,

$$Z = \int \frac{dp\,dq}{(2\pi\hbar)^{3N}} \exp\left(-\frac{\beta}{2m}\sum p_i^2\right) = V^N \left(\frac{2\pi m}{\beta}\right)^{\frac{3N}{2}} \left(\frac{1}{2\pi\hbar}\right)^{3N}, \qquad (5.112)$$

whereas an explicit computation with discrete levels gives (see Exercise 5.8)

$$Z_1 = V^N \left(\frac{2\pi m}{\beta}\right)^{\frac{3N}{2}} \left(\frac{1}{2\pi\hbar}\right)^{3N} + \mathcal{O}\left(\frac{\beta}{2m}\frac{\hbar^2\pi^2}{V^{2/3}}\right), \qquad (5.113)$$

which shows that the continuum approximation is valid for $\lambda_T \equiv (2\pi\hbar^2/mk_BT)^{1/2} \ll L$. This condition is always satisfied because the volume of the system is always taken to be macroscopic, even while the quantum effects are being discussed. The quantity λ_T is called the thermal wavelength and represents the de Broglie wavelength of a nonrelativistic particle of energy k_BT.

The concept of the grand partition function, introduced in Section 5.3, can also be generalised to quantum theory in a straightforward manner. Corresponding to Eq. (5.38),

$$P(N, E) = \mathcal{G}^{-1} \exp \beta(\mu N - E) = \exp \beta(\Phi + \mu N - E), \qquad (5.114)$$

$$\Phi = -T \ln\left[\sum_{N,E} \exp \beta(\mu N - E)\right] = -\beta^{-1} \ln \mathcal{G} = -PV. \qquad (5.115)$$

From the thermodynamic potential $\Phi = -PV$ we can compute all the relevant quantities. To begin with, the mean values for N and E are given by

$$\langle N \rangle = \sum_{E,N=0}^{\infty} \frac{N}{\mathcal{G}} e^{\beta(\mu N - E)} = \frac{1}{\beta}\left(\frac{\partial \ln \mathcal{G}}{\partial \mu}\right)_\beta = -\left(\frac{\partial \Phi}{\partial \mu}\right)_\beta, \qquad (5.116)$$

$$\langle E \rangle = \sum_{N,E} \frac{z^N}{\mathcal{G}} E e^{-\beta E} = \left(\frac{\partial \ln \mathcal{G}}{\partial(-\beta)}\right)_z = -\beta\left(\frac{\partial \Phi}{\partial \beta}\right)_z. \qquad (5.117)$$

Further, our formalism also allows us to compute the fluctuations around the mean values. (We saw in Section 5.4 a simple example of this phenomena when we computed the fluctuations in the energy from statistical mechanics). Starting from

$$\langle N \rangle = e^{\beta \Phi} \sum_N N e^{\beta \mu N} \sum_k e^{-\beta E_k} \qquad (5.118)$$

and differentiating with respect to μ, we get

$$\frac{\partial \langle N \rangle}{\partial \mu} = e^{\beta \Phi} \sum_N \beta N^2 e^{\beta \mu N} \sum_k e^{-\beta E_k} + e^{\beta \Phi} \beta \frac{\partial \Phi}{\partial \mu} \sum_N N e^{\beta \mu N} \sum_k e^{-\beta E_k}$$

$$= \beta \langle N^2 \rangle + \beta \frac{\partial \Phi}{\partial \mu} \langle N \rangle = \beta(\langle N^2 \rangle - \langle N \rangle^2) = \beta(\Delta N)^2. \qquad (5.119)$$

Thus the fluctuations can be determined from the knowledge of the mean occupation number as a function of the chemical potential.

Exercise 5.8
Numerical estimates for a quantum ideal gas: (1) Consider an ideal gas of N molecules confined inside a cubical box of size L with a total energy E. Estimate the value of $n = (n_x^2 + n_y^2 + n_z^2)^{1/2}$ for a typical macroscopic system at room temperature. What is the relative energy-level spacing $(\Delta E/E)$ for this system? (2) Prove Eq. (5.113).

5.8 Partition Function for Bosons and Fermions

We now compute explicitly the physical variables in quantum-statistical mechanics, taking into account the discrete nature of the energy levels and the indistinguishability of the particles.

Let the single-particle energy level for the system be the $\{\epsilon_j\}$. If there are n_1 particles with energy ϵ_1, n_2 with ϵ_2, ..., etc., then the total energy of this configuration is $E = \sum \epsilon_j n_j$ and this configuration is completely specified by the set of integers $\{n_j\}$. These integers are constrained by the relation $\sum n_j = N$. Further, if the system is made of fermions, then each of the n_j must be 0 or 1. To find the quantum partition function under such conditions, we have to evaluate the sum

$$Z_Q(\beta, N) = {\sum_{\{n_j\}}}' \exp\left(-\beta \sum_j n_j \epsilon_j\right), \tag{5.120}$$

over all configurations $\{n_j\}$ subject to the constraint $\sum n_i = N$. (The subscript Q is for quantum.) The prime on the summation is a reminder of this constraint, which makes the sum difficult to evaluate. To circumvent this difficulty we use the grand partition function $\mathcal{G}_Q(z, \beta)$ defined by

$$\mathcal{G}_Q(z, \beta) = \sum_{N=0}^{\infty} e^{\beta \mu N} Z_Q(\beta, N) \equiv \sum_{N=0}^{\infty} z^N Z_Q(\beta, N), \quad z \equiv e^{\beta \mu}. \tag{5.121}$$

Given $\mathcal{G}_Q(z, \beta)$, the partition function Z_Q can be obtained as the coefficient of z^N in its expansion. To evaluate $\mathcal{G}_Q$, note that it can be written as

$$\mathcal{G}_Q(z, \beta) = \sum_{N=0}^{\infty} z^N {\sum_{\{n_j\}}}' \exp\left(-\beta \sum_j n_j \epsilon_j\right) = \sum_{N=0}^{\infty} {\sum_{\{n_j\}}}' \prod_j [z \exp(-\beta \epsilon_j)]^{n_j}. \tag{5.122}$$

The double summation is first over the set $\{n_j\}$, satisfying the constraint $\sum n_i = N$, and then over all possible values of N. But this is equivalent to the summation over all possible values in the set $\{n_j\}$ without any constraint. Therefore

$$\mathcal{G}_Q(z, \beta) = \sum_{\{n_j\}} \prod_j [z \exp(-\beta \epsilon_j)]^{n_j}, \tag{5.123}$$

where the prime on the summation is dropped because n_j's are no longer required for satisfying any constraint. Rewriting this expression as

$$\mathcal{G}_Q(z, \beta) = \sum_{n_1, n_2, \dots} [(ze^{-\beta \epsilon_1})^{n_1} (ze^{-\beta \epsilon_2})^{n_2} \cdots]$$

$$= \left[\sum_{n_1} (ze^{-\beta \epsilon_1})^{n_1}\right]\left[\sum_{n_2} (ze^{-\beta \epsilon_2})^{n_2}\right] \cdots \prod_j \left[\sum_n (ze^{-\beta \epsilon_j})^n\right], \tag{5.124}$$

we can easily find the final result: In the case of Bose particles n can take the values $0, 1, 2, \dots$, and the sum n is a geometric progression, giving

$$\mathcal{G}_Q^B = \prod_j \frac{1}{(1 - ze^{-\beta \epsilon_j})}. \tag{5.125}$$

In the case of fermions, $n = 0$ or 1 and

$$\mathcal{G}_Q^F(z, \beta) = \prod_j (1 + ze^{-\beta \epsilon_j}). \tag{5.126}$$

The partition function for an N-particle system is the coefficient of z^N in these expressions; but we will not need the explicit form for Z_Q.

For comparison, let us consider the corresponding result in classical statistical mechanics. Any state specified by the $\{n_j\}$ is counted only once in quantum theory; in contrast, permutation of particles is treated as a new state in classical statistical mechanics so that the energy E can be achieved in

$$g = {}^N C_{n_1} {}^{N-n_1} C_{n_2} \cdots = \frac{N!}{n_1! n_2! \cdots} = \frac{N!}{\prod_i n_i!} \tag{5.127}$$

ways in the classical limit of quantum statistical mechanics. Because permutation of the integers n_j is treated as producing distinct states, we need to evaluate Eq. (5.120) with the factor of Eq. (5.127) put in. We should also divide the final partition function by $N!$ for reasons discussed before (see Section 5.5). Hence $\mathcal{G}_C$ is given by

$$\begin{aligned}
\mathcal{G}_C(z, \beta) &= \sum_{N=0}^{\infty} \frac{z^N}{N!} \sideset{}{'}\sum_{\{n_j\}} \frac{N!}{\prod_j n_j!} \exp\left(-\beta \sum_j \epsilon_j n_j\right) \\
&= \sum_{N=0}^{\infty} \frac{1}{N!} \sideset{}{'}\sum_{\{n_j\}} \frac{N!}{\prod_j n_j} \prod_i (z e^{-\beta \epsilon_i})^{n_i} \\
&= \sum_{N=0}^{\infty} \frac{1}{N!} \sideset{}{'}\sum_{\{n_j\}} \frac{N!}{n_1! \, n_2! \cdots} x_1^{n_1} x_2^{n_2} \cdots,
\end{aligned} \tag{5.128}$$

where $x_i = z e^{-\beta \epsilon_i}$. This sum over $\{n_j\}$ with the constraint $\sum n_i = N$ is just the multinomial expansion for the quantity $(x_1 + x_2 + x_3 + \cdots)^N$. Therefore

$$\mathcal{G}_C(z, \beta) = \sum_{N=0}^{\infty} \frac{1}{N!} \left(z \sum_j e^{-\beta \epsilon_j}\right)^N = \exp\left(z \sum_j e^{-\beta \epsilon_j}\right). \tag{5.129}$$

This is the final answer for the case of (classical) Maxwell–Boltzmann statistics. Of course, in this case, the partition function itself can be expressed in closed form.

The thermodynamic potential $\Phi = -PV$ is related to $\mathcal{G}$ by $\Phi = -\beta^{-1} \ln \mathcal{G}$. For the three cases, usually called *Fermi–Dirac* [Eq. (5.126)], *Bose–Einstein* [Eq. (5.125)] or *Maxwell–Boltzmann* [Eq. (5.129) considered above] systems, Φ is given in a unified manner as

$$\Phi = -\frac{\beta^{-1}}{a} \sum_{\epsilon} \ln\left[1 + a e^{\beta(\mu - \epsilon)}\right] = -PV, \tag{5.130}$$

with $a = 1, -1$, or 0 for Fermi–Dirac, Bose–Einstein, or Maxwell–Boltzmann systems, respectively. Using Eq. (5.118), we get the mean number of particles as

$$\langle N \rangle = -\left(\frac{\partial \Phi}{\partial \mu}\right)_\beta = \frac{\beta^{-1}}{a} \sum_\epsilon \frac{a\beta e^{\beta(\mu-\epsilon)}}{\left(1 + a e^{\beta(\mu-\epsilon)}\right)} = \sum_\epsilon \frac{1}{\left(e^{\beta(\epsilon-\mu)} + a\right)} \quad (5.131)$$

and the mean energy as

$$\langle E \rangle = \beta\left(\frac{\partial \Phi}{\partial \beta}\right)_z = -\frac{1}{a} \sum_\epsilon \frac{-a\epsilon e^{\beta(\mu-\epsilon)}}{\left(1 + a e^{\beta(\mu-\epsilon)}\right)} = \sum_\epsilon \frac{\epsilon}{\left(e^{\beta(\epsilon-\mu)} + a\right)}. \quad (5.132)$$

These expressions have a fairly simple interpretation. They show that the *average* number of particles with an energy ϵ is given by

$$n(\epsilon) = \frac{1}{\left[e^{\beta(\epsilon-\mu)} + a\right]}, \quad (5.133)$$

where $\langle N \rangle = \sum n(\epsilon)$ and $\langle E \rangle = \sum \epsilon n(\epsilon)$. These two relations determine the two parameters β and μ in terms of $\langle N \rangle$ and $\langle E \rangle$, which may be set equal to the *actual* number of particles N and energy E of the system when the fluctuations are negligible.

All average thermodynamic properties of the system can be computed directly from expression (5.133). Because $n(\epsilon) \geq 0$ for all ϵ, it follows that $z \geq 0$. Further, for bosons, $(z^{-1}e^{\beta\epsilon} - 1) > 0$ for all ϵ, implying that $z \leq 1$. Hence z is bounded by

$$0 \leq z \leq 1 \quad \text{(bosons)}, \qquad 0 \leq z \leq \infty \quad \text{(fermions)}. \quad (5.134)$$

It is also clear that $z \ll 1$ corresponds to the classical limit, in which

$$n(\epsilon) \approx z e^{-\beta\epsilon} \quad \text{(bosons and fermions)}. \quad (5.135)$$

Hence significant quantum effects occur at the other end of the range: for bosons around $z \lesssim 1$ and for fermions when $z \to \infty$.

The fluctuations in the number of particles brings out the difference among the three cases. Using Eq. (5.131), we can write the total fluctuation as

$$(\Delta N)^2 = \frac{1}{\beta} \frac{\partial \langle N \rangle}{\partial \mu} = \sum_\epsilon \frac{e^{\beta(\epsilon-\mu)}}{\left(e^{\beta(\epsilon-\mu)} + a\right)^2} \equiv \sum_\epsilon (\Delta n)^2_\epsilon, \quad (5.136)$$

where Δn_ϵ is the fluctuation in the occupancy of the energy level ϵ. This can be rearranged to give

$$\left(\frac{\Delta n_\epsilon}{n_\epsilon}\right)^2 = e^{\beta(\epsilon-\mu)} = \left(\frac{1}{n_\epsilon} - a\right). \quad (5.137)$$

In the Boltzmann case ($a = 0$) the fluctuations are Poissonian and $\Delta n_\epsilon/n_\epsilon$ has the $1/\sqrt{n_\epsilon}$ scaling. In the case of bosons ($a = -1$) fluctuations are enhanced with $(\Delta n_\epsilon)^2 = n_\epsilon + n_\epsilon^2$. The term proportional to n_ϵ arises from the particle nature of the quanta, whereas n_ϵ^2 can be interpreted as being due to the wave nature of the quanta. (We shall see this more clearly in Section 5.11 when we study photons.) For bosons, these terms add up. In contrast, for fermions, $(\Delta n_\epsilon)^2 = n_\epsilon - n_\epsilon^2$, where the wavelike contribution acts with an opposite sign.

So far, we have not taken into account the spin of the particle. We can do this by multiplying the expressions for N, E, P, etc., by an extra factor g_s; for electrons, with two spin states, $g_s = 2$.

The summation over ϵ in Eqs. (5.131), (5.132), etc., can be replaced with integration even for Fermi and Bose systems, as explained in Section 5.7 [see discussion around Eq. (5.113)]. In all these expressions, the energy $\epsilon(p) = \sqrt{p^2c^2 + m^2c^4} - mc^2$ depends on only the magnitude p of the momentum, allowing the replacement d^3p with $4\pi p^2 dp$. Further, in expression (5.130) for pressure, rewritten as

$$\frac{PV}{T} = \int \frac{Vd^3p}{(2\pi\hbar)^3} g_s \ln\left[1 + aze^{-\beta\epsilon(p)}\right], \tag{5.138}$$

we can do an integration by parts and use the relation $p(d\epsilon/dp) = p^2c^2/(\epsilon + mc^2)$ to obtain

$$P = g_s \int_0^\infty \frac{4\pi p^2 \, dp}{(2\pi\hbar)^3} \left(\frac{1}{z^{-1}e^{\beta\epsilon} + a}\right)\left(\frac{1}{3}p\frac{d\epsilon}{dp}\right)$$

$$= g_s \int_0^\infty \frac{4\pi p^2 \, dp}{(2\pi\hbar)^3} \left(\frac{1}{z^{-1}e^{\beta\epsilon} + a}\right)\left[\frac{1}{3}\frac{c^2p^2}{\epsilon(p) + mc^2}\right]. \tag{5.139}$$

This equation and the corresponding results of Eqs. (5.131) and (5.132) for N/V and E/V, rewritten as

$$\frac{N}{V} = g_s \int_0^\infty \frac{4\pi p^2 \, dp}{(2\pi\hbar)^3} \frac{1}{z^{-1}e^{\beta\epsilon} + a}, \quad \frac{E}{V} = g_s \int_0^\infty \frac{4\pi p^2 \, dp}{(2\pi\hbar)^3} \frac{\epsilon}{z^{-1}e^{\beta\epsilon} + a}, \tag{5.140}$$

form the basic set of relations in quantum statistical mechanics. They give the pressure, energy, and total number of particles of the system in terms of the parameters μ and β or, equivalently, in terms of $z = e^{\beta\mu}$ and β. By eliminating these parameters between any two of the expressions, we can obtain all the relevant thermodynamical information.

In these expressions, energy $\epsilon(p)$ has been defined as $[(p^2c^2 + m^2c^4)^{1/2} - mc^2]$ so that $\epsilon \to 0$ as $p \to 0$, and hence the total energy E of the system does not include the rest energy of particles. If we use the expression $(p^2c^2 + m^2c^4)^{1/2}$ for $\epsilon(p)$, then the chemical potential should be changed from μ to $\mu + mc^2$.

One general result can be obtained immediately if the energy scales as a power of momentum: $\epsilon \propto p^\alpha$, as in the case of nonrelativistic ($\alpha = 2$) or extreme relativistic ($\alpha = 1$) particles. Then $p(d\epsilon/dp) = \alpha\epsilon$ and the expression for pressure gives

$$P = g_s \int_0^\infty \frac{4\pi p^2 dp}{(2\pi\hbar)^3} \left(\frac{1}{z^{-1}e^{\beta\epsilon} + a}\right)\left(\frac{1}{3}\alpha\epsilon\right) = \frac{\alpha}{3}\left(\frac{E}{V}\right) \quad \text{(for } \epsilon \propto p^\alpha\text{)}.$$

$$(5.141)$$

Thus, for nonrelativistic particles, pressure is two-thirds of the energy density, whereas for extreme relativistic particles it is one-third of the energy density. This result was obtained in Section 5.5.

Exercise 5.9
Another approach to quantum statistics: Consider a system with energy levels ϵ_j and degeneracies g_j. Any given microstate can be specified by giving the number of particles that is occupying these energy levels; i.e., by specifying the occupation numbers n_j. Obtain the number of different ways in (1) classical distinguishable particles, (2) bosons, (3) fermions can be distributed in these energy levels subject to the constraint that the total number of particles $\sum n_j = N$ and total energy $\sum n_j \epsilon_j = E$ are fixed. Show that the configuration that can be achieved in the maximum number of ways is the one that satisfies Eq. (5.133).

5.9 Fermions

In general, if the energy is not a simple power of the momentum, then the integrals in Eqs. (5.139) and (5.140) are nontrivial. We can express them in a more tractable form by using the parameter $q = \beta mc^2$ and changing the integration variable to $y = \beta\epsilon$. Then

$$\frac{N}{V} = \frac{g_s}{2\pi^2}\left(\frac{\hbar}{mc}\right)^{-3} \int_0^\infty \frac{dy}{q}\left(\frac{2y}{q}\right)^{1/2}\left(1 + \frac{y}{2q}\right)^{1/2}\left(1 + \frac{y}{q}\right)\frac{1}{z^{-1}e^y + 1},$$

$$(5.142)$$

$$\frac{E}{V} = \frac{g_s}{2\pi^2}\left(\frac{\hbar}{mc}\right)^{-3} mc^2 \int_0^\infty \frac{dy}{q}\left(\frac{2y}{q}\right)^{1/2}\left(\frac{y}{q}\right)\left(1 + \frac{y}{2q}\right)^{1/2}\left(1 + \frac{y}{q}\right)\frac{1}{z^{-1}e^y + 1}.$$

$$(5.143)$$

The nature of the result depends on the relative values of the two parameters z and q. We can identify four separate regimes of behaviour for this system.

When $z \ll 1$, the system is nearly classical and the quantum corrections can be obtained as a power-series expansion in z. Further, if $q \gg 1$, we are in the nonrelativistic regime, whereas if $q \ll 1$ we are in the ultrarelativistic regime.

If $z \gg 1$, quantum effects of the fermionic system are quite significant and cannot be treated as small perturbation. In this case, if we further have $q \ll 1$, then we are in the extreme relativistic regime. However, it turns out that the opposite condition $q \gg 1$ may not ensure nonrelativistic behaviour for a quantum Fermi gas. This limit is somewhat more subtle and has to be handled separately.

We start with the classical limit for which $z \ll 1$. In this case, both relativistic and nonrelativistic limits can be handled fairly easily and only the nonrelativistic limit is of particular interest. In Subsection (5.9.2) below, we deal with the $z \gg 1$ limit in which quantum-mechanical effects are important. In this case, both relativistic and nonrelativistic limits are of astrophysical interest.

5.9.1 Classical Limit: $z \ll 1$

In the classical limit, $z \ll 1$, so that $(z^{-1}e^y) \gg 1$ even for $y = 0$; then we can approximate the distribution function by

$$\frac{1}{z^{-1}e^y + 1} = \frac{ze^{-y}}{1 + ze^{-y}} \cong ze^{-y}(1 - ze^{-y}), \tag{5.144}$$

retaining just the two leading terms in the Taylor expansion. We can easily evaluate the integrals in the two limits of $q \gg 1$ [which corresponds to nonrelativistic (NR) particles] and $q \ll 1$ [which corresponds to extreme relativistic (ER) particles] by using the standard result

$$\int_0^\infty x^\alpha e^{-nqx} dx = (nq)^{-(\alpha+1)} \Gamma(\alpha + 1) \tag{5.145}$$

to give

$$\frac{N}{V} = \frac{g_s}{2\pi^2} \left(\frac{\hbar}{mc} \right)^{-3} \begin{cases} zq^{-3/2} \sqrt{\frac{\pi}{2}} [1 - (z/2^{3/2})] & \text{(NR)} \\ zq^{-3} 2[1 - (z/2^3)] & \text{(ER)} \end{cases}, \tag{5.146}$$

$$\frac{E}{V} = \frac{g_s}{2\pi^2} \left(\frac{\hbar}{mc} \right)^{-3} mc^2 \begin{cases} zq^{-5/2} \frac{3}{2} \sqrt{\frac{\pi}{2}} [1 - (z/2^{5/2})] & \text{(NR)} \\ zq^{-4} 6[1 - (z/2^4)] & \text{(ER)} \end{cases}. \tag{5.147}$$

Solving for z in terms of $n = (N/V)$ by using the first expression and substituting for z in the second, we get, in the nonrelativistic limit,

$$E = \frac{3}{2} PV = \frac{3}{2} Nk_B T \left[1 + \frac{1}{2} \left(\frac{N}{g_s V} \right) \left(\frac{\pi \hbar^2}{mk_B T} \right)^{3/2} \right], \tag{5.148}$$

where we have also used Eq. (5.141) to express pressure in terms of energy density. This result shows that the pressure exerted by a fermion gas on the walls of the container is larger than that of an ideal gas – which is understandable from the fact that fermions tend to avoid clustering compared with an ideal gas.

In this nonrelativistic limit, the integrals in Eqs. (5.139) and (5.140) can be expressed in a more formal way as follows. Introducing the thermal wavelength $\lambda_T = (2\pi\hbar^2/mk_BT)^{1/2}$ and a set of functions defined by

$$f_n(z) \equiv \frac{1}{\Gamma(n)} \int_0^\infty \frac{y^{n-1}}{z^{-1}e^y + 1} \, dy, \qquad (5.149)$$

we find that the integrals for number density, pressure, and energy become

$$\frac{N}{V} = \frac{g_s}{\lambda_T^3} f_{3/2}(z), \qquad \frac{PV}{T} = \frac{g_s V}{\lambda_T^3} f_{5/2}(z), \qquad \frac{E}{V} = \left(\frac{3}{2}k_B T\right)\frac{g_s}{\lambda_T^3} f_{5/2}(z). \qquad (5.150)$$

These expressions are valid for nonrelativistic particles, both in the classical ($z \ll 1$) and the quantum-mechanical ($z \gg 1$) limits. In the classical limit, $f_n(z)$ can be expanded in a Taylor series as

$$f_n(z) = \sum_{k=0}^\infty (-1)^k \frac{z^{k+1}}{(k+1)^n} = z - \frac{z^2}{2^n} + \frac{z^3}{3^n} - \cdots . \qquad (5.151)$$

Using Eqs. (5.150), we get

$$\frac{n\lambda_T^3}{g_s} = f_{3/2}(z) = \left[z - \frac{z^2}{2^{3/2}} + \mathcal{O}(z^3)\right]. \qquad (5.152)$$

Inverting this expression for z, we obtain to the same order of accuracy:

$$z = \left(\frac{n\lambda_T^3}{g_s}\right) + \frac{1}{2^{3/2}}\left(\frac{n\lambda_T^3}{g_s}\right)^2 + \mathcal{O}\left[\left(\frac{n\lambda_T^3}{g_s}\right)^3\right]. \qquad (5.153)$$

Because $z = e^{\beta\mu}$, this equation also gives the corrections to the chemical potential that are due to the quantum nature of the particle. By using the Taylor series in Eqs. (5.151) and (5.150), we can compute the corrections to classical behaviour to any required order of accuracy.

These expressions also provide an estimate of the validity of the approximations. The condition $z \ll 1$ translates to $n\lambda_T^3 \ll 1$. Because $n^{-1/3}$ represents the mean interparticle separation, this condition implies that classical approximation is valid whenever the mean interparticle distance is far larger than the thermal wavelength λ_T of the particles. In addition, nonrelativistic approximations assume that $k_B T \ll mc^2$.

The behaviour of a classical relativistic fermion gas can be obtained by use of the corresponding expressions. Because this is not of much relevance to astrophysics, we do not discuss it here.

5.9.2 Quantum Limit: $z \gg 1$

It is clear from the previous discussion that quantum effects will be important at high densities and low temperatures. (In this limit, the system is said to be degenerate.) To study these effects, we need an approximate expressions for the integrals in Eq. (5.140) for $z \gg 1$. The nature of this approximation can be understood from the following consideration.

For a fermionic system with a given number density and energy density, the chemical potential will depend on temperature. Let us consider a fermionic gas in the limit of $T \to 0$ and assume that the chemical potential has a nonzero, finite limit when $T \to 0$. We denote this limiting value by the symbol ϵ_F, called the Fermi energy. (The claim that chemical potential has a finite, nonzero limit as $T \to 0$ needs to be proved; we will do this later; see Eq. (5.171).) Consider now the limit of the distribution function,

$$n(\epsilon) = \frac{1}{\{\exp[\beta(\epsilon - \mu)] + 1\}}, \tag{5.154}$$

as $\beta \to \infty$ and $\mu \to \epsilon_F$. For all $\epsilon > \epsilon_F$, the argument of the exponential is positive and hence the exponential rises without limit as $\beta \to \infty$; thus $n(\epsilon) \to 0$ for all $\epsilon > \epsilon_F$ when $T \to 0$. On the other hand, for all $\epsilon < \epsilon_F$, the argument of the exponential is negative and hence the exponential goes to zero as $\beta \to \infty$; thus $n(\epsilon) \to 1$ for all $\epsilon < \epsilon_F$ when $T \to 0$.

We thus reach the conclusion that, at zero temperature, the fermionic system has all the levels up to Fermi energy ϵ_F fully occupied, with higher energy levels remaining unoccupied. In all the integrals, we can replace the distribution function $n(\epsilon)$ with unity and limit the integration to the range $(0, \epsilon_F)$ in energy, or, equivalently, to a range $(0, p_F)$ in momentum, where p_F is the Fermi momentum corresponding to the energy ϵ_F. In this limit, we can evaluate all relevant integrals without having to resort to any further approximation. With a dimensionless variable $x = (p_F/mc)$, the integrals become

$$n = \frac{N}{V} = \frac{g_s}{(2\pi\hbar)^3} \frac{4\pi}{3} p_F^3 = \frac{g_s}{6\pi^2} \left(\frac{\hbar}{mc}\right)^{-3} x^3, \tag{5.155}$$

$$P = \frac{g_s}{48\pi^2} \left(\frac{\hbar}{mc}\right)^{-3} (mc^2) F(x), \tag{5.156}$$

$$E = \frac{g_s}{48\pi^2} \left(\frac{\hbar}{mc}\right)^{-3} (mc^2) G(x), \tag{5.157}$$

with

$$F(x) = x(x^2 + 1)^{1/2}(2x^2 - 3) + 3\sinh^{-1} x, \tag{5.158}$$

$$G(x) = 8x^3\left[(1 + x^2)^{1/2} - 1\right] - F(x). \tag{5.159}$$

From expression (5.155) for number density, we get $x = (3/4\pi g_s)^{1/3}(n\lambda_c^3)^{1/3}$, where $\lambda_c = (h/mc)$. Using this in Eqs. (5.156) and (5.157), we can express the pressure and the energy in terms of the number density, thereby completely solving the problem.

Both the nonrelativistic and the extreme relativistic limits are of interest in this case. When $x = (p_F/mc) \ll 1$ all the fermions have momenta much less than mc and the system will be nonrelativistic. On the other hand, when $x = (p_F/mc) \gtrsim 1$, fermions near the top of the occupied levels have relativistic energies and the system should be treated as relativistic. Note that whether a system is relativistic or nonrelativistic is now determined by the Fermi momentum and *not* by the temperature. In fact, we are considering the zero-temperature limit of the system.

In these two limits, the functions $F(x)$ and $G(x)$ can be expanded in a Taylor series, giving

$$F(x) = \begin{cases} \dfrac{8}{5}x^5 - \dfrac{4}{7}x^7 + \cdots & (x \ll 1) \\[2mm] 2x^4 - 2x^2 + \cdots & (x \gg 1) \end{cases}, \tag{5.160}$$

$$G(x) = \begin{cases} \dfrac{12}{5}x^5 - \dfrac{3}{7}x^7 + \cdots & (x \ll 1) \\[2mm] 6x^4 - 8x^3 + \cdots & (x \gg 1) \end{cases}. \tag{5.161}$$

Using these results and expressing the pressure in terms of number density, we get

$$P = \left(\frac{g_s\pi}{6}\right)\left(\frac{mc^2}{\lambda_c^3}\right)\begin{cases} (8/5)\left[n\lambda_c^3(3/4\pi g_s)\right]^{5/3} & (x \ll 1) \\[2mm] 2\left(3n\lambda_c^3/4\pi g_s\right)^{4/3} & (x \gg 1) \end{cases}. \tag{5.162}$$

The two limiting forms simplify for the case of $g_s = 2$, giving

$$P = \begin{cases} \left[(3\pi^2)^{2/3}/5\right](\hbar^2/m)n^{5/3} \\[2mm] \left[(3\pi^2)^{1/3}/4\right](\hbar c)n^{4/3} \end{cases}. \tag{5.163}$$

This result, which is of astrophysical importance, shows that the pressure of a quantum Fermi system scales as $P \propto \rho^{5/3}$ in the nonrelativistic limit and as $P \propto \rho^{4/3}$ in the extreme relativistic limit.

We now turn to the task of justifying the above approximation from the original integrals in Eqs. (5.139) and (5.140). For this, we have to evaluate integrals of the form

$$I = \int_0^\infty \frac{f(y)\,dy}{z^{-1}e^y + 1} \tag{5.164}$$

in the limit of $z \gg 1$. By a series of transformations, we can express this integral as

$$I = \int_0^\infty \frac{f(y)\,dy}{z^{-1}e^y + 1} \equiv \int_0^\infty \frac{f(y)\,dy}{e^{y-\xi} + 1} = \int_{-\xi}^\infty \frac{f(\xi + x)\,dx}{e^x + 1}$$

$$= \int_0^\infty \frac{f(\xi + x)\,dx}{e^x + 1} + \int_0^\xi \frac{f(\xi - x)\,dx}{e^{-x} + 1}$$

$$= \int_0^\infty \frac{f(\xi + x)\,dx}{e^x + 1} + \int_0^\xi f(\xi - x)\,dx\left\{1 - \frac{1}{e^x + 1}\right\}$$

$$= \int_0^\infty \frac{f(\xi + x)\,dx}{e^x + 1} - \int_0^\xi \frac{f(\xi - x)}{e^x + 1}dx + \int_0^\xi f(\xi - x)\,dx, \quad (5.165)$$

where $z = e^\xi$. This expression is exact. We now use the conditions $z \gg 1$ and $\xi \gg 1$ to set the upper limit of the second integral to infinity. Then

$$I \cong \int_0^\xi f(y)\,dy + \int_0^\infty \frac{dx}{e^x + 1}[f(\xi + x) - f(\xi - x)]$$

$$= \int_0^\xi f(y)\,dy + 2f'(\xi)\int_0^\infty \frac{x\,dx}{e^x + 1} + \frac{1}{3}f'''(\xi)\int_0^\infty \frac{x^3dx}{e^x + 1} + \cdots +$$

$$= \int_0^\xi f(y)\,dy + \frac{\pi^2}{6}f'(\xi) + \mathcal{O}[f'''(\xi)], \quad (5.166)$$

where we have used the result

$$\int_0^\infty \frac{x\,dx}{e^x + 1} = \frac{\pi^2}{12} \quad (5.167)$$

and the Taylor series expansion for $f(\xi \pm x)$. Note that the integrals in relation (5.166) are strongly suppressed by the exponential factor for $x \gg 1$; because $\xi \gg 1$, Taylor expansion in x/ξ is valid in the region where the integral contributes most. The leading term of the approximate expression shows that we can approximate the integral by taking the contribution of $f(y)$ in the range $(0, \xi)$.

Exercise 5.10
Integrals: Show that, for $x > 1$,

$$\int_0^\infty \frac{z^{x-1}dz}{e^z + 1} = (1 - 2^{1-x})\Gamma(x)\zeta(x), \qquad \int_0^\infty \frac{z^{x-1}dz}{e^z - 1} = \Gamma(x)\zeta(x), \quad (5.168)$$

$$\int_0^\infty \frac{dz}{e^z + 1} = \ln 2, \quad (5.169)$$

where $\zeta(x)$ is the zeta function.

Consider now the integral for number density of nonrelativistic fermions when $z \gg 1$. By using the exact expression in Eq. (5.140) and approximating $f_n(z)$ by using relation (5.166) we find that

$$\frac{N}{g_s V} \lambda_T^3 \equiv \left(\frac{n\lambda_T^3}{g_s}\right) = f_{3/2}(z) \cong \frac{4}{3\sqrt{\pi}}(\ln z)^{3/2} + \frac{\pi^{3/2}}{6}(\ln z)^{-1/2}. \quad (5.170)$$

To leading-order,

$$\beta\mu \equiv \ln z = \beta\frac{\hbar^2}{2m}\left(\frac{6\pi^2}{g_s}n\right)^{2/3}, \quad (5.171)$$

showing that the chemical potential μ reaches a finite limit as $z \to \infty$, as assumed above. Equation (5.171) also gives the expression for the Fermi energy $\epsilon_F = (\hbar^2/2m)(6\pi^2 n/g_s)^{2/3}$ which matches with the result from Eq. (5.155) for nonrelativistic particles.

By using relation (5.166), we can also calculate the higher-order corrections systematically. Taking the leading-order solution for $\ln z$ to be $\ln z_0 = \beta\epsilon_F$, we find that the higher-order corrections are given by an expansion

$$\ln z \cong \ln z_0\left[1 - \frac{\pi^2}{12}\frac{1}{(\ln z_0)^2} + \cdots\right] = \beta\epsilon_F\left[1 - \frac{\pi^2}{12}\left(\frac{k_B T}{\epsilon_F}\right)^2 + \cdots\right]. \quad (5.172)$$

By using similar expansions,

$$f_{5/2} = \frac{1}{\Gamma(5/2)}\left[\frac{2}{5}(\ln z)^{5/2} + \frac{\pi^2}{4}(\ln z)^{1/2} + \cdots\right],$$
$$f_{3/2} = \frac{1}{\Gamma(3/2)}\left[\frac{2}{3}(\ln z)^{3/2} + \frac{\pi^2}{12}\frac{1}{(\ln z)^{1/2}} + \cdots\right], \quad (5.173)$$

we find that

$$E = \frac{3}{5}N\epsilon_F\left[1 + \frac{5\pi^2}{12}\left(\frac{k_B T}{\epsilon_F}\right)^2 + \cdots\right], \quad (5.174)$$

$$P = \frac{2}{3}\frac{E}{V} = \frac{2}{5}\left(\frac{N}{V}\right)\epsilon_F\left[1 + \frac{5\pi^2}{12}\left(\frac{k_B T}{\epsilon_F}\right)^2 + \cdots\right]. \quad (5.175)$$

These corrections also indicate the range of validity of the approximations. The zero-temperature approximation on which the results in Eqs. (5.155)–(5.162) were based is valid as long as $k_B T \ll \epsilon_F$. Because the Fermi energy can be very high for certain systems, this condition is satisfied even when the temperature is large by conventional standards. For example, the estimate in Exercise 5.11 shows that the Fermi energy of electrons in a sodium metal is of the order of a few electron volts, corresponding to temperatures higher than 10^4 K. Thus the electron gas in a metal can be approximated as a zero-temperature Fermi gas for $T \lesssim 10^4$ K. Similar features occur in the astrophysics of white dwarfs and neutron stars.

A similar analysis can be performed in the case of relativistic Fermi particles as well. In general, the corrections to number density, pressure, and energy that are due to finite temperature can be expressed as

$$n = \frac{4\pi g_s}{3}\left(\frac{h}{mc}\right)^{-3} x^3 \left(1 + \pi^2 \frac{1+2x^2}{2x^4}\frac{1}{q^2} + \cdots\right), \qquad (5.176)$$

$$P = \frac{\pi g_s}{6}\left(\frac{h}{mc}\right)^{-3} mc^2 F(x)\left[1 + 4\pi^2 \frac{x(1+x^2)^{1/2}}{F(x)}\frac{1}{q^2} + \cdots\right], \qquad (5.177)$$

$$E = \frac{\pi g_s}{6}\left(\frac{h}{mc}\right)^{-3} mc^2 G(x)$$

$$\times \left[1 + 4\pi^2 \frac{(1+3x^2)(1+x^2)^{1/2}-(1+2x^2)}{xG(x)}\frac{1}{q^2} + \cdots\right], \qquad (5.178)$$

where $x = (p_F/mc)$, $q = \beta mc^2$, and $F(x)$, $G(x)$ are given by Eqs. (5.158) and (5.159). The accuracy of these approximation is again determined by the smallness of the correction terms. In general, these conditions reduce to $k_B T \ll \epsilon_F$.

From the expression for energy, we can obtain the specific heat at constant density (or volume, if the number is fixed). Using the condition $(\partial n/\partial T)_\rho = 0$ in Eq. (5.176), we can find the variation of x with respect to temperature. With this expression, straightforward algebra gives the specific heat as

$$C_V = k_B \left(\frac{k_B T}{mc^2}\right)\frac{1}{\lambda_c^3}\frac{1}{3}x(1+x^2)^{1/2}. \qquad (5.179)$$

This is the specific heat for unit volume; to obtain the specific heat for unit mass, C_V should be divided by the mass density ρ. The fact that $C_V \propto T$ for $k_B T \ll \epsilon_F$ is obvious from Eq. (5.174). When heat is added to the system, only a fraction of fermions $n(k_B T/\epsilon_F)$ acquires the energy $(k_B T)$; hence the temperature-dependent part of E scales as $n(k_B T)^2/\epsilon_F$, leading to $C_V = (\partial E/\partial T) \simeq k_B n$ $(k_B T/\epsilon_F)$. This expression has the same form as Eq. (5.179) when Eqs. (5.171) and (5.176) are used to determine n and ϵ_F.

For future reference, the various exponents that are appropriate for a degenerate Fermi gas are mentioned here. In any realistic situation the total pressure of a system will be contributed by both degenerate electrons and the ions. Exponents such as χ_T in Eqs. (5.83) are defined as logarithmic derivatives and hence cannot be separated into the exponents of the constituents. Even though the degeneracy pressure P_e of electrons will be much greater than that of ions (P_I), the same is not true regarding their temperature derivative. Equation (5.175) shows that $(\partial P_e/\partial T)_\rho \propto T$, whereas for an ideal gas of ions, $(\partial P_I/\partial T)_\rho = (N_A k_B \rho/\mu_I)$, where N_A is the Avogadro number and μ_I is the mean molecular weight of ions introduced in Eq. (5.93). Clearly, at low enough temperatures, χ_T is contributed essentially by ions and we have $\chi_T \approx (N_A k_B \rho T / \mu_I P_e)$. The density exponent, on the other hand, is dominated by degeneracy pressure and we get

$$\chi_\rho \to \frac{\rho}{P_e}\left(\frac{\partial P_e}{\partial \rho}\right)_T \to \begin{cases} 5/3 & \text{nonrelativistic} \\ 4/3 & \text{relativistic} \end{cases}. \tag{5.180}$$

The ion specific heat (per mole) is given by

$$C_V = \frac{3N_A k_B}{2\mu_I} = \frac{1.247 \times 10^8}{\mu_I}\ \text{erg g}^{-1}\,\text{K}^{-1} = \frac{0.012}{\mu_I}\text{J kg}^{-1}\,\text{K}^{-1} \tag{5.181}$$

and dominates over electronic specific heat at low temperatures. From these results, we can easily find that $C_P = C_{V,\text{ion}}$, $\Gamma_3 - 1 = (2/3)$, $\Gamma_1 = \chi_\rho$, and $\nabla_{\text{ad}} = (2/3\chi_\rho)$.

Exercise 5.11
Estimate of degeneracy effects: (1) Estimate the thermal wavelength for hydrogen at room temperature. Hence show that $n\lambda^3 \ll 1$ for such a gas at room temperature and pressure. (2) Estimate the corresponding quantity $n\lambda_e^3$ for the electrons in a sodium metal at room temperature and show that, in this case, $n\lambda_e^3 \gg 1$.

Exercise 5.12
For long winter evenings II: Prove Eqs. (5.146), (5.147), (5.150), (5.155)–(5.159), and (5.176)–(5.180), supplying the intermediate steps.

5.10 Bosons

Let us next consider the corresponding situation for the bosons. We again start with the assumption that the replacement

$$\sum_\epsilon \to \int d^3 p \tag{5.182}$$

is possible. This will lead us to a difficulty, the resolution of which exhibits an interesting physical phenomenon. With this replacement, for nonrelativistic particles, Eqs. (5.139) and (5.140) reduce to

$$\frac{N}{V}\lambda_T^3 = n\lambda_T^3 = G_{3/2}(z), \quad \frac{P}{T} = \frac{1}{\lambda_T^3}G_{5/2}(z), \quad E = \frac{3}{2}PV, \tag{5.183}$$

where we have set the spin degeneracy factor $g_s = 1$ for simplicity and defined the functions

$$G_n(z) = \frac{1}{\Gamma(n)}\int_0^\infty \frac{y^{n-1}dy}{z^{-1}e^y - 1}. \tag{5.184}$$

Note that $G_n(z)$ is an increasing function of z. Therefore $G_n(z) < G_n(1)$, as $z \leq 1$ for bosons. In particular, because

$$G_{3/2}(1) = \frac{1}{\Gamma(3/2)} \int_0^\infty \frac{y^{1/2}dy}{e^y - 1} = \zeta[3/2] \cong 2.612, \qquad (5.185)$$

it follows that the density must satisfy the constraint

$$\frac{N}{V}\lambda_T^3 = G_{3/2}(z) \leq G_{3/2}(1) = 2.61 \qquad (5.186)$$

or

$$\frac{N}{V}\left(\frac{2\pi}{mk_BT}\right)^{3/2}\hbar^3 < 2.61, \qquad (5.187)$$

which is equivalent to the condition $T > T_c$, where

$$k_BT_c \equiv \left[\frac{1}{\zeta(3/2)}\frac{N\hbar^3}{V}\right]^{2/3}\left(\frac{2\pi}{m}\right) = \frac{2\pi}{(2.61)^{2/3}}\left(\frac{N}{V}\right)^{2/3}\frac{\hbar^2}{m} \cong 3.3\frac{\hbar^2}{m}\left(\frac{N}{V}\right)^{2/3}.$$

$$(5.188)$$

The question arises as to what happens to the system when the temperature is lowered below T_c.

The above difficulty has its roots in the continuum approximation, which was introduced in expression (5.182). In this approximation, the measure $4\pi p^2 dp$ gives zero weight to the $p=0$ ground state. For $T < T_c$, a finite fraction of the particles will occupy this ground state, which requires that the $p=0$ state be treated separately. (This phenomenon is called Bose–Einstein condensation.) Thus, if $T < T_c$, there will be a two population of particles. The excited states ($p > 0$) can contain N_{ex} particles determined by Eqs. (5.183),

$$\frac{N_{ex}}{V} = \left(\frac{T}{T_c}\right)^{3/2}\left(\frac{N}{V}\right), \qquad (5.189)$$

and the remaining particles will be contained in the ground state (with $p=0$),

$$\frac{N_g}{V} = \left[1 - \left(\frac{T}{T_c}\right)^{3/2}\right]\left(\frac{N}{V}\right). \qquad (5.190)$$

A more formal way of deriving this result is to write the first equation in Eqs. (5.183) as

$$n\lambda_T^3 = G_{3/2}(z) + n(\epsilon = 0)\lambda_T^3 = G_{3/2}(z) + \frac{\lambda_T^3}{V}\left(\frac{z}{1-z}\right), \qquad (5.191)$$

where we have used the fact that

$$n(\epsilon = 0) = \left(\frac{1}{V}\frac{1}{z^{-1}e^{\beta\epsilon} - 1}\right)_{\epsilon=0} = \frac{1}{V}\left(\frac{z}{1-z}\right). \qquad (5.192)$$

Given a value of $n\lambda_T^3$, we have to solve this equation for z and take the limit of $N \to \infty$, $V \to \infty$ with finite (N/V). Note that, except very near $z \approx 1$, the V^{-1} factor will kill the second term; near $z \approx 1$, however, this term will dominate over $G_{3/2}(z)$. It is therefore clear that, to the accuracy of $\mathcal{O}(\lambda_T^3/V)$, the right-hand side of Eq. (5.191) is given by

$$\mathcal{F}(z) \equiv G_{3/2}(z) + \frac{\lambda_T^3}{V}\left(\frac{z}{1-z}\right) = \begin{cases} G_{3/2}(z) & [z < 1 - \mathcal{O}(\lambda_T^3/V)] \\ (\lambda_T^3/V)[z/(1-z)] & (z \approx 1) \end{cases}.$$

(5.193)

The solution to the equation $n\lambda_T^3 = \mathcal{F}(z)$ is very well approximated by

$$z = \begin{cases} 1 & [\text{for } n\lambda_T^3 \ge G_{3/2}(1)] \\ \text{root of } G_{3/2}(z) = n\lambda_T^3 & [\text{for } n\lambda_T^3 \le G_{3/2}(1)] \end{cases}.$$

(5.194)

Thus the behaviour is normal for $n\lambda_T^3 \le G_{3/2}(1)$, i.e., for $T > T_c$. For $T < T_c$, z becomes 1. The occupations of ground and excited states are therefore given by

$$\frac{N_g}{N} = \begin{cases} 1 - (T/T_c)^{3/2} & [\text{for } n\lambda_T^3 \ge G_{3/2}(1)] \\ 0 & [\text{for } n\lambda_T^3 \le G_{3/2}(1)] \end{cases},$$

(5.195)

$$\left(\frac{N_{\text{ex}}}{N}\right) = \begin{cases} (T/T_c)^{3/2} & [\text{for } n\lambda_T^3 \ge G_{3/2}(1)] \\ (1/n\lambda_T^3)\, G_{3/2}(z) & [\text{for } n\lambda_T^3 \le G_{3/2}(1)] \end{cases}.$$

(5.196)

The pressure for the system will be

$$\frac{P}{T} = \frac{1}{\lambda_T^3}G_{5/2}(z) = \begin{cases} (1/\lambda_T^3)\, G_{5/2}(1) & [\text{for } n\lambda_T^3 \ge G_{3/2}(1)] \\ (1/\lambda_T^3)\, G_{5/2}(z) & [\text{for } n\lambda_T^3 \le G_{3/2}(1)] \end{cases}.$$

(5.197)

[In the calculation of P/T, the ground state contributes a term $V^{-1}\ln(1-z)$; as $V \to \infty$, this term is zero for both phases. It is obviously zero for $T > T_c$; for $T < T_c$, $(1-z)$ is $\mathcal{O}(\lambda_T^3/V)$ and $V^{-1}\ln V$ vanishes for large V.] This calculation shows that, for $T < T_c$, the pressure is independent of the volume. Because $E = (3/2)PV$ we get

$$E = \frac{3}{2}\frac{VT}{\lambda_T^3}\begin{cases} G_{5/2}(1) & \text{for } T < T_c \\ G_{5/2}(z) & \text{for } T > T_c \end{cases}.$$

(5.198)

For $T < T_c$, $E \propto T^{5/2}$ and hence the specific heat is proportional to $T^{3/2}$. It can be shown that the specific heat is continuous at $T = T_c$ but its derivative is discontinuous (see Exercise 5.13).

Exercise 5.13
Specific heat of a low-temperature Bose system: Calculate the specific heat C_V for the above system as a function of temperature. Show that the specific heat is continuous at $T = T_c$ with a value $[C_V(T_c)/Nk_B] = (15/4)[\zeta(5/2)/\zeta(3/2)] \approx 1.92$. Next compute the

derivative of the specific heat $(\partial C_V / \partial T)$ as a function of temperature. Show that this quantity is discontinuous across $T = T_c$, with the discontinuity being given by $3.665(Nk_B/T_c)$.

5.11 Statistical Mechanics of the Electromagnetic Field

The basic formalism of statistical mechanics does not depend on the explicit form of the Hamiltonian; any Hamiltonian system with a large number of degrees of freedom can be studied with this formalism. We now develop the statistical mechanics of a system that is quite different from a gas of particles, viz., the electromagnetic field confined inside a cubical box of size L, the walls of which are kept at a temperature T. The electromagnetic field can exchange energy with the material making up the walls and is governed by the Hamiltonian

$$H = \frac{1}{8\pi} \int d^3\mathbf{x}(\mathbf{E}^2 + \mathbf{B}^2). \tag{5.199}$$

To study the statistical mechanics of this system we should integrate $\exp(-\beta H)$ over the phase space of the system and obtain the partition function $Z(\beta)$. Therefore, to begin with, we need to express the Hamiltonian in terms of the independent degrees of freedom q_i and the canonically conjugate momenta p_i.

This issue has already been tackled in Chap. 4, Subsection 4.5.1, where it was shown that the Hamiltonian for the electromagnetic field can be expressed in the form

$$H = \sum_{\alpha=1}^{2} \sum_{\mathbf{k}} \frac{1}{2}\left(P_{\mathbf{k}\alpha}^2 + k^2 Q_{\mathbf{k}\alpha}^2\right) = \sum_{\alpha} \sum_{\mathbf{k}} H_{\mathbf{k}\alpha}, \tag{5.200}$$

where α is the polarisation index, taking the value of 1 or 2 for each $\mathbf{k}$. (It is convenient to work with units $c = 1$, which we have adopted.)

The statistical mechanics of the electromagnetic field is therefore identical to that of these oscillators and we have to compute only the quantum-mechanical partition function for this system. Because the oscillators do not interact with each other (except for the weak coupling through the material of the walls, which allows them to exchange energy) the partition function will be the product

$$Z = \prod_{\mathbf{k}\alpha} Z_{\mathbf{k}\alpha}, \tag{5.201}$$

where $Z_{\mathbf{k}\alpha}$ is the partition function of the individual oscillator. The mean value of the energy will be

$$E = -\frac{\partial}{\partial \beta} \ln Z = \sum_{\mathbf{k}\alpha} E_{\mathbf{k}\alpha}, \quad E_{\mathbf{k}\alpha} = -\frac{\partial}{\partial \beta} \ln Z_{\mathbf{k}\alpha}. \tag{5.202}$$

Let us calculate $Z_{\mathbf{k}\alpha}$ first classically and then quantum mechanically. Classically, we have

$$Z_{k\alpha}^{\text{cla}} = \int \frac{dp_{k\alpha}\, dq_{k\alpha}}{(2\pi\hbar)} \exp\left[-\frac{\beta}{2}\left(p_{k\alpha}^2 + k^2 q_{k\alpha}^2\right)\right]$$

$$= \left(\frac{1}{2\pi\hbar}\right)\left(\frac{2\pi}{\beta}\right)\frac{1}{|\mathbf{k}|} = \frac{1}{\beta\hbar\omega_k} = \frac{k_B T}{\hbar\omega_{\mathbf{k}}}, \tag{5.203}$$

giving

$$E_{k\alpha} = \frac{1}{\beta} = k_B T. \tag{5.204}$$

This result is understandable from the equipartition and virial theorems (see Exercise 5.6). Equipartition of energy implies that each degree of freedom has $(1/2)k_B T$ of kinetic energy. Virial theorem, when applied to a quadratic potential, shows that kinetic and potential energies are equal on the average. Hence the total energy is $k_B T$ per oscillator.

Quantum mechanically, the energy levels of the oscillator are given by $\hbar\omega_k(n + 1/2)$, and the partition function becomes

$$Z_{k\alpha} = \sum_{n=0}^{\infty} \exp\left[-\beta\hbar\omega_k\left(n + \frac{1}{2}\right)\right] = e^{-\frac{1}{2}\beta\hbar\omega_k}\left(\frac{1}{1 - e^{-\beta\hbar\omega_k}}\right). \tag{5.205}$$

[It is easy to verify that, at high temperatures, this result goes over to the expression in Eq. (5.203).] The logarithm of the partition function is

$$\ln Z_{k\alpha} = -\frac{1}{2}\beta\hbar\omega_k - \ln(1 - e^{-\beta\hbar\omega_k}), \tag{5.206}$$

giving the mean energy as

$$E_{k\alpha} = \frac{1}{2}\hbar\omega_k + \frac{\hbar\omega_k e^{-\beta\hbar\omega_k}}{1 - e^{-\beta\hbar\omega_k}} = \frac{1}{2}\hbar\omega_k + \frac{\hbar\omega_k}{(e^{\beta\hbar\omega_k} - 1)}. \tag{5.207}$$

The first term is the quantum ground-state energy of the oscillator that is a constant that is independent of the temperature. We drop this term as no other satisfactory method has been found to deal with it. Because it is independent of the temperature, it will not contribute to specific heat and other exponents.

The second term in this expression is identical to the energy of a Bose gas of particles with $\epsilon_k = \hbar\omega_k$ and $\mu = 0$. Hence the electromagnetic field behaves as though it is made of a collection of bosons with zero chemical potential. The number of particles (photons) with frequency ω_k is $n(\omega) = (e^{\beta\hbar\omega} - 1)^{-1}$.

To find the total energy, we have to sum over all $\mathbf{k}$, α. Because α does not appear in the expressions, summing over α is the same as multiplying by a factor 2. To sum over $\mathbf{k}$, we can consider the limit of $V \to \infty$, making the spacing between successive **k** values smaller and smaller. Then, with the usual replacement

$$\sum_{k,\alpha} = 2\sum_{k} = 2\int \frac{V d^3 k}{(2\pi)^3}, \qquad (5.208)$$

the energy density will become

$$\frac{E}{V} = \int_0^\infty \frac{8\pi k^2 \, dk}{8\pi^3} E(k) = \int_0^\infty \frac{k^2 \, dk}{\pi^2} E(k) = \int_0^\infty \frac{\omega^2 \, d\omega}{\pi^2} E(\omega), \quad (5.209)$$

with

$$E(k) = E(\omega) = \begin{cases} k_B T, & \text{(classical)} \\ \hbar\omega(e^{\beta\hbar\omega} - 1)^{-1} & \text{(quantum)} \end{cases}. \qquad (5.210)$$

[The ground-state energy ($\hbar\omega/2$) has been dropped in the quantum result.] The classical result leads to an unacceptable divergence,

$$\left(\frac{E}{V}\right)_{\text{class}} = \int_0^\infty \frac{\omega^2 \, d\omega}{\pi^2} (k_B T) \to \lim_{\Lambda \to \infty} (k_B T \Lambda^3), \qquad (5.211)$$

which is temperature dependent. This will make other thermodynamical quantities such as C_V diverge as well. This divergence arises because, in classical theory, each degree of freedom has the energy $k_B T$ that is due to equipartition and this system has infinite number of degrees of freedom.

Quantum mechanically, *this* particular divergence is cured. The term [$\hbar\omega/(e^{\beta\hbar\omega} - 1)$] – which goes over to $k_B T$ in the $\hbar \to 0$ limit – ensures convergence when $\hbar \neq 0$. The temperature-dependent term in quantum theory is

$$\left(\frac{E}{V}\right)_{\text{quantum}} = \int_0^\infty \frac{\omega^2 \, d\omega}{\pi^2} \frac{\hbar\omega}{e^{\beta\hbar\omega} - 1}$$

$$= \frac{(k_B T)^4}{\pi^2 (\hbar c)^3} \int_0^\infty \frac{x^3 dx}{e^x - 1} = \frac{\pi^2}{15} \frac{(k_B T)^4}{(c\hbar)^3} \equiv aT^4, \qquad (5.212)$$

where $a = (\pi^2/15)(k_B^4/c^3\hbar^3)$ is called Stefan's constant. (We have temporarily reintroduced the c factor.) The energy density in the frequency range $d\omega$,

$$dU_\omega = \frac{2(4\pi\omega^2) \, d\omega}{(2\pi)^3} (\hbar\omega) n(\omega) = \frac{\hbar\omega^3}{\pi^2} (e^{\beta\hbar\omega} - 1)^{-1} \, d\omega, \qquad (5.213)$$

is called the Planck spectrum.

For $\hbar\omega \ll k_B T$ and $\hbar\omega \gg k_B T$ the expression for $n(\omega) = (e^{\beta\hbar\omega} - 1)^{-1}$ has simple asymptotic forms. The $\hbar\omega \ll k_B T$ limit corresponds to the long-wavelength classical ($\hbar \to 0$) regime of the radiation. The equipartition of energy suggests that each mode (having two polarisation states) should have energy $\epsilon_\omega = 2(k_B T/2) = k_B T$ or $n_\omega = \epsilon_\omega/\hbar\omega = k_B T/\hbar\omega$. The quantity $n(\omega)$ does reduce to $k_B T/\hbar\omega$ in this limit. The $\hbar\omega \gg k_B T$ limit corresponds to the regime in which photons behave as particles. In such a case we expect $n_\omega = \exp(-\hbar\omega/k_BT)$, based on Boltzmann statistics. (In this limit, $n_\omega \ll 1$ and quantum statistical effects are ignorable; hence we get Boltzmann statistics rather than Bose–Einstein statistics.) Again, this is what we obtain from $n(\omega)$ in this limit. From our general result of Eq. (5.137), we also know that the fluctuations in the number of photons (or mean energy) are

$$(\Delta n)^2 \equiv \left(\frac{\Delta E}{\hbar\omega}\right)^2 = n^2 + n. \tag{5.214}$$

If photons were to be interpreted as particles, then we would expect $(\Delta n)^2 \simeq \bar{n}$, giving the usual Poisson fluctuations of $(\Delta n/n) \simeq n^{-1/2}$. For this to occur it is necessary that $\bar{n} \gg \bar{n}^2$, that is, $\bar{n} \ll 1$, which requires that $\beta\hbar\omega \gg 1$. On the other hand, if $\beta\hbar\omega \ll 1$ then $\bar{n} \gg 1$ and $(\Delta n)^2 \simeq \bar{n}^2$, which characterises the wavelike fluctuations. Thus we may think of photons as particles when $\hbar\omega \gg k_BT$ and as waves when $\hbar\omega \ll k_BT$. The limiting domains $\hbar\omega \ll k_BT$ and $\hbar\omega \gg k_BT$ are called the Rayleigh–Jeans regime and the Wien's regime, respectively. The corresponding limits of the Planck spectrum, $U_\omega \simeq (\omega^2/\pi^2)k_BT$ and $U_\omega \simeq (\hbar\omega^3/\pi^2)$ $\exp(-\beta\hbar\omega)$, are called the Rayleigh-Jeans spectrum and the Wien's spectrum, respectively.

The pressure that is due to radiation is given by $P_{\text{rad}} = (1/3)aT^4$. When a system contains matter and radiation at some temperature T, the total pressure and energy that are due to the unit mass of material are given by

$$P = \frac{\rho N_A k_B T}{\mu} + \frac{aT^4}{3} = P_{\text{gas}} + P_{\text{rad}}, \quad E = \frac{3N_A k_B T}{2\mu} + \frac{aT^4}{\rho}, \tag{5.215}$$

where N_A is the Avogadro number and μ is the molecular weight. The various thermodynamic exponents for a system made of radiation and matter can be easily determined from the above relation. Denoting the ratio between gas pressure to total pressure as $r \equiv (P_{\text{gas}}/P)$, we find that $\chi_\rho = r$, $\chi_T = 4 - 3r$, and

$$C_V = \frac{3N_A k_B}{2\mu}\left(\frac{8 - 7r}{r}\right), \tag{5.216}$$

$\Gamma_3 - 1 = (2/3)[(4 - 3r)/(8 - 7r)]$, $\Gamma_1 = r + (4 - 3r)(\Gamma_3 - 1)$, $(\Gamma_2/(\Gamma_2 - 1)) = [(32 - 24r - 3r^2)/2(4 - 3r)]$, and, finally, $\gamma = \Gamma_1/r$. When $r \to 1$ we get the ideal gas values of $\chi_\rho = \chi_T = 1$ and $\Gamma_1 = \Gamma_2 = \Gamma_3 = \gamma = (5/3)$; when $r \to 0$ we get the values corresponding to radiation with $\chi_\rho = 0$, $\chi_T = 4$ and $\Gamma_1 = \Gamma_2 = \Gamma_3 = (4/3)$. Note that $\gamma = (\Gamma_1/\chi_\rho)$ diverges in this case. For a mixture of radiation and ideal gas, the values are intermediate to the limiting values.

Exercise 5.14
For long winter evenings III: Verify the statements made in the text regarding a system made of radiation and ideal gas.

Exercise 5.15
Lattice vibrations in a solid: The quantisation of a system of harmonic oscillators (each labelled by a three-dimensional vector **k**) occurs in several other contexts. For example, the vibration of the ions in a solid lattice can be thought of as representing the vibrations of a large number of harmonic oscillators. If there are N atoms in a solid, then the total degrees of freedom for the system will be $3N$. What is more, the frequency of vibration in this case cannot be arbitrarily high because the corresponding wavelength of vibration cannot be smaller than the lattice spacing. The dispersion relation for the oscillators can be written as $k = (\omega/u_l)$ for the longitudinal vibrations and $k = (\omega/u_t)$ for the two transverse vibrations, where u_l and u_t are the speeds of propagation of waves in the longitudinal and the transverse directions, respectively. Let ω_m be the maximum frequency of vibration chosen in such a way that the total degrees of freedom is $3N$. (1) Show that $\omega_m = \bar{u}(6\pi^2 N/V)^{1/3}$, with $(3/\bar{u}^3) = (2/u_t^3) + (1/u_l^3)$. (2) Show that the total energy of the system is given by $E = N\epsilon_0 + 3Nk_BT D(\Theta/T)$, where ϵ_0 is the total ground-state energy of the system, $k_B\Theta = \hbar\omega_m$, and

$$D(x) = \frac{3}{x^3} \int_0^x \frac{z^3\, dz}{e^z - 1}. \tag{5.217}$$

(3) Show that the corresponding specific heat is given by the expression $C = 3Nk_B$ $[D(\Theta/T) - (\Theta/T)D'(\Theta/T)]$. Obtain the limiting forms of the specific heat for $T \ll \Theta$ and $T \gg \Theta$. In particular, show that at low temperatures $C = (12Nk_B\pi^4/5)(T/\Theta)^3$. Explain this result physically. (We have seen earlier in Section 5.9 that the electronic specific heat of metals scale as $C \propto T$; so the total specific heat of ions and electrons together varies as $C = a_1T + a_2T^3$ at low temperatures.)

Exercise 5.16
Doppler effect on temperature: An observer is moving with speed v through a region containing isotropic blackbody radiation of temperature T. Show that the intensity of radiation measured by this observer, in a direction that makes angle θ with respect to the direction of motion, will be Planckian in form but with a Doppler-shifted temperature:

$$T' = T\left[\frac{\sqrt{1 - (v^2/c^2)}}{1 - (v/c)\cos\theta}\right]. \tag{5.218}$$

Exercise 5.17
Energy fluctuations for waves: A classical radiation field in thermal equilibrium may be thought of as made up of a superposition of electromagnetic waves with random phases. The mean energy of a system will be proportional to $\langle \mathbf{E}^2 \rangle$, where $\mathbf{E}$ is the electric field and the averaging is over the random phases. The fluctuations in the energy will be proportional to $[\langle \mathbf{E}^4 \rangle - \langle \mathbf{E}^2 \rangle^2]$. Compute this fluctuation and show that this corresponds to the n^2 term in Eq. (5.214).

Exercise 5.18
Temporal coherence of thermal radiation: The thermal spectrum can be thought of as arising because of a randomly fluctuating electromagnetic field with specific stochastic

properties. Let any one component of the electric field be denoted by a Gaussian random variable $E(t)$, with $\langle E(t) \rangle = 0$ and $\langle E(t)E(t') \rangle = C(t - t')$, where $C(t - t')$ is the temporal correlation function. Relate C to a Fourier transform of the Planck spectrum and obtain its explicit form.

5.12 Ionisation and Pair-Creation Equilibria

The key condition that governs the equilibria of multicomponent systems is the equality of temperature, pressure, and chemical potential, which makes the total entropy a local maxima when varied with respect to energy, volume, or number of particles. We now apply this principle for the study of three important processes in astrophysics. The first one deals with the ionisation of hydrogen at high temperatures and the second one deals with production of electron–positron pairs at high temperatures. The last one summarises the corresponding case for equilibria of different nuclear species.

5.12.1 Ionisation Equilibrium for Hydrogen

A neutral hydrogen atom, in the ground state, contains an electron and a proton bound together with an energy of $E_0 = -13.6$ eV, which is equivalent to a temperature of $T_I \simeq 1.578 \times 10^5$ K. At temperatures $T \simeq T_I$, a certain fraction of the atoms will be ionised, thereby leading to a composition made of protons, electrons, and neutral hydrogen atoms. Let n_p, n_e, and n_H denote the number densities of protons, electrons, and neutral hydrogen atoms, respectively, with $n_p = n_e$. Further, let the total number density be $n = n_p + n_H$. The ionisation fraction is defined to be the ratio $y \equiv [n_p/(n_p + n_H)] = (n_p/n) = (n_e/n)$. We are interested in determining y as a function of the temperature T and density n.

When the reaction $p + e \Leftrightarrow H$ reaches equilibrium, the total chemical potential on the left-hand side should be equal to that on the right-hand side, giving $\mu_e + \mu_p = \mu_H$. Assuming that all the components behave as an ideal gas, the chemical potential is related to the number density and temperature by

$$z = e^{\beta\mu} = \frac{n\lambda_T^3}{g_s} = \frac{n\hbar^3}{g_s}\left(\frac{2\pi}{mk_BT}\right)^{3/2} \tag{5.219}$$

[see Eq. (5.153)], which can be rewritten as

$$n = g_s\left(\frac{mk_BT}{2\pi}\right)^{3/2}\frac{1}{\hbar^3}e^{\beta\mu}. \tag{5.220}$$

In this particular case, we have

$$n_e = \frac{2}{\hbar^3}\left(\frac{m_e k_B T}{2\pi}\right)^{3/2} e^{\beta\mu_e}, \quad n_p = \frac{2}{\hbar^3}\left(\frac{m_p k_B T}{2\pi}\right)^{3/2} e^{\beta\mu_p}, \quad (5.221)$$

$$n_H = \frac{4}{\hbar^3}\left[\frac{(m_p + m_e)k_B T}{2\pi}\right]^{3/2} e^{\beta\mu_H} e^{\beta|E_0|}. \quad (5.222)$$

The extra factor $\exp\beta|E_0| = \exp(-\beta E_0)$ in n_H arises because of the binding energy of the hydrogen atom relative to free electrons and photons. The g_s factors for the three species are determined as follows: The ground state of a hydrogen atom has two possibilities for electron spin and two for proton, giving $g_s = 4$; at the temperatures of interest, we may treat all these states as having the same energy. The proton and the electron have two possible spin orientations, giving $g_s = 2$ for each of them.

Forming the combination $(n_e n_p / n_H)$ and using the conservation of chemical potentials, we get

$$\frac{n_e n_p}{n_H} = \left(\frac{k_B T}{2\pi}\right)^{3/2} \frac{1}{\hbar^3}\left(\frac{m_e m_p}{m_e + m_p}\right)^{3/2} e^{-\beta|E_0|} \simeq \left(\frac{m_e k_B T}{2\pi\hbar^2}\right)^{3/2} e^{-\beta|E_0|},$$

$$(5.223)$$

where we have used the fact that $m_e \ll m_p$. Using the definition of the ionisation fraction y, we can write as

$$\frac{y^2}{1 - y} = \frac{1}{n}\left(\frac{m_e k_B T}{2\pi\hbar^2}\right)^{3/2} e^{-\beta|E_0|} \quad (5.224)$$

(this relation is called Saha's equation). This quadratic equation in y can be easily solved to give y in terms of n and T. For pure hydrogen, this expression becomes

$$\frac{y^2}{1 - y} = \frac{4.01 \times 10^{-9}}{\rho} T^{3/2} \exp\left(-\frac{1.578 \times 10^5}{T}\right), \quad (5.225)$$

where ρ is in grams per cubic centimeter and T is in Kelvin. (The numerical coefficient is 4.01×10^{-6} if SI units are used.) To understand the temperature dependence of ionisation better, let us determine the temperature at which half the atoms are ionised [i.e., $y = (1/2)$] as a function of the density. Setting $y = 1/2$ gives the following curve in the ρ–T plane:

$$\rho_{1/2}(T) = 0.51 \text{ g cm}^{-3}\left(\frac{T}{1.6 \times 10^5 \text{ K}}\right)^{3/2} \exp\left(-\frac{1.578 \times 10^5 \text{ K}}{T}\right), \quad (5.226)$$

which is shown in Fig. 5.1.

We see that the temperature changes by a factor of only 3 or so when the density changes by ~ 6 orders of magnitude. Thus, for a wide range of interesting densities, $y = 1/2$ corresponds to a temperature of $\sim 10^4$. It should be stressed that

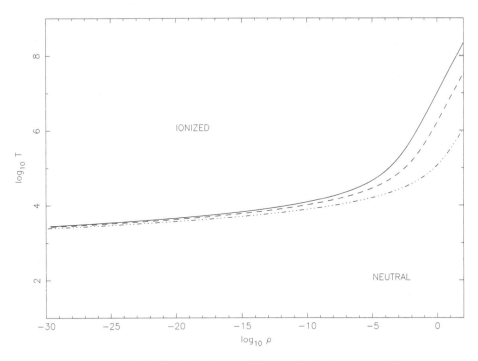

Fig. 5.1. The $\rho-T$ relation in ionisation equilibrium for hydrogen gas. The curves are obtained for different percentages of ionisation, corresponding to $y = 0.9$ (solid curve), 0.5 (dashed curve), and 0.05 (dashed–dotted curve).

significant ionisation (50% in the above case) occurs at a temperature that is well below the temperature corresponding to the binding energy E_0. This is because the prefactor to the exponential in Eq. (5.224), $(1/n)(m_e k_B T/2\pi\hbar^2)^{3/2} = (n\lambda_T^3)^{-1}$, is quite large; in fact, the quantity $(n\lambda_T^3)^{-1}$ must necessarily be large for an ideal-gas approximation to be valid. The temperature at which half the atoms are ionised, $T_{1/2}$, is of the order of $k_B T_{1/2} \approx |E_0|/10$.

The analysis leading to Eq. (5.224) can be easily generalised to more complex atomic systems. In general, the right-hand side of Eq. (5.224) will be multiplied by $2Z_{i+1}/Z_i$, where Z_n is the partition function for the atom with n electrons removed. The same correction, with $Z_{i+1} = 1$, will take care of the fact that the neutral hydrogen atom can exist in different energy states $E_n = (-E_0/n^2)$. The sum over all n in $Z = \sum \exp(-\beta E_n)$ formally diverges, but in physical situations, high-n states are modified by other processes. In a solid, for example, the Coulomb interaction of nearby atoms contribute $\Delta E \simeq e^2 n^{1/3}$, where n is the number density. This provides a cutoff at $|E_k| \simeq e^2 n^{1/3}$, when $e^2 n^{1/3} \gg k_B T$. At high temperatures the cutoff arises from a process called Debye shielding, which corrects the energy levels by $\Delta E \simeq e^2/\lambda_D$ with $\lambda_D^2 = (k_B T/4\pi e^2 n)$. [This will be discussed in Chap. 12; see Eq. (12.33).] Then the cutoff is determined by $|E_k| \simeq e^2/\lambda_D$ if $k_B T \gg (e^2/\lambda_D)$.

When the possibility of ionisation is taken into account, the equation of state for the system relating the total pressure to the total density and temperature becomes quite complicated. For a partially ionised system, the pressure and energy are given by

$$P = n(1 + y)k_B T, \quad E = \frac{3}{2}n(1 + y)k_B T + n|E_0|y. \tag{5.227}$$

(In the case of hydrogen, this relation is obvious from the expressions for pressure, energy, and ionisation factor.) The corresponding differential relation is

$$\frac{dP}{P} = \left[1 + \mathcal{D}\left(\frac{3}{2} + \frac{|E_0|}{k_B T}\right)\right]\frac{dT}{T} + (1 - \mathcal{D})\frac{d\rho}{\rho}, \tag{5.228}$$

where

$$\mathcal{D}(y) = \frac{y(1 - y)}{(2 - y)(1 + y)}. \tag{5.229}$$

This allows us to compute all the relevant exponents introduced in Section 5.6. As y approaches 0 or 1, $\mathcal{D} \to 0$ and all the Γ's reach the ideal-gas value of 5/3. This is to be expected because a fully neutral or fully ionised gas will behave as an ideal gas. However, when the temperature is near the half-ionisation point, Γ_3 drops rapidly from the ideal-gas value (of 5/3) to a much lower value. The physical reason for this rapid drop is the following: If an ideal gas is compressed adiabatically, then (because $\Gamma_3 = 5/3$), $P \propto \rho^{5/3}$, $T \propto P/\rho \propto \rho^{2/3}$, and the gas initially heats up. Because ionisation is exponentially sensitive to temperature, the heating up of the gas accelerates the ionisation. Consequently, the energy supplied in adiabatic compression goes into ionisation and not to the thermal motion of gas molecules. Thus, in the presence of ionisation, T does not increase as rapidly as $\rho^{2/3}$ and the value of Γ_3 is less than its value for an ideal gas.

Exercise 5.19
For long winter evenings IV: Evaluate the exponents for partially ionised gas and verify the statements given above.

In the discussion above, we considered the ionisation of hydrogen, assuming that it exists either as a neutral atom in the ground state or as a proton and an electron. In general, the neutral atom could also be in an excited state rather than in the ground state. The fraction of hydrogen atoms in the energy level E_n is the product of the fraction of the neutral hydrogen atoms and the relative probability of the atom's being found in the excited state with energy E_n compared with being in the ground state. Because the latter probability is given by the standard Boltzmann factor, we can easily compute the fraction of hydrogen atoms in any excited state. For example, consider the fraction of neutral hydrogen atoms in the $n = 2$ state. This is given by

$$\frac{n_2}{n} = \frac{n_2}{n_{\text{neutral}}} \frac{n_{\text{neutral}}}{n} \cong \frac{n_2}{(n_1 + n_2 + \cdots)}(1 - y) \cong \frac{n_2}{n_1}(1 - y)$$

$$= \frac{g_2}{g_1} e^{-\Delta E/k_B T}[1 - y(n_e, T)], \tag{5.230}$$

where ΔE is the energy difference between the states $n = 2$ and $n = 1$, $g_2 = 2n^2 = 8$ is the degeneracy of $n = 2$ state, $g_1 = 2$, and we have assumed that $n_2, n_3, \ldots,$ $\ll n_1$. Writing $(1/n) = (y/n_e)$ in Eq. (5.224) we can solve for $(1 - y)$ and get

$$1 - y = \left[1 + \frac{1}{n_e} \left(\frac{m_e k_B T}{2\pi\hbar^2} \right)^{3/2} e^{-|E_0|/k_B T} \right]^{-1}. \tag{5.231}$$

Substituting Eq. (5.231) into Eq. (5.230), we find that

$$n_2 = 4ne^{-\Delta E/k_B T}\left[1 + \frac{1}{n_e} \left(\frac{m_e k_B T}{2\pi\hbar^2} \right)^{3/2} e^{-|E_0|/k_B T} \right]^{-1}. \tag{5.232}$$

This expression shows that n_2 has a maximum around a temperature of $\sim 10^4$ K. At lower temperatures, the Boltzmann factor reduces the probability of finding the hydrogen atom in the excited state; at higher temperatures, there is a paucity of neutral atoms and hence the fraction comes down. We will see in Vol. II that this idea finds practical application in the study of stellar physics.

Our analysis can also be easily generalised for elements heavier than hydrogen. Other things being equal, atoms with low ionisation potentials will be relatively more ionised at any given temperature. Among the common elements, helium and neon are hardest to ionise and have binding energies in excess of 20 eV; (hydrogen, carbon, nitrogen, oxygen, fluorine, phosphorus, sulfur, chlorine and argon) are the next in line with binding energies in the range of 10–20 eV, and the easiest to ionise are (lithium, sodium, magnesium, aluminium, potassium, calcium, and silicon), etc., with binding energies of ~ 5 eV. In solar atmosphere ($T = 6000$ K), for example, the exponential factors $\exp(-|E_0|/k_B T)$ for sodium and hydrogen have a relative value of $\sim 10^7$, giving $N(\text{Na}^+)/N(\text{H}^+) \approx 10^7$. This more than compensates for the relative abundance of sodium to hydrogen (which is $\sim 10^{-6}$) in solar atmospheres and can produce significant effects.

Finally, note that elements with atomic number greater than unity can have multiple ionisation states. In helium, for example, one electron can be removed (leaving the other electron bound to the atom), thereby producing a He^+ ion. The complete dissociation of helium, removing both the electrons, will give a doubly ionised He^{++} ion. Equation (5.223) can be easily generalised to take into account multiple ionisation states. We get in its place the relation

$$\frac{n_e n_i}{n_0} = \frac{2Z_i}{Z_0} \left(\frac{m_e k_B T}{2\pi\hbar^2} \right)^{3/2} e^{-\beta|E_0|}, \tag{5.233}$$

where n_0 is the number density of neutral atoms in the ground state, n_i is the number density of atoms in the ith ionisation state, and Z_i and Z_0 are the partition functions for the ion in the ith ionisation state and for the neutral atom, respectively.

A similar analysis can also be used to study the ionisation equilibrium in much more complicated cases having more than one species. As an example, the procedure for the case of a hydrogen–helium mixture with weight fractions X and Y is briefly mentioned. In this case, there are three ionisation energies, corresponding to hydrogen ($\chi_H^0 = 13.6$ eV), neutral helium ($\chi_{He}^0 = 24.6$ eV), and singly ionised helium ($\chi_{He}^1 = 54.4$) eV. The corresponding fractional number densities of different species of particles are taken to be

$$x_H^0 = \frac{n_H^0}{n_H}, \quad x_H^1 = \frac{n_H^1}{n_H}, \quad x_{He}^0 = \frac{n_{He}^0}{n_{He}}, \quad x_{He}^1 = \frac{n_{He}^1}{n_{He}}, \quad x_{He}^2 = \frac{n_{He}^2}{n_{He}},$$

(5.234)

with the conditions

$$x_H^0 + x_H^1 = 1, \quad x_{He}^0 + x_{He}^1 + x_{He}^2 = 1. \tag{5.235}$$

The number of electrons e per atom is given by

$$e = \left[X x_H^1 + \frac{1}{4} Y \left(x_{He}^1 + 2 x_{He}^2 \right) \right] \mu, \tag{5.236}$$

where μ is the molecular weight introduced in Eq. (5.92). Saha's ionisation equation now gives the three constraints

$$\frac{x_H^1}{x_H^0} \frac{e}{e+1} = K_H^0, \quad \frac{x_{He}^1}{x_{He}^0} \frac{e}{e+1} = K_{He}^0, \quad \frac{x_{He}^2}{x_{He}^1} \frac{e}{e+1} = K_{He}^1, \tag{5.237}$$

with

$$K_i^r = \frac{2 Z_{r+1}}{Z_r} \left(\frac{k_B T}{P_{\text{gas}}} \right) \left(\frac{m_e k_B T}{2\pi\hbar^2} \right)^{3/2} \exp\left[-\left(\chi_i^r / k_B T \right) \right]. \tag{5.238}$$

[We have written the $(1/n)$ factor of Saha's equation as $(k_B T / P_{\text{gas}})$ which is a convenient parameterisation used very often.] Given the values of X, Y, P_{gas}, and T, the six Eqs. (5.235), (5.236), and (5.237) determine the six unknown quantities $x_H^0, x_H^1, x_{He}^0, x_{He}^1, x_{He}^2$, and e. These are coupled algebraic equations that have to be solved numerically.

The qualitative behaviour of the solution is as follows: At low temperatures, both hydrogen and helium remain neutral. As the temperature is increased, hydrogen becomes ionised and helium becomes singly ionised. When the temperature increases still further, the fraction of singly ionised helium reaches a maximum and starts decreasing and simultaneously the fraction of doubly ionised helium starts to increase. At still higher temperatures, both hydrogen and helium will be completely ionised.

5.12.2 Pair Creation

The annihilation of an electron and a positron will lead to the emission of photons; if the photons are energetic enough, it is also possible to produce electron–positron pairs out of the radiation. We now discuss the equilibrium properties of the reaction $e^- + e^+ \Longleftrightarrow \gamma$. In particular we are interested in the number density of electron–positron pairs as a function of temperature.

Because photons have zero chemical potential [see the discussion following Eq. (5.207)], the conservation of chemical potentials gives $\mu_+ + \mu_- = 0$. Treating electrons and positrons as ideal gas, we can use Eq. (5.220) to relate the chemical potential to the number densities $n_\pm$, with μ replaced with $\mu - mc^2$ to take into account the change in the rest mass. Taking the product of number densities and using the conservation law for the chemical potential, we get

$$n_+ n_- = 4\left(\frac{m k_B T}{2\pi\hbar^2}\right)^3 \exp(-2mc^2/k_B T). \qquad (5.239)$$

If n_0 is the number density of electrons in the absence of pair production, then $n_- = n_+ + n_0$. This gives another relation between n_+ and n_-. The n_+ can be determined by solving these two relations simultaneously, and we get:

$$n_+ = n_- - n_0 = -\frac{1}{2}n_0 + \left[\frac{1}{4}n_0^2 + \frac{1}{2}\left(\frac{\hbar}{mc}\right)^{-6}\left(\frac{k_B T}{\pi mc^2}\right)^3 e^{-2mc^2/k_B T}\right]^{1/2}. \qquad (5.240)$$

To understand the temperature dependence of pair production, we can plot a curve in the T–ρ plane corresponding to $n_+ = n_0$. This curve is shown in Fig. 5.2.

As in the case of ionisation, the temperature at which $n_+ = n_0$ varies very little when the density changes by several orders of magnitude. Most of the pair production takes place around 10^8–10^9 K, which is lower than the temperature $T_0 = (m_e c^2/k_B) \simeq 5 \times 10^9$ K, corresponding to the rest-mass energy of the electron. This occurs for the same reason as in the case of ionisation, discussed before.

The above analysis, in fact, is valid only for $k_B T \ll mc^2$ because the quantum effects have been ignored. When $k_B T \gtrsim mc^2$, there will be a copious production of pairs and $n_\pm \gg n_0$. In that case, charge neutrality requires that $n_+ \approx n_-$. Because the chemical potential is uniquely determined by the number density, this implies that $\mu_- \approx \mu_+$. Combined with the conservation law $\mu_- + \mu_+ = 0$,

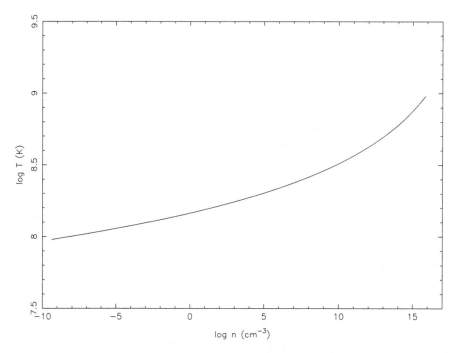

Fig. 5.2.   The curve relates the density and the temperature at which the number density of positrons produced is equal to the original number density of electrons.

$\mu_{\pm} \approx 0$. Then the number densities become

$$n_+ = n_- = \frac{1}{\pi^2 \hbar^3} \int_0^\infty \frac{p^2 dp}{e^{\beta \epsilon} + 1}, \quad \epsilon = \sqrt{p^2 c^2 + m^2 c^4}. \tag{5.241}$$

When $k_B T \gtrsim mc^2$, we can set $\epsilon \cong pc$ and obtain

$$n_+ = n_- \cong \frac{3\zeta(3)}{2\pi^2} \left(\frac{k_B T}{\hbar c} \right)^3. \tag{5.242}$$

This number is comparable with the number density of photons at this temperature and the pair production is a dominant feature.

5.12.3 Nuclear Statistical Equilibrium

The arguments that were used to derive the conditions of statistical equilibrium in the case of ionisation and pair production are fairly general and can be used to describe a wide class of equilibrium reactions. For example, when several different nuclei are in statistical equilibrium because of nuclear reactions, the number density of any particular species with atomic number Z, atomic weight

A, and mass $M(Z, A)$ will be given by

$$n_i \equiv n(Z, A) = \frac{g(Z, A)A^{3/2}}{2^A \theta^{A-1}} n_p^Z n_n^{A-Z} \exp\left[\frac{Q(Z, A)}{k_B T}\right], \qquad (5.243)$$

where

$$Q(Z, A) \equiv c^2[Zm_p + (A - Z)m_n - M(Z, A)], \quad \theta \equiv \left(\frac{m_H k_B T}{2\pi\hbar^2}\right)^{3/2}, \qquad (5.244)$$

$g(Z, A)$ is the nuclear partition function, and n_p and n_n give the number density of protons and neutrons, respectively. These are, in turn, determined by the conservation of charge and baryon number expressed in the form

$$\sum_i n_i A_i = \frac{\rho}{m_H}, \quad \sum_i n_i Z_i = \left(\frac{\rho}{m_H}\right) Y_e, \qquad (5.245)$$

where Y_e is the mean number of electrons per baryons. This result will be of use in the study of stellar remnants in Vol. II.

5.13 Time Evolution of Distribution Functions

The discussion so far has been confined to the steady-state behaviour of systems. We next consider some general features of the time evolution of a macroscopic systems. The most powerful approach to this problem is through the distribution function introduced in Chap. 3. Our aim will be to obtain an equation for the evolution of the distribution function and to develop useful approximation schemes for solving it. This topic will be further elaborated on in Chaps. 6, 8, and 10.

In the absence of creation or destruction of particles or scattering between the particles, the distribution function is constant along the phase-space trajectory of any fiducial particle. Let the phase-space trajectory of a fiducial particle be denoted by a parameterised curve $x^i(s)$, $p^\alpha(s)$, where s any well-defined parameter. (For massive particles, s could be taken to be the proper time.) Here $x^i = (ct, \mathbf{x})$ is a four vector and $p^\alpha = (\mathbf{p})^\alpha$ are the components of the momentum three vector; the fourth component, $p^0 = (E_\mathbf{p}/c)$, is the energy of the particle (divided by c) and is given in terms of p^α by $E_\mathbf{p} = (c^2 p^2 + m^2 c^4)^{1/2}$. The conservation of the distribution function along the fiducial trajectory implies that

$$\begin{aligned}
0 = \frac{df}{ds} &= \frac{dx^i}{ds}\frac{\partial f}{\partial x^i} + \frac{dp^\alpha}{ds}\frac{\partial f}{\partial p^\alpha} = \frac{E_\mathbf{p}}{mc^2}\frac{\partial f}{\partial t} + \frac{\mathbf{p}}{m}\cdot\frac{\partial f}{\partial \mathbf{x}} + \frac{d\mathbf{p}}{ds}\cdot\frac{\partial f}{\partial \mathbf{p}} \\
&= \frac{E_\mathbf{p}}{mc^2}\left(\frac{\partial f}{\partial t} + \mathbf{v}\cdot\frac{\partial f}{\partial \mathbf{x}} + \frac{mc^2}{E_\mathbf{p}}\frac{d\mathbf{p}}{ds}\cdot\frac{\partial f}{\partial \mathbf{p}}\right).
\end{aligned} \qquad (5.246)$$

Using the relation $(E_{\mathbf{p}}/mc^2)\,ds = dt$, we can rewrite this equation as

$$\frac{\partial f}{\partial t} + \mathbf{v}\cdot\frac{\partial f}{\partial \mathbf{x}} + \frac{d\mathbf{p}}{dt}\cdot\frac{\partial f}{\partial \mathbf{p}} = 0, \tag{5.247}$$

which is same as the corresponding nonrelativistic equation. In the presence of sources or sinks for the particles or if there are collisions (scattering) between particles, then the right-hand side of Eq. (5.247) will not be zero and will contain terms that will represent creation, destruction, and scattering of the particles.

The above analysis applies equally well to zero-mass particles travelling with the velocity of light. Such particles, when not affected by systematic external forces, will have $\dot{\mathbf{p}} = 0$ and $\mathbf{v} = c\hat{\mathbf{k}}$, where $\hat{\mathbf{k}}$ is a unit vector in the direction of propagation. The distribution function for such zero-mass particles satisfies the equation

$$\frac{\partial f}{\partial t} + c\hat{\mathbf{k}}\cdot\frac{\partial f}{\partial \mathbf{x}} = \mathcal{C}, \tag{5.248}$$

where $\mathcal{C}$ describes creation, destruction, and scattering of such particles, if present.

The distribution function f and the occupation number n are proportional to each other, with $n = (2\pi\hbar)^3 f$, because, by definition, $dN \equiv n[d^3\mathbf{x}d^3\mathbf{p}/(2\pi\hbar)^3] \equiv f d^3\mathbf{x}d^3\mathbf{p}$. The equations for f can be easily converted to give equations for n whenever needed.

Given the time evolution of the distribution function and the definitions for different moments introduced in Chap. 3, Section 3.6, we can write the evolution equations that are satisfied by the various moments. To do this, it is convenient to go back to Eq. (5.246) and write it as

$$u^i\frac{\partial f}{\partial x^i} + F^\alpha\frac{\partial f}{\partial p^\alpha} = \mathcal{C}, \tag{5.249}$$

where u^i is the four velocity, F^α are the spatial components of the four force, $F^i = dp^i/ds = (E_{\mathbf{p}}/mc^2)(dp^i/dt)$, and $\mathcal{C}$ denotes processes that violate the conservation of the distribution function. This equation can be rewritten as

$$\frac{\partial}{\partial x^i}(u^i f) + \frac{\partial}{\partial p^\alpha}(F^\alpha f) = \mathcal{C} \tag{5.250}$$

in all physically relevant situations. It is obvious that u^i can be brought inside the integral sign as x^i and $u^i = p^i/m$ are independent variables. The force F^α can be brought inside the integral sign if it depends on only x^i; the only important exception is in the case of electromagnetism, in which $F^i = (q/m)F^i{}_k u^k = (q/m^2)F^i{}_k p^k$.

But even in this case the result is true, because

$$F^\alpha \frac{\partial f}{\partial p^\alpha} = \frac{\partial}{\partial p^\alpha}(F^\alpha f) - f \frac{\partial F^\alpha}{\partial p^\alpha} = \frac{\partial}{\partial p^\alpha}(F^\alpha f) - f F^\alpha_\mu \frac{\partial p^\mu}{\partial p^\alpha}$$

$$= \frac{\partial}{\partial p^\alpha}(F^\alpha f) - f F^\alpha_\mu \delta^\mu_\alpha = \frac{\partial}{\partial p^\alpha}(F^\alpha f), \tag{5.251}$$

where we have used the fact that for the antisymmetric tensor $F^\mu_\mu = 0$.

We now multiply Eq. (5.250) by $m(d^3 \mathbf{p}/E_\mathbf{p})$ and integrate over the momentum space. Denoting the result of integration on of the right-hand side by $\mathcal{C}_1$, we get

$$\frac{\partial}{\partial x^i} \left(\int \frac{d^3 \mathbf{p}}{E_\mathbf{p}} p^i f \right) + m \int \frac{d^3 \mathbf{p}}{E_\mathbf{p}} \frac{\partial}{\partial p^\alpha}(F^\alpha f) = \mathcal{C}_1. \tag{5.252}$$

The first term on left can be written as $(\partial J^i / \partial x^i)$ by use of the definition of J^i in Eq. (3.43). The integral in the second term becomes, on integration by parts and with $(\partial E_\mathbf{p}/\partial p^\alpha) = v^\alpha$,

$$\int \frac{d^3 \mathbf{p}}{E_\mathbf{p}} \frac{\partial}{\partial p^\alpha}(F^\alpha f) = \int d^3 \mathbf{p} \frac{\partial}{\partial p^\alpha} \left(\frac{F^\alpha f}{E_\mathbf{p}} \right) + \int \frac{d^3 \mathbf{p}}{E_\mathbf{p}} \left(\frac{v^\alpha F_\alpha}{E_\mathbf{p}} \right) f. \tag{5.253}$$

The first term on the right vanishes on integration for any reasonable f. Using $F^\alpha = (dp^\alpha/ds) = (E_\mathbf{p}/mc^2)(dp^\alpha/dt)$ in the second term, we get

$$m \int \frac{d^3 \mathbf{p}}{E_\mathbf{p}} \frac{\partial}{\partial p^\alpha}(F^\alpha f) = \frac{1}{c^2} \int \frac{d^3 \mathbf{p}}{E_\mathbf{p}} \left(\mathbf{v} \cdot \frac{d\mathbf{p}}{dt} \right) f = \int \frac{d^3 \mathbf{p}}{E_\mathbf{p}} \left[\frac{d(E_\mathbf{p}/c^2)}{dt} f \right]. \tag{5.254}$$

Putting these results together, we find that

$$\frac{\partial J^i}{\partial x^i} + \int \frac{d^3 \mathbf{p}}{E_\mathbf{p}} \left[f \frac{d(E_\mathbf{p}/c^2)}{dt} \right] = \mathcal{C}_1. \tag{5.255}$$

The second term on the left-hand side is a relativistic effect that associates a mass density $E_\mathbf{p}/c^2$ with energy $E_\mathbf{p}$ and vanishes in the nonrelativistic limit of $c \to \infty$. This term also vanishes in the extreme relativistic limit, in which the external force is ignorable. In either case, we get $(\partial J^i / \partial x^i) = \mathcal{C}_1$.

In a similar manner, we can multiply Eq. (5.250) by p^j and integrate over $(d^3 p/E_\mathbf{p})$ to obtain an equation describing the evolution of T^{ij}. The form of this equation depends on the nature of the force field F^α. When there is no external force, the equation reduces to

$$\frac{\partial T^{ij}}{\partial x^i} = \mathcal{C}^j, \tag{5.256}$$

where $\mathcal{C}^j$ denotes the moment of the right-hand side.

In several physical situations the collisional terms will satisfy standard conservation laws and will not act as a source or sink for particles or energy. Then we will get $(\partial J^i/\partial x^i) = 0; (\partial T^{ij}/\partial x^i) = 0$. Although these equations are exact, they do not form a closed system. This is a general feature that arises whenever moments of the evolution equation for f are taken. The time derivative of the moment calculated with n powers of momenta will be determined by the derivatives of the moment with $(n + 1)$ powers of momenta; the system will never close with a finite number of moment equations.

To make any progress, we have to determine the form of the stress tensor T^{ik} and the flux vector J^i by some external, physical requirements. If we can thereby express these quantities in terms of other variables – such as density, pressure, velocity, etc. – we will have a closed set of equations that we can solve. When collisions are effective, i.e., when the mean free path l is far less than the scale R of the system, we can do this in a systematic manner by expanding the distribution function in perturbation series, with each term being lower than the preceding one by the small factor l/R, that is, we use the ansatz

$$f = f_0 + f_1 + f_2 \cdots \tag{5.257}$$

in the transport equation where $f_n = \mathcal{O}[(l/R)^n]$ and solve the equation by standard perturbation methods order by order. This will be discussed in detail in Chap. 8.

In practise, we are more interested in the form of T^{ik} and J^i corresponding to the above series than in the actual form of f_n's. Once we know the form of T^{ik} and J^i, we can close the moment equations and work with them rather than with the distribution function. In particular, moment equations are much easier to adapt for different physical situations with different boundary conditions, etc. Given J^i and T^{ab}, Eqs. (5.255), (5.256) will govern the evolution of several bulk physical parameters of the system such as the energy density, momentum flux, etc. In Exercises 5.20 and 5.21, we discuss the form of J^i and T^{ab} and the resulting equations for two important cases that will be studied in detail in Chap. 8.

Exercise 5.20
Equations of motion for an ideal fluid: The stress tensor and the current for an ideal fluid were obtained in Chap. 3, Section 3.6 to be $T_b^a = (p + \rho)u^a u_b - p\delta_b^a$ and $J^a = nu^a$. Determine the equations of motion for the fluid from the conservation law for these quantities. Show that the nonrelativistic limit of these equations reduces to

$$\frac{\partial \rho}{\partial t} + \nabla \cdot (\rho \mathbf{u}) = 0, \quad \rho \left[\frac{\partial \mathbf{u}}{\partial t} + (\mathbf{u} \cdot \nabla)\mathbf{u} \right] \cong -\nabla p \tag{5.258}$$

and the condition that the entropy of the fluid elements does not change during the flow.

Exercise 5.21
Stress tensor for a nonideal fluid: T^{ik} and J^i of the ideal fluid were obtained in Chap. 3, with the gradients in temperature, number density, and bulk velocity being ignored. In the next order of approximation, in which these gradients are taken into account, we expect T_{ik} and J_i will contain terms that are proportional to the gradients $(\partial T/\partial x^i)$ and $(\partial u_k/\partial x^i)$. We write these expressions, correct to linear order in the gradients, as

$$T_{ik} = wu_i u_k - p\eta_{ik} + \tau_{ik}, \quad J_i = nu_i + h_i, \tag{5.259}$$

where $w = (p + \rho)$ is the enthalpy. In a relativistic theory, because all energy fluxes involve equivalent mass fluxes, it is necessary to define h_i, etc., more precisely as follows. In the proper rest frame of the fluid element we shall demand that (1) the momentum of the element be zero and (2) the energy be expressible in terms of other thermodynamic variables in the same functional form as in the absence of dissipative processes. This requires that, in the proper frame, $\tau^{0i} = 0$, which can be written in four-dimensional form as $\tau_{ik}u^i = 0$. Similarly, we demand that $h_i u^i = 0$ so that in the rest frame n^0 is same as the proper number density n. Using this condition and the form of the expressions in Eqs. (5.259) show that

$$\frac{\partial}{\partial x^i}\left(su^i - \frac{\mu}{T}h^i\right) = -h^i \frac{\partial}{\partial x^i}\left(\frac{\mu}{T}\right) + \frac{\tau_i^k}{T}\frac{\partial u^i}{\partial x^k}. \tag{5.260}$$

The left-hand side is the divergence of the entropy current $[su^i - (\mu/T)h^i]$, which was zero in the absence of dissipative terms. In the presence of dissipation, the entropy has to increase and the right-hand side of Eq. (5.260) should be positive. Further, because we are computing the first-order corrections, the quantities τ_{ik} and h_i must be linear in the gradients $(\partial u^i/\partial x^k)$ and $[\partial(\mu/T)/\partial x^k]$. Taking $\tau_{ab} = M_{abik}(\partial u^i/\partial x_k)$, $h^i = N^{ik}[\partial(\mu/T)/\partial x^k]$, substituting into Eq. (5.260) and using the conditions $\tau_{ik}u^i = 0$, $h_i u^i = 0$ along with the positivity of right-hand side, determine the forms of τ_{ik} and h_i:

$$\tau^{ik} = -c\eta\left(\frac{\partial u_i}{\partial x^k} + \frac{\partial u_k}{\partial x^i} - u_k u^l \frac{\partial u_i}{\partial x^l} - u_i u^l \frac{\partial u_k}{\partial x^l}\right) - c\left(\zeta - \frac{2}{3}\eta\right)\frac{\partial u^l}{\partial x^l}(g_{ik} - u_i u_k),$$
$$\tag{5.261}$$

$$h^i = \frac{\kappa}{c}\left(\frac{nT}{w}\right)^2\left[\frac{\partial}{\partial x^i}\left(\frac{\mu}{T}\right) - u_i u^k \frac{\partial}{\partial x^k}\left(\frac{\mu}{T}\right)\right], \tag{5.262}$$

where the coefficients η and ζ describe viscosity – arising from velocity gradients – and κ describes thermal conduction – arising from the temperature gradient.

5.14 Evolution under Scattering

Let us next consider a more complicated situation in which the distribution function changes because of the existence of a nonzero term on the right-hand side of Eq. (5.249). This can arise, in general, because of two kinds of processes: (1) The presence of sources or sinks can lead to creation or destruction of the particles or (2) scattering between particles can populate or depopulate different regions of momentum space.

Of these, the first type of term is usually important in radiative processes when photons can be absorbed or emitted and will be discussed in Chap. 6. The scattering (or collision) of the particles occurs for both material particles and for photons. Collisions between particles could be either between particles of the same kind (say, molecules of a gas colliding with each other) or it could be between particles of different kind (say, between photons and electrons). The nature of the scattering will also depend on whether the forces between the particles are of short range (like intermolecular forces) or of long range (as in a plasma containing charged particles). In the case of photons, the most important process is the scattering of photons by charged particles.

The study of collisions in gas molecules will be taken up in Chap. 8, and the case of long-range interactions will be discussed in Chaps. 9 and 10. Here we discuss the collision between photons and charged particles in order to illustrate how the collision term C affects the evolution.

The specific process we study is the scattering between electrons and photons. A general scattering event will be between a photon of momentum $\mathbf{k}'$ and an electron of momentum $\mathbf{p}'$, leading to a photon of momentum $\mathbf{k}$ and electron of momentum $\mathbf{p}$. Quantum mechanically, this scattering should be thought of as absorption of a photon (with momentum $\mathbf{k}'$) and an electron (with momentum $\mathbf{p}'$) followed by the emission of a photon (with momentum $\mathbf{k}$) and an electron (of momentum $\mathbf{p}$). In this case, the gain in the number of photons that is due to scattering can be written as

$$\left[\frac{dn(\mathbf{k})}{dt}\right]_{\text{gain}} = \int \frac{V d^3\mathbf{p}}{(2\pi\hbar)^3} \int \frac{V d^3\mathbf{p}'}{(2\pi\hbar)^3} \int \frac{V d^3\mathbf{k}'}{(2\pi\hbar)^3}$$

$$\times \mathcal{R}(\mathbf{p}', \mathbf{k}'; \mathbf{p}, \mathbf{k}) n_e(\mathbf{p}')[1 - n_e(\mathbf{p})] n(\mathbf{k}')[1 + n(\mathbf{k})]. \quad (5.263)$$

The factor $n(n + 1)$ arises because of the probability of the absorption of a photon of momentum $\mathbf{k}'$ and emission of a photon of momentum $\mathbf{k}$. The proportionality to $n_e(1 - n_e)$ arises for a similar reason, taking into account the fermionic nature of electrons (see the discussion in Chap. 4, Subsection 4.5.2). The quantity $\mathcal{R}$ represents the rate for this scattering event and has the dimension of inverse time (s^{-1}). The integral over $\mathbf{p}$ and $\mathbf{p}'$ takes care of the density of states for the initial and the final states of the electron and the integral over $\mathbf{k}'$ takes care of the photon distribution function. The corresponding term for loss of photons with momentum $\mathbf{k}$ is given by

$$\left[\frac{dn(\mathbf{k})}{dt}\right]_{\text{loss}} = \int \frac{V d^3\mathbf{p}}{(2\pi\hbar)^3} \int \frac{V d^3\mathbf{p}'}{(2\pi\hbar)^3} \int \frac{V d^3\mathbf{k}'}{(2\pi\hbar)^3}$$

$$\times \mathcal{R}(\mathbf{p}, \mathbf{k}; \mathbf{p}', \mathbf{k}') n_e(\mathbf{p})[1 - n_e(\mathbf{p}')] n(\mathbf{k})[1 + n(\mathbf{k}')], \quad (5.264)$$

so that the net change that is due to scattering becomes

$$
\left[\frac{dn(\mathbf{k})}{dt}\right]_{\text{scatt}} = \int \frac{V d^3\mathbf{p}}{(2\pi\hbar)^3} \int \frac{V d^3\mathbf{p}'}{(2\pi\hbar)^3} \int \frac{V d^3\mathbf{k}'}{(2\pi\hbar)^3} \mathcal{R}(\mathbf{p}', \mathbf{k}'; \mathbf{p}, \mathbf{k})
$$

$$
\times \{n_e(\mathbf{p}')[1 - n_e(\mathbf{p})]n(\mathbf{k}')[1 + n(\mathbf{k})]
$$

$$
- n_e(\mathbf{p})[1 - n_e(\mathbf{p}')]n(\mathbf{k})[1 + n(\mathbf{k}')]\}, \tag{5.265}
$$

as time reversibility implies that $\mathcal{R}(\mathbf{p}, \mathbf{k}; \mathbf{p}', \mathbf{k}') = \mathcal{R}(\mathbf{p}', \mathbf{k}'; \mathbf{p}, \mathbf{k})$. The evolution of the number distribution will be governed by the equation $(dn/dt) = (dn/dt)_{\text{scat}}$. Before we simplify this expression further, it is interesting to ask about the steady-state distribution for a system undergoing such a scattering. Clearly, steady state will be reached when the expression in the braces in Eq. (5.265) vanishes, which gives the relation

$$
\frac{n_e(\mathbf{p})}{1 - n_e(\mathbf{p})} \frac{n(\mathbf{k})}{1 + n(\mathbf{k})} = \frac{n_e(\mathbf{p}')}{1 - n_e(\mathbf{p}')} \frac{n(\mathbf{k}')}{1 + n(\mathbf{k}')}. \tag{5.266}
$$

The left-hand side depends on the momentum, energy, etc., of the initial state, and the right-hand side depends on corresponding quantities after the scattering. It is clear that, if the logarithm of either side is a linear function of additive conserved quantities in the collision, then this equation will be automatically satisfied. Hence we conclude that, in steady state,

$$
\ln \frac{n_e(\mathbf{p})}{1 - n_e(\mathbf{p})} + \ln \frac{n(\mathbf{k})}{1 + n(\mathbf{k})} = \begin{cases} \text{linear function of} \\ \text{additive, conserved quantities .} \\ \text{in the collision} \end{cases} \tag{5.267}
$$

Because the relevant additive constants are energy and momentum, we must have,

$$
\ln \frac{n_e(\mathbf{p})}{1 - n_e(\mathbf{p})} + \ln \frac{n(\mathbf{k})}{1 + n(\mathbf{k})} = -\beta[\epsilon(\mathbf{p}) + E(\mathbf{k})] - \mathbf{U} \cdot [\mathbf{p} + \mathbf{k}] + \text{constant},
$$

$$
\tag{5.268}
$$

where β and $\mathbf{U}$ are constants. Equating the variables corresponding to the electron and the photon separately leads to two equations,

$$
\ln \frac{n_e(\mathbf{p})}{1 - n_e(\mathbf{p})} = -\alpha_1 - \beta\epsilon(\mathbf{p}) - \mathbf{U} \cdot \mathbf{p}, \quad \ln \frac{n(\mathbf{k})}{1 + n(\mathbf{k})} = -\alpha_2 - \beta E(\mathbf{k}) - \mathbf{U} \cdot \mathbf{k},
$$

$$
\tag{5.269}
$$

where α_1 and α_2 are constants. This gives the steady-state distribution functions

$$
n_e(\mathbf{p}) = \frac{1}{[\exp(\alpha_1 + \beta\epsilon + \mathbf{U} \cdot \mathbf{p}) + 1]}, \tag{5.270}
$$

$$
n(\mathbf{k}) = \frac{1}{[\exp(\alpha_2 + \beta E + \mathbf{U} \cdot \mathbf{k}) - 1]}. \tag{5.271}
$$

These are the Fermi–Dirac and Bose–Einstein distribution functions in a coordinate frame with bulk velocity $\mathbf{U}$. In the rest frame, with $\mathbf{U} = 0$, the distribution functions are specified by the constants α_1, α_2, and β. The constants α_1 and α_2 are determined by the total number of electrons and photons, respectively (which cannot change when the system evolves only through scattering), and β is determined by the *total* energy of the system. The quantity β^{-1} is proportional to the equilibrium temperature of the system, which is clearly the same for photons and electrons in steady state. Note that pure scattering drives the photons to a Bose–Einstein distribution and *not* to Planck distribution, because processes like emission and absorption – which can change the number of photons – have been ignored.

To do explicit computations with Eq. (5.265), we need the form of $\mathcal{R}$. In general, when a formal relativistically invariant form is needed, $\mathcal{R}$ is expressed in the form

$$\mathcal{R} = |M|^2 \left(\frac{c}{\hbar}\right) \left[\frac{(2\pi\hbar)^3}{V}\right]^3 \left(\frac{c}{2\epsilon_\mathbf{p}}\right) \left(\frac{c}{2E_\mathbf{k}}\right) \left(\frac{c}{2\epsilon_{\mathbf{p}'}}\right) \left(\frac{c}{2E_{\mathbf{k}'}}\right) \delta_D[\Delta\mathbf{P}]\delta_D[\Delta E/c],$$

(5.272)

where the different factors have the following interpretation: The two Dirac delta functions conserve total momentum and energy (with $\Delta\mathbf{P}$ and ΔE denoting the difference in momenta and energy between initial and final states). Each particle in the initial and the final state gets a factor $c/2E$ from normalisation. The factor $(2\pi\hbar)^3/V$ arises because of the manner in which the phase volume is normalised. Finally, $c/\hbar$ arises from the calculation of the transition rate if the matrix element $|M|^2$ is taken to be dimensionless. It is easy to verify that, in this normalisation, $\mathcal{R}$ has the dimension of per second. Substituting Eq. (5.272) into Eq. (5.265) and eliminating the momentum delta function by integrating over d^3p, we find that the final expression becomes

$$\left[\frac{dn(\mathbf{k})}{dt}\right]_{\text{scatt}} = \int d^3\mathbf{p}' \, d^3\mathbf{k}' \left(\frac{c}{2\epsilon_{\mathbf{p}'}}\right) \left(\frac{c}{2\epsilon_\mathbf{p}}\right) \left(\frac{c}{2E_{\mathbf{k}'}}\right) \left(\frac{c}{2E_\mathbf{k}}\right) \delta_D$$

$$\times [\Delta U/c] \left(\frac{c}{\hbar}\right) |M|^2 \{n_e(\mathbf{p}')[1 - n_e(\mathbf{p})]n(\mathbf{k}')[1 + n(\mathbf{k})]$$

$$- n_e(\mathbf{p})[1 - n_e(\mathbf{p}')]n(\mathbf{k})[1 + n(\mathbf{k}')]\}, \qquad (5.273)$$

where $\mathbf{p}$ is treated as a function of $(\mathbf{p}', \mathbf{k}', \mathbf{k})$ by use of momentum conservation and the term inside the braces is the same as that in Eq. (5.265).

For many other applications, it is, however, better to recast Eq. (5.265) in a different form, as follows: We begin by noting that, because the collisions conserve momentum, the integration over $\mathbf{p}$ can be omitted with the understanding that $\mathbf{p} = \mathbf{p}' + \mathbf{k}' - \mathbf{k}$ in $\mathcal{R}$. Further, because the collisions conserve energy, one more degree of freedom can be eliminated. It is convenient to retain the integration over $\mathbf{p}'$ and over two angles (θ, ϕ) that define the direction of scattering in a suitable frame. In that case, the integrations in Eq. (5.265) can be expressed in terms of a suitably defined scattering cross section $\sigma(\theta, \phi)$ and incident flux as

$$\int \frac{V d^3 \mathbf{p}}{(2\pi\hbar)^3} \int \frac{V d^3 \mathbf{p'}}{(2\pi\hbar)^3} \int \frac{V d^3 \mathbf{k'}}{(2\pi\hbar)^3} \mathcal{R}(\mathbf{p'}, \mathbf{k'}; \mathbf{p}, \mathbf{k}) \cdots$$

$$= \int \frac{V d^3 \mathbf{p'}}{(2\pi\hbar)^3} \int \frac{d\Omega}{V} \sigma(\theta, \phi) |\mathbf{v'_{rel}}| \cdots, \tag{5.274}$$

where $|\mathbf{v}|_{rel}$ is the relative velocity of scattering particles. Note that $(\sigma v / V)$ has the dimensions of per second, the same as that of $\mathcal{R}$. With this modification, Eq. (5.265) is very general and can be applied to many different cases. We now describe two simple examples.

As a first example, consider the scattering of photons by nonrelativistic, non-degenerate $(n_e \ll 1)$ electrons. In this case, we can set $(1 - n_e) \approx 1$ and $v_{rel} = c$ to get

$$\left(\frac{dn}{dt}\right)_{scatt} = \int \frac{V d^3 \mathbf{p'}}{(2\pi\hbar)^3} \int \frac{d\Omega}{V} \sigma(\Omega) c \{ n_e(\mathbf{p'}) n(\mathbf{k'})[1 + n(\mathbf{k})]$$

$$- n_e(\mathbf{p}) n(\mathbf{k})[1 + n(\mathbf{k'})] \} = \int d^3 \mathbf{p'} \, d\Omega \sigma(\Omega) c \{ n(\mathbf{k'})[1 + n(\mathbf{k})] f_e(\mathbf{p'})$$

$$- n(\mathbf{k})[1 + n(\mathbf{k'})] f_e(\mathbf{p}) \} \tag{5.275}$$

where we have denoted the phase-space density of electrons as $f_e \equiv n_e/(2\pi\hbar)^3$. This expression will be used in Chap. 6 to study the scattering of electrons and photons.

As a second example, consider molecular collisions in a nonrelativistic, classical gas of identical particles for which it is possible to simplify the equation still further: (1) To begin with, we are now interested in the collisions between molecules of the same kind, in contrast to collisions between particles of different types (viz., electrons and photons). It is therefore adequate to use a single distribution function rather than two separate ones; we set $n_e = n$ in our equation, where n now refers to the phase-space density of molecules. (2) All stimulated emission terms of the form $1 \pm n$ can be approximated by unity when occupation numbers are small and quantum degeneracy effects are ignorable. (3) We can assume that molecules are point particles with mass m and are nonrelativistic. It is then convenient to work with the centre of mass and relative coordinates and define the scattering cross section in this frame. Let the centre-of-mass momentum be $\mathbf{P}$ and the relative momentum be $\mathbf{q}$ in the initial state. Because molecules are approximated as nonrelativistic point particles, it is easy to show that the Jacobian from $d^3\mathbf{p'}d^3\mathbf{k'}$ to $d^3\mathbf{P}d^3\mathbf{q}$ is unity. With these simplifications, we will get

$$\left(\frac{dn}{dt}\right)_{scatt} = \int \frac{V d^3 \mathbf{p'}}{(2\pi\hbar)^3} \int \frac{d\Omega \, \sigma(\Omega)}{V \quad m} |\mathbf{p'} - \mathbf{k'}| \{ n(\mathbf{p'}) n(\mathbf{k'}) - n(\mathbf{p}) n(\mathbf{k}) \}. \tag{5.276}$$

Writing the phase-space density, viz., the distribution function as $f = n/(2\pi\hbar)^3$ and changing the variables of integration to simplify the notation, we find that this equation becomes

$$\left(\frac{df}{dt}\right)_{scatt} = \int d^3\mathbf{p}_1 d\Omega \sigma(\Omega)|\mathbf{v} - \mathbf{v}_1|(f'f_1' - ff_1). \qquad (5.277)$$

Here $f = f(\mathbf{x}, \mathbf{p}, t)$, $f_1 = f(\mathbf{x}, \mathbf{p}_1, t)$, $f' = f(\mathbf{x}, \mathbf{p}', t)$, and $f_1' = f(\mathbf{x}, \mathbf{p}_1', t)$. The collision described by $\sigma(\Omega)$ changes the initial momenta $(\mathbf{p}, \mathbf{p}_1)$ to final momenta $(\mathbf{p}', \mathbf{p}_1')$, conserving the total momentum and energy. This form will be used in Chap. 8 to study fluids.

6

Radiative Processes

6.1 Introduction

In this chapter we use the formalism developed in Chaps. 4 and 5 to study several radiative processes. The results obtained here will be needed extensively in the study of several astrophysical systems in Vols. II and III.

6.2 Macroscopic Quantities for Radiation

The discussion in Chap. 4 concentrated on the microscopic properties of emission and absorption of photons by quantum-mechanical systems. In many astrophysical contexts, we are interested in radiation processes that take place in bulk matter to which several atoms contribute independently. In the study of such processes, it is convenient to define macroscopically useful quantities in terms of the formalism developed in Chap. 4.

The key quantity in any macroscopic description of a collection of particles is the distribution function $f(x^i, \mathbf{p})$, introduced in Chap. 3, Section 3.6. If the number of photons in a phase-space volume $d^3x d^3p$ is dN, then we have

$$
\begin{aligned}
dN &= f(x^i, \mathbf{p}) \, d^3\mathbf{x} \, d^3\mathbf{p} = f[x^i, (h\nu/c)\hat{\mathbf{k}}] \, d^3\mathbf{x} \, d^3\mathbf{p} \\
&= n(x^i, \mathbf{p}) \frac{d^3\mathbf{x} \, d^3\mathbf{p}}{(2\pi\hbar)^3} = n[x^i, (h\nu/c)\hat{\mathbf{k}}] \frac{d^3\mathbf{x} \, d^3\mathbf{p}}{(2\pi\hbar)^3},
\end{aligned} \tag{6.1}
$$

where n is the number of photons in a particular quantum state labelled by the wave vector $\mathbf{k}$ and momentum $(h\nu/c)\hat{\mathbf{k}}$, where $\hat{\mathbf{k}}$ is the unit vector in the direction of propagation. In conformity with astronomical practise, we are now using the frequency $\nu = (\omega/2\pi)$. The energy-momentum tensor corresponding to this distribution function is (see Chap. 3, Section 3.6)

$$
T^{ab}(x^i) = \int \frac{d^3\mathbf{p}}{E(p)} c^2 p^a p^b f(x^i, \mathbf{p}). \tag{6.2}
$$

The integration over p in $d^3p = p^2\,dp\,d\Omega$ can be converted into an integration over ν by use of $p = (h\nu/c)$. Defining the symbol $\hat{k}^a = k^a/k^0$, where k^a is the wave vector of the photons, we find that T^{ab} becomes

$$T^{ab}(x^i) = \int \frac{h^4\nu^3}{c^3} \hat{k}^a\hat{k}^b \, f(x^i, \nu, \hat{\mathbf{k}}) \, d\nu\, d\Omega. \qquad (6.3)$$

This expression suggests defining a quantity (called specific intensity of radiation) by

$$I_\nu(x^i, \hat{\mathbf{k}}) = (h^4\nu^3/c^2)f = (h\nu^3/c^2)n \qquad (6.4)$$

so that the energy-momentum tensor becomes

$$T^{ab}(x^i) = \frac{1}{c} \int d\nu\, d\Omega\, \hat{k}^a\hat{k}^b I_\nu(x^i, \hat{\mathbf{k}}). \qquad (6.5)$$

Note that $\hat{k}^a$ (which is *not* a four vector) has the four components $(1, \hat{\mathbf{k}})$. Because $T^{00} = (dE/dV)$ is the energy per unit volume, it is clear that $I_\nu = (c\,dE/dV\,d\nu\,d\Omega) = (dE/dt\,dA\,d\nu\,d\Omega)$ is the energy flowing per unit area per second per unit frequency range into a solid angle $d\Omega$. The units for I_ν are ergs per square centimeters per seconds per hertz per steradians; because the $dt\,d\nu$ combination is dimensionless, this would also be dimensionally the same as ergs per square centimeters per steradians, although such a form is somewhat misleading.

So far we have not bothered to indicate the polarisation of the photons explicitly. In general we should have a distribution function, energy-momentum tensor, intensity, etc., for each of the two degrees of polarisation, and summation over polarisations will require that the expressions be multiplied by a factor of 2. Therefore, in general,

$$I_\nu = \sum_{\alpha=1}^{2} I_{\nu\alpha} = \sum_{\alpha=1}^{2} n_{\alpha\nu}\frac{h\nu^3}{c^2} = n_\nu\left(\frac{2h\nu^3}{c^2}\right), \qquad (6.6)$$

where the last expression is valid for an unpolarised beam.

Relation (6.4) shows that the specific intensity of radiation contains the same amount of information as the density of photons in phase space or the distribution function in phase space. It is also clear from Eq. (6.6) that any process that conserves n_ν (like propagation through loss-free optical systems) conserves specific intensity. Historically, however, the theory of radiative processes was developed in terms of the intensity, as this is a macroscopically well-defined, observable quantity that was known even before the quantisation of the electromagnetic field was understood. Following this tradition, the probabilities for absorption, emission, etc., can be expressed directly in terms of intensities. With

$n_\nu = (c^2/2h\nu^3)I_\nu$, Eqs. (4.137) and (4.138) become

$$\left(\frac{dP}{dt}\right)_{abs} = \left(\frac{2q^2}{mc^3}\right)\omega^2 f_{12}\left(\frac{c^2}{2h\nu^3}I_\nu\right) = 4\pi^2\left(\frac{q^2}{mc}\right)\left(\frac{f_{12}}{h\nu}\right)I_\nu \equiv BI_\nu, \quad (6.7)$$

$$\left(\frac{dP}{dt}\right)_{emi} = \left(\frac{2q^2}{mc^3}\right)\left(1 + \frac{c^2}{2h\nu^3}I_\nu\right)\omega^2 f_{12} \equiv A + BI_\nu, \quad (6.8)$$

with

$$A \equiv 4\pi^2\left(\frac{q^2}{mc}\right)\left(\frac{2\nu^2}{c^2}\right)f_{12}; \quad B \equiv 4\pi^2\left(\frac{q^2}{mc}\right)\left(\frac{f_{12}}{h\nu}\right). \quad (6.9)$$

(We have denoted the oscillator strength by the symbol f_{12} rather than by f_{fi}.) These equations relate the emission and absorption rates of radiation to the intensity by use of two constants, A and B. These constants are called Einstein's A, B coefficients. Because the transitions are between two definite quantum states 1 and 2, these coefficients also depend on the quantum states we are considering. [It is obvious from Eq. (6.9) that f_{12} depends explicitly on the quantum state and hence A and B also has this dependence. We indicate this dependence explicitly through the notation A_{12}, etc., only when needed.] Using the definition of bound–bound scattering cross section (see Chap. 4, Section 4.5), we can express σ_{bb} in terms of B:

$$\sigma_{bb} = \frac{\pi q^2}{mc}f_{12}\delta(\nu - \nu_{12}) = \left(\frac{B}{4\pi}\right)h\nu\delta(\nu - \nu_{12}) = \frac{Bh\nu}{4\pi}\phi_\nu. \quad (6.10)$$

It follows from the definitions that $(A/B) = (2h\nu^3/c^2) = (\hbar\omega^3/2\pi^2c^2)$. This particular ratio between the Einstein's coefficients is merely a historic accident arising from the fact that we chose to use intensity as a fundamental variable rather than the photon number. Note that the ratio is precisely the one between intensity and photon number in Eq. (6.6); it has no deep or fundamental significance.

The intensity I_c of radiation at which the stimulated and the spontaneous emission rates are equal is given by $BI_c = A$. The energy flux contained in a small frequency interval $d\omega$ in such a beam will be $I_c d\omega = (A/B)d\omega = (\hbar\omega^3/2\pi^2c^2)d\omega$. For visible light, with $\omega \approx 3 \times 10^{15}$ Hz and $d\omega \approx 2\pi \times 10^{10}$ Hz, we get $I_c d\omega \approx 10$ W cm$^{-2} \approx 10^5$ W m^{-2}. Conventional light sources do not have such high intensities (see Exercise 6.2) whereas a laser beam does.

Several macroscopically useful quantities can now be defined in terms of A and B. To begin with, we note that the probability per second for an atom in an excited state 2 to decay spontaneously to a lower energy state 1 is A_{21}. Therefore the number of decays per second per unit volume of material is $n_2 A_{21}$, where n_2 is the number of excited atoms per unit volume. Each decay produces the energy $h\nu\delta(\nu - \nu_{12})d\nu \equiv h\nu\phi_\nu d\nu$ in the frequency range $d\nu$, in which the direction of the emitted photon is equally likely to be anywhere within the total solid angle

4π. Therefore, to obtain the rate of emission of energy per unit solid angle, we have to divide by 4π. Putting these together, we obtain

$$\left(\frac{dE}{d^3x\,dt\,dv\,d\Omega}\right)_{\text{spon.emi}} \equiv j_v = \frac{n_2 A_{21}}{4\pi}hv\delta(v-v_{12}) = \frac{n_2 A_{21}hv}{4\pi}\phi_v, \quad (6.11)$$

which gives the amount of energy emitted from unit volume of matter per second in the frequency interval dv per solid angle. Clearly, this emissivity j_v is a macroscopically well-defined quantity. The emissivity per atom is

$$J_v \equiv \frac{j_v}{n_2} = \left(\frac{dE}{dt\,dv\,d\Omega}\right)_{\text{atom}} = \frac{A_{21}hv}{4\pi}\phi_v. \quad (6.12)$$

We shall see in Chap. 7 that the line profile function ϕ_v can be approximated as $\phi_v \approx (\Delta v)^{-1}$, where Δv is the width of a spectral line in realistic cases. Because A_{21} can be calculated from theory, this allows the computation of emissivity for any given process.

Let us next consider the absorption of radiation by the atom. Let the number of atoms at the energy level E_1 in a small volume d^3x be $n_1 d^3x$. The fraction of those that will make a transition to upper level, absorbing the energy $hv\phi_v\,dv$ in a small time interval dt, is $B_{12}I_v\,dt(d\Omega/4\pi)$. So the net energy absorbed is

$$dE_{\text{abs}} = (n_1 d^3x)(B_{12}I_v\,dt)(d\Omega/4\pi)(hv\phi_v\,dv) \equiv \alpha_v I_v d^3x\,dt\,d\Omega\,dv, \quad (6.13)$$

where second equation defines the absorption coefficient α_v. Comparing the two expressions, we get

$$\alpha_v = n_1 B_{12}hv\phi_v(4\pi)^{-1} = n_1\sigma_{bb}(v) \quad (6.14)$$

where the second equality follows from Eq. (6.10). The α_v has the dimension of inverse length. Note that the absorption coefficient α_v is defined by the relation

$$\alpha_v I_v = \left(\frac{dE}{d^3x\,dt\,dv\,d\Omega}\right)_{\text{abs}}. \quad (6.15)$$

This is because the net absorption is proportional to the intensity of radiation shining on the matter; to define a useful coefficient that is dependent on only the properties of matter, it is necessary to factor out the intensity I_v on the left-hand side. On the other hand, the rate of spontaneous emission of energy is independent of the intensity of radiation and hence the emissivity is defined directly by Eq. (6.11).

The coefficients j_v and α_v describe spontaneous emission and absorption, respectively. What remains is the induced (or stimulated) emission, which is the second term in Eq. (6.8). Even though this is an emission process, it is proportional to the amount of radiation intensity falling on the matter. Therefore, to arrive at a useful coefficient for induced emission α_v^{ind}, we will take a clue from the absorption coefficient and define this quantity by the relation

$$\alpha_\nu^{\text{ind}} I_\nu = \left(\frac{dE}{d^3x \, dt \, d\nu \, d\Omega} \right)_{\text{ind}}. \tag{6.16}$$

The analysis for α_ν^{ind} is identical except for the replacement of $n_1 B_{12}$ with $n_2 B_{21}$, which is equal to $n_2 B_{12}$. Thus $\alpha_\nu^{\text{ind}} = n_2 B_{12} h\nu \phi_\nu (4\pi)^{-1}$.

We next define a quantity called the net absorption coefficient as $\alpha_\nu^{\text{net}} = \alpha_\nu - \alpha_\nu^{\text{ind}}$. This is the net absorption because it takes into account the true absorption and subtracts out the induced emission, so this will represent the change in the radiation at the frequency ν that is due to absorption and induced emission combined. Explicitly,

$$\alpha_\nu^{\text{net}} = \frac{h\nu}{4\pi} \phi_\nu (n_1 - n_2) B_{12} = (n_1 - n_2)\sigma_{bb}(\nu) = (n_1 - n_2)\frac{\pi q^2}{mc} f_{12}\phi_\nu. \tag{6.17}$$

In many respects, we can think of induced emission as negative absorption. The quantity α_ν^{net} and the emissivity j_ν together serve as useful macroscopically defined quantities that govern the radiative properties of matter.

The j_ν and different α_ν's are proportional to the number density of atoms participating in the emission or absorption process. In general, we can write $\alpha = n\sigma$, where n is the number density of agents involved in the process and σ is a suitable cross section. Astronomers prefer to work with quantities defined for unit *mass* rather than with number density. In that case, we introduce a quantity κ_ν, called opacity, with the definition

$$\alpha = n\sigma \equiv \rho\kappa, \tag{6.18}$$

where ρ is the mass density of the material; κ is in units of square centimeters per gram.

These definitions can provide useful insights into macroscopic radiative processes. As an example, note that the ratio $j_\nu/\alpha_\nu^{\text{net}}$ can be expressed in terms of the ratios A/B and n_2/n_1 as follows:

$$\frac{j_\nu}{\alpha_\nu^{\text{net}}} = \frac{A_{21}n_2}{(n_1 - n_2)B_{21}} = \frac{(A/B)_{21}}{(n_1/n_2) - 1} = \frac{2h\nu^3}{c^2}\frac{1}{(n_1/n_2) - 1}. \tag{6.19}$$

The ratio is entirely fixed by the ratio of atoms at the two levels n_1/n_2 and can be calculated from the state of the matter distribution. Suppose we are dealing with a situation in which the population of atoms at different energy levels is known. Such a system will, in general, both absorb and emit radiation. Equation (6.19) shows that if, for example, the emission properties of the system and the distribution of atoms are known, then we can find the absorption properties. This principle will be used several times in subsequent sections.

Some general results can be obtained with the above formalism when the system has reached a steady state. Consider the transitions between two particular energy levels, labelled 1 and 2, with an energy difference $h\nu$. Let n_1 and n_2 denote the number of atoms in these two levels and let $n(\nu)$ denote the number of photons at the frequency ν corresponding to the energy-level difference between these two states. The system is said to be in steady state with respect to these two energy levels if n_1, n_2, and $n(\nu)$ do not change with time. This requires that the number of atoms making a downward transition per second, $n_2(dP_{\text{emi}}/dt)$, be balanced by the number of atoms making the upward transition per second, $n_1(dP_{\text{abs}}/dt)$. Then,

$$\frac{n_1}{n_2} = \frac{(dP_{\text{emi}}/dt)}{(dP_{\text{abs}}/dt)} = \frac{1 + n(\nu)}{n(\nu)} = \frac{1}{n(\nu)} + 1, \tag{6.20}$$

giving

$$n(\nu) = \frac{1}{(n_1/n_2) - 1}. \tag{6.21}$$

This is a general condition for the existence of steady state between radiation at frequency ν and matter distribution with level populations n_2 and n_1. Note that, in the steady state, $n(\nu)$ is completely determined by the population of matter at different energy levels. We can define a temperature T_{12} for two energy levels E_1 and E_2 by the relation

$$\frac{n_2}{n_1} \equiv \exp - \frac{(E_2 - E_1)}{k_B T_{12}} = \exp\left(-\frac{h\nu}{k_B T_{12}}\right). \tag{6.22}$$

If matter is in thermodynamic equilibrium, T_{12} will be the thermodynamic temperature of the system and will be the same for all levels E_1 and E_2. In that case, Eq. (6.21) shows that

$$n(\nu) = \frac{1}{e^{h\nu/k_B T} - 1}. \tag{6.23}$$

We have thus succeeded in (re)deriving the distribution function for the number of photons with frequency ν when radiation is in equilibrium with matter that has a temperature T. The fact that, the probability for emission of radiation scaled as $1 + n$ was crucial in obtaining this formula. If there is no induced emission and $1 + n$ is replaced with 1 in Eq. (6.20), then the formula for $n(\nu)$ in Eq. (6.21) will not have -1 in the denominator and will read as $n(\nu) = (n_2/n_1)$. In thermal equilibrium, this will give $n(\nu) = \exp(-h\nu/k_B T)$, which is what we would have expected from the application of Boltzmann statistics to photons. Equation (6.23) reaffirms the facts that photons do not obey Boltzmann statistics and that their thermal-equilibrium properties are instead governed by a different formula.

The spectrum in Eq. (6.23) was obtained earlier in Chap. 5 from the requirement of statistical equilibrium for electromagnetic field. The current analysis shows that the same result can be obtained from (1) the laws of absorption and emission of radiation and (2) the thermal-equilibrium condition for matter, thereby establishing the overall consistency of the formalism.

Given the number density of photons in thermal equilibrium, we can compute several other useful quantities. To begin with, the intensity I_ν in thermal equilibrium is

$$I_\nu = \frac{2h\nu^3}{c^2} n = \frac{2h\nu^3}{c^2(e^{\beta h\nu} - 1)} \equiv B_\nu, \qquad (6.24)$$

with $\beta = (k_B T)^{-1}$. This intensity (also called the Planckian spectrum or black-body spectrum) plays a vital role in the physics of radiation. The two limiting forms of the Planck spectrum for $h\nu \gg k_B T$ and $h\nu \ll k_B T$ are called Wien's spectrum and Rayleigh–Jeans spectrum, respectively. The limiting forms for the intensity are

$$B_W(\nu, T) = \frac{2h\nu^3}{c^2} e^{-h\nu/k_B T}, \quad B_{RJ}(\nu, T) = \frac{2\nu^2}{c^2} k_B T, \qquad (6.25)$$

with the first form being valid for $(\nu/\text{GHz}) \ll 21[T(\text{K})]$. (These two regimes correspond to cases in which photons behave as particles or waves, as described in Chap. 5, Section 5.11.) The fact that, in the Rayleigh–Jeans regime, intensity is proportional to temperature is often used to define a quantity called brightness temperature by the relation

$$T_b = \frac{c^2}{2k_B} \frac{1}{\nu^2} I_\nu = \frac{\lambda^2}{2k_B} I_\nu, \qquad (6.26)$$

where I_ν is the intensity of the source. If the source is approximately thermal and the frequency range satisfies $(\nu/\text{GHz}) \ll 21[T(\text{K})]$, then T_b will be approximately constant and will give the thermodynamic temperature of the source.

Very often, we are also interested in the energy density U_ν of radiation when it is in equilibrium with matter at some temperature T. From the definitions

$$U_\nu = \frac{dE}{d^3x \, d\nu} = \frac{dE}{d\nu \, dA(c \, dt)}, \quad I_\nu = \frac{dE}{d\nu \, dt \, dA \, d\Omega}, \qquad (6.27)$$

it follows that

$$cU_\nu = \frac{dE}{d\nu \, dA \, dt} = \int d\Omega \frac{dE}{d\nu \, dA \, dt \, d\Omega} = 4\pi I_\nu \qquad (6.28)$$

for any isotropic radiation. Therefore

$$U_\nu \, d\nu = \frac{4\pi}{c} I_\nu d\nu = \frac{8\pi h}{c^3} \frac{\nu^3}{e^{\beta h\nu} - 1} d\nu = \frac{\hbar}{\pi^2 c^3} \frac{\omega^3}{e^{\beta \hbar\omega} - 1} d\omega, \qquad (6.29)$$

which is the same result as obtained in Eq. (5.213). The total energy density is

$$U = \int_0^\infty d\nu U_\nu = \frac{8\pi h}{c^3} \left(\frac{1}{\beta h}\right)^4 \int_0^\infty \frac{dx \, x^3}{e^x - 1} = \frac{\pi^2 (k_B T)^4}{15(\hbar c)^3}. \qquad (6.30)$$

The quantity U_ν represents energy per unit volume per unit frequency range and is in units of ergs per cubic centimeter per hertz. The integrated energy density U in units of ergs per cubic centimeter. The flux in any direction, integrated over all frequencies, is $F \equiv (c/4\pi)U \equiv \sigma T^4$, where $\sigma \equiv (\pi/60)(k_B^4/\hbar^3 c^2)$; F is in units of ergs per seconds per square centimeters per steradians. The number density of photons in thermal equilibrium is given by

$$n_{\text{tot}} = \int_0^\infty dv \left(\frac{U_\nu}{h\nu}\right) = \int_0^\infty \frac{8\pi}{c^3} \frac{v^2 dv}{e^{\beta h\nu} - 1} = \frac{2\zeta(3)}{\pi^2} \left(\frac{k_B T}{\hbar c}\right)^3, \quad (6.31)$$

where $\zeta(3) \approx 1.202$ is the Riemann zeta function of the order of 3. Note that the ratio $(U/n_{\text{tot}}) \approx 2.7 k_B T$ is determined entirely by the temperature of matter. An arbitrary radiation field will, of course, not have any definite value for the ratio U/n_{tot}. If we start with such a radiation field at some initial time $t = t_i$, let it interact with matter, and eventually come into equilibrium, then this ratio has to reach the value given above. In other words, when radiation and matter come into equilibrium, photons will be emitted or absorbed in such a fashion as to finally bring about this ratio. This necessarily assumes that emission and absorption have played a significant role in bringing about the steady state. There can be other situations in which steady state is brought about by, say, scatterings in which the photon number cannot change. We have seen in Chap. 5, Section 5.14, that the steady-state distribution of photons will be a Bose–Einstein distribution in such a case.

Exercise 6.1
Aspects of Planck spectrum: (1) Show that the intensity of thermal radiation treated as a function of wavelength reaches a maximum at $\lambda_{\max}$, where $(\lambda_{\max}/\text{cm})(T/K) = 0.29$. Similarly, the intensity, treated as a function of frequency, reaches a maximum at $\nu_{\max}$, where $(\nu_{\max}/\text{GHz}) = 58.8(T/K)$. These are not the same, in the sense that $\nu_{\max} \neq (c/\lambda_{\max})$. A more relevant procedure is to study the maximum of $\mathcal{E}_\nu \equiv \nu U_\nu$, which gives the energy density per unit logarithmic interval in frequency. For what values of ν and λ does $\mathcal{E}_\nu$ reach a maximum? (2) Compute $n_{\max} = n(\nu_{\max})$. What does this suggest regarding the classical nature (or otherwise) of photons near the peak of the Planck spectrum?

The relation that we obtained earlier in Eq. (6.19), connecting the emissivity and absorption coefficient, takes a specific form when matter is in thermal equilibrium. Then $(n_1/n_2) = \exp(\beta h\nu)$, and Eq. (6.19) becomes

$$\alpha_\nu^{\text{net}} = \frac{c^2}{2h\nu^3} \left(e^{h\nu/k_B T} - 1\right) j_\nu = j_\nu B_\nu^{-1}(T). \quad (6.32)$$

In practical situations, the emissivity of radiation at a frequency ν will be determined by some detailed physical process. This result shows that if matter is in thermal equilibrium, then the frequency dependence of the absorption coefficient can be determined by this formula. Then Eq. (6.17) becomes

$$\alpha_\nu^{net} = (n_1 - n_2) \frac{\pi q^2}{mc} f_{12}\phi_\nu = \frac{\pi q^2}{mc} f_{12}\phi_\nu \, n_1 \left(1 - e^{-h\nu/k_B T}\right). \quad (6.33)$$

This expression has the limiting forms

$$\alpha_\nu^{net} = \begin{cases} n_1 \left(\dfrac{\pi q^2 f_{12}}{mc}\right)\left(\dfrac{h\nu}{k_B T}\right)\phi_\nu & \text{(for } h\nu \ll k_B T) \\[2mm] n_1 \left(\dfrac{\pi q^2 f_{12}}{mc}\right)\phi_\nu & \text{(for } h\nu \gg k_B T) \end{cases}. \quad (6.34)$$

To give a flavour of the formalism based on α_ν and j_ν, let us consider a simple representation of the propagation of radiation in a medium that is emitting and absorbing radiation. We ignore all scattering effects and also any spatial dependence of the processes. Then, when the radiation travels a distance dl, the intensity I_ν will increase by the amount $j_\nu dl$ because of the emission of photons; the absorption of the photons will cause the intensity to decrease by the net amount $-\alpha_\nu I_\nu dl$ during the same time. (We have dropped the superscript net on α_ν^{net} for simplicity.) Hence the rate of change of intensity can be described by a differential equation of the form

$$\frac{dI_\nu}{dl} = j_\nu - \alpha_\nu I_\nu \quad (6.35)$$

if scattering is ignored. This has the solution

$$I_\nu = \frac{j_\nu}{\alpha_\nu}\left(1 - e^{-\alpha_\nu l}\right), \quad (6.36)$$

which satisfies the boundary condition $I_\nu = 0$ at $l = 0$. If the path length of the light through the source is R, then the intensity at the edge of the source is

$$I_\nu(R) = \frac{j_\nu}{\alpha_\nu}(1 - e^{-\alpha_\nu R}) = \begin{cases} j_\nu R & \text{(if } \alpha_\nu R \ll 1) \\ (j_\nu/\alpha_\nu) & \text{(if } \alpha_\nu R \gg 1) \end{cases}. \quad (6.37)$$

The limiting forms show an interesting feature. Note that $\alpha = n\sigma$ has the dimensions of inverse length and α^{-1} can be interpreted as a mean free path for the process governed by the cross section σ. When $\alpha_\nu R \ll 1$, the mean free path is much larger than the source size and the absorption is not very effective. (The source is then said to be optically thin.) In that case, we directly observe from the source the emissivity j_ν, unhindered by absorption processes. In the other extreme limit, when $\alpha_\nu R \gg 1$ (called optically thick), the observed intensity is the ratio between the emissivity and the absorption coefficient of the source. In particular, if the emission arises from matter in thermal equilibrium, then this ratio is fixed to be B_ν, completely independent of the details of the process. Thus an optically thick medium in thermal equilibrium always emits the blackbody spectrum irrespective of the specific process involved in the emissivity j_ν. In that case, Eq. (6.37) also shows that $I_\nu = B_\nu[1 - \exp(-\alpha_\nu R)]$, which is always less than B_ν. No material system in thermal equilibrium can emit at a given frequency ν more intensity than a blackbody can emit at the same frequency.

Radiation emitted from a gas cloud in which matter is in thermal equilibrium, but is optically thin to the frequency at which observations are made, is of considerable interest. In this case, we have from Eq. (6.37)

$$I_\nu(R) \approx j_\nu R = B_\nu(T) R \, \alpha_\nu^{\text{net}} \approx B_\nu(T) R \frac{\pi q^2}{mc} f_{12} \phi_\nu n_1 \left(1 - e^{-h\nu/k_B T}\right), \quad (6.38)$$

with the limiting forms

$$I_\nu(R) = \begin{cases} B_\nu(T) n_1 R \left(\dfrac{\pi q^2 f_{12}}{mc}\right)\left(\dfrac{h\nu}{k_B T}\right) \phi_\nu & \text{(for } h\nu \ll k_B T) \\[4mm] B_\nu(T) n_1 R \left(\dfrac{\pi q^2 f_{12}}{mc}\right)\phi_\nu & \text{(for } h\nu \gg k_B T) \end{cases}. \quad (6.39)$$

In the first case, which corresponds usually to radio frequencies, the observed brightness temperature of the source is

$$k_B T_B(\nu) \simeq k_B T \alpha_\nu^{\text{net}} R \simeq (n_1 R)\left(\frac{\pi q^2}{mc}\right) f_{12} h\nu \, \phi(\nu) \quad (h\nu \ll k_B T). \quad (6.40)$$

(Eq. (6.38) also gives, more directly, the first relation above.) Determining $T_B(\nu)$ will help us to determine the factors that affect the linewidth $\phi(\nu)$ as well as the column density $n_1 R$.

Finally, we comment on the Lorentz transformation properties of the various quantities that we have introduced. From the definition of intensity I_ν in terms of the photon occupation number it is clear that $I_\nu \propto \nu^3 n$. Because n is Lorentz invariant it follows that I_ν/ν^3 is invariant. The change of intensity dI_ν that is due to emission can be written as $dI_\nu \propto \rho J_\nu dt$; so $(dI_\nu/\nu^3) \propto (\rho J_\nu/\nu^2)(dt/\nu)$. Both dt and ν are zeroth components of four vectors and hence the ratio (dt/ν) is Lorentz invariant. A more formal way of deriving this result is to note that $h\nu$ represents the energy of the photon and write

$$\frac{dt}{E} = \frac{d^4 x}{d^3 \mathbf{x} E} = \left[\frac{d^4 x}{(d^3 \mathbf{x}\, d^3 \mathbf{p})}\right]\left[\frac{d^3 p}{E}\right], \quad (6.41)$$

in which we have multiplied both the denominator and the numerator by $d^3 x d^3 p$. From our results in Chap. 3, Section 3.6, we know that $d^4 x$, $d^3 x d^3 p$, and $(d^3 p/E)$ are all individually Lorentz invariant; hence dt/E is Lorentz invariant. It follows that $(\rho J_\nu/\nu^2)$ is Lorentz invariant. Similarly, $dI_\nu \propto \rho \kappa_\nu I_\nu dt$, giving $(dI_\nu/\nu^3) \propto \rho \kappa_\nu \nu (dt/\nu)(I_\nu/\nu^3)$. Because (dt/ν), (dI_ν/ν^3), etc., are Lorentz invariant, it follows that $\rho \kappa_\nu \nu$ is Lorentz invariant.

Exercise 6.2
Numerical estimates for light: Estimate the total intensity $I_\nu d\nu$ (in watts per square centimeters), the amplitude of the electric field (in volts per centimeters), and the number density of photons in the logarithmic band $n_\nu(d\nu/\nu)$ for (a) a mercury lamp, (b) a laser.

Exercise 6.3
Aspects of intensity: (1) Show that the intensity of an extended source (like a brightly illuminated wall) is independent of the distance to it. (2) A spherical source of radiation (with radius R) has a uniform intensity I_ν. Show that the total flux of radiation from the source at a distance r from the center of the source will be

$$S_\nu \equiv \int_\Omega I_\nu \cos\theta \, d\Omega = \pi I_\nu \left(\frac{R}{r}\right)^2. \tag{6.42}$$

More generally, a spherical source of radius R at a distance D will subtend a solid angle $d\Omega = (2\pi R \sin\theta)(R d\theta / D^2)$. The total flux received from the source will then be

$$S_\nu = \int_0^{\pi/2} I_\nu \frac{2\pi R^2}{D^2} \sin\theta \cos\theta \, d\theta. \tag{6.43}$$

Exercise 6.4
More radiative transfer: When j_ν and α_ν vary along the path, we can still integrate Eq. (6.35) in terms of an optical depth τ_ν defined by the relation $d\tau_\nu \equiv -\alpha_\nu dl$. In that case, show that the solution to Eq. (6.35) is given by

$$I_\nu(l) = I_\nu(0)e^{-\tau_\nu(l)} + \int_0^{\tau_\nu(l)} B_\nu[T(\tau)]e^{-\tau} \, d\tau \tag{6.44}$$

if $j_\nu = \alpha_\nu B_\nu(T)$ locally within the source. Of particular importance is the case in which the medium is isothermal with spatially constant temperature. Then this result becomes

$$I_\nu(l) = I_\nu(0)e^{-\tau_\nu(l)} + B_\nu(T)\left[1 - e^{-\tau_\nu(l)}\right]. \tag{6.45}$$

In terms of the brightness temperature introduced in Eq (6.26), we can write this result as

$$T_b(l) = T_b(0)e^{-\tau_\nu(l)} + T\left[1 - e^{-\tau_\nu(l)}\right]. \tag{6.46}$$

Describe qualitatively what this equation implies for a source viewed through a medium in thermal equilibrium for different ranges of the parameters.

Exercise 6.5
Radiation from a spherical source I: A spherical cloud of radius R and temperature T is a distance d from Earth, with $d \gg R$. It emits thermal radiation at the rate $P(\nu) = [dE/(dt d\nu dV)]$. (1) What is the brightness of the cloud as measured from Earth as a function of the distance b from the centre of the cloud? (2) What is the effective temperature of the cloud and the flux F_ν measured at Earth from the entire cloud? Give the answers for both optically thin and optically thick limits. [Answers: Consider the optically thin case first. (1) Because $j_\nu = P_\nu/4\pi$, the intensity is

$$I_\nu(b) = \int j_\nu(z) \, dz = \frac{P_\nu}{2\pi}\sqrt{R^2 - b^2}. \tag{6.47}$$

(2) The total power is $L = (4/3)\pi R^3 P$, where P is the total power integrated over all frequencies. Equating this to $L = 4\pi R^2 \sigma T_{\text{eff}}^4$, we get $T_{\text{eff}} = (PR/3\sigma)^{1/4}$. The flux F_ν can be determined from the energy conservation $4\pi d^2 F_\nu = (4/3)\pi R^3 P_\nu$, giving $F_\nu = P_\nu(R^3/3d^2)$. In the optically thick case, $I_\nu = B_\nu(T)$, independent of b. The effective temperature is clearly $T_{\text{eff}} = T$. The flux at the surface is $\pi B_\nu(T)$, giving $F_\nu(d) = \pi(R/d)^2 B_\nu(T)$.]

Exercise 6.6

Radiation from a spherical source II: A spherical cloud of radius R is a distance d from Earth with $d \gg R$. It emits photons at a uniform rate $\Gamma = [dN/(dt\,dV)]$. A detector on Earth has a beam width of half angle $\Delta\theta$ and an effective area ΔA. Assume that the photons are propagating through an optically thin medium. (1) What is the observed intensity towards the centre of the cloud if the source is completely resolved? (2) What is the average intensity once the source is completely unresolved? [Answers: (1) Integrating along the line through the centre gives $I_0 = (\Gamma/4\pi)2R = (R\Gamma/2\pi)$ photons per unit area per second per solid angle. (2) The average intensity is the total flux divided by the solid angle intercepted by the detector $\Delta\Omega_{\text{Det}} = \pi(\Delta\theta_{\text{det}})^2$. Because the flux is $F = (L/4\pi d^2) = (R^3\Gamma/3d^2)$, the mean intensity is $\bar{I} = [R^3\Gamma/3\pi d^2(\Delta\theta_{\text{Det}})^2]$. This can be written as $(\bar{I}/I_0) = (2/3)(\Delta\theta_s/\Delta\theta_{\text{Det}})^2$, where $\Delta\theta_s = (R/d)$ is the angular size of the source.]

Exercise 6.7

Radiation from a relativistic sphere: Suppose that material is ejected from a point P in the form of a spherical shell moving with speed v. Every point on the surface of the ejected shell of material is assumed to emit radiation with a luminosity (in the instantaneous rest frame) $L_0(\nu) = k\nu^{-\alpha}$. Let Q be a point on the sphere, with PQ making an angle θ with respect to the line connecting the observer with P. The radiation reaching the observer from Q will be Doppler shifted by the factor

$$(1 + z) \equiv \frac{1 + [v/c]\cos\theta}{\sqrt{1 - [v/c]^2}}. \tag{6.48}$$

Show that the observed flux density is

$$S = S_0(1 + z)^{-(3+\alpha)} \tag{6.49}$$

where $S_0 \equiv L_0(\nu)/4\pi R^2$ is the flux density in the nonrelativistic limit.

Exercise 6.8

Astrophysical masers: Consider a system with two levels and populations N_1 and N_2 in the upper and the lower levels. The equation for radiative transfer involving just these two levels can be approximated by

$$\frac{dI}{ds} = h\nu_0[B(N_2 - N_1)I + A], \tag{6.50}$$

where A and B are the relevant Einstein coefficients. Assume that there exists some pumping mechanism that populates the upper level at a rate R and that collisions can produce transition between the two states without production of photons at a rate C.

Write down and integrate the equation for radiative transfer under such conditions. In particular, show that (1) if $C \gg BI$, the specific intensity grows exponentially as the beam travels through such a region. (2) If $C \ll BI$ then the specific intensity becomes saturated and varies linearly with s. Explain physically the origin of this effect. [Hint: Show that $I(s)$ is determined by the transcendental equation $I + (C/B) \ln I = h\nu Rs$; the asymptotic forms are easy to obtain from this.]

6.3 Absorption and Emission in the Continuum Case

The discussion so far dealt with emission and absorption of radiation between two energy levels E_1 and E_2, which were assumed to be discrete. No significantly new complications arise when these results are generalised to a situation in which the energy levels form a continuum. We now work out some simple generalisations of the formulas for α_ν, j_ν, etc., for the continuum case in order to illustrate the changes.

To fix the ideas, let us consider the emission of radiation by a charged particle, say, an electron, that is *not* bound in an atom. Classically, the electron will emit radiation when it follows an accelerated trajectory that changes its momentum. Quantum mechanically, the initial and the final states of the electron (when it is a free particle) can be thought of as the eigenstates of momentum and energy, with eigenvalues $\mathbf{p}$ and $E(p)$. These states are labelled by a continuous parameter $\mathbf{p}$ or E. (In more complicated situations, such as electron scattering in a Coulomb field, say, it will not be possible to diagonalise both momentum and energy; then the stationary solutions to the Schrodinger equation will be parameterised by the energy E, which can take a continuous range of values.) In either case, we can find the quantity $(d\epsilon/dt\,d\nu) = P(\nu, E)$ which is the energy radiated per second per frequency interval at frequency ν by an electron with energy E. Given the distribution function $f(\mathbf{p})$ for the electrons, we can also compute the number of electrons per unit volume with energy between E and $E + dE$ as

$$n(E)\,dE \equiv f(\mathbf{p})\frac{d^3p}{dE}\,dE = 4\pi p^2 f(p)\frac{dp}{dE}\,dE, \qquad (6.51)$$

where $p = p(E)$ and the second equality is for an isotropic distribution $f(\mathbf{p}) = f(p)$. Because each electron emits the energy $d\epsilon = P(\nu, E)\,dt\,d\nu$, the total emissivity over all solid angles is

$$4\pi j_\nu = \int d^3p\, f(p) P(\nu, E) = \int_0^\infty dE\, 4\pi p^2 \left(\frac{dp}{dE}\right) f(E) P(\nu, E) \qquad (6.52)$$

or

$$j_\nu = \frac{d\epsilon}{d\Omega\, d\nu\, dt\, d^3x} = \int_0^\infty dE\, n(E) P(\nu, E). \qquad (6.53)$$

To determine the coefficients of absorption and induced emission, we relate $P(v, E)$ to the A, B coefficients as follows:

$$P(v, E_2) = \left\{\begin{matrix} \text{energy radiated per second, at frequency } v \\ \text{by an electron of energy } E_2 \end{matrix}\right\}$$

$$= hv \int dE_1 \, A_{21}\phi_{21}(v) = \left(\frac{2hv^3}{c^2}\right) hv \int dE_1 \, B_{21}\phi_{21}(v). \quad (6.54)$$

The first relation follows from the definition of A_{21}, and the second relation arises from the fact that $A = (2hv^3/c^2)B$. The net absorption coefficient α_v, on the other hand, is defined to be

$$\rho\kappa_v = \alpha_v = \frac{hv}{4\pi} \int dE_1 \, dE_2 \, [n(E_1)B_{12} - n(E_2)B_{21}]\phi_{21}(v) \quad (6.55)$$

[see Eq. (6.17)]. The integral is over all states E_1 and E_2 but $\phi_{21}(v)$ will ensure that only states with $v = v_{12} = (E_2 - E_1)/h$ contribute significantly. The first term represents true absorption, and the second one represents stimulated emission. This quantity can be expressed in terms of P by use of Eq. (6.54). The second term on the right-hand side of Eq. (6.55) is

$$-\frac{hv}{4\pi} \int dE_2 \, n(E_2) \int dE_1 \, B_{21}\phi_{21}(v)$$

$$= -\frac{hv}{4\pi} \int dE_2 \, n(E_2)\left(\frac{c^2}{2hv^3}\right)\frac{P(v, E_2)}{hv}$$

$$= -\frac{c^2}{8\pi hv^3} \int dE_2 \, n(E_2)P(v, E_2), \quad (6.56)$$

and first term is

$$-\frac{hv}{4\pi} \int dE_1 \, dE_2 \, n(E_1)B_{12}\phi_{21}$$

$$= \frac{hv}{4\pi} \int dE_2 \, n(E_2 - hv) \int dE_1 \, B_{21}\phi_{21}$$

$$= \frac{c^2}{8\pi hv^3} \int dE_2 \, n(E_2 - hv)P(v, E_2). \quad (6.57)$$

Therefore

$$\alpha_v = \frac{c^2}{8\pi hv^3} \int dE_2\{n(E_2 - hv) - n(E_2)\}P(v, E_2)$$

$$= \frac{c^2}{8\pi hv^3} \int d^3p_2[f(p_2') - f(p_2)]P(v, E_2), \quad (6.58)$$

where p_2' is the momentum corresponding to energy $E_2 - hv$. Equations (6.58) and (6.53) [or Eq. (6.52)] give the relation we were seeking. These equations represent the emissivity and net absorption coefficient in terms of the distribution function of the electrons as well as the rate of energy emission P in an individual process. In thermal equilibrium, $f(p) \propto \exp[-\beta E(p)]$ and

$$f(p_2') - f(p_2) = f(p_2)(e^{\beta h v} - 1) \tag{6.59}$$

so that

$$\alpha_v = \frac{c^2}{8\pi h v^3}(e^{\beta h v} - 1)\int d^3 p_2 f(p_2) P(v, E_2)$$

$$= \frac{c^2}{8\pi h v^3}(e^{\beta h v} - 1)(4\pi j_v) = \frac{c^2}{2h v^3}(e^{\beta h v} - 1) j_v. \tag{6.60}$$

This is, of course, the same result obtained [see Eq. (6.32)] for the discrete case.

6.4 Scattering of Electromagnetic Radiation

An electromagnetic wave that is incident upon a charged particle will make it oscillate. Such an oscillating charged particle will be moving in an accelerated trajectory and hence will radiate electromagnetic waves in all directions. The net effect can be thought of as scattering of the incident electromagnetic radiation by a charged particle. Let us study the details of this process.

Consider an electromagnetic wave with the electric field $\mathbf{E} = \mathbf{E}_0 \cos(\mathbf{k} \cdot \mathbf{x} - \omega t)$ that is incident upon a charged particle. Assuming that the resultant motion of the charge is nonrelativistic, we can retain the electric force $q\mathbf{E}$ but ignore the magnetic force $(q/c)\mathbf{v} \times \mathbf{B}$. We can also ignore the $\mathbf{k} \cdot \mathbf{x}$ term in $\cos(\mathbf{k} \cdot \mathbf{x} - \omega t)$ to the same order of accuracy (because $kx \simeq kvt$ and $\omega t = kct$) and write the equation of motion for the charge as

$$m\ddot{\mathbf{x}} = \left(\frac{m}{q}\right)\ddot{\mathbf{d}} = q\mathbf{E}, \tag{6.61}$$

where $\mathbf{d} = q\mathbf{x}$ is the dipole moment. Such an oscillating charge will radiate energy at the same frequency as that of the incident wave. The intensity radiated in some direction $\hat{\mathbf{n}}'$ can be determined from the general formula [see Eq. (4.38)]

$$\frac{dE}{dt d\Omega} = \frac{1}{4\pi c^3}(\ddot{\mathbf{d}} \times \hat{\mathbf{n}}')^2 = \frac{q^4}{4\pi m^2 c^3}(\mathbf{E} \times \hat{\mathbf{n}}')^2 = \frac{q^4}{4\pi m^2 c^3} E^2 \sin^2\theta, \tag{6.62}$$

where θ is the direction between $\mathbf{E}$ and $\hat{\mathbf{n}}'$. Because the incident flux is $S = (cE^2/4\pi)$, the scattering cross section is

$$\left(\frac{d\sigma}{d\Omega}\right) \equiv \frac{1}{S}\left(\frac{dE}{dt d\Omega}\right) = \left(\frac{q^2}{mc^2}\right)^2 \sin^2\theta \equiv r_0^2 \sin^2\theta, \quad r_0 \equiv \frac{q^2}{mc^2}. \tag{6.63}$$

The total scattering cross section for electrons with $q = e$, obtained by integration over all solid angles, is

$$\sigma_T = \frac{8\pi}{3}\left(\frac{e^2}{mc^2}\right)^2 = 6.7 \times 10^{-25} \text{ cm}^2 = 6.7 \times 10^{-29} \text{ m}^2 \qquad (6.64)$$

and is called the Thomson scattering cross section.

The result in Eq. (6.63) is valid for radiation polarised along a specific direction $\hat{\mathbf{e}} = (\mathbf{E}/E)$. For unpolarised radiation we have to average $\sin^2\theta = 1 - (\hat{\mathbf{n}}' \cdot \hat{\mathbf{e}})^2$ over all $\hat{\mathbf{e}}$ perpendicular to the direction of propagation $\hat{\mathbf{n}} = (\mathbf{k}/k)$. We can do this by noting that the average of $e_a e_b$ is

$$\langle e_a e_b \rangle = \frac{1}{2}\left(\delta_{ab} - \frac{k_a k_b}{k^2}\right), \qquad (6.65)$$

and hence

$$\langle \sin^2\theta \rangle = 1 - n'_a n'_b \langle e_a e_b \rangle = \frac{1}{2}[1 + (\hat{\mathbf{n}} \cdot \hat{\mathbf{n}}')^2]. \qquad (6.66)$$

Therefore the scattering cross section for unpolarised radiation is

$$\left(\frac{d\sigma}{d\Omega}\right)_{\text{unpol}} = \frac{1}{2}r_0^2[1 + (\hat{\mathbf{n}} \cdot \hat{\mathbf{n}}')^2] = \frac{1}{2}r_0^2(1 + \cos^2\Theta), \qquad (6.67)$$

where Θ is the scattering angle between the incident $\hat{\mathbf{n}}$ and $\hat{\mathbf{n}}'$ directions.

Exercise 6.9

Polarisation in Thomson scattering: Estimate the Stokes parameters and degree of polarisation of the scattered light if the incident light is unpolarised.

As an example of Thomson scattering in a slightly more complicated circumstance, let us consider the cross section for scattering of a linearly polarised wave by a charged particle that is acted on by three different forces: (1) a harmonic restoring force $-\omega_0^2 x$ (as we saw in Chap. 4, Section 4.5.3, the quantum theoretical effects can be mimicked by such a harmonic restoring force), (2) the force that is due to the incident electromagnetic wave, and (3) the radiation reaction force. The equation of motion for the charged particle will then be

$$\ddot{\mathbf{x}} + \omega_0^2\mathbf{x} = \frac{q}{m}\mathbf{E}_0 e^{-i\omega t} + \frac{2q^2}{3mc^3}\dddot{\mathbf{x}}. \qquad (6.68)$$

Because the damping and driving forces are usually small compared with the restoring harmonic force, we can replace $\dddot{\mathbf{x}}$ with $-\omega_0^2\dot{\mathbf{x}}$. Then the equation of motion is

$$\ddot{\mathbf{x}} + \gamma\dot{\mathbf{x}} + \omega_0^2\mathbf{x} = \frac{q}{m}\mathbf{E}_0 e^{-i\omega t}, \qquad (6.69)$$

where $\gamma = (2q^2/3mc^3)\omega_0^2$. Solving this equation, we get

$$\mathbf{x} = \frac{q}{m}\mathbf{E}_0\frac{e^{-i\omega t}}{\omega_0^2 - \omega^2 - i\omega\gamma}. \tag{6.70}$$

Obtaining the dipole moment from this expression and finding the energy radiated by using the Larmor's formula, as before, we find that the total scattering cross section turns out to be

$$\sigma = \frac{8\pi}{3}\left(\frac{q^2}{mc^2}\right)^2\frac{\omega^4}{\left(\omega^2 - \omega_0^2\right)^2 + \omega^2\gamma^2}. \tag{6.71}$$

This expression has three interesting limiting forms: (1) When $\omega \gg \omega_0$, the restoring force is irrelevant and the charged particle behaves as though it is free. In this case, $\sigma \approx \sigma_T$. (2) In the opposite limit, when $\omega \ll \omega_0$, the cross section is $\sigma(\omega) \propto \omega^4$. In this case, the electric field appears nearly static and the dipole moment produced by the incident field is describable in terms of a static polarisability. This gives rise to ω^4 dependence and such a scattering is called Rayleigh scattering. (3) When $\omega \approx \omega_0$, Eq. (6.71) reduces to

$$\sigma \cong \frac{8\pi}{3}\left(\frac{q^2}{mc^2}\right)^2\frac{\omega_0^2}{4(\omega - \omega_0)^2 + \gamma^2}$$

$$= \frac{2\pi^2q^2}{mc}\left[\frac{(\gamma/2\pi)}{(\omega - \omega_0)^2 + (\gamma/2)^2}\right] \quad (\omega \cong \omega_0). \tag{6.72}$$

In arriving at the second equality we have used $\gamma = (2q^2/3mc^3)\omega_0^2$ and grouped the terms such that the function in the brackets is normalised to unity when integrated over ω. This function has the Lorentzian shape of a standard resonance curve with a width determined by the damping term γ. Integration over $\nu = \omega/2\pi$ gives

$$\int_0^\infty \sigma(\nu)\,d\nu = \frac{\pi q^2}{mc} = 2.65 \times 10^{-2} \text{ cm}^2 \text{ Hz} = 2.65 \times 10^{-6} \text{ m}^2 \text{ Hz}. \tag{6.73}$$

This result is valid in quantum theory as well, provided the right-hand side is multiplied by the oscillator strength f_{ij} [see Eq. (4.136), Chap. 4]. The cross section at resonance, $\omega = \omega_0$ is $\sigma(\omega_0) = (3/4\pi)\lambda_0^2$, where λ_0 is the wavelength of the radiation. Also note that $\Delta\omega \simeq \gamma \simeq (r_0/c)\omega_0^2$ whereas the corresponding width in wavelength is $\Delta\lambda \cong (c/\omega_0^2)\Delta\omega \simeq r_0$.

In a plasma, Thomson scattering of photons by the charged particle will give the photons a mean free path of $l_T = (n_e\sigma_T c)^{-1}$, where n_e is the electron density. This will be the main source of scattering of photons in a fully ionised plasma. [We can ignore scattering by the protons because the Thomson cross section for protons is lower by the factor $(m_e/m_p)^2 \simeq 10^{-6}$.] Note that $\sigma_T \ll \sigma_{matter}$, where $\sigma_{matter} \simeq 10^{-15}$ cm^2 is the scattering cross section of matter that is due to molecular collisions. Hence the photon mean free path is usually much longer than the matter mean free path. The opacity κ is in units of square centimeters per gram and can be interpreted as the cross section per unit mass of the material. The corresponding opacity in the case of Thomson scattering by electrons is given by

$$\kappa_{es} = \frac{n_e}{\rho}\sigma_T = \left(\frac{\sigma_T}{2m_H}\right)(1 + X_H) \cong 0.2(1 + X_H) \text{ cm}^2 \text{ g}^{-1}$$

$$\cong 0.02(1 + X_H) \text{ m}^2 \text{ Kg}^{-1}, \tag{6.74}$$

where X_H is the mass function of hydrogen and we have used Eq. (5.95), Chap. 5, to estimate (n_e/ρ).

Thomson scattering does not change the frequency of the radiation. If we think of radiation of frequency ω as made of photons with energy $\hbar\omega$, then Thomson scattering is a scattering between a photon and, say, an electron. In general, such a scattering will transfer both momentum and energy. However, as we shall see in Section 6.5 (also see Exercise 4.5, Chap. 4), the amount of energy transferred from the photon to the electron is $\Delta E \simeq (\hbar\omega/mc^2)\hbar\omega$. For this quantity to be negligible, we must have $\hbar\omega \ll mc^2$. If this condition is violated, in the frame in which the electron was at rest before scattering, then the scattering cross section is *not* given by Eq. (6.67). Quantum electrodynamical corrections modify the formula for the scattering cross section (for unpolarised radiation) to

$$\left(\frac{d\sigma}{d\Omega}\right)_{\text{unpol}} = \frac{1}{2}r_0^2\left(\frac{\omega_1}{\omega}\right)^2\left(\frac{\omega}{\omega_1} + \frac{\omega_1}{\omega} - \sin^2\Theta\right), \tag{6.75}$$

where ω and ω_1 are the frequencies of the incident and the scattered photons. [When $\omega \cong \omega_1$, this reduces to Eq. (6.67).] The total cross section for scattering is modified to

$$\sigma_{KN} = \frac{3}{8}\sigma_T\frac{1}{\epsilon}\left\{\left[1 - \frac{2(\epsilon + 1)}{\epsilon^2}\right]\ln(2\epsilon + 1) + \frac{1}{2} + \frac{4}{\epsilon} - \frac{1}{2(2\epsilon + 1)^2}\right\}, \tag{6.76}$$

where $\epsilon = (\hbar\omega/mc^2)$. This is called the Klein–Nishina cross section. When $\epsilon \ll 1$, this reduces to $\sigma_{KN} \approx \sigma_T(1 - 2\epsilon) \approx \sigma_T$. In the other limit, when $\epsilon \gg 1$, the scattering cross section is given by $\sigma_{KN} \approx (3/8)(\sigma_T/\epsilon)(\ln 2\epsilon + 0.5)$, which decreases roughly as $(1/\epsilon)$.

The cross section σ_T also plays a key role in determining the annihilation and creation of electron–positron pairs at high energies and several related phenomenon. Considering the importance of processes leading to $e^+ - e^-$ pairs in high-energy astrophysics, we briefly summarise several relevant features of relativistic pair production. There exists number of mechanisms that can produce $e^\pm$ pairs, such as $\gamma + \gamma \rightarrow e^+ + e^-; \gamma + e^\pm \rightarrow e^\pm + e^+ + e^-; \gamma + Z \rightarrow Z + e^+ + e^-; e^\pm + e^\pm \rightarrow e^\pm + e^\pm + e^+ + e^-;$ and $e^\pm + Z \rightarrow e^\pm + Z + e^+ + e^-$. Of these, the $\gamma\gamma$ production has a cross section of the order of σ_T. Those involving a charged particle and a photon scale as $\alpha\sigma_T$, where $\alpha = (e^2/\hbar c)$ is the fine-structure constant and those with two charged particles scale as $\alpha^2 \sigma_T$. Clearly, the photon–photon pair production is the most dominant phenomenon. This process occurs provided the photons have sufficient total energy $(2m_e c^2)$ in the centre-of-mass frame of the pair that is produced. They should also be in the zero-spin state in that frame and electron and positron are created in a singlet-spin state. (Because of angular-momentum conservation, the other initial state for photons, with spin $2\hbar$, cannot lead to pair production.) The cross section for pair production in the centre-of-mass frame is given by

$$\sigma_{\gamma\gamma}(s) = \frac{3}{8}\frac{\sigma_T}{s}\left[\left(2 + \frac{2}{s} - \frac{1}{s^2}\right)\cosh^{-1}s^{1/2} - \left(1 + \frac{1}{s}\right)\left(1 - \frac{1}{s}\right)^{1/2}\right], \quad (6.77)$$

where $s = (E/m_e c^2)^2$ and E is the centre-of-mass energy for the photons. The two limits of this process are

$$\sigma_{\gamma\gamma}(s) \rightarrow \frac{3}{8}\sigma_T \begin{cases} \sqrt{s-1} & (s-1 \ll 1) \\ [\ln(4s) - 1]/s & (s \gg 1) \end{cases}. \quad (6.78)$$

The inverse process corresponds to the annihilation of electron and positron with the emission of a pair of photons. Once again the initial and the final states must have zero total angular momentum. The cross section for this process can be obtained from the above by use of the principle of detailed balance. In the limit of very slow and very rapid motion of the $e^\pm$ in the centre-of-mass frame, the annihilation cross section is given by

$$\sigma_{\text{ann}} = \frac{3}{16}\sigma_T \begin{cases} (1 + \beta_{\text{CM}}^2)/\beta_{\text{CM}} & (\alpha \ll \beta_{\text{CM}} \ll 1) \\ [2\ln(2\gamma_{\text{CM}}) - 1]/\gamma_{\text{CM}} & (\beta_{\text{CM}} \gg 1) \end{cases}, \quad (6.79)$$

where $\beta = (v/c)$ and γ is the Lorentz factor of the electron, all evaluated in the centre-of-mass frame. Since $\sigma_{\text{ann}} \propto \beta^{-1}$ in the nonrelativistic limit, the annihilation-rate coefficient $\alpha_{\text{ann}} = (v\sigma_{\text{ann}}) \approx (3/8)(\sigma_T c)$ is independent of the velocity in the nonrelativistic limit. The spectrum of emitted photons in the centre-of-mass frame is a two-photon line at $\gamma_{\text{CM}} m_e c^2$. In the nonrelativistic limit, the lab frame spectrum is close to a delta function at $m_e c^2 = 511$ keV. The cooling rate that is due to this process is $\Lambda_{\text{ann}} = (2m_e c^2)\alpha_{\text{ann}} = (3/4)\sigma_T m_e c^3$.

In the discussion so far, we have assumed that the scattering was by some well-defined microscopic charged particles whose internal structure is not of relevance. The situation is different when we study the scattering of radiation, for example, by dust grains of size of $\sim 0.1\ \mu$m. The detailed theory of such a scattering is fairly complicated, and analytical results are available in only some limiting cases. The relevant information is briefly summarised below.

A dust grain of radius a will have a geometrical cross section πa^2; the actual cross section for scattering can be conveniently parameterised as $Q\pi a^2$, where Q takes into account both the true scattering as well as any absorption of radiation by the dust grain. With the two separate Q factors denoted by Q_{sca} and Q_{abs}, the net effect (often called extinction) will depend on $Q_{ext} = Q_{abs} + Q_{sca}$. A dust grain along a line of sight to an astronomical source will remove radiation from the source by a factor that depends on Q_{ext}. It is also conventional to define a quantity called an albedo by the ratio (Q_{sca}/Q_{ext}). Detailed study shows that the key parameter that determines the efficiency of scattering is the ratio $x = (2\pi a/\lambda)$ between the size of the grain a and the wavelength λ of the radiation. Both Q_{ext} and Q_{sca} increase with x as a power law x^{α} for $x \ll 1$; for $x \gg 1$, both Q_{abs} and Q_{sca} approach unity and $Q_{ext} \to 2$; for small spherical particles, it is possible to obtain analytic expressions in terms of the complex reflective index m of the material. We find that, for $x \ll 1$,

$$Q_{sca} = \frac{8}{3}x^4 \mathrm{Re}\left\{\left(\frac{m^2 - 1}{m^2 + 1}\right)^2\right\}, \quad Q_{abs} = -4x\mathrm{Im}\left(\frac{m^2 - 1}{m^2 + 1}\right). \tag{6.80}$$

The scattering has the Rayleigh form $\propto \lambda^{-4} \propto \omega^4$ in this limit.

It must be stressed that the above theoretical result assumes a refractive index that is independent of wavelength, which is unrealistic. Hence, for real material, the x dependence is often quite different from what is predicted by the theory given above. It is found that, in general, $Q \propto x^{\alpha}$ with $\alpha \approx 1$–2. In the case of large grains ($x \gg 1$), an empirical formula is available that describes the scattering cross section $S(\theta)$ and is given by

$$S(\theta) = [1 - g^2][1 + g^2 - 2g\cos\theta]^{-3/2}, \tag{6.81}$$

where g is a constant characterising the dust grain, and this equation has been parameterised such that $g = \langle\cos\theta\rangle$. Note that when g is close to 0, scattering is isotropic, whereas if g is close to unity, $S(\theta)$ peaks in the forward direction. This function, however, does not describe correctly scattering by small ($\lesssim 0.05$-μm) particles nor the backscattering by larger particles. Outside this regime, this formula is useful. In the astronomical context, scattering by carbon and silicates is of primary importance. The absorption efficiency for silicate grains of radius 0.1 μm peaks near 9.7 μm and at 8 μm. For carbon, there is a broad peak in the absorption near 2100 Å. These results will be of interest in the study of extinction of light by interstellar grains.

Exercise 6.10
Scattering by a polarisable sphere: A plane-polarised wave of wavelength λ is scattered by a solid sphere of radius a, where $\lambda \gg a$. Show that the scattering cross section is

$$\sigma = \pi a^2 Q_{scatt}, \tag{6.82}$$

with

$$Q_{scatt} = \frac{8(ka)^4}{3(1 + 3/4\pi\alpha)^2}, \tag{6.83}$$

where α is the polarisability of the material.

Exercise 6.11
X-ray halo that is due to scattering: X-rays from a source located at distance D from the observer are scattered by dust grains around the line of sight. Show that, to the lowest order of approximation, the intensity of scattered light at an angle α away from the line of sight to the source will be given by

$$I(\alpha) = \frac{f_X D}{\sin \alpha} \int \int \frac{d\sigma}{d\Omega}(\phi, a) n_d(a) \, da \, d\phi, \qquad (6.84)$$

where $n_d(a) \, da$ is the number density of interstellar dust grains with radius a, $(d\sigma/d\Omega)(\phi, a)$ is the scattering cross section at a scattering angle ϕ for a dust grain of radius a, and f_X is the observed x-ray flux from the source.

6.5 Radiation Drag on a Charged Particle

When the Thomson scattering is a valid approximation, the photon transfers all its *momentum* to the electron but a negligible amount of *energy*. Hence this scattering must be accompanied by a force acting on the particle, which can also be viewed as follows: When an electromagnetic wave hits a charged particle, it makes the particle oscillate and radiate. The radiation of energy will lead to a damping force on the particle. This drag force can be obtained by averaging the force in Eq. (4.78), over one period of the wave. The first term with $\dot{\mathbf{E}}$ averages to zero, and the second term gives

$$\langle \mathbf{f} \rangle = \frac{2}{3} \left(\frac{q^2}{mc^2} \right)^2 \langle E^2 \rangle \hat{\mathbf{n}} = \frac{8\pi}{3} \left(\frac{q^2}{mc^2} \right)^2 \frac{\langle E^2 \rangle}{4\pi} \hat{\mathbf{n}} = \sigma_T U \hat{\mathbf{n}}, \qquad (6.85)$$

where $U \hat{\mathbf{n}}$ is the flux of radiation. Note that the mean force that is due to this process is *quadratic* in an electric field whereas the instantaneous force acting on the charged particle is *linear* in the electric field.

The same result can be obtained more elegantly from Eq. (4.86). In a frame in which the charge is at rest, $u^i = (1, 0, 0, 0)$ and $g^i = (\gamma \mathbf{f} \cdot \mathbf{v}, \gamma \mathbf{f}) = (0, \mathbf{f})$. From Eq. (4.86) we get

$$g^i = \sigma_T [T^{i0} - T^{00} u^i] = (0, \sigma_T U \hat{\mathbf{n}}), \qquad (6.86)$$

which agrees with Eq. (6.85).

In the above discussion, we considered a *single* plane wave that scattered by a charged particle. A more complicated situation arises when a charged particle with velocity $\mathbf{v}$ is moving through a region of space containing an isotropic bath of radiation with energy density U_{rad}. The charged particle will be constantly scattering the electromagnetic waves that make up the radiation bath. Because the charge has a nonzero velocity $\mathbf{v}$, this scattering will be anisotropic and hence the momentum transfer to the charged particle will be in the direction opposite to the velocity. (The reason for the drag force is easier to understand if we think of radiation as made of photons. Clearly, the charged particle will be hit by more photons in the front than in the back and thus will experience a drag force.)

The simplest way to calculate the drag is to use Eq. (4.86). Using $T^{ab} = U_{rad}$ dia$(1, 1/3, 1/3, 1/3)$ for an isotropic radiation bath (see Chap. 3, Section 3.6) and $u^i = (\gamma, \gamma \mathbf{v})$, we get

$$T^{ab} u_a u_b = U_{rad} \gamma^2 \left(1 + \frac{1}{3} v^2 \right), \quad T^{ab} u_b = \left(U_{rad} \gamma, -\frac{1}{3} U_{rad} \gamma \mathbf{v} \right). \quad (6.87)$$

This gives, on use of Eq. (4.86),

$$g^i = \left(-\frac{4}{3} \sigma_T U_{rad} \gamma^3 v^2, -\frac{4}{3} \sigma_T U_{rad} \gamma^3 \mathbf{v} \right) = (\gamma \mathbf{f} \cdot \mathbf{v}, \gamma \mathbf{f}), \quad (6.88)$$

that is,

$$\mathbf{f} = -\frac{4}{3} \sigma_T U_{rad} \gamma^2 \left(\frac{\mathbf{v}}{c} \right), \quad -\mathbf{f} \cdot \mathbf{v} = \frac{4}{3} \sigma_T U_{rad} \gamma^2 \left(\frac{v^2}{c^2} \right) c. \quad (6.89)$$

This result is valid for any isotropic radiation field with energy density U_{rad}.

The work done by this drag force, $\mathbf{f}_{drag} \cdot \mathbf{v} = -(4/3) \sigma_T U_{rad} \gamma^2 (v/c)^2$, will reduce the kinetic energy and hence the velocity of the charged particle. This loss of energy by the particle will appear as net gain of energy of radiation given by

$$\frac{dE}{dt} = \frac{4}{3} \sigma_T U_{rad} \gamma^2 \left(\frac{v}{c} \right)^2 c. \quad (6.90)$$

Thus a charged particle, moving relativistically through a radiation bath, can transfer its kinetic energy to the radiation. This process, called inverse Compton scattering, will be discussed in greater detail in Section 6.6.

Exercise 6.12
Alternative derivation: The radiation drag on a particle, Eq. (6.90), can be obtained directly as follows. Treat a radiation bath as equivalent to randomly fluctuating $\mathbf{E}$ and $\mathbf{B}$ fields with $\langle \mathbf{E} \rangle = \langle \mathbf{B} \rangle = 0$; $\langle \mathbf{E}^2 \rangle = \langle \mathbf{B}^2 \rangle = 4\pi U_{rad}$. Evaluate the mean Lorentz force on the charge and show that the mean power radiated by the charge is

$$\left(\frac{dE}{dt} \right)_{rad} = \sigma_T c \gamma^2 \left(1 + \frac{v^2}{3c^2} \right) U_{rad}. \quad (6.91)$$

Argue that the radiation absorbed by the charge is $(dE/dt)_{abs} = \sigma_T c U_{rad}$. Hence calculate the net energy transfer from the charge to the radiation bath and obtain Eq. (6.90).

Finally, we mention some corrections to Eq. (6.90), for later reference. The above result uses the Thomson scattering cross section in the rest frame of the charged particle. If the photon energy, in the rest frame of the charge, is comparable with mc^2, then it is necessary to use the Klein–Nishina cross section [Eq. (6.76)]. In that case, the above equation is modified to

$$\frac{dE}{dt} = \frac{4}{3}\sigma_T U_{\text{rad}}\gamma^2 c \left(\frac{v}{c}\right)^2 \left(1 - \frac{63}{10}\frac{\gamma\langle E^2\rangle}{mc^2\langle E\rangle} + \cdots\right), \tag{6.92}$$

where $\langle E\rangle$, etc., denote the mean value of the energy of the photon, etc. For a Planckian distribution of photons, the correction term is approximately $150(\gamma k_B T/mc^2)$. The above formula, of course, assumes that the correction term is small, i.e., $\gamma k_B T \ll mc^2$. In the extreme relativistic limit, the corresponding result becomes

$$\frac{dE}{dt} \simeq \frac{3}{8}\sigma_T m^2 c^5 \int_{(mc^2/\gamma)}^{\infty} \frac{dE}{E} n_{\text{ph}}(E)\left(\ln\frac{4E\gamma}{mc^2} - \frac{11}{6}\right), \tag{6.93}$$

where $n_{\text{ph}}(E)$ is the spectrum of photons. For a Planckian spectrum this reduces to

$$\frac{dE}{dt} \simeq \frac{\sigma_T}{16}\frac{(mck_B T)^2}{\hbar^3}\left(\ln\frac{4\gamma k_B T}{mc^2} - 1.981\right). \tag{6.94}$$

These results are of importance in the study of active galactic nuclei in Vol. III.

6.6 Compton Scattering and Comptonisation

In Sections 6.4 and 6.5 we considered the scattering of a plane wave by a charged particle (which was originally at rest) and the interaction of a moving charged particle with a radiation bath. We next discuss the case of a *collection* of charged particles (say, electrons in a plasma) interacting with a radiation field. It is now convenient – and necessary – to think of the radiation field as made of a collection of photons. The interaction between the electrons and the photons can cause net energy transfer to either particle and we need to take into account different possibilities.

Consider a plasma embedded in a radiation field of temperature T_{rad}. The scattering of photons by the electrons in the plasma will continuously transfer energy between the two components. The high-energy photons with $m_e v^2 \ll \hbar\omega \ll m_e c^2$ will transfer energy to the low-energy electrons, but will gain energy from the high-energy electrons (with $\hbar\omega \ll m_e v^2$). (It is assumed that $\hbar\omega \ll m_e c^2$, so that the quantum electrodynamical effects, like pair production, are negligible.) In thermal equilibrium, the net transfer of energy will be zero. But if the electron temperature T_e is very different from the photon temperature, there can be a net transfer of energy. When $T_e \gg T_{\text{rad}}$, the electrons cool (on the average) by transferring energy to photons. This process, inverse Compton scattering, will cause the spectrum of the photons to be distorted. On the other hand, if $T_{\text{rad}} \gg T_e$, the energy will be transferred (on the average) from the photons to the electrons and this process is called *Compton scattering*. In astrophysical applications, inverse Compton scattering plays a more important role than Compton scattering, essentially because it can serve as a mechanism for generating high-energy photons. We shall discuss both the processes but will concentrate more on inverse Compton scattering.

We begin by considering the scattering of a photon by an electron that was originally at rest. Let the initial and the final four momenta of the photon be $k_i^a = (\hbar\omega_i/c)[1, \mathbf{n}_i]$ and $k_f^a = (\hbar\omega_f/c)[1, \mathbf{n}_f]$, respectively, and those of the electron be $p_i^a = (mc, 0)$ and $p_f^a = (E/c, \mathbf{p})$, respectively. The conservation of momentum and energy is expressed by the equation $p_i^a + k_i^a = p_f^a + k_f^a$. Squaring this equation and using the components to eliminate the final electron momentum (see Exercise 4.5, Chap. 4), we get

$$\frac{\omega_f}{\omega_i} = \left[1 + \left(\frac{\hbar\omega_i}{m_e c^2}\right)(1 - \cos\theta)\right]^{-1}, \tag{6.95}$$

where $\cos\theta = (\mathbf{n}_i \cdot \mathbf{n}_f)$. When $\hbar\omega_i \ll m_e c^2$, we can expand the expression in the bracket in a Taylor series and obtain

$$\frac{\omega_f - \omega_i}{\omega_i} = \frac{\Delta E_\gamma}{E_\gamma} = -\left(\frac{\hbar\omega_i}{m_e c^2}\right)(1 - \cos\theta). \tag{6.96}$$

To find the mean energy transfer, this expression should be averaged over θ. In the rest frame of the electron, the scattering has front–back symmetry, making $\langle\cos\theta\rangle = 0$. Hence the average energy lost by the photon per collision is

$$\langle\Delta E_\gamma\rangle = -\left(\frac{\hbar\omega_i}{m_e c^2}\right)\hbar\omega_i. \tag{6.97}$$

Let us next consider the average energy gained by the photon field from the charged particle. We saw in Section 6.6 that the net addition of energy to the photon field is given by [see Eq. (6.90)]

$$P = \frac{4}{3}\sigma_T c U_{\text{rad}}\gamma^2\left(\frac{v}{c}\right)^2. \tag{6.98}$$

The mean number of photons scattered per second is $N_c = (\sigma_T c n_{\text{rad}}) = (\sigma_T c U_{\text{rad}}/\hbar\omega_i)$, where $\hbar\omega_i$ is the average energy of the photon defined by $\hbar\omega_i = (U_{\text{rad}}/n_{\text{rad}})$. Hence the average energy gained by the photon in one collision is

$$\langle\Delta E_\gamma\rangle = \frac{P}{N_c} = \frac{4}{3}\gamma^2\left(\frac{v}{c}\right)^2\hbar\omega_i = \frac{4}{3}\gamma^2\left(\frac{v}{c}\right)^2\langle E_\gamma\rangle. \tag{6.99}$$

In the relativistic limit, $(\Delta E_\gamma/E_\gamma) \approx (4/3)\gamma^2 \gg 1$, and this process can be a source of high-energy photons. For example, if $\gamma = 10^3$ (which is possible in some astrophysical cases), this process can convert radio photons to UV photons, far-infrared photons to x rays, and optical photons to gamma rays. It should, however, be noted that the scattering cross section σ_T has to be modified when $E_\gamma \gtrsim m_e c^2$. The modified Klein–Nishina cross section in expression

(6.78) decreases for $E_\gamma > m_e c^2$ and the inverse Compton process ceases to be effective.

When $v \ll c$, the energy gain by photons per collision is $\langle \Delta E/E \rangle \simeq (4 k_B T_e / m_e c^2)$ as $\langle m v^2 \rangle \cong 3 k_B T_e$. Combining this with Eq. (6.97) we find that the mean fractional energy change of photons, per collision, is

$$\left\langle \frac{\Delta E_\gamma}{E_\gamma} \right\rangle = -\frac{\langle \hbar \omega \rangle}{m_e c^2} + \frac{4 k_B T_e}{m_e c^2} = \frac{4 k_B T_e - \langle E_\gamma \rangle}{m_e c^2}. \tag{6.100}$$

If $4 k_B T_e > \langle E_\gamma \rangle$, the net energy transfer is from electrons to photons (inverse Compton scattering), and if $4 k_B T_e < \langle E_\gamma \rangle$, the net energy transfer is from photons to electrons. We may say that, in a typical collision between an electron and a photon, the electron energy changes by $E^2/m_e c^2$ and photon energy changes by $(4 k_B T_e / m_e c^2) E$.

The process described above acts as a major source of cooling for relativistic plasma as well as a mechanism for producing high-energy photons. The time scale for Compton cooling of an individual relativistic particle is

$$t_{cc} \simeq \frac{\gamma m_e c^2}{P} \simeq 4 \times 10^{-3} \gamma^{-1} \beta^{-2} \left(\frac{T_R}{10^6 \, \text{K}} \right)^{-4} \text{s}, \tag{6.101}$$

where T_R is the radiation temperature. If electrons are nonrelativistic with temperature T_e, this time scale is

$$t_{cc} \simeq \frac{k_B T_e}{P} \simeq \frac{1}{n_\gamma \sigma_T c} \left(\frac{m_e c^2}{k_B T_R} \right) = 1.3 \times 10^{-3} \left(\frac{T_R}{10^6 \, \text{K}} \right)^{-4} \text{s}. \tag{6.102}$$

As the energy is progressively transferred from the electrons to the photons, through repeated scattering, the mean energy will increase towards $4 k_B T_e$, when the net transfer will cease.

When electrons and photons coexist in a region of size l, the repeated scattering of photons by the electrons will distort the original spectrum of the photons. The mean free path of the photon that is due to Thomson scattering is $\lambda_\gamma = (n_e \sigma_T)^{-1}$. If the size of the region l is such that $(l/\lambda_\gamma) \gg 1$, then the photon will undergo several collisions in this region; but if $(l/\lambda_\gamma) \lesssim 1$ then there will be few collisions. It is convenient to define an optical depth $\tau_e \equiv (l/\lambda_\gamma) = (n_e \sigma_T l)$ so that $\tau_e \gg 1$ implies strong scattering.

If $\tau_e \gg 1$, then the photon goes through $N_s (\gg 1)$ collisions in travelling a distance l. From standard random-walk arguments, we have $N_s^{1/2} \lambda_\gamma \simeq l$ so that $N_s = (l/\lambda_\gamma)^2 = \tau_e^2$. On the other hand, if $\tau_e \lesssim 1$, then $N_s \simeq \tau_e$; therefore an estimate for the number of scatterings is $N_s \simeq \max(\tau_e, \tau_e^2)$. The average fractional change in the photon energy, per collision, is given by $4(k_B T_e / m_e c^2)$. Hence the condition for a significant change of energy is

$$1 \simeq N_s \left(\frac{4 k_B T_e}{m_e c^2} \right) = \left(\frac{4 k_B T_e}{m_e c^2} \right) \max \left(\tau_e, \tau_e^2 \right). \tag{6.103}$$

Defining a parameter y (called the Compton y parameter) by

$$y = \frac{k_B T_e N_s}{m_e c^2} = \left(\frac{k_B T_e}{m_e c^2}\right) \max\left(\tau_e, \tau_e^2\right), \qquad (6.104)$$

the condition for significant scattering becomes $y \simeq 1/4$.

The optical depth of the region in which this process is significant is given by relation (6.103). Because $l = \tau_e/n_e\sigma_T$, the size of the region in which this process will be important is $l = (\tau_e/n_e\sigma_T)$ and the corresponding time scale is $t_c \simeq (l/c)$. Explicitly,

$$t_c = \frac{l}{c} = \begin{cases} (n_e\sigma_T c)^{-1}(m_e c^2/4k_B T_e) & (\text{for } \tau_e \ll 1) \\ (n_e\sigma_T c)^{-1}(m_e c^2/4k_B T_e)^{1/2} & (\text{for } \tau_e \gg 1) \end{cases}. \qquad (6.105)$$

A more precise condition for repeated scattering to change the spectrum of the radiation field can be obtained as follows: The change in the energy of a typical photon after a single scattering is given by the factor $(\mathcal{E}'/\mathcal{E}) = (1 + 4k_B T_e/m_e c^2)$, with $k_B T_e \ll m_e c^2$. After N_s scatterings, the energy change is by the factor

$$\left(\frac{\mathcal{E}'}{\mathcal{E}}\right) = \left(1 + \frac{4k_B T_e}{m_e c^2}\right)^{N_s} \simeq \exp\left(\frac{4k_B T_e N_s}{m_e c^2}\right) = \exp(4y), \qquad (6.106)$$

where we have used Eq. (6.104). Suppose that the initial mean frequency of the radiation field is ω_i with $\hbar\omega_i \ll k_B T_e$. The energy gain by photons (usually called Comptonisation) goes on till the mean energy of the photons raises to $4k_B T_e$. The critical optical depth needed for this is determined by

$$\frac{\mathcal{E}'}{\mathcal{E}} = \left(\frac{4k_B T_e}{\hbar\omega_i}\right) = \exp\left[4\left(\frac{k_B T_e}{m_e c^2}\right)\tau_{\text{crit}}^2\right], \qquad (6.107)$$

giving

$$\tau_{\text{crit}} = \left[\left(\frac{m_e c^2}{4k_B T_e}\right)\ln\left[\frac{4k_B T_e}{\hbar\omega_i}\right]\right]^{1/2}. \qquad (6.108)$$

When the optical depth of the region is comparable with τ_{crit}, the spectrum of the photons will evolve because repeated scattering. Such an evolution is described by an equation called Kompaneets equation, which we shall discuss in Section 6.7.

Finally, note that when a 10-keV photon scatters a free electron, the recoil energy is ~ 200 eV, which is much larger than the typical ionisation potential of valence electrons in atoms. Therefore even bound electrons (except the inner shell electrons of high-Z atoms) behave as though they are free in hard-x-ray scattering.

6.7 Kompaneets Equation

To study these effects more systematically, we need a differential equation that determines the evolution of the photon distribution function as the scattering proceeds. We now derive this equation, called Kompaneets equation.

Let us assume that the medium is reasonably homogeneous over the length scales of interest and that the changes in the number $n(\omega)$ of photons of frequency ω occur only because of scattering. Then the evolution equation for photon number density is [see Eq. (5.275)]

$$\frac{\partial n(\omega)}{\partial t} = \int d^3p \int d\Omega \left(\frac{d\sigma}{d\Omega}\right) c\{n(\omega)[1+n(\omega')]N(E) - n(\omega')[1+n(\omega)]N(E')\}$$

(6.109)

where $(d\sigma/d\Omega)$ is the electron–photon scattering cross section, $n(\omega)$ is the photon distribution function, and $N(E)$ is the electron distribution function. The rate of scattering of photons from frequency ω to frequency ω' by electrons of energy E is described by the term

$$\int d^3p N(E) \int c \left(\frac{d\sigma}{d\Omega}\right) d\Omega\{n(\omega)[1+n(\omega')]\}.$$

(6.110)

The proportionality to $n(\omega)$ and $N(E)$ is obvious; the $[1+n(\omega')]$ term takes into account the stimulated emission effects. Strictly speaking, we should also include a factor $[1 - N(E)]$ to take into account the fermion nature of electrons. This is ignored because electrons are assumed to be nondegenerate with $N(E) \ll 1$. The quantity $d\sigma/d\Omega$ is the differential scattering cross section of Eq. (6.63) for Thomson scattering, and the integration over d^3p takes into account all the electrons with energy $E = p^2/2m$. Similarly, the scattering of photons from ω' to ω is described by the term

$$\int d^3p N(E') \int c \left(\frac{d\sigma}{d\Omega}\right) d\Omega\{n(\omega')[1+n(\omega)]\}.$$

(6.111)

We take the distribution of electrons as given by the Boltzmann distribution with $N(E) \propto \exp(-E/k_B T_e)$. Then Eq. (6.109) is an integrodifferential equation for the unknown quantity $n(\omega, t)$ and is difficult to solve. It can, however, be converted into a differential equation when the frequency change Δ (corresponding to the energy transfer $\hbar\Delta$) in each encounter is small. We now describe this technique, which, incidentally, is applicable to many other similar situations.

To approximate the integrodifferential equation in Eq. (6.109) by a differential equation, we expand $n(\omega') = n(\omega + \Delta)$ and $N(E') = N(E - \hbar\Delta)$ in a Taylor series in Δ, retaining, up to quadratic order,

$$n(\omega') = n(\omega) + \frac{\hbar\Delta}{k_B T_e} \frac{\partial n}{\partial x} + \frac{1}{2}\left(\frac{\hbar\Delta}{k_B T_e}\right)^2 \frac{\partial^2 n}{\partial x^2} + \cdots, \qquad (6.112)$$

$$N(E') = N(E) + \hbar\Delta \frac{\partial N}{\partial E} + \frac{1}{2}\hbar^2\Delta^2 \frac{\partial^2 N}{\partial E^2} + \cdots$$

$$= N(E) + \frac{\hbar\Delta}{k_B T_e} N(E) + \frac{1}{2}\left(\frac{\hbar\Delta}{k_B T_e}\right)^2 N(E) \ldots, \qquad (6.113)$$

where $x \equiv (\hbar\omega/k_B T_e)$ and we have used $N(E) \propto \exp(-E/k_B T_e)$. Substituting these expansions into the original equation, we get

$$\frac{\partial n}{\partial t} = \frac{\hbar}{k_B T_e}\left[\left(\frac{\partial n}{\partial x}\right) + n(n+1)\right] I_1$$

$$+ \frac{1}{2}\left(\frac{\hbar}{k_B T_e}\right)^2 \left[\frac{\partial^2 n}{\partial x^2} + 2(1+n)\frac{\partial n}{\partial x} + n(n+1)\right] I_2, \qquad (6.114)$$

with

$$I_1 = \int d^3p\, d\Omega \left(\frac{d\sigma}{d\Omega}\right) cN(E)\Delta, \qquad (6.115)$$

$$I_2 = \int d^3p\, d\Omega \left(\frac{d\sigma}{d\Omega}\right) cN(E)\Delta^2. \qquad (6.116)$$

To proceed further, we need an estimate for Δ in the individual scattering. The conservation of energy and momentum in the electron–photon scattering can be expressed by

$$\hbar\omega + \frac{p^2}{2m} = \hbar\omega' + \frac{p'^2}{2m}, \qquad \frac{\hbar\omega}{c}\hat{\mathbf{n}} + \mathbf{p} = \frac{\hbar\omega}{c}\hat{\mathbf{n}}' + \mathbf{p}'. \qquad (6.117)$$

Solving for $\mathbf{p}'$, squaring, and substituting into the first equation leads to a quadratic equation on $\Delta \equiv \omega' - \omega$. When $(\Delta/\omega) \ll 1$, we can ignore the Δ^2 term in this equation and solve for Δ, obtaining

$$\hbar(\omega' - \omega) \equiv \hbar\Delta \cong -\frac{\hbar\omega c\mathbf{p}\cdot(\hat{\mathbf{n}} - \hat{\mathbf{n}}') + \hbar^2\omega^2(1 - \hat{\mathbf{n}}\cdot\hat{\mathbf{n}}')}{mc^2 + \hbar\omega(1 - \hat{\mathbf{n}}\cdot\hat{\mathbf{n}}') - c\mathbf{p}\cdot\hat{\mathbf{n}}'}. \qquad (6.118)$$

The second term in the numerator is a correction to the first of $\mathcal{O}(\hbar\omega/mcv) \simeq \mathcal{O}(v/c)$; similarly the second and the third terms in the denominator are small corrections to mc^2. Hence, to lowest order,

$$\hbar\Delta = \hbar(\omega' - \omega) \cong -\left(\frac{\hbar\omega}{mc}\right)\mathbf{p}\cdot(\hat{\mathbf{n}} - \hat{\mathbf{n}}'). \qquad (6.119)$$

The integration for I_2 is now straightforward. Using $\Delta = -(\omega/mc)\mathbf{p}\cdot(\hat{\mathbf{n}} - \hat{\mathbf{n}}')$, we get

$$
I_2 = \left(\frac{\omega}{mc}\right)^2 \int c\left(\frac{d\sigma}{d\Omega}\right) d\Omega \int d^3 p N(E) [\mathbf{p} \cdot (\hat{\mathbf{n}} - \hat{\mathbf{n}}')]^2
$$

$$
= \left(\frac{\omega}{mc}\right)^2 \int c\left(\frac{d\sigma}{d\Omega}\right) d\Omega \int_0^\infty 2\pi p^2 dp(\sin\psi \, d\psi) N(p) p^2 |\hat{\mathbf{n}} - \hat{\mathbf{n}}'|^2 \cos^2\psi
$$

$$
= \frac{1}{3}\left(\frac{\omega}{mc}\right)^2 \int c\left(\frac{d\sigma}{d\Omega}\right) d\Omega |\hat{\mathbf{n}} - \hat{\mathbf{n}}'|^2 \int_0^\infty 4\pi p^2 dp [N(p) p^2]. \tag{6.120}
$$

Because $N(p) \propto \exp(-p^2/2mk_BT_e)$, the p integral gives $\langle p^2 \rangle n_e = 2mn_e \langle p^2/2m \rangle = 3k_BT_e mn_e$, where n_e is the number density of electrons. The angular integration leads to

$$
\int d\Omega \left(\frac{d\sigma}{d\Omega}\right) |\hat{\mathbf{n}} - \hat{\mathbf{n}}'|^2 = \int d\Omega \frac{1}{2} r_0^2 (1 + \cos^2\theta)(2 - 2\cos\theta)
$$

$$
= r_0^2 \int d\Omega (1 + \cos^2\theta) = r_0^2 4\pi \left(1 + \frac{1}{3}\right) = 2\sigma_T. \tag{6.121}
$$

Putting everything together,

$$
I_2 = 2\left(\frac{\omega}{mc}\right)^2 (k_BT_e)(mc)(n_e\sigma_T) = \frac{2n_e\sigma_T(k_BT_e)^3}{\hbar^2 mc} x^2. \tag{6.122}
$$

Let us next consider I_1. In the lowest order, $\Delta \propto \mathbf{p} \cdot (\hat{\mathbf{n}} - \hat{\mathbf{n}}')$ and at this order I_1 will be zero when integrated over all $\mathbf{p}$. Thus, to obtain the nonzero contribution to I_1 we need to expand Eq. (6.118) to a higher order in v/c. We can, however, determine I_1 by an indirect procedure. Note that, by definition, I_1 is the energy-transfer rate divided by the mean energy (k_BT_e) of the electrons. If ΔE is the mean energy transfer per collision, then $I_1 = (\Delta E/k_BT_e)(\sigma_T n_e c)$, as the rate of collisions is $\sigma_T n_e c$. From the discussion in Section 6.6 [see Eq. (6.100)] we know that $\Delta E = (\hbar\omega/mc^2)(4k_BT_e - \hbar\omega)$. Therefore

$$
I_1 = \left(\frac{k_BT_e}{mc^2}\right) \sigma_T n_e x(4 - x). \tag{6.123}
$$

With the form of I_2 and I_1, Eq. (6.114) becomes

$$
\left(\frac{mc^2}{k_BT_e}\right) \frac{1}{n_e\sigma_T c} \frac{\partial n}{\partial t} = \frac{1}{x^2} \frac{\partial}{\partial x}\left[x^4\left(\frac{\partial n}{\partial x} + n + n^2\right)\right]. \tag{6.124}
$$

The combination that appears in the denominator of the left-hand side can be written in the form

$$
\left(\frac{k_BT_e}{mc^2}\right) n_e\sigma_T ct = \left(\frac{k_BT_e}{mc^2}\right) n_e\sigma_T l = \left(\frac{k_BT_e}{mc^2}\right) \tau_e = y, \tag{6.125}
$$

which is essentially the Compton y parameter, defined in Section 6.6. [The last equality follows from Eq. (6.104) in the limit of weak scattering.] So the final result is

$$\frac{\partial n}{\partial y} = \frac{1}{x^2}\frac{\partial}{\partial x}\left[x^4\left(\frac{\partial n}{\partial x} + n + n^2\right)\right].$$

(6.126)

This equation is called Kompaneets equation. The solution to this equation describes the evolution of the photon spectrum that is due to repeated weak scattering with a nonrelativistic thermal bath of electrons. Except in special cases (discussed below), the equation has to be solved numerically.

Exercise 6.13
Filling in the details: Prove Eq. (6.118).

Exercise 6.14
Fokker–Planck equation: The derivation of Kompaneets equation illustrates a general technique that is of value in many different situations. Consider a random variable Y with a probability $P(y_1, t_1 \mid y_2, t_2)$ of having a value y_2 at time t_2 if it starts with value y_1 at time t_1. We will assume that the random process governing the evolution is stationary and depends on only $t_2 - t_1$ so that $P(y_1, t_1 \mid y_2, t_2) = P(y_1, 0 \mid y_2, t_2 - t_1)$. (1) Show that this requires the condition

$$P(y_1 \mid y_3, t_3) = \int P(y_1 \mid y_2, t_2) \, P(y_2 \mid y_3, t_3 - t_2) \, dy_2,$$

(6.127)

where the 0 in $P(y_1, 0 \mid y_2, t_2 - t_1)$ is omitted for simplicity of notation. (2) Use the above result to compare the probabilities at times t and $t + \Delta t$ and show that, in the limit of $\Delta t \to 0$, the probability satisfies the equation

$$\frac{\partial}{\partial t}P(y_0 \mid y, t) = \sum_{n=1}^{\infty}\frac{(-1)^n}{n!}\frac{\partial^n}{\partial y^n}[M_n(y) \, P(y_0 \mid y, t)],$$

(6.128)

with

$$M_n(y) \equiv \lim_{\Delta t \to 0}\frac{1}{\Delta t}\int (y' - y)^n P(y \mid y', \Delta t) \, dy'.$$

(6.129)

In several physical contexts, only $M_1 \equiv A$ and $M_2 \equiv B$ will have nonzero limits, with the rest of the moments vanishing as $\Delta t \to 0$. In that case, Eq. (6.128) reduces to the Fokker–Planck equation

$$\frac{\partial}{\partial t}P = -\frac{\partial}{\partial y}[A(y) \, P] + \frac{1}{2}\frac{\partial^2}{\partial y^2}[B(y) \, P]$$

(6.130)

where $P(y_0 \mid y, t)$ is treated as a function of y and t with fixed y_0. (3) Convince yourself that the derivation of Kompaneets equation is physically same as that of Fokker–Planck equation above.

Let us now rederive some of the preceding results, by using Eq. (6.126). To begin with, the steady-state solution to this equation [with $(\partial n/\partial y) = 0$] is determined by

$$\frac{\partial n}{\partial x} = -n(n+1). \tag{6.131}$$

This can be integrated to give

$$n = [\exp(x - x_0) - 1]^{-1}, \tag{6.132}$$

which is the Bose–Einstein distribution with nonzero $\mu = \beta^{-1}x_0$. The reason for the Bose–Einstein distribution rather than the Planck distribution has already been discussed in Chap. 5, Section 5.14. In the case of Comptonisation, the scattering between electrons and photons cannot change the total number of the photons but can change the mean energy. Therefore the final configuration cannot be a Planck spectrum because of the constraints on both number and energy. Hence the final distribution of the photons, undergoing repeated scattering with electrons, will be a Bose distribution. The β and μ of the distribution will be determined by the total number and energy of the photons. When $\mu \gg 1$, we will have $n \ll 1$ and $n(x) \propto \exp(-x)$. This spectrum is the same as Wien's spectrum.

Kompaneets equation has a particularly simple solution when both n and n^2 are ignored compared with $(\partial n/\partial x)$. Then we obtain the equation

$$\frac{\partial n}{\partial y} \simeq \frac{1}{x^2} \frac{\partial}{\partial x}\left[x^4 \frac{\partial n}{\partial x}\right] = \frac{1}{x^3} \frac{\partial}{\partial(\ln x)}\left[x^3 \frac{\partial n}{\partial(\ln x)}\right]. \tag{6.133}$$

It is convenient to transform coordinates from $(y, q = \ln x)$ to (y, z), where $z = q + 3y$. Noting that

$$\begin{aligned}
dn &= \left(\frac{\partial n}{\partial y}\right)_q dy + \left(\frac{\partial n}{\partial q}\right)_y dq = \left(\frac{\partial n}{\partial y}\right)_q dy + \left(\frac{\partial n}{\partial q}\right)_y (dz - 3dy) \\
&= \left[\left(\frac{\partial n}{\partial y}\right)_q - 3\left(\frac{\partial n}{\partial q}\right)_y\right] dy + \left(\frac{\partial n}{\partial q}\right)_y dz \\
&= \left(\frac{\partial n}{\partial y}\right)_z dy + \left(\frac{\partial n}{\partial q}\right)_y dz,
\end{aligned} \tag{6.134}$$

we can transform relation (6.133) to the simple form:

$$\left(\frac{\partial n}{\partial y}\right)_q = \left(\frac{\partial n}{\partial y}\right)_z + 3\left(\frac{\partial n}{\partial q}\right)_y = \left(\frac{\partial^2 n}{\partial z^2}\right)_y + 3\left(\frac{\partial n}{\partial q}\right)_y \tag{6.135}$$

or

$$\left(\frac{\partial n}{\partial y}\right)_z = \left(\frac{\partial^2 n}{\partial z^2}\right)_y. \tag{6.136}$$

This is a diffusion equation with the solution

$$n(z, y) = \int_{-\infty}^{+\infty} dz' \left(\frac{1}{4\pi y}\right)^{1/2} n(z', 0) \exp\left[-\frac{(z - z')^2}{4y}\right], \quad (6.137)$$

where $n(z', 0)$ is the initial distribution. By transforming back to x, y and writing $z' = \ln \mu$, we find that the solution in terms of the original variables is

$$n(x, y) = \frac{1}{(4\pi y)^{1/2}} \int_0^{\infty} \frac{d\mu}{\mu} n(\mu, 0) \exp\left[-\frac{1}{4y}\left(3y + \ln\frac{x}{\mu}\right)^2\right]. \quad (6.138)$$

As a special case, suppose that the initial photon spectrum is approximated adequately by the Rayleigh–Jeans limit of the Planck spectrum, with $n(x, 0) = x^{-1}$. Then Eq. (6.138) reduces to

$$n(x, y) = \frac{1}{(4\pi y)^{1/2}} \int_0^{\infty} \frac{d\mu}{\mu^2} \exp\left[-\frac{1}{4y}\left(3y - \ln\frac{\mu}{x}\right)^2\right]. \quad (6.139)$$

Substituting $p = \ln(\mu/x)$ into Eq. (6.139), we can evaluate the integral by elementary means and we find that

$$n(x, y) = x^{-1} e^{-2y}. \quad (6.140)$$

Because the effective temperature is proportional to the number density of photons at the Rayleigh–Jeans end, the temperature is *reduced* by a factor e^{-2y}. This result will be of importance in studying the interaction of a cosmic microwave background with a hot gas in galaxy clusters in Vol. III.

Let us next consider the rate of change of the total energy of the radiation with respect to y. The total energy of the photons is

$$E_{\text{pho}} = \int_0^{\infty} 2n(\omega)\hbar\omega \frac{4\pi\omega^2}{c^3} d\omega = \frac{8\pi (k_B T)^4}{\hbar^3 c^3} \int_0^{\infty} n(x)x^3 \, dx. \quad (6.141)$$

Its rate of change is determined by

$$\frac{\hbar^3 c^3}{8\pi (k_B T_e)^4} \frac{dE_{\text{pho}}}{dy} = \int_0^{\infty} dx x^3 \frac{\partial n}{\partial y} = \int_0^{\infty} dx x^3 \left\{\frac{1}{x^2}\frac{\partial}{\partial x}\left[x^4\left(\frac{\partial n}{\partial y} + n + n^2\right)\right]\right\}. \quad (6.142)$$

In many of the astrophysically interesting situations, $n \ll 1$, so that we can ignore n^2 compared with n. Integrating expression (6.142) by parts, we get

$$\frac{\hbar^3 c^3}{8\pi (k_B T_e)^4} \frac{dE_{\text{pho}}}{dy} = 4 \int_0^{\infty} nx^3 \, dx - \int_0^{\infty} nx^4 \, dx. \quad (6.143)$$

If $n(x)$ is significant mostly for $x \ll 1$, then the first integral dominates over the second and

$$\frac{dE_{\text{pho}}}{dy} \cong 4E_{\text{pho}} \tag{6.144}$$

or

$$E_{\text{pho}}(t) \cong E_{\text{pho}}(0)\exp 4y = E_{\text{pho}}(0)\exp\left(\frac{t}{t_{\text{comp}}}\right), \tag{6.145}$$

with

$$t_{\text{comp}} = \left(\frac{mc^2}{4k_B T_e}\right)\frac{1}{(n_e\sigma_T c)}. \tag{6.146}$$

In this limit, the optical depth is $\tau_e \approx (n_e\sigma_T)(ct)$ and the result from relation (6.145) is the same as that obtained in Eq. (6.106).

Finally, we note that Eq. (6.143) can also be used to study the change in the electron energy. Because $E_{\text{pho}} + E_e$ should be a constant, $dE_e/dy = -dE_{\text{pho}}/dy$. Let us assume, for the sake of definiteness, that $n(x) = n_0\exp(-x/\alpha)$. The parameter α measures the ratio of photon temperature to electron temperature. For this case,

$$4\int_0^\infty nx^3\,dx - \int_0^\infty nx^4\,dx = -24\alpha^4 n_0(\alpha - 1), \tag{6.147}$$

so that

$$\frac{dE_{\text{elec}}}{dt} = -\frac{dE_{\text{pho}}}{dt} = \frac{8}{3}E_{\text{ele}}\left(\frac{E_{\text{pho}}\sigma_T}{mc}\right)(\alpha - 1). \tag{6.148}$$

With $E_{\text{elec}} = (3/2)k_B T_e$, $E_{\text{pho}} = aT^4$, Eq. (6.148) becomes

$$\frac{dT_e}{dt} = \frac{8}{3}\left(\frac{\sigma_T}{mc}\right)(aT^4)(T - T_e). \tag{6.149}$$

This (approximate) equation governs the cooling of the electron gas that is due to a transfer of energy to the photons.

It often happens in astrophysical systems that the plasma is not fully ionised but has a fractional ionisation of x_e. For a hydrogen plasma, we may define $x_e = n_e/(n_p + n_H) = n_p/(n_p + n_H)$, where n_e, n_p, and n_H denote the number densities of electrons, protons, and hydrogen atoms, respectively. The collisions between e, p, and H will maintain a common temperature for matter T_m; the transfer of energy between photons and matter, however, is mainly due to Thomson scattering of electrons by photons. Then, in energy-balance Eq. (6.149), the left-hand side will be multiplied by $(n_e + n_p + n_H) = (n_p + n_H)(1 + x_e)$ and the right-hand side will be multiplied by $n_e = (n_p + n_H)x_e$. This changes the equation to

$$\frac{dT_m}{dt} = \frac{8}{3}\left(\frac{x_e}{1 + x_e}\right)\left(\frac{\sigma_T}{m_e c}\right)(aT^4)(T - T_m). \tag{6.150}$$

If the gas is not hydrogen, we need to multiply the right-hand side further by the molecular weight μ. The corresponding cooling time is

$$
t_{\text{cool}} = \left(\frac{1}{T_m}\frac{dT_m}{dt}\right)^{-1} = \frac{3}{8}\left(\frac{T}{T_m} - 1\right)^{-1}\left(\frac{1 + x_e}{x_e}\right)\left(\frac{m_e c}{\mu \sigma_T}\right)\left(\frac{1}{aT^4}\right)
$$
$$
\approx \left(\frac{mc^2}{k_B T_e}\right)\left(\frac{1}{n_\gamma \sigma_T c}\right). \tag{6.151}
$$

The second equality is valid if $T_m \gg T$, $aT^4 \approx n_\gamma k_B T_e$, $x_e \approx 1$, and $\mu \approx 1$.

The Kompaneets equation does not take into account either the input of photons or the escape of photons from the region under consideration. In some cases, it is necessary to model a situation in which a photon source is present and the photons escape out of the region. Usually there will be a source of photons with low energies that will be scattered to higher energies through Comptonisation and will eventually escape. With the presence of a source and an escape route, it is possible to have steady-state solutions with power-law spectra. To see how this possibility arises, let us consider a simple model in which we add to the Kompaneets equation two terms: (1) a source of soft photons $Q(x)$ with $Q(x)$ vanishing for $x > x_s$ with $x_s \ll 1$; (2) a term $-[n/\max(\tau_e, \tau_e^2)]$ that represents the escape of photons. In the case of steady state, Kompaneets equation now becomes

$$
0 = \left(\frac{k_B T_e}{mc^2}\right)\frac{1}{x^2}\frac{\partial}{\partial x}\left[x^4\left(\frac{\partial n}{\partial x} + n + n^2\right)\right] + Q(x) - \frac{n}{\max\left(\tau_e, \tau_e^2\right)}. \tag{6.152}
$$

It is easy to see that for $x \gg 1$, the approximate solution is still $n(x) \propto \exp(-x)$. On the other hand, for $x_s \ll x \ll 1$ we can ignore the $n(n+1)$ term and obtain an approximate power-law solution given by

$$
n(x) \propto x^m, \quad m = -\frac{3}{2} \pm \sqrt{\frac{9}{4} + \frac{4}{y}}. \tag{6.153}
$$

We need to take the positive root in the above solution for $y \gg 1$, which leads to the low-frequency limit of Wien's law. On the other hand, for $y \ll 1$, the negative root is appropriate. For $y \approx 1$, it is necessary to take the linear combination of the solutions and no single power law is possible. It is easy to see that the input energy is significantly amplified when $y \gtrsim 1$, as to be expected.

The Kompaneets equation derived above is also based on the assumption that the distribution of electrons is given by a Boltzmann distribution. In principle, the derivation can be repeated along the same lines for any other distribution of electrons. Of particular relevance to astrophysics is the case in which a power-law distribution of electrons with $N(E)\,dE = CE^{-p}\,dE$ scatters with photons with a distribution $n_{\text{in}}(\epsilon)$. In general, this is a complicated problem; but if the plasma is thin, we may assume that each photon is scattered only once against the electron distribution. It is then fairly straightforward to show that the resulting spectrum of photons (after a single scattering) is given by

$$n_{\text{out}}(\epsilon) = A\epsilon^{-(p-1)/2}, \tag{6.154}$$

where A is a constant given by

$$A = \frac{3}{8}(3C\sigma_T c)\frac{2^p(p^2 + 4p + 11)}{(p+3)^2(p+1)(p+5)}\int_0^\infty x^{(p-1)/2}n_{\text{in}}(x)\,dx. \tag{6.155}$$

Note that this is a power-law spectrum that is independent of the initial photon spectrum.

6.8 Equations of Radiative Transport

In Section 6.2, a coefficient of absorption, a coefficient for induced emission, and an emissivity to describe spontaneous emission were introduced. In many contexts, we also have to take into account scattering of radiation by matter, which can be handled by the scattering cross section introduced in Section 6.4 and further developed in Section 6.7. The subject of radiative transport deals with the issue of solving for the intensity distribution in a given physical context that is described by the coefficients for absorption, emission, and scattering. Considering the complexity of this subject, we approach it in two steps. First, we consider situations in which scattering is not important; then we take into account complications introduced by scattering.

In Chap. 5, Section 5.14, we obtained an equation for the evolution of the distribution function of massless particles. Because the frequency of the photon does not change during the free propagation [described by the left-hand side of Eq. (5.248)], the same equation can be used to describe the *intensity* of radiation, which is proportional to f at a given frequency, giving

$$\frac{1}{c}\frac{\partial I_\nu}{\partial t} + \hat{\mathbf{k}}\cdot\frac{\partial I_\nu}{\partial \mathbf{x}} = C_\nu, \tag{6.156}$$

where C_ν represents processes that change the intensity $I_\nu(x^i, \hat{\mathbf{k}})$ along the path of a ray. From the discussion in Section 6.2, we know that C_ν can be written as

$$C = \frac{\rho j_\nu}{4\pi} - \rho\kappa_\nu^{\text{abs}}I_\nu + \rho\kappa_\nu^{\text{ind}}I_\nu + \cdots. \tag{6.157}$$

The first term represents spontaneous emission and j_ν is emissivity per unit mass with the dimensions of energy emitted per second per frequency interval per unit mass. Multiplied by ρ (which has dimensions of mass per unit volume) and divided by 4π, this term has the dimensions of $(dE/dt\,dV\,d\nu\,d\Omega)$, which is the same as the dimensions of rate of change of intensity per unit length. [In Section 6.2 we denoted by j_ν what we are denoting here by $(\rho j_\nu/4\pi)$.] The second term represents the absorption of photons; we have used the standard definitions $\alpha = n\sigma = \rho\kappa$, where all quantities are in units of inverse centimeters and κ is in square centimeters per gram. The third term represents induced emission with the same conventions. The $\cdots$ in Eq. (6.157) indicate scattering terms, which have been ignored for the present and will be discussed later.

Even after the scattering term is ignored, Eq. (6.157) is fairly difficult to handle self-consistently. In particular, the induced emission term depends on the direction of photon, as induced emission produces a photon in the same state as that of the original photon. To make progress, it is usually assumed that the induced emission term can be ignored compared with the spontaneous emission, which is valid if the mean occupation number of photons, in the frequency range we are interested in, is small compared with unity. In this case, we can approximate Eq. (6.157) by retaining only the first two terms and further assuming that j_ν and κ_ν^{abs} are independent of $\hat{\mathbf{k}}$.

To extract the physics out of this equation, we use the moments defined in Eq. (6.5). Because the frequency remains constant, it is convenient to redefine the moments with *only* angular integrations as

$$cT^{ab}(x^i, \nu) = \int d\Omega \hat{k}^a \hat{k}^b I_\nu(x^i, \hat{\mathbf{k}}), \quad \hat{k}^a \equiv (1, \hat{\mathbf{k}}). \tag{6.158}$$

It is conventional to give separate symbols U_ν, $\mathbf{F}_\nu$, and $P_\nu^{\alpha\beta}$ for the time–time, time–space, and space–space parts of this moments by defining

$$cU_\nu(x^i) = cT^{00} = \int d\Omega I_\nu(x^i, \hat{\mathbf{k}}), \tag{6.159}$$

$$[\mathbf{F}_\nu(x^i)]^\alpha = cT^{0\alpha} = \left[\int d\Omega \, \hat{k}^\alpha I_\nu(x^i, \hat{\mathbf{k}}) \right], \tag{6.160}$$

$$cP_\nu^{\alpha\beta}(x^i) = cT^{\alpha\beta} = \int d\Omega \hat{k}^\alpha \hat{k}^\beta I_\nu(x^i, \hat{\mathbf{k}}). \tag{6.161}$$

The quantities U_ν, $\mathbf{F}_\nu$, and $P^{\alpha\beta}$ are called the energy density, radiative flux, and (three-dimensional) radiative stress tensor, respectively. (We will omit the subscript ν on $P_\nu^{\alpha\beta}$ to simplify notation.) The time evolution of these quantities can be determined from the general result for the evolution of the moments of the distribution function [see Eq. (5.256)]:

$$\frac{\partial T^{ik}}{\partial x^i} = C^k, \tag{6.162}$$

where C^k is the result of taking the moment of the right-hand side of Eq. (5.256). This equation obviously remains valid even when moments are defined with only angular integration. In our case, the left-hand side is a four vector with components

$$\frac{\partial T^{ik}}{\partial x^i} = \left[\frac{\partial U_\nu}{\partial t} + \nabla \cdot \mathbf{F}_\nu, \quad \frac{1}{c} \frac{\partial F^\alpha(\nu)}{\partial t} + c \frac{\partial P^{\alpha\beta}(\nu)}{\partial x^\beta} \right]. \tag{6.163}$$

On the right-hand side, because j_ν and κ_ν^{abs} have no dependence on $\hat{\mathbf{k}}$, we obtain the four vector with components

$$C^k = \left[\rho\left(j_\nu - c\kappa_\nu^{abs}U_\nu\right),\ -\rho\kappa_\nu^{abs}\mathbf{F}_\nu\right]. \tag{6.164}$$

Identifying the terms, we get

$$\frac{\partial U_\nu}{\partial t} + \nabla \cdot \mathbf{F}_\nu = \rho\left(j_\nu - c\kappa_\nu^{abs}U_\nu\right), \tag{6.165}$$

$$\frac{1}{c}\frac{\partial F^\alpha}{\partial t} + c\frac{\partial P^{\alpha\beta}}{\partial x^\beta} = -\rho\kappa^{abs}F^\alpha. \tag{6.166}$$

In Eq. (6.166) the subscript ν, indicating the frequency dependence, has been suppressed throughout for simplicity of notation.

These equations have a fairly straightforward interpretation. Equation (6.165) shows that the change in the energy density is due to three separate processes: the radiative flux, gain that is due to emission, and loss that is due to absorption. Equation (6.166) describes the force balance in the medium in terms of the rate of change of momentum flux.

In a material medium, the rate of loss of momentum by the radiation should be compensated for by the rate of gain of momentum by matter. Therefore Eq. (6.166) also allows us to determine the radiative force per unit volume acting on the matter. Because the net momentum flow of photons is $1/c$ times the net energy flow, the radiative force is given by the right-hand side of Eq. (6.166) divided by c. Integrated over all frequencies, the net radiative force is given by

$$\mathbf{f}_{rad} = \frac{\rho}{c}\int_0^\infty \kappa_\nu\mathbf{F}_\nu\,d\nu. \tag{6.167}$$

[Because F has the dimension of energy per unit area per unit time, the quantity $(\rho\kappa F/c)$ has the dimensions of force per unit volume.]

Exercise 6.15

Eddington luminosity: Consider a cloud of ionised gas that is accreting to a luminous central object with a luminosity L. The radiation from the central object will exert a force on the in-falling gas through Thomson scattering. Use Eq. (6.167) to show that this force is given by $f_{rad} = (\sigma_T/\mu_e)(L/4\pi cr^2)$, where the opacity is due only to Thomson scattering. The ratio between this outward force and the inward gravitational force that is due to the central object is $(f_{rad}/f_{grav}) = (L/L_E)$, where

$$L_E = \frac{4\pi cGM\mu_e}{\sigma_T} = 1.51 \times 10^{38}\,\mu_e\left(\frac{M}{M_\odot}\right)\ \text{erg s}^{-1} \tag{6.168}$$

is called the Eddington luminosity. Clearly, the radiative force will prevent accretion of a plasma when the luminosity approaches the Eddington luminosity. [Note that the time scale to radiate away an amount Mc^2 (comparable with the rest-mass energy of the central source) at the rate L_E is

$$t_E = \frac{\sigma_T c}{4\pi G m_p} \cong 4 \times 10^8 \text{ yr,}$$

which is expressible entirely in terms of fundamental constants. It may be noted that the above argument leading to $L_{Edd} = (4\pi GMc/\kappa)$ is quite general and is independent of the actual source of opacity κ. The numerical value obtained above is applicable for fully ionised hydrogen and can be easily changed for any other fully ionised gas. Another extreme example is given by a gas of solar abundance in which all the refractory elements have condensed into dust grains of radius $r_d \approx 0.1 \ \mu$m. If the radiation is at wavelengths shorter than r_d, opacity will be dominated by dust absorption and scattering and we will get $\kappa = (\sigma_{dust}/\mu_{dust} m_{dust})$, where $\sigma_d \approx \pi r_d^2 \approx 3 \times 10^{-10}$ cm^2, m_d is the mean mass per dust grain $= (4\pi/3)m_{atom}(r_d/r_{atom})^3 \approx 3 \times 10^{10} \ m_p$ (if $m_{atom} \approx 6 \ m_p$), and $(\mu_d m_d)^{-1} = (n_{dust}/\rho_{total}) = (\rho_{dust}/m_d \rho_{total}) = (Z/m_d)$. Taking $Z = 0.02$ we get $\kappa = 120 \text{ cm}^2\text{g}^{-1}$ and, correspondingly, $L_{Edd} \approx 4.2 \times 10^{35}(M/M_\odot)$ erg s^{-1}. For radiation with $\lambda \gtrsim r_d$, the opacity κ will decrease substantially and the radiation will not interact efficiently with dust, increasing L_{Edd} significantly.]

In general, Eqs. (6.165) and (6.166) are as intractable as the original set. In fact, taking a finite number of moments of the original equation will never give a closed set of equations. It is obvious that we do not have an equation for the time rate of change of $P^{\alpha\beta}$. Any attempt to obtain this by taking the next moment will lead to an appearance of a still higher-order moment in the equations. Thus equations like Eqs. (6.165) and (6.166) are useful only if some further approximations can be made. We now discuss one such approximation that is very useful.

There are several situations in astrophysics in which the mean free path for the photon, l_ν, is much smaller than the size R over which macroscopic parameters like temperature, etc., vary in the system under consideration. In such a case, it is reasonable to assume that the intensity at any given location would have reached a local thermal equilibrium with matter, giving $I_\nu \approx B_\nu(T)$, $U_\nu \approx (4\pi/c)B_\nu(T)$, and $P^{\alpha\beta} = P\delta^\alpha_\beta$; $P_\nu \approx (U_\nu/3)$. The radiation flux cannot, of course, be computed from this expression, as it will vanish for an isotropic intensity distribution. Radiation flux arises only because there are small spatial gradients in the temperature. To estimate the flux, we find the relative values of various terms in the equations, thereby obtaining a useful approximation.

In both equations, the time-derivative terms are very small and ignorable in most astrophysical contexts. For example, in Eq. (6.165), the time derivative is of the order of U_ν/t_{evl}, where t_{evl} is the time scale for the evolution of the astrophysical system, say, a star. The second term on the right-hand side is of the order of $(\rho\kappa_\nu)(cU_\nu) \approx (cU_\nu/l_\nu)$. The ratio between these two terms is $(l_\nu/ct_{evl}) \ll 1$. The situation is identical in Eq. (6.166) as well. It follows that, in order to satisfy Eq. (6.166), near equality must exist between the second term on the left-hand side and the right-hand side. This gives

$$F^\alpha \approx -\frac{c}{\rho\kappa}\frac{\partial P^{\alpha\beta}}{\partial x^\beta} \qquad (6.169)$$

and allows us to estimate the flux as $F_\nu \approx (cU_\nu)(l_\nu/R)$, which is a factor of l_ν/R lower than the flux in the absence of opacity that is due to any matter.

Having estimated F, we can use it in Eq. (6.165); we see that the second term on the left-hand side is $(F_\nu/R) \approx (l_\nu/R^2)U_\nu c$. The ratio of this term to the second term on the right-hand side is $(l_\nu/R)^2 \ll 1$. Therefore neither of the two terms on the left-hand side can match the second term on the right-hand side of Eq. (6.165). This equation can be satisfied only if the two terms on the right-hand side balance each other. Thus we get the second relation

$$U_\nu = \frac{\rho j_\nu}{c\kappa_\nu\rho} = \frac{\rho j_\nu}{c\alpha_\nu} = \frac{4\pi}{c}B_\nu(T), \qquad (6.170)$$

where the last equality follows from Eq. (6.32), which is valid whenever matter is in thermal equilibrium. [We stress that the definition of j_ν has changed slightly; what was called j_ν there is denoted as $(\rho j_\nu/4\pi)$ here.]

We thus conclude that if $R \gg l_\nu$ and $ct_{\rm evl} \gg l_\nu$, then the solution to the moment equations can be expressed as Eq. (6.169) and Eq. (6.170). In this case, the radiative flux is given by

$$F^\alpha \cong -\frac{c}{\rho\kappa}\frac{\partial}{\partial x^\beta}\left(\frac{1}{3}\delta_\beta^\alpha U_\nu\right) = -\frac{4\pi}{3\rho\kappa_\nu}\left(\frac{\partial B_\nu}{\partial T}\right)\frac{\partial T}{\partial x^\alpha}. \qquad (6.171)$$

The flux is proportional to the gradient of the temperature with a coefficient that depends on the properties of matter and temperature. This is a general feature of diffusion phenomena, and the approximation can be called the radiative diffusion approximation. Integrating this relation over all frequencies and defining a quantity called the Roseland mean opacity by

$$\frac{1}{\kappa_R} \equiv \frac{\int_0^\infty (1/\kappa_\nu)(\partial B_\nu/\partial T)\,d\nu}{\int_0^\infty (\partial B_\nu/\partial T)\,d\nu} \qquad (6.172)$$

we can write the final result as

$$\mathbf{F}_{\rm rad} = -\frac{c}{3\rho\kappa_R}\nabla(aT^4), \qquad (6.173)$$

where we have also used the relation

$$\int_0^\infty \frac{\partial B_\nu}{\partial T}\,d\nu = \frac{d}{dT}\int_0^\infty B_\nu(T)\,d\nu = \frac{d}{dT}\left(\frac{caT^4}{4\pi}\right). \qquad (6.174)$$

The corresponding radiative force acting on the unit volume of matter can be found from Eq. (6.167) and we get $\mathbf{f}_{\rm rad} = -(1/3)\nabla(aT^4) = -\nabla P$, which is just the gradient of the radiation pressure. These results are extensively used (for example) in the study of stellar interiors.

We now turn to the discussion of scattering, which has been ignored so far. Let us first consider scattering that does not change the frequency of the photon, but only the direction, and write the relevant scattering cross section as

$$\frac{d\sigma(\hat{\mathbf{k}} \to \hat{\mathbf{k}})}{d\Omega} = \sigma\phi(\hat{\mathbf{k}}, \hat{\mathbf{k}}'), \tag{6.175}$$

where σ is the total scattering cross section and ϕ carries the angular dependence. Clearly, ϕ is symmetric in its arguments, dimensionless, and is normalised so that the integral over the solid angle is unity. Then the gain in the number of photons with wave vector $\mathbf{k}$ that is due to scattering from all other $\mathbf{k}'$ is given by

$$\left[\frac{dn(\mathbf{k})}{dt}\right]_{\text{gain}} = \int d\Omega'[N_e\sigma\phi(\hat{\mathbf{k}}, \hat{\mathbf{k}}')c]n(\mathbf{k}')[1 + n(\mathbf{k})]. \tag{6.176}$$

Here N_e is the number density of scatterers (the subscript e is for electrons, but it could be any charged particle) and $N_e\sigma\phi c$ gives the number of scatterings per second for a single photon. The multiplication by $n(\mathbf{k}')[1 + n(\mathbf{k}')]$ arises because scattering of radiation should be thought of as an absorption of a photon in the state labelled by $\mathbf{k}$ followed by emission into a state labelled by $\mathbf{k}'$. The probability for absorption scales as $n(\mathbf{k}')$, and the probability for emission scales as $[1 + n(\mathbf{k})]$.

The loss of photons in the state labelled by $\mathbf{k}$ is given by a similar term:

$$\left[\frac{dn(\mathbf{k})}{dt}\right]_{\text{loss}} = \int d\Omega' N_e\sigma\phi(\hat{\mathbf{k}}, \hat{\mathbf{k}}')cn(\mathbf{k})[1 + n(\mathbf{k}')]. \tag{6.177}$$

Subtracting the loss from the gain, we find that the net rate of change of photon number is given by

$$\frac{dn(\mathbf{k})}{dt} = \int d\Omega' N_e\sigma\phi(\hat{\mathbf{k}}, \hat{\mathbf{k}}')c[n_{\mathbf{k}'}(1 + n_{\mathbf{k}}) - n_{\mathbf{k}}(1 + n_{\mathbf{k}'})]$$

$$= \int d\Omega' N_e\sigma\phi(\hat{\mathbf{k}}, \hat{\mathbf{k}}')c[n(\mathbf{k}') - n(\mathbf{k})], \tag{6.178}$$

as the $n(\mathbf{k}')n(\mathbf{k}')$ term cancels out. Hence we would have arrived at the correct result for this simple case (of scattering with no change of frequency) even if we did not include the $(1 + n)$ factor. However, such coincidences do not occur for more complicated cases.

Because the angular integration is over only ϕ in the second term on the right-hand side of Eq. (6.178), we get

$$\frac{1}{c}\left[\frac{dn(\mathbf{k})}{dt}\right]_{\text{scattering}} = N_e\sigma\int d\Omega'\phi(\hat{\mathbf{k}}, \hat{\mathbf{k}}')n(\mathbf{k}') - N_e\sigma n(\mathbf{k})\int d\Omega'\phi(\hat{\mathbf{k}}, \hat{\mathbf{k}}')$$

$$= -\rho\kappa_\nu^{\text{sca}}n(\mathbf{k}) + \rho\kappa_\nu^{\text{sca}}\int d\Omega'\phi(\hat{\mathbf{k}}, \hat{\mathbf{k}}')n(\mathbf{k}'). \tag{6.179}$$

In arriving at this result, we have used the normalisation condition on ϕ and defined the scattering opacity by the standard procedure: $\rho\kappa = n\sigma$. Because the photon number is proportional to the intensity (at the fixed frequency we are working with), the same equation holds for intensity as well. Taking into account all the emission, absorption, and scattering (without frequency change), Eq. (6.156) becomes modified to

$$\frac{1}{c}\frac{\partial I_\nu}{\partial t} + \hat{\mathbf{k}} \cdot \nabla I_\nu = \frac{1}{4\pi}\rho j_\nu - \rho\kappa_\nu^{\text{abs}} I_\nu - \rho\kappa_\nu^{\text{sca}} I_\nu + \rho\kappa_\nu^{\text{sca}} \int d\Omega' \phi(\hat{\mathbf{k}}, \hat{\mathbf{k}}') I_\nu(\hat{\mathbf{k}}').$$
(6.180)

The above derivation of Eq. (6.179) shows the origin of the two scattering terms clearly but is somewhat heuristic. A more formal way of obtaining the same equation is as follows: We obtained the general form of the equation describing scattering between electrons and photons in Chap. 5, Section 5.14 [see Eq. (5.273)]. This equation should reduce to Eq. (6.178) under suitable approximations. To see how this comes about, let us consider Eq. (5.273) in the Thomson scattering limit in the mean rest frame of electrons. In this case, the energy of the photon or the electron does not change. Then we can set, ignoring the energy transfer to electron in the mean rest frame of the electron,

$$\epsilon_{\mathbf{p}} = \epsilon_{\mathbf{p}'} \cong mc^2, \quad E_{\mathbf{k}} = E_{\mathbf{k}'}, \quad \delta_{\text{D}}(\Delta U/c) = \delta_{\text{D}}\left[\frac{\hbar}{c}(\omega_{\mathbf{k}} - \omega_{\mathbf{k}'})\right],$$

$$d^3\mathbf{k} = k'^2 dk' d\Omega'.$$
(6.181)

Writing $k' = (\hbar\omega/c)$, we find that Eq. (5.273) becomes

$$\left(\frac{dn}{dt}\right)_{\text{scatt}} = \left(\frac{1}{4mc}\right)^2 \frac{c}{\hbar} \int d^3\mathbf{p}' \int d\omega' d\Omega' |M|^2 \delta_{\text{D}}(\omega - \omega')$$
$$\times \{n_e(\mathbf{p}')[1 - n_e(\mathbf{p})]n(\mathbf{k}')[1 + n(\mathbf{k})] - n_e(\mathbf{p})$$
$$\times [1 - n_e(\mathbf{p}')]n(\mathbf{k})[1 + n(\mathbf{k}')]\},$$
(6.182)

where we have set $\omega = \omega'$ inside the integral, making use of the presence of the Dirac delta function. In the absence of a significant change in momentum, the terms $n_e(1 - n_e)$ involving electron distribution functions is the same in both the terms within the braces. Taking this out as a common factor, we see that the n^2 term of the photon distribution cancels out. In most practical situations, $n_e \ll 1$ and we can write the number density of electrons (that is, number of electrons per unit space volume) as

$$N_e = \int \frac{d^3\mathbf{p}'}{(2\pi\hbar)^3} n_e(\mathbf{p}').$$
(6.183)

With these simplifications, we get

$$\left(\frac{dn}{dt}\right)_{\text{scatt}} = \frac{c}{\hbar}\left(\frac{1}{4mc}\right)^2 \int d^3\mathbf{p}' \int d\Omega' |M|^2 n_e(\mathbf{p}')[n(\mathbf{k}') - n(\mathbf{k})]$$
$$= \frac{\pi^3}{2}\left(\frac{\hbar}{mc}\right)^2 cN_e \int d\Omega' |M|^2 [n(\mathbf{k}') - n(\mathbf{k})].$$
(6.184)

In the case of Thomson scattering, with our normalisation conventions, the matrix element can be written as

$$|M|^2 = \frac{16}{3\pi^2} \left(\frac{e^2}{\hbar c}\right)^2 \phi(\hat{\mathbf{k}}, \hat{\mathbf{k}}'), \quad \phi(\hat{\mathbf{k}}, \hat{\mathbf{k}}') = \frac{3}{16\pi}[1 + (\hat{\mathbf{k}} \cdot \hat{\mathbf{k}}')^2], \quad (6.185)$$

where ϕ has been normalised to unity for angular integration. This gives the final result

$$\left(\frac{dn}{dt}\right)_{\text{scatt}} = \sigma_T(cN_e) \int d\Omega' \phi(\hat{\mathbf{k}}, \hat{\mathbf{k}}')[n(\mathbf{k}') - n(\mathbf{k})]$$

$$= -c(\sigma_T N_e)n(\mathbf{k}) - c(\sigma_T N_e) \int d\Omega' \phi(\hat{\mathbf{k}}, \hat{\mathbf{k}}')n(\mathbf{k}'). \quad (6.186)$$

This matches with our earlier result [Eq. (6.179)] and illustrates the assumptions that were involved in it.

It is easy to work out how our moment equations become modified in the presence of scattering. To get the first moment Eq. (6.165) we need to integrate the right-hand side of the evolution equation over all the solid angles $d\Omega$. In this case, the two scattering terms will precisely cancel, as

$$-\int d\Omega I_\nu(\hat{\mathbf{k}}) + \int d\Omega \int d\Omega' \phi(\hat{\mathbf{k}}, \hat{\mathbf{k}}')I_\nu(\hat{\mathbf{k}}')$$

$$= -\int d\Omega I_\nu(\hat{\mathbf{k}}) + \int d\Omega' I_\nu(\hat{\mathbf{k}}') = 0 \quad (6.187)$$

because of the normalisation of ϕ. Thus our first moment Eq. (6.165) remains unchanged in the presence of scattering (without frequency change).

To obtain the second moment equation, we have to multiply the right-hand side of Eq. (6.156) by $\hat{\mathbf{k}}$ and integrate over $d\Omega$. The result now will depend on the angular dependence of the scattering determined by ϕ. In most cases, the scattering will have a forward–backward symmetry (as in Thomson scattering, for example) and ϕ will satisfy the further condition that

$$\int \phi(\hat{\mathbf{k}}, \hat{\mathbf{k}}')(\hat{\mathbf{k}}\, d\Omega \quad \text{or} \quad \hat{\mathbf{k}}'\, d\Omega') = 0. \quad (6.188)$$

In this case, the last term on the right-hand side of Eq. (6.186) will not contribute to the first moment and we will get

$$\frac{1}{c}\frac{\partial F^\alpha}{\partial t} + c\frac{\partial P^{\alpha\beta}}{\partial x^\beta} = -\rho\left(\kappa_\nu^{\text{abs}} + \kappa_\nu^{\text{sca}}\right)F^\alpha, \quad (6.189)$$

which differs from Eq. (6.166) only by the replacement of κ_ν^{abs} with $\kappa_\nu^{\text{tot}} \equiv \kappa_\nu^{\text{abs}} + \kappa_\nu^{\text{sca}}$. Thus, in the case of scattering (with forward–backward symmetry and no frequency change, as in Thomson scattering), all our earlier conclusions follow with the opacity given by κ_ν^{tot}. In particular, the solution based on radiative diffusion approximation (6.171) and the expression for radiative force, etc., continue to be valid.

Finally let us consider the expression for the radiative flux in the diffusion approximation, which was obtained above from a formal argument involving thermal equilibrium. Although it is clear from the modification of Eqs. (6.166)–(6.189) that all the results will continue to hold in the presence of scattering, it is worthwhile to rederive the diffusion approximation more directly. This is particularly important because it highlights a very general technique in dealing with the equation that describes the evolution of distribution function. We now illustrate this technique when the photon number is changing purely because of scattering (without frequency change and possessing forward–backward symmetry) by using basic Eq. (6.178), written in the form

$$
\frac{dn(x^i, \mathbf{k})}{dt} = \frac{\partial n(x^i, \mathbf{k})}{\partial t} + c\hat{\mathbf{k}} \cdot \frac{\partial n(x^i, \mathbf{k})}{\partial \mathbf{x}}
$$
$$
= -\sigma N_e c n(x^i, \mathbf{k}) + N_e c \sigma \int d\Omega' \phi(\hat{\mathbf{k}}, \hat{\mathbf{k}}') n(x^i, \mathbf{k}'). \quad (6.190)
$$

In steady state the time derivative $(\partial n / \partial t) = 0$; an exact solution to this equation is still impossible because of the integration over $d\Omega'$. We obtain an approximate solution of the form $n = n_0 + n_1$, where n_0 is the equilibrium term and $n_1 \approx (l_\nu / R) n_0$, where $l_\nu = (N_e \sigma)^{-1}$ is the mean free path that is due to scattering and R is the scale over which n_0 changes. This approximation will be valid to first-order accuracy in l_ν / R. Substituting this ansatz into Eq. (6.190), we get

$$
c\hat{\mathbf{k}} \cdot \frac{\partial n_0}{\partial \mathbf{x}} + c\hat{\mathbf{k}} \cdot \frac{\partial n_1}{\partial \mathbf{x}} = \left[-\sigma N_e c n_0 + N_e c \sigma \int d\Omega' \phi(\hat{\mathbf{k}}, \hat{\mathbf{k}}') n_0 \right]
$$
$$
+ \left\{ -\sigma N_e c n_1 + N_e c \sigma \int d\Omega' \phi(\hat{\mathbf{k}}, \hat{\mathbf{k}}') n_1 \right\}. \quad (6.191)
$$

We now assume that the equilibrium distribution $n_0(\mathbf{x}, \mathbf{k}) = n_0(\mathbf{x})$ is independent of $\mathbf{k}$. Given this isotropy, n_0 can be pulled out of the integral in the second term within the brackets of Eq. (6.191). The integral over ϕ gives unity and the terms inside the brackets cancel each other. On the left-hand side, the first term will dominate over the second and hence we drop the second term. This leads to the equation

$$
c\hat{\mathbf{k}} \cdot \frac{\partial n_0(\mathbf{x})}{\partial \mathbf{x}} = -\sigma N_e c n_1(\mathbf{x}, \mathbf{k}) + \sigma N_e c \int d\Omega' \phi(\hat{\mathbf{k}}, \hat{\mathbf{k}}') n_1(\mathbf{x}, \mathbf{k}'). \quad (6.192)
$$

The $\mathbf{k}$ dependence of the left-hand side is of the form $\hat{\mathbf{k}} \cdot \mathbf{A}(\mathbf{x})$ and hence we must have $n_1(\mathbf{x}, \mathbf{k})$ of the form $n_1(\mathbf{x}, \mathbf{k}) = \mathbf{B}(\mathbf{x}, \nu) \cdot \hat{\mathbf{k}}$. On substituting this expression into Eq. (6.192), the integral over $d\Omega'$ vanishes because of the forward–backward symmetry of the scattering, giving

$$
c\hat{\mathbf{k}} \cdot \frac{\partial n_0(\mathbf{x})}{\partial \mathbf{x}} = -\sigma N_e c \mathbf{B}(\mathbf{x}) \cdot \hat{\mathbf{k}}. \quad (6.193)
$$

This allows us to determine $\mathbf{B}(\mathbf{x})$. More importantly, we find that

$$n_1(\mathbf{x}, \mathbf{k}) = \mathbf{B}(\mathbf{x}) \cdot \hat{\mathbf{k}} = -\left(\frac{1}{N_e \sigma}\right)(\hat{\mathbf{k}} \cdot \nabla n_0). \tag{6.194}$$

Obviously, $(n_1/n_0) = \mathcal{O}(l_\nu/R)$, justifying the original claim and the approximation; the result also shows that the first-order perturbation to the distribution function (which is proportional to the photon number) is driven by the gradient of the equilibrium solution, which happens to be a very general feature.

The energy flux carried by the photons is given by the time–space component of the stress tensor in Eq. (6.160). The equilibrium part, n_0, does not contribute to flux because it is isotropic and we are left with

$$
\begin{aligned}
F^\alpha = c T^{0\alpha} &= \int d\Omega \, \hat{k}^\alpha I_{\nu 1} = \int d\Omega \, \hat{k}^\alpha \left(\frac{h\nu^3}{c^2} n_1\right) \\
&= \int d\Omega \, \hat{k}^\alpha \left[-\frac{1}{N_e \sigma} \hat{k}^\beta \frac{\partial}{\partial x^\beta} \left(\frac{h\nu^3}{c^2} n_0\right)\right] \\
&= -\frac{1}{N_e \sigma} \frac{\partial}{\partial x^\beta} \int d\Omega \, \hat{k}^\alpha \hat{k}^\beta I_{\nu 0} = -\frac{c}{N_e \sigma} \frac{\partial P^{\alpha\beta}}{\partial x^\beta}.
\end{aligned}
\tag{6.195}
$$

The second equality follows from Eq. (6.160), the third from Eq. (6.4), the fourth from Eq. (6.194), and the last one from Eq. (6.161). This is identical to relation (6.169) as $N_e \sigma = \rho \kappa$.

So far, we have not assumed any specific form for n_0 except that it is independent of $\mathbf{k}$. In most of the practical situations, n_0 can be taken to be the equilibrium distribution at the local temperature $T(\mathbf{x})$; then the spatial dependence of n_0 will arise through the spatial dependence of the temperature and we can write $P^{\alpha\beta} = (1/3)aT^4\delta^{\alpha\beta}$; the radiative flux and the force on matter will now be given by Eq. (6.173).

The procedure described above is of very general validity and can be used to derive first-order corrections to the distribution functions in several contexts. Quite generically, these corrections will be driven by the gradients of parameters appearing in the equilibrium distribution and can be computed in a manner identical to that described above.

Exercise 6.16

Combined scattering and absorption: When scattering is present, the effective optical depth of a medium for absorption of radiation is increased. Show that, to the lowest order, the net optical depth can be taken to be $\tau_* \approx \sqrt{\tau_a(\tau_a + \tau_s)}$, where τ_a and τ_s are the individual optical depths for absorption and scattering, respectively.

6.9 Bremsstrahlung

As the next example of radiative process used extensively in astrophysics, we consider the radiation emitted by a plasma that is due to the scattering of electrons and ions. A simple derivation of this effect is provided first and then extra complications are discussed.

6.9.1 Classical Bremsstrahlung

In a plasma, electrons are constantly accelerated during their collisions with ions, which leads to the emission of radiation by the plasma, called (thermal) *bremsstrahlung*. To study the properties of bremsstrahlung, we begin by considering an individual scattering event between an electron and an ion in which an electron of velocity v and impact parameter b is scattered by an ion. The acceleration experienced by the electron is $a \simeq (Ze^2/mb^2)$ and lasts for a time $(2b/v)$. From the expression for radiated power [see Eq. (4.69)] we find the total energy emitted to be

$$\frac{dE}{d\omega} = \frac{2}{3\pi}\frac{e^2}{c^3}\left(\frac{Ze^2}{mb^2}\frac{2b}{v}\right)^2 = \frac{8}{3\pi}\frac{Z^2e^6}{m^2c^3}\left(\frac{1}{vb}\right)^2. \tag{6.196}$$

Because the acceleration lasted for a time $(2b/v)$, there will be very little power at frequencies $\omega > \omega_{max} \simeq (v/2b)$.

This is the amount of radiation emitted in a single collision. If the number densities of ions n_i and electrons n_e are the same and given by n, then the total amount of energy emitted per unit volume per second that is due to all collisions with an impact parameter in the range $(b, b + db)$ will be $[(n_i n_e v)(2\pi b db)(dE/d\omega)] = [n^2 v(2\pi b db)(dE/d\omega)]$. So

$$\left(\frac{dE}{dV\,d\omega\,dt}\right)_{total} = n^2 v db\left(\frac{dE}{d\omega}\right)(2\pi b) = \frac{16Z^2e^6n^2}{3m^2c^3v}\frac{1}{b}\,db. \tag{6.197}$$

Integrating over b with the limits b_1 and b_2, we get

$$\left(\frac{dE}{dV\,d\omega\,dt}\right) = \frac{16Z^2e^6n^2}{3m^2c^3v}\ln\left(\frac{b_2}{b_1}\right). \tag{6.198}$$

The limits of integration b_2 and b_1 need to be determined by some physical consideration.

We can fix the value of b_2 by noting that most of the radiation is at frequencies $\omega < (v/2b)$. Therefore we must have $b < (v/\omega)$ and we can set $b_2 = (v/\omega)$. The lower limit is somewhat more difficult to determine. Note that the entire analysis is based on the assumption that the deflection in each encounter is small. This is justifiable because each logarithmic interval in the impact parameter contributes the same amount to the radiation, as is evidenced by the logarithmic dependence on b. By the same token, such an analysis, based on weak scattering, will break down below an impact parameter $b_{1,c}$ where

$$\frac{Ze^2}{b_{1,c}} = \frac{1}{2}mv^2, \tag{6.199}$$

giving $b_{1,c} = (2Ze^2/mv^2)$. On the other hand, quantum effects are important if the angular momentum involved in the trajectory is of the order of $\hbar$, that is, if $mvb_{1,Q} \simeq \hbar$ or when $b_{1,Q} \simeq (\hbar/mv)$. Which of these two expressions should be used for b_1 depends on the velocity of the particle. The ratio between these two is

$$\frac{b_{1,Q}}{b_{1,c}} \simeq \frac{v\hbar}{Ze^2} = \left(\frac{1}{Z\alpha}\right)\left(\frac{v}{c}\right), \tag{6.200}$$

where $\alpha = (e^2/\hbar c) \simeq 10^{-2}$. Thus for particles with $(v/c) \gtrsim 10^{-2}Z$ we should use $b_{1,Q}$; for particles with $(v/c) \lesssim 10^{-2}Z$ we should use $b_{1,c}$. Therefore

$$\left(\frac{b_2}{b_1}\right) \simeq \begin{cases} (mv^3/Ze^2\omega) & \text{(for low velocities)} \\ (mv^2/\hbar\omega) & \text{(for high velocities)} \end{cases}. \tag{6.201}$$

Because this ratio appears only inside the logarithm, its exact numerical value is not very important. In summary, we can express the result as

$$\left(\frac{dE}{dV\,d\omega\,dt}\right) = \frac{16Z^2e^6n^2}{3m^2c^3v}\ln\left(\frac{mv^3}{Ze^2\omega}\right) \quad \left(\frac{v\hbar}{Ze^2} \ll 1, \omega \ll \frac{mv^3}{Ze^2}\right)$$

$$= \frac{16Z^2e^6n^2}{3m^2c^3v}\ln\left(\frac{mv^2}{\hbar\omega}\right) \quad \left(\frac{v\hbar}{Ze^2} \ll 1, \omega \ll \frac{mv^2}{\hbar}\right). \tag{6.202}$$

This is the amount of energy emitted per unit volume per second per unit frequency range if all the electrons have the velocity v. In a plasma, the electrons have a distribution of velocities and Eq. (6.202) needs to be averaged over the velocity distribution. However, before doing that, we provide a more rigorous derivation of the bremsstrahlung emission, both in the classical and the quantum-mechanical context.

Classically, bremsstrahlung arises because of the accelerated trajectory of an electron in the presence of an ion. For the free electrons, the exact trajectory is a hyperbola in the x–y plane [see Chap. 2, Eqs. (2.115)], which can be written in a parametric form as

$$x = a(\epsilon - \cosh\xi), \quad y = a\sqrt{\epsilon^2 - 1}\sinh\xi, \quad t = \sqrt{\frac{ma^3}{Ze^2}}(\epsilon\sinh\xi - \xi),$$

$$\tag{6.203}$$

where

$$a = \frac{Ze^2}{2\epsilon} = \frac{Ze^2}{mv^2}, \quad \epsilon = \sqrt{1 + b^2\left(\frac{mv^2}{Ze^2}\right)^2}. \tag{6.204}$$

In these expressions, v, b, and ϵ denote the initial velocity, impact parameter, and the eccentricity of the orbit, respectively. It is also convenient to define a dimensionless variable μ by

$$\mu \equiv \frac{Ze^2\omega}{mv^3}. \tag{6.205}$$

To calculate the radiation emitted in the dipole approximation, we need to evaluate the Fourier transform of the dipole moment, which in turn requires computation of the Fourier transform $x(\omega)$, $y(\omega)$ of the trajectory $x(t)$, $y(t)$. This is best done by relating the Fourier transform of $x(t)$ to that of $\dot{x}(t)$ by $\dot{x}(\omega) = -i\omega x(\omega)$ and writing

$$x(\omega) = -\frac{\dot{x}(\omega)}{i\omega} = \frac{i}{\omega} \int_{-\infty}^{\infty} \dot{x}(t)\, e^{-i\omega t}\, dt = \frac{i}{\omega} \int_{-\infty}^{\infty} e^{-i\omega t(\xi)}\left(\frac{dx}{d\xi}\right)d\xi$$

$$= -\frac{ia}{\omega} \int_{-\infty}^{\infty} d\xi\,(\sinh\xi)\,[\exp - i\omega t(\xi)]. \tag{6.206}$$

In arriving at the third inequality, we have used the relation that $\dot{x}\,dt = dx$; it is then convenient to convert the variable of integration to ξ, which is done in the last step. We can evaluate the integral by using the definition of Hankel functions,

$$\int_{-\infty}^{\infty} e^{p\xi - ix\sinh\xi}\, d\xi = i\pi H_p^{(1)}(ix) \tag{6.207}$$

to give

$$x(\omega) = \frac{\pi a}{\omega} H_{i\mu}^{(1)'}(i\mu\epsilon), \quad y(\omega) = -\frac{\pi a\sqrt{\epsilon^2 - 1}}{\omega\epsilon} H_{i\mu}^{(1)}(i\mu\epsilon). \tag{6.208}$$

From the general result for dipole radiation, we can find the spectral distribution of energy:

$$\frac{dE}{d\omega} = \frac{2}{3\pi}\frac{e^2\omega^4}{c^3}[|x(\omega)|^2 + |y(\omega)|^2]$$

$$= \frac{2\pi}{3}\frac{Z^2e^6}{m^2c^3}\frac{\omega^2}{v^4}\left[|H_{i\mu}^{(1)'}(i\mu\epsilon)|^2 - \left(1 - \frac{1}{\epsilon^2}\right)|H_{i\mu}^{(1)}(i\mu\epsilon)|^2\right]. \tag{6.209}$$

We next must multiply this expression by $(n^2 v)(2\pi b\,db)$ and integrate over all impact parameters b in the range $(0, \infty)$. From Eq. (6.204), we see that $b\,db = a^2\epsilon\,d\epsilon$. We can perform the resulting integral over ϵ by using the identity for Hankel functions,

$$z\left[Z_p'^2 + \left(\frac{p^2}{z^2} - 1\right)Z_p^2\right] = \frac{d}{dz}(zZ_pZ_p'), \tag{6.210}$$

leading to

$$\left(\frac{dE}{dV\,d\omega\,dt}\right) = \left(\frac{4\pi^2 Z^2 e^6 n^2}{3m^2 c^3 v}\right)\left(\frac{Ze^2\omega}{mv^3}\right)\left|H_{i\mu}^{(1)}(i\mu)\right| H_{i\mu}^{(1)'}(i\mu). \quad (6.211)$$

This result is finite, showing that the logarithmic divergence in Eq. (6.198) is an artifact of the approximations that were used.

To obtain the asymptotic limits of this result, we need the approximation for the Hankel functions in various limiting cases. For $\mu \ll 1$, the Hankel functions are

$$H_{i\mu}^{(1)'}(i\mu) \cong H_0^{(1)'}(i\mu), \quad H_0^{(1)}(i\mu) \cong \frac{2}{i\pi}\ln\left(\frac{2}{\gamma\mu}\right) \quad (\mu \ll 1), \quad (6.212)$$

where γ is the exponential of Euler's constant; $\gamma \approx 1.78$. On the other hand, for $\mu \gg 1$,

$$H_{i\mu}^{(1)}(i\mu) \cong -\frac{i}{\pi\sqrt{3}}\Gamma\left(\frac{1}{3}\right)\left(\frac{6}{\mu}\right)^{1/3},$$

$$H_{i\mu}^{(1)'}(i\mu) = \frac{1}{\pi\sqrt{3}}\Gamma\left(\frac{2}{3}\right)\left(\frac{6}{v}\right)^{2/3} \quad (\mu \gg 1). \quad (6.213)$$

Using these results, we can easily show that the asymptotic forms of Eq. (6.211) are

$$\left(\frac{dE}{d\omega\,dt\,dV}\right) \cong \left(\frac{16}{3}\frac{Z^2 e^6 n^2}{m^2 c^3 v}\right)\ln\left(\frac{2}{\gamma}\frac{mv^3}{Ze^2\omega}\right) \quad \left(\omega \ll \frac{mv^3}{Ze^2}\right)$$

$$\cong \frac{16 Z^2 e^6 n^2}{3m^2 c^3 v}\left(\frac{\pi}{\sqrt{3}}\right) \quad \left(\omega \gg \frac{mv^3}{Ze^2}\right). \quad (6.214)$$

For low ω this expression matches with the result in Eq. (6.202) except for a numerical factor inside the logarithm. The present analysis also gives the result for high ω, provided that classical considerations are valid; this limit was not available in the previous analysis.

6.9.2 Quantum Bremsstrahlung

We next outline the results in the quantum-mechanical case [corresponding to the second case in Eq. (6.202)]. To do this, it is necessary to compute the matrix element $\mathbf{d}_{fi}$ between the initial and the final states. We assume that these states are described by momentum eigenstates of the form $\exp(i\mathbf{p} \cdot \mathbf{x})$ and $\exp(i\mathbf{p}' \cdot \mathbf{x})$, where $\mathbf{p}$ and $\mathbf{p}'$ denote the initial and the final momenta of the electron. To evaluate the matrix element, we note that the equation of motion for the charged particle in the Coulomb field gives

$$\ddot{\mathbf{d}} = e\ddot{\mathbf{x}} = -\frac{Ze^2}{m}\nabla\left(\frac{1}{r}\right). \tag{6.215}$$

From this it follows that

$$\langle\mathbf{p}'|\,\mathbf{d}\,|\mathbf{p}\rangle = -\frac{1}{\omega^2}\langle\mathbf{p}'|\,\ddot{\mathbf{d}}\,|\mathbf{p}\rangle = \frac{Ze^2}{m\omega^2}\langle\mathbf{p}'|\nabla\left(\frac{1}{r}\right)|\mathbf{p}\rangle$$

$$= \left(\frac{Ze^2}{m\omega^2}\right)\left(\frac{4\pi i\mathbf{q}}{q^2}\right), \quad \mathbf{q} = \mathbf{p}' - \mathbf{p}, \quad 2m\omega = p^2 - p'^2. \tag{6.216}$$

We can now compute the probability for an electron with an initial momentum $\mathbf{p} = m\mathbf{v}$ to emit a photon of wave vector $\mathbf{k}$ and end up with a final momentum $\mathbf{p}' = m\mathbf{v}'$ by using our results in Chap. 4, Section 4.5. The final result is

$$\frac{d\sigma_{\mathbf{k}\mathbf{p}'}}{d\omega\,d\Omega_{\mathbf{k}}\,d\Omega_{\mathbf{p}'}} = \frac{Z^2e^6}{\pi^2}\left(\frac{v'}{v}\right)\frac{(\mathbf{e}\cdot\mathbf{q})(\mathbf{e}^*\cdot\mathbf{q})}{q^4\omega}, \quad \mathbf{q} = \mathbf{p}' - \mathbf{p} = \hbar\mathbf{k}. \tag{6.217}$$

Summation over polarisations gives a factor $\sin^2\Theta$, where Θ is the angle between $\mathbf{k}$ and $\mathbf{q}$. Integrating over the directions of the photon leads to

$$\frac{d\sigma}{d\theta\,d\omega} = \frac{16}{3}\frac{Z^2e^6}{m^2}\frac{v'}{v}\frac{1}{\omega}\frac{\sin\theta}{v^2 + v'^2 - 2vv'\cos\theta}, \tag{6.218}$$

where θ is the scattering angle. Finally, integrating over θ will give the differential cross section for the emission of a photon of frequency ω as

$$\frac{d\sigma}{d\omega} = \frac{16}{3}\frac{Z^2e^6}{m^2v^2\omega}\ln\frac{v+v'}{v-v'}$$

$$= \frac{16}{3}\frac{Z^2e^6}{m^2v^2\omega}\ln\left\{\frac{1 + [1 - (2\omega/mv^2)]^{1/2}}{1 - [1 - (2\omega/mv^2)]^{1/2}}\right\}. \tag{6.219}$$

To find the amount of energy emitted per unit time per unit volume of plasma we first multiply this differential cross section first by $\hbar\omega$ (to get the energy emitted), then by the flux nv, and finally by another factor n to take into account the number of scattering centres per unit volume. This gives the quantum-mechanical result

$$\left(\frac{dE}{dt\,dV\,d\omega}\right) = \omega n^2 v\frac{d\sigma}{d\omega} = \frac{16}{3}\frac{Z^2e^6n^2}{m^2vc^3}\ln\left\{\frac{1 + [1 - (2\omega/mv^2)]^{1/2}}{1 - [1 - (2\omega/mv^2)]^{1/2}}\right\}. \tag{6.220}$$

Note that this has the same prefactor as in relation (6.214) or Eq. (6.202). The limiting forms of this expression are of interest. When $\hbar\omega \ll mv^2$, we get

$$\left(\frac{dE}{dt\,d\omega\,dV}\right) \simeq \frac{16}{3}\frac{Z^2e^6n^2}{m^2c^3v}\ln\left(\frac{2mv^2}{\hbar\omega}\right) \quad (\hbar\omega \ll mv^2), \tag{6.221}$$

which matches with the second limit in Eq. (6.202) except for an unimportant factor 2 within the logarithm. Further, when $\hbar\omega \to (1/2)mv^2$, the expression in Eq. (6.220) vanishes; this is to be expected because, quantum mechanically, an electron with energy $mv^2/2$ cannot emit a photon with energies $\hbar\omega > mv^2/2$. Finally, it must be noted that the actual wave functions describing the electron in the presence of a nucleus are not plane waves and hence our analysis is valid only when $(Ze^2/\hbar v) \ll 1$ and $(Ze^2/\hbar v') \ll 1$.

Exercise 6.17
Quantum bremsstrahlung: Prove Eq. (6.217) by using the formalism developed in Chap. 4, Section 4.5.

6.9.3 Thermal Bremsstrahlung

Having obtained the expression for bremsstrahlung emission in an encounter with velocity v, we now turn to averaging the result over the velocity distribution. This is conventionally done as follows: We first note that all our results can be written in the form

$$\left(\frac{dE}{dV\,d\omega\,dt}\right) = \frac{16Z^2e^6n^2}{3m^2c^3}\frac{1}{v}\ln\left(\frac{b_2}{b_1}\right) \equiv \frac{16\pi\,Z^2e^6n^2}{3\sqrt{3}m^2c^3}\frac{1}{v}g_{ff}(v,\omega) \quad (6.222)$$

where we have defined the quantity $g_{ff}(v,\omega)$, called the *Gaunt factor*:

$$g_{ff}(v,\omega) = \frac{\sqrt{3}}{\pi}\ln\left(\frac{b_2}{b_1}\right). \quad (6.223)$$

Detailed tables of Gaunt factors are available for actual astrophysical situations. In averaging expression (6.222), most of the contributions will come from the $\langle(1/v)\rangle$ factor, as we can ignore the weak dependence of g_{ff} on v. For a plasma in thermal equilibrium, electrons will have a Maxwellian distribution of velocities. Because an electron needs to have a minimum energy $\frac{1}{2}mv_{min}^2 \simeq \hbar\omega$ to emit a photon of energy $\hbar\omega$, the averaging of $1/v$ will lead to a factor

$$\left\langle\frac{1}{v}\right\rangle = \left(\frac{m}{2\pi k_B T}\right)^{3/2}\int_{v_{min}}^{\infty}\left(\frac{1}{v}\right)4\pi v^2\,dv\left(\exp-\frac{mv^2}{2k_B T}\right)$$

$$= \sqrt{\frac{2m}{\pi k_B T}}\exp\left(-\frac{\hbar\omega}{k_B T}\right). \quad (6.224)$$

Hence

$$j(\omega) = \left(\frac{dE}{dV\,dt\,d\omega}\right) \simeq \frac{16\pi\,Z^2e^6n^2}{3\sqrt{3}m^2c^3}\left(\frac{2m}{\pi k_B T}\right)^{1/2}\exp\left(-\frac{\hbar\omega}{k_B T}\right) \propto n^2 T^{-1/2}$$

$$(6.225)$$

if we ignore the Gaunt factor. The last proportionality follows for $\hbar\omega \ll k_B T$. This quantity $j(\omega)$, called the specific emissivity of the plasma, represents the amount of energy emitted by unit volume of plasma per second per unit frequency range. To obtain the volume emissivity we have to integrate $j(\omega)$ over all the frequencies:

$$J = \left(\frac{dE}{dV\,dt}\right) = \int_0^\infty d\omega\, j(\omega) = \frac{16Z^2 e^6 n^2}{3m^2 c^3 \hbar}\left(\frac{2mk_B T}{\pi}\right)^{1/2} \propto n^2 T^{1/2}. \tag{6.226}$$

Numerically,

$$4\pi j_\nu = 6.8 \times 10^{-38} n^2 T^{-1/2} e^{-h\nu/k_B T}\, \bar{g}_{ff}\ \text{erg cm}^{-3}\ \text{s}^{-1}\ \text{Hz}^{-1}, \tag{6.227}$$

$$4\pi J \simeq 1.4 \times 10^{-27} n^2 T^{1/2} \bar{g}_B\ \text{erg cm}^{-3}\ \text{s}^{-1} \tag{6.228}$$

if all variables are expressed in cgs units. (The numerical factor in both equations should be divided by 10 if SI units are used.) In these expressions, $\bar{g}_{ff}$ and $\bar{g}_B$ represent the velocity-averaged Gaunt factor and the frequency-averaged Gaunt factor, respectively. For most of the relevant range, $\bar{g}_{ff}$ varies between 1 and 5 and can be set to unity for order-of-magnitude estimates. Similarly $\bar{g}_B(T)$, which is a frequency average of $\bar{g}_{ff}(\omega)$, varies between 1.1 and 1.5. Choosing a value of $\bar{g}_B \approx 1.2$ usually gives an accuracy of $\sim$20%.

For reference, we give the fitting formulas for the Gaunt factors, which are useful in the radio-frequency domain. The thermally averaged Gaunt factor in radio frequencies is given by

$$\bar{g}_{ff}(T,\omega) = \frac{\sqrt{3}}{\pi} \ln\left[\left(\frac{2k_B T}{\gamma m}\right)^{3/2} \frac{2m}{\gamma Z e^2 \omega}\right], \quad (10^2\ \text{K} \lesssim T \lesssim 8.9 \times 10^5\ \text{K}), \tag{6.229}$$

$$\bar{g}_{ff}(T,\omega) = \frac{\sqrt{3}}{\pi} \ln\left(\frac{4k_B T}{\hbar\omega\gamma}\right), \quad (T \gtrsim 8.9 \times 10^5\ \text{K}),$$

where $\gamma \approx 1.78$ is the exponential of Euler's constant. Substituting numbers, we get

$$\bar{g}_{ff}(T,\nu) \cong 11.69 + 0.83 \ln\left(\frac{T}{10^4\ \text{K}}\right) - 0.55 \ln\left(\frac{\nu}{100\ \text{MHz}}\right), \tag{6.230}$$

where $\nu = (\omega/2\pi)$.

The emission of bremsstrahlung by a plasma acts as a cooling mechanism. The time scale for cooling can be estimated as $t_{\text{cool}} \simeq (E_{\text{thermal}}/J)$, where $E_{\text{thermal}} \simeq nk_B T$ is the thermal energy per unit volume. So $t_{\text{cool}} \propto (nk_B T/n^2 T^{1/2}) \propto T^{1/2} n^{-1}$.

In the calculations performed above, we have ignored the scattering of electrons by electrons and ions by ions. The reason for concentrating on electron–ion collisions is the following: For a system of nonrelativistic charges with the same charge-to-mass ratio (q/m), the second derivative of the dipole moment,

$$\ddot{\mathbf{d}} = \sum q_i \ddot{\mathbf{x}}_i = \sum \left(\frac{q_i}{m_i}\right) m_i \ddot{\mathbf{x}}_i = \frac{q}{m} \sum \dot{\mathbf{p}}_i, \qquad (6.231)$$

vanishes because of momentum conservation. Hence the dipole radiation from an electron–electron collision and an ion–ion collision will be subdominant to the effects from electron–ion collisions as long as the system is nonrelativistic.

6.9.4 Free–Free Absorption

The process of bremsstrahlung is also called free–free emission, as it involves emission of radiation by a charged particle in a free (unbound) motion. The inverse of this process is called free–free absorption, which corresponds to electrons absorbing photons in the presence of ions. It is possible to obtain information about free–free absorption from the expressions that have been obtained for free–free emission. To do this, it is convenient to characterise this process by a frequency-dependent free–free absorption cross section $\sigma_{\mathrm{ffa}}(\omega)$ and a corresponding absorption coefficient $\alpha_{ff} \equiv n\sigma_{\mathrm{ffa}}$. From our general result in Eq. (6.32), we know that $\alpha_{ff} = n\sigma_{\mathrm{ffa}} = (j_\nu/B_\nu)$, where B_ν is the blackbody intensity. By using Eq. (6.225) we find that

$$\alpha_{ff}(\omega) = \frac{j(\omega)}{B(\omega)} = n\sigma_{\mathrm{ffa}} = \left(\frac{32\pi^3}{3}\right)^{1/2}\left(\frac{Z^2 e^2}{\hbar c}\right)\left(\frac{mc^2}{k_B T}\right)^{1/2}$$
$$\times \left(\frac{nc^3}{\omega^3}\right)(n\sigma_T)(1 - e^{-\beta\hbar\omega}). \qquad (6.232)$$

For $(\hbar\omega/k_B T) \lesssim 1$, this gives $\sigma_{\mathrm{ffa}} \propto (nT^{-1/2}/\omega^3)(\omega/T) \propto (n/\omega^2 T^{3/2})$. The time scale for free–free absorption is $t_{ff} = (n\sigma_{\mathrm{ffa}}c)^{-1}$. This is given by

$$t_{ff} \approx \frac{3}{8}\left(\frac{\pi}{2}\right)^{1/2}\frac{(mk_B T)^{1/2}}{Z^2 e^6 n^2}\frac{\omega^3}{\hbar^2 c^3 (2\pi)^3}(1 - e^{-\hbar\omega/k_B T})^{-1}$$
$$\propto T^{1/2} n^{-2}(1 - e^{-\hbar\omega/k_B T})^{-1}. \qquad (6.233)$$

The process of free–free absorption is important, for example, in stellar physics. In stellar interiors, most of the opacity arises because of photons being absorbed by fully ionized hydrogen and helium. In this case $\hbar\omega \approx kT$ and the effective free–free absorption cross section scales with temperature as $nT^{-7/2}$. Hence the mean free path scales as $l \propto T^{7/2} n^{-2}$. Writing $\hbar\omega \cong k_B T$ and replacing n^2 with $n_e n_i$, where n_e and n_i are the number densities of electrons and ions, respectively, we find that Eq. (6.232) leads to the opacity called

Kramer's opacity:

$$\kappa_{ff} \equiv \frac{\alpha_{ff}}{\rho} \cong \left(\frac{32\pi^3}{3}\right)^{1/2} \left(\frac{Z^2 e^2}{\hbar c}\right) \left(\frac{mc}{\hbar}\right)^{1/2} \left(\frac{\hbar c}{k_B T}\right)^{7/2} \left(\frac{n_e n_i}{\rho}\right) \sigma_T$$

$$\simeq 4 \times 10^{-24} \left(\frac{Z^2 n_e n_i}{\rho T^{3.5}}\right) \text{cm}^2 \text{ g}^{-1} \simeq 4 \times 10^{-25} \left(\frac{Z^2 n_e n_i}{\rho T^{3.5}}\right) \text{m}^2 \text{ kg}^{-1}.$$

$$(6.234)$$

When n_e and n_i are written in terms of mean molecular weights μ_e and μ_i and an average nuclear charge $\bar{Z}$ is introduced, this becomes

$$\kappa_{ff} \simeq 10^{23} \text{ cm}^2 \text{ g}^{-1} \left(\frac{\rho \bar{Z}^2}{\mu_e \mu_i}\right) T^{-3.5}$$

$$\cong 4 \times 10^{22} \text{ cm}^2 \text{ g}^{-1} (X + Y)(1 + X)\rho T^{-3.5}, \qquad (6.235)$$

where the second expression, with X and Y denoting the mass fraction of hydrogen and helium, respectively, is a good approximation in stellar interiors. (The numerical factor should be divided by 10 if SI units are used.) We will need this result in Vol. II.

In the above calculations, we have assumed that all the radiation that is emitted by the plasma escapes from the system. This cannot, of course, be quite true as the cross section for free–free absorption increases as ω^{-2} at low frequencies. Hence the low-frequency radiation emitted by the accelerated electrons will be reabsorbed within the plasma. This process, called self-absorption, will distort the low-frequency radiation emitted by the plasma. When $\hbar\omega \ll k_B T$, the absorption coefficient in Eq. (6.232) scales as $\alpha_\nu \propto T^{-3/2} \nu^{-2}$. If the size of the region emitting bremsstrahlung is R, then, at low frequencies, $\alpha_\nu R \gg 1$ and the intensity will be that of thermal radiation with $B_\nu \propto \nu^3 (e^{h\nu/k_B T} - 1)^{-1} \propto T\nu^2$. At intermediate and high frequencies, $\alpha_\nu R \ll 1$, and we have $I_\nu \propto j_\nu$, which is independent of ν at intermediate scales and decreases as $\exp(-h\nu/k_B T)$ at high frequencies.

For future reference, the free–free absorption coefficient α_ν^{ff} is given in the radio-frequency regime. Assuming that $\bar{g}_{ff} \approx 10$, we get from Eq. (6.232), including a Gaunt factor,

$$\alpha_\nu^{ff} = 5.5 \times 10^{-2} \left(\frac{n_e}{1 \text{ cm}^{-3}}\right)^2 \left(\frac{T}{10^4 \text{ K}}\right)^{-3/2} \left(\frac{\nu}{100 \text{ MHz}}\right)^{-2} \text{kpc}^{-1}. \quad (6.236)$$

Equivalently, we can say that $\alpha_\nu R = 1$ at a critical frequency ν_c, where

$$\nu_c \cong 23 \text{ MHz} \left(\frac{n_e}{1 \text{ cm}^{-3}}\right) \left(\frac{T}{10^4 \text{ K}}\right)^{-3/4} \left(\frac{R}{1 \text{ kpc}}\right)^{1/2}. \qquad (6.237)$$

The bremsstrahlung emission received from an optically thin plasma along any given direction clearly depends on the integral of n_e^2 over the line of sight.

Because of this reason, it is often convenient to define a quantity called the emission measure by

$$\frac{\text{EM}}{\text{pc cm}^{-6}} = \int_0^{s/\text{pc}} \left(\frac{n_e}{\text{cm}^{-3}}\right)^2 d\left(\frac{s}{\text{pc}}\right). \tag{6.238}$$

With this definition, the optical depth for free–free absorption along this line of sight can be written as

$$\tau_\nu = \int_0^s \alpha_\nu \, ds = 3.014 \times 10^{-2} \left(\frac{T}{\text{K}}\right)^{-3/2} \left(\frac{\nu}{\text{GHz}}\right)^{-2} \left(\frac{\text{EM}}{\text{pc cm}^{-6}}\right) \langle g_{ff} \rangle, \tag{6.239}$$

where the Gaunt factor is given by relation (6.230).

These results will be useful in Vols. II and III. For a more precise estimate of optical depth τ_ν, we use a better power-law fit to the Gaunt factor, and the numerical fit can be expressed as

$$\tau_\nu = 8.24 \times 10^{-2} T^{-1.35} \nu^{-2.1} \int n_e^2 ds = 8.24 \times 10^{-2} T^{-1.35} \nu^{-2.1} \text{EM}, \tag{6.240}$$

where T is measured in degrees Kelvin, ν in gigahertz, and EM, the emission measure, is measured in cm^{-6} pc.

Exercise 6.18

Bremsstrahlung from a spherical source: A spherical cloud of fully ionised hydrogen is undergoing collapse isothermally at a temperature T_0 and radius $R(t)$. The mass of the cloud is M_0. (1) Find the total bremsstrahlung luminosity of the sphere when it is optically thin. Take the Gaunt factor as $g \approx 1.2$. (2) At some stage during the contraction, the cloud becomes optically thick. What is the luminosity after it has become optically thick? (3) Determine the critical radius at which the transition from thin to thick emission occurs in terms of the mass and the temperature of the cloud. [Answers: All answers are in cgs units. (1) $\mathcal{L}_{\text{thin}} = 1.6 \times 10^{20} M_0^2 T_0^{1/2} R^{-3}(t)$. (2) $\mathcal{L}_{\text{thick}} = 7.1 \times 10^{-4} T_0^4 R^2(t)$. (3) The conditions that the transitions can be determined by equating the two luminosities found in (1) and (2). This gives $R(t_0) \approx 4.7 \times 10^4 M_0^{2/5} T_0^{-7/10}$.]

6.10 Synchrotron Radiation: Basics

Charges spiraling in a magnetic field at relativistic speeds will emit radiation called synchrotron radiation. This topic, although conceptually fairly straight-forward, is mathematically complicated. First, a simple approach aimed at explaining the process is provided, and then a more rigorous derivation is given.

We saw in Chap. 3, Section 3.8, that a charged particle in a constant magnetic field moves in a circular trajectory in the plane perpendicular to **B**. The angular velocity is

$$\omega = \frac{cqB}{\mathcal{E}} = \frac{qB}{mc}\sqrt{1 - \frac{v^2}{c^2}} = \frac{qB}{mc}\left(\frac{1}{\gamma}\right) = 17.6\gamma^{-1}\,\text{MHz}\left(\frac{B}{1\,\text{G}}\right), \quad (6.241)$$

where $\mathcal{E}$ is the energy of the particle. If $\mathbf{v} \cdot \mathbf{B} = 0$, then the particle will move in a circular path of radius

$$r_B = \frac{v}{\omega} = \frac{mcv}{qB}\gamma. \quad (6.242)$$

If $\mathbf{v} \cdot \mathbf{B}/B \equiv v_\parallel \neq 0$, then the motion is a superposition of a circular motion perpendicular to $\mathbf{B}$ and linear motion along $\mathbf{B}$.

The energy radiated by such a particle can be found from general formula (4.73). Using $\mathbf{E} = 0$, $(\mathbf{v} \times \mathbf{B})^2 = v^2 B^2 \sin^2 \alpha$, where α (called the pitch angle) is the angle between $\mathbf{B}$ and $\mathbf{v}$, we get

$$\frac{d\mathcal{E}}{dt} = \frac{2}{3}\left(\frac{q^2}{m}\right)^2 \gamma^2 v^2 B^2 \sin^2 \alpha = 2\left(\frac{B^2}{8\pi}\right)\sigma_T \gamma^2 v^2 \sin^2 \alpha. \quad (6.243)$$

If we average over all possible angles α by using $\langle \sin^2 \alpha \rangle = (2/3)$ we get

$$\frac{d\mathcal{E}}{dt} = \frac{4}{3}(\sigma_T c \gamma^2 \beta^2)\left(\frac{B^2}{8\pi}\right) = \frac{4}{3}(\sigma_T c \gamma^2 \beta^2)U_B, \quad (6.244)$$

where $U_B = (B^2/8\pi)$ is the energy density in the magnetic field. We can find the time scale for energy loss by this process by dividing the energy of the electron (γmc^2) by $(d\mathcal{E}/dt)$; this gives

$$t_{\text{sync}} \equiv \frac{\gamma mc^2}{(d\mathcal{E}/dt)} \simeq 5 \times 10^8 \text{s}(\gamma^{-1}B_{\text{gauss}}^{-2}) \cong 10^{10} \text{ yr}\left(\frac{B}{10^{-6}\,\text{G}}\right)^{-2}\left(\frac{\mathcal{E}}{\text{GeV}}\right)^{-1} \quad (6.245)$$

for the extreme relativistic case with $\beta \approx 1$.

Let us next consider the spectrum of synchrotron radiation from a highly relativistic charged particle. From the discussion in Chap. 4, Subsection 4.3.2, it is known that most of the radiation is emitted in the forward direction into a cone with opening angle

$$\Delta\theta \simeq \left(1 - \frac{v^2}{c^2}\right)^{1/2} \simeq \gamma^{-1} = \frac{mc^2}{\mathcal{E}}. \quad (6.246)$$

This fact has two implications for the frequency spectrum of radiation emitted in the ultrarelativistic case: (1) An observer will receive the radiation only when the direction of observation is within a cone of angular width $\Delta\theta$. (2) The electric field $E(t)$ at any time t (at the point of observation) can depend on the angle θ only through the combination $\gamma\theta$ [see relation (4.53)]. The time of arrival of radiation, t, can be related to the angle θ in the following way: Consider radiation emitted at two instances, $t_1 = 0$ and $t_2 = t$, when the charged particle is at positions $\theta_1 = 0$ and $\theta_2 = \theta$. Clearly $r_B\theta = v(t_2 - t_1) = vt$, giving $t = (r_B\theta/v)$. The arrival time of the pulses at the observer will be less by the amount of time $(r_B\theta/c)$ taken by the radiation to travel the extra distance $r_B\theta$. Therefore $t_{obs} \simeq t - r_B\theta/c = (r_B\theta/v)(1 - v/c)$. Hence

$$\theta(t_{obs}) \simeq \frac{vt_{obs}}{r_B} \frac{1}{(1 - v/c)} \simeq \frac{2vt_{obs}}{r_B}\gamma^2, \tag{6.247}$$

giving

$$\gamma\theta \simeq t_{obs}\frac{2v}{r_B}\gamma^3 = t_{obs}(2\omega_B\gamma^3 \sin\alpha) \equiv \frac{4}{3}\omega_c t_{obs}. \tag{6.248}$$

In arriving at the second equality, we have used $v = \omega_B r_B \sin\alpha$, where the $\sin\alpha$ factor takes into account the fact that only the component of B perpendicular to v contributes to the radiation. The last equality defines ω_c:

$$\omega_c = \frac{3}{2}\left(\frac{qB\sin\alpha}{mc}\right)\gamma^2 \cong 10^2 \text{ MHz} \left(\frac{B}{10^{-6}\,\text{G}}\right)\left(\frac{\mathcal{E}}{\text{GeV}}\right)^2. \tag{6.249}$$

The numerical factor $3/2$ in the definition of ω_c is chosen so as to make future analysis simple.

Because the electric field $\mathcal{E}$ at the observed location is a function of only $\gamma\theta \propto \omega_c t_{obs}$, its Fourier transform

$$E(\omega) = \int_{-\infty}^{\infty} E(t)e^{-i\omega t}\,dt = \int_{-\infty}^{+\infty} E(\omega_c t)e^{-i\omega t}\,dt$$

$$= \frac{1}{\omega_c}\int_{-\infty}^{+\infty} E(q)\exp\left[-i\left(\frac{\omega}{\omega_c}\right)q\right]dq = E(\omega/\omega_c) \tag{6.250}$$

depends on ω only through the ratio (ω/ω_c). Hence, the energy radiated per orbital period, $(d\mathcal{E}/d\omega)(1/T) \equiv (d\mathcal{E}/dt\,d\omega)$, also depends on ω only through some function $F(\omega/\omega_c)$.

These features of synchrotron radiation can be understood qualitatively as follows. If the charged particle is moving nonrelativistically, its trajectory will be a periodic function and the characteristic frequency will be the cyclotron frequency $\omega_{cyc} \approx (qB/mc)$. In that case, the particle will emit cyclotron radiation essentially at this frequency. As its speed increases, three effects come into play. First, the higher harmonics of ω_{cyc} start contributing to the spectrum with a relative strength that depends on the powers of v/c. Second, the rotation frequency starts decreasing as $\omega_B \propto \gamma^{-1}$. Finally, the radiation becomes confined to a cone with opening angle $\Delta\theta \approx \gamma^{-1}$ and the observer will be able to see the radiation only at intervals $\Delta t \propto \gamma^{-3}\omega_B^{-1}$. In the frequency domain, we would expect to see a series of spikes at ω_B (and its harmonics) with a cutoff at a frequency of

$\omega_c \propto (1/\Delta t) \propto \omega_B \gamma^{-3}$. When γ becomes large, the harmonics tend to be closely spaced, with each contribution becoming broader because of the distribution of γ and pitch angle. The spectrum will therefore appear continuous, with a maximum around ω_c. In fact, most of the energy will be emitted at frequencies close to $\omega_c \propto \omega_B \gamma^3 \propto B\gamma^2 \propto B\mathcal{E}^2$, where $\mathcal{E}$ is the energy of the charged particle. It follows that $\mathcal{E} \propto \omega_c^{1/2} B^{-1/2}$. Numerically,

$$\mathcal{E} \approx 4 v_c^{1/2} B^{-1/2} \text{ erg} \tag{6.251}$$

if all quantities are in cgs units. By using the relation between the energy and synchrotron frequency, we can express t_{sync} in Eq. (6.245) as

$$t_{\text{sync}} \cong 5 \times 10^8 \text{ yr} \left(\frac{B}{10^{-6} \text{ G}} \right)^{-3/2} \left(\frac{v}{1 \text{ GHz}} \right)^{-1/2}, \tag{6.252}$$

where $v = (\omega/2\pi)$.

Synchrotron radiation will be linearly polarised with a high degree of polarisation. To see this, consider the radiation emitted by a single electron circling in a magnetic field. When the electron is viewed in the orbital plane, the radiation is 100% linearly polarised with the electric vector oscillating perpendicular to the magnetic field. When viewed along the direction of the magnetic field, the radiation is circularly polarised. (If it is right-handed circular polarisation when viewed from the top, it will be left-handed circular polarisation when seen from the bottom.) Consider now the relativistic motion of the electron that beams the radiation within a narrow cone in the direction of motion. The radiation is now mostly confined to the plane of the orbit and, in any realistic situation, there will be a bunch of charged particles with a distribution of pitch angles. In that case, the two components of the circular polarisation will effectively cancel out whereas the linear polarisation will survive to a significant extent. Hence the radiation is expected to be linearly polarised to a high degree.

Finally we discuss a radiative process called curvature radiation, which is closely related to synchrotron process. Consider a charged particle moving under the influence of a magnetic field with gently curving field lines. The motion of the charge will be a spiral around the field line with a drift along the field line if r_L is much smaller than the curvature radius R of the field line. The drift along the field line can be locally approximated as a motion in a circle of radius R and will – in turn – lead to synchrotron radiation, which is called curvature radiation. We shall have occasion to use this concept in Vol. II in the study of pulsars.

6.11 Synchrotron Radiation: Rigorous Results

Let us now work out the rigorous theory of synchrotron radiation. We take the plane of orbit to be the x–y plane, with the origin at the centre of the orbit. The magnetic field is taken to be along the negative z axis. The other geometrical

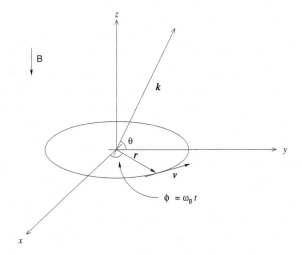

Fig. 6.1.   The geometry for synchrotron radiation.

information is given in Fig. 6.1. At any given time t, $\phi = \omega_B t$ is the angle between the radius vector of the particle and the x axis.

6.11.1 Angular Distribution of Radiation

To determine the angular distribution of radiation, taking into account one full orbital cycle of the particle, we must integrate expression (4.54) over one orbital period. However, the integrand in Eq. (4.54) is a function of the retarded time t_R and hence we use the result

$$dt = \frac{\partial t}{\partial t_R} dt_R = \left(1 - \frac{v_R}{c}\right) dt_R \tag{6.253}$$

to do the integration. In this particular case, it is also convenient to express the acceleration in terms of the velocity and magnetic field as

$$\mathbf{w} = \frac{q}{mc}\sqrt{1 - \frac{v^2}{c^2}}\mathbf{v} \times \mathbf{B}. \tag{6.254}$$

Converting the integration over retarded time to 1 over the angle $\phi = \omega_B t$, we get

$$\frac{dE}{d\Omega} = \frac{q^4 B^2 v^2}{8\pi^2 m^2 c^5}\left(1 - \frac{v^2}{c^2}\right)\int_0^{2\pi} \frac{\left(1 - \frac{v^2}{c^2}\right)\sin^2\theta + \left(\frac{v}{c} - \cos\theta\cos\phi\right)^2}{\left(1 - \frac{v}{c}\cos\theta\cos\phi\right)^5} d\phi. \tag{6.255}$$

This integration can be carried out by elementary means, and the result is

$$\frac{dE}{d\Omega} = \frac{q^4 B^2 v^2 (1 - \beta^2)}{8\pi^2 m^2 c^5} \left[\frac{2 + \beta^2 \cos^2 \theta}{(1 - \beta^2 \cos^2 \theta)^{5/2}} - \frac{(1 - \beta^2)(4 + \beta^2 \cos^2 \theta) \cos^2 \theta}{4(1 - \beta^2 \cos^2 \theta)^{7/2}} \right],$$

(6.256)

where $\beta = v/c$. This expression can be used to find the ratio $\mathcal{R}$ between the intensity of radiation perpendicular to the plane of the orbit $[\theta = (\pi/2)]$ and the intensity of radiation in the plane of orbit $[(\theta = 0)]$:

$$\mathcal{R} = 2 \left(1 - \frac{v^2}{c^2} \right)^{5/2} \left(1 + \frac{3}{4} \frac{v^2}{c^2} \right)^{-1}.$$

(6.257)

As $v \to c$, this ratio vanishes, showing that most of the radiation is confined to the plane of the orbit.

6.11.2 Spectral Distribution of Radiation

Let us next consider the spectral distribution of the radiation. For this, we use Eq. (4.47) that, in this case, becomes

$$\mathbf{A}_n = q \frac{e^{ikR_0}}{cR_0} \oint e^{i(\omega_B nt - \mathbf{k} \cdot \mathbf{r})} d\mathbf{r},$$

(6.258)

where the integration is over the trajectory of the particle. Using the relations $x = r \cos \omega_B t$, $y = r \sin \omega_B t$, and

$$\mathbf{k} \cdot \mathbf{r} = kr \cos \theta \sin \phi = (nv/c) \cos \theta \sin \phi,$$

(6.259)

where $k = (n\omega_B/c) = (nv/cr)$ for the nth harmonic, we find the Fourier components of the x component of the vector potential as

$$A_{xn} = -\frac{qv}{2\pi c R_0} e^{ikR_0} \int_0^{2\pi} e^{in[\phi - (v/c) \cos \theta \sin \phi]} \sin \phi \, d\phi.$$

(6.260)

This integral can be expressed in terms of the derivatives of a Bessel function:

$$A_{xn} = -\frac{iqv}{cR_0} e^{ikR_0} J_n' \left(\frac{nv}{c} \cos \theta \right).$$

(6.261)

Similarly we find

$$A_{yn} = \frac{q}{R_0 \cos \theta} e^{ikR_0} J_n \left(\frac{nv}{c} \cos \theta \right).$$

(6.262)

Given the vector potential, the intensity in any harmonic can be obtained from the general relations given in Section 4.3. In this particular case, we use Eq. (4.48) and

$$|\mathbf{A} \times \mathbf{k}|^2 = A_x^2 k^2 + A_y^2 k^2 \sin^2 \theta$$

(6.263)

to get

$$\frac{dI_n}{d\Omega} = \frac{n^2 q^4 B^2}{2\pi c^3 m^2}\left(1 - \frac{v^2}{c^2}\right)\left[\tan^2\theta \, J_n^2\left(\frac{nv}{c}\cos\theta\right) + \frac{v^2}{c^2}J_n'\left(\frac{nv}{c}\cos\theta\right)\right].$$

(6.264)

To determine the total intensity of radiation emitted at a frequency $\omega = n\omega_B$, this expression has to be integrated over all angles. Unfortunately this integration cannot be carried out in closed form. However, by a series of transformations, it can be written as

$$I_n = \frac{2q^4 B^2\left(1 - \frac{v^2}{c^2}\right)}{m^2 c^2 v}\left[\frac{nv^2}{c^2}J_{2n}'\left(\frac{2nv}{c}\right) - n^2\left(1 - \frac{v^2}{c^2}\right)\int_0^{(v/c)} J_{2n}(2n\xi)\, d\xi\right].$$

(6.265)

Exercise 6.19
Angular and spectral distribution of synchrotron radiation: Obtain Eqs. (6.255)–(6.257) and (6.261)–(6.265).

Some amount of simplification occurs if we consider the ultrarelativistic limit in which most of the radiation will arise from large n. In that case, we can use the asymptotic formula for the Bessel functions:

$$J_{2n}(2n\xi) \cong \frac{1}{\sqrt{\pi}\, n^{1/3}}\Phi[n^{2/3}(1 - \xi^2)],$$

(6.266)

where $\Phi(t)$ is the Airy function, defined as

$$\Phi(t) = \frac{1}{\sqrt{\pi}}\int_0^\infty \cos\left(\frac{\xi^3}{3} + \xi t\right) d\xi.$$

(6.267)

Then Eq. (6.265) becomes

$$I_n = -\frac{2}{\sqrt{\pi}}\left(\frac{q^4 B^2}{mc\mathcal{E}}\right)u^{1/2}\left[\Phi'(u) + \frac{u}{2}\int_u^\infty \Phi(u)\, du\right], \quad u = n^{2/3}\left(\frac{mc^2}{\mathcal{E}}\right)^2.$$

(6.268)

When $u \to 0$, the second term within the bracket vanishes and the first term becomes

$$\Phi'(0) = -\frac{1}{\sqrt{\pi}}\int_0^\infty \xi \sin\frac{\xi^3}{3}\, d\xi = -\frac{1}{\sqrt{\pi}\, 3^{1/3}}\int_0^\infty x^{-1/3}\sin x\, dx$$

$$= -\frac{3^{1/6}\Gamma(2/3)}{2\sqrt{\pi}} \simeq -0.46.$$

(6.269)

Hence, for $u \ll 1$, we get

$$I_n \cong 0.52 \frac{q^4 B^2}{m^2 c^3} \left(\frac{mc^2}{\mathcal{E}}\right)^2 n^{1/3}, \quad 1 \ll n \ll \left(\frac{\mathcal{E}}{mc^2}\right)^3. \qquad (6.270)$$

On the other hand, for $u \gg 1$, we can use the asymptotic form of the Airy function given by

$$\Phi(t) \approx \frac{1}{2t^{1/4}} \exp\left(-\frac{2}{3} t^{3/2}\right) \qquad (6.271)$$

to obtain

$$I_n = \frac{q^4 B^2}{2\sqrt{\pi} m^2 c^3} \left(\frac{mc^2}{\mathcal{E}}\right)^{5/2} n^{1/2} \exp\left[-\frac{2}{3} n \left(\frac{mc^2}{\mathcal{E}}\right)^3\right], \quad n \gg \left(\frac{\mathcal{E}}{mc^2}\right)^3. \qquad (6.272)$$

It is obvious therefore that the spectrum has a maximum for $n \approx \gamma^3$. The main part of the radiation is concentrated in the region of frequencies for which

$$\omega \approx \omega_B \left(\frac{\mathcal{E}}{mc^2}\right)^3 = \left(\frac{qB}{mc}\right) \left(\frac{\mathcal{E}}{mc^2}\right)^2. \qquad (6.273)$$

These values of ω are very large compared with the spacing ω_B between neighbouring harmonics; therefore we can think of the spectrum as continuous and introduce – in place of I_n – a differential quantity

$$dI = I_n \, dn = I_n \frac{d\omega}{\omega_B}. \qquad (6.274)$$

Doing this, reexpressing the Airy function in terms of Macdonald's function by using the relation

$$\Phi(t) \approx \sqrt{t/3\pi} \, K_{1/3}\left(\frac{2}{3} t^{3/2}\right), \qquad (6.275)$$

and using the formulas

$$K_{\nu-1}(x) - K_{\nu+1}(x) = -\frac{2\nu}{x} K_\nu, \quad 2K_\nu'(x) = -K_{\nu-1}(x) - K_{\nu+1}(x), \qquad (6.276)$$

$$\Phi'(t) = -\frac{1}{\sqrt{3\pi}} K_{2/3}\left(\frac{2}{3} t^{3/2}\right),$$

we can finally express the spectral distribution as

$$\frac{dI}{d\omega} = \frac{\sqrt{3} q^3 B}{2\pi mc^2} F\left(\frac{\omega}{\omega_c}\right), \quad F(\xi) = \xi \int_\xi^\infty K_{5/3}(z) \, dz, \qquad (6.277)$$

where ω_c is defined in Eq. (6.249). For $\xi \lesssim 0.01$, $F(\xi) \cong 2.13 \xi^{1/3}$, and for $\xi \gtrsim 20$, $F(\xi) \cong (\pi \xi/2)^{1/2} e^{-\xi}$ within 5% accuracy.

6.11.3 Radiation from a Power-Law Spectrum of Electrons

All the above results can be easily generalised to a situation in which the radiation is emitted by a collection of charged particles with a distribution in energy. In this case, the total radiation is given by

$$P_{\text{tot}}(\omega) = \int_0^\infty \left[\frac{dI(\mathcal{E}, \omega)}{d\omega} \right] n(\mathcal{E}) \, d\mathcal{E}. \tag{6.278}$$

If $n(\mathcal{E})d\mathcal{E} = C\mathcal{E}^{-p}d\mathcal{E}$, then the integral over the Macdonald functions in Eqs. (6.277) can be performed in closed form to give

$$P_{\text{tot}}(\omega) = \frac{\sqrt{3}q^3 C B \sin\alpha}{2\pi mc^2(p+1)} \Gamma\left(\frac{3p-1}{12}\right) \Gamma\left(\frac{3p+19}{12}\right)$$

$$\times \left(\frac{m^3 c^5 \omega}{3q B \sin\alpha}\right)^{-(p-1)/2} \propto \omega^{-s} B^{1+s}, \tag{6.279}$$

with $p = 2s + 1$. This scaling has a simple interpretation: Synchrotron power radiated by a relativistic particle of energy $\mathcal{E}$ in a magnetic field B scales as $B^2\gamma^2 \propto B^2\mathcal{E}^2$ [see Eq. (6.244)]. The power radiated by a set of particles with a spectrum $n(\mathcal{E})$ is

$$P \propto \begin{pmatrix} \text{power radiated} \\ \text{by a particle of} \\ \text{energy } \mathcal{E} \end{pmatrix} \times \begin{pmatrix} \text{number of} \\ \text{particles with} \\ \text{energy } \mathcal{E} \end{pmatrix} \times \begin{pmatrix} \text{Jacobian} \\ \text{from } \mathcal{E} \text{ to } \nu \end{pmatrix} \propto (B^2\mathcal{E}^2)n(\mathcal{E})\frac{d\mathcal{E}}{d\nu}. \tag{6.280}$$

Because $\mathcal{E} \propto \nu^{1/2} B^{-1/2}$ [see Eq. (6.251)], it follows that $(d\mathcal{E}/d\nu) \propto B^{-1/2}\nu^{-1/2}$; writing $n(\mathcal{E}) \propto \mathcal{E}^{-p} \propto \nu^{-p/2} B^{p/2}$, we get

$$P \propto B^2\, \nu B^{-1}\, \nu^{-p/2} B^{p/2} \nu^{-1/2} B^{-1/2} \propto \nu^{-\frac{1}{2}(p-1)} B^{\frac{1}{2}(p-1)} \propto \nu^{-s} B^{1+s}, \tag{6.281}$$

with $p = 2s + 1$.

We can obtain the total intensity of emission along a line of sight through an optically thin source of size L by multiplying this expression by L. Numerically, this result can be expressed as

$$I(\nu) = 0.933\, a(s) C L (B \sin\alpha)^{s+1} \left(\frac{6.26 \times 10^9}{\nu/\text{GHz}}\right)^s \text{ Jy rad}^{-2}, \tag{6.282}$$

where $s = (p - 1)/2$ and

$$a(s) = \left(\frac{2^s}{s+1}\right) \Gamma\left(\frac{3s+1}{6}\right) \Gamma\left(\frac{3s+11}{6}\right). \tag{6.283}$$

The above discussion assumes that the magnetic field is homogeneous and regular. In the other extreme, when the field is randomly oriented, the quantity

$(\sin\alpha)^{s+1}$ has to be averaged over all angles. With

$$\frac{1}{2}\int_0^{\pi/2}\sin^{s+1}\alpha\,\sin\alpha\,d\alpha = \frac{\pi}{2}\frac{\Gamma\left(\frac{s+3}{2}\right)}{\Gamma\left(\frac{s+4}{2}\right)}, \tag{6.284}$$

the result for random orientation of the magnetic field becomes

$$I(\nu) = 13.5\,b(s)\,CLB^{s+1}\left(\frac{6.26\times10^9}{\nu/\text{GHz}}\right)^s\,\text{Jy rad}^{-2}, \tag{6.285}$$

with

$$b(s) = 2^{s-4}\sqrt{\frac{3}{\pi}}\frac{\Gamma\left(\frac{3s+1}{6}\right)\Gamma\left(\frac{3s+11}{6}\right)\Gamma\left(\frac{s+3}{2}\right)}{(s+1)\Gamma\left(\frac{s+4}{2}\right)}. \tag{6.286}$$

The nature of polarisation also changes when there is a collection of particles. Even though the radiation from each particle is elliptically polarised, for a distribution of particles that vary smoothly over the pitch angle, the elliptical component will cancel out because we will obtain equal contributions from both sides of the line of sight. A certain degree of linear polarisation will remain, which can be computed from the above formulas through a series of mathematical manipulations. Here merely the final result is given. The degree of polarisation for a set of particles, all with single energy E but a smooth distribution across the pitch angle, is given by

$$\rho(\omega) = \frac{G(x)}{F(x)}, \quad G(x) = x K_{2/3}(x), \quad x = \frac{\omega}{\omega_c}, \tag{6.287}$$

and F is defined in Eqs. (6.277). For a power-law distribution, this reduces to $\rho = 3(p+1)/(3p+7)$.

6.11.4 Synchrotron Self-Absorption

The above discussion assumes tacitly that all the photons that are emitted by a synchrotron system escape the system. In reality, of course, some amount of photons will be absorbed by the system and we would like to know how the absorption affects the net spectrum of photons that is observed from a synchrotron source. To work out this effect, we must first compute the synchrotron (self-) absorption coefficient α_ν for a power-law distribution of electrons. If the electrons are ultrarelativistic, then $\mathcal{E}\simeq pc$ and

$$f(p)d^3p = 4\pi p^2 f(p)\,dp = 4\pi(\mathcal{E}/c)^2(d\mathcal{E}/c)f(\mathcal{E}/c) = N(\mathcal{E})\,d\mathcal{E}, \tag{6.288}$$

i.e., $N(\mathcal{E})\propto f(\mathcal{E})\mathcal{E}^2$. Then the absorption coefficient α_ν becomes [see Eq. (6.58)]

$$\alpha_\nu = \frac{c^2}{8\pi h\nu^3}\int d\mathcal{E}\,P(\nu,\mathcal{E})\mathcal{E}^2\left[\frac{N(\mathcal{E}-h\nu)}{(\mathcal{E}-h\nu)^2} - \frac{N(\mathcal{E})}{\mathcal{E}^2}\right]. \tag{6.289}$$

When $h\nu \ll \bar{\mathcal{E}}$, where $\bar{\mathcal{E}}$ is the typical energy of the electron, we can expand $N(\mathcal{E} - h\nu)$ in a Taylor series in $h\nu$ and obtain

$$\alpha_\nu = -\frac{c^2}{8\pi\nu^2} \int d\mathcal{E} \, P(\nu, \mathcal{E}) \mathcal{E}^2 \frac{\partial}{\partial\mathcal{E}} \left[\frac{N(\mathcal{E})}{\mathcal{E}^2}\right]. \tag{6.290}$$

For a power-law distribution $N(\mathcal{E}) = C\mathcal{E}^{-p}$,

$$-\mathcal{E}^2 \frac{d}{d\mathcal{E}}\left(\frac{N}{\mathcal{E}^2}\right) = (p+2)C\mathcal{E}^{-(p+1)} = \frac{(p+2)N(\mathcal{E})}{\mathcal{E}} \tag{6.291}$$

$$\alpha_\nu = \frac{(p+2)c^2}{8\pi\nu^2} \int d\mathcal{E} \, P(\nu, \mathcal{E}) \frac{N(\mathcal{E})}{\mathcal{E}}. \tag{6.292}$$

From our earlier discussion, we know that $P(\nu, \mathcal{E}) \propto F(\nu/\nu_c) \propto F(\nu/\mathcal{E}^2)$; using this result and introducing a variable $x = (\nu/\mathcal{E}^2)$, we find that α_ν becomes

$$\alpha_\nu \propto \frac{1}{\nu^2} \int \frac{\mathcal{E}}{\mathcal{E}} N(\mathcal{E}) F(\nu/\mathcal{E}) \propto \frac{1}{\nu^2} \int \frac{dx}{x} \left(\frac{\nu}{x}\right)^{p/2} F(x) \propto \nu^{-\frac{1}{2}(p+4)}. \tag{6.293}$$

Thus $\alpha_\nu R \gg 1$ at low frequencies and $\alpha_\nu R \ll 1$ at high frequencies. When $\alpha_\nu R \gg 1$, the observed intensity at low frequencies is

$$I_\nu \propto \left(\frac{j_\nu}{\alpha_\nu}\right) \propto \frac{\nu^{-(p-1)/2}}{\nu^{-(p+4)/2}} \propto \nu^{5/2}, \tag{6.294}$$

independent of p. When $\alpha_\nu R \ll 1$, the observed intensity is $I_\nu \propto j_\nu \propto \nu^{-(p-1)/2}$. Note that low-frequency synchrotron intensity has a $\nu^{2.5}$ behaviour in contrast to a thermal spectrum, which has ν^2 behaviour. It is also clear from the preceding discussion that synchrotron radiation will peak at some intermediate frequency.

The $\nu^{5/2}$ dependence has a simple physical interpretation. We know that the low-frequency thermal radiation has the form $B_\nu \approx (2\nu^2/c^2)k_B T$, in which the first factor comes from phase space and the second from the mean energy of the photons. In the case of a synchrotron source, the phase-space factor remains the same but $k_B T$ should be replaced with the mean energy ϵ of a synchrotron electron emitting radiation at the frequency ν. From $\epsilon = \gamma m_e c^2$ and $\nu = \gamma^2 \nu_L$, where ν_L is the Larmor frequency [see Eq. (6.249)], we get

$$I_\nu \approx \left(\frac{2\nu^2}{c^2}\right)\left(\frac{\nu}{\nu_L}\right)^{1/2} m_e c^2 \propto B_0^{-1/2} \nu^{5/2}. \tag{6.295}$$

Exercise 6.20

Compact radio sources and Compton catastrophe: Consider a spherical source of radius R_s located at a distance r subtending an angle $\theta_s \approx (R_s/r)$. The source has a magnetic field B, energy density of photons U_{ph}, and emits photons by synchrotron as well as possibly by inverse Compton scattering.

(1) Show that the synchrotron flux F_ν from this source will be $F_\nu \approx (1/3)\rho j_\nu \theta_s^2$ if it is optically thin and $F_\nu = \pi B_\nu \theta_s^2$ if it is optically thick. The corresponding brightness temperature of an angularly resolved source will be $T_B = (c^2/2\pi k_B)(F_\nu/\nu^2\theta_s^2)$. The maximum brightness temperature will occur roughly at ν_m where the synchrotron radiation peaks.

(2) Argue that the ratio between inverse Compton and synchrotron luminosities of the source will be given approximately by

$$\frac{L_C}{L_S} \approx \frac{8\pi \nu_m F_{\nu m}}{\theta_s^2 c B_0^2}. \tag{6.296}$$

Expressing $F_{\nu m}$ in terms of the brightness temperature, show that $F_{\nu m} \propto T_B^5 B_0^2 \theta_s^2$, where the brightness temperature is evaluated at the maximum frequency. Hence conclude that $(L_C/L_S) \propto T_B^5 \nu_m$.

(3) Include the proportionality constants for an order-of-magnitude evaluation and obtain

$$\frac{L_C}{L_S} \approx \left[\frac{T_B}{10^{12} \text{ K}}\right]^5 \left(\frac{\nu_m}{10^{8.5} \text{ Hz}}\right). \tag{6.297}$$

Hence argue that if the brightness temperature greatly exceeds 10^{12} K at radio frequencies, the inverse Compton losses will be very high. For a more general version of this bound, prove that if

$$k_B T_B > \left(\frac{\lambda_m}{r_0}\right)^{1/5} m_e c^2, \tag{6.298}$$

where λ_m is the wavelength at maximum frequency and r_0 is the classical electron radius, then Compton luminosity will greatly exceed the synchrotron luminosity.

(4) Argue that if Compton losses should not be too large, then the angular size of a synchrotron source of total flux F_ν is bounded by $\theta > 10^{-3}$ arc sec $(F_\nu/\text{Jy})^{1/2} (\nu_m/\text{GHz})^{-9/10}$.

6.12 Photoionisation

Two other radiative processes of vital importance in astrophysics are *ionisation* and *recombination*. In ionisation, a sufficiently energetic photon induces the electron to make a transition from a bound state to a free state. In recombination, a free electron is bound to an ion, emitting a photon. We now compute the cross sections for these processes, starting with the ionisation of a hydrogen atom.

Consider the process in which an electron in a hydrogen atom absorbs the photon and makes a transition from the $n = 1$ ground state to a free particle state. Let the rate of such transitions be dP/dt when the number of incident photons is N_γ. The cross section σ_{bf} (*bf* stands for bound–free) for photoionisation is defined by the relation

$$\left(\frac{dP}{dt}\right) dN_{\text{elec}} = \left(\frac{dP}{dt}\right) \times \frac{V d^3 p_e}{(2\pi\hbar)^3} \equiv \sigma_{bf}(\omega) \left[\frac{N_\gamma}{V} c\right] \left[\frac{V d^3 p_\gamma}{(2\pi\hbar)^3}\right], \tag{6.299}$$

where $dN_{\text{elec}} = [Vd^3p_e/(2\pi\hbar)^3]$ is the density of states available for the final electron. The flux of incident photons is $(N_\gamma/V)c$ times the density of states dN_γ for photons: $dN_\gamma = [Vd^3p_\gamma/(2\pi\hbar)^3] = 4\pi V(\omega/c)^2d(\omega/c)/(2\pi)^3$. This flux, multiplied by the cross section σ_{bf}, should give the rate for a process – here the rate of ionisation. Because the free electron is produced by the process, we should also include a density of state factor for the electron. Thus

$$\left(\frac{dP}{dt}\right)\frac{Vd^3p_e}{(2\pi\hbar)^3} = \sigma_{bf}(\omega)cn(\omega)\frac{4\pi(\omega^2/c^2)d(\omega/c)}{(2\pi)^3}, \qquad (6.300)$$

with $n = (N_\gamma/V)$. To evaluate $\sigma_{bf}(\omega)$, we need to compute (dP/dt), which essentially involves computing the matrix element of $e^{i\mathbf{k}\cdot\mathbf{x}}\mathbf{p}$ between the initial and the final states of the electron. The initial state is the $n = 1$ ground state of a hydrogen atom with energy $-E_0$. Assume, for simplicity, that the final state is a plane-wave state. Then the wave function for the initial state is

$$\phi_i(\mathbf{x}) = \left(\frac{1}{\pi a_0^3}\right)^{1/2}\exp\left(-\frac{|\mathbf{x}|}{a_0}\right), \qquad a_0 = \frac{\hbar^2}{mq^2}, \qquad (6.301)$$

and the final state can be approximated as a plane-wave,

$$\phi_f(\mathbf{x}) = \left(\frac{1}{V}\right)^{1/2}\exp(-i\mathbf{k}_e\cdot\mathbf{x}), \qquad (6.302)$$

with momentum $\mathbf{p}_e = \hbar\mathbf{k}_e$. (We denote the electronic charge by q rather than e in this section for future convenience.) The matrix element of $e^{i\mathbf{k}\cdot\mathbf{x}}\mathbf{p} = -i\hbar e^{i\mathbf{k}\cdot\mathbf{x}}\nabla$ between ϕ_i and ϕ_f can be evaluated easily with the result

$$\int d^3x\,\exp[i(\mathbf{k}-\mathbf{k}_e)\cdot\mathbf{x}-|\mathbf{x}|/a_0]$$

$$= 2\pi\int_0^\infty dr\,re^{-r/a_0}\int_{-1}^{+1}d\mu e^{ir\mu|\mathbf{k}-\mathbf{k}_e|} = \frac{8\pi a_0^3}{\left(1+a_0^2|\mathbf{k}-\mathbf{k}_e|^2\right)^2} \qquad (6.303)$$

to give

$$|\langle\phi_f|e^{i\mathbf{k}\cdot\mathbf{x}}\hat{\mathbf{e}}\cdot\mathbf{p}|\phi_i\rangle|^2 = 64\pi\frac{\hbar^2 a_0^3}{V}(\mathbf{k}_e\cdot\hat{\mathbf{e}}_\alpha)^2\left(1+a_0^2|\mathbf{k}-\mathbf{k}_e|^2\right)^{-4}. \qquad (6.304)$$

Substituting this expression into our general formula for the rate of absorption [see Eq. (4.126)],

$$\left(\frac{dP_\alpha}{d\Omega\,dt\,d\omega}\right) = \left(\frac{\omega}{2\pi\hbar c^3}\right)n_{k\alpha}\frac{q^2}{m^2}|\langle E_f|e^{i\mathbf{k}\cdot\mathbf{x}}\hat{\mathbf{e}}_\alpha\cdot\mathbf{p}|E_i\rangle|^2\delta(\omega-\omega_{fi}), \qquad (6.305)$$

we find that

$$\left(\frac{dP_\alpha}{d\Omega\,dt\,d\omega}\right) = 32\left(\frac{q}{m}\right)^2\left(\frac{\hbar a_0^3}{Vc^3}\right)\frac{(\omega n_{k\alpha})(\mathbf{k}_e\cdot\hat{\mathbf{e}}_\alpha)^2}{\left(1+a_0^2|\mathbf{k}-\mathbf{k}_e|^2\right)^4}\delta(\omega-\omega_{fi}). \qquad (6.306)$$

To proceed further we have to integrate this expression over $d\omega$ and $d\Omega$ and sum over the polarisation states. Taking $\mathbf{k}$ along the z axis, $\hat{\mathbf{e}}_\alpha$ along the x axis, and $\mathbf{k}_e$ along the direction specified by (θ, ϕ), we have the relations

$$\mathbf{k}_e \cdot \hat{\mathbf{e}}_\alpha = k_e \sin\theta \cos\phi, \quad |\mathbf{k} - \mathbf{k}_e|^2 = k^2 + k_e^2 - 2kk_e \cos\theta. \tag{6.307}$$

Further, because we expect that $\hbar k \ll m_e c$ and $\hbar k_e \ll m_e c$ in the nonrelativistic limit, we can write

$$1 + |\mathbf{k} - \mathbf{k}_e|^2 \cong a_0^2 \frac{2m\omega}{\hbar}, \tag{6.308}$$

where we have used the fact that

$$k_e^2 = \frac{(2mE_f)}{\hbar^2} = \frac{2m}{\hbar^2}\left(\hbar\omega_{fi} - \frac{q^2}{2a_0}\right) = \frac{2mck}{\hbar} - \frac{1}{a_0^2}. \tag{6.309}$$

The angular integrations can now be performed with the result

$$\int_0^{2\pi} d\phi \int_0^\pi \sin^2\theta \cos^2\phi \sin\theta \, d\theta \, d\phi = \frac{4\pi}{3}, \tag{6.310}$$

giving

$$\left(\frac{dP}{dt}\right) = \left(\frac{8\pi}{3}\right)\left(\frac{a_0^3}{V}\right)\left(\frac{q^2}{\hbar c}\right)\left(\frac{\hbar k_e}{mc}\right)^2\left(\frac{\hbar}{ma_0^2}\right)^4\frac{n(\omega)}{\omega^3}. \tag{6.311}$$

The cross section can now be identified from definition (6.299); we get

$$\sigma_{bf}(\omega) = \frac{8\pi}{3}\left(\frac{q^2}{mc}\right)\left(\frac{\hbar}{ma_0^2}\right)^4 \frac{1}{\omega^5}(a_0 k_e)^3, \tag{6.312}$$

with

$$k_e^2 = \frac{2m}{\hbar^2}\left(\hbar\omega - \frac{q^2}{2a_0}\right). \tag{6.313}$$

We have assumed in our analysis that the final wave function of the electron is that of a free particle. This requires that the final energy of the electron be far greater than q^2/a_0 or, equivalently, $\hbar\omega \gg q^2/a_0$. In that case $k_e^2 \simeq (2m\omega/\hbar)$ and the above result becomes

$$\sigma_{bf} = \frac{2^8}{3}(\pi a_0^2)\left(\frac{q^2}{\hbar c}\right)\left(\frac{E_0}{\hbar\omega}\right)^{7/2}, \quad E_0 = \frac{mq^4}{2\hbar^2}. \tag{6.314}$$

Strictly speaking, the final state of the electron is not a plane-wave state but rather a scattering state in the $(-q^2/r)$ potential with an energy $E_p = (p^2/2m)$. To do a more precise computation, we need to evaluate the matrix element between the initial bound state and the final scattering state. We now indicate how this can be done.

The final scattering state can be described as a superposition of partial waves with all values of l. However, because the photon has an angular momentum of one unit and the electron is initially in an s state with zero angular momentum, the final state of the electron can be only a p state. Hence, in the standard expansion of the scattering state,

$$\Psi_p = \frac{1}{p}\left(\frac{\pi}{2}\right)^{1/2}\sum_{l=0}^{\infty} i^l(2l+1)e^{-i\delta_l}R_{pl}(r)P_l(\hat{\mathbf{n}}\cdot\hat{\mathbf{n}}_1), \qquad (6.315)$$

where $R_{pl}(r)$ is the radial part of the scattering state, $\hat{\mathbf{n}} = (\mathbf{p}/p)$, and $\hat{\mathbf{n}}_1 = (\mathbf{r}/r)$, we need to retain only the $l=1$ state. Omitting the unimportant phase factors, we can therefore write the final state as

$$\Psi_p = \frac{3}{p}\left(\frac{\pi}{2}\right)^{1/2}R_{p1}(r)(\hat{\mathbf{n}}\cdot\hat{\mathbf{n}}_1). \qquad (6.316)$$

In this case, the relevant matrix element is

$$\hat{\mathbf{e}}_\alpha\cdot\mathbf{p}_{fi} = \frac{3}{\sqrt{2}}\left(\frac{1}{a_0}\right)^{5/2}\frac{1}{p}\int d\Omega_1\int r^2 dr(\hat{\mathbf{n}}_1\cdot\hat{\mathbf{n}})(\hat{\mathbf{n}}_1\cdot\hat{\mathbf{e}}_\alpha)e^{-r/a_0}R_{p1}(r)$$

$$= \frac{2^{3/2}}{pa_0^{5/2}}(\hat{\mathbf{n}}\cdot\hat{\mathbf{e}}_\alpha)\int_0^\infty r^2 dr\, e^{-r/a_0}R_{p1}(r). \qquad (6.317)$$

The radial part of the wave function in a scattering state is

$$R_{p1} = \frac{2}{3a_0}\left[\frac{1+\mu^2}{\mu(1-e^{-2\pi\mu})}\right]^{1/2}pr\, e^{-ipr}F(2+i\mu,4,2ipr), \qquad (6.318)$$

where F is the hypergeometric function and $\mu\equiv(q^2/\hbar c)(c/v)$. To evaluate the integral we need two results:

$$\int_0^\infty dz\, e^{-\lambda z}z^{\gamma-1}F(\alpha,\gamma,kz) = \Gamma(\gamma)\lambda^{\alpha-\gamma}(\lambda-k)^{-\alpha} \qquad (6.319)$$

$$\left(\frac{\mu+i}{\mu-i}\right)^{i\mu} = e^{-\mu\cot^{-1}\mu}. \qquad (6.320)$$

Using these, we find that

$$\hat{\mathbf{e}}_\alpha\cdot\mathbf{p}_{fi} = \frac{2^{7/2}\pi\mu^3(\hat{\mathbf{n}}\cdot\hat{\mathbf{e}}_\alpha)}{p^{1/2}(1+\mu^2)^{3/2}}\frac{e^{-2\mu\cot^{-1}\mu}}{(1-e^{-2\pi\mu})^{1/2}}. \qquad (6.321)$$

From the energy conservation we have

$$\hbar\omega = \frac{p^2}{2m}(1+\mu^2), \qquad (6.322)$$

or $\mu=[(\hbar\omega/E_0)-1]^{-1/2}$. The rest of the analysis proceeds exactly as before and we get

$$\sigma_{bf} = \frac{2^9\pi^2}{3}\left(\frac{q^2}{\hbar c}\right)a_0^2\left(\frac{E_0}{\hbar\omega}\right)^4\frac{e^{-4\mu\cot^{-1}\mu}}{1-e^{-2\pi\mu}}. \qquad (6.323)$$

This result is valid for both $\hbar\omega \simeq E_0$ as well as $\hbar\omega \gg E_0$ as long as $\hbar\omega \ll mc^2$. When $\hbar\omega \gg E_0$, that is, when $\mu \ll 1$,

$$\frac{e^{-4\mu \cot^{-1}\mu}}{1 - e^{-2\pi\mu}} \simeq \frac{1}{2\pi\mu} = \frac{1}{2\pi}\left(\frac{q^2}{\hbar c}\right)^{-1}\frac{(2m\hbar\omega)^{1/2}}{mc} = \frac{1}{2\pi}\left(\frac{\hbar\omega}{E_0}\right)^{1/2} \qquad (6.324)$$

and the cross section reduces to expression (6.314):

$$\sigma_{bf} = \frac{2^8\left(\pi a_0^2\right)}{3}\left(\frac{q^2}{\hbar c}\right)\left(\frac{E_0}{\hbar\omega}\right)^{7/2} = \left(\frac{128\pi}{3}\right)\left(\frac{q^2}{mc^2}\right)\left(\frac{c}{\omega_0}\right)\left(\frac{\omega_0}{\omega}\right)^{7/2}, \qquad (6.325)$$

where $\omega_0 = (E_0/\hbar)$. On the other hand, when $\hbar\omega > E_0$ (large μ) we get

$$\frac{e^{-4\mu \cot^{-1}\mu}}{1 - e^{-2\pi\mu}} \simeq e^{-4} \qquad (6.326)$$

(where e is the base of the natural logarithm) and

$$\sigma_{bf} = \left(\frac{2^9}{3e^4}\right)\left(\pi^2 a_0^2\right)\left(\frac{q^2}{\hbar c}\right). \qquad (6.327)$$

Results (6.325) and (6.327) show that the exact ω dependence of σ_{bf} is complicated and has different limiting forms for $\hbar\omega \approx E_0$ and $\hbar\omega \gg E_0$. In general, it is conventional to write the result for σ_{bf} as

$$\sigma_{bf}(\omega) = \frac{8\pi}{3\sqrt{3}}\left(\frac{q^2}{\hbar c}\right)^5\left(\frac{mc^4}{\hbar\omega^3}\right)g_{bf}(\omega), \qquad (6.328)$$

where $g_{bf}(\omega)$ is again called a Gaunt factor. It varies slowly with ω and goes as $\omega^{-1/2}$ if $\hbar\omega \gg E_0$. If $g_{bf}(\omega)$ is nearly constant, $\sigma_{bf}(\omega) \propto \omega^{-3}$; such an approximation is often made within various astrophysical contexts.

A similar analysis can be performed for photoionisation from any level n of the hydrogen atom. The result in Eq. (6.328) is multiplied by the factor n^{-5}. The result can be expressed conveniently in the form

$$\sigma_{bf}(\omega, n) = n\sigma_0\left(\frac{\omega_n}{\omega}\right)^3 g_n(\omega)$$

for frequencies $\omega > \omega_n$, with $\hbar\omega_n = E_0/n^2$, and the constant σ_0 is given by

$$\sigma_0 \equiv \frac{64}{3\sqrt{3}}\left(\frac{q^2}{\hbar c}\right)\pi a_0^2.$$

The same considerations are applicable to other atoms. In general, photoionisation can act as one of the dominant mechanisms for the absorption of photons, for example in stellar interiors. The bound–free opacity can, in general, be expressed in the form $\kappa_{bf} = \sum(N_n/\rho)\sigma_{bf,n}$ where $N_n \propto n^2$ is the number density of atoms in the nth state for a particular species and $\sigma_{bf,n}$ is the bound–free cross section from the level n for that particular species. The summation is over all energy levels and all species of atoms. It is clear that at any given temperature, only atoms that are remaining neutral will contribute significantly. Further, because $\sigma_{bf,n} \propto \omega^{-3} n^{-5} \propto n^6 n^{-5} \propto n$ and $N_n \propto n^2$, the combination $N_n \sigma_{bf,n}$ grows as n^3; hence the major contribution to κ_{bf} arises from highly excited states of atoms that still remain neutral at a given temperature. In stellar interiors, such a contribution arises mainly from atomic species that are heavier than hydrogen and helium (called metals by astrophysicists). A reasonable approximation for κ_{bf} in this context is

$$\kappa_{bf} \approx 4 \times 10^{25} Z(1+X)\rho T^{-3.5} \text{ cm}^2 \text{ g}^{-1},$$

$$\approx 4 \times 10^{24} Z(1+X)\rho T^{-3.5} \text{ m}^2 \text{ kg}^{-1} \tag{6.329}$$

where X and Z are hydrogen and the metal mass fraction, respectively. Because $\hbar\omega \simeq k_B T$ at the peak of the Planck spectrum, the scaling $\sigma_{bf} \propto \omega^{-3.5}$ gives the bound–free opacity $\kappa_{bf} \propto T^{-3.5}$. This formula is reasonable for temperatures above 10^4 K.

If the hydrogen gas is immersed in a high-energy radiation field with intensity I_ν, then the rate of photoionisation from a level i to a continuum state, formally denoted by k, is given by

$$R_{ik} = \int_{\nu_0}^{\infty} d\nu \, d\Omega \left[\left(\frac{dE}{dt dA \, d\Omega \, d\nu} \right) \frac{1}{h\nu} \right] \sigma_{bf}(\nu) = 4\pi \int_{\nu_0}^{\infty} d\nu \left[\frac{\sigma_{bf}(\nu) I_\nu}{h\nu} \right],$$

$$\tag{6.330}$$

with

$$\sigma_{bf}(\nu) \simeq \frac{mq^{10}}{3\pi^2 \sqrt{3} c\hbar^6} \frac{1}{n^5 \nu^3} = \frac{2.815 \times 10^{29} Z^4}{n^5 \nu^3} \text{ cm}^2 = \frac{2.815 \times 10^{25} Z^4}{n^5 \nu^3} \text{ m}^2.$$

$$\tag{6.331}$$

This follows from the definition of photoionisation cross section $\sigma_{bf}(\nu)$ in Eq. (6.328) with $\omega = 2\pi\nu$, and we have (1) included the n^{-5} factor but (2) ignored the Gaunt factor for simplicity. Assuming that $h\nu_0 \gg k_B T$ and that the radiation field is Planckian, we have $I_\nu \approx (2h\nu^3/c^2) \exp(-h\nu/k_B T)$, so that

$$R_{ik} = \frac{8}{3\sqrt{3}\pi} \left(\frac{q^2}{\hbar c} \right)^5 \left(\frac{mc^2}{\hbar} \right) \int_{\nu_0}^{\infty} \frac{e^{-h\nu/k_B T}}{\nu n^5} d\nu$$

$$= \frac{8}{3\sqrt{3}\pi} \left(\frac{q^2}{\hbar c} \right)^5 \left(\frac{mc^2}{\hbar} \right) \frac{E_1(h\nu_0/k_B T)}{n^5}$$

$$\cong \frac{8}{3\sqrt{3}\pi} \left(\frac{q^2}{\hbar c} \right)^5 \left(\frac{mc^2}{\hbar} \right) \left(\frac{k_B T}{h\nu_0} \right) \frac{1}{n^5} \exp\left(-\frac{h\nu_0}{k_B T} \right) \quad (h\nu_0 \gg k_B T),$$

$$\tag{6.332}$$

where E_1 is the exponential integral and we have used the asymptotic form $E_1(z) \approx (e^{-z}/z)$ for $z \gg 1$.

The process described above involves splitting the hydrogen atom into a proton and a free electron by the absorption of a photon. The inverse process will correspond to the recombination of a free electron and a proton to form a bound state of a hydrogen atom with the emission of a photon. The cross section for this process, σ_{rec}, can be related to the cross section for photoionisation σ_{bf} by the following argument. In equilibrium, the reaction $e^- + p \to H + \gamma$ should be balanced by the reaction $H + \gamma \to e^- + p$. This means that

$$p_i^2 g_i \sigma_{i \to f} = p_f^2 g_f \sigma_{f \to i}, \tag{6.333}$$

where p_i and p_f are the momenta of relative motion of the particles and g_i and g_f take into account the degeneracies of the initial and the final states, respectively. Let us first consider direct recombination to the ground state of the hydrogen atom. Then $g_i = 2 \times 2$ for electron and proton spins and $g_f = 2 \times 2 \times 2$ for the electron, proton, and photon spins. Therefore

$$\sigma_{rec} = \left(\frac{g_f}{g_i}\right)\left(\frac{p_f}{p_i}\right)^2 \sigma_{bf} = \frac{2(\hbar k)^2}{(mv)^2}\sigma_{bf} = \frac{2(\hbar \omega)^2}{(mv)^2}\left(\frac{\sigma_{bf}}{c^2}\right), \tag{6.334}$$

where v is velocity of the incident electron and ω is the frequency of the emitted photon. When the temperature of the electron gas is much lower than the ionisation potential of the hydrogen atom, we can write

$$\hbar \omega = \frac{\hbar^2 k_e^2}{2m} + \frac{mq^2}{\hbar^2 a_0} \simeq \frac{mq^2}{\hbar^2 a_0} = \frac{m^2 q^4}{\hbar^4}. \tag{6.335}$$

In this case, by using Eq. (6.327) in Eq. (6.334), we find that

$$\sigma_{rec} = \left(\frac{2^{10}\pi^2}{3e^4}\right)\left(\frac{q^2}{\hbar c}\right)\left(\frac{a_0^2}{m^2 c^2 v_e^2}\right)\frac{1}{E_0^2}, \tag{6.336}$$

where E_0 is the binding energy of hydrogen.

Note that $v\sigma_{rec}n_e n_i = v\sigma_{rec}n_i^2$ will represent the rate of recombination per unit volume of matter, where v is the relative velocity of electrons with respect to ions and n_i is the density of ions or electrons. (This quantity is in units of per cubic centimeter per second.) Therefore, with a system of electrons with a distribution of velocities, the really relevant quantity that regulates recombination is the mean value of $v\sigma_{rec}$. This suggests that we define a quantity called the coefficient of radiative recombination by the relation $\alpha = \langle v_e \sigma_{rec}\rangle$, where v_e is the speed of the electrons. In thermal equilibrium, the averaging is done over a Maxwellian distribution with temperature T. With this definition, the rate of change of number density of electrons that is due to recombination will be $-\alpha n_e n_i$, where n_e and n_i are the electron and the ion number densities, respectively. Averaging $\sigma_{rec}v_e$ by using the result that, $\langle v_e^{-1}\rangle = (2m/\pi k_B T)^{1/2}$

for a Maxwellian distribution, we find that

$$\alpha \equiv \langle \sigma_{\text{rec}} v_e \rangle = \left(\frac{2^{10} \pi^{3/2}}{3 e^4} \right) \left(\frac{q^2}{\hbar c} \right)^3 \left(\frac{a_0^3 E_0}{\hbar} \right) \left(\frac{E_0}{k_B T} \right)^{1/2}$$

$$\simeq 35 \left(\frac{q^2}{\hbar c} \right)^3 \left(\frac{a_0^3 E_0}{\hbar} \right) \left(\frac{E_0}{k_B T} \right)^{1/2}. \tag{6.337}$$

Numerically,

$$\alpha = \langle \sigma_{\text{rec}} v \rangle \approx 1.4 \times 10^{-13} \left(\frac{k_B T}{1 \, \text{eV}} \right)^{-1/2} \, \text{cm}^3 \, \text{s}^{-1}$$

$$\approx 1.4 \times 10^{-19} \left(\frac{k_B T}{1 \, \text{eV}} \right)^{-1/2} \, \text{m}^3 \, \text{s}^{-1}. \tag{6.338}$$

In the above analysis, we have considered the recombination of an electron and proton directly to the ground state of the hydrogen atom. The discussion can be easily generalised when the recombination produces an atom in the nth excited state. The relation between recombination and photoionisation cross sections, given in Eq. (6.337), now becomes modified to

$$\sigma_{\text{rec}}(n) = \left(\frac{g_f}{g_i} \right) \left(\frac{\hbar \omega}{c m_e v} \right)^2 \sigma_{bf}(\omega, n),$$

with $g_f / g_i = 2n^2$ and $\hbar \omega = (m_e^2 v^2 / 2) + \hbar \omega_n$. The recombination rate is determined by the integral over the Maxwellian distribution $f_e(v)$ for the electrons:

$$\alpha_n = \int_0^\infty \sigma_{\text{rec}}(n) v f_e(v) 4\pi v^2 \, dv = n^2 \lambda_T^3 e^{\hbar \omega_n / k_B T} I(n),$$

where $\lambda_T = (2\pi \hbar^2 / m_e k_B T)^{1/2}$ is the thermal wavelength of the electron and

$$I(n) \equiv \frac{8\pi}{c^2} n \sigma_0 v_n^3 E_1(x_n), \quad x_n \equiv \frac{\hbar \omega_n}{k_B T},$$

where E_1 is the exponential integral:

$$E_1(x) \equiv \int_x^\infty \frac{e^{-t}}{t} \, dt. \tag{6.339}$$

The final result can be reexpressed in the form

$$\alpha_n = \left(\frac{\sigma_0 c}{\sqrt{\pi}} \right) \left(\frac{q^2}{\hbar c} \right)^3 x_n^{3/2} e^{x_n} E_1(x_n), \quad x_n \equiv \frac{\hbar \omega_n}{k_B T},$$

where $(\sigma_0 c / \sqrt{\pi})(q^2 / \hbar c)^3 \simeq 5.2 \times 10^{-14}$ cm^3 s^{-1} is the order of magnitude.

The total recombination rate is obtained by the summation of this expression over all bound states. In practical situations the recombination to the level $n = 1$ can be ignored as it produces a photon that is sufficiently energetic to ionise another hydrogen atom immediately. Thus it is really necessary to sum the re-combination coefficient only from $n = 2$ in order to obtain the total recombination coefficient:

$$\alpha = \sum_{n=2}^{\infty} \alpha_n.$$

We can estimate such sums by using the approximate formula

$$\sum_{n=a}^{b} F(n) \approx \frac{1}{2}[F(b) + F(a)] + \int_{a}^{b} F(n)\, dn.$$

Straightforward integration with $a = 2$ and $b \to \infty$ gives the estimate

$$\alpha = \left(\frac{q^2}{\hbar c}\right)^3 \left(\frac{\sigma_0 c}{\sqrt{\pi}}\right) x_2^{1/2} \left[x_2 e^{x_2} E_1(x_2)\left(\frac{1}{2} + \frac{1}{x_2}\right) + \gamma + \ln x_2\right],$$

where $\gamma = 0.577$ is Euler's constant and $x_2 = (h\nu_2/k_B T) = 39\,500\ \text{K}/T$.

The radiative recombination can serve as a cooling mechanism for the gas because each recombination removes the kinetic energy of the captured electron from the gas. To leading order, α now scales as $x_2^{3/2} \propto T^{-3/2}$ [compared with Eq. (6.337), which gives $\alpha \propto T^{-1/2}$] so that the rate of energy loss that is due to recombination is $\alpha(k_B T) \propto T^{-1/2}$. More formally, we can obtain the net energy-loss rate by first multiplying the recombination cross section by the electron energy flux $n_e \nu f(\nu)(m\nu^2/2)$ and the density of ions n_i and then summing over all velocities and the excited states. This result is conventionally expressed in the form

$$\epsilon_\nu\, d\nu = \frac{8}{3}\left(\frac{2\pi}{3}\right)^{1/2} \frac{Z^2 q^6}{m^2 c^3}\left(\frac{m}{k_B T}\right) n_i n_e Q\, d\nu$$

$$\approx 5.4 \times 10^{-39} Z^2 \frac{n_i n_e}{T^{1/2}} Q\, d\nu\ \text{erg}\ \text{s}^{-1}\ \text{cm}^{-3}\ \text{Hz}^{-1}\ \text{rad}^{-2}, \quad (6.340)$$

where ϵ_ν is the amount of energy lost per second per unit volume of plasma in unit frequency range and unit solid angle. (The numerical factor should be divided by 10 if SI units are used.) The coefficient Q can be expressed as

$$Q = [g(\nu, T) + f(\nu, T)]\exp\left(-\frac{h\nu}{k_B T}\right), \quad (6.341)$$

where $f(\nu, T)$ takes into account the effects of recombination cooling and $g(\nu, T)$ is the Gaunt factor. In general, these quantities have to be evaluated numerically. At temperatures greater than $\sim 10^7$ K, bremsstrahlung dominates the cooling whereas near 10^4–10^6 K, line radiation from the gas can be the main source of cooling. If we write the net cooling rate as $C = (dE/dt\, dV) = n_e n_p \Lambda(T)$, then the quantity $\Lambda(T)$ (called the radiative cooling function) is well approximated by

$$\Lambda(T) \approx \begin{cases} 2.5 \times 10^{-27} \, T^{1/2} \; \text{erg cm}^3 & (\text{for } T > 10^7 \, \text{K}) \\ 6 \times 10^{-19} \, T^{-0.6} \; \text{erg cm}^3 & [\text{for } T \simeq (2 \times 10^5 - 10^7) \, \text{K}]. \end{cases} \quad (6.342)$$

(The numerical factor should be multiplied by 10^{-13} if SI units are used.)

6.13 Collisional Ionisation

It is also possible for an electron in an atom to make a transition from level i to j during a collision of the atom with a target particle. The rate of excitation C_{ij} (and deexcitation C_{ji}) are defined in terms of a collisional cross section $\sigma_{ij}(v)$ by the relations

$$C_{ij} = \int_{v_0}^{\infty} n_e(v) \, v \, \sigma_{ij}(v) \, dv, \quad C_{ji} = \int_0^{\infty} n_e(v) \, v \, \sigma_{ji}(v) \, dv, \quad (6.343)$$

where $(1/2)mv_0^2 = E_{ij}$ determines the minimum velocity of the target particle needed to induce the transition. For a Maxwellian distribution of particles, we can easily transform the resulting integrals into giving

$$C_{ij} = \left(\frac{2N_e}{k_B T}\right) \left(\frac{2}{\pi m k_B T}\right)^{1/2} \int_{E_{ij}}^{\infty} \sigma_{ij}(E) e^{-E/k_B T} \, E \, dE, \quad (6.344)$$

(where N_e is the number of electrons per unit volume) and a corresponding expression for σ_{ji}. By an analysis similar to that which led to Eq. (6.337), we can relate C_{ij} to C_{ji} as

$$C_{ji} = \left(\frac{g_i}{g_j}\right) C_{ij} \exp(E_{ij}/k_B T). \quad (6.345)$$

An important case of collisional excitation of ions by electrons corresponds to those transitions that are radiatively forbidden. The determination of σ_{ij} from fundamental considerations is fairly complicated; however, in a vast majority of cases, $\sigma \propto \lambda^2$, where $\lambda = (\hbar/mv)$ is the de Broglie wavelength of the particle. Because $v^2 \propto E$, the cross section scales as $\sigma_{ij} \propto (1/E)$ and lower-energy particles are deflected more in the collision. In this case, the cross section can be written in the form

$$\sigma_{ij} \simeq \pi \lambda^2 \simeq (\pi \hbar^2 / 2mE). \quad (6.346)$$

Motivated by this scaling, we define a collision strength $\Omega(i, j)$ through the relation

$$\sigma_{ij}(E) = \frac{\pi \hbar^2}{2mE} \frac{\Omega(i, j)}{g_i}. \quad (6.347)$$

Typically, $\Omega(i, j)$ varies between 0.1 and 0.3. From Eq. (6.344) we get

$$C_{ij} = N_e \sqrt{\frac{2\pi}{mk_BT}} \frac{\hbar^2}{m} \frac{\Omega(i, j)}{g_i} e^{-E_{ij}/k_BT}, \tag{6.348}$$

$$C_{ji} = N_e \left(\frac{2\pi}{mk_BT}\right)^{1/2} \left(\frac{\hbar^2}{m}\right) \frac{\Omega(i, j)}{g_j}$$

$$= 8.6 \times 10^{-6}\,\mathrm{s}^{-1} \left(\frac{N_e}{\mathrm{cm}^{-3}}\right) \left(\frac{T}{1\,\mathrm{K}}\right)^{-1/2} \frac{\Omega(i, j)}{g_j}. \tag{6.349}$$

In the case of radiatively permitted transitions, the collisional cross section in the dipole approximation is usually parameterised in the form

$$\sigma_{ij}(E) \cong \frac{8\pi}{\sqrt{3}} \left(\frac{E_0}{E}\right) \left(\frac{E_0 f}{E_{ij}}\right) \pi a_0^2 G, \tag{6.350}$$

where f is the oscillator strength for the relevant levels and G is a numerical constant. The corresponding numerical values in this case are

$$C_{ij} \simeq 3.9 N_e \left(\frac{k_BT}{E_{ij}}\right)^{5/3} \frac{f}{T^{3/2}} \exp(-E_{ij}/k_BT). \tag{6.351}$$

Similar results hold for the collisional ionisation of neutral atoms with the coefficient 3.9 replaced with 2.16 and all variables are in cgs units. Finally, the corresponding result for collisional ionisation under the same approximations is

$$C_{ik} \simeq 10^9 N_e \left(\frac{k_BT}{I}\right) \exp(-I/k_BT) \frac{\sigma_0}{T^{1/2}}, \tag{6.352}$$

where σ_0 is the photoionisation cross section at the threshold. It must be stressed that these are fairly crude approximations that could be trusted to within only an order of magnitude.

7

Spectra

7.1 Introduction

The analysis of the spectra of astrophysical systems provides valuable information about their composition and dynamics. The purpose of this brief chapter is to introduce some basic concepts of atomic and molecular spectroscopy that are needed to appreciate the role played by spectra in astrophysics. The ideas developed in this chapter will be used in the study of stellar atmospheres, the interstellar medium (Vol. II), and in extragalactic astronomy (Vol. III).

7.2 Width of Spectral Lines

When a system makes a transition between two discrete energy levels E_2 and E_1 emitting a single photon, the frequency of the photon should be equal to $\omega = (E_2 - E_1)/\hbar$. Such a transition should lead to a sharp spectral line of infinite intensity and zero width. In reality, the frequency of the photon that is emitted is not precisely determined and the observed spectral line will have a finite width and intensity. The nature of the width of the spectral line contains important information about the state of the physical system.

The finite width of the spectral line can arise because of several reasons, among which three particular processes are of importance in astrophysics. To begin with, all energy levels (except the ground state) have a finite intrinsic width, that is, the energy of an excited state can be ascertained within only a finite accuracy ΔE_2 around a mean value E_2. This is because all excited states have a nonzero probability per second $\mathcal{P}$ for making a spontaneous transition to lower energy levels. The quantity $\tau = (1/\mathcal{P})$ represents the lifetime of the excited state. Because the excited state has a finite lifetime, an uncertainty principle predicts that the energy of the excited state will be uncertain by the amount $\Delta E_2 \approx (\hbar/\tau)$. When a transition occurs from level 2 to the ground state, the frequency of the emitted photon will be uncertain by the amount $\Delta \omega \approx (\Delta E_2/\hbar) \approx (1/\tau)$. In the spectral line, this will lead to a finite width; this width, which is inherently quantum mechanical, is called the natural linewidth.

The second process that contributes to the width of a spectral line is the motion of the system emitting the photon. Because of Doppler shift, the observed frequency of the photon will be different from the rest frequency. The mean, bulk, flow velocity of the system will shift the mean frequency to a different value. If the atoms or molecules emitting the radiation have a distribution in velocities around the mean velocity, the Doppler shift of the observed frequency will also have a distribution directly related to the velocity distribution. This will lead the observed frequency's being spread around a mean frequency, depending on the velocity dispersion of the system.

The third source for the width of the spectral line arises because of collisions between the atoms and molecules. Because collisions – which are intrinsically random by nature – can influence the radiative transitions, the frequency of the emitted photon will be randomly distributed around a mean value because of collisional effects. The width in this case will be governed by the frequency of the collisions.

In most astrophysical contexts, these processes act independently and hence contribute separately to the total width of the spectral line. It is therefore possible to analyse them individually and combine their effects together in the end.

7.2.1 Natural Width of Spectral Lines

When the probability for a spontaneous radiative transition was derived in Chap. 4, Section 4.5, the energy levels were treated as well defined with zero width. To see the effect of a finite width of energy levels on the frequency of emitted photon, we repeat that derivation along a different line.

Let Ψ be the wave function of the system made up of the atom and the electromagnetic field and let $H = H_0 + V$ be the Hamiltonian of the system, with V representing the interaction between the atom and the field. To calculate the amplitude for radiative transitions, we seek a solution of Schrodinger's equation (in units with $\hbar = 1$),

$$i\frac{\partial \Psi}{\partial t} = (H_0 + V)\Psi, \tag{7.1}$$

in the form of an expansion in terms of the unperturbed states, $\psi_n^{(0)}(t) \equiv \Psi_n^{(0)} e^{-i E_n t}$, of the system

$$\Psi = \sum_n a_n(t)\psi_n^{(0)}(t) = \sum_n a_n(t)e^{-i E_n t}\Psi_n^{(0)}. \tag{7.2}$$

Substituting Eq. (7.2) into Eq. (7.1), we find that the time-dependent coefficients

$a_n(t)$ satisfy the equation

$$i\frac{da_n}{dt} = \sum_m \langle n|V|m\rangle a_m \exp[i(E_n - E_m)t]. \tag{7.3}$$

At the initial instant, the system is assumed to be in an excited state $|2\rangle$ with energy E_2. Let the final state be the one in which the system has an energy E_1 and the electromagnetic field has a photon of frequency ω. This state is denoted by the symbol $|\omega, 1\rangle$. With this boundary condition, it follows that at $t = 0$ we must have

$$a_2 = 1, \quad a_m = 0 \quad \text{for } |m\rangle \neq |2\rangle. \tag{7.4}$$

The solution to Eq. (7.3) with this initial condition will give the probability amplitude for the system to make a transition from 2 to 1 with an emission of photon in the frequency range $(\omega, \omega + d\omega)$, with the corresponding probability of $dP = |a_{\omega 1}(t)|^2 \, d\omega$. We can obtain the form of $a_{\omega 1}(t)$ by solving Eq. (7.3) perturbatively; for example, we obtain the lowest-order solution by integrating Eq. (7.3) after replacing the a_n's on the right-hand side with the values in Eqs. (7.4). Usually we are interested in the probability for transition in the asymptotic limit as $t \to \infty$. Because the amplitudes vary with time at the characteristic time scale $\omega^{-1} = (E_2 - E_1)/\hbar$, the asymptotic limit is attained for $t \gg \hbar/(E_2 - E_1)$.

This approach, however, requires modification when we take into account the fact that the excited level has a finite lifetime, say, $t_{\text{decay}} = (1/\Gamma)$. If the energy levels have zero width, then ω^{-1} is the *only* relevant time scale in the problem and the asymptotic limit of $t \to \infty$ can be replaced with the condition $\omega t \gg 1$. If the excited energy level E_2 has a finite width Γ and corresponding lifetime t_{decay}, the above approach is valid only for $t \ll t_{\text{decay}}$. In other words, the results obtained in Chap. 4 are valid for time scales with $\omega^{-1} \ll t \ll \Gamma^{-1}$. Because the widths of the energy levels are small, $\Gamma \ll \omega$ and this interval is fairly large.

For time scales comparable with $1/\Gamma$, it is necessary to take into account the fact that there is a nonzero amplitude for the excited state to decay spontaneously. We can do this by modifying the conditions in Eqs. (7.4) and assuming that the function $a_2(t)$ decreases in time according to the law

$$a_2(t) = \exp\left(-\frac{1}{2}\Gamma t\right). \tag{7.5}$$

Integrating Eq. (7.3) with this assumption will take into account the finite width of the excited state. The integration, of course, can be done perturbatively, as above.

It is obvious that even in this case, only the term with $|m\rangle = |2\rangle$ contributes on the right-hand side of Eq. (7.3). Hence the equation for $a_{\omega 1}(t)$ becomes

$$i\frac{da_{\omega 1}}{dt} = \langle \omega, 1|V|2\rangle e^{i(E_1+\omega-E_2)t} a_2 = \langle \omega, 1|V|2\rangle \exp\left[i(\omega - \omega_{12})t - \frac{1}{2}\Gamma t\right],$$

$$\tag{7.6}$$

where $\omega_{12} = E_2 - E_1$. Integrating this equation with the condition $a_{\omega 1}(0) = 0$ gives the amplitude as

$$a_{\omega 1} = i \langle \omega, 1|V|2 \rangle \frac{1 - \exp[i(\omega - \omega_{12})t - (1/2)\Gamma t]}{\omega - \omega_{12} + (1/2)i\Gamma}. \tag{7.7}$$

The squared modulus of this amplitude gives the probability for the transition. For time scales with $\omega_{12}^{-1} \ll t \ll \Gamma^{-1}$, we can set $\Gamma \approx 0$. The squared modulus will then give a factor $\sin^2[(\omega - \omega_{12})t/2]/[(\omega - \omega_{12})/2]^2$. Using the limit of this expression as $t \to \infty$, we can rederive a finite rate of transition obtained in Chap. 4.

We are, however, interested in the limit $\Gamma t \gg 1$. In this case, the exponential term in the numerator of Eq. (7.7) vanishes, and the corresponding rate of transition becomes

$$d\mathcal{R} = |a_{\omega 1}|^2 d\omega \cong |\langle \omega, 1|V|2 \rangle|^2 \frac{d\omega}{(\omega - \omega_{12})^2 + (1/4)\Gamma^2}. \tag{7.8}$$

Because the width $\Gamma \ll \omega$, we can put $\omega = \omega_{12}$ in the matrix element. Then the quantity $2\pi|\langle \omega, 1|V|2 \rangle|^2$ is the probability per unit time for the emission of the photon; we ignore the width of the energy level. This is precisely the result obtained in Chap. 4. Comparison of the expression obtained above with Eq. (4.123) of shows that the Dirac delta function $\delta_D(\omega - \omega_{12})$ is replaced with the spectral line shape ϕ_ω with the form

$$\phi_\omega = \frac{\Gamma}{(\omega - \omega_{12})^2 + (\Gamma/2)^2}, \tag{7.9}$$

which is a Lorentzian. (Note that the integral of ϕ over $\nu = (\omega/2\pi)$ gives unity in this normalisation.) This result gives the spreading of the spectral line that is due to the finite lifetime of the energy level.

In the present analysis we have ignored complications arising from the polarisation of photons, density of states, etc. But these dependences are the same as those that were obtained in Chap. 4, even in the presence of a finite width to the energy level. Further, $|1\rangle$ was taken to be the ground state with zero width. In the case of transitions between two levels, each of finite width, it is easy to show that the above formula is still applicable, with Γ replaced with the sum of the widths.

It only remains to determine the value of Γ. From the definition of the decay rate in Eq. (7.5) it is clear that Γ is the rate of spontaneous transition from level 2 to 1, which was determined in Chap. 4, Section 4.5, to be proportional to $(q^2\omega_{12}^2/mc^3)f_{12}$. If we use the model of a damped harmonic oscillator (discussed in Chap. 6, Section 6.4) we can take

$$\Gamma = \frac{2q^2}{3mc^3}\omega_{12}^2 f_{12}. \tag{7.10}$$

For electric dipole transitions with $f_{12} \approx 1$, a good estimate will be $\Gamma \approx (q^2\omega_{12}^2/mc^3)$.

Given the line-profile function ϕ_ω, we can obtain the bound–bound absorption cross section as

$$\sigma_{bb} = \frac{\pi q^2}{mc} f_{12}\phi_\omega \simeq \frac{\pi(q^2/mc)\Gamma f_{12}}{(\omega - \omega_{12})^2 + (\Gamma/2)^2}. \tag{7.11}$$

This expression shows that the absorption cross section at the centre of the line, at $\omega = \omega_{12}$, is given by $\sigma_{bb}(\omega = \omega_{12}) = (4\pi q^2/mc\Gamma)f_{12} = (3/2\pi)\lambda_{12}^2$, where λ_{12} is the wavelength of the radiation involved; thus the absorption cross section at the centre of a naturally broadened line is of the order of the square of the wavelength involved.

7.2.2 Doppler Width of Spectral Lines

The observed frequency ω and the rest frame frequency ω_0 of a photon, emitted by an atom moving with a velocity v_z along the line of sight, are related by

$$\omega \cong \omega_0 \left(1 + \frac{v_z}{c}\right) \tag{7.12}$$

if $v_z \ll c$ (see Chap. 3, Subsection 3.12.1). In astrophysical situations the emission usually arises from a collection of atoms with a Maxwellian distribution of velocities (corresponding to some temperature T) around some bulk velocity $\bar{v}$. In that case, the probability of observing a photon with frequency ω can be related to the probability of the atom's having a velocity $v_z = (c/\omega_0)(\omega - \omega_0)$. Because v_z and ω are linearly related, the Jacobian of the transformation is a constant and we get

$$\mathcal{P}(\omega) \propto \exp\left(-\frac{1}{2}\frac{M[v_z - \bar{v}]^2}{k_B T}\right)\Bigg|_{v_z = c\left[\frac{\omega - \omega_0}{\omega_0}\right]} \propto \exp\left(-\frac{1}{2}\frac{Mc^2}{k_B T}\left\{\frac{\omega - \bar{\omega}}{\omega_0}\right\}^2\right)$$

$$\equiv \frac{1}{(\Delta\omega)\sqrt{\pi}} \exp\left[-\frac{(\omega - \bar{\omega})^2}{(\Delta\omega)^2}\right], \tag{7.13}$$

where M is the mass of the atom in which the transition occurs and $\bar{\omega} = \omega_0[1 + (\bar{v}/c)]$. In the last equality, we have fixed the proportionality constant by normalising $\mathcal{P}(\omega)$ and using $\Delta\omega \lesssim \bar{\omega}$. It follows that the line profile now is a Gaussian peaked at $\bar{\omega}$ with a width given by

$$(\Delta\omega)^2 = \omega_0^2\left(\frac{2k_B T}{Mc^2}\right), \quad \frac{\Delta\omega}{\omega_0} = \sqrt{\frac{2k_B T}{Mc^2}} \cong 4 \times 10^{-5}\left(\frac{T}{10^4\,\mathrm{K}}\right)^{1/2}\left(\frac{M}{m_H}\right)^{-1/2}. \tag{7.14}$$

We obtain the corresponding full width at half maximum (FWHM) by multiplying this result by $2(\ln 2)^{1/2}$. Note that the line profile arising from a thermal distribution of velocities is quite different from the natural line profile derived in Subsection (7.2.1) above. The natural line profile – being a Lorentzian – falls slowly compared with the Gaussian and hence will dominate away from the resonance frequency.

The absorption cross section corresponding to this line profile, analogous to the one in Eq. (7.11), is

$$\sigma = \frac{\pi q^2}{mc} \frac{f_{12}}{\omega_0} \left(\frac{Mc^2}{2\pi k_B T}\right)^{1/2} \exp\left[-\left(\frac{Mc^2}{2k_B T \omega_0^2}\right)(\omega - \bar{\omega})^2\right]. \quad (7.15)$$

The cross section at the centre of the line is given by $\sigma(\omega = \bar{\omega}) = (\pi q^2 f/m\omega_0)$ $(M/2\pi k_B T)^{1/2}$. The optical depth τ_0 at the line centre is $\tau_0 = NL\sigma = N_c\sigma$, where N_c is the column density of the absorbing atoms. Numerically $\tau_0 = (N_c\lambda f)$ $(q^2/m_e c^2)(c/b) \cong 1.5 \times 10^{-2}(N_c\lambda f/b)$, with $b \equiv (2\pi k_B T/M)^{1/2}$ if all quantities are in cgs units.

Exercise 7.1

Gaussian core and Lorentzian wings: Estimate at how far away from ω_0 the Lorentzian width becomes comparable with the Gaussian width at a given temperature.

The argument that leads to Doppler broadening from thermal motions can also be generalised for other kinds of random velocity fields that could be present in the fluid. In particular, it may be possible to describe turbulent motion of the fluid in terms of a suitably defined mean-square turbulent velocity dispersion v_{tur}^2. Then the total mean-square velocity of thermal and turbulent motion will determine the Doppler linewidth of the spectrum.

Exercise 7.2

Information from Doppler broadening: Three lines (H-α, OII, and OIII) were detected from a gaseous region with the wavelength $\lambda = (6563, 3729, 5007)$Å and linewidths $\Delta\lambda = (0.374, 0.154, 0.207)$Å. Assume that the linewidths are due to a combination of turbulent velocities and thermal Doppler broadening. What is the temperature and turbulent velocity distribution of the line-emitting region? [Answer: When both thermal and turbulent effects are present, $(\Delta\lambda/\lambda)^2 = (\Delta v/v)^2 = (v_{\text{turb}}/c)^2 + (2k_B T/Mc^2)$. In this particular case, $M = Am_p$, where A is the atomic weight of the element, being 1 for hydrogen and 16 for oxygen. $(\Delta\lambda/\lambda)^2 = (3.25, 1.71, 1.71) \times 10^{-9}$ for the three lines. Noting that $(\Delta\lambda/\lambda)^2$ is a linear function of $(1/A)$, we can solve for T and v_{turb}^2. We get $(1 - 1/16)(2k_B T/m_p c^2) = (3.25 - 1.71) \times 10^{-9}$, leading to $T = 8820$ K. Substituting into any one line gives $v_{\text{turb}} = 12$ km s^{-1}.]

7.2.3 Collisional Broadening of Spectral Lines

In Section (7.2.1) above an atomic system undergoing transition between two energy levels differing in frequency by $\omega_{12} \equiv \omega_0$ and having a net width Γ was described. The radiation from such a system can be modelled semiclassically in terms of an electric field that varies with time as $E(t) \propto \exp[i\omega_0 t - (\Gamma t/2)]$.

The Fourier transform of this electric field will lead to a Lorentzian peaked at ω_0 with a width Γ. This is precisely the line profile obtained in Eq. (7.9) above.

This profile will be modified if the atoms emitting the radiation are undergoing random collisions. The primary effect of these collisions will be to change the phase of the electric field of the radiation in a random manner, so that the electric field will now become

$$E(t) = E_0 e^{i\omega_0 t - \Gamma t/2} e^{i\phi(t)}, \tag{7.16}$$

where $\phi(t)$ is a random function of time such that

$$\langle e^{i[\phi(t_1)-\phi(t_2)]}\rangle = \begin{cases} 0 & \text{(if a collsion occurs during } t_1 < t < t_2) \\ 1 & \text{(if no collision occurs during } t_1 < t < t_2) \end{cases}. \tag{7.17}$$

This is equivalent to assuming that each collision changes the phase completely randomly so that the mean value vanishes if a collision has taken place. It follows that

$$\langle e^{i(\phi(t_1)-\phi(t_2))}\rangle = \text{probability that no collision occurs during } t_1 < t < t_2$$
$$= \exp(-\nu_c|t_1 - t_2|), \tag{7.18}$$

where ν_c^{-1} is the mean time between collisions. In arriving at the last equality, we have modelled collisions as a random Poisson process.

To find the line profile in the presence of collisions, we have to take the Fourier transform of Eq. (7.16) and compute its mean value. The average of the Fourier transform is given by [with the notation $\phi_1 = \phi(t_1)$, $\phi_2 = \phi(t_2)$]

$$\langle |E(\omega)|^2\rangle = |E_0|^2 \left\langle \int_0^\infty dt_1 e^{-\Gamma t_1/2 + i\phi_1 + i(\omega_0-\omega)t_1} \int_0^\infty dt_2 e^{-\Gamma t_2/2 - i\phi_2 - i(\omega_0-\omega)t_2}\right\rangle$$

$$= |E_0|^2 \int_0^\infty dt_1 dt_2 e^{-(\Gamma/2)(t_1+t_2)}\langle e^{i(\phi_1-\phi_2)}\rangle e^{i(\omega_0-\omega)(t_1-t_2)}. \tag{7.19}$$

Changing the variables to $v \equiv t_1 + t_2$ and $u \equiv t_1 - t_2$ and using Eq. (7.18), we find that the integral becomes

$$\langle |E_\omega|^2\rangle = \frac{|E_0|^2}{2}\int_{-\infty}^\infty du \int_{v=|u|}^\infty dv e^{-\Gamma v/2} e^{-\nu_c|u|} e^{i(\omega_0-\omega)u}$$

$$= \frac{|E_0|^2}{\Gamma}\int_{-\infty}^\infty du e^{-\nu_c|u| + i(\omega_0-\omega)u} e^{-\Gamma|u|/2}. \tag{7.20}$$

This result shows that the effect of collisions is merely to change the natural width of the line by

$$\Gamma/2 \rightarrow \Gamma/2 + \nu_c = \Gamma'/2.$$

The line profile remains Lorentzian with an increased width.

When all the three sources of broadening are present, the net line profile takes a fairly complicated form. We can determine it by taking the Lorentzian profile that is due to natural and collisional broadening and convolving with the probability distribution for the velocities. If an atom has a velocity v_z along the line of sight, then its line profile will be given by $\phi_L[\omega - \omega_0(1 + v_z/c)]$, where ϕ_L is the Lorentzian profile with a total width $\Gamma'/2 = \Gamma/2 + v_c$. Because the probability for the velocity to be v_z is given by a Maxwellian distribution at some temperature T, the net line profile is proportional to the integral

$$\phi_{\text{tot}}(\omega) \propto \int \phi_L \left(\omega - \omega_0 \left[1 + \frac{v_z}{c} \right] \right) \mathcal{P}(v_z) \, dv_z$$

$$\propto \int dv_z \frac{\exp(-Mv_z^2/2k_BT)}{\left[(\omega - \omega_0 - [\omega_0 v_z/c])^2 + (\Gamma'/2)^2 \right]}. \tag{7.21}$$

For simplicity, we have assumed that there is no bulk flow, i.e., $\bar{v} = 0$. With a series of simple transformations, the integral in expression (7.21) can be expressed in the form

$$\phi_{\text{tot}}(\omega) = \frac{2}{\sqrt{\pi}} \frac{a}{(\Delta\omega)_D} \int_{-\infty}^{\infty} dq \frac{e^{-q^2}}{(q-u)^2 + a^2}, \tag{7.22}$$

where

$$u = \left(\frac{\omega - \omega_0}{\omega_0} \right) \left(\frac{c}{v_0} \right), \qquad a = \left(\frac{\Gamma'}{4\omega_0} \right) \left(\frac{c}{v_0} \right),$$

$$v_0 = \sqrt{\frac{2k_BT}{M}}, \qquad (\Delta\omega)_D = \omega_0 \left(\frac{v_0}{c} \right). \tag{7.23}$$

The constant in front of the integral in Eq. (7.22) is chosen such that $\phi(\omega)$ is normalised to unity with respect to integration over $\nu = (\omega/2\pi)$. The function on the right-hand side of Eq. (7.22) is called the Voigt function and is plotted in Fig. 7.1 for different values of a. The asymptotic forms of this function can be determined as follows: Writing $q = xu$ and changing the variable of integration to x, we get

$$\phi = \frac{2}{\sqrt{\pi}} \frac{1}{(\Delta\omega)_D} \left(\frac{a}{u} \right) \int_{-\infty}^{\infty} dx \frac{e^{-x^2u^2}}{(x-1)^2 + (a/u)^2}. \tag{7.24}$$

For $u \gg 1$, most of the contribution to the integral comes from a region around the origin $(x = 0)$ of width $\Delta x \simeq (1/u) \ll 1$. Hence we can set $(x-1)^2 \approx 1$ in the denominator and perform the integration to get

$$\phi \cong \frac{2}{\Delta\omega_D} \frac{a}{u^2 + a^2} \quad (u \gg 1), \tag{7.25}$$

which is a Lorentzian. To obtain the limiting form for small u, we transform the

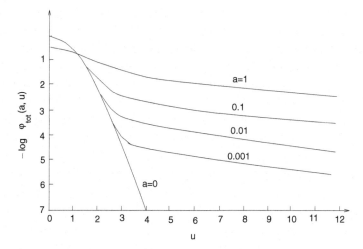

Fig. 7.1.   The Voigt profile as a function of u for different values of a.

integral to the form

$$\phi = \frac{2}{\sqrt{\pi}} \frac{a}{(\Delta\omega)_D} \int_{-\infty}^{\infty} dy \, \frac{e^{-(u+y)^2}}{y^2 + a^2}$$

$$= \frac{2}{\sqrt{\pi}} \frac{a}{(\Delta\omega)_D} \int_{-\infty}^{\infty} dy \int_0^{\infty} d\mu \, e^{-(u+y)^2 - \mu(y^2 + a^2)}. \qquad (7.26)$$

Performing the y integration first, we get

$$\phi = \frac{a}{\pi^{3/2}} e^{-u^2} \int_0^{\infty} \frac{d\lambda}{(1+\lambda)^{1/2}} e^{-\lambda a^2 - [u^2/(1+\lambda)]}. \qquad (7.27)$$

The $\exp(-\lambda a^2)$ suppresses the integral for all $\lambda \gtrsim a^{-2}$. So if $u^2 \lesssim (1 + a^{-2})$, we can ignore the u dependence of the integral and get $\phi \propto \exp(-u^2)$. Normalising, we get

$$\phi \cong \frac{2\sqrt{\pi}}{(\Delta\omega)_D} e^{-u^2} \qquad \left(u^2 \ll 1 + \frac{1}{a^2}\right), \qquad (7.28)$$

which is a Gaussian. Thus near the origin the function is dominated by the Gaussian and varies as $\exp(-u^2)$, whereas near the edges it is dominated by the Lorentzian and varies as u^{-2}. A good approximation to the Voigt profile with correct asymptotic limits is given by the sum of relations (7.25) and (7.28):

$$\phi = \left(\frac{2\sqrt{\pi}}{\Delta\omega_D}\right) \left(e^{-u^2} + \frac{a}{\sqrt{\pi} u^2}\right). \qquad (7.29)$$

This profile is used extensively in the study of spectral lines from astrophysical systems.

All the results of Chap. 6 involving the line-profile function ϕ_ν can now be recast in terms of Doppler, Lorentzian, or Voigt profiles, depending on the context. In particular, the maximum intensity of the line [at the line centre, which is infinite for a line of zero width with $\phi_\nu = \delta_D(\nu - \nu_{12})$] is determined by $\phi_\nu(\nu_{12})$. For the Doppler and the Lorentzian profiles, these are given by

$$\phi(\nu_{12}) = \begin{cases} \left(\dfrac{\ln 2}{\pi}\right)^{1/2} \left(\dfrac{2}{\Delta\nu_D}\right) \\[2ex] \dfrac{2}{\pi}\left(\dfrac{1}{\Delta\nu_L}\right) \end{cases} . \tag{7.30}$$

To a good approximation, $\phi(\nu_{12}) \approx (\Delta\nu)^{-1}$. The results regarding the absorption and emission of radiation by systems in thermal equilibrium [see Eqs. (6.33) and (6.39)] now become

$$\alpha^{\text{net}}(\nu_{12}) = \left(\frac{\pi q^2 f_{12}}{mc}\right)n_1\left(\frac{1}{\Delta\nu}\right)\left(1 - e^{-h\nu/k_B T}\right) = \begin{cases} \left(\dfrac{\pi q^2 f_{12}}{mc}\right)\left(\dfrac{1}{\Delta\nu}\right)\left(\dfrac{h\nu_{12}}{k_B T}\right)n_1 \\[2ex] \left(\dfrac{\pi q^2 f_{12}}{mc}\right)\left(\dfrac{1}{\Delta\nu}\right)n_1 \end{cases} \tag{7.31}$$

in the case of the absorption coefficient (where the two cases are for $h\nu \ll k_B T$ and $h\nu \gg k_B T$, respectively) and

$$I_\nu(R) \cong \begin{cases} B_\nu(T)\,n_1\,R\,\left(\dfrac{\pi q^2 f_{12}}{mc}\right)\left(\dfrac{h\nu_{12}}{k_B T}\right)\left(\dfrac{1}{\Delta\nu}\right) & (h\nu \ll k_B T) \\[2ex] B_\nu(T)\,n_1\,R\,\left(\dfrac{\pi q^2 f_{12}}{mc}\right)\left(\dfrac{1}{\Delta\nu}\right) & (h\nu \gg k_B T) \end{cases} \tag{7.32}$$

in the case of optically thin emission. At radio frequencies, the equivalent brightness temperature at the line centre will be

$$T_B(\nu_{12}) = \left(\frac{n_1 R}{\Delta\nu}\right)\left(\frac{\pi q^2 f_{12}}{mc}\right)\left(\frac{h\nu_{12}}{k_B T}\right) \quad (h\nu \ll k_B T). \tag{7.33}$$

Exercise 7.3
Numerical estimate: Estimate the natural, Doppler, collisional width of a typical spectral line arising from solar atmosphere.

7.3 Curve of Growth

It is clear that the shape of the spectral line contains useful information about the physical characteristics of the system that is emitting or absorbing the radiation. To extract this information in a methodical form, astronomers use two concepts, called the equivalent width of a line and the curve of growth. We will now introduce these concepts.

For the sake of definiteness, let us consider an absorption line with a line profile given by the Voigt function $\phi(\omega)$ centred at the frequency ω_0. The intensity of this line at a frequency ω can be expressed as $I(\omega) = I_0 \exp[-\tau(\omega)]$, where I_0 is the intensity at the centre of the line and $\tau(\omega) = NL\sigma(\omega)$ is the optical depth of the medium for absorption, where N gives the number density of the absorbers and L is the effective optical path length. Because $\sigma(\omega)$ is proportional to the line-profile function [see Chap. 4, Eq. (4.135)] we can write the intensity as $I(\omega) = I_0 \exp[-C\phi(\omega)]$, where C is a constant independent of ω but proportional to the number density of the absorbing atoms. The equivalent width of the line is defined by

$$W[N] = \int dv \left[\frac{I_0 - I(v)}{I_0} \right] = \int dv [1 - e^{-C\phi(v)}], \tag{7.34}$$

where the integral is taken over the line. [This is a measure of the total amount of intensity removed from the vicinity of the line; hence it is independent of the spectral resolution achieved by the detector as long as the lines are distinct. In observational astronomy, it is conventional to define this integral over $v = (\omega/2\pi)$ rather than over ω.] Geometrically the equivalent width gives the width of a rectangular spectral line with the same area as the actual spectral line when both the lines are normalised to unity at large frequencies. The variations of the equivalent width with the constant C (which is proportional to the concentration of absorbing atoms) is of considerable interest and is called the curve of growth. Because the form of $\phi(\omega)$ is complicated, the integral for $W(C)$ cannot be done in closed form. However, the general behaviour of $W(C)$ can be ascertained along the following lines.

Consider first the case in which the concentration of absorbers is low so that $\tau_v = C\phi(v) \ll 1$ at the relevant frequencies. In that case, the exponential can be expanded in Taylor series and we get

$$W[N] \cong \int dv \, \tau_v = NL \int \sigma_v \, dv \cong NL(\pi q^2 f/mc), \tag{7.35}$$

where we have used Eq. (4.136). The curve of growth $W(N)$ is proportional to N for a low concentration of absorbers. When the optical depth is large, $\tau \gg 1$ near $v = v_0$; away from the line centre, the optical depth will decrease with a Gaussian profile in the region dominated by Doppler broadening and with a Lorentzian profile in the region dominated by collisional or natural broadening. In either case, the integrand $[1 - \exp(-\tau_v)]$ in Eq. (7.34) acts as a step function, being nearly unity for $\tau_v \gg 1$ and vanishing for $\tau_v \ll 1$. Hence the equivalent width can be approximated as $W(N) \approx (2\Delta)$, where Δ is the deviation from the line centre to the frequency at which $\tau_v = NL\sigma_v = 1$. This quantity Δ can now be determined from the form of σ_v given by Eq. (7.15) for Doppler broadening and Eq. (7.11) for natural broadening. In the former case, an elementary calculation gives

$$W[N] \approx 2c_1 [\ln(NLc_2/c_1)]^{1/2}, \quad c_1 = v_0 \left(\frac{2\pi k_B T}{Mc^2} \right)^{1/2}, \quad c_2 = \left(\frac{\pi q^2 f}{mc} \right),$$

(7.36)

which shows that W varies as $\sqrt{\ln N}$ when Doppler broadening is dominant. A corresponding calculation with Eq. (7.11) approximated as $\sigma_v \approx c_2 \Gamma / [2\pi (v - v_{12})]^2$, far away from the line centre, gives

$$W[N] \approx \frac{(NLc_2\Gamma)^{1/2}}{\pi},$$

(7.37)

which is proportional to $\sqrt{N}$. In summary, the curve of growth $W(N)$ varies as N for low values of N, as $(\ln N)^{1/2}$ for moderate N, and as $N^{1/2}$ for large N.

Exercise 7.4

Saddle-point evaluation of curve of growth: Evaluate the integral of Eq. (7.34) in the saddle-point limit and show that W varies as $\sqrt{\ln N}$ in the Doppler regime and as $\sqrt{N}$ in the regime dominated by the natural linewidth.

In astronomical spectroscopy, we are often interested in determining the quantity N that contains information about the abundance of absorbing or emitting atoms of a particular species from the knowledge of the spectral line. From the spectral line, we can determine the equivalent width W and try to read off N from the curve of growth. (In practice, N cannot be varied for a given source. However, because only the combination Nf occurs in the equivalent width, we can determine W as a function of f for several different lines with different f. Fitting these observed values and treating N as an unknown parameter allows us to determine N.) This can be done fairly accurately for low values of N, where $W \propto N$. At moderate values of N, W changes very little for large variation in N because the dependence is logarithmic. This implies that when we attempt to determine N from W, a small error in W can lead to a large error in N and the result will be inaccurate. These concepts are used in the study of stellar atmospheres and intergalactic medium.

We have defined the equivalent width as an integral over the spectral line in frequency space, because of which $W(N)$ has the dimension of inverse time. In conventional spectroscopy, the equivalent width is defined as an integral over the line in the wavelength space, in which dv of Eq. (7.34) is replaced with $(cd\lambda/\lambda^2)$. If we denote the integral of $d\lambda$ over the line by W_λ, then it is easy to show that $W_\lambda \approx (\lambda^2/c)W(N)$. In the linear regime, this will give

$$\frac{W_\lambda}{\lambda} \approx (NLf\lambda) \left(\frac{\pi q^2}{m_e c^2} \right) = 8.85 \times 10^{-18} f \left(\frac{\lambda}{1000 \text{ Å}} \right) \left(\frac{N_c}{1 \text{ cm}^{-2}} \right), \quad (7.38)$$

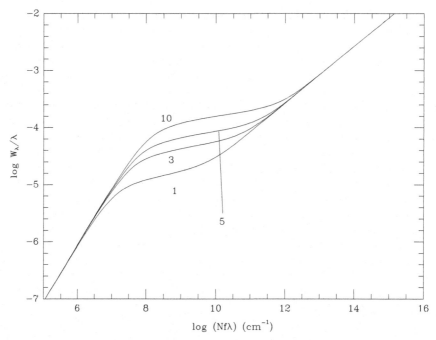

Fig. 7.2.   Equivalent width as a function of column density obtained by exact numerical integration for a few values of Doppler width b in kilometres per second.

where $N_c \equiv NL$ is the column density. Correspondingly, in the flat portion of the curve of growth we get

$$\frac{W_\lambda}{\lambda} \approx \frac{2b}{c} \quad b = \left(\frac{2\pi k_B T}{M}\right)^{1/2} \tag{7.39}$$

if the logarithmic dependence on the column density is ignored. Finally, in the wings of the line,

$$\frac{W_\lambda}{\lambda} \approx \left(\frac{N_c \lambda^2 f \Gamma}{c}\right)^{1/2} \left(\frac{q^2}{m_e c^2}\right)^{1/2}. \tag{7.40}$$

Figure 7.2 gives (W_λ/λ) as a function of $N f \lambda$ for different values of the Doppler velocity b. As to be expected, the linear part and the wings are independent of b whereas the in-between logarithmic part varies with b.

Exercise 7.5

Practice with equivalent width: The following data are based on absorption lines of a cloud that is assumed to be fitted by a single curve of growth. For the four lines from HI, NI(a), NI(b), and CII with $\lambda = 1216, 1201, 1200,$ and $1336\,\text{Å}$, respectively, the equivalent widths are $W_\lambda = 8.96, 0.077, 0.083,$ and $0.0018\,\text{Å}$. The oscillator strengths and widths for these transitions are $f = 0.416, 0.0442, 0.0884,$ and 0.0227 and

$\Gamma = 6.27, 4.09, 4.10, 4.97 \times 10^8$ Hz. Determine on which part of the curve of growth these lines fall. How can we use the data to estimate the column densities of hydrogen, nitrogen, and chlorine? [Answer: From the data, we compute W_λ/λ for HI, NI(a), NI(b), and ClII to be $7.37 \times 10^{-3}, 6.41 \times 10^{-5}, 6.92 \times 10^{-5}$, and 1.35×10^{-6}, respectively. The NI(a) and NI(b) lines differ by almost a factor of 2 in the f values but have nearly same W_λ/λ. Hence they must be in the flat part of the curve. HI has a much larger W_λ/λ and hence will be on the damping wing of the line, whereas ClII, having a smaller W_λ/λ, must be on the linear part of the curve of growth. This allows us to determine the column densities of ClII and HI easily. For ClII, $W_\lambda/\lambda \approx 8.85 \times 10^{-13} N_c \lambda f$ (in cgs), giving $N_c \approx 5 \times 10^{12}$ cm^{-2}. For HI, using the fact that $(W_\lambda/\lambda) \propto \lambda(N_c f \Gamma)^{1/2}$ and plugging in the constants appropriately gives $N_c \approx 1.5 \times 10^{20}$ cm^{-2}. Because NI lies in the flat part of the curve, it is difficult to determine its column density accurately, but the basic procedure is as follows. We first note that NI(a) and NI(b) form a doublet with f values differing approximately by factor of 2. If τ_0 is the centre optical depth of the 1201-Å line, it follows that $W(2\tau_0)/W(\tau_0) \approx (6.92/6.41) \approx 1.079$. Given the curve of the growth line, this equation can be solved to estimate τ_0 and the velocity b. By using Fig. 7.2, we estimate $b \approx 5$ km s^{-1}. When the width of the line is essentially due to Doppler spread, the optical depth at the centre of the line is related to b through the equation $\tau_0 \approx 1.5 \times 10^{-2}(N_c \lambda f/b)$. Using the known values of τ_0 and b, we get $N_c \lambda f$ and thus determine N_c. In this case we obtain $N_c \approx 3 \times 10^{15}$ cm^{-2}.]

7.4 Atomic-Energy Levels

Because the radiation emitted in a transition between two energy levels has a characteristic frequency related to the energy difference between the levels, it is important to have a clear understanding of the different energy levels of the system in order to use the observed spectrum as a diagnostic. This requires understanding of the energy levels of the atomic and the molecular systems and the properties of the transitions between different energy levels in this system. We now address these questions.

7.4.1 Energy Levels in the Nonrelativistic Theory

The simplest of all such systems is the hydrogen atom, which we begin with. Similar considerations are applicable to the energy levels of more complex atoms whenever the electron that is under consideration can be assumed to move in a Coulomb field of the nuclei. In the lowest order of approximation, the interaction potential coupling the electron to the nucleus is $V_{Coul}(r) = -(Ze^2/r)$. In this case the energy levels are the expression

$$E_n = -Z^2 \frac{m_e e^4}{2n^2\hbar^2} \equiv -\frac{Z^2 \mathcal{R}}{n^2}, \tag{7.41}$$

where $\mathcal{R} = (m_e e^4/2\hbar^2) \approx 13.6$ eV. Though $Z = 1$ for hydrogen, we explicitly retain the Z dependence because many ions, such as He^+, have a hydrogenlike structure.

A quantum state of the electron in this approximation is given by three integers, l, m, and n. The fact that the energy levels are independent of m is a consequence of spherical symmetry, but the fact that E_n's are independent of l is a feature special to Coulomb interaction that arises because of the existence of an extra conserved operator in quantum theory, which is analogous to the Runge–Lenz vector in the classical Kepler problem (see Chap. 2, Section 2.5).

The transitions between different energy levels in the hydrogen atom are usually described by a special terminology. All transitions to $n = 1$ are said to form a series of spectral lines called the Lyman series, transitions to $n = 2$ are called the Balmer series, and transitions to $n = 3$ are called the Paschen series. The transition is denoted by a subscript α when the principle quantum number changes by unity, by β when it changes by 2, etc. For example, the Lyman α line is produced in a transition from $n = 2$ to $n = 1$ and has a wavelength of $L_\alpha = 1216$ Å; the Balmer α line (also called the H_α line) corresponds to the transition between $n = 3$ and $n = 2$ and has a wavelength of $H_\alpha = 6563$ Å. The H_β line arises because of a transition between $n = 4$ and $n = 2$ (with $H_\beta = 4861$ Å), the H_γ line arises in the transition between $n = 5$ and $n = 2$ (with $H_\gamma = 4340$ Å), and the H_δ line arises from $n = 6$ to $n = 2$ (with $H_\delta = 4101$ Å). These lines, H_α, H_β, H_γ, and H_δ, that belong to the Balmer series lie in the visible part of the spectrum and hence are of special importance. The transition rate $A_{nn'}$ for a hydrogen atom varies around 10^8 s^{-1} for low-order strong transitions but can be quite small for large values of n. For example, the recombination line of hydrogen between $n = 110$ and $n' = 109$ occurs at a wavelength of 6 cm with a transition probability of 0.38 s^{-1}.

The series of spectral lines described above, of course, correspond to transitions between discrete energy levels. It is also possible for the hydrogen atom to be ionised with the absorption of a photon, causing the electron to make a transition from a discrete level to the continuum. For example, the ionisation energy for the $n = 2$ level is 3.4 eV and the corresponding wavelength is 3647 Å. Therefore a transition from $n = 2$ to the continuum can absorb all photons with wavelengths shorter than 3647 Å. Similarly, the capture of a free electron by a proton to the level $n = 2$ will lead to the emission of photons at wavelengths shorter than 3647 Å. This emission, called the Balmer continuum, will exist at all wavelengths short of 3647 Å. Similarly, we have the Lyman continuum starting at 912 Å and the Paschen continuum starting at 8204 Å.

Line radiation will also be emitted when an electron and ion recombine to form a bound neutral atom, a process that corresponds to a transition between $n = \infty$ and some finite value of n. This process leads to a series of lines, usually called recombination lines, that arise because the electron cascades down to lower energy levels after forming an excited state of the atom. The characteristics of the recombination lines depend on the physical conditions rather critically and – in particular – on whether the material can be considered to be in local thermodynamic equilibrium. These issues will be discussed in Vols. II and III within the appropriate astrophysical contexts.

One variant of recombination line is the radio recombination line, which arises because of transitions between high-n states of hydrogenlike atoms. Lines corresponding to $n + 1 \to n$ are strongest and are called α lines; those for transitions $n + 2 \to n$ are called β lines, etc. In the standard notation used by astronomers, H 109 α is a line corresponding to the transition $110 \to 109$ of hydrogen, and He 127 β corresponds to the transition $129 \to 127$ of helium. The α transitions with $n > 160$ in most of the abundant elements produce lines in the radio band with $\lambda > 1$ cm. For such high values of n, the transitions rates can be determined by a semiclassical analysis. The rate of transition for $n + 1 \to n$ for large n is (see Exercise 7.6)

$$A_{n+1,n} \simeq \frac{1}{3} \frac{me^{10}}{c^3 \hbar^6} \frac{1}{n^5} \simeq \frac{5.36 \times 10^{10}}{n^5} \, \text{s}^{-1}. \tag{7.42}$$

In this case, the linewidth is essentially determined by Doppler broadening and the shape of the line is a Gaussian, with the full Doppler width at half intensity given by

$$\frac{\Delta \nu}{\nu} = \frac{2}{c} \left[2 \left(\frac{k_B T_e}{M} + v_t^2 \right) \ln 2 \right]^{1/2}, \tag{7.43}$$

where v_t^2 is the velocity dispersion that is due to turbulence, if it is present. For such a line shape, the value of line profile at the centre of the line will be [see Eq. (7.30)]

$$\phi(0) = \left(\frac{\ln 2}{\pi} \right)^{1/2} \frac{2}{\Delta \nu}. \tag{7.44}$$

Given the number densities N_e and N_i of electrons and ions, respectively, in a given temperature T_e, we can compute the optical depth τ_L (at the centre of the line) from a plasma in which such radio recombination lines are produced. This calculation is straightforward, and the result can be expressed in terms of the emission measure of the plasma as

$$\tau_L = 1.92 \times 10^3 \left(\frac{T_e}{K} \right)^{-5/2} \left(\frac{\text{EM}}{\text{cm}^{-6} \, \text{pc}} \right) \left(\frac{\Delta \nu}{\text{kHz}} \right)^{-1}, \tag{7.45}$$

where EM is the emission measure of the plasma defined as

$$\text{EM} = \int N_e(s) N_i(s) \, ds = \int \left[\frac{N_e(s)}{\text{cm}^{-3}} \right]^2 d \left(\frac{s}{\text{pc}} \right) \tag{7.46}$$

if $N_e = N_i$. Because $\tau_L \ll 1$ in most contexts, the brightness temperature of the line [see Chap. 6, relation (6.40)] is given by $T_L \approx T_e \alpha_\nu R \simeq T_e \tau_L$. Numerically

$$T_L = 1.92 \times 10^3 \left(\frac{T_e}{K} \right)^{-3/2} \left(\frac{\text{EM}}{\text{cm}^{-6} \, \text{pc}} \right) \left(\frac{\Delta \nu}{\text{kHz}} \right)^{-1}. \tag{7.47}$$

Such a plasma will also be emitting continuum radiation, which can be modelled as a thermal bremsstrahlung, as described in Chap. 6, Section 6.9. For a pure hydrogen plasma, we can obtain the brightness temperature of the line compared with that of the continuum by dividing the above expression by the corresponding results for thermal bremsstrahlung; this is given by

$$
\left(\frac{T_L}{T_c}\right) \simeq 2.3 \times 10^4 \left(\frac{\Delta \nu}{\text{kHz}}\right)^{-1} \left(\frac{\nu}{\text{GHz}}\right)^{2.1} \left(\frac{T_e}{\text{K}}\right)^{-1.15}, \tag{7.48}
$$

where we have used the fitting function in Eq. (6.240). The recombination line is often used as a good diagnostic of astrophysical plasmas.

Exercise 7.6
Semiclassical transition rates: The change in the dipole moment in a transition $n + 1 \rightarrow n$ can be approximated (for $n \gg 1$) to be $ea_n/2$, where a_n is the radius of the nth orbital. Using this expression and classical radiation formula, obtain the result in relation (7.42).

Exercise 7.7
Filling in the blanks: (1) Derive Eqs. (7.45)–(7.48) when the plasma is in thermal equilibrium. (2) When the plasma has an optical depth τ_c for the continuum and τ_L for the line radiation, show that the brightness temperature of the line alone is given by $T_L = T_e e^{-\tau_c}(1 - e^{-\tau_L})$. What is the optimum frequency at which we could attempt to detect the line radiation?

Exercise 7.8
Radio recombination lines: (1) Determine the frequency of the H $109\,\alpha$ line and find a β and γ line of hydrogen recombination with nearby frequencies. (2) Radio recombination lines can allow us to determine the ratio of helium to hydrogen abundance in HII regions if we measure, say, the $109\,\alpha$ line from recombination of ionised hydrogen and from the recombination of singly ionised helium. Determine the frequency difference between them and find the equivalent Doppler velocity shift corresponding to this frequency difference. Will it be easy to separate these lines in a spectrum? [Answers: (1) For hydrogenlike recombination lines, $\nu = 6.58 \times 10^9$ Hz $(\Delta n/n^3)$. For $n = 109$, $\Delta n = 1$, this gives $\nu = 5007$ MHz. To find a β line close to this frequency, we look for an integer n_1 such that $(2/n_1^3) = (1/109^3)$, giving $n_1 = 137$. This is a H $137\,\beta$ line with frequency $\nu = 5005$ MHz. Similarly, for a γ line, we take $(3/n_2^3) = (1/109^3)$ giving $n_2 = 157$. This is a H $157\,\gamma$ line with a frequency $\nu = 4955$ MHz. (2) If the nuclear mass is Am_p, the reduced mass μ for a hydrogenlike configuration will be $\mu = Am_p m_e/(m_e + Am_p) = m_e[1 + (m_e/Am_p)]^{-1}$. Because the energy levels scale as $E_n = (1/2)\mu c^2 (Z\alpha)^2$, the energy levels will change slightly with A. For the $109\,\alpha$ line of hydrogen and helium, we get $\Delta \nu = \nu_{\text{He}} - \nu_H \approx 2.07$ MHz. The corresponding Doppler velocity $v = c(\Delta \nu/\nu) \approx 124$ km s^{-1}. At a temperature of ~ 10 K, the Doppler velocities are $b \approx 12$ km s^{-1}. Because $v \gg b$ the lines are easily resolved.]

There is another important complication in the case of a hydrogen atom that has an astrophysical significance. A neutral hydrogen atom does have a residual electromagnetic field in its vicinity because the electron in a given quantum state cannot completely cancel the electric field of the proton. The residual electric field is attractive and is, in fact, strong enough to allow the *neutral* hydrogen atom to capture another electron and become a negative hydrogen ion. It is possible to have a bound state made up of two electrons and a proton, making a H^- ion with a binding energy of ~0.75 eV. (The presence of an extra electron polarises the hydrogen atom, leading to a residual electric field in the vicinity that can, in turn, bind the electron.) The bound–free absorption from this state gives rise to a continuous opacity at wavelengths shorter than 16 502 Å, corresponding to the energy 0.75 eV. This can be a very significant source of opacity in cool stars such as the Sun.

7.4.2 Fine Structure of Energy Levels

The actual potential felt by the electron in the hydrogen atom, however, is not just $V_{Coul}(r)$, for several reasons. The electron carries a spin $\hbar/2$ and a magnetic moment associated with it, that lead to corrections to the Hamiltonian over and above the nonrelativistic expression $H_{nr} = (p^2/2m) + V_{Coul}$. The expression for kinetic energy, $p^2/2m$, also picks up higher-order corrections from the full relativistic expression for kinetic energy. The correct theory of an electron moving in the hydrogen atom can be obtained from the Dirac equation describing the electron that automatically takes into account the relativistic corrections as well as the effects that are due to spin. The results of the relativistic theory are briefly described below.

In the nonrelativistic theory, a level with principal quantum number n can have the orbital quantum number l ranging through the values $l = 0, 1, \ldots, n-1$. This holds true even in the relativistic theory. The total angular momentum of the electron j will be the sum of the orbital angular momentum l and the spin angular momentum $s = \pm(1/2)$. Hence for any given state labelled by n and l, j can take the values $j = l \pm (1/2)$. For historical reasons, states with $l = 0, 1, 2, \ldots$, are denoted by the letters S, P, D, $F \ldots$. With this notation, a quantum state is usually specified in the form $n^{2s+1}A_j$, where A stands for any of the letters S, P, D, F, etc. For example, the state $3\,^2P_{3/2}$ will have $n = 3$, $l = 1$ (because it is an P state), $j = 3/2$, and $s = 1/2$ (because $2s + 1 = 2$). The energy levels in the full relativistic theory are given by

$$E_{nj}^{exact} = mc^2[(1 + Z\alpha\{n - j - 1/2 + [(j + 1/2)^2 - Z^2\alpha^2]^{1/2}\}^{-1})^{-1/2} - 1],$$

$$(7.49)$$

where $\alpha = (e^2/\hbar c) \approx 10^{-2}$ is the fine-structure constant. The energy levels now depend on both n and j – in contrast to the nonrelativistic theory in which the energy levels depended on only n. Because $Z\alpha \ll 1$ for hydrogen, we can expand this expression in a Taylor series in $Z\alpha$ to obtain the lowest-level corrections to the energy. This gives the energy levels as

$$E_{nj} \equiv E_n + \Delta E_{nj} = E_n \left[1 + \frac{(Z\alpha)^2}{n^2} \left(\frac{n}{j+1/2} - \frac{3}{4} \right) \right],$$ (7.50)

where the second term in the brackets gives the fractional corrections to the energy levels and is seen to be of $\mathcal{O}(\alpha^2)$.

The above expression shows that the relativistic corrections and the spin of the electron introduce a splitting in the original energy levels. (This splitting is said to produce a fine structure to the energy levels, and a simple order-of-magnitude estimate of this phenomenon was given in Chap. 1, Subsection 1.4.1.) For example, consider the state with $n = 2$ and $l = 1$, which has a energy of 3.4-eV in the nonrelativistic case. With the spin of the electron taken into account, this state can have the total angular momentum $j = 1/2$ or $j = 3/2$. These two states, denoted by $2\,{}^2P_{1/2}$ and $2\,{}^2P_{3/2}$, differ in their j values and will have different energies in the correct relativistic theory. From Eq. (7.50) we find the energies of these levels as

$$\Delta E_{j-1/2} = -\frac{5}{64}\alpha^2 \mathcal{R} = -5.66 \times 10^{-5} \text{ eV},$$

$$\Delta E_{j-3/2} = -\frac{1}{64}\alpha^2 \mathcal{R} = -1.1 \times 10^{-5} \text{ eV}.$$ (7.51)

Thus the relativistic corrections splits the $n = 2$ energy level into two levels separated in energy by 4.5×10^{-5} eV. The transitions from the $n = 3$ to the $n = 2$ states can now produce spectral lines that differ slightly in wavelength because the final state of the electron could be either of the two energy levels. Because the H_α line has a wavelength of 6563 Å, the typical separation of wavelength between the two lines will be approximately $10^{-5} \times 6563$ Å $= 0.06$ Å. This is an extremely tiny separation and is observable only at high resolution. For more complicated atoms, however, the fine structure is easily visible; for example, the sodium D lines are ~ 6 Å apart.

The fine structure that is due to relativistic corrections leads to energy levels that depend on only n and j. Hence we expect that states such as, say, $2\,{}^2S_{1/2}$ and $2\,{}^2P_{1/2}$ will have the same energy even after these corrections are taken into account. Quantum field-theoretic corrections to the interactions of electrons and photons, which are neglected in the Dirac equation for the single electron, change the energy of these two levels because of phenomena called radiative corrections. The order of magnitude of these corrections is $\Delta E \approx (mc^2)Z^4\alpha^5 \ln(1/\alpha)$. The $2\,{}^2S_{1/2}$ level becomes higher than the $2\,{}^2P_{1/2}$ level, with the splitting corresponding to a frequency of ~ 1057 MHz. This splitting is called the Lamb shift.

The following point should be noted. In the Taylor expansion of Eq. (7.49) the next-order correction will be of the order of $(mc^2)(Z\alpha)^6$. The ratio of this term to the radiative correction is approximately $(Z^2\alpha)/\ln(1/\alpha)$, which is a small number for hydrogen. In other words, the exact expression in Eq. (7.49) is *not* of significance for second and higher orders because the field-theoretic corrections dominate at these orders.

7.4.3 *Hyperfine Structure of Energy Levels*

The fine structure of energy levels discussed above arises because of the coupling of the electron spin with its orbital motion as well as from the relativistic corrections to the expression for kinetic energy. A further splitting of energy levels of a hydrogen atom occurs because the spin of the electron also couples to the spin of the nucleus.

The electron and the proton have magnetic moments given by

$$\mathbf{m} = -\frac{g_e e}{2m_e c}\mathbf{s}, \quad \mathbf{M} = -\frac{g_p e}{2m_p c}\mathbf{S}, \tag{7.52}$$

where $g_e \approx 2.0023$, $g_p \approx 5.586$, and $\mathbf{s}$ and $\mathbf{S}$ denote the spins of the electron and the proton, respectively. The magnetic moment of the proton will generate a magnetic field

$$\mathbf{B} = \frac{3(\mathbf{M} \cdot \hat{\mathbf{k}})\hat{\mathbf{k}} - \mathbf{M}}{|\mathbf{x}|^3} \tag{7.53}$$

at the location of the electron, which will couple to the magnetic moment of the electron through an interaction Hamiltonian of the form $H_{\text{spin-spin}} = -\mathbf{m} \cdot \mathbf{B}$. In the presence of such a coupling, neither spin will be conserved, and it is convenient to work with states labelled by the total spin $\mathbf{F} = \mathbf{S} + \mathbf{s}$. The contribution to the energy that is due to the spin–spin coupling can be computed in a straightforward manner by use of quantum-mechanical perturbation theory. The result is

$$E_m = g_e g_p \alpha^4 \left(\frac{m_e}{m_p}\right)(m_e c^2)\left\{\frac{1}{3}\left[F(F+1) - \frac{3}{2}\right]\right\}$$

$$\equiv E_{ss}\left\{\frac{1}{3}\left[F(F+1) - \frac{3}{2}\right]\right\}. \tag{7.54}$$

This shows that the energy of the system is different for the state $F = 1$ (corresponding to parallel spins) and the state $F = 0$ (corresponding to antiparallel spins). We get $E(F = 1) = (E_{ss}/6)$ and $E(F = 0) = -(E_{ss}/2)$. This splitting of the ground state of the hydrogen atom that is due to spin–spin coupling is called hyperfine splitting. The transition between the two levels leads to an emission of a photon with frequency $\omega = (2/3)(E_{ss}/\hbar)$. The wavelength of this transition is ~ 21 cm and is in the radio band. The so-called 21-cm line from the neutral hydrogen has been used extensively in astrophysics.

The hyperfine transition involves the flipping of the spin of the interacting particles that is due to the magnetic coupling. By an analysis similar to that performed in Chap. 4, we can show that the transition rate is given by

$$A_{if} = \frac{4\omega^3}{3\hbar c^3} |\langle f | \mathbf{M} + \mathbf{m} | i \rangle|^2, \tag{7.55}$$

where $\mathbf{M} + \mathbf{m}$ is the total magnetic moment of the electron–proton system. (This formula is completely analogous to the dipole radiation formula derived in Chap. 4.) The 21-cm line in hydrogen atom arises because of a transition from the $F = 1$ state to the $F = 0$ state. For $F = 1$, the component of the angular-momentum can take the values -1, 0, or 1 along a specific axis, thereby introducing a threefold degeneracy. The transition rate from any of these three upper levels is given by

$$A_{10} = \frac{4\omega^3}{3\hbar c^3} (\mu_p + \mu_e)^2 = \frac{g_e}{12} \left(1 + \frac{g_p m_e}{g_e m_p} \right)^2 \left(\frac{r_e \omega}{c} \right) \left(\frac{\hbar \omega}{m_e c^2} \right) \omega, \tag{7.56}$$

where $r_e = (e^2/m_e c^2)$ is the classical electron radius and ω is the transition frequency corresponding to the wavelength of 21 cm. The numerical value of the transition rate is approximately 2.9×10^{-15} s^{-1}. Correspondingly, the lifetime of the $F = 1$ state against such a decay is large, approximately 3×10^{14} s. The low transition rate and long lifetime is characteristic of magnetic dipole transitions. In spite of this, the 21-cm line has been extensively observed primarily because of the high abundance of neutral hydrogen in the universe. An order-of-magnitude estimate of this result was given in Chap. 1, Subsection 1.4.1.

The analysis of 21-cm emission from any bulk system containing neutral hydrogen is fairly straightforward because, for any realistic astrophysical temperatures, $\hbar \omega \ll k_B T$ so that $\exp(-\hbar \omega / k_B T) \simeq 1$. The populations of atoms in the upper and the lower levels are then determined only by the statistical weights, and the ratio between them gives $(n_u/n_d) = g_u/g_d = 3$. The 21-cm absorption coefficient is given by

$$\kappa_\nu = B_{ij} h \nu N_i \phi_\nu (1 - e^{-h\nu/k_B T}) \cong A_{ji} \frac{g_j}{g_i} \frac{c^2 h}{8\pi} \frac{\phi_\nu}{\nu} \frac{N_i}{k_B T}, \tag{7.57}$$

where the upper level is j, the lower level is i with population N_i, ϕ_ν is the line profile, and we have used the fact that $h\nu \ll k_B T$. Because the 21-cm line falls at radio wavelength, it is conventional to relate the flux to a brightness temperature [see Eq. (6.26)]. The brightness temperature in this particular case will be given by $T_B = T(1 - \exp[-\tau_\nu]) \approx T\tau_\nu$ for optical depth $\tau_\nu = \kappa_\nu \rho R$, corresponding to a cloud of size R along the line of sight. The brightness temperature, integrated over the line profile, will be

$$\int T_B d\nu = T\rho R \int \kappa_\nu d\nu \cong T\rho R A_{ji} \frac{g_j}{g_i} \frac{c^2 h}{8\pi k} \frac{N_i}{T\nu_0} \int \phi_\nu d\nu$$

$$= \frac{c^2 h}{8\pi k} A_{ji} \frac{g_j}{g_i} \frac{1}{\nu_0} \frac{N_i}{N_{tot}} (N_{tot} \rho R), \tag{7.58}$$

where ν_0 is the frequency corresponding to the 21-cm line and the expression within the parentheses gives the column density of the gas. For the 21-cm line, $N_j/N_i = 3$, $N_i/N_{tot} = 1/4$. By using this we can express the column density $C = N_{tot}\rho R$ in terms of the integrated line profile:

$$C = \frac{8\pi k}{c^2 h}\frac{4}{3A_{ji}}\nu_0 \int T_B dv = 3.9 \times 10^{14} \int T_B dv \text{ cm}^{-2}$$

$$= 3.9 \times 10^{18} \int T_B dv \text{ m}^{-2}. \tag{7.59}$$

The value of the integral depends on the broadening mechanism. If the line is Doppler broadened and optically thin, then $T_B(\nu) = T_B(0)\exp\{-[(\nu - \nu_0)/\Delta\nu_D]^2\}$, giving

$$\int T_B dv = T_B(0)\sqrt{\pi}\,(\Delta\,\nu_D). \tag{7.60}$$

The relevant equations are somewhat different in the case of emission from a cloud that is not spatially resolved by the instrument. In that case, it is only the total flux from the cloud that is measurable. If a spherical cloud, located at a distance d, has a radius R and is optically thin and homogeneous, then the total flux received from the cloud will be

$$F_\nu = \frac{1}{3}\frac{R^3}{d^2} A_{ji}\frac{g_j}{g_i} h\,\nu\phi_\nu \left(\frac{N_i}{N_{tot}}\right) N_{tot}\rho. \tag{7.61}$$

The rest of the analysis can be performed as above.

Exercise 7.9

Practice with 21-cm observations: (1) A neutral hydrogen cloud located at $d = 30$ pc emits 21-cm radiation. If the observed flux is $f = 4.5 \times 10^{-15}$ erg cm^{-2} s^{-1}, what is the mass of the neutral hydrogen in the cloud? (2) Along some direction in the galaxy, 21-cm emission is seen in a narrow line N centred at a radial velocity of 100 km s^{-1} and a broad line B centred at a velocity of 50 km s^{-1}. The brightness temperature measurements across the lines give a central brightness of 25.5 K and a FWHM of 5 km s^{-1} for N and a central brightness of 6.2 K and a FWHM of 34 km s^{-1} for B. There is a bright radio source with brightness temperature $T_0 = 10^9$ K in the same direction. The absorption of this source shows that the difference in the brightness temperature of the background source and the brightness of the source with absorption is $\sim 10^{8.5}$ for N and $10^{5.9}$ for B. Determine (i) the Doppler velocity widths, (ii) the optical depth at the line centre, (iii) the kinetic temperature of the absorbing systems, and (iv) the column densities. [Answers: (1) The luminosity of the source is $4\pi d^2 f = 4.85 \times 10^{26}$ erg s^{-1}. Equating this to $L_{21cm} = (3/4)N_H A_{21}h\nu$ with $A_{21} = 2.869 \times 10^{-15}$ s^{-1}, we get $N_H = 2.39 \times 10^{58}$; this corresponds to $M_H = m_p N_H \approx 20 M_\odot$. (2) (i) Because b is related to the FWHM by $0.5 = \exp[-(\text{FWHM}/2b)^2]$, we have $b = \text{FWHM}/(4\ln 2)^{1/2}$. Given FWHM $= (5, 34)$ km s^{-1} for (N, B) we get $b = (3, 20)$ km s^{-1}, respectively. (ii) Because T_0 is much higher than the gas temperature, we can write the brightness temperature as

$T_B = T_0 e^{-\tau}$. This gives $\log(1 - e^{-\tau}) = \log(T_0 - T_B) - \log T_0 = (-0.5, -3.1)$ for N, B. Hence the optical depths are $\tau = (0.38, 8 \times 10^{-4})$ for N, B. (iii) If T_k is the kinetic temperature, then at the line centre, $T_{B,\text{centre}} = T_k(1 - e^{-\tau_c})$. Using the known values of $T_{B,\text{centre}}$ and τ_c, we get $T_k = (80, 8000)$ K for N, B. (iv) The central optical depth is given by $\tau_c = 5.49 \times 10^{-14}\,(N_H/\sqrt{\pi}bT_k)$ when all quantities are in cgs units. This gives $N_H = 3.23 \times 10^{18}\tau_{\text{centre}}(b/1 \text{ km s}^{-1})(T_k/1 \text{ K})$. Substituting the known values, we get $N_H = (3, 4) \times 10^{20}$ cm^{-2} for N, B.]

7.4.4 X-Ray Emission from Atoms

Although the energies associated with valence-shell transitions of neutral atoms lead to lines in the UV region, inner-shell transitions lead to lines in the x-ray regime of the spectra. This is most easily seen from the fact that, for a hydrogenlike atom with a nuclear charge Z, the energy levels scale as $E \approx (1/2)(Z\alpha)^2 m_e c^2$. The K-shell electrons, which are relatively unshielded from the nucleus, have the ionisation potentials that scale as Z^2; for carbon, we have 36×13.6 eV $= 490$ eV, for oxygen 64×13.6 eV $= 870$ eV, and for iron 676×13.6 eV $= 9.19$ keV. These rough estimates turn out to be reasonably realistic, and inner-shell ionisation can lead to x-ray spectra.

When a K-shell electron is ejected through any of the several possible mechanisms (collision with a charged particle or energetic neutrino; absorption of photon; recoil in a Compton scattering with a photon) the vacancy created is filled in one of the two ways. In the first process, called Auger ionisation, an outer-shell electron falls to the K shell and the released energy kicks out another outer-shell electron from the atom. The energy available for this process scales as $E \propto Z^2$, and hence the phase volume scales as $p^3 \propto E^{3/2} \propto Z^3$. The second process involves an outer electron that fills the K-shell hole, emitting a photon that carries off the energy. In this case, the phase space for the photon increases in proportion to $p^3 \propto E^3 \propto Z^6$. The ratio of these two processes scales as Z^3, with Auger ionisation dominating lower-Z elements and photon emission (usually called x-ray fluorescence) dominating for higher-Z elements. The transition occurs around the iron group elements; for iron, the ratio is 0.34.

The transitions that cause the emission of x-ray lines also act as a main source of opacity in the x-ray band. The photoionisation cross section just above the threshold is $\sigma \approx (2\pi)^{3/2}\alpha a^2$ [see Eq. (6.327)], where the orbital radius a varies as Z^{-1}. Hence the K-shell photoionisation scales down as Z^{-2}. To find the opacity of a gas with a given proportion of different elements, we must sum over the individual opacities, weighing each one by the relative abundance of the elements. For a neutral gas with solar abundance, the net absorption scales with photon energy E roughly as E^{-3} all the way from 13.6 eV to ~ 10 keV. When the gas is ionised, the opacity drops but *not* as a gradual function of ionisation. The behaviour of the K-shell electrons does not change significantly until most of the outer electrons are stripped. Thus, for example, opacity in the range of

700 eV to 10 keV is not affected until carbon and nitrogen lose all but two electrons. Similarly, opacities in the kilo-electron-volt band are affected only after oxygen, silicon, sulfur, neon, and argon are stripped down to K shells. To a good degree of approximation, opacity in the kilo-electron-volt range is given by $\tau = (N_H/10^{21.5}\ \text{cm}^{-2})(E/1\ \text{keV})^{-\alpha}$, with $\alpha \approx 2.6 - 3$. These facts play a key role in several astrophysical contexts.

7.5 Selection Rules

The radiative transitions between different energy levels are caused by the interaction of the system with the electromagnetic field. All such transitions are governed by matrix elements of the form $\langle f|\mathcal{O}|i \rangle$, where i and f refer to the initial and the final energy levels, respectively, and $\mathcal{O}$ is an operator constructed from the dynamical variables of the radiating system. If this matrix element vanishes between the two states i and f, then the probability for such a transition is zero and we may say that the transition is forbidden. By studying all pairs of initial and final states and different forms of operators $\mathcal{O}$, we can arrive at a set of selection rules for transitions that are allowed. Such a study is important in classifying the spectra of matter.

It should be stressed that the vanishing or otherwise of the matrix element depends on both the quantum states and the operator $\mathcal{O}$. It is possible that the matrix element vanishes when $\mathcal{O}$ is approximated, to leading order, by some expression $\mathcal{O}_0$. Then the transition cannot take place at the leading order of the approximation. However, it is quite possible that higher-order terms in the expansion of $\mathcal{O}$ will have nonvanishing matrix elements, thereby allowing the transition to take place with lower probability. In such a case, the lifetime of the excited state will tend to be longer.

To illustrate these ideas in a concrete form, let us study the selection rules for radiation in the electric dipole approximation discussed in Chap. 4. The initial and the final states may be taken to be characterised by the integers n, l, and m and n', l', and m' for a hydrogenlike atom. The electric dipole transitions are governed by the matrix elements of the form $\langle f|\mathbf{e}_\alpha \cdot \mathbf{x}|i \rangle$, which can be separated into matrix elements of the form $\langle f|z|i \rangle$ and $\langle f|x \pm iy|i \rangle$. For the electric dipole transition to occur, at least one of these matrix elements has to be nonvanishing. The first case corresponds to the emission or absorption of radiation propagating in the x–y plane and polarised in the z direction (i.e., $\mathbf{e}_\alpha \cdot \mathbf{x} = z$), and the second case corresponds to the interaction with radiation propagating in the z direction and circularly polarised (i.e., $\mathbf{e}_\alpha \propto \hat{\mathbf{x}} \pm i\hat{\mathbf{y}}$). The matrix element of z is given by the integrals of the form

$$\langle f|z|i \rangle \propto \left[\int_0^\infty r\mathcal{R}_{n'l'}\mathcal{R}_{nl}\,dr \right]\left[\int_{-1}^{+1} \mu P_{l'}^{|m'|}(\mu)P_l^{|m|}(\mu)\,d\mu \right]\left[\int_0^{2\pi} e^{i(m-m')\phi}\,d\phi \right],$$

$$(7.62)$$

where $\mu = \cos\theta$. The integral over r cannot vanish because the expansion of $r\mathcal{R}_{nl}$ in terms of radial wave functions will have nonvanishing coefficients with all other eigenfunctions. The integral over ϕ, however, will vanish unless $m' = m$, i.e., unless $\Delta m = 0$. When this condition is satisfied, we have to check whether the integral over μ vanishes or not. Using the recursion relation

$$\mu P_l^{|m|}(\mu) = \frac{1}{2l+1}\left[(l-|m|)P_{l+1}^{|m|}(\mu) + (l+|m|)P_{l-1}^{|m|}(\mu)\right], \qquad (7.63)$$

in expression. (7.62), we can easily see that the integral vanishes unless $\Delta l = \pm 1$. Thus the matrix element of z vanishes unless $\Delta l = \pm 1$ and $\Delta m = 0$.

When these conditions are not satisfied and $\langle f|z|i\rangle = 0$, dipole radiation can still arise if the matrix element of $x \pm iy$ is nonvanishing. Writing $x \pm iy = r(1 - \mu^2)^{1/2}e^{\pm i\phi}$, we see that the ϕ integral vanishes unless $\Delta m = \pm 1$. When this condition is satisfied, we have to check whether the μ integral, which is of the form

$$I = \int_{-1}^{+1} (1-\mu^2)^{1/2} P_l^{|m'|}(\mu) P_l^{|m|}(\mu)\, d\mu, \qquad (7.64)$$

is nonzero. Making use of the recurrence relation

$$(1-\mu^2)^{1/2} P_l^{|m|}(\mu) = \frac{1}{2l+1}\left[P_{l-1}^{|m'|}(\mu) - P_{l+1}^{|m'|}(\mu)\right], \qquad (7.65)$$

(which is valid for $|m'| < |m|$), we see that the integral is nonzero only when $\Delta l = \pm 1$. We can use an analogous recurrence relation for the case $|m'| > |m|$ to arrive at the same conclusion.

The preceding discussion illustrates the general procedure that is followed to obtain the selection rules. In this particular case, however, the rule $\Delta l = \pm 1$ could have been arrived at more easily by the following argument. The dipole operator $\mathbf{d} = q\mathbf{x}$ has odd parity under space reversal. On the other hand, the spherical harmonic Y_{lm} has the parity $(-1)^l$. Hence the matrix element of the dipole operator will be nonvanishing only if Δl is an odd integer. That integer must be equal to unity because, in a single-photon emission, the angular momentum can change only by unity.

When the electric dipole radiation is forbidden, the transition can still proceed through other routes. For example, it may be recalled that the dipole approximation is based on the Taylor expansion of the quantity $\exp(i\mathbf{k}\cdot\mathbf{x})$ in which only the first term is retained. In the next leading order, the transition will be governed by the matrix element of the form $k^a e_\alpha^b \langle f|x_a p_b|i\rangle$. The tensor $x_a p_b$ can be expressed as a sum of symmetric and antisymmetric parts; the matrix element of the symmetric part can be related by a series of simple algebraic transformations to the matrix element of the electric quadrupole operator $Q_{ab} \equiv -e(3x_a x_b - x^2 \delta_{ab})$. The matrix elements of this operator vanish unless $\Delta l = 0$ or ± 2 and $\Delta m = 0, \pm 1, \pm 2$. These are the selection rules for the electric quadrupole transitions.

The antisymmetric part of $x_a p_b$ can be related to the angular momentum $\mathbf{L} = \mathbf{x} \times \mathbf{p}$ and hence to the magnetic-moment operator $\mathbf{M} = -(e/2m_e c)\mathbf{L}$. The transitions proceeding through the antisymmetric part are called the magnetic dipole transitions. By an analysis similar to that given above, it can be shown that the selection rules now are $\Delta l = 0$ and $\Delta m = 0, \pm 1$.

It is possible that transitions between certain states are strictly forbidden even when all the terms in the perturbation expansion are taken into account. As an example, consider the transition from a $2s$ state of a hydrogen atom to a $1s$ state. Both these states have spherically symmetric wave functions and hence all multipole moments have vanishing matrix elements between these states. This is analogous to the classical result that the charge distribution with a spherically symmetric motion cannot emit electromagnetic radiation. Such a transition, however, can proceed through two-photon emission.

The analysis described so far does not take into account electronic spin, fine structure, and other complications. When the fine structure of energy levels is taken into account, the selection rule for electric dipole transitions become $\Delta l = \pm 1$ and $\Delta j = 0, \pm 1$. For example, in the case of a Balmer series (corresponding to transitions $n \to 2$), the following seven transitions are allowed by the selection rules: $np_{1/2}\text{--}2s_{1/2}$, $np_{3/2}\text{--}2s_{1/2}$, $ns_{1/2}\text{--}2p_{1/2}$, $ns_{1/2}\text{--}2p_{3/2}$, $nd_{3/2}\text{--}2p_{1/2}$, $nd_{3/2}\text{--}2p_{3/2}$, $nd_{5/2}\text{--}2p_{3/2}$. However, because levels $ns_{1/2}$ and $np_{1/2}$ coincide as well as levels $np_{3/2}$ and $nd_{3/2}$, each Balmer transition leads to only five distinct lines. Further, because the energy differences of a multiplet rapidly decrease with increasing n, the fine-structure splitting of a spectral line in a transition between two levels is essentially dominated by the fine structure of the lower level. Each line of a Balmer series therefore essentially consists of a doublet or, more precisely, two groups of closely spaced lines.

For hyperfine transitions, the electric dipole selection rules continue to be valid. In addition, F must satisfy the condition $\Delta F = 0, \pm 1$, with the transition $F = 0 \to F = 0$ remaining forbidden.

Exercise 7.10

Practice with selection rules: Which of the following radiative transitions are forbidden and why? (1) $(2p)^2 \, {}^3P_0 \to 2p3s \, {}^3P_0$; (2) $2p3p \, {}^3S_1 \to 2p4d \, {}^3D_2$; (3) $2p3s \, {}^1P_1 \to 2p3p \, {}^1P_1$; (4) $2p3p \, {}^3D_1 \to 3p4d \, {}^3F_2$; (5) $2p \, {}^2P_{1/2} \to 3d \, {}^2D_{5/2}$. [Answers: (1) Forbidden because it is a $J = 0 \to J = 0$ transition. (2) Forbidden because $|\Delta L| > 1$. (3) Allowed. (4) Allowed. (5) Forbidden because $|\Delta J| > 1$.]

7.6 Energy Levels of Diatomic Molecules

Just like atoms, molecules also possess discrete energy levels, and transitions between these levels can lead to spectral lines that are characteristic of the molecule. The structure of the molecules is more complicated than that of atoms but the

basic principle behind the analysis is the same. We confine our attention to diatomic molecules, which have the simplest structure.

The Hamiltonian of a diatomic molecule with two nuclei A and B and two
valence electrons 1 and 2 can be written as the sum $H = H_{AB} + H_{ele}$, where H_{AB}
is the sum of the kinetic-energy operators of the two nuclei and H_{ele} represents
the kinetic energy and the potential energy of interaction of the electrons:

$$H_{AB} \equiv -\frac{\hbar^2}{2M_A}\nabla_A^2 - \frac{\hbar^2}{2M_B}\nabla_B^2, \quad H_{ele} \equiv -\frac{\hbar^2}{2m_e}(\nabla_1^2 + \nabla_2^2) + V. \quad (7.66)$$

We can determine the energy levels of the system by solving the Schroedinger
equation, $H\phi = E\phi$. Because exact solutions to such many-body systems are
impossible to obtain, it is necessary to resort to some approximation scheme.

In this particular case, we can readily obtain such an approximation (called the
Born–Oppenheimer approximation) by noting the fact that the ratio of the masses
of electrons and nuclei is quite small. Hence we can ignore the back reaction
of electrons on the nuclear wave function in the lowest order of approximation.
As far as the nuclei are concerned, it is necessary to take into account their
rotational and vibrational energy levels. The valence electrons may be thought
of as moving in an effective potential produced by the nuclei and dependent on
the separation $\mathbf{R}$ between them.

Mathematically, such an approximation can be implemented along the following lines. The wave function for the system is taken to be of the form
$\phi \approx \phi_{ele}(\mathbf{x}_1, \mathbf{x}_2, R)\phi_{AB}(\mathbf{x}_A, \mathbf{x}_B)$, where $\mathbf{x}_1, \mathbf{x}_2$ are the coordinates of the electrons
and $\mathbf{x}_A, \mathbf{x}_B$ are the coordinates of the nuclei and $\mathbf{R} = \mathbf{x}_A - \mathbf{x}_B$. In the Schroedinger
equation

$$\phi_{AB} H_{ele}\phi_{ele} + H_{AB}\phi_{ele}\phi_{AB} = E\phi_{ele}\phi_{AB}, \quad (7.67)$$

we can expand the action of H_{AB} in the form

$$H_{AB}\phi_{ele}\phi_{AB} = \phi_{ele}H_{AB}\phi_{AB} - \frac{\hbar^2}{M_A}\nabla_A\phi_{ele}\cdot\nabla_A\phi_{AB}$$
$$- \frac{\hbar^2}{M_B}\nabla_B\phi_{ele}\cdot\nabla_B\phi_{AB} + \phi_{AB}H_{AB}\phi_{ele}. \quad (7.68)$$

The last term on the right-hand side is smaller than the first term by the ratio of the
masses of electron and the nuclei, m_e/m. The second term is proportional to the
mean value of the relative velocity of the molecules, which is again subdominant
to the first term when the back reaction of the electrons on the nuclear motion is
ignored. The Born–Oppenheimer approximation consists of ignoring the second
and the third terms on the right-hand side and taking $H_{AB}\phi_{ele}\phi_{AB} \approx \phi_{ele}H_{AB}\phi_{AB}$.
Using this approximation in the Schroedinger equation and dividing throughout

by the wave function ϕ, we get

$$\frac{1}{\phi_{\text{ele}}} H_{\text{ele}} \phi_{\text{ele}} = -\frac{1}{\phi_{AB}} H_{AB} \phi_{AB} + E. \tag{7.69}$$

To the same order of approximation, we can find a consistent solution to this equation by requiring that each side be equal to a constant $E(R)$ that depends on only the relative separation of the nuclei. This leads to the pair of equations

$$H_{\text{ele}} \phi_{\text{ele}} = E_{\text{ele}}(R) \phi_{\text{ele}}, \tag{7.70}$$

$$H_{AB} \phi_{AB} + E_{\text{ele}}(R) \phi_{AB} = E \phi_{AB}, \tag{7.71}$$

which describe the diatomic molecule.

We can further simplify Eq. (7.71) by noting that the Hamiltonian H_{AB} can be expressed as the sum of the terms representing the kinetic energies of the centre-of-mass motion and the relative velocity:

$$H_{AB} = -\frac{\hbar^2}{2M} \nabla_{\text{CM}}^2 - \frac{\hbar^2}{2\mu} \nabla_R^2, \tag{7.72}$$

where M is the total mass and μ is the reduced mass of the ions. The wave function ϕ_{AB} can therefore be written as the product $\phi_{AB} = \phi_{\text{trans}}(\mathbf{X}_{\text{CM}}) \phi_{\text{int}}(\mathbf{R})$, with the total energy E expressed as the sum $E_{\text{trans}} + E_{\text{int}}$, where E_{trans} represents the overall kinetic energy of the centre-of-mass motion of the molecule and E_{int} is the energy, that is due to the relative motion of the nuclei. The translational kinetic energy is clearly unimportant in the study of internal transitions of the molecules. The really nontrivial part of the physics is contained in the wave function $\phi_{\text{int}}(\mathbf{R})$, which satisfies the equation

$$-\frac{\hbar^2}{2\mu} \nabla_R^2 \phi_{\text{int}} + E_{\text{ele}}(R) \phi_{\text{int}} = E_{\text{int}} \phi_{\text{int}}. \tag{7.73}$$

Introducing the standard (R, θ, φ) coordinate system, we can expand the Laplacian as

$$\nabla_R^2 = \frac{1}{R^2} \left[\frac{\partial}{\partial R} \left(R^2 \frac{\partial}{\partial R} \right) - \frac{L^2}{\hbar^2} \right], \tag{7.74}$$

where $\mathbf{L}^2$ is the angular-momentum operator. Separating out the angular and radial parts of the wave function in the form

$$\phi_{\text{int}}(\mathbf{R}) = \frac{1}{R} Z_{\text{vib}}(R) Y_{Jm}(\theta, \varphi), \tag{7.75}$$

we see that the radial part satisfies the equation

$$-\frac{\hbar^2}{2\mu} \frac{d^2 Z_{\text{vib}}}{dR^2} + E_{\text{ele}}(R) Z_{\text{vib}} = \left[E_{\text{int}} - \frac{J(J+1)\hbar^2}{2\mu R^2} \right] Z_{\text{vib}}. \tag{7.76}$$

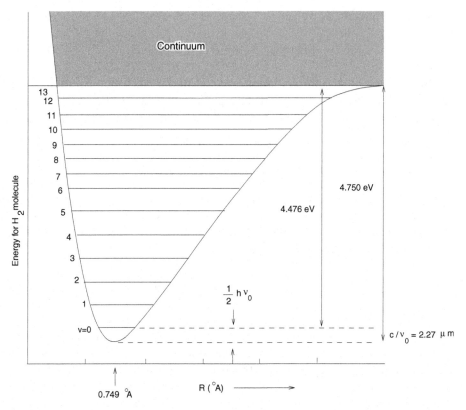

Fig. 7.3.   Vibrational potential energy for a diatomic molecule of hydrogen.

To proceed further, it is necessary to ascertain the form of $E_{\text{ele}}(R)$ that was introduced as a separation of variables in the original equation. Physically, it is clear that this quantity $E_{\text{ele}}(R)$ is the potential energy between the two nuclei at separation R. The exact shape of this potential is fairly complicated and is shown in Fig. 7.3. It is well approximated by the form

$$E_{\text{ele}}(R) = D_e\big[e^{-2\alpha(R-R_0)} - 2e^{-\alpha(R-R_0)}\big], \qquad (7.77)$$

where R_0, D_e, and α are constants. This potential is attractive at large distances, is repulsive near the origin, and has a minimum of $-D_e$ at $R = R_0$. The quantity D_e corresponds to the energy that needs to be supplied to the molecule in order to dissociate it. Near the minimum of the potential, it can be Taylor expanded as a harmonic oscillator potential with

$$E_{\text{ele}}(R) = E_{\text{ele}}(R_0) + \frac{\mu}{2}\omega_0^2(R - R_0)^2 + \cdots, \qquad (7.78)$$

where R_0 is the location of the minimum of the potential and $\mu\omega_0^2 = E_{\text{ele}}''(R_0) = D_e\alpha^2$. In this harmonic approximation, the energy eigenvalue appearing on the right-hand side of Eq. (7.76) must have the form $\hbar\omega_0[v + (1/2)]$, where $v =$ 0, 1, 2, ..., is called the vibrational quantum number. The total internal energy of the system is now given by

$$E_{int} = E_{ele}(R_0) + \left(\frac{\hbar^2}{2\mu R_0^2}\right) J(J+1) + \hbar\omega_0\left(v + \frac{1}{2}\right). \tag{7.79}$$

The above result should be thought of as the lowest-order approximation to the energy levels of the system. A more accurate expression for the energy levels is given by

$$E_v = \hbar\omega_0\left[\left(v + \frac{1}{2}\right) - \beta\left(v + \frac{1}{2}\right)^2\right], \tag{7.80}$$

where $\beta = (\hbar\omega/4D_e)$. Further, the above analysis approximates the rotational contribution to the potential $[J(J+1)\hbar^2/2\mu R^2]$ at $R = R_0$ while arriving at the final expression. More precisely, this term should be moved to the left-hand side of Eq. (7.76) and the total potential {which is the sum $E_{ele}(R) + [J(J+1) \hbar^2/2\mu R^2]$} should be Taylor expanded about the equilibrium position. In this case the equilibrium position shifts to the location

$$R_1 \simeq R_0 + \frac{\hbar^2}{2\mu} \frac{J(J+1)}{\alpha^2 R_0^3 D_e}, \tag{7.81}$$

because the rotational motion "stretches" the molecule slightly. This couples the rotational and the vibrational motion and the energy levels in Eqs. (7.79) and (7.80) are modified to

$$\bar{E}_{s,v,J} = -D_e + \hbar\omega_0\left[\left(v + \frac{1}{2}\right) - \beta\left(v + \frac{1}{2}\right)^2\right] + \frac{\hbar^2}{2\mu R_0^2}J(J+1)$$

$$- a\left(v + \frac{1}{2}\right)J(J+1) - bJ^2(J+1)^2, \tag{7.82}$$

where

$$a = \frac{3\hbar^3\omega_0}{4\mu\alpha R_0^3 D_e}\left(1 - \frac{1}{\alpha R_0}\right), \quad b = \frac{\hbar^4}{4\mu^2\alpha^2 R_0^6 D_e}. \tag{7.83}$$

The above analysis shows that the vibrational and the rotational energies of the nucleus can be described in terms of the two integers J and v. The complete solution of the problem requires solving Eq. (7.70) to obtain the full wave function. From a practical point of view, this is unnecessary as far as the energy levels are concerned. The expression for the energy obtained above can be used to describe most of the transitions that take place in a diatomic molecule.

Exercise 7.11

Specific heat of diatomic molecules: The energy levels of a diatomic molecule, characterised by the two integers v and J, are given by

$$\epsilon_{vJ} = \epsilon_0 + \hbar\omega\left(1 + \frac{1}{2}\right) + \frac{\hbar^2 J(J+1)}{2I}. \tag{7.84}$$

Calculate the partition function for a system of diatomic molecules by summing over v
and integrating over J (which is a reasonable approximation to summing over J if the
rotational levels are closely spaced). Compute the mean energy and the specific heat of
these molecules at low and high temperatures.

7.7 Aspects of Diatomic Spectra

Because the state of a diatomic molecule depends on the electronic state as well
as the rotational and the vibrational state of the nuclei, the spectral lines from the
molecule can arise because of different kinds of transition. Of these, the electronic
transitions require the maximum amount of energy followed by the vibrational
transitions and the rotational transitions. Because of this wide difference in the
energy scales, it is possible to discuss these transitions separately.

7.7.1 Rotational Transitions

At the lowest order of approximation, radiative transitions are caused by the
coupling between the electric field of the radiation and the dipole moment
operator of the atomic system in the form $H = \mathbf{E} \cdot \mathbf{d}$. The dipole moment
of the molecule can be expressed as the sum of the electronic and the nuclear
dipole moments, $\mathbf{d} = \mathbf{d}_{el} + \mathbf{d}_{nucl}$. The electronic dipole moment, in turn, is the
sum of the dipole moments of all the electrons in the system; similarly, the nu-
clear dipole moment is given by $\mathbf{d}_{nucl} = e(Z_A \mathbf{X}_A + Z_B \mathbf{X}_B)$. It is obvious that
$\mathbf{d}_{nucl}$ vanishes for homonuclear molecules such as H_2, O_2, etc.

Let us consider the rotational transitions in which the initial and the final
states have the same electronic wave function ϕ_{el} and vibrational wave function
$Z_{vib}(R)$. The transitions are now governed by the matrix element of the form

$$\langle \Psi_f | H | \Psi_i \rangle = \int_0^{2\pi} d\varphi \int_0^\pi Y_{Jm}^*(\theta, \varphi) \langle Z_{vib} | H_D | Z_{vib} \rangle Y_{J'm'}(\theta, \varphi) \sin\theta \, d\theta, \tag{7.85}$$

where the matrix element $\langle Z_{vib} | H_D | Z_{vib} \rangle$ is in turn given by integrations over
the electronic and the vibrational wave functions:

$$\langle Z_{vib} | H_D | Z_{vib} \rangle \equiv -\int_0^\infty \mathbf{E}_\omega \cdot \mathbf{D}(R, \theta, \varphi) |Z_{vib}(R)|^2 \, dR, \tag{7.86}$$

with

$$\mathbf{D}(R, \theta, \varphi) \equiv \int (\mathbf{d}_{ele} + \mathbf{d}_{nucl}) |\phi_{ele}(\mathbf{X}_1, \mathbf{X}_2, \ldots, \mathbf{X}_N; R)|^2 d^3 X_1 d^3 X_2 \cdots d^3 X_N,$$

(7.87)

where the integration is over all the electronic coordinates. The quantity $\mathbf{D}$ is the average of the dipole moment over the electronic states of the system and is usually called the permanent dipole moment of the molecule. This again vanishes for homonuclear molecules because of the symmetry of electronic distribution. It follows that pure rotational spectra cannot arise at the lowest order of approximation for the homonuclear molecules.

Assuming that the system has a permanent dipole moment, it is again obvious from symmetry that it should be oriented along the axis connecting the two nuclei. In that case the interaction Hamiltonian can be written as

$$\mathbf{E}_\omega \cdot \mathbf{D} = |\mathbf{E}_\omega| D(R) \cos \theta,$$

(7.88)

where θ is the angle between the direction of the electric-field vector (which may be taken to be the z axis) and $\mathbf{R}$. In the lowest-order approximation, we can replace $D(R)$ with $D(R_0) \equiv D_0$, called the reduced dipole moment. Then the matrix element becomes

$$\langle \Psi_f | H | \Psi_i \rangle = -|\mathbf{E}_\omega| D_0 \int_0^{2\pi} d\varphi \int_0^\pi \cos \theta Y_{Jm}^*(\theta, \varphi) Y_{J'm'}(\theta, \varphi) \sin \theta \, d\theta.$$

(7.89)

This expression will be nonvanishing only if $\Delta J = \pm 1$, $\Delta m = 0$. The second condition $\Delta m = 0$ will be replaced with $\Delta m = \pm 1$ if the electric field is taken to be circularly polarised. These are the selection rules for the purely rotational transitions along with the requirement $D_0 \neq 0$. A transition from a level $J + 1$ to level J can proceed from any sublevel m' of $J + 1$ to any other sublevel m of J that satisfies the selection rules. To take this into account, it is convenient to introduce a transition dipole moment by the relation

$$|\mathbf{D}_{J+1,J}|^2 \equiv \sum_{m=-J}^{J} D_0^2 \left| \oint Y_{Jm}^* \cos \theta Y_{J+1,m'} \, d\Omega \right|^2 = \frac{J+1}{2J+3} D_0^2.$$

(7.90)

The rate of spontaneous transition can now be expressed as

$$A_{J+1,J} = \frac{4\omega^3}{3\hbar c^3} |\mathbf{D}_{if}|^2 = \frac{4(J+1)\omega^3}{3(2J+3)\hbar c^3} D_0^2.$$

(7.91)

In summary, the simplest form of rotational transitions occur in heteronuclear molecules. The most abundant heteronuclear molecule is CO, for which the fundamental transition $J = 1$ to $J = 0$ has a rate $A_{10} \approx 6 \times 10^{-8}$ s^{-1}. In the simplest context, the rotational spectrum produces a series of lines with the frequency varying as $\hbar\omega = (J\hbar^2/\mu R_0^2)$; that is, the lines are equally spaced in the frequency space. For CO, $\mu = 6.859$ amu $= 1.11 \times 10^{-23}$ gm and

$R_0 = 1.13 \times 10^{-8}$ cm. The lowest transition frequency ($J = 1 \rightarrow 0$) is 115 GHz, corresponding to a wavelength of 2.6 mm. The next-higher transitions have frequencies of 230 GHz ($J = 2 \rightarrow 1$), 345 GHz ($J = 3 \rightarrow 2$), etc.

7.7.2 Vibrational–Rotational Spectrum

We next consider transitions in which both the vibrational and the rotational quantum numbers change but the electronic state remains the same. The corresponding transition element is given by

$$\langle \Psi_f | H | \Psi_i \rangle = -|\mathbf{E}_\omega| \int_0^{2\pi} d\varphi \int_0^\pi Y_{Jm}(\theta, \varphi) \langle v | D | v' \rangle \cos \theta \, Y_{J'm'}(\theta, \varphi) \sin \theta \, d\theta,$$

(7.92)

where

$$\langle v | D | v' \rangle \equiv -N_v N_{v'} \int_{-\infty}^\infty H_v(x/x_0) D(R) H_{v'}(x/x_0) e^{-x^2/x_0^2} \, dx,$$

(7.93)

where $x = (R - R_0)$, $x_0 = (\hbar/\mu\omega_0)^{1/2}$, H_v is the Hermite polynomial, and N_v is the normalisation constant. To be consistent with our approximation, we Taylor expand $D(R)$, retaining only the leading-order term in x

$$D(R) = D_0 + \left(\frac{dD}{dR} \right)_0 x + \cdots .$$

(7.94)

The first term D_0 gives purely rotational transitions with $v' = v$. The second term will have a nonvanishing matrix element, provided that $\Delta v = \pm 1$. This gives the selection rule for vibrational transitions at the leading order [along with the requirement $(dD/dR) \neq 0$ at $R = R_0$].

The rotational part of the transition still gives the selection rule $\Delta J = \pm 1$. There is, however, one further complication that needs to be mentioned. Let Λ denote the component of the electronic angular momentum along the internuclear axis. If $\Lambda = 0$, then $\Delta J = \pm 1$ must be obeyed for the transition to take place. However, if $\Lambda \neq 0$ then it is possible to have the transition proceed even if $\Delta J = 0$. In describing the spectral lines these two cases should be dealt with separately.

In a transition between the levels v, J and v', J', with $v' = v \pm 1$, we obtain a series of spectral lines corresponding to $\Delta J = +1$ or $\Delta J = -1$. The first set is called the R branch and corresponds to the frequencies

$$\omega = \omega_0 + 2(J + 1)B/\hbar \quad (R \text{ branch}),$$

(7.95)

where $J = 0, 1, 2, \ldots$, and $B = (\hbar^2/2\mu R_0^2)$. The transitions corresponding to $\Delta J = -1$ are called P branch with frequencies

$$\omega = \omega_0 - 2JB/\hbar, \quad (P \text{ branch}),$$

(7.96)

with $J = 1, 2, 3, \ldots$, etc. There will be no line at ω_0 if $\Lambda = 0$ as $\Delta J = 0$ is forbidden.

The above expressions give a set of equally spaced lines in frequencies. When the corrections to the energy levels arising from anharmonicity, etc., are taken into account, the lines of the R branch approach closer but those of the P branch spread apart for larger values of J.

7.7.3 Electronic–Vibrational–Rotational Transitions

This case can be analysed along lines similar to those of Subsection 7.7.2. When the electronic configuration changes, the term proportional to $\mathbf{d}_{\mathrm{nucl}}$ drops out and the angular dependence of $\mathbf{d}_{\mathrm{ele}}$ gives the selection rules $\Delta J = 0, \pm 1$ and $\Delta m = 0, \pm 1$. The $\Delta J = 0$ transition is not allowed if $\Lambda = 0$ in both initial and final states. The transition from $J = 0$ to $J = 0$, however, is forbidden for all values of Λ. It should be noted that electronic transitions do not require that the molecule possess a permanent dipole moment.

The actual rate of transitions is governed by the matrix element of the form

$$\mathbf{D}_{if} \propto \int Z_{\mathrm{vib}, f}^*(R) Z_{\mathrm{vib}, i}(R) \, dR, \tag{7.97}$$

called the Franck–Condon factor. In principle, we now have no restrictions on Δv. However, the probability for transition will be quite different for transitions involving different values of Δv.

In summary, the spectra of diatomic molecules arise from three different factors. For a given value of electronic transition, $n' \rightarrow n$, there exist different possible choices for Δv that lead to a vibrational fine structure. For each choice of $v' \rightarrow v$, the different possible choices $\Delta J = 0, \pm 1$ lead to a still finer rotational structure. For emission spectra, $\Delta J = +1$ is called the P branch, $\Delta J = 0$ is called the Q branch, and $\Delta J = -1$ is called the R branch. Convention with the reverse sign is used for absorption spectra. The R branch starts off at higher frequencies than the Q branch, which in turn begins at higher frequencies than the P branch.

Exercise 7.12

Pure rotation spectra: Show that a diatomic molecule will lead to nearly pure rotation spectra in the temperature range

$$\frac{\hbar^2}{m_p a_0^2} \ll \hbar k_B T \ll \frac{\hbar^2}{m_p a_0^2} \left(\frac{m_p}{m_e}\right)^{1/2}. \tag{7.98}$$

Because the moment of inertia of the molecule is different in the initial and the final states of the electronic transition, the frequency spacing between successive spectral lines will not be monotonic. For example, if $B' > B$ in the transition, then the R branch ($J' = J + 1$) is given by

$$v_R = v_{\text{ele,vib}} + h^{-1}[2B + (3B - B')J - (B' - B)J^2]. \tag{7.99}$$

This function increases with J, reaches a maximum, and then turns around for larger values of J. In contrast, the corresponding formulas for the Q branch ($J' = J$) and the P branch ($J' = J - 1$ with $J \geq 1$) are monotonic and are given by

$$v_Q = v_{\text{ele,vib}} - h^{-1}(B' - B)J(J + 1), \tag{7.100}$$

$$v_P = v_{\text{ele,vib}} - h^{-1}[(B' + B)J - (B' - B)J^2]. \tag{7.101}$$

This leads to a fairly complex structure of spectral lines in a diatomic molecule.

8
Neutral Fluids

8.1 Introduction

We use the concepts developed in some of the previous chapters (especially Chap. 5) to study the physical processes in fluids in this chapter. The emphasis is on aspects of fluid mechanics that are of relevance in astrophysics. This chapter will be needed in Chaps. 9 and 10 and in several sections of Vols. II and III.

8.2 Molecular Collisions and Evolution of the Distribution Function

At the microscopic level, a fluid can be thought of as a collection of molecules. Ignoring the internal structure of the molecules, we can specify the state of any molecule by giving its position $\mathbf{x}$ and momentum $\mathbf{p}$. Let $dN = f(\mathbf{x}, \mathbf{p}, t) \, d^3\mathbf{x} d^3\mathbf{p}$ denote the number of molecules in a phase volume $d^3\mathbf{x} d^3\mathbf{p}$ at time t. We are interested in the form and evolution of this distribution function.

The distribution function changes because of two kinds of physical processes. Macroscopic force fields, such as the gravitational field, can exert forces on the molecules and influence their motion. Such a force, $\mathbf{F}_{sm}(\mathbf{x}, t) \equiv -\nabla U_{sm}(\mathbf{x}, t)$, will vary smoothly over the microscopic scales and can be derived from a suitable potential U_{sm}. A particular molecule will also experience the force $\mathbf{F}_{coll}$ that is due to collision with another molecule whenever it is close to another molecule. At a fundamental level, collisions arise because of two molecules interacting by means of the intermolecular force. However, because this force is of very short range, we use the term collision, which suggests a contact interaction. (We also ignore the possibility of three or more molecules colliding simultaneously because the probability for such events is very small under the conditions we are interested in.) The collisions are assumed to be uncorrelated in space and time and conserve the total momentum and energy of the colliding particles. The dynamics of such a collision can be completely characterised by a differential cross section $\sigma(\Omega)$, where Ω denotes the two angular coordinates θ and ϕ.

The equations describing the evolution of the distribution function under such circumstances have already been developed in Chap. 5, Section 5.14, where it was shown that

$$\frac{\partial f}{\partial t} + \mathbf{v} \cdot \nabla f - \nabla U_{sm} \cdot \frac{\partial f}{\partial \mathbf{p}} = C[f], \tag{8.1}$$

with

$$C[f] = \int |\mathbf{v} - \mathbf{v}_1|(f' f_1' - f f_1)\sigma(\Omega) \, d\Omega \, d^3\mathbf{p}_1. \tag{8.2}$$

Here $f = f(\mathbf{x}, \mathbf{p}, t)$, $f_1 = f(\mathbf{x}, \mathbf{p}_1, t)$, $f' = f(\mathbf{x}, \mathbf{p}', t)$, and $f_1' = f(\mathbf{x}, \mathbf{p}_1', t)$. The collision described by $\sigma(\Omega)$ changes the initial momenta $(\mathbf{p}, \mathbf{p}_1)$ to final momenta $(\mathbf{p}', \mathbf{p}_1')$, conserving the total momentum and energy. In the absence of collisions, the evolution of distribution function is described by Eq. (8.1), with the right-hand side set to zero.

In deriving Eq. (8.1) we have assumed that a single collision can change the momentum of the particle by an arbitrary amount. Hence $C[f]$ involves an integral over $\mathbf{p}_1$ and Eq. (8.1) is an integrodifferential equation that is extremely difficult to solve. Progress can be made only with further simplifying assumptions and the validity of such assumptions depends on the physical context. The various possibilities are now briefly described.

(1) When the momentum (or energy) transfer in a single collision is small, we can expand the collision term in a suitable Taylor series in the momentum (or energy) transfer and obtain a differential equation. This was done in Chap. 6, Section 6.7, to derive the Kompaneets equation. A similar approach will again be used in Chaps. 9 and 10 to obtain the effect of long-range electromagnetic and gravitational interactions. Because differential equations are easier to handle than integrodifferential equations, this approach allows us to make some progress.

(2) Given a collisional process with a cross section of σ, the collisional mean free path is $l \approx (N\sigma)^{-1}$, where N is the space density of particles. If $l \gg R$, where R is the typical scale of the system, the collisions are ignorable and we can set the right-hand side of Eq. (8.1) to zero. In the case of photons, this corresponds to the optically thin situation, which is comparatively easy to handle. This is because, for photons, there is no force term on the left-hand side of Eq. (8.1) that represents the large-scale external field (except when gravitational bending of light needs to be taken into account, as we shall see in Vol. III). In other physical situations with $l \gg R$, e.g., in galactic dynamics, we have to handle the gravitational field of the particles in a self-consistent manner, and this again makes the equations difficult to handle. We shall discuss this issue in Chap. 10.

(3) If $l \ll R$, then collisions are very effective and we may expect some kind of local statistical equilibrium to prevail. In the context of photons this

corresponds to the optically thick limit, which we handled in Chap. 6, Section 6.8, by taking the moments of the evolution equation. This is the most hopeful procedure when $l \ll R$.

In the case of astrophysical fluids, all the above situations arise in one context or the other. Most of the discussion in this chapter deals with the situation in item (3) above. When $l \ll R$, we have some hope of closing the moment equations by using some physically well-motivated criteria to solve them. As a first step, we derive the relevant moment equations.

We obtained the results of taking the moments by using the first and the second powers of four momenta in Chap. 3, Section 3.6, and in Chap. 5, Section 5.14. Because the collision between two molecules conserves the four momenta, the moments of the right-hand side of Eq. (8.1) will vanish. We also assume that no external force field is present on the fluid. (The equations can be easily modified to take external force fields into account.) In this case, the equations reduce to

$$\frac{\partial J^{\alpha}}{\partial x^{\alpha}} = 0, \quad \frac{\partial T^{\alpha\beta}}{\partial x^{\alpha}} = 0, \tag{8.3}$$

where

$$J^{\alpha}(x^{\mu}) = c \int \frac{d^3\mathbf{p}}{E(\mathbf{p})} p^{\alpha} f(x^{\mu}, \mathbf{p}), \quad T^{\alpha\beta}(x^{\mu}) = c^2 \int \frac{d^3\mathbf{p}}{E(\mathbf{p})} p^{\alpha} p^{\beta} f(x^{\mu}, \mathbf{p}). \tag{8.4}$$

(In this chapter we use the Greek letters α, β, etc., to denote four-dimensional quantities and they take values 0, 1, 2, 3; the Roman letters i, k, etc., denote the spatial components and run over 1, 2, 3. This change of convention from Chap. 3 is due to the fact that we are concerned predominantly with three-dimensional vectors etc. in this chapter. Within the context of three-dimensions, the raising and the lowering of an index does not change the sign and hence we do not worry about the positioning of the indices in three-dimensional vectors, tensors, etc.) By taking the Newtonian limit of these equations, we will be able to obtain the moment equations for a nonrelativistic fluid. The components of J^{α} are

$$J^{\alpha} = \left[\int d^3\mathbf{p} f, \int d^3 p \left(\frac{\mathbf{v}}{c} \right) f \right] \equiv \{N(x^{\mu}), N(x^{\mu})[\mathbf{V}(x^{\mu})/c)]\}, \tag{8.5}$$

where we have defined the number density $N(x^{\mu})$ and the bulk velocity field

$$\mathbf{V}(x^{\mu}) = \langle \mathbf{v} \rangle = \frac{\int d^3\mathbf{p}\mathbf{v} f}{\int d^3\mathbf{p} f} \tag{8.6}$$

by averaging over the momentum distribution of the particles; in arriving at the spatial components, we have also used the relation $(\mathbf{p}/E) = (\mathbf{v}/c^2)$ and defined the symbol $\langle \cdots \rangle$ to denote the averaging over the distribution function. The vector field $\mathbf{V}(\mathbf{x}, t)$ represents the macroscopic flow velocity of the fluid at a location $\mathbf{x}$ at time t. This velocity field is the weighted average of the microscopic velocities of the molecules at the location $\mathbf{x}$ at time t. The conservation law for J^α now becomes

$$\frac{\partial N}{\partial t} + \nabla \cdot (N\mathbf{V}) = 0 = \frac{\partial \rho}{\partial t} + \nabla \cdot (\rho \mathbf{V}), \tag{8.7}$$

where we have defined the mass density $\rho = Nm$, where m denotes the mass of the fluid particles. This is the standard continuity equation for the fluid and represents the conservation of mass in the smooth fluid limit.

Let us next consider the components of $T^{\mu\nu}$ in the nonrelativistic limit. The time–time part is

$$T^{00} = c^2 \int \frac{d^3\mathbf{p}}{E(\mathbf{p})} p^0 p^0 f(\mathbf{p}, x^i) \cong \int d^3\mathbf{p} f mc^2 \left(1 + \frac{1}{2}\frac{v^2}{c^2}\right)$$

$$= \rho c^2 + \int d^3 p f \left(\frac{1}{2}mv^2\right) \equiv \rho c^2 + N\bar{\epsilon}, \tag{8.8}$$

where we have expanded $\gamma = [1 - (v/c^2)]^{-1/2}$ in a Taylor series, retaining the first nontrivial term and defining the average kinetic energy as $\bar{\epsilon}$. The space–time part becomes

$$T^{0i} = T^{i0} = c^2 \int \frac{d^3\mathbf{p}}{E(\mathbf{p})} p^0 p^i f = \int d^3\mathbf{p} f \, mcv^i \left(1 + \frac{1}{2}\frac{v^2}{c^2}\right) = \left(\rho c\mathbf{V} + \frac{1}{c}\mathbf{q}\right)^i, \tag{8.9}$$

where we have defined the quantity

$$\mathbf{q} \equiv \int d^3\mathbf{p} \left(\frac{1}{2}mv^2\right)\mathbf{v} f. \tag{8.10}$$

The first term in Eq. (8.9) represents the flux of rest-mass energy that is due to the bulk flow and the second term represents the flux of kinetic energy. Finally, for the space–space part, we get

$$T^{ik} = c^2 \int \frac{d^3\mathbf{p}}{E(\mathbf{p})} p^i p^k f \cong \int d^3\mathbf{p} \, mv^i v^k f, \tag{8.11}$$

which may be thought of as a three-dimensional stress tensor. Substituting these expressions into the time component of the conservation law $(\partial T^{\mu 0}/\partial x^\mu) = 0$ gives

$$\frac{1}{c}\frac{\partial}{\partial t}(\rho c^2 + N\bar{\epsilon}) + \nabla \cdot \left(\rho c\mathbf{V} + \frac{\mathbf{q}}{c}\right) = 0, \tag{8.12}$$

which becomes, on use of continuity equation (8.7),

$$\frac{\partial(N\bar{\epsilon})}{\partial t} + \nabla \cdot \mathbf{q} = 0. \tag{8.13}$$

This relation ensures conservation of energy. The quantity $N\bar{\epsilon}$ is the kinetic-energy density in the fluid and the vector field $\mathbf{q}$ represents the kinetic-energy flux.

For future use, we note that $N\bar{\epsilon}$ can be expressed in a different form, separating out the bulk flow energy and kinetic energy of random motion. Writing $\mathbf{v} = \mathbf{V} + \mathbf{u}$ in the definition of $N\bar{\epsilon}$ in Eq. (8.8), we get

$$
N\bar{\epsilon} = N\left\langle \frac{1}{2}m(\mathbf{V}+\mathbf{u})^2 \right\rangle = \frac{1}{2}\rho V^2 + \frac{1}{2}Nm\langle u^2 \rangle = \rho\left(\frac{1}{2}V^2 + \epsilon\right)
$$

$$
= \frac{1}{2}\rho V^2 + \frac{3}{2}NT, \tag{8.14}
$$

where ϵ is internal energy per unit mass of the fluid. (We are using energy units to measure the temperature; in conventional units T will be replaced with $k_B T$.)

The spatial part of the conservation law $(\partial T^{\mu j}/\partial x^\mu) = 0$ becomes

$$
\frac{1}{c}\frac{\partial}{\partial t}\left(\rho c\mathbf{V} + \frac{\mathbf{q}}{c}\right)^i + \frac{\partial T^{ik}}{\partial x^k} = 0. \tag{8.15}
$$

The second term within the parentheses is of higher order in $1/c$ and hence can be dropped, giving

$$
\frac{\partial}{\partial t}(\rho V^i) + \frac{\partial T^{ik}}{\partial x^k} = 0. \tag{8.16}
$$

where the space–space part of the stress tensor has to be taken from Eq. (8.11). This is the force equation in the fluid limit. The quantity T^{ik} represents the stress tensor of the fluid; the gradient of the stress tensor gives rise to the acceleration of the fluid element. Alternatively, we can think of Eq. (8.16) as the momentum conservation equation. The quantity $\rho\mathbf{V}$ denotes the momentum density in the fluid, and the gradient of the stress tensor determines how this quantity changes.

Before proceeding further with the analysis of these equations, we derive an important relation that is valid for *any* system that satisfies Eqs. (8.7) and (8.16). This result, known as the tensor virial theorem, is independent of the specific form of T^{ik} given in Eq. (8.11) and will be used in later sections with different forms of T^{ik} arising from different physical processes. The tensor virial theorem depends on one preliminary result, which we derive first: For any function $F(\mathbf{x}, t)$, we claim that

$$
\frac{d}{dt}\int_V \rho F d^3\mathbf{x} = \int_V \rho \frac{dF}{dt} d^3\mathbf{x}, \tag{8.17}
$$

where the integration is over an entire region at the boundaries of which various physical quantities vanish. The proof is straightforward:

$$
\frac{d}{dt}\int_V \rho F d^3\mathbf{x} = \int_V \frac{\partial}{\partial t}(\rho F) d^3\mathbf{x} = \int_V \left\{\rho\frac{\partial F}{\partial t} + F[-\nabla\cdot(\rho\mathbf{V})]\right\} d^3\mathbf{x}
$$

$$
= \int_V \left[\rho\frac{\partial F}{\partial t} - \nabla\cdot(F\rho\mathbf{V}) + \rho\mathbf{V}\cdot\nabla F\right] d^3\mathbf{x}
$$

$$
= \int_V \rho\left(\frac{dF}{dt}\right) d^3\mathbf{x}. \tag{8.18}
$$

In the first line, Eq. (8.7) is used to express $\partial \rho / \partial t$ as a divergence; in arriving at the last line the divergence term has been removed by Gauss theorem and the other two terms have been combined to give the total time derivative of F. Let us next define the second moment of the mass distribution by

$$I^{ab} \equiv \int_V \rho x^a x^b d^3 \mathbf{x}. \tag{8.19}$$

Consider the second time derivative of I^{ab}, which can be transformed as follows:

$$\frac{d^2 I^{ab}}{dt^2} = \frac{d}{dt} \int_V \rho \frac{d}{dt}(x^a x^b) d^3 \mathbf{x} = \frac{d}{dt} \int_V \rho V^i \frac{\partial}{\partial x^i}(x^a x^b) d^3 \mathbf{x}$$

$$= \frac{d}{dt} \int_V \rho (V^a x^b + V^b x^a) d^3 \mathbf{x} = \int_V \frac{\partial}{\partial t}(\rho V^a x^b + \rho V^b x^a) d^3 \mathbf{x}. \tag{8.20}$$

One of the time derivatives has been pulled into the integral sign by the result obtained in Eq. (8.18) with $F = x^a x^b$. To convert this equation into more manageable form, we find an expression for $[\partial(\rho V^a x^b)/\partial t]$ in terms of derivatives of T^{ik}. Note that we can write

$$\frac{\partial(T^{jl} x^k)}{\partial x^l} = T^{jk} + x^k \frac{\partial T^{jl}}{\partial x^l} = T^{jk} - x^k \frac{\partial}{\partial t}(\rho V^j). \tag{8.21}$$

Because $\mathbf{x}$ and t are independent variables, this equation implies the relation

$$\frac{\partial}{\partial t}(\rho x^k V^j) = T^{jk} - \frac{\partial}{\partial x^l}(T^{jl} x^k). \tag{8.22}$$

Expression (8.20) can now be written as

$$\frac{d^2 I^{ab}}{dt^2} = \int_V \left[(T^{ab} + T^{ba}) - \frac{\partial}{\partial x^l}(T^{al} x^b + T^{bl} x^a) \right] d^3 \mathbf{x}. \tag{8.23}$$

The term with the spatial derivative can be converted by Gauss theorem to a surface integral that vanishes, thereby providing the relation

$$\frac{d^2 I^{ab}}{dt^2} = 2 \int T^{ab} d^3 \mathbf{x}. \tag{8.24}$$

This result is called the tensor virial theorem and is extremely useful in many astrophysical contexts. We can obtain a scalar virial theorem from this expression by taking the trace of both sides:

$$\frac{d^2 I}{dt^2} = 2 \int T d^3 \mathbf{x}, \quad I \equiv I^a_a = \int \rho |\mathbf{x}|^2 d^3 \mathbf{x}, \quad T \equiv T^a_a. \tag{8.25}$$

If the system is quasi periodic or steady, then the time average of the left-hand side will vanish, implying that the time average of the stress tensor integrated over the system should be zero.

Let us return to the dynamics of the fluid. Equations (8.7), (8.13), and (8.16) determine the nature of fluid flow in the nonrelativistic limit. Although these equations are exact, they do not form a closed system. As emphasised above, this is a general feature that arises whenever moments of the evolution equation are taken. The time derivative of the moment calculated with n powers of momenta will be determined by the derivatives of the moment with $n + 1$ powers of momenta; the system will never close with a finite number of moment equations.

To make any progress, we have to determine the form of the stress tensor T^{ik} and the energy-flux vector q^i by some external, physical requirements. If we manage to express these quantities in terms of other variables, Such as density, pressure, velocity, etc., we will have a closed set of equations that we can solve. When collisions are effective, i.e, when $l \ll R$, we can do this in a systematic manner by expanding the distribution function in perturbation series, with each term being lower than the preceding one by the small factor l/R, that is, we use the ansatz

$$f = f_0 + f_1 + f_2 \cdots \tag{8.26}$$

in Eq. (8.1), where $f_n = \mathcal{O}[(l/R)^n]$, and solve the equation by standard perturbation methods order by order. (This was done in Chap. 6, Section 6.8, to study radiative transfer.)

In practice, we are more interested in the form of T^{ik} and q^i corresponding to the above series than in the actual form of f_n's. This has already been done in Chap. 5, Section 5.13, for both ideal and viscous fluids. By taking the nonrelativistic limits of the fully relativistic expressions in Eqs. (5.261) and (5.262) we can determine T^{ik} and q^i. Once we know the form of T^{ik} and q^i, we can close the moment equations and work with them rather than with the distribution function. Let us consider the ideal and viscous fluid equations arising from the corresponding forms for T^{ik} and q^i.

8.3 Stress Tensor for an Ideal Fluid

The space–space part of the four-dimensional energy-momentum tensor, obtained in Chap. 3, Section 3.6, Eqs. (3.55), becomes, when $v \ll c$ and $P \ll \rho c^2$,

$$T_{ik} = \rho V_i V_k + P \delta_{ik}. \tag{8.27}$$

Similarly, for **q** we find

$$\mathbf{q} = \left[\frac{1}{2} V^2 + w \right] \rho \mathbf{V}, \tag{8.28}$$

where $w = \epsilon + (P/\rho)$ is the heat function (enthalpy) per unit mass of the fluid. The fact that energy flux is $\rho \mathbf{V}(V^2/2 + w)$ rather than $\rho \mathbf{V}(V^2/2 + \epsilon)$ has a simple interpretation. Using $w = \epsilon + P/\rho$, we can write the flux of energy through a surface as

$$-\int \rho \mathbf{V} \left(\frac{1}{2} V^2 + \epsilon \right) \cdot \hat{\mathbf{n}} \, dS - \int P \mathbf{V} \cdot \hat{\mathbf{n}} \, dS. \qquad (8.29)$$

The first term is the bulk kinetic energy and internal energy transported through the surface, and the second term is the work done by the pressure forces on the fluid. Because these equations express T_{ik} and q_i in terms of other variables ρ, $\mathbf{V}$, and P, we can now use the moment equations to study the evolution of the fluid.

Substitution of these expressions into Eqs. (8.7), (8.13), and (8.16) will lead to the equations of motion for a gas in which all dissipative processes are ignored. We get the conservation laws for mass and momentum from Eqs. (8.7) and (8.16):

$$\frac{\partial \rho}{\partial t} + \nabla \cdot (\rho \mathbf{V}) = \frac{\partial \rho}{\partial t} + (\mathbf{V} \cdot \nabla)\rho + \rho \nabla \cdot (\mathbf{V}) = \frac{d\rho}{dt} + \rho \nabla \cdot \mathbf{V} = 0, \quad (8.30)$$

$$\frac{\partial \mathbf{V}}{\partial t} + (\mathbf{V} \cdot \nabla)\mathbf{V} = \frac{d\mathbf{V}}{dt} = -\frac{\nabla P}{\rho}. \qquad (8.31)$$

Further, such an ideal fluid has no dissipative processes and hence the fluid elements do not exchange heat during the motion. It follows that the entropy per fluid particle $s \equiv (S/N)$ will also be conserved for each fluid element. This gives

$$\frac{ds}{dt} = \frac{\partial s}{\partial t} + \mathbf{V} \cdot \nabla s = 0. \qquad (8.32)$$

This equation is equivalent to the conservation law for heat flux [Eq. (8.13)] and is also the nonrelativistic version of Eq. (5.260), but it is often easier to use it in this form. We now need to solve for the five variables, ρ, $\mathbf{V}$, P, and we have five equations in Eqs. (8.30)–(8.32). The internal energy per unit mass can be expressed in terms of temperature as $\epsilon = (3/2)(T/\mu m_H)$ and the temperature, in turn, can be expressed in terms of pressure and density by use of the equation of state.

The system described by these equations represents what is known as an ideal fluid, which we shall study in Section 8.6 below. However, first we proceed to the next order of approximation and determine the corrections to the stress tensor and energy flux that are due to the existence of gradients in temperature, number density, and bulk velocity.

8.4 Stress Tensor for a Viscous Fluid

We obtain T^{ik} and q^i of the ideal fluid after ignoring contributions from the gradients in temperature, number density, and bulk velocity. In the next order of approximation, in which these gradients are taken into account, we expect T_{ik} and q_i to contain terms that are proportional to the gradients $(\partial T/\partial x^i)$ and

$(\partial V_k/\partial x^i)$. We write these expressions, correct to linear order in the gradients, as

$$T_{ik} = \rho V_i V_k + P\delta_{ik} - \sigma_{ik}, \quad q_i = \left(\frac{1}{2}\rho V^2 + w\right)V_i + Q_i. \quad (8.33)$$

We can determine these corrections σ_{ik} and Q_i by solving Eq. (8.1) in the next order of perturbation to determine f_1 and then computing T^{ik} and q^i by using $f = f_0 + f_1$. However, we can obtain these corrections more simply by taking the nonrelativistic limit of the expressions obtained in Chap. 5, Exercise 5.21. For example, the spatial components of the four-dimensional energy-momentum tensor obtained in Exercise 5.21 in the nonrelativistic limit is given by

$$\sigma_{ik} = 2\eta\left(V_{ik} - \frac{1}{3}\delta_{ik}\nabla\cdot\mathbf{V}\right) - \zeta\delta_{ik}\nabla\cdot\mathbf{V}. \quad (8.34)$$

Although this result is correct, the manner in which we have obtained it (by taking the nonrelativistic limit of the fully relativistic expression) hides the physical origin of different terms. In view of this fact, we provide an alternative derivation of this expression and the one for q_i.

Let us consider the correction Q^i first. In an ideal fluid, the flux of energy is described by $\mathbf{q}$, given in Eq. (8.28). In the presence of temperature and velocity gradients we expect an additional flow of energy because the random motion of molecules at different locations will now be different. To the lowest order at which we are working, we can expand q_i as a Taylor series in the gradients of temperature and velocity and retain only the lowest-order terms. This will contribute to $\mathbf{q}$ a term of the form $-\kappa\nabla T$ (where κ is called the thermal conductivity) from the temperature gradient in addition to terms arising from velocity gradients. The latter contribution can be related to σ_{ik}. Because the energy density contributed by velocity gradients is already included in $-\sigma_{ik}$, the corresponding energy flux along the ith direction must be $-V^k\sigma_{ik}$. Hence the net energy flux must have the form

$$Q_i = -\kappa\frac{\partial T}{\partial x^i} - V^k\sigma_{ik}. \quad (8.35)$$

Let us next consider the contributions to T_{ik} arising from a spatial variation of $\mathbf{V}$, which are a bit more involved. To begin with, it is clear that we can again expand T_{ik} in a Taylor series and retain only the terms linear in $(\partial V_i/\partial x^k)$. We can write this gradient as a sum of symmetric and antisymmetric tensors:

$$\frac{\partial V_i}{\partial x^k} = V_{ik} + \frac{1}{2}\left(\frac{\partial V_i}{\partial x^k} - \frac{\partial V_k}{\partial x^i}\right), \quad (8.36)$$

where

$$V_{ab} \equiv \frac{1}{2}\left(\frac{\partial V_a}{\partial x_b} + \frac{\partial V_b}{\partial x_a}\right) \quad (8.37)$$

is the symmetric part.

We do not, however, expect any viscous dissipation when the entire fluid is in a state of uniform rigid rotation with angular velocity $\mathbf{\Omega}$, for which $\mathbf{V} = (\mathbf{\Omega} \times \mathbf{x})$. In this case, $V_{ik} = 0$ but the antisymmetric part does not vanish. It follows that T_{ik} should not have any contribution from the antisymmetric part if there should be no viscous dissipation in a state of uniform rotation. Hence T_{ik} must be a linear combination of V_{ik} and $\delta_{ik}(\nabla \cdot \mathbf{V})$, which are the only two symmetric, independent, second-rank tensors that are linear in the gradients of $\mathbf{V}$. Such a linear combination can be written as $\alpha V_{ik} + \beta \delta_{ik}(\nabla \cdot \mathbf{V})$. It is, however, conventional to separate out from V_{ik} its trace and rewrite the expression in the form

$$\sigma_{ik} = 2\eta \left(V_{ik} - \frac{1}{3}\delta_{ik}\nabla \cdot \mathbf{V} \right) - \zeta \delta_{ik}\nabla \cdot \mathbf{V}. \tag{8.38}$$

Thus the viscous-energy tensor is parameterised by two coefficients of viscosity, η and ζ. The term proportional to ζ characterises compressive motions of a fluid element; the term involving η characterises shearing motions, which does not change the volume of the fluid element. To linear order in the gradients of $\mathbf{V}$ and T no other contributions can arise to T_{ik}, and our system is completely determined by the three constants η, ζ, and κ.

The actual values of the coefficients of viscosity and heat conduction cannot be calculated by the above approach. To do this we have to substitute our ansatz Eq. (8.26) into Eq. (8.1) and determine explicitly the form of f_1. By solving the resulting equation we will be able to determine the form of these coefficients. (This was done in Chap. 6, Section 6.8 to calculate the photon flux that is due to the temperature gradient. The same approach works here as well.) We merely quote here the final result. If the angular scattering cross section is $(d\sigma/d\theta)$ for molecular collisions, then the thermal conductivity κ and viscosity η are given by

$$\kappa = \frac{75}{64}(m^4\beta^4)\left(\frac{m\beta}{2\pi}\right)^{1/2} \int_0^\pi d\theta \int_0^\infty dv \left(\frac{d\sigma}{d\theta}\right)(v^7 \sin^2 \theta) \exp\left(-\frac{1}{2}\beta m v^2\right),$$
$$\tag{8.39}$$

and $\eta = (4m/15)\kappa$, where $\beta = T^{-1}$. For a monatomic gas with no internal degrees of freedom, compressional motion cannot dissipate energy internally and hence $\zeta = 0$. In general, η plays a more important role than ζ. If the molecules are approximated as hard spheres of diameter d, then the above expressions reduce to

$$\kappa = \left(\frac{75}{64d^2\sqrt{\pi}}\right)\left(\frac{T}{m}\right)^{1/2}, \qquad \eta = \left(\frac{4m}{15}\right)\kappa. \tag{8.40}$$

The results are fairly insensitive to the explicit form of the interaction and change only slightly when higher-order perturbations are taken into account.

An order-of-magnitude derivation of the transport coefficients can be provided along the following lines: Consider a gaseous system in local thermodynamic equilibrium in which the particles are moving randomly in all directions. Let ΔA

be a small area in any plane perpendicular to the y axis. The number of particles crossing this area ΔA, in unit time, either from above or from below will be $(nv\Delta A/6)$. (The factor 6 arises because, on the average, one third of particles will be travelling along the y axis of which half will be travelling downwards.) The particles collide, on the average, within one mean free path $l=(n\sigma)^{-1}$. Hence the particles coming from above will transport properties of the gas at the point $y+l$, whereas the particles going from below will carry properties of the gas at the point $y-l$. If the properties of the gas vary along the y axis, then this transport of particles (or photons) can lead to interesting physical effects.

Consider the case in which the density of the particles varies along the y axis. The *net* number ΔN of particles that crosses the area ΔA per second will be $(1/6)\Delta Av[n(y+l)-n(y-l)]\approx(1/3)\Delta Avl$. The flux of particles along the y axis, $J_y\equiv(\Delta N/\Delta A)$, is conventionally expressed as $J_y=D(\partial n/\partial y)$, where D is called the coefficient of diffusion. A comparison with the expression obtained above shows that $D=(1/3)(vl)$.

If the temperature $T(y)$ of the gas varies along the y axis then the average energy of the particle $\epsilon(y)$ will vary along the y axis by $(\partial\epsilon/\partial y)=(\partial\epsilon/\partial T)(\partial T/\partial y)=C_V(\partial T/\partial y)$, where $C_V=3/2$ is the specific heat per particle. The transport of the particles will also carry thermal energy Q across the surface because of the temperature gradient. We can define the thermal conductivity κ by the relation $Q=\kappa(\partial T/\partial y)$. Comparing with the previous expression, we find that

$$\kappa = \frac{1}{3}(nvC_Vl) = \frac{vC_V}{3\sigma} = \frac{1}{2}\frac{v}{\sigma}. \tag{8.41}$$

Last, there is also a net momentum transfer per second across the area ΔA by the amount

$$\frac{\Delta P}{\Delta t} \simeq \frac{1}{6}(nv\Delta A)\left[2ml\left(\frac{\partial u_x}{\partial y}\right)\right]. \tag{8.42}$$

It follows that there will be a net force on the area ΔA in the xz plane when a velocity gradient exists along the y axis. We can express this viscous force per unit area $F_x=\Delta P/\Delta t\Delta A$ by the relation $F_x\equiv\eta(\partial u_x/\partial y)$, where η is the coefficient of viscosity. Comparing the expressions, we find

$$\eta = \frac{1}{3}mnlv \simeq \frac{mv}{3\sigma} = \left(\frac{2m}{3}\right)\kappa. \tag{8.43}$$

If we take $\sigma=(\pi d^2)$, then κ computed in Eq. (8.41) is 0.42 times the exact value and η computed in Eq. (8.43) is 1.04 times the exact value.

The above analysis shows that the gradients in density, temperature, and velocity field lead to the diffusion of particles, energy, and viscous dissipation. The expressions obtained above for particle and energy transport are quite general and can be used in different contexts when appropriate values for v and l are supplied.

8.5 Equations of Motion for the Viscous Fluid

The analysis in Section 8.4 shows that the flow of fluids can be described in terms of some well-defined macroscopic variables and certain coefficients that describe dissipative processes. The basic variables are (1) The macroscopic velocity $\mathbf{v}(\mathbf{x}, t)$, (2) the density $\rho(\mathbf{x}, t)$, and (3) the pressure $p(\mathbf{x}, t)$. All the other thermodynamic quantities are determined once ρ and p are given. [The velocity field and the pressure have been denoted so far by capital letters $\mathbf{V}(\mathbf{x}, t)$ and $P(\mathbf{x}, t)$ to avoid confusion with microscopic velocity $\mathbf{v}$ and momentum $\mathbf{p}$. Because the latter quantities will not be needed any longer we use the lowercase letters $\mathbf{v}(\mathbf{x}, t)$ and $p(\mathbf{x}, t)$ to denote fluid velocity and pressure, respectively, from now on.] The dissipative processes can be characterised by the coefficient of thermal conduction κ and the coefficients of viscosity η and ζ.

By substituting the form for T^{ik} and q^i into Eqs. (8.7), (8.13), and (8.16), we can obtain the macroscopic equations governing the flow of the fluid. Of these, the mass conservation equation has the same form for ideal and nonideal fluids:

$$0 = \frac{\partial \rho}{\partial t} + \frac{\partial}{\partial x^i}(\rho v^i) = \frac{\partial \rho}{\partial t} + v^i \frac{\partial \rho}{\partial x^i} + \rho(\nabla \cdot \mathbf{v}) \qquad (8.44)$$

or

$$\frac{d\rho}{dt} + \rho(\nabla \cdot \mathbf{v}) = 0. \qquad (8.45)$$

To obtain the force equation (also called the Euler equation) for a viscous fluid, we substitute the expression for T_{ik},

$$T_{ik} = \rho v_i v_k + p\delta_{ik} - 2\eta \left[v_{ik} - \frac{1}{3}\delta_{ik}(\nabla \cdot \mathbf{v}) \right] - \zeta \delta_{ik} \nabla \cdot \mathbf{v}, \qquad (8.46)$$

into Eq. (8.16). This gives

$$\frac{\partial}{\partial t}(\rho v_i) + \frac{\partial}{\partial x_k}(\rho v_i v_k) + \frac{\partial p}{\partial x^i}$$

$$= \eta \left[\frac{\partial}{\partial x_k}\left(\frac{\partial v_i}{\partial x^k}\right) + \frac{\partial}{\partial x_k}\left(\frac{\partial v_k}{\partial x^i}\right) \right] - \frac{2}{3}\eta \frac{\partial}{\partial x^i}(\nabla \cdot \mathbf{v}) + \zeta \frac{\partial}{\partial x^i}(\nabla \cdot \mathbf{v})$$

$$= \eta \nabla^2 v_i + \left(\frac{1}{3}\eta + \zeta\right) \frac{\partial}{\partial x^i} \nabla \cdot \mathbf{v}. \qquad (8.47)$$

Expanding the first two terms on the left-hand side and using the continuity equation to eliminate the $v_i(\partial \rho/\partial t)$ term leads to

$$\rho \frac{d\mathbf{v}}{dt} = -\nabla p + \eta \nabla^2 \mathbf{v} + \left(\zeta + \frac{1}{3}\eta\right) \nabla(\nabla \cdot \mathbf{v}). \qquad (8.48)$$

This is called the Navier-Stokes equation.

To derive the generalisation of Eq. (8.32) to viscous fluids, we start with the equation for energy flux [Eq. (8.13)] and substitute $N\bar{\epsilon} = \rho(v^2/2 + \epsilon)$ and the expression for the heat flux,

$$q_a = \rho\left(\frac{1}{2}v^2 + w\right)v_a - v^b\sigma_{ab} - \kappa\frac{\partial T}{\partial x_a}, \tag{8.49}$$

to obtain

$$\frac{\partial}{\partial t}\left[\rho\left(\frac{1}{2}v^2 + \epsilon\right)\right] = -\frac{\partial}{\partial x^a}\left[\rho v^a\left(\frac{1}{2}v^2 + w\right) - v_b\sigma^{ab} - \kappa\frac{\partial T}{\partial x_a}\right]. \tag{8.50}$$

The left-hand side denotes the rate of change of energy density; the two contributions to the energy density are from steady motion (with velocity **v**) and random motion [with an internal energy $\rho\epsilon = (3/2)NT$]. The energy flux on the right-hand makes three contributions: the first term represents the flow of kinetic energy and work done against pressure. The second term is the flux of energy that is due to internal friction. The last term is the heat flux that is due to conduction. We now reexpress this equation in a more convenient form.

Taking the dot product of the Euler equation with **v** and using the continuity equation, we get

$$\frac{\partial}{\partial t}\left(\frac{1}{2}\rho v^2\right) + \frac{\partial}{\partial x^k}\left(\frac{1}{2}\rho v^2 v^k\right) = -v^i\frac{\partial p}{\partial x^i} + v^i\frac{\partial\sigma_{ki}}{\partial x_k}. \tag{8.51}$$

Subtraction of this equation from Eq. (8.50) gives, after some rearrangement,

$$\frac{\partial}{\partial t}(\rho\epsilon) + \nabla\cdot(\rho\epsilon\mathbf{v}) = -p\nabla\cdot\mathbf{v} + \nabla\cdot(\kappa\nabla T) + \sigma^{ik}\frac{\partial v_i}{\partial x^k}. \tag{8.52}$$

Using the continuity equation, we can rewrite this as

$$\rho\frac{d\epsilon}{dt} + p\nabla\cdot\mathbf{v} = \nabla\cdot(\kappa\nabla T) + \sigma^{ik}\frac{\partial v_i}{\partial x^k}. \tag{8.53}$$

However, because

$$p\nabla\cdot\mathbf{v} = -(p/\rho)(d\rho/dt) = p\rho(d\rho^{-1}/dt) = \rho p\frac{d\mathcal{V}}{dt}, \tag{8.54}$$

where $\mathcal{V} \equiv \rho^{-1}$ is the specific volume, the left-hand side of Eq. (8.53) is

$$\rho\left(\frac{d\epsilon + pd\mathcal{V}}{dt}\right) = \rho T\frac{ds}{dt} = \rho T\left[\frac{\partial s}{\partial t} + (\mathbf{v}\cdot\nabla)s\right] \tag{8.55}$$

where s is the entropy density. Using this result, we obtain

$$\rho T\frac{ds}{dt} = \rho T\left[\frac{\partial s}{\partial t} + (\mathbf{v}\cdot\nabla)s\right] = \nabla\cdot(\kappa\nabla T) + \sigma^{ik}\frac{\partial v_i}{\partial x^k}. \tag{8.56}$$

This equation shows how the entropy of the fluid element increases because of dissipative processes. Obviously we recover Eq. (8.32) if $\kappa = 0$ and $\sigma_{ik} = 0$.

In general both the pressure and the density (or specific volume) of the fluid element will change as it moves along. Among these, the change in pressure is negligible if the fluid velocity is small compared with the velocity of sound (this will be discussed in detail in Section 8.8); the density variations will depend on temperature gradients as well and cannot, in general, be ignored. Hence we may think of the derivatives on the left-hand side of Eq. (8.56) as taken at constant pressure rather than at constant volume. In that case, we can write

$$\left(\frac{\partial s}{\partial t}\right) = \left(\frac{\partial s}{\partial T}\right)_P \left(\frac{\partial T}{\partial t}\right) = \frac{C_P}{T}\left(\frac{\partial T}{\partial t}\right), \quad \nabla s = \left(\frac{\partial s}{\partial T}\right)_P \nabla T = \frac{C_P}{T}\nabla T,$$

(8.57)

leading to the final equation

$$\rho C_P \left(\frac{dT}{dt}\right) = \sigma^{ik}\frac{\partial v_i}{\partial x^k} + \nabla\cdot(\kappa\nabla T).$$

(8.58)

In summary, the basic equations for fluid flow, including viscous dissipation and heat conduction, are

$$\frac{d\rho}{dt} + \rho\nabla\cdot\mathbf{v} = 0,$$

(8.59)

$$\rho\frac{d\mathbf{v}}{dt} = -\nabla p + \eta\nabla^2\mathbf{v} + \left(\zeta + \frac{1}{3}\eta\right)\nabla(\nabla\cdot\mathbf{v}),$$

(8.60)

$$\rho T\left(\frac{ds}{dt}\right) = \sigma^{ik}\frac{\partial v_i}{\partial x^k} + \nabla\cdot(\kappa\nabla T),$$

(8.61)

where

$$\frac{d}{dt} = \frac{\partial}{\partial t} + \mathbf{v}\cdot\nabla$$

(8.62)

and s is the entropy density. The temperature $T(\mathbf{x}, t)$ can be expressed in terms of $p(\mathbf{x}, t)$ and $\rho(\mathbf{x}, t)$ by use of the equation of state. These five equations will allow us to determine the five variables ρ, p, and $\mathbf{v}$.

From our derivation, it is clear that these equations conserve energy and momentum of the fluid. It is also of interest to ask how the entropy evolves under the above equations. To begin with, it can be verified by simple algebra that the derivatives of the velocity satisfy the identity

$$\sigma^{ik}\frac{\partial v_i}{\partial x^k} = \frac{1}{2}\eta\left(2V_{ik} - \frac{2}{3}\delta_{ik}\nabla\cdot\mathbf{v}\right)^2 + \zeta(\nabla\cdot\mathbf{v})^2.$$

(8.63)

Using this expression, the continuity equation, and Eq. (8.56), we find

$$\frac{\partial(\rho s)}{\partial t} = \rho\frac{\partial s}{\partial t} + s\frac{\partial\rho}{\partial t} = -s\nabla\cdot(\rho\mathbf{v}) - \rho(\mathbf{v}\cdot\nabla)s + \frac{1}{T}\nabla\cdot(\kappa\nabla T)$$

$$+ \frac{\eta}{2T}\left(2V_{ik} - \frac{2}{3}\delta_{ik}\nabla\cdot\mathbf{v}\right)^2 + \frac{\zeta}{T}(\nabla\cdot\mathbf{v})^2.$$

(8.64)

This equation describes the change of entropy of a fluid element with time. On the right-hand side the first two terms can be combined to give $-\nabla \cdot (\rho s \mathbf{v})$. Further, using

$$\frac{1}{T} \nabla \cdot (\kappa \nabla T) = \nabla \cdot \left(\frac{\kappa \nabla T}{T}\right) + \kappa \left(\frac{\nabla T}{T}\right)^2 \qquad (8.65)$$

in Eq. (8.64) and integrating the equation over a large volume of the fluid, we get

$$\frac{d}{dt} \int \rho s \, d^3x = \int \kappa \left(\frac{\nabla T}{T}\right)^2 d^3x + \int \frac{\eta}{2T} \left(2 V_{ik} - \frac{2}{3} \delta_{ik} \nabla \cdot \mathbf{v}\right)^2 d^3x$$
$$+ \int \frac{\zeta}{T} (\nabla \cdot \mathbf{v})^2 \, d^3x. \qquad (8.66)$$

(It is assumed that surface integrals vanish when taken over a surface at large distance.) The first term on the right-hand side gives the entropy increase arising from thermal conduction, and the other two terms describe the entropy increase arising from viscous dissipation.

In the above analysis, we have been using standard results of equilibrium thermodynamics in a context in which the system is only in *local* thermodynamic equilibrium and velocity and temperature gradients are present. Strictly speaking, the thermodynamic relations need to be modified in such a case. The quantities ρ, ϵ, and $\mathbf{v}$, being defined through microscopic (primitive) variables, continue to retain their validity. The entropy $s = s(\rho, \epsilon)$, however, will not be the true thermodynamic entropy in the strict sense. However, because the thermodynamic entropy should be a maximum in equilibrium, it cannot change when the variables deviate from their equilibrium values to first-order. In other words, to the first-order accuracy in the gradients of T and $\mathbf{v}$, the entropy used above may be taken to be the thermodynamic entropy.

Similar remarks apply to pressure. If we take the pressure for a viscous fluid $p = p(\rho, \epsilon)$ to be of the same form as in thermal equilibrium, such a pressure will *not* give the correct normal component of force on a surface element. There is a correction to the pressure term proportional to $\nabla \cdot \mathbf{v}$ that arises when velocity gradients are present. We see this most directly by noting that the T_{ik} for a viscous fluid has two contributions that are proportional to δ_{ik}: one term of the form $p\delta_{ik}$, where p is the thermodynamic pressure, and another term $\zeta(\nabla \cdot \mathbf{v})\delta_{ik}$ that arises from velocity gradients.

It should be noted that the form of corrections to the ideal fluid, to the lowest order in the gradients of macroscopic variables, is completely determined by η, ζ, and κ. No other terms are permitted in this order of approximation in the description of the fluid.

The structure of Eqs. (8.59), (8.60), and (8.61) allow easy generalisation to more complicated situations. For example, consider an astrophysical context in which radiative forces, heating, and cooling of gas as well as gravitational field are present. The modifications of these equations to incorporate these effects will be as follows: (1) the equation for mass conservation, Eq. (8.59), continues to hold even in this context. (2) In force equation (8.60), we should add on the right-hand side $-\rho\nabla\phi$ to represent the gravitational force and $\mathbf{f}_{rad}$ to represent the radiation force. (The expression for $\mathbf{f}_{rad}$ was derived in Chap. 6, Section 6.8.) (3) To the right-hand side of Eq. (8.61) we have to add the work done by gravitational and radiative forces $(-\rho\mathbf{v}\cdot\nabla\phi + \mathbf{v}\cdot\mathbf{f}_{rad})$. In addition, we should add a term $\Gamma - \Lambda$, where Γ represents the volumetric gain of energy that is due to heating and Λ represents the volumetric loss of energy that is due to cooling.

The components of the gravitational force $\mathbf{F} = -\nabla\phi$ can also be expressed in the form $(\partial T^{ik}/\partial x^i)$, where the gravitational stress tensor is given by

$$T^a{}_b = \frac{1}{4\pi G}\left[\partial^a\phi\,\partial_b\phi - \frac{1}{2}\delta^a_b(\partial^i\phi\,\partial_i\phi)\right],\tag{8.67}$$

with $\partial_a \equiv (\partial/\partial x^a)$. The integral of the trace, $T = T^a{}_a$, of this quantity gives the gravitational potential energy of matter:

$$\int_{\mathcal{V}} d^3\mathbf{x}\,T = \int_{\mathcal{V}} d^3\mathbf{x}\,T^a{}_a = -\frac{1}{8\pi G}\int_{\mathcal{V}}(\partial^a\phi\,\partial_a\phi)\,d^3\mathbf{x}$$
$$= -\frac{1}{8\pi G}\int_{\mathcal{V}}[\partial_a(\phi\,\partial^a\phi) - \phi\nabla^2\phi]\,d^3\mathbf{x} = \frac{1}{2}\int\rho\phi\,d^3\mathbf{x}.\tag{8.68}$$

The last equality follows from converting the divergence by use of Gauss theorem and $\nabla^2\phi = 4\pi G\rho$. For an ideal self-gravitating fluid, the corresponding integral over the *total* stress tensor will give

$$\int d^3\mathbf{x}\,T_{tot} = \int d^3\mathbf{x}[T_{gas} + T_{grav}] = \int d^3\mathbf{x}\left[\rho v^2 + 3p + \frac{1}{2}\rho\phi\right]$$
$$\equiv 2K + U + 3\int_{\mathcal{V}} d^3\mathbf{x}\,p,\tag{8.69}$$

where K is the kinetic energy and U is the gravitational potential energy. If the system is in a steady state, then the scalar virial theorem [Eqs. (8.25)] shows that this expression should vanish. Let us explore briefly some consequences of this result.

(1) Denoting the third term as $3pV$, where V is the effective volume of the system defined by this relation, the virial theorem in steady state can be written as $2K + U + 3pV = 0$. We apply this relation to a (nearly) spherical configuration of fluid with mass M, radius R, and temperature T in pressure equilibrium with the surroundings that exert a pressure p. The gravitational potential energy of such a configuration will be $U_{gr} = -\alpha(GM^2/R)$, where α is a numerical constant possibly of the order of unity. The kinetic energy can be written as $K = (3/2)Mu^2$, where $u^2 = (k_BT/m)$ is the square of the typical velocity. The virial theorem now

gives the relation

$$p = \frac{1}{4\pi}\left(\frac{3Mu^2}{R^3} - \alpha\frac{GM^2}{R^4}\right). \tag{8.70}$$

In the absence of gravity ($G = 0$), this equation gives $pV = Mu^2$, which is merely the ideal-gas law. In this particular case, this relation may be interpreted as giving the balance between the external pressure p required for keeping a cloud of volume V and temperature T confined in some region. In the presence of gravitational field, the situation changes qualitatively and we obtain the relation

$$p + \frac{\alpha}{4\pi}\frac{GM^2}{R^4} = \frac{3Mu^2}{4\pi R^3}. \tag{8.71}$$

This shows that the amount of pressure required for keeping the cloud confined is *less* in the presence of gravity as self-gravity will help in keeping the cloud together. As we increase the value of M, a point is reached when p vanishes. For larger values of M, the pressure p will become negative, indicating an instability in the system. The critical value of mass is given by

$$M_{\text{crit}} = \frac{3}{\alpha}\left(\frac{u^2 R}{G}\right) = \frac{3}{\alpha}\left(\frac{k_B T R}{mG}\right). \tag{8.72}$$

Clouds of a given size R and temperature T will collapse under self-gravity if $M > M_{\text{crit}}(T, R)$. We use this result in Vols. II and III in the study of star formation and galaxy formation.

(2) In many astrophysical contexts, we must deal with gaseous systems in which the mean kinetic energy K and the total internal energy of the gas U_{int} are related by $K = (3/2)(\gamma - 1)U_{\text{int}}$, where γ is the ratio of the specific heats [see the discussion preceding Eq. (5.96) in Chap. 5]. For such systems we can draw some general conclusions by invoking virial theorem (8.25), written in the form

$$\frac{1}{2}\frac{d^2 I}{dt^2} = 2K + U_{\text{gr}} = U_{\text{gr}} + 3(\gamma - 1)U_{\text{int}}$$
$$= 3(\gamma - 1)E - (3\gamma - 4)U_{\text{gr}}, \tag{8.73}$$

where $E = U_{\text{gr}} + U_{\text{int}}$ is the total energy of the system. We have now ignored the integral over external pressure by assuming that the system is isolated. In steady state, the time derivative must vanish, implying that

$$E = \frac{3\gamma - 4}{3(\gamma - 1)}U_{\text{gr}}, \tag{8.74}$$

which gives a simple relation between the total energy and the gravitational potential energy U_{gr}. For a bound system, we require that $E < 0$; because $U_{\text{gr}} < 0$ it is necessary that $\gamma > 4/3$. The system will be unstable when $\gamma < 4/3$. Some insight into the nature of instability can be obtained by rewriting Eq. (8.73) in the form

$$E = \frac{\gamma - (4/3)}{\gamma - 1} U_{\mathrm{gr}} + \frac{(d^2 I/dt^2)}{6(\gamma - 1)}. \tag{8.75}$$

When $\gamma > 4/3$ and $E < 0$, the last term can vanish and the system will be in equilibrium. If some physical process now drives γ towards $4/3$ with E still negative, the last term cannot vanish and we must have $2E \approx (d^2 I/dt^2)$ when $\gamma \approx 4/3$. Because $E < 0$, this implies that $(d^2 I/dt^2) < 0$. Usually such a condition would imply a collapse of the system.

When radiative effects are significant and matter and radiation are strongly coupled, we also have to modify the equations of radiative transport derived in Chap. 6, Section 6.8. The transport equation for the energy density of radiation will now be

$$\frac{\partial U_{\mathrm{rad}}}{\partial t} + \nabla \cdot [\mathbf{F}_{\mathrm{rad}} + (U_{\mathrm{rad}} + P_{\mathrm{rad}})\mathbf{v}] = -\mathbf{v} \cdot \mathbf{f}_{\mathrm{rad}} - \Gamma + \Lambda. \tag{8.76}$$

The structure of this equation is obvious from the fact that whatever appears as a source for matter should appear as a sink for radiation and vice versa. The flux term now contains radiative enthalpy flux $(U_{\mathrm{rad}} + P_{\mathrm{rad}})\mathbf{v}$, carried by the bulk motion of the matter and the energy flux $\mathbf{F}_{\mathrm{rad}}$ that is due to the diffusion of photons. To the same order of approximation, we can take $\mathbf{f}_{\mathrm{rad}} = -\nabla P_{\mathrm{rad}}$. In this case, Eq. (8.76) simplifies to

$$\frac{\partial U_{\mathrm{rad}}}{\partial t} + \nabla \cdot [\mathbf{F}_{\mathrm{rad}} + U_{\mathrm{rad}}\mathbf{v}] = -P_{\mathrm{rad}} \nabla \cdot \mathbf{v} - \Gamma + \Lambda. \tag{8.77}$$

All the quantities in the above equation are integrated over frequencies.

8.6 Flow of Ideal Fluids

We rapidly overview in this section several aspects of fluid flow, starting with ideal fluids. The equations developed here will be used repeatedly in the coming sections.

The simplest (and most drastic) approximation that we can make in the study of fluids is to set $\kappa = 0$, $\eta = 0$, and $\zeta = 0$, thereby ignoring all dissipative processes. Such a fluid is called ideal. We have seen above [see Eq. (8.32)] that, for an ideal fluid,

$$\frac{ds}{dt} = \frac{\partial s}{\partial t} + \mathbf{v} \cdot \nabla s = 0, \quad s = s(\rho, p). \tag{8.78}$$

This implies that s is constant for a given fluid element but does not necessarily imply that all fluid elements have the same entropy. We next rewrite Euler's equation in a more convenient form. Using the identity

$$(\mathbf{v} \cdot \nabla)\mathbf{v} = \frac{1}{2}\nabla v^2 - \mathbf{v} \times (\nabla \times \mathbf{v}), \tag{8.79}$$

we find that Euler's equation becomes

$$\frac{\partial \mathbf{v}}{\partial t} - (\mathbf{v} \times \boldsymbol{\Omega}) = -\frac{1}{2}\nabla v^2 - \frac{1}{\rho}\nabla p - \nabla \phi, \tag{8.80}$$

where $\boldsymbol{\Omega} = \nabla \times \mathbf{v}$ is the vorticity and we have introduced the gravitational potential ϕ in order to keep the discussion general. Taking the curl of this equation we find that

$$\frac{\partial \boldsymbol{\Omega}}{\partial t} = \nabla \times (\mathbf{v} \times \boldsymbol{\Omega}) - \nabla \times \left(\frac{1}{\rho}\nabla p\right) = \nabla \times (\mathbf{v} \times \boldsymbol{\Omega}) + \frac{1}{\rho^2}(\nabla \rho \times \nabla p). \tag{8.81}$$

This equation, along with Eq. (8.78) and the continuity equation, provides a set of five equations for the five variables ρ, p, and $\mathbf{v}$. (Note that s is treated as a given function of ρ and p and $\boldsymbol{\Omega}$ is given in terms of $\mathbf{v}$.) These equations form a fairly formidable set and cannot be solved in closed form without further approximations. Several such approximations are used in astrophysics; some of them are listed below and their important consequences discussed.

Exercise 8.1

Kinematics of fluid flow: The general pattern of fluid flow is characterised by the velocity field $v_i(x^k)$. This exercise explores different types of fluid flow and introduces some standard jargon. (1) Decompose the quantity $\partial_k v_i \equiv (\partial v_i / \partial x^k)$ into the form

$$\partial_k v_i \equiv S_{ki} + \frac{1}{3}\theta \delta_{ki} - \frac{1}{2}\epsilon_{ikl}\Omega^l, \tag{8.82}$$

where S_{ki} is a symmetric, traceless, second-rank tensor, $\theta \equiv \nabla \cdot \mathbf{v}$, and $\boldsymbol{\Omega} \equiv \nabla \times \mathbf{v}$. Prove that this decomposition is unique. (2) Show that θ (called expansion) produces a volume change of the fluid element without rotation or deformation; S_{ik} (called shear) deforms the fluid element without volume change or rotation; $\boldsymbol{\Omega}$ (called vorticity) rotates the fluid element without volume change or deformation. (3) Expressing the stress tensor of Eq. (8.46) in terms of the variables S_{ij} and θ, show that equation of motion (8.16) becomes

$$\rho\left[\frac{\partial v^i}{\partial t} + v^j\frac{\partial v^i}{\partial x^j}\right] = -\frac{\partial p}{\partial x^i} - \rho\frac{\partial \phi}{\partial x^i} + \frac{\partial}{\partial x^i}(\zeta\theta) + 2\frac{\partial}{\partial x^j}(\eta S^{ij}), \tag{8.83}$$

where we have also added the gravitational force on the right-hand side. Similarly, show that Eq. (8.64) for entropy generation can be written in the form

$$T\left[\frac{\partial}{\partial t}(\rho s) + \nabla \cdot (\rho s \mathbf{v})\right] - T\nabla \cdot \left(\frac{\kappa \nabla T}{T}\right) = \zeta\theta^2 + 2\eta S_{jk}S^{jk} + \frac{\kappa}{T}(\nabla T)^2. \tag{8.84}$$

Exercise 8.2

Frozen vector fields: Consider any vector field $\mathbf{A}(\mathbf{x}, t)$ that has zero divergence, $\nabla \cdot \mathbf{A} = 0$, as the magnetic field or the vorticity. As the fluid flows, this vectorial quantity will change in space and time. The field lines of a vector field are the curves drawn in such a way that the tangent to the curve is in the direction of the vector field at every point. One particular situation of interest arises when the fluid moves in such a way that the field lines of **A** are frozen into the fluid; in other words, each fluid particle remains attached to the same field line as time goes on. The fluid particle, of course, can move up and down the field line like beads in a string, but they cannot move from one field line to another. In this case, show that **A** must satisfy

$$\frac{\partial \mathbf{A}}{\partial t} - \nabla \times (\mathbf{v} \times \mathbf{A}) = \frac{d\mathbf{A}}{dt} - (\mathbf{A} \cdot \nabla)\mathbf{v} + (\nabla \cdot \mathbf{v})\mathbf{A} = 0. \tag{8.85}$$

8.6.1 Barotropic Fluids

The structure of the last term in Eq. (8.81) shows that if p can be expressed as a function of ρ alone, then ∇p and $\nabla \rho$ will be parallel to each other, making this term vanish. Fluids in which pressure can be expressed as a function of density alone are called barotropic. (We shall see specific examples of barotropic fluids later in Chap. 10.) For a barotropic fluid, Eq. (8.81) becomes

$$\frac{\partial \mathbf{\Omega}}{\partial t} = \nabla \times (\mathbf{v} \times \mathbf{\Omega}). \tag{8.86}$$

From our general result in Exercise 8.2, it follows that the flux of $\mathbf{\Omega}$, viz., the quantity

$$\Gamma \equiv \int \mathbf{\Omega} \cdot d\mathbf{s} = \int (\nabla \times \mathbf{v}) \cdot d\mathbf{s} = \oint \mathbf{v} \cdot d\mathbf{l}, \tag{8.87}$$

called the circulation, is conserved. (In this context, conservation means that the value of the circulation defined through some contour C has the same value as the contour moves through the fluid.) Equivalently, the vorticity $\mathbf{\Omega}$ is frozen on to the fluid. This result is important because, very often, boundary conditions might ensure that $\mathbf{\Omega} = 0$ at an initial instant for a fluid element; then the conservation implies that $\mathbf{\Omega}$ continues to remain zero as the fluid flows.

8.6.2 Steady Flows of Ideal Fluids

The equations of motion for the ideal fluid take still simpler forms if more assumptions are made. A flow is called steady if $(\partial/\partial t)(\text{anything}) = 0$. In such a case, it is convenient to introduce the concept of streamlines, which are the integral curves of the velocity field determined by the equations

$$\frac{dx}{v_x} = \frac{dy}{v_y} = \frac{dz}{v_z}. \tag{8.88}$$

The tangent to the streamline at any point gives the direction of the velocity at that point. In steady flow, streamlines do not change with time and coincide with the paths of the fluid particle. (This, of course, does not happen in nonsteady flow: The tangent to the streamline gives the direction of the velocity of fluid particles at different spatial locations at a given time, whereas the tangent to the actual path of particles gives the velocities at different times.) In the case of a steady flow of an ideal fluid, both $(\partial s/\partial t)$ and (ds/dt) vanish, implying that $v^i(\partial s/\partial x^i) = 0$, that is, s remains constant along each streamline. Further, along any streamline,

$$dw = d\epsilon + pd\left(\frac{1}{\rho}\right) + \frac{dp}{\rho} = Tds - pd\left(\frac{1}{\rho}\right) + pd\left(\frac{1}{\rho}\right) + \frac{dp}{\rho}$$

$$= Tds + \frac{dp}{\rho} = \frac{dp}{\rho}, \tag{8.89}$$

as entropy is constant $(ds = 0)$ along any streamline. Now consider Euler's equation (8.80), which can be written in the form

$$\frac{1}{2}\nabla v^2 - (\mathbf{v} \times \mathbf{\Omega}) = -\frac{1}{\rho}\nabla p - \nabla\phi, \tag{8.90}$$

because $(\partial \mathbf{v}/\partial t) = 0$. Let $\mathbf{l}$ be a unit tangent vector to any streamline. Taking the dot product of the above equation with $\mathbf{l}$, we find that the first term on the left-hand side gives the rate of change of $(1/2)v^2$ along the streamline; the second term is perpendicular to $\mathbf{v}$ (and hence to $\mathbf{l}$) and thus makes no contribution. The first term on the right-hand side becomes dw when evaluated along the streamline and the second term gives $d\phi$ along the streamline. It follows that

$$B \equiv \left(\frac{1}{2}v^2 + w + \phi\right) \tag{8.91}$$

is a constant along a streamline. The constancy of B is called the Bernoulli equation. Note that, for a perfect gas, $w = (\gamma p/\rho)(\gamma - 1)^{-1}$ [see Eq. (5.97)] and for an incompressible liquid, with $\gamma \to \infty$, we have $w = (p/\rho)$.

The above analysis has used only the fact that s is constant along each streamline. Such a flow may be called adiabatic in the sense that the fluid element does not exchange heat with its surroundings. If we further assume that $s(\rho, p) = \text{constant}$ for *all* fluid elements globally, then we have an *isoentropic* flow. Given the condition $s(\rho, p) = \text{constant}$, we can express p as a function of ρ and the flow will necessarily be barotropic and the circulation Γ will also be conserved.

A special case of steady flow corresponds to hydrostatics in which there is no actual fluid flow, i.e., $\mathbf{v} = 0$. Then the pressure gradient balances the gravitational force and the condition of equilibrium becomes

$$\nabla p = -\rho\nabla\phi, \quad \nabla^2\phi = 4\pi G\rho, \tag{8.92}$$

where the second equation assumes that the fluid is self-gravitating, i.e., the gravitational force is due to only the mass density of the fluid. Combining these equations, we get $\nabla \cdot (\nabla p/\rho) = -4\pi G\rho$. This equation follows from the requirement of mechanical equilibrium alone and we have not assumed the existence of thermal equilibrium. Further, taking the curl of the first equation in

Eqs. (8.92), we find that $\nabla \phi \times \nabla \rho = 0$; i.e., contours of constant density coincide with equipotential surfaces, allowing us to write $\rho = \rho(\phi) = -(dp/d\phi)$. This, in turn, implies that the pressure difference between two equipotential surfaces depends on only the potential difference between these two surfaces. Therefore the isobars also coincide with equipotential surfaces.

For a nonrotating body in thermal equilibrium, p and ρ will be spherically symmetric and Eqs. (8.92) combine to give

$$\frac{1}{r^2}\frac{d}{dr}\left(\frac{r^2}{\rho}\frac{dp}{dr}\right) = -4\pi G\rho. \tag{8.93}$$

If the system is further assumed to be barotropic, we will have a relation $p = p(\rho)$ and Eq. (8.93) can be immediately integrated. One possible choice, which has applications in several astrophysical contexts, is given by the polytropic equation of state for which $p \propto \rho^{(n+1)/n}$. We shall discuss the solutions for polytropic spheres in Chap. 10.

8.6.3 Irrotational, Isoentropic Flow of Ideal Fluids

Consider next the situation in which the flow is not necessarily steady but is irrotational (i.e., $\nabla \times \mathbf{v} = 0$) and isoentropic. In this case we can set $\mathbf{v} = \nabla \psi$, where ψ may be thought of as a potential for the velocity field. This kind of flow can arise in any barotropic fluid if $\Omega = 0$ at the boundary. (It must be stressed, however, that this is an idealisation while we are considering flow past material bodies as we have not taken the boundary condition on the material correctly. In general, a thin boundary layer will form around the body in which the dynamics will be lot more complicated.) Setting $\Omega = 0$ and $\mathbf{v} = \nabla\psi$ in the Euler equation for an ideal fluid (8.80), we get

$$\nabla\left(\frac{\partial\psi}{\partial t} + \frac{1}{2}v^2\right) = -\frac{\nabla p}{\rho} - \nabla\phi. \tag{8.94}$$

When the flow is isoentropic, $ds = 0$ and Eq. (8.89) gives $(\nabla p/\rho) = \nabla w$, where w is the enthalpy. Then,

$$\frac{\partial\psi}{\partial t} + \frac{1}{2}v^2 + w + \phi = f(t), \tag{8.95}$$

where $f(t)$ is an arbitrary function of time. This function can be set to zero without any loss of generality as we can always add to ψ an arbitrary function of time without changing the relation $\mathbf{v} = \nabla\psi$. We thus get

$$\frac{\partial\psi}{\partial t} + \frac{1}{2}v^2 + w + \phi = 0, \tag{8.96}$$

which is the generalisation of Bernoulli equation (8.91) for a time-dependent situation.

8.6.4 Incompressible, Irrotational Flow of Ideal Fluids

Finally we consider the most drastic (and most tractable) case in which we take $\rho = $ constant. This condition is equivalent to assuming that the fluid is incompressible and reduces the continuity equation to $\nabla \cdot \mathbf{v} = 0$. Further, because $\rho = $ constant can be taken as a barotropic condition, the circulation is conserved and we also have

$$\frac{\partial(\nabla \times \mathbf{v})}{\partial t} = \nabla \times [\mathbf{v} \times (\nabla \times \mathbf{v})]. \tag{8.97}$$

Thus the equations for fluid flow can be expressed entirely in terms of velocity alone.

If the flow is further assumed to be irrotational, then $\boldsymbol{\Omega} = 0$ and we can set $\mathbf{v} = \nabla \psi$. In this case, the condition $\nabla \cdot \mathbf{v} = 0$ shows that ψ satisfies Laplace's equation $\nabla^2 \psi = 0$.

8.7 Flow of Viscous Fluids

We next consider briefly some aspects of viscous flow. Although this is a more realistic description of fluids, it is mathematically more intractable. Fortunately viscous effects are small in several astrophysical contexts, and hence the description outlined here will be needed only in special cases. For a viscous fluid, we start with equation of motion (8.48) written in the form

$$\left(\frac{\partial \mathbf{v}}{\partial t} + \mathbf{v} \cdot \nabla\right)\mathbf{v} = -\frac{\nabla p}{\rho} + \frac{\eta}{\rho}\nabla^2 \mathbf{v} + \left(\frac{\zeta}{\rho} + \frac{1}{3}\frac{\eta}{\rho}\right)\nabla(\nabla \cdot \mathbf{v}). \tag{8.98}$$

Taking the curl of this equation and using the vector identity

$$\nabla \times (\mathbf{v} \cdot \nabla)\mathbf{v} = -\nabla \times [\mathbf{v} \times (\nabla \times \mathbf{v})], \tag{8.99}$$

we get

$$\frac{\partial \boldsymbol{\Omega}}{\partial t} - \nabla \times (\mathbf{v} \times \boldsymbol{\Omega}) = \frac{\eta}{\rho}\nabla^2 \boldsymbol{\Omega} + \frac{1}{\rho^2}(\nabla \rho) \times \left[\nabla p - \eta\nabla^2 \mathbf{v} - \left(\zeta + \frac{1}{3}\eta\right)\nabla\theta\right].$$
$$\tag{8.100}$$

On the right-hand side the first term represents the diffusion of vorticity $\boldsymbol{\Omega}$ and the second term arises because of density gradients. This generalises Eq. (8.81) for the ideal fluid.

8.7.1 Incompressible Flow of Viscous Fluids

If the fluid is incompressible, implying that $\rho = $ constant and $\nabla \cdot \mathbf{v} = 0$, we obtain

$$\frac{\partial \boldsymbol{\Omega}}{\partial t} - \nabla \times (\mathbf{v} \times \boldsymbol{\Omega}) = \frac{\eta}{\rho}\nabla^2 \boldsymbol{\Omega}. \tag{8.101}$$

By expanding $\nabla \times (\mathbf{v} \times \mathbf{\Omega})$ and using $\nabla \cdot \mathbf{v} = 0$, we can write this as

$$\frac{\partial \mathbf{\Omega}}{\partial t} + (\mathbf{v} \cdot \nabla)\mathbf{\Omega} - (\mathbf{\Omega} \cdot \nabla)\mathbf{v} = \frac{\eta}{\rho}\nabla^2 \mathbf{\Omega} \tag{8.102}$$

or

$$\frac{d\mathbf{\Omega}}{dt} - (\mathbf{\Omega} \cdot \nabla)\mathbf{v} = \frac{\eta}{\rho}\nabla^2 \mathbf{\Omega}. \tag{8.103}$$

To relate the pressure to the velocity field, we take the divergence of Eq. (8.80). This gives, on use of $\nabla \cdot \mathbf{v} = 0$,

$$\nabla \cdot [(\mathbf{v} \cdot \nabla)\mathbf{v}] = \frac{\partial}{\partial x^i}\left(v^k \frac{\partial v^i}{\partial x^k}\right) = \frac{\partial v^k}{\partial x^i}\frac{\partial v^i}{\partial x^k} = \frac{\partial^2}{\partial x^i \partial x^k}(v^i v^k) = -\frac{1}{\rho}\nabla^2 p. \tag{8.104}$$

Given Eq. (8.101) and $\nabla \cdot \mathbf{v} = 0$, we can determine the velocity field. Then we can determine p by integrating Eq. (8.104). This provides the complete solution to the flow of incompressible viscous fluid.

8.7.2 Scaling Relations in Viscous Flows

It is possible to obtain some *general* results regarding the steady, incompressible viscous flow of fluids past solid bodies – a problem that is of relevance in several practical situations. In analysing such a flow, we introduce a quantity called the Reynolds number, R, which is defined as

$$R = \frac{\rho u l}{\eta} \equiv \frac{u l}{\nu}, \tag{8.105}$$

where l is a characteristic scale in the problem (which could, for example, be related to the dimension of the solid body moving through the fluid), u is the characteristic velocity, and $\nu \equiv (\eta/\rho)$.

In the study of incompressible viscous fluids, the only dimensional parameter that appears in the equation is $\nu = (\eta/\rho)$, which has the dimension L^2/T. Consider now the motion of a body, characterised by the size l and velocity u, through this fluid. The boundary conditions will introduce the variables l and u into the problem. The only dimensionless quantity that can be formed from l, ν, and u is $R = (ul/\nu)$. By reexpressing the equations of motion in terms of the ratios $(\mathbf{v}/u)$, (x/l), and (ut/l), we can reduce the equations to a dimensionless form that involves only the Reynolds number R. Further, in the steady state, the solution will have no time dependence. It follows that the solutions to the equations of motion will depend on only dimensionless variables and R.

In particular, the velocity and the pressure must scale in proportion to u and ρu^2, which are the only two quantities with the appropriate dimensions. Hence we must have

$$\mathbf{v} = u\mathbf{f}_1\left(\frac{\mathbf{x}}{l}, R\right), \quad p = \rho u^2 f_2\left(\frac{\mathbf{x}}{l}, R\right), \tag{8.106}$$

where $\mathbf{f}_1$ and f_2 are functions that can be determined only by solving the equations of fluid flow. Similarly the only combination having the dimensions of force is $\rho u^2 l^2$. Hence the force acting on the solid body must have the form $f = \rho u^2 l^2 f_3(R)$. These relations express the following important fact: Bodies of geometrically similar shape, differing only in the overall size, will induce flows that are self-similar if the Reynolds number R remains constant, that is, if we solve the problem with a given Reynolds number for a body of some size, then we can find the solution for a body of different size (but the same shape) by rescaling the original solution, provided the Reynolds number is same.

Exercise 8.3

Conditions for incompressibility and adiabaticity: Two of the most important approximations discussed above – and invoked in fluid mechanics frequently – are incompressibility and adiabaticity. It is therefore important to determine the conditions under which these assumptions are valid. The conditions for incompressibility, for example, can be determined by estimating the fractional change in the density ($\Delta\rho/\rho$) from the laws of thermodynamics and the equations of fluid motion and demanding that ($\Delta\rho/\rho) \ll 1$. (1) Show that this leads to the condition

$$\frac{\Delta\rho}{\rho} \lesssim \frac{v}{c_s}\left(\frac{l/\tau}{c_s}\right) + \frac{v^2}{c_s^2} + \frac{\Delta\phi}{c_s^2} + \left(\frac{\eta+\zeta}{\rho l v}\right)\left(\frac{v}{c_s}\right)^2 + \left(\frac{\beta}{C_P}\right)\left(\frac{\kappa}{\rho v}\right)\left(\frac{\Delta T}{l}\right)$$

$$+ \left(\frac{\eta+\zeta}{\rho l v}\right)\left(\frac{v}{c_s}\right)^2\left(\frac{\beta c_s^2}{C_P}\right), \tag{8.107}$$

where $c_s^2 = (\partial p/\partial\rho)_s$, $\beta = -(1/\rho)(\partial\rho/\partial T)_p$, $C_P = T(\partial s/\partial T)_p$, and (l, τ) are the characteristic length and time scales over which the fluid properties vary and $\Delta\phi$ denotes a typical drop in the gravitational potential. Interpret physically the constraint arising from each of the terms on the right-hand side. (2) Show that the conditions for adiabaticity are

$$\frac{\Delta T}{l} \ll \frac{\rho c_s^2 v}{\kappa}, \quad \frac{\rho l v}{\eta} \gg \frac{v^2}{c_s^2}. \tag{8.108}$$

Note that incompressibility implies adiabaticity, although the converse is not necessarily true.

8.8 Sound Waves

We now proceed to the study of *compressible* fluids and begin by investigating the propagation of waves through such fluids. A compressible fluid can support small-amplitude oscillations that are usually called sound waves. To obtain the equations governing the propagation of sound, we have to linearise the fluid equation around a background solution, which is characterised by a fluid at rest ($v_0 = 0$) with density ρ_0 and pressure p_0.

To begin with, we ignore dissipative effects and treat the fluid as ideal. Also, because the oscillations are expected to be small, the velocity will be small and the quadratic term in velocity in Euler's equation can be ignored. Taking $p = p_0 + p'$, $\rho = \rho_0 + \rho'$, and $\mathbf{v} = \mathbf{v}'$, we linearise the equations in p', ρ' and $\mathbf{v}'$. This gives

$$\frac{\partial \rho'}{\partial t} + \rho_0 \nabla \cdot \mathbf{v}' = 0, \tag{8.109}$$

$$\frac{\partial \mathbf{v}'}{\partial t} + \frac{1}{\rho_0} \nabla p' = 0. \tag{8.110}$$

To proceed further we need a relation between p' and ρ'. Because all motion in an ideal fluid can be taken to be adiabatic to the lowest order of approximation, we may set $p' = (\partial p/\partial \rho)_s \rho'$, where the derivative is taken at constant entropy. Substituting for ρ' in terms of p', we can convert equation of continuity to

$$\frac{\partial p'}{\partial t} + \rho_0 \left(\frac{\partial p}{\partial \rho} \right)_s \nabla \cdot \mathbf{v}' = 0. \tag{8.111}$$

These equations can be solved by the ansatz that $\mathbf{v}'$ is irrotational with $\mathbf{v}' = \nabla \psi$. Then Eq. (8.110) gives $p' = -\rho_0 (\partial \psi/\partial t)$; using this in Eq. (8.111) we find that ψ satisfies the wave equation

$$\frac{\partial^2 \psi}{\partial t^2} - c_s^2 \nabla^2 \psi = 0, \tag{8.112}$$

where $c_s^2 \equiv (\partial p/\partial \rho)$. For an ideal gas in adiabatic flow, with $p \propto \rho^\gamma \propto \rho^{5/3}$ (see Chap. 5, Section 5.6), $c_s = (5p/3\rho)^{1/2} \propto \rho^{1/3}$. (Hereafter we drop the subscript 0 on ρ_0, etc., for simplicity.)

The solutions to wave equation (8.112) for waves propagating towards the positive x axis can be taken in the form $\psi(x, t) = f(x - c_s t)$. It is clear that $\mathbf{v}' = \nabla \psi$ is also along the x axis. Hence the waves are longitudinal, with the displacement along the direction of propagation. For $\psi = f(x - c_s t)$ we have $v' = (\partial \psi/\partial x) = f'$ and $p' = -\rho(\partial \psi/\partial t) = \rho c_s f'$. Hence $v' = (p'/\rho c_s)$. By using $p' = c_s^2 \rho$, we can write this as $v' = c_s(\rho'/\rho)$. The condition $\rho' \ll \rho$ implies that $v' \ll c_s$, that is, the flow must be slow compared with the sound speed.

For the temperature variation that is due to a sound wave, we have $T' = (\partial T/\partial p)_s p'$; by using the thermodynamic relation $(\partial T/\partial p)_s = (\partial V/\partial S)_p (T/C_P)$ $(\partial V/\partial T)_p$ (see Table 5.1, Chap. 5) and the relation $v' = (p'/\rho c_s)$, we find

$$T' = \frac{c_s \beta T}{C_P} v', \tag{8.113}$$

where $\beta = V^{-1}(\partial V/\partial T)_p$ is the coefficient of thermal expansion.

The sound waves, in general, allow the propagation of perturbations in various physical quantities along the fluid. There are, however, two exceptions to this rule: The entropy and vorticity of the fluid are not propagated with the speed of sound. This follows from the fact that, in an ideal fluid, entropy and circulation are strictly conserved in each fluid element. They do not move relative to the fluid; relative to a fixed system of coordinates, they move with a velocity appropriate to each point in the fluid.

We have assumed that the gas is ideal and dissipative processes are not in operation. In the presence of viscosity and heat conduction, the amplitude of the sound waves will be damped and the energy will be dissipated as heat. The propagation of small perturbations in a fluid will also become modified in the presence of radiative processes that can heat and cool the gas. This will be discussed later in Section 8.11.

8.9 Supersonic Flows

The analysis in Section 8.8 assumed that the unperturbed fluid was at rest, i.e., $\mathbf{v}_0 = 0$. Let us now consider a case in which the fluid is moving with a steady, constant velocity $\mathbf{v}$ when seen in the lab frame K with coordinates $\mathbf{x}$. We introduce another frame K' with coordinates $\mathbf{x}'$ that is comoving with the fluid. Our analysis in Section 8.8 is applicable in K' in which the fluid is at rest, so that, in K', the physical quantities representing the sound wave will vary as $\exp[i(\mathbf{k} \cdot \mathbf{x}' - kc_s t)]$. Because $\mathbf{x}' = \mathbf{x} - \mathbf{v}t$, it follows that the variables will appear to oscillate as $\exp\{i[\mathbf{k} \cdot \mathbf{x} - (kc_s + \mathbf{k} \cdot \mathbf{v})t]\}$ in K. Thus the frequency in K is $\omega = kc_s + \mathbf{k} \cdot \mathbf{v}$ and the phase velocity of propagation of the disturbance is

$$\frac{\partial \omega}{\partial \mathbf{k}} = \mathbf{v} + c_s \hat{\mathbf{n}}, \tag{8.114}$$

where $\hat{\mathbf{n}}$ is the unit vector in the direction of propagation. Note that this result is valid even when $v > c_s$, as no specific assumptions were made in this regard.

This result has a simple interpretation: If a gas in steady motion, with velocity $\mathbf{v}$, is perturbed at some point O then the perturbation is propagated in all directions with the speed of sound c_s relative to the gas. With respect to a fixed coordinate system, the velocity of propagation of the disturbance has two components: The perturbations are carried along with the gas flow with velocity $\mathbf{v}$ and are propagated in the rest frame of the gas with a velocity $c_s \hat{\mathbf{n}}$, where $\hat{\mathbf{n}}$ is a unit vector in any arbitrary direction. Thus the net velocity of propagation of the disturbance is $\mathbf{v} + c_s \hat{\mathbf{n}}$.

We can obtain all possible values of this vector by placing one end of the vector $\mathbf{v}$ at O and drawing a sphere with radius c_s from its tip (see Fig. 8.1). We see from the figure that very different conclusions emerge depending on whether $v < c_s$ or $v > c_s$. If $v < c_s$ then $\mathbf{v} + c_s \hat{\mathbf{n}}$ can take all possible directions and the perturbation can propagate in all possible directions. The disturbance that starts

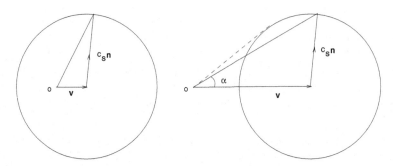

Fig. 8.1.   Propagation of a disturbance in a fluid moving with velocty v. When $v < c$, the disturbance can propagate in all directions with respect to O. When $v > c$, it can propagate only into a conical region in the forward direction.

from any point O in a gas can eventually reach any other point in the gas if the gas is moving *subsonically*.

On the other hand, if $v > c_s$, the direction of the vector $\mathbf{v} + c_s \hat{\mathbf{n}}$ can lie only within a cone with vertex at O with a semivertical angle α, where $\sin \alpha = (c_s/v)$. Thus a disturbance starting at O does *not* reach all the points downstream if the gas is flowing *supersonically*.

This fact has far-reaching consequences in the study of fluid mechanics. As a simple example, consider an object moving through a fluid with some speed v. In the rest frame of the body, the fluid will be moving around the object. When $v < c_s$, the presence of the body affects the fluid flow at all finite distances both upstream and downstream, as the signals originating from the region near the body can propagate ahead of the body and influence the fluid. However, if the speed of the body is supersonic, then the effect of the body can be felt only downstream (that is, behind the body). This is because the signals propagating with the speed of sound are lagging behind the body (which is moving supersonically). In the rest frame of the body, we may say that the fluid in the upstream region flows "blindly" onto the body, without being cognizant of the existence of the body.

Because subsonic and supersonic flows have very different characteristics, the transition between the two kinds of flow is of special significance. This transition, from subsonic to supersonic flow, can occur either continuously or discontinuously. In this section and the next we consider examples of continuous transitions. The discontinuous transition is closely related to shock waves and will be studied in Sections 8.11 and 8.12.

8.9.1 de Laval nozzle

To understand some of the peculiarities of the supersonic flow, as well as the *continuous* transition from subsonic to supersonic flow, we shall study a simple example known as the de Laval nozzle. Let us consider the steady adiabatic flow through a tube with a variable cross-sectional area A. The conservation of mass implies that $\rho v A = \text{constant}$ or $d(\rho v)/\rho v = -(dA/A)$. The Euler equation for the one-dimensional steady flow becomes

$$\mathbf{v} \cdot \frac{d\mathbf{v}}{dt} = v\frac{dv}{dt} = -\frac{1}{\rho}\mathbf{v} \cdot \nabla p = -\frac{1}{\rho}\frac{dp}{dt} \tag{8.115}$$

because $(\partial p/\partial t) = 0$. This implies that

$$v dv = -\frac{1}{\rho} dp = -\frac{1}{\rho}\left(\frac{\partial p}{\partial \rho}\right)_s d\rho = -\frac{c_s^2(\rho)}{\rho} d\rho, \tag{8.116}$$

or, equivalently, $d\rho/\rho = -v(dv/c_s^2)$. [We have used the definition $c_s^2 = (\partial p/\partial \rho)$ in these relations; of course, this quantity c_s^2 itself is now a variable, being dependent on the local density.] We can use this equation to relate v and A:

$$-\frac{dA}{A} = \frac{d(\rho v)}{\rho v} = -v\frac{dv}{c_s^2} + \frac{dv}{v} = \left(1 - \frac{v^2}{c_s^2}\right)\frac{dv}{v}. \tag{8.117}$$

Combining all these relations, we get

$$\frac{dA}{A} = \frac{c_s^2}{v^2}\left(1 - \frac{v^2}{c_s^2}\right)\frac{d\rho}{\rho} = \frac{p}{\rho v^2}\left(1 - \frac{v^2}{c_s^2}\right)\frac{dp}{p} = -\left(1 - \frac{v^2}{c_s^2}\right)\frac{dv}{v}. \tag{8.118}$$

These equations show that the characteristics of the flow are very different depending on whether $v > c_s$ or $v < c_s$. Suppose that the flow is from left to right. When $v < c_s$ and the area A increases to the right (subsonic diffuser), the speed v decreases but ρ and p increase. If, on the other hand, A decreases to the right (subsonic nozzle) then v increases and ρ and p decrease. This is what we normally expect; in a constricted region v is larger and p is lower.

The behaviour is exactly the opposite in the case of supersonic flow with $v > c_s$. In this case, an increase in area (supersonic diffuser) leads to an increase in v and a decrease in ρ and p; similarly a decrease in A (supersonic nozzle) leads to a decrease in v and an increase in ρ and p. When the area decreases, ρ increases quite a lot, requiring that the speed *also* decrease in order to keep $\rho v A$ constant. It is clear that the high compressibility and the supersonic flow go hand in hand in this case.

Consider now the flow of gas through a tube in which A decreases, reaches a minimum, and increases again. The flow begins subsonically and the speed of flow increases as the area decreases. From Eq. (8.117), we see that $(dv/dx) > 0$ for $(dA/dx) < 0$ as long as $v < c_s$; further $v = c_s$ occurs only at the point where $(dA/dx) = 0$. Hence two possibilities arise; (1) $v < c_s$ at the minimum-area point (throat). In that case velocity increases, reaches a maximum, and decreases again, with $v < c_s$ everywhere. This is the normal situation in which the velocity of flow is largest at the narrowest point of the tube and the flow is everywhere subsonic. (2) The second, and more interesting, possibility is the one with $v = c_s$ at the throat; then the flow begins subsonically, reaches sonic speed at the narrowest part of the tube, and makes a transition to supersonic flow to the right; then – on the right side – we have v increasing with increasing area.

To see these effects explicitly, we consider the flow of an ideal polytropic gas through a tube of variable cross section in which the cross-sectional area decreases to a minimum value and increases again. For an ideal polytropic gas, $c_s^2 = (\gamma p/\rho) \propto T$ and $w = (\gamma p/\rho)(\gamma - 1)^{-1} = c_s^2(\gamma - 1)^{-1}$ [see Eq. (5.97)]. The adiabatic condition $\rho T^{1/(1 - \gamma)} = \text{constant}$ now becomes

$$\rho = \rho_0 \left(\frac{c_s}{c_0} \right)^{2/(\gamma - 1)}. \tag{8.119}$$

The conservation laws for the fluid flow through a region of varying cross-sectional area

$$\rho v A = \dot{M}, \quad \frac{1}{2} v^2 + w = w_0, \tag{8.120}$$

can now be written as

$$v c_s^{2/(\gamma - 1)} = k \frac{\dot{M}}{A}, \quad v^2 + \frac{2}{(\gamma - 1)} c_s^2 = \frac{2}{(\gamma - 1)} c_0^2, \tag{8.121}$$

where k is an unimportant constant and c_0 is the speed of sound at the location $v = 0$. The second equation of Eqs. (8.121) defines a quadratic curve in the v–c_s plane (see the solid curve in Fig. 8.2). The dashed curves are based on the first equation of Eqs. (8.121) for different values of A. The intersection of the unbroken and the dashed curves (corresponding to area A) will give the flow velocity at the location with cross section A. The flow begins around the point P and moves up along the quadratic curve as the area of the cross section decreases. If the area of cross section decreases sufficiently to reach a value $A_{\min}$ before the flow reaches the line $v = c_s$, supersonic motion does *not* occur. On the other hand, if $A_{\min}$ is reached at $v = c_s$ the flow can become supersonic. Thus, to achieve supersonic motion, we have to adjust $\dot{M}$ and $A_{\min}$ such that $A_{\min}$ is reached when $v = c_s$. The condition for this can be easily worked out to be

$$\dot{M} = \left(\frac{2}{\gamma + 1} \right)^{\frac{(\gamma+1)}{2(\gamma-1)}} \rho_0 c_0 A_{\min} \equiv \dot{M}_{\text{crit}}. \tag{8.122}$$

This shows that if the flow has to make a transition from subsonic to supersonic, then the mass flux cannot be arbitrary but is constrained to have a very specific value.

Exercise 8.4

Booster rockets: Rockets work most efficiently when the hot gas produced by the burning of the fuel goes out through the de Laval nozzle. In the space shuttle, for example, a

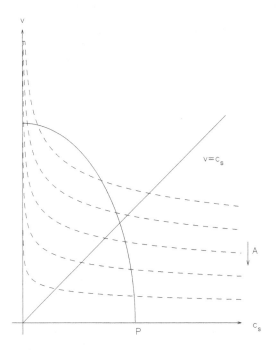

Fig. 8.2.   Flow of a gas through a region of variable cross section; see text for discussion.

thrust of $T \approx 2 \times 10^{12}$ dyn is produced by burning the fuel at a rate of $\dot{M} \approx 10^8$ gm s^{-1} for $\sim 10^2$ s. The de Laval nozzle is such that the exit area is approximately four times the minimum area. Estimate (1) the exit speed V_{exit}, (2) the Mach number $\mathcal{M}_{\text{exit}} \equiv (V_{\text{exit}}/c_s)$ at exit, (3) the exit pressure, and (4) the stagnation pressure in the combustion region. Assume (somewhat unrealistically) steady, isoentropic flow of ideal gas. [Hint: From the first equation of Eqs. (8.121), $A^{-1} \propto \mathcal{M} c_s^{(\gamma + 1)/(\gamma - 1)}$, where $\mathcal{M} = v/c_s$; using it in the second of Eqs. (8.121), we can relate (A/A_{min}) to $\mathcal{M}$.]

   In this particular case, the variation in the area of cross section and the initial conditions (which started the flow) have governed the dynamics. A very similar pattern of flow arises when fluid accretes to a central object, influenced by its gravitational field. We discuss this phenomenon – called spherical accretion – next.

8.9.2 Spherical Accretion

A simple example in which many of the concepts developed in Subsection 8.9.1 finds application is in the accretion of gas to a central object. Consider a spherical object of mass M and radius R embedded inside a gas cloud. We expect the gas to flow towards the central mass because of gravitational attraction. In a highly idealised situation, we may consider the flow to be radially inwards at some steady rate $\dot{M}$. Let us further assume that the gas is ideal and polytropic with an index γ, where $1 < \gamma < 5/3$.

For a steady spherically symmetric flow, the continuity equation

$$\frac{1}{r^2}\frac{\partial}{\partial r}(r^2 \rho v) = 0 \tag{8.123}$$

shows that $\rho v r^2$ should be a constant, where v is the radial component of the velocity. We set $\dot{M} = -4\pi r^2 \rho v$ to be the constant accretion rate. The Euler equation, with a force term arising from the gravitational field of the central object, is

$$v\frac{\partial v}{\partial r} + \frac{1}{\rho}\frac{\partial p}{\partial r} + \frac{GM}{r^2} = 0. \tag{8.124}$$

To complete the system of equations we add the polytropic relation $p \propto \rho^\gamma$. We now have three equations for the three variables p, ρ, and v.

Writing $(\partial p/\partial r) = (\partial p/\partial \rho)(\partial \rho/\partial r) = c_s^2(\partial \rho/\partial r)$ in Euler equation (8.124) and using

$$\frac{1}{\rho}\frac{\partial \rho}{\partial r} = -\frac{1}{vr^2}\frac{\partial}{\partial r}(vr^2) \tag{8.125}$$

[which can be obtained from Eq. (8.123)], we can rewrite the Euler's equation as

$$v\frac{\partial v}{\partial r} - \frac{c_s^2}{vr^2}\frac{\partial}{\partial r}(vr^2) + \frac{GM}{r^2} = 0. \tag{8.126}$$

This can be rearranged into a more convenient form,

$$\frac{1}{2}\left(1 - \frac{c_s^2}{v^2}\right)\frac{\partial v^2}{\partial r} = -\frac{GM}{r^2}\left(1 - \frac{2c_s^2 r}{GM}\right), \tag{8.127}$$

which allows us to understand several features of accretion.

To begin with, note that c_s^2 will tend to some constant value at very large distances from the central object. (In general, of course, c_s^2 depends on ρ and hence on r.) Therefore the factor $(1 - 2c_s^2 r/GM)$ will be negative definite at large r, making the right-hand side of Eq. (8.127) positive at large r. On the left-hand side, we want $\partial v^2/\partial r$ to be negative because we expect the gas to be at rest at large distances and flow with an acceleration as it approaches the central body. Hence $v^2 < c_s^2$ at large r, i.e., the flow is subsonic at large distances. As the gas approaches the central object, the factor $(1 - 2c_s^2 r/GM)$ will tend to zero at some critical radius r_c. Using $c_s^2 = \gamma p/\rho = \gamma kT$, we can estimate this critical radius as

$$r_c = \frac{GM}{2c_s^2(r_c)} \simeq 7.5 \times 10^{13}\left(\frac{T}{10^4 K}\right)^{-1}\left(\frac{M}{M_\odot}\right)\text{cm}. \tag{8.128}$$

This radius is much larger than the size of typical central masses encountered in astrophysics. Hence we must consider the flow at $r < r_c$ as well. Repeating the above analysis for $r < r_c$, we can conclude that $v^2 > c_s^2$. Hence the flow must

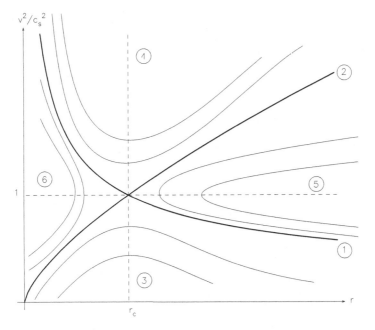

Fig. 8.3. The fluid velocity in spherical accretion in different cases. See text for discussion.

make a transition from subsonic to supersonic speeds at $r = r_c$. Further, at $r = r_c$, we must have either $v^2 = c_s^2$ or $(\partial v^2/\partial r) = 0$

The solutions can now be classified based on their behaviour near $r = r_c$ and are shown schematically in Fig. 8.3: (1) $v^2 = c_s^2$ at $r = r_c$, $v^2 \to 0$ as $r \to \infty$, $v^2 < c_s^2$ for $r > r_c$, and $v^2 > c_s^2$ for $r < r_c$. This is the relevant solution that describes accretion that starts out as subsonic at large distances and becomes supersonic near the compact object. (2) $v^2 = c_s^2$ at $r = r_c$, $v^2 \to 0$ as $r \to 0$, $v^2 > c_s^2$ for $r > r_c$, and $v^2 < c_s^2$ for $r < r_c$. This represents a time-reversed situation to spherical accretion involving a wind flowing from the compact object. The flow is subsonic near the object and becomes supersonic at large distances. (3) $(\partial v^2/\partial r) = 0$ at $r = r_c$, and $v^2 < c_s^2$ for all r. (4) $(\partial v^2/\partial r) = 0$ at $r = r_c$, and $v^2 > c_s^2$ for all r. Solutions (3) and (4) are either subsonic or supersonic in the entire range and hence cannot describe accretion onto a compact object of sufficiently small radius. (5) $(\partial v^2/\partial r) = \infty$ when $v^2 = c_s^2$, and $r > r_c$ always. (6) $(\partial v^2/\partial r) = \infty$ when $v^2 = c_s^2$, and $r < r_c$ always. These two solutions are unphysical because they give two possible values for v^2 at any given r. Thus we conclude that only the type 1 solution is relevant to the problem of spherical accretion onto a compact object.

To proceed further, we integrate Eq. (8.124) by using $(\partial p/\partial \rho) = \gamma(p/\rho) = c_s^2$ to obtain

$$\frac{v^2}{2} + \frac{c_s^2}{(\gamma - 1)} - \frac{GM}{r} = \text{constant}. \tag{8.129}$$

Because we must have $v^2 \to 0$ as $r \to \infty$, the constant must be $c_s^2(\infty)/(\gamma - 1)$. On the other hand $v^2(r_c) = c_s^2(r_c)$ and $(GM/r_c) = 2c_s^2(r_c)$. Using these relations, we can relate $c_s(r_c)$ to $c_s(\infty)$, obtaining

$$c_s^2(r_c)\left(\frac{1}{2} + \frac{1}{\gamma - 1} - 2\right) = \frac{c_s^2(\infty)}{(\gamma - 1)} \tag{8.130}$$

or

$$c_s(r_c) = c_s(\infty)\left[\frac{2}{(5 - 3\gamma)}\right]^{1/2}, \tag{8.131}$$

provided that $1 < \gamma < 5/3$. (The cases of $\gamma = 1$ and $\gamma \geq 5/3$ require separate discussion; see Exercises 8.6 and 8.7.) Because $c_s^2 \propto \rho^{\gamma - 1}$, we also get

$$\rho(r_c) = \rho(\infty)\left[\frac{c_s(r_c)}{c_s(\infty)}\right]^{2/(\gamma - 1)}. \tag{8.132}$$

Finally, because $r^2 \rho v$ is a constant, we can write $\dot{M} = 4\pi r_c^2 \rho(r_c)c_s(r_c)$; by using this we can relate the accretion rate $\dot{M}$ to the conditions of the gas at infinity. Straightforward algebra gives

$$\dot{M} = \pi G^2 M^2 \frac{\rho(\infty)}{c_s^3(\infty)}\left(\frac{2}{5 - 3\gamma}\right)^{(5 - 3\gamma)/2(\gamma - 1)}. \tag{8.133}$$

Numerically, this corresponds to the accretion rate

$$\dot{M} = 1.4 \times 10^{11} \text{g s}^{-1}\left(\frac{M}{M_\odot}\right)^2\left[\frac{\rho(\infty)}{10^{-24}\text{g cm}^{-3}}\right]\left[\frac{c_s(\infty)}{10 \text{ km s}^{-1}}\right]^{-3} \tag{8.134}$$

if $\gamma = 1.4$. To complete the solution we find $v(r)$ in terms of $c_s(r)$ from the conservation of $\dot{M}$. This gives

$$-v = \frac{\dot{M}}{4\pi r^2 \rho(r)} = \frac{\dot{M}}{4\pi r^2 \rho(\infty)}\left[\frac{c_s(\infty)}{c_s(r)}\right]^{2/(\gamma - 1)}. \tag{8.135}$$

Substituting this relation into Eq. (8.129) gives an algebraic equation for $c_s^2(r)$. Because this equation has fractional exponents it has to be solved numerically. Given $c_s^2(r)$, Eq. (8.135) determines $v(r)$ and the relation $\dot{M} = r^2 \rho(r)v(r)$ fixes $\rho(r)$. This completes the solution to our problem. We see that there is a steady flow of matter to the central object that starts out subsonically but makes a transition to supersonic flow at a critical radius.

It may appear strange that the flow rate $\dot{M}$ is uniquely fixed at a value dependent only on M, $c_s(\infty)$, and $\rho(\infty)$, implying that a gas can flow to a compact object at only a critical rate. (We found a similar result in Subsection 8.9.1 when the flow made a transition from subsonic to supersonic flow in the de Laval nozzle.) This can be understood as follows: Using the relations $\dot{M} = -4\pi \rho v r^2$,

$c_s^2(r) = c^2(\infty)[\rho/\rho(\infty)]^{\gamma-1}$ and Eq. (8.129), we can estimate how $v(r)$ and $\rho(r)$ change when $\dot{M}$ changes slightly. At fixed r, we have,

$$\frac{\delta \dot{M}}{\dot{M}} = \left(1 - \frac{v^2}{c_s^2}\right)\frac{\delta v}{v}, \qquad \frac{\delta \rho}{\rho} = -\frac{v^2}{c_s^2}\frac{\delta v}{v}. \qquad (8.136)$$

As we increase v (at any r), $\dot{M}$ increases until $v = c_s$; then, as v increases, $\dot{M}$ decreases. Hence, at any given r, there is a maximum permitted flow rate, that occurs when $v^2 = c_s^2$. The maximum value is

$$|\dot{M}_{max}(r)| = 4\pi\rho c_s r^2 = 4\pi r^2 c(\infty)\rho(\infty)\left(\frac{c_s}{c_\infty}\right)\left[\frac{\rho}{\rho(\infty)}\right]$$

$$= 4\pi r^2 c(\infty)\rho(\infty)\left[\frac{c_s}{c(\infty)}\right]^{\frac{\gamma+1}{\gamma-1}}$$

$$= 4\pi r^2 c(\infty)\rho(\infty)\left[\frac{2(\gamma-1)}{(\gamma+1)}\left(\frac{GM}{r^2 c(\infty)} + \frac{1}{\gamma-1}\right)\right]^{\frac{\gamma+1}{2(\gamma-1)}}. \qquad (8.137)$$

The last equality follows from Eq. (8.129) {with $v = c_s$ and the constant on the right-hand side set to $[c^2(\infty)/(\gamma-1)]$}. For $1 < \gamma \leq 5/3$, this function has the following behaviour: It decreases as $r^{-(5-3\gamma)/2(\gamma-1)}$ for small r, reaches a minimum value $\dot{M}_{crit}$ at $r = r_c$, and increases as r^2 for large r. Because we must have $|\dot{M}| < |\dot{M}_{max}|$ at all r, it follows that $\dot{M} > \dot{M}_{crit}$ is forbidden. Further, for $\dot{M} < \dot{M}_{crit}$, the flow is subsonic everywhere. To have a flow that makes the transition from subsonic to supersonic, we must have exactly $\dot{M} = \dot{M}_{crit}$. This is the reason for the critical flow rate.

Note that as $\gamma \to 5/3$, the rate $\dot{M}$ drops to zero because the sonic point moves closer to the origin, whereas as $\gamma \to 1$, the rate $\dot{M}$ diverges because the sonic point moves to very large radius. The fact that $\dot{M}$ vanishes for a strictly adiabatic gas [with $c_s(\infty) \neq 0$] shows that for a realistic transonic accretion flow it is necessary to have some radiative cooling.

A similar procedure can be used to analyse isothermal accretion. The time-reversed solution will allow us to describe winds from astronomical bodies. These issues will be discussed in Vol. II.

Exercise 8.5

Accretion onto a moving star: Consider a star moving with a speed V with respect to the interstellar medium that has a density ρ_∞ and speed of sound c_∞. Give a qualitative argument as to why the accretion rate in this case can be written in the form

$$\dot{M} = \lambda_* 4\pi\rho_\infty\frac{(GM)^2}{(V^2 + c_s^2)^{3/2}}, \qquad (8.138)$$

where λ_* is a coefficient of the order of unity.

Exercise 8.6
Isothermal accretion: (1) Show that the equations governing the spherical accretion can
be written in dimensionless form as

$$x^2 yz = \lambda, \qquad \frac{z^2}{2} + H(y) - \frac{1}{x} = 0, \tag{8.139}$$

where

$$x = \left(rc_\infty^2/GM\right), \qquad y = (\rho/\rho_\infty), \qquad z = |v|/c_\infty, \tag{8.140}$$

$$H(y) = \ln y \quad (\text{for } \gamma = 1); \qquad H(y) = \frac{\gamma}{\gamma - 1}(y^{\gamma-1} - 1) \quad (\text{for } \gamma \neq 1). \tag{8.141}$$

(2) Manipulate these equations to give

$$\left(z - \frac{1}{z}\right) dz = \left(\frac{2}{x} - \frac{1}{x^2}\right) dx \tag{8.142}$$

in the isothermal case with $\gamma = 1$. This implies that the sonic transition occurs when
$z = 1$. Show that this requires λ to have the critical value $\lambda_c = (e^{3/2}/4) \approx 1.12$. How does
the velocity field look for $\lambda > \lambda_c$, $\lambda < \lambda_c$, and $\lambda = \lambda_c$? Describe the flow in physical terms.
   (3) Redo the analysis in the text in terms of x, y, and z and show that the sonic point
is determined by the condition

$$H(y) - \frac{3}{2}yH'(y) = 0. \tag{8.143}$$

Hence conclude that

$$y = e^{3/2} \quad (\text{for } \gamma = 1), \quad \text{and} \quad y^{\gamma-1} = \frac{2}{5 - 3\gamma} \quad (\text{for } 1 \neq \gamma \leq 5/3). \tag{8.144}$$

Now study the limit of $\gamma = 5/3$ and explain physically the impossibility of steady supersonic accretion for $\gamma \geq 5/3$.
   (4) The time-reversed solution to the accretion problem can describe a stellar wind
flowing out from a star. We are now looking for solutions that start with small speeds near
the star, accelerate through the sonic point, and acquire large speeds far away. Describe
the isothermal wind in detail, parameterising the solutions by the density of fluid at the
sonic point. Can this solution be physically reasonable?

Exercise 8.7
Nonadiabatic spherical accretion: When heating and cooling are present, the adiabatic
condition is replaced with

$$\left(\frac{\partial}{\partial t} + v\frac{\partial}{\partial r}\right)\left[\frac{v^2}{2} + \frac{\gamma}{\gamma - 1}\frac{p}{\rho} - \left(1 - \frac{L}{L_E}\right)\frac{GM}{r}\right] = H - C, \tag{8.145}$$

where L_E is the Eddington luminosity included to take into account the radiation pressure
(see Exercise 6.15, Chap. 6) and $H - C$ is the net heating in units of erg per second per
gram. Explain why this equation is correct and state the assumptions involved in it.

Analyse this case and show that the sonic point now satisfies the equation

$$2r_c = \left(1 - \frac{L}{L_E}\right)\frac{GM}{c_s^2} - \frac{r_c^2}{c_s^3}(\gamma - 1)(H - C). \tag{8.146}$$

8.10 Steepening of Sound Waves

In the analysis of sound waves in Section 8.8 it was assumed that the amplitude of the disturbance was small. In that case, the solutions were functions of $(x \pm c_s t)$ and the wave propagated without any change in its profile. Because velocity v, density ρ, and pressure p were all functions of only $(x \pm c_s t)$, we could express all these variables in terms of each other.

When the amplitude of the perturbation is not small, these relations no longer hold. In particular, because c_s is proportional to $\rho^{1/3}$, regions of the fluid with higher density will be travelling faster. This will necessarily distort the shape of the wave and cause it to steepen. Such a distortion of shape cannot be understood in linearised theory, and we need to study the exact equations.

In the absence of shock waves, the flow of an ideal fluid is adiabatic. If the gas is homogeneous at some initial instant, then $s = $ constant and the flow is isoentropic, allowing us to express the pressure in terms of the density alone. For a sound wave propagating along the x axis, all physical quantities depend on only x and t and we can take the velocity field to be $v_x = v(x, t)$, $v_y = v_z = 0$. In this case, the equations can be integrated in closed form if we further assume that the velocity can be expressed as a function of the density. Let us consider such a flow with a velocity $v_x = v(\rho)$ that depends on only the density ρ. In this case, the continuity and the Euler equations,

$$\frac{\partial \rho}{\partial t} + \frac{\partial(\rho v)}{\partial x} = 0, \qquad \frac{\partial v}{\partial t} + \frac{v\partial v}{\partial x} + \frac{1}{\rho}\frac{\partial p}{\partial x} = 0, \tag{8.147}$$

can be transformed into the form

$$\frac{\partial \rho}{\partial t} + \frac{d(\rho v)}{d\rho}\frac{\partial \rho}{\partial x} = 0, \qquad \frac{\partial v}{\partial t} + \left(v + \frac{1}{\rho}\frac{dp}{dv}\right)\frac{\partial v}{\partial x} = 0. \tag{8.148}$$

We now use the relation $(\partial x/\partial t)_\rho = -[(\partial \rho/\partial t)/(\partial \rho/\partial x)]$, etc., to obtain from these equations the partial derivatives

$$\left(\frac{\partial x}{\partial t}\right)_\rho = \frac{d(\rho v)}{d\rho} = v + \rho\frac{dv}{d\rho}, \qquad \left(\frac{\partial x}{\partial t}\right)_v = v + \frac{1}{\rho}\frac{dp}{dv}. \tag{8.149}$$

However, because $\rho = \rho(v)$, it follows that $(\partial x/\partial t)_\rho = (\partial x/\partial t)_v$. Equating the expressions for these two and simplifying, we find that

$$v = \pm\int\frac{c_s}{\rho}d\rho = \pm\int\frac{dp}{\rho c_s}. \tag{8.150}$$

Substituting this result back into the expression for $(\partial x / \partial t)_v$ gives $(\partial x / \partial t)_v = v \pm c_s(v)$. Integrating, we find that

$$x = t[v \pm c_s(v)] + f(v), \qquad (8.151)$$

where $f(v)$ is an arbitrary function of v to be fixed by the initial conditions. This relation determines v as an implicit function of x and t. Because all other quantities, such as ρ, p, etc., can be expressed in terms of v, we have completely solved the problem.

As a specific example, consider the case of a polytropic gas for which the adiabaticity condition is equivalent to $\rho T^{1/(1-\gamma)} = \text{constant}$. Because $c_s \propto \sqrt{T}$, it follows that

$$\rho = \rho_0 \left(\frac{c_s}{c_0} \right)^{2/(\gamma-1)}. \qquad (8.152)$$

By using this relation in Eq. (8.150), we can determine $c_s(v)$:

$$c_s = c_0 \pm \frac{1}{2}(\gamma - 1)v. \qquad (8.153)$$

Hence

$$\rho = \rho_0 \left[1 \pm \frac{1}{2}(\gamma - 1)\frac{v}{c_0} \right]^{2/(\gamma-1)}, \qquad p = p_0 \left[1 \pm \frac{1}{2}(\gamma - 1)\frac{v}{c_0} \right]^{2\gamma/(\gamma-1)}. \qquad (8.154)$$

Substituting these results into our general relation (8.151), we get

$$x = t \left[\pm c_0 + \frac{1}{2}(\gamma + 1)v \right] + f(v). \qquad (8.155)$$

This relation can be rewritten in a more convenient form as

$$v = F \left\{ x - \left[\pm c_0 + \frac{1}{2}(\gamma + 1)v \right] t \right\}, \qquad (8.156)$$

where F is another arbitrary function.

The wave solution obtained here and described by Eq. (8.151) is different qualitatively from the one obtained in Section 8.8 for small amplitudes. In Eq. (8.151) we may think of the disturbance as propagating for two reasons: the fluid is moving with velocity v while the disturbance is moving with respect to the fluid with velocity $c_s(v)$. However, because c_s depends on the density of gas, the net velocity of propagation $v \pm c_s$ is different at different points. In general, the velocity of propagation increases with density. Thus high-density regions propagate faster and overtake low-density regions, giving rise to a density profile that is changing progressively, as shown in Fig. 8.4. This leads to the conclusion that, after a finite duration of time, the density is bound to become a multiple-valued function of

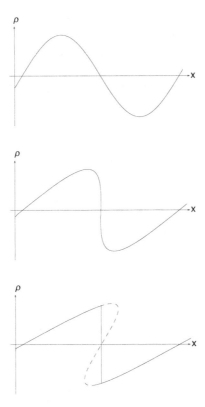

Fig. 8.4. The occurrence of discontinuities during the propagation of the sound wave is illustrated. Because the speed of sound increases with density, high-density regions move faster and overtake low-density regions. This will eventually lead to a density configuration, indicated by the dashed curve, in which the density becomes multiple valued. Such a situation is physically unacceptable and is avoided by the formation of strong discontinuity in the medium. The resulting density profile is indicated by the solid curve.

x. Because this is physically impossible, a sharp discontinuity (shock) formed in the fluid.

The location of this discontinuity can be easily ascertained. Let t_0 be the time of formation of discontinuity. At the time $t = t_0$, the graph of velocity v as a function of x becomes vertical at some location x_0, i.e., the derivative $(\partial x/\partial v)_t$ vanishes. It is also clear that this point should be a point of inflexion for the function $x(v)$, thereby requiring that $(\partial^2 x/\partial v^2)_t = 0$. Hence the location and the time of formation of shock wave are determined by the conditions

$$\left(\frac{\partial x}{\partial v}\right)_t = 0, \quad \left(\frac{\partial^2 x}{\partial v^2}\right)_t = 0. \tag{8.157}$$

For a polytropic gas we are considering, we can use Eq. (8.155) to write this condition more explicitly. We get

$$t = -\frac{2f'(v)}{(\gamma + 1)}, \qquad f''(v) = 0. \tag{8.158}$$

Given $v = v(x, t)$ these equations can be solved to give (x_c, t_c), which is the location of the shock.

This condition is modified in the case in which the discontinuity is formed at a boundary between moving gas and stationary gas. Then the second derivative f'' need not vanish; but we need to satisfy $(\partial x/\partial t)_v = 0$ at the surface where $v = 0$. Therefore, in this case,

$$t_c = -\frac{2f'(0)}{(\gamma + 1)}. \tag{8.159}$$

Let us illustrate the propagation of such a shock wave by an example. Assume that the gas located in a semi-infinite cylindrical pipe ($x > 0$) is terminated by a piston at one end $x = 0$, at $t \leq 0$. The piston begins to move at $t = 0$ with a uniformly accelerated velocity $v_{\text{pist}} = \pm at$. Consider first the case in which the piston moves away from the gas with $v_{\text{piston}} = -at$. This sends a rarefaction wave through the gas. Because the gas and the piston must have same velocity at the surface of the piston, we must have $v = -at$ at $x = -\frac{1}{2}at^2$ for all $t > 0$. This condition, when used in Eq. (8.155), gives

$$f(-at) = -c_0 t + \frac{1}{2}\gamma at^2 = \left(\frac{c_0}{a}\right)(-at) + \frac{1}{2}\frac{\gamma}{a}(-at)^2, \tag{8.160}$$

showing that the function f has the form

$$f(z) = \left(\frac{c_0}{a}\right) z + \frac{1}{2}\frac{\gamma}{a}z^2. \tag{8.161}$$

Using this result again in Eq. (8.155), we get

$$x - \left[c_0 + \frac{1}{2}(\gamma + 1)v\right]t = f(v) = \frac{c_0}{a}v + \frac{1}{2}\frac{\gamma}{a}v^2. \tag{8.162}$$

Solving, we find the velocity as a function of x and t as

$$v = \frac{1}{\gamma}\left(\left\{\left[c_0 + \frac{1}{2}(\gamma + 1)at\right]^2 - 2a\gamma(c_0 t - x)\right\}^{1/2} - \left[c_0 + \frac{1}{2}(\gamma + 1)at\right]\right). \tag{8.163}$$

This solution gives the change in the velocity of flow in a region between the piston [at $x = -(1/2)at^2$] and the forward front (at $x = c_s t$). The gas at $x > c_s t$ is undisturbed. The velocity of flow is everywhere to the negative x axis (like that of the piston) and decreases monotonically to the right. The pressure and the density increase monotonically to the right.

If the piston moves to the right with a velocity $v_{\text{piston}} = +at$, we can obtain the corresponding solution by changing the sign of a in the expression. In this case, a compression wave propagates into the gas and a shock front is formed at the time determined by Eq. (8.159); we find that $t = 2c_0/a(\gamma + 1)$.

8.11 Shock Waves

Shocks arise in several astrophysical contexts in which there is a supersonic flow of gas relative to a solid obstacle. In the laboratory frame, this could arise either because of an object moving through a fluid with supersonic velocities or because gas flowing with supersonic velocities suddenly encounters an obstacle. Because of the boundary conditions near the solid body, the flow has to change from supersonic to subsonic somewhere near the object. Further, this information cannot reach upstream where the flow is supersonic. Hence the pattern of flow changes very rapidly over a short region, resulting in a shock wave. In the frame in which the gas is at rest, we may consider the solid body as moving with supersonic speed and pushing the gas. This necessarily leads to steepening of the density profile at some place ahead of the body. We consider some of the properties of the shock waves in this section.

We idealise the surface of discontinuity as a plane with gas flow occurring perpendicularly to it, with the flow being supersonic on the left side. It is convenient to study the situation in a frame in which the shock front is at rest. Let p_1, ρ_1, and v_1 denote the state of the gas to the left of the shock front, and let p_2, ρ_2, and v_2 give the corresponding values on the right side. The physical quantities on the two sides of the shock front are related by the conservation of mass, energy, and momentum. The conservation of mass gives

$$\rho_1 v_1 = \rho_2 v_2 \equiv j. \tag{8.164}$$

The conservation of energy flux, taken in conjunction with the above equation, implies that

$$w_1 + \frac{1}{2}v_1^2 = w_2 + \frac{1}{2}v_2^2. \tag{8.165}$$

Finally, the conservation of momentum flux gives

$$p_1 + \rho_1 v_1^2 = p_2 + \rho_2 v_2^2. \tag{8.166}$$

These three equations relate the set (p_2, ρ_2, v_2) to p_1, ρ_1, and v_1. The enthalpy w can be expressed in terms of other variables for a polytropic gas with index γ:

$$w = \frac{\gamma}{\gamma - 1}\frac{p}{\rho} = \frac{\gamma}{\gamma - 1}k_B T. \tag{8.167}$$

These equations can be solved by straightforward algebraic manipulations. It is convenient to write the solution expressing the ratios (ρ_2/ρ_1), (v_2/v_1), etc., in terms of the Mach number $M_1 \equiv (v_1/c_s)$ as one side. We find that

$$\frac{\rho_2}{\rho_1} = \frac{(\gamma + 1)M_1^2}{(\gamma + 1) + (\gamma - 1)(M_1^2 - 1)} = \frac{v_1}{v_2}, \tag{8.168}$$

$$\frac{p_2}{p_1} = \frac{(\gamma + 1) + 2\gamma(M_1^2 - 1)}{(\gamma + 1)}, \tag{8.169}$$

$$\frac{T_2}{T_1} = \frac{[(\gamma + 1) + 2\gamma(M_1^2 - 1)][(\gamma + 1) + (\gamma - 1)(M_1^2 - 1)]}{(\gamma + 1)^2 M_1^2}. \tag{8.170}$$

Because we have assumed that the flow is supersonic on side 1, we have $M_1 > 1$; it is clear from the above equations that $p_2 > p_1$, $\rho_2 > \rho_1$, $v_2 < v_1$, and $T_2 > T_1$. The strongest shock that we can have corresponds to the limit $M_1 \to \infty$. In this case we find that (ρ_2/ρ_1) tends to the finite value $(\gamma + 1)/(\gamma - 1)$. In the same limit, the pressure and temperature ratios diverge.

The derivation of the junction conditions use the conservation of mass, momentum, and energy. Of these, the first two are fairly well satisfied in a wide variety of contexts. There are, however, situations in which energy conservation may not be applicable at a shock front. First, energy can be added to the flow if chemical or nuclear reactions occur around the shock front. Second, if the gas is heated to a very high temperature, it will lose energy through radiation (some aspects of radiative shocks will be discussed later towards the end of this section). Third, the energy may be conducted away far upstream by particles moving with superthermal speeds so as to preheat the incoming gas. This can thicken the shock front so that it cannot be approximated as a discontinuity. When these effects do not occur and the shock discontuinity satisfies conservation of mass, momentum, and energy, the shock is called adiabatic.

In the above discussion, we have assumed that the fluid is ideal on either side of the shock front. In reality the shock front will have a small thickness over which the physical parameters will change rapidly. Because the gradients are large, dissipative processes will be important in the transition region that will determine the thickness of this region.

This thickness can be estimated as follows. Taking into account the viscous and the thermal effects, we can write the conservation laws for the flux of mass, momentum, and energy as

$$\rho v = \text{constant}, \tag{8.171}$$

$$\rho v^2 + p - \left(\frac{4}{3}\eta + \zeta\right)\frac{\partial v}{\partial x} = \text{constant}, \tag{8.172}$$

$$\rho v^2 \left(\frac{1}{2}v^2 + \epsilon + \frac{p}{\rho}\right) - \left(\frac{4}{3}\eta + \zeta\right)v\frac{\partial v}{\partial x} - \kappa\frac{\partial T}{\partial x} = \text{constant}. \tag{8.173}$$

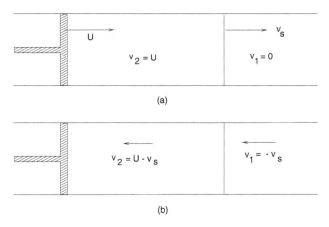

(a)

(b)

Fig. 8.5. Motion of gas in a tube with a moving piston: (a) description in the lab frame, (b) description in the rest frame of the shock front.

In arriving at the above equations, we have used the fact that the component σ_{xx} of the viscous tensor is $\sigma_{xx} = [(4/3)\eta + \zeta](\partial v/\partial x)$. In the shock layer, η and ζ will be of comparable magnitude or η will dominate over ζ; therefore we can set $\zeta = 0$ while making an order-of-magnitude estimate. Further, the momentum flux that is due to the friction in Eq. (8.172), given by the term $(4/3)\,\eta(\partial v/\partial x)$, must have the same order of magnitude as the other two terms, in particular ρv^2. If the velocity changes by Δv over a region Δx, then $\eta(\partial v/\partial x) \simeq \rho v(\Delta v/\Delta x)$. Equating this to ρv^2, we find that $\Delta x \simeq v(\Delta v/v^2)$. For a strong shock, we may take $\Delta v \simeq v$; using this and noting that $v \simeq lv$, where l is the mean free path, we find that $\Delta x \simeq l$. In other words, a strong shock has a thickness that is comparable with the mean free path.

We can obtain the same result from Eq. (8.173) by noting that $\kappa(\partial T/\partial x)$ must be of the same order as $\rho v(p/\rho) \simeq nk_B T v$. This gives $\Delta x \simeq (\kappa/nk_B v)(\Delta T/T)$. Using $\Delta T \simeq T$ and $\kappa \simeq nk_B vl$, we again find that $\Delta x \simeq l$.

As an example of a shock front, let us assume again that the gas in a semi-infinite cylindrical pipe is terminated by the piston at one end (see Fig. 8.5). At $t = 0$, the piston begins to move towards the positive x axis with a constant speed U. As the piston begins to move, a shock wave is formed in front of the piston and moves towards the positive x direction. At $t = 0$, the shock and the piston are coincident but at $t > 0$ the shock is ahead of the piston and will be moving with speed v_s. Thus, for $t > 0$, the gas is divided into two regions; region 1 ahead of the shock wave where gas pressure is equal to its initial value p_1 and a region 2 that lies between the shock front and the piston where the pressure is p_2. In region 2, the gas moves with constant velocity equal to that of the piston, U. Hence the difference in velocity between regions 1 and 2 is also equal to U.

We now consider the motion in the frame in which the shock front is at rest and the gas has velocities $v_1 = -v_s$ and $v_2 = U - v_s = U + v_1$ in the two regions. The velocity difference between the two regions $v_1 - v_2$ is, of course, the same in all frames and is given by U. The conservation laws (8.164)–(8.166) can be rearranged to give

$$\frac{\rho_1}{\rho_2} = \frac{p_1(\gamma + 1) + p_2(\gamma - 1)}{p_1(\gamma - 1) + p_2(\gamma + 1)},$$

$$j^2 = \left(\frac{p_2 - p_1}{\rho_2 - \rho_1}\right)\rho_1\rho_2, \tag{8.174}$$

$$\frac{\gamma}{\rho_1\rho_2}(\rho_2 - \rho_1) = \frac{1}{\sqrt{\rho_1\rho_2}}[(p_2 - p_1)(\rho_2 - \rho_1)]^{1/2}.$$

In our case we find that

$$v_1 - v_2 = U = \frac{(p_2 - p_1)}{\sqrt{\rho_1}}\left[\frac{2}{(\gamma - 1)p_1 + (\gamma + 1)p_2}\right]^{1/2}. \tag{8.175}$$

This equation can be rewritten as a quadratic equation for the pressure ratio (p_2/p_1) in terms of p_1 and U:

$$\left(\frac{p_2}{p_1}\right)^2 - \left(\frac{p_2}{p_1}\right)\left[2 + (\gamma + 1)\frac{\rho_1 U^2}{2p_1}\right] + \left[1 - \frac{(\gamma - 1)\rho_1 U^2}{2p_1}\right] = 0. \tag{8.176}$$

The solution can be expressed in terms of the velocity of sound, $c_1 = \sqrt{\gamma p_1/\rho_1}$, in the undisturbed gas:

$$\frac{p_2}{p_1} = 1 + \frac{\gamma(\gamma + 1)U^2}{4c_1^2} + \frac{\gamma U}{c_1}\left[1 + \frac{(\gamma + 1)^2 U^2}{16c_1^2}\right]^{1/2}. \tag{8.177}$$

With this ratio known, the conservation laws provide the expression for $|v_1|$ as

$$|v_1| = \frac{\gamma + 1}{4}U + \left[c_1^2 + \frac{(\gamma + 1)^2 U^2}{16}\right]^{1/2}. \tag{8.178}$$

Note that these results are valid in the frame in which the shock front is at rest. In the frame in which the undisturbed gas is at rest, the shock front moves with a speed equal to $v_{\text{shock}} = v_1$. Therefore expression (8.178) also gives the velocity of the shock front in the undisturbed medium. Given the conditions of the undisturbed medium (ρ_1, p_1, T_1) and the speed of piston U, Eq. (8.177) determines p_2; with (p_2/p_1) known, Eqs. (8.174) determine ρ_2. Equation (8.178) relates the speed of the piston to the speed of the shock front; specifying either one of them is enough to fix all other parameters. For $U \gg c_1$, we have:

$$\frac{p_2}{p_1} = \frac{\gamma(\gamma + 1)}{2}\left(\frac{U}{c_1}\right)^2, \quad v_{\text{shock}} \cong \frac{1}{2}(\gamma + 1)U. \tag{8.179}$$

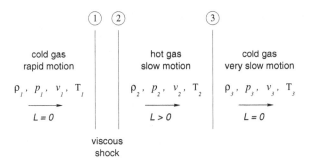

Fig. 8.6. Shock front in conjunction with radiative cooling.

Exercise 8.8
Filling in details: Prove Eqs. (8.168)–(8.173).

The shocks discussed above correspond to the simple case of a fluid in which no radiative processes or ionisation is taking place. A much richer and complicated set of phenomena can arise when we take radiative processes into account. We now discuss some of these effects briefly.

Let us first consider the effect of a shocked gas going out of thermal equilibrium and thus initiating the process of radiative cooling. We assume that any radiation from the shocked gas occurs under optically thin conditions and hence the gas can be described by a net cooling function $\mathcal{L}(\rho, T) = (\Lambda - \Gamma)/\rho$, where Λ is the volumetric cooling rate and Γ is the corresponding heating rate. In such a case, if we view the phenomena in the frame in which a shock front is at rest, we will expect to see the following (see Fig. 8.6). At the left extreme is the cold gas with parameters ρ_1, p_1, v_1, and T_1 moving rapidly to the right. This gas is in thermal equilibrium and thus $\mathcal{L} = 0$ in this region. The gas is shocked viscously in a narrow layer between 1 and 2 (see Fig. 8.6). In the immediate downstream (which is the middle region in the figure), the gas has the parameters ρ_2, p_2, v_2, and T_2. The density here has been compressed at most only by a factor $(\gamma + 1)/(\gamma - 1)$, and the gas is moving subsonically with respect to the viscous shock front. However, the temperature of the gas would have increased by a large factor, thereby throwing the gas out of thermal equilibrium, making $\mathcal{L}$ a large positive quantity. When the gas cools radiatively, it will be compressed by the pressure of the ambient medium and its density will increase. Because the mass flux ρv has to remain constant, an increase in ρ leads to a decrease in flow velocity. Slowly the gas cools enough to come back to thermal equilibrium with $\mathcal{L} = 0$ and we reach the region 3 with parameters ρ_3, p_3, v_3, and T_3.

The detailed structure of such shock fronts can be obtained from our governing equations. For a steady flow, we have the conditions

$$\rho v = \text{constant}, \quad \rho v^2 + p = \text{constant}, \tag{8.180}$$

and the equation for energy flux, which can be written as

$$\rho v \frac{d}{dx}\left[\frac{1}{(\gamma-1)}\frac{p}{\rho}\right] = -p\frac{dv}{dx} - \rho\mathcal{L}. \tag{8.181}$$

Moving ρv inside the derivative and simplifying, we get,

$$\frac{v}{(\gamma-1)}\frac{dp}{dx} + \frac{\gamma}{(\gamma-1)}p\frac{dv}{dx} = -\rho\mathcal{L}. \tag{8.182}$$

Now, by using $dp = -\rho v dv$ [which is obtained from the second equation of Eqs. (8.180)] and defining the adiabatic sound speed as $c_s^2 = (\gamma p/\rho)$, we get

$$\frac{1}{(\gamma-1)}(c_s^2 - v^2)\frac{dv}{dx} = -\mathcal{L}(\rho, T). \tag{8.183}$$

This equation, along with Eqs. (8.180) and the equation of state $p = (\rho/m)k_B T$, can be solved for $\rho(x)$, $p(x)$, $v(x)$, and $T(x)$. It is, however, easy to see the qualitative behaviour of the solution in the middle regime. Here, because $\mathcal{L} > 0$ and $v < c_s$, Eq. (8.183) implies that v will decrease with x. This raises ρ (because ρv is a constant) and also slightly raises the pressure p. The large increase in ρ coupled with a small increase in p leads to a decrease in T. This decreases the net cooling rate, and eventually $\mathcal{L}$ goes back to zero.

In general, the detailed structure of gas in the middle and the right regions has to be determined by the solution of Eq. (8.183). In many circumstances, however, the final temperature of the gas T_3 is not very different from the initial temperature T_1. In this case, we can write (and solve) matching conditions between region 1 and region 3, assuming that $T_1 = T_3$, without worrying about the detailed structure in region 2. (This is known as isothermal shock.) A simple calculation shows that

$$v_1 v_3 = c_T^2, \quad \frac{\rho_3}{\rho_1} = \left(\frac{v_1}{c_T}\right)^2, \tag{8.184}$$

where $c_T = (k_B T/m)^{1/2}$ is the isothermal sound speed. We note that, unlike the nonradiative shock, the isothermal shock has no limiting value for the density contrast and ρ_3 can have arbitrarily high values. In realistic circumstances, magnetic fields can prevent the formation of very high densities, which we shall discuss in Chap. 9.

Exercise 8.9
Matching the isothermal shock: Prove Eqs. (8.184) and explain in physical terms why there is no limiting value for density in this case.

8.12 Sedov Solution for Strong Explosions

Another example in which shock-wave conditions occur is around an "explosion." An explosion can release a large amount of energy E into an ambient gas in a time scale that is small. This will result in the propagation of a spherical shock wave centred at the location of the explosion. The resulting flow pattern can be determined everywhere if we make the following assumptions: (1) The explosion is idealised as one that releases energy E instantaneously at the origin. The shock wave is taken to be so strong that the pressure p_2 behind the shock is far greater than the pressure p_1 of the undisturbed gas. (2) We neglect the original energy of the ambient gas in comparison with the energy E that it acquires as a result of the explosion. (3) We assume that the gas flow is governed by the equations appropriate for an adiabatic polytropic gas with index γ.

For a strong shock, $p_2 \gg p_1$ and we can ignore p_1. In the same limit, $(\rho_2/\rho_1) \simeq (\gamma+1)/(\gamma-1)$ and hence ρ_2 is completely specified by ρ_1. Thus the strong explosion is entirely characterised by the total energy E and the density ρ_1 of the unperturbed gas. Let the radius of the shock front at time t be $R(t)$. Using E, ρ_1, and t, we can find only one quantity with the dimension of length, viz., $(Et^2/\rho_1)^{1/5}$. Hence we *must* have

$$R(t) = R_0 \left(\frac{Et^2}{\rho_1} \right)^{1/5}, \tag{8.185}$$

where R_0 is a constant to be determined by the solution of the equations of motion. All other physical quantities can now be determined in terms of the expression for $R(t)$. The speed of the shock wave with respect to the undisturbed gas is

$$u_1 = \frac{dR}{dt} = \frac{2R}{5t} = \frac{2}{5} R_0 E^{1/5} \rho_1^{-1/5} t^{-3/5}. \tag{8.186}$$

We can determine the pressure p_2, density ρ_2, and the speed of shock propagation $v_2 = u_2 - u_1$ (relative to a fixed coordinate system at the back of the shock) by using our junction conditions (8.179) across the shock front:

$$v_2 = \frac{2}{(\gamma+1)} u_1, \quad \rho_2 = \rho_1 \frac{(\gamma+1)}{(\gamma-1)}, \quad p_2 = \frac{2}{(\gamma+1)} \rho_1 u_1^2. \tag{8.187}$$

The density remains constant while v_2 and p_2 decrease as $t^{-3/5}$ and $t^{-6/5}$, respectively.

These scaling laws can be given a qualitative interpretation along the following lines. During the time interval t since the explosion, the mass swept up by the fluid will be $\sim \rho_1 R(t)^3$ and the fluid velocity will be about $R(t)/t$. Therefore the kinetic energy of the swept up gas will be $\rho_1 R^5 t^{-2}$. On the other hand, the internal energy density will be of the order of $\rho_1 \dot{R}^2$, giving the total internal energy inside the expanding shock as $\rho_1 \dot{R}^2 R^3$. Equating either of these energies to the total energy E will lead to the scaling relation in Eq. (8.185).

To proceed further we must explicitly integrate the equations of motion and determine the velocity $v(r, t)$, pressure $p(r, t)$, and density $\rho(r, t)$ of the gas behind the shock. The relevant equations are

$$\frac{\partial v}{\partial t} + v \frac{\partial v}{\partial r} = -\frac{1}{\rho} \frac{\partial p}{\partial r}, \quad \frac{\partial \rho}{\partial t} + \frac{\partial (\rho v)}{\partial r} + \frac{2\rho v}{r} = 0, \tag{8.188}$$

$$\left(\frac{\partial}{\partial t} + v \frac{\partial}{\partial r} \right) \ln \left(\frac{p}{\rho^\gamma} \right) = 0. \tag{8.189}$$

Equation (8.189) – the conservation of entropy – replaces the equation of energy conservation. Instead of p we can equivalently use the variable $c_s^2 = \gamma p/\rho$.

We look for a self-similar solution to these equations, with all variables depending essentially on the quantity $\xi \equiv [r/R(t)]$ with suitable scaling. More precisely, we invoke the ansatz

$$v = \frac{2r}{5t} V(\xi), \quad \rho = \rho_1 G(\xi), \quad c_s^2 = \frac{4\gamma^2}{25t^2} Z(\xi) \tag{8.190}$$

with the boundary conditions

$$V(1) = \frac{2}{(\gamma + 1)}, \quad G(1) = \frac{(\gamma + 1)}{(\gamma - 1)}, \quad Z(1) = \frac{2\gamma(\gamma - 1)}{(\gamma + 1)^2}. \tag{8.191}$$

It is possible to obtain a relation between Z and V that will simplify the integration of the equations. Because we have ignored p_1/ρ_1 compared with E, it follows that the total energy of the gas contained within a sphere bounded by the shock is a constant equal to E. Further, because the flow is self-similar, the energy of gas inside any sphere of smaller radius that follows the trajectory $\xi = \text{constant}$ must also remain constant. There is an amount of energy $4\pi r^2 \rho v(w + \frac{1}{2}v^2) \, dt$ that flows out of a sphere of radius r in time dt. On the other hand, the volume of the sphere increases by $4\pi r^2 v_n dt$ during this time, where $v_n = 2r/5t$ is the radial velocity of this sphere. The energy of gas inside the extra volume is $4\pi r^2 \rho v_n (\epsilon + \frac{1}{2}v^2) \, dt$. Equating the two energies and simplifying the expression, we obtain the result

$$Z = \frac{\gamma(\gamma - 1)(1 - V)V^2}{2(\gamma V - 1)}. \tag{8.192}$$

Given this relation, it is fairly straightforward (although tedious) to integrate Eqs. (8.188) and (8.189). The second and third equations can be written as

$$\frac{dV}{d \ln \xi} - (1 - V) \frac{d \ln G}{d \ln \xi} = -3V, \tag{8.193}$$

$$\frac{d \ln Z}{d \ln \xi} - (\gamma - 1) \frac{d \ln G}{d \ln \xi} = -\frac{5 - 2V}{1 - V}. \tag{8.194}$$

Eliminating $[d(\ln G)/d \ln \xi]$ between these equations and using Eq. (8.192) to express $[dZ/d \ln \xi]$ in terms of $[dV/d \ln \xi]$, we can determine $(dV/d \ln \xi)$ as a function of V alone:

$$\frac{1}{V}\frac{dV}{d \ln \xi} = \frac{\gamma(1 - 3\gamma)V^2 + (8\gamma - 1)V - 5}{\gamma(\gamma + 1)V^2 - 2(\gamma + 1)V + 2}. \tag{8.195}$$

This equation allows integration in closed form. A straightforward integration with suitable boundary conditions gives

$$\xi^5 = \left[\frac{2}{(\gamma + 1)V}\right]^2 \left\{\frac{\gamma + 1}{7 - \gamma}[5 - (3\gamma - 1)V]\right\}^{n_1} \left[\frac{\gamma + 1}{\gamma - 1}(\gamma V - 1)\right]^{n_2}, \tag{8.196}$$

where

$$n_1 = -\frac{13\gamma^2 - 7\gamma + 12}{(3\gamma - 1)(2\gamma + 1)}, \qquad n_2 = \frac{5(\gamma - 1)}{2\gamma + 1}. \tag{8.197}$$

We can also determine $G(V)$ in a similar way and obtain

$$G = \frac{\gamma + 1}{\gamma - 1}\left[\frac{\gamma + 1}{\gamma - 1}(\gamma V - 1)\right]^{n_3} \left\{\frac{\gamma + 1}{7 - \gamma}[5 - (3\gamma - 1)V]\right\}^{n_4} \left[\frac{\gamma + 1}{\gamma - 1}(1 - V)\right]^{n_5}, \tag{8.198}$$

with

$$n_3 = \frac{3}{2\gamma + 1}, \qquad n_4 = -\frac{n_1}{2 - \gamma}, \qquad n_5 = -\frac{2}{2 - \gamma}. \tag{8.199}$$

The solutions are shown in Fig. 8.7.

This provides the complete solution to our problem. The constant R_0 can now be determined by the condition that the total energy – which is the integral of $\rho(v^2/2 + \epsilon)$ – should be E. Writing $\epsilon = w - P/\rho = c_s^2/(\gamma - 1) - c_s^2/\gamma = c_s^2/[\gamma(\gamma - 1)]$, we get

$$E = \int_0^R 4\pi r^2 \, dr\rho\left[\frac{1}{2}v^2 + \frac{c_s^2}{\gamma(\gamma - 1)}\right], \tag{8.200}$$

which reduces to

$$R_0^5 \frac{16\pi}{25} \int_0^1 G\left[\frac{1}{2}V^2 + \frac{Z}{\gamma(\gamma - 1)}\right]\xi^4 \, d\xi = 1. \tag{8.201}$$

The integral needs to be evaluated numerically; for a gas with $\gamma = 7/5$, we get $R_0 = 1.033$. The shock solution described here has application in the study of supernova explosions, as we shall see in Vol. II.

8.13 Fluid Instabilities

Fluid motion exhibits a rich variety of instabilities, depending on the physical contexts. These instabilities play a vital role in astrophysical flows and are often

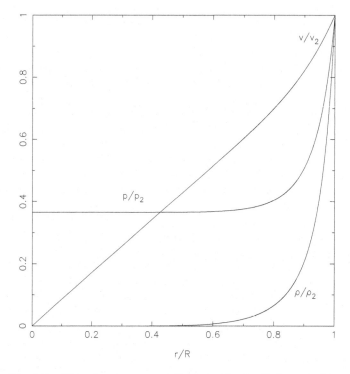

Fig. 8.7.   Velocity, pressure, and density for the Sedov solution.

quite complex in nature. In this section, a summary is given of some of these instabilities and their physical importance is discussed.

We confine our attention to linear stability analysis, which is performed along the following lines: We start with a system that is in equilibrium and perturb the relevant variables slightly. When the perturbations are small, all the relevant equations can be linearised in terms of the perturbed quantities. For such a linear system, we can attempt solutions of the form $\exp i(\omega t + \mathbf{k} \cdot \mathbf{x})$. The presence of an imaginary part to ω signals instability.

We begin with the simplest of all instabilities, which arises because of the buoyancy effects.

8.13.1 Rayleigh–Taylor Instability

Consider two different fluids of densities ρ and ρ', with the fluid of density ρ' resting on top with a constant gravitational field $\mathbf{g}$ acting downwards. To keep the geometry simple, we use a coordinate system in which the z axis is pointing vertically upwards and the x–y plane is the equilibrium surface separating the two liquids. We also assume that both layers of fluid have finite heights – h for the bottom fluid and h' for the top one. The fluid is expected to be at rest on the top and the bottom planes (see Fig. 8.8).

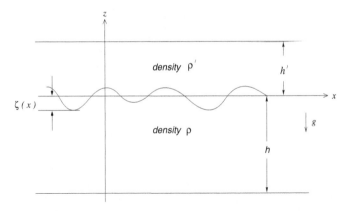

Fig. 8.8.   Geometry for Rayleigh–Taylor instability.

Let the interface between the two fluids be distorted slightly. We are interested in knowing whether such a distortion can lead to an instability and mixing of fluids or whether it will result in a stable wave propagation. It turns out that the situation is unstable when a denser liquid is located on top of a lighter liquid with a gravitational field directed downwards. This instability is called the Rayleigh–Taylor instability.

In studying this phenomenon, we can clearly confine our attention to perturbations that are independent of the y axis. Let $\zeta(x)$ denote the displacement of the surface in the z direction at the coordinate x. Because we are interested in the linear limit, we can ignore the $\mathbf{v} \cdot \nabla \mathbf{v}$ term in the fluid equations, which implies that we have a potential flow with $\mathbf{v} = \nabla \psi$. We further assume that the fluid is incompressible, allowing us to write $w = p/\rho$ in Eq. (8.96). Ignoring the v^2 term, this equation gives the pressure as

$$p = -\rho g z - \rho \frac{\partial \psi}{\partial t}, \tag{8.202}$$

where we have used a gravitational potential term $\phi = gz$. Because the pressure has to be continuous at the surface of separation, we must satisfy the condition

$$\rho g \zeta + \rho \frac{\partial \psi}{\partial t} = \rho' g \zeta + \rho' \frac{\partial \psi'}{\partial t} \quad \text{[at } z = \zeta(x)\text{]}, \tag{8.203}$$

which can be solved to give

$$\zeta = \frac{1}{g(\rho - \rho')} \left(\rho' \frac{\partial \psi'}{\partial t} - \rho \frac{\partial \psi}{\partial t} \right). \tag{8.204}$$

Moreover, the z component of the velocity must be the same for each of the liquids at the surface of separation. Because $\mathbf{v} = \nabla\psi$, this gives the condition

$$\frac{\partial \psi}{\partial z} = \frac{\partial \psi'}{\partial z} \quad [\text{at } z = \zeta(x)]. \tag{8.205}$$

Finally, we note that the z component of the velocity is equal to the time derivative of the displacement ζ at the lowest order of approximation. Hence we can write $v_z = (\partial\psi/\partial z) = (\partial\zeta/\partial t)$. Substituting this into Eq. (8.204) we get

$$g(\rho - \rho')\frac{\partial \psi}{\partial z} = \rho'\frac{\partial^2 \psi'}{\partial t^2} - \rho\frac{\partial^2 \psi}{\partial t^2}. \tag{8.206}$$

We now must determine the form of ψ and ψ'. For the incompressible fluid we are studying, ψ must satisfy the Laplace equation (see Subsection 8.6.4) and – because of our assumption – it must be independent of the y coordinate. When the ansatz $\psi = f(z)\cos(kx - \omega t)$ is substituted into the Laplace equation, it follows that $f(z) \propto \exp(\pm kz)$. Further, the velocity field has to vanish at the top and the bottom surfaces, that is, at $z = h'$ and $z = -h$. We can easily verify that the solution with the correct boundary conditions is given by

$$\psi = A \cosh k(z + h) \cos(kx - \omega t)$$
$$\psi' = B \cosh k(z - h') \cos(kx - \omega t). \tag{8.207}$$

Substituting these into Eq. (8.206), we get two linear equations for A and B. Equating the ratio for A/B from these two equations, we get the dispersion relation

$$\omega^2 = \frac{kg(\rho - \rho')}{\rho \coth kh + \rho' \coth kh'}, \tag{8.208}$$

which determines the properties of the perturbation. This relation contains several pieces of useful information.

For the equilibrium to be stable, we need $\omega^2 > 0$, which requires that $\rho > \rho'$. In this case, the dispersion relation describes gravity waves propagating in the interface between two liquids. The following limiting forms of this dispersion relation are of interest. First, if we take $\rho' = 0$, we get $\omega^2 = gk \tanh kh$, which is the dispersion relation for gravity waves propagating on the surface of a fluid of depth h. When the depth is very large ($kh \gg 1$), this reduces to $\omega^2 \approx gk$, whereas for $kh \ll 1$ we get $\omega^2 = k^2 gh$. These two limits are called deep-water waves and shallow-water waves. Note that shallow-water waves are nondispersive (i.e., $\omega \propto k$, making the phase velocity $v_{ph} \equiv \omega/k$ the same as the group velocity $v_g \equiv d\omega/dk$) whereas deep-water waves are not.

Second, when $\rho' \neq 0$, we still have three limiting cases. For $kh \gg 1$ and $kh' \gg 1$ (with both liquids very deep),

$$\omega^2 = \frac{kg(\rho - \rho')}{\rho + \rho'}, \tag{8.209}$$

whereas for $kh \ll 1$, $kh' \ll 1$ (long waves),

$$\omega^2 = k^2 \frac{g(\rho - \rho')hh'}{\rho h' + \rho' h}, \tag{8.210}$$

and, if $kh \gtrsim 1$ and $kh' \ll 1$,

$$\omega^2 = k^2 g h'(\rho - \rho')/\rho. \tag{8.211}$$

Finally, our result shows that the equilibrium is unstable if $\rho < \rho'$, that is, if the heavier liquid is residing on top of the lighter liquid. This is the Rayleigh–Taylor instability. In general, Rayleigh-Taylor instability occurs whenever $d\rho/dz$ is positive and the gravity is acting downwards. In such a configuration, the system can lower the potential energy by moving the heavier liquid downwards this is the physical origin of this instability.

8.13.2 Kelvin–Helmholtz Instability

The Rayleigh–Taylor instability is driven by the gravitational field; it is because of the presence of a vertical gravitational field that the energy of the system will be lower if the heavier fluid is at the bottom. There is another instability that occurs at the two-fluid interface that is independent of the gravitational field. This arises whenever there is a velocity gradient to the flow along the direction that separates the fluids. We now study this instability, which is called the Kelvin–Helmholtz instability.

Consider two layers of an incompressible fluid with velocities $\mathbf{v}_1$ and $\mathbf{v}_2$. We assume that one layer is attempting to "slide" on top of another along the common surface that they share. We consider a small region of the fluid on both sides of the surface of separation with fluid velocities tangential to the surface; the surface itself is assumed to be a plane. This kind of a flow – which occurs very often in nature, e.g., wind blowing on the surface of a lake – happens to be unstable to small perturbations. We first ignore the gravitational field to avoid any effects that arise from buoyancy; the role of gravity is discussed at the end.

Consider a perturbation in which the surface of discontinuity is slightly distorted from the planar shape. We show that, in an ideal incompressible fluid, such a perturbation will grow and we determine the time scale for these instabilities to grow. It is convenient to work in a frame of reference in which the fluid velocity is zero on one side of the separation and $\mathbf{v}$ on the other side. Let the direction of $\mathbf{v}$ be the x axis and normal to the surface be along the z axis. If the surface of discontinuity receives a small perturbation in which all quantities (the displacement of the surface in the z direction, pressure, and velocity) are periodic functions of the form $\exp[i(kx - \omega t)]$, we can linearise the equations in the small perturbation and denote the deviations by primed quantities. If the dispersion relation between ω and k leads to an imaginary part for ω, then the perturbations will lead to an instability.

The linearised equations for the perturbed quantities $\mathbf{v}'$ and p' for an incompressible ideal fluid are

$$\nabla \cdot \mathbf{v}' = 0, \qquad \frac{\partial \mathbf{v}'}{\partial t} + (\mathbf{v} \cdot \nabla)\mathbf{v}' = -\frac{\nabla p'}{\rho}. \tag{8.212}$$

Because the original velocity is along the x axis, $(\mathbf{v} \cdot \nabla)\mathbf{v}' = v(\partial \mathbf{v}'/\partial x)$. Taking the divergence of both sides of Euler equation, we find that p' must satisfy the Laplace equation: $\nabla^2 p' = 0$. Because the variation along the x axis is fixed, let us try an ansatz of the form $p' = f(z)\exp[i(kx - \omega t)]$. This leads us to the equation

$$\frac{d^2 f}{dz^2} = k^2 f \tag{8.213}$$

with the solution $f \propto \exp(\pm kz)$. Assuming that the region under consideration (side 1, say) corresponds to positive values of z, the solution with proper behaviour for large z is

$$p_1' \propto e^{-kz} e^{i(kx - \omega t)}. \tag{8.214}$$

From Eqs. (8.212) we see that this corresponds to the velocity distribution

$$v_z' = \frac{k p_1'}{i\rho_1(kv - \omega)}. \tag{8.215}$$

Consider next the displacement of the surface $l(x, t)$. Because the velocity component normal to the surface is equal to the rate of displacement of the surface itself, we can write $(dl/dt) = v_z'$, correct to the linear order. Therefore

$$\frac{\partial l}{\partial t} = v_z' - v\frac{\partial l}{\partial x}, \tag{8.216}$$

where the value of v_z' must be taken on the surface. Using $l \propto \exp[i(kx - \omega t)]$ in this equation, we find that $v_z' = il(kv - \omega)$; Substituting into Eq. (8.215), we get

$$p_1' = -\frac{l\rho_1}{k}(kv - \omega)^2. \tag{8.217}$$

On side 2, similar considerations apply except that we must set $v = 0$ and use the e^{kz} solution instead of e^{-kz} solution. This gives

$$p_2' = \frac{l\rho_2}{k}\omega^2. \tag{8.218}$$

If the liquids on both sides are the same, then $\rho_1 = \rho_2$; but the above analysis is applicable even when two different fluids are in contact with each other with $\rho_1 \neq \rho_2$. To obtain the dispersion relation we need to use only the fact that $p_1' = p_2'$ at the surface of separation. This leads to the quadratic equation $\rho_1(kv - \omega)^2 = -\rho_2\omega^2$ in ω with the solution

$$\omega = kv\frac{\rho_1 \pm i\sqrt{\rho_1\rho_2}}{\rho_1 + \rho_2}. \tag{8.219}$$

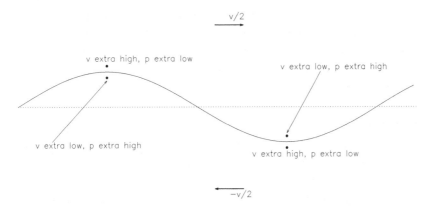

Fig. 8.9.   Intuitive way of understanding Kelvin–Helmholtz instability. The perturbation in the figure will grow because of the pressure difference.

The frequency ω is complex with a positive imaginary part. Hence the fluid is unstable to shearing flow of the kind described above. When $\rho_1 = \rho_2$, the imaginary part of ω is given by Im $\omega = (kv/2)$. The inverse of this quantity gives the time scale for the growth of instability. Alternatively, we can consider waves with real ω that lead to an imaginary part in k, corresponding to unbounded spatial growth. These two descriptions are equivalent and illustrate a general result for any type of wave with a dispersion relation $\omega(k)$ that is an analytic function. Let us suppose that a solution to this dispersion relation describes a wave with real $k = k_R$ that grows in time at a growth rate $\omega_I(k_R)$, where the subscripts denote real and imaginary parts. Then there will exist another solution that describes a wave with real frequency ω and an imaginary wave vector such that the spatial growth of amplitude occurs at a rate $k_I(\omega_R)$. The Cauchy–Riemann equation for the analytic functions relates these two modes by means of $(\partial \omega_I/\partial k_I) = (\partial \omega_R/\partial k_R) = v_g$, the group velocity.

A simple, intuitive way of understanding this instability is illustrated in Fig. 8.9. This figure shows the interface between the two liquids, at some moment of time, in a frame of reference moving with velocity $\mathbf{v}/2$. In this frame, the upper liquid moves with velocity $\mathbf{v}/2$ and the lower liquid with velocity $-\mathbf{v}/2$. Suppose that the perturbation is such that the velocity is slightly higher on top of a crest and slightly lower below a trough. Because of Bernouli's theorem, the pressure will be lower in regions of higher velocity and vice versa. This will cause the perturbation to grow.

In the above discussion, we have not taken into account the effect of gravitational field. If there is a vertical gravitational field acting on the liquids and if the liquid on top is lighter, buoyancy effects can work against Kelvin–Helmholtz instability. Consider a gas that has a density that decreases with increasing height (that is, $d\rho/dz < 0$) in a gravitational field g that is acting downwards. The flow of the gas is along the x axis, and $v_x = v_x(z)$ is a function of z. In general, such a velocity gradient will tend to produce Kelvin–Helmholtz instability. However, as we saw in Subsection 8.13.1, the condition $d\rho/dz < 0$ ensures that buoyancy forces will try to stabilise any vertical motion of the fluid. When both the buoyancy forces *and* the velocity gradient are present, instability can occur only if the velocity gradient is sufficiently large. The condition for stability under the above circumstances can be obtained by similar analysis (see Exercise 8.10), and the result can be expressed in terms of a quantity called Richardson's number, which is defined as

$$\mathcal{R}i \equiv -\frac{g}{\rho}\frac{(d\rho/dz)}{(dU/dz)^2} = -\frac{g(\partial\rho/\partial s)_P(ds/dz)}{\rho(dU/dz)^2}. \tag{8.220}$$

For stability, $\mathcal{R}i$ should be greater than $1/4$. It is clear from the definition of $\mathcal{R}i$ that there is a competition between the shear that occurs in the denominator and the buoyancy that occurs in the numerator, and the stability is decided by which of the these two factors dominate.

Exercise 8.10
Richardson criteria: Repeat the stability analysis in the presence of a gravitational field and obtain the criterion mentioned above.

Exercise 8.11
Rotational instability: Consider an axisymmetric fluid distribution rotating in steady state about the z axis under the action of gravity and pressure gradient. Let $\Omega(R)$ be the angular velocity of the fluid at a radius R from the axis. Show that the fluid motion is unstable if the quantity $|\Omega(R)R^2|$ decreases outwards.

8.13.3 Thermal Instability

A different kind of instability, which occurs essentially in optically thin gases subjected to radiative processes, is called thermal instability. The basic idea behind this instability is quite simple, although its actual realisation is often accompanied by several complications.

Consider a gas of density ρ and temperature T that is subjected to certain internal heating and cooling processes. Let the *net* heat loss per unit mass be $\mathcal{L}(\rho, T)$. In thermal equilibrium, we have the condition $\mathcal{L}(\rho, T) = 0$, which will be a simple connected curve in the ρ–T plane (see Fig. 8.10). Further, let us assume that the gas is ideal with an equation of state $p \propto \rho T$ and that it is in pressure equilibrium with the surroundings. This implies that we can treat p as an externally specified constant. It could happen that the $p = $ constant line intersects the $\mathcal{L} = 0$ line at several points. Figure 8.10 illustrates the situation in which there are three intersections, A, B, and C, in some relevant range. At any one of these three points, the blob of gas is in both mechanical and thermal equilibrium. However, not all these points constitute stable equilibrium configurations. Suppose that a small blob of gas is perturbed away from the $\mathcal{L} = 0$

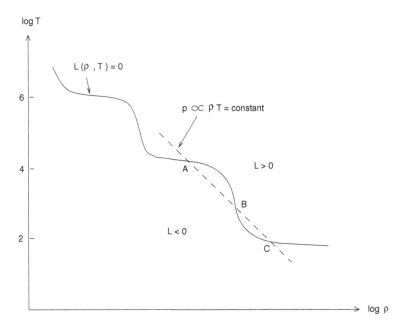

Fig. 8.10. Graphical illustration of thermal instability.

curve, but along the $p =$ constant line. The constancy of p is based on the idea that mechanical equilibrium with the surroundings can be achieved in a time scale shorter than the time scale for thermal equilibrium. If the blob was originally at A and if we displace it slightly to lower temperatures and higher densities along the $p =$ constant curve (shown by the dashed line in the figure), then we see that it enters a region where $\mathcal{L} < 0$. In this region, heating exceeds cooling and the blob will heat up and expand back towards A. On the other hand, if we do the same with a blob that was originally at B, it will end up at a region with $\mathcal{L} > 0$ where the cooling exceeds heating, driving it further to still lower temperatures. A similar analysis shows that C is stable. Thus, in the simple configuration discussed above, A and C are stable equilibrium configurations, but B is not. Such thermal instabilities play a vital role in the study of interstellar media.

The nature of the curve in Fig. 8.10 arises because different physical processes dominate the heating and cooling in a gas at different regions of the temperature–density plane. At temperatures of 10 to 100 K, collisions excite the rotational levels of molecules and the fine-structure levels of atoms. The excited atom returns back to the ground state, emitting a photon. Such collisional excitation followed by radiative deexcitation has a cooling rate that depends sensitively on the temperature T. This is because the fraction of particles with energy E exceeding a threshold energy drops off as $\exp(-E/k_B T)$. For such a cooling to balance any source of external heating, T need to change only a little within this range and can still cause ρ to change a lot. But as T approaches and exceeds

$\sim 10^3$ K, the Boltzmann factor approaches unity for the energies relevant for rotational and fine-structure excitations. By now almost all the particles will be participating in the inelastic collision and hence we need large changes in T to compensate for small changes in ρ if cooling has to balance heating. When T approaches 10^4 K, there will be enough particles in the Maxwellian distribution with energies sufficient to collisionally excite the low-lying electronic states of common elements and – once again – a small increase in T is sufficient to balance large decreases in ρ. The same pattern repeats as T approaches and exceeds 10^5 K; the temperature sensitivity is restored at 10^6 K when inner-shell elements such as oxygen and iron contribute to radiative cooling. Above 10^7 K, line cooling becomes inefficient and free–free emission becomes the dominant process.

We can construct a simple quantitative model for thermal instability by perturbing the fluid equations in the presence of heating and cooling. Ignoring gravitation and viscosity, we take the relevant equations to be

$$\frac{\partial \rho}{\partial t} + \nabla \cdot (\rho \mathbf{v}) = 0, \tag{8.221}$$

$$\frac{\partial \mathbf{v}}{\partial t} + \frac{1}{2}\nabla(|\mathbf{v}|^2) + (\nabla \times \mathbf{v}) \times \mathbf{v} = -\frac{1}{\rho}\nabla p, \tag{8.222}$$

$$T\left(\frac{\partial s}{\partial t} + \mathbf{v} \cdot \nabla s\right) = -\mathcal{L}(\rho, T), \tag{8.223}$$

where

$$\rho\mathcal{L} \equiv \Lambda - \Gamma, \tag{8.224}$$

$$s = C_V \ln(p\rho^{-\gamma}) + \text{constant}, \quad p = \mathcal{R}\rho T, \tag{8.225}$$

and $\mathcal{R} = (k_B/m_H) = C_P - C_V$ and $\gamma = (C_P/C_V)$. The equilibrium state corresponds to $\rho = \rho_0$, $T = T_0$, and $\mathbf{v} = 0$, with $\mathcal{L}(\rho_0, T_0) = 0$. We now consider small perturbations ρ_1, T_1, and $\mathbf{v}_1$ and linearise the equations in the perturbed quantities. An analysis similar to that in Section 8.8 will now give

$$\frac{\partial^2 \rho_1}{\partial t^2} - \nabla^2 p_1 = 0, \tag{8.226}$$

$$\frac{C_V}{p_0}\frac{\partial p_1}{\partial t} - \frac{C_P}{\rho_0}\frac{\partial \rho_1}{\partial t} = \left(\frac{\partial \mathcal{L}}{\partial T}\right)_p \left(\frac{\rho_1}{\rho_0}\right) - \left(\frac{\partial \mathcal{L}}{\partial T}\right)_\rho \left(\frac{p_1}{p_0}\right). \tag{8.227}$$

Taking the Laplacian of Eq. (8.227) and using Eq. (8.226) to eliminate $\nabla^2 p_1$, we obtain the final equation as

$$\frac{\partial}{\partial t}\left(\frac{\partial^2 \rho_1}{\partial t^2} - c_s^2\nabla^2\rho_1\right) = N_P c_s^2\nabla^2\rho_1 - N_V\frac{\partial^2\rho_1}{\partial t^2}, \tag{8.228}$$

where

$$N_P \equiv \frac{1}{C_P}\left(\frac{\partial \mathcal{L}}{\partial T}\right)_P, \quad N_V \equiv \frac{1}{C_V}\left(\frac{\partial \mathcal{L}}{\partial T}\right)_\rho. \quad (8.229)$$

To solve Eq. (8.228) we use the ansatz $\rho_1 \propto \exp(\mu t + i\mathbf{k} \cdot \mathbf{x})$ and obtain the cubic dispersion relation

$$\mu^3 + N_V\mu^2 + k^2 c_s^2 \mu + N_P k^2 c_s^2 = 0. \quad (8.230)$$

The exact solution to this equation is somewhat complicated. However, asymptotic limits can be easily obtained. In the limit of short wavelengths, we have the three roots

$$\mu = -N_P, \quad \mu = \pm ikc_s + \frac{1}{2}(N_P - N_V). \quad (8.231)$$

This shows that we have a growing instability if $N_P < 0$, that is, when $(\partial \mathcal{L}/\partial T)_P < 0$. This is the criterion that was illustrated in Fig. 8.10. This mode occurs at constant pressure; small blobs of gas will tend to maintain pressure equilibrium with their surrounding when a misbalance in the heating or the cooling rate occurs. The second root corresponds to sound waves with damping that is due to radiative processes if $N_V > N_P$. For an ideal gas, it is easy to show that

$$N_P - N_V = \left(\frac{1}{C_P} - \frac{1}{C_V}\right)\left(\frac{\partial \mathcal{L}}{\partial T}\right)_s = -(\gamma - 1)\rho\left(\frac{\partial \mathcal{L}}{\partial P}\right)_s, \quad (8.232)$$

which is negative.

In the limit of very long wavelengths, the three roots are given by

$$\mu = -N_V, \quad \mu = \pm ika_{\text{eff}}, \quad (8.233)$$

where

$$a_{\text{eff}} = \left(\frac{N_P}{N_V}\right)^{1/2} c_s. \quad (8.234)$$

This will lead to an instability if $N_V < 0$, that is, $(\partial \mathcal{L}/\partial T)_\rho < 0$. Very large blobs may not change their volume and density appreciably by the pressure of the ambient medium. In practical situations, the first kind of instability seems to be of greater relevance.

Exercise 8.12

Simpler picture of thermal instability: Consider a neutral hydrogen gas that is stationary and in pressure equilibrium with the surroundings so that $P = n_H k_B T = $ constant. Let the heating rate be $H = \Gamma_0 n_H$, where Γ_0 is a constant and the cooling rate $L = n_H^2 \Lambda(T)$. Suppose that the temperature of the gas is now perturbed slightly by an amount δT. Show that the perturbation evolves as $\delta T \propto \exp(-t/t_{\text{cool}})$ and find an expression for t_{cool}. [Answers: The equation governing the heat loss in this particular case is given by $n_H(d/dt)(3k_BT/2) - k_BT(dn_H/dt) = H - L$. With $n_HT = \text{constant}$, this reduces to $(5/2)n_H(dk_BT/dt) = H - L$. Writing $T = T_0 + \delta T$ and expanding the right-hand side in a Taylor series, we get $(d\delta T/dt) = ((2/5)(1/n_Hk_B)\{[\partial(H - L)/\partial T]\}_p)_{T_0}\delta T$. This clearly has exponential solutions of the form $\delta T \propto \exp(-t/t_{\text{cool}})$, with $t_{\text{cool}} = -((2/5)(1/n_Hk_B)\{[\partial(H - L)/\partial T]\}_p)_{T_0}^{-1}$. For the form of H and L given in the question this reduces to $t_{\text{cool}} = (5/2)[k_BT_0/n_H\Lambda(T_0)(d\ln\Lambda/d\ln T - 1)_{T_0}]$.]

8.14 Conduction and Convection

A fluid can be in mechanical equilibrium, exhibiting no macroscopic motion, even when it is not in thermal equilibrium. For fluids in gravitational fields, this essentially requires a balance between pressure gradients and gravity. Such a fluid can exhibit a temperature gradient that would cause a flux of heat to be transported from the high-temperature region to the low-temperature region. From our discussion in Section 8.5, we know that the temperature distribution is governed by Eq. (8.58). The absence of fluid motion necessarily implies that the temperature differences are small; hence we can take the conductivity, heat capacity, and density to be approximately constant. In that case, Eq. (8.58) reduces to

$$\frac{dT}{dt} = \chi\nabla^2 T, \qquad (8.235)$$

where $\chi = (\kappa/\rho C_P)$. This is a diffusion equation that describes the flow of heat that is due to thermal conduction without any mechanical motion.

Because this equation shows that the conductive flux is proportional to the temperature gradient, we can define conductivity κ and a conductive opacity κ_{cond} (in analogy with radiative opacity) by writing

$$F_{\text{cond}} \equiv -\frac{4ac}{3\kappa_{\text{cond}}\rho}T^3\frac{dT}{dr} \equiv -\kappa\frac{dT}{dr}. \qquad (8.236)$$

The definition of conductivity κ is identical to the one introduced in Eq. (8.35). The conductivity is related to opacity by $\kappa = (4acT^3/3\rho\kappa_{\text{cond}})$. When both radiative and conductive transports of heat exist, the total opacity κ_{tot} is related to the individual opacities by $\kappa_{\text{tot}} = (\kappa_{\text{rad}}\kappa_{\text{cond}})/(\kappa_{\text{rad}} + \kappa_{\text{cond}})$.

The conductivity for an ionised gas is mostly contributed by electrons and can be expressed in the form $\kappa = n_eC_Vv_e\lambda/3$, where C_V is the specific heat at constant volume, v_e is the typical electron velocity, and λ is the mean free path [see Eq. (8.41)]. Using the relations $\lambda \approx (\sigma n_i)^{-1}$, where n_i is number density of ions, and $\sigma = \pi r_0^2$, where $r_0 = (2Ze^2/m_ev^2)$ is the typical distance of the closest approach between electrons and ions, we can estimate the conductivity for a gas in which electron–ion scattering dominates the heat transfer. We then get

$$\kappa = \frac{1}{3}\left(\frac{m_e^2}{4\pi Z^2e^4}\right)\left(\frac{n_e}{n_i}\right)(C_Vv_e^5).$$

To proceed further, we need to esimate $(C_V v_e^5)$, which depends on whether the electron gas is classical or degenerate. For a classical electron gas, $C_V = (3/2)k_B$ and $v_e^2 = (3k_B T/m_e)$, giving

$$\kappa \simeq \left(\frac{9\sqrt{3}}{8\pi}\right)\left(\frac{m_e c^2}{Ze^2}\right)^2 \left(\frac{\mu_i}{\mu_e}\right) k_B c \left(\frac{k_B T}{m_e c^2}\right)^{5/2}$$

$$\approx 10^3 \text{erg s}^{-1} \text{cm}^{-1} \text{K}^{-1}\left(\frac{\mu_i}{\mu_e}\right)\left(\frac{T}{10^4 \text{ K}}\right)^{5/2},$$
(8.237)

where μ_e and μ_i are the electronic and ionic molecular weights, respectively, defined so that $n_e = (\rho/m_p)\mu_e^{-1}$ and $n_i = (\rho/m_p)\mu_i^{-1}$. This result is valid for nondegenerate electrons. (The numerical factor 10^3 becomes 10^{-2} if SI units are used.)

For a system made of degenerate fermions, however, the thermal conductivity is different, because the scattering processes are different when all the energy levels $\epsilon < \epsilon_F$ are filled. In general, the Coulomb scattering of electrons by ions will change the momentum of the electrons from some initial value to final value. In a completely degenerate system, such a scattering is possible only if the final energy level is not filled. Hence only the electrons near the Fermi surface, having momenta near $m_e v_F = p_F \propto (\rho/\mu_e)^{1/3}$, can participate in the scattering; further, $C_V \simeq k_B(k_B T/\epsilon_F)$, giving $C_V v_e^5 \simeq k_B(k_B T/\epsilon_F)(2\epsilon_F/m_e)^{5/2} \simeq 6\pi^2(\hbar^3 k_B^2/m_e^4)n_e T$. The corresponding conductivity that is due to electrons in degenerate matter will be

$$\kappa = \frac{\pi}{2}\left(\frac{\hbar^3 k_B^2}{Z^2 m_p m_e^2 e^4}\right)\left(\frac{\mu_i}{\mu_e^2}\right)\rho T.$$
(8.238)

Note that the corresponding $\chi = (\lambda v/3)$ is given by $\chi = (m_e/Ze^2)^2(v_F^5/12\pi n_i)$. This result, as we shall see in Vol. II, has important implications for the conduction of heat by white dwarfs.

Exercise 8.13
Thermal instability with thermal conduction: (1) As an extension of the Exercise 8.12, consider the situation in which the gas was originally in equilibrium with a constant temperature T_0, density n_0, and a thermal conductivity κ. Let the temperature be perturbed by an amount $\delta T \propto \exp[i(\mathbf{k}\cdot\mathbf{x} - \omega t)]$, keeping the pressure constant spatially and temporally. Derive the dispersion relation for the perturbation and determine the stability criteria. (2) In the relevant range of temperature, $\Lambda(T)$ can be approximated as $\Lambda(T) \simeq \Lambda_0(T/10^6 \text{ K})^{-2.7}$, where $\Lambda_0 \approx 8 \times 10^{23}$ in cgs units. What is the longest wavelength perturbation that is stable in an interstellar gas with $T = 10^6$ K and number density of electrons $n_e = 0.003$ cm^{-3}? [Answers: (1) The equation governing the temperature now becomes $(5/2)n_0 k_B(dT/dt) = (H - L) + \nabla\cdot(\kappa\nabla T)$. Perturbing T as suggested in the question leads to $(5/2)n_0 k_B(\partial\delta T/\partial t) = -(5/2)(n_0 k_B\delta T/t_{\text{cool}}) + \nabla\cdot(\kappa\nabla\delta T)$. Using $\nabla\cdot(\kappa\nabla\delta T) = -\kappa k^2\delta T$ and $(\partial\delta T/\partial t) = -i\omega\delta T$, we get the dispersion

relation $i\omega = t_{cool}^{-1} + (2/5)(k^2\kappa/n_0k_B)$. In the absence of conductivity, we found that $t_{cool} < 0$ will induce an instability. However, now small wavelength perturbations can be stable because of finite thermal conductivity (even when $t_{cool} < 0$), provided that $k^2 > (5/2)(n_0k_B/\kappa)(-1/t_{cool})$. (2) The maximum wavelength $\lambda_{max} = (2\pi/k_{min})$ is given by $\lambda_{max} = 2\pi(2\kappa/5n_0k_B)^{1/2}(-t_{cool})^{1/2}$. From $\Lambda(10^6$ K$) = 8 \times 10^{-23}$ and $(d \ln \Lambda/d \ln T) \approx -2.7$, we get the cooling time scale $t_{cool} \approx -4 \times 10^{14}$ s. For $T = 10^6$ K, $n = 0.003$, the thermal conductivity is approximately 6×10^8 in cgs units. This shows that $\lambda_{max} \approx 1$ kpc.]

Let us next consider the question as to whether such a mechanical equilibrium with a temperature gradient is stable. It turns out that the equilibrium is stable only when the temperature gradient is not too large. If this is not the case, then fluid motions will occur that tend to mix the fluid in such a way as to equalise the temperature. Such a fluid flow is called convection.

One simple condition for the absence of convection is the following. Let us consider a fluid column located in a gravitational field. Let a blob with specific volume $V(p, s)$ move from a height z to a height $z + \xi$, where the pressure is p'. On displacement, its specific volume becomes $V(p', s)$. This motion of the blob will be stable and the resulting force will drive it back to the original location, provided the fluid element is heavier than the fluid it displaces in its new position. The specific volume of the latter is $V(p', s')$, where s' is the equilibrium entropy at the height $z + \xi$. For stability we must have $V(p', s') - V(p', s) > 0$. Writing $s' - s = \xi(ds/dz)$ and using the fact that $(\partial V/\partial s)_p = (T/C_P)(\partial V/\partial T)_p > 0$, we obtain the condition that $(ds/dz) > 0$, that is, the entropy must increase with height for stability. This condition can be translated into a condition on the temperature gradient by

$$\frac{ds}{dz} = \left(\frac{\partial s}{\partial T}\right)_p \frac{dT}{dz} + \left(\frac{\partial s}{\partial p}\right)_T \frac{dp}{dz} = \frac{C_P}{T}\frac{dT}{dz} - \left(\frac{\partial V}{\partial T}\right)_p \frac{dp}{dz} > 0, \quad (8.239)$$

where we have used the identity $(\partial V/\partial T)_p = -(\partial s/\partial p)_T$; see Table 5.1. Using the equilibrium condition $(dp/dz) = -g\rho = -(g/V)$, we obtain

$$-\frac{dT}{dz} < \frac{g\beta T}{C_P}, \quad (8.240)$$

where $\beta = (\partial \ln V/\partial T)_p$ is the thermal-expansion coefficient. If the temperature decreases with a gradient whose magnitude is larger than the right-hand side of this equation, the equilibrium is unstable to convective flow.

In a more general setting – for example, in a star with gravitational force decreasing with radius r – the condition $(ds/dz) > 0$ of Eq. (8.239) can be stated as follows. The fluid will be unstable if the entropy decreases outward from the centre. For an ideal gas, with $s \propto \ln[pT^{\gamma/(\gamma-1)}]$ [see Equation (5.96)], $(\partial s/\partial T)_p = [\gamma/(\gamma - 1)]T^{-1}$, and $(\partial s/\partial p)_T = p^{-1}$, the condition $(ds/dr) > 0$

for convection to occur can be written as

$$\left|\frac{d \ln T}{d \ln r}\right| > \left(\frac{\gamma - 1}{\gamma}\right)\left|\left(\frac{d \ln p}{d \ln r}\right)\right| = \nabla_{\text{ad}}\left|\left(\frac{d \ln p}{d \ln r}\right)\right|, \qquad (8.241)$$

where we have used the result $\nabla_{\text{ad}} = (\gamma - 1)/\gamma$, valid for an ideal gas. This shows that convection occurs if $\nabla \equiv |[d(\ln T)/d(\ln p)]| > \nabla_{\text{ad}}$.

The criterion for convection $\nabla > \nabla_{\text{ad}}$ is valid only when chemical composition is homogeneous. If that is not the case, we will get a contribution $(\Delta \rho/\rho) = (\partial \ln \rho/\partial \ln \mu)(\Delta \mu/\mu)$, where μ is interpreted as the mean mass per particle. The stability will now be ensured if $\nabla < [\nabla_{\text{ad}} + (\phi/\delta)\nabla_{\mu}]$, where $\delta \equiv -(\partial \ln \rho/\partial \ln T)_{p,\mu}$, $\phi \equiv (\partial \ln \rho/\partial \ln \mu)_{T,p}$, and $\nabla_{\mu} = (\partial \ln \mu/\partial \ln p)_{\text{ext}}$. [If heavier particles are located at smaller r in a star, for example, then $(d\mu/dr) < 0$ and we obtain a stabilising influence against convection.] More about this aspect is discussed in Vol. II.

When convection occurs, it can be a rapid source of transport of energy flux in a fluid. In spite of the fact that such a convective flux plays an important role in several astrophysical contexts, we do not yet have a fundamental theory to describe this phenomenon. A naive model is summarised below for the convective transport of energy; called the mixing-length theory, it is used extensively in several astrophysical contexts. This model assumes that a, hot, rising-fluid element travels a distance l before releasing the excess heat energy to the surrounding fluid. This length scale l, called the mixing length, is usually written as $l = \alpha H_p$, where H_p is the characteristic scale height of pressure variation defined through $H_p \equiv -(\partial \ln p/\partial r)^{-1}$ and α is a numerical constant. After travelling a distance of the order of l, the excess heat flow per unit volume from the fluid element to its surroundings will be $\delta q = \rho C_p \delta T$, where δT is the difference in the temperature between the rising-fluid element and the surrounding gas, given by

$$\delta T = \left(\frac{dT}{dr}\bigg|_{\text{ad}} - \frac{dT}{dr}\bigg|_{\text{act}}\right)l = \delta\left(\frac{dT}{dr}\right)l. \qquad (8.242)$$

The corresponding convective heat flux (amount of energy carried per unit area per unit time) is given by $F_c = \delta q \, \bar{v}_c = (C_p \delta T)\rho \, \bar{v}_c$, where $\bar{v}_c$ is the typical velocity of the fluid element. We can estimate this quantity by equating the kinetic energy gained by the fluid element in travelling a distance l to the work done by the buoyancy force over a distance of the order of l. The force acting on the fluid element is $f_{\text{net}} = -g\delta\rho$; because the fluid element is in pressure equilibrium throughout its motion, we can express $\delta\rho$ in terms of δT by using the relation

$$0 = \delta p = \frac{p}{\rho}\delta\rho + \frac{p}{T}\delta T. \qquad (8.243)$$

This gives the net force as $f_{\text{net}} = \rho g(\delta T/T)$. The difference δT is zero initially and has some value δT_f just before the fluid element mixes with the surrounding gas. Therefore the average force acting throughout the motion may be taken to be $(1/2)\rho g(\delta T_f / T)$. Equating the work done by this force over a distance l to the average kinetic energy of the fluid element, $(1/2)\rho\langle v^2\rangle$, we can estimate the mean convective speed $\bar{v}_c$ of the fluid element. We assume that $\bar{v}_c^2 = \alpha'\langle v^2\rangle$, where α' is another numerical factor of the order of unity. Then we get

$$\bar{v}_c = \left(\frac{\alpha' g}{T}\right)^{1/2}\left[\delta\left(\frac{dT}{dr}\right)\right]^{1/2} l = \alpha'^{1/2}\alpha\left(\frac{T}{g}\right)^{1/2}\left(\frac{k_B}{\mu m_H}\right)\left[\delta\left(\frac{dT}{dr}\right)\right]^{1/2},$$

$$(8.244)$$

where we have used the fact that $l = \alpha H_p = \alpha(p/\rho g)$. This corresponds to a convective flux

$$F_c = (C_P \rho \bar{v}_c)\delta T = \alpha^2 \alpha'^{1/2}\rho\, C_P\left(\frac{k_B}{\mu m_H}\right)^2\left(\frac{T}{g}\right)^{3/2}\left[\delta\left(\frac{dT}{dr}\right)\right]^{3/2}. \quad (8.245)$$

This expression shows that, although F_c is not very sensitive to α', it does depends strongly on the unknown parameter α and on the differences in temperature gradients. In spite of this fact, this theory provides reasonable agreement with observations in modelling stellar structures with α in the range 0.5–3.

The following point needs to be emphasised regarding the convective flux; the process of convection is quite efficient even when the temperature gradient exceeds the adiabatic gradient by a very small amount. Once convection is initiated it is very efficient in transporting energy and leads to a temperature gradient that is very close to the adiabatic gradient. We shall have occasions to use these facts in Vol. II.

The above analysis did not take viscosity into account. Viscosity in the fluid, in general, works against fluid motion. Thus convective fluid motion can actually occur only when the temperature gradient is large enough to overcome the effects of viscosity. The conditions for this, by use of a simple model called Rayleigh–Bernard convection, are briefly described.

Consider a fluid confined between two planes separated by a vertical distance of h with gravity acting downwards. The planes are kept at different temperatures with temperature decreasing with height. Under the circumstances we expect the fluid heated at the bottom to become lighter and rise to the top, setting off convection, provided it can overcome the viscous effects. Let us first consider the static solution to equations of fluid mechanics that we obtain by setting all the time derivatives to zero. In the case of an incompressible fluid, the equations have five unknown functions $\mathbf{v}$, (p/ρ), and T and three parameters ν, χ, βg, where β is the thermal-expansion coefficient and χ is the conductivity. The boundary conditions will involve the length scale h and the temperature difference Θ. From these parameters, we can form two dimensionless parameters that are called the Prandtl number, $P \equiv (\nu/\chi)$, and the Rayleigh number, $\mathcal{R} \equiv (\beta g \Theta h^3/\nu\chi)$. All other quantities can be expressed as suitably scaled combinations:

$$\mathbf{v} = (v/h)\mathbf{f}(\mathbf{r}/h, \mathcal{R}, P), \quad T = \Theta f(\mathbf{r}/h, \mathcal{R}, P). \tag{8.246}$$

As the Rayleigh number is gradually increased, the temperature difference dominates over viscosity and convection will set in at some critical Rayleigh number, $\mathcal{R}_{\text{crit}}$. Naively, we would have expected this transition to occur when $\mathcal{R}$ is of the order of unity; in this particular case, $\mathcal{R}_{\text{crit}}$ turns out to be ~ 1708, that is, a fluid of height h and a downward temperature gradient A become unstable when

$$\frac{\beta g A h^3}{v \chi} > 1708. \tag{8.247}$$

The critical Rayleigh number is much larger than unity (and is ~ 2000) because the equations governing the instability turn out to be sixth order and even numerical coefficients like $\pi^6 \approx 10^3$ can make a significant difference in the estimate. When $\mathcal{R} > \mathcal{R}_{\text{crit}}$, there is a steady convection flow that is periodic in the x–y plane. The whole space between the planes is divided into identical cells, and the fluid moves in closed paths in each cell without migrating from one cell to another.

8.15 Turbulence

The equations for viscous flow can, in principle, be solved when the flow is steady for any given boundary condition. (It is assumed that the boundary conditions and the external fields, like gravity, do not depend on time.) Such solutions exist for all values of the Reynolds number $R = (UL/v)$, where U and L are the characteristic velocity and the length scales, respectively, of the flow and $v \equiv (\mu/\rho)$ is the kinematic viscosity. Such formal solutions, however, may not actually be realised in nature because they could be unstable. When the steady flow is unstable, any small perturbation will grow in time and destroy the original characteristics of the flow.

As one possible example of such instabilities, let us consider again the Kelvin–Helmholtz instability discussed in Subsection 8.13.2. The analysis in that subsection shows instabilities will grow near any region with a strong velocity gradient in the fluid. The growth of such an instability will further increase the velocity gradient and the flow pattern will become very complicated and time dependent.

The mathematical investigation of the instability is extremely difficult, and the general theoretical problem of stability of steady flow around finite bodies has not yet been solved. It is certain that steady flow is stable for sufficiently small Reynolds numbers. Experiments in fluid mechanics suggest that when the Reynolds number R increases, we eventually reach a critical value R_{crit} beyond which the flow becomes unstable; we may say that turbulence sets in at $R = R_{\text{crit}}$. The value of R_{crit} depends on the nature of the flow and the boundary condition and seems to be in the general range 10–100. This issue is of tremendous practical importance in, say, estimating the drag force that acts on a material body moving through the fluid. For example, the drag force per unit length f_{drag} acting on a cylinder of diameter d and moving with speed U through a fluid of density ρ can be expressed in the form $f_{\text{drag}} = C_{\text{drag}}(R)(\rho U^2 d/2)$, where $C_{\text{drag}}(R)$, called the drag coefficient, is a function of only the Reynolds number.

Though a highly turbulent flow cannot be analysed quantitatively from the basic equations of fluid mechanics, we can, however, make some progress by introducing reasonable physical assumptions regarding the behaviour of energy in the turbulent flow. We assume that, in the mean rest frame of the fluid, the turbulent velocity is homogeneous and isotropic and can be described in terms of coexisting eddies of different sizes. Let λ be the size of a generic eddy and let v_λ and ϵ_λ denote the typical change in the velocity (across the eddy) and energy per unit mass contained in the eddy, respectively. At very large scales, it seems reasonable to assume that viscosity does not play any crucial role and that the energy that is injected into the system is merely transferred to the next level of smaller eddies. The rate at which energy can cascade from larger scales to smaller scales is given by

$$\dot{\epsilon} \simeq \left(\frac{1}{2}v_\lambda^2\right)\left(\frac{v_\lambda}{\lambda}\right) \simeq \frac{v_\lambda^3}{\lambda}. \tag{8.248}$$

The first factor gives the energy per unit mass and unit volume in the eddies of scale λ and the second factor $(\lambda/v_\lambda)^{-1}$ gives the inverse time scale at which the transfer of energy occurs at this scale. [This effect arises because of the nonlinear term $(\mathbf{v} \cdot \nabla)\mathbf{v}$ in the Euler equation.] In steady state, the energy injected into the system at large-scale L has be transferred to some small scale λ_s, at which it can be dissipated as viscous heat. For the intermediate scales $\lambda_s \ll \lambda \ll L$, the energy cannot be accumulated and must be transferred at a constant rate $\dot{\epsilon}$ all the way down to the smallest scales; it then follows that v_λ^3/λ is a constant in the intermediate scales, allowing us to write

$$v_\lambda \approx U\left(\frac{\lambda}{L}\right)^{1/3}, \tag{8.249}$$

where U is the velocity at some macroscopically large-scale L. Note that the largest eddies carry larger velocity but the smallest ones carry larger vorticity, $\Omega \approx v_\lambda/\lambda \propto \lambda^{-2/3}$.

To estimate the smallest-scale λ_s at which this process stops, we have to calculate the energy dissipation that is due to viscosity and equate it to $\dot{\epsilon}$. The above analysis shows that $v_\lambda = (\dot{\epsilon}\lambda)^{1/3}$. If the shear σ at some scale λ is of the order of v_λ/λ, the viscous heating rate at scale λ will be

$$h_\lambda \simeq \eta\sigma^2 \simeq \eta\left(\frac{v_\lambda}{\lambda}\right)^2 \simeq \eta\dot{\epsilon}^{2/3}\lambda^{-4/3}. \tag{8.250}$$

At large scales h is negligible compared with $\dot{\epsilon}$, and hence the same amount of energy will be transmitted from scale to scale. As we reach smaller scales, viscous heating becomes important and when $h_\lambda \simeq \rho\dot{\epsilon}$, at some scale $\lambda = \lambda_s$, the cascading of energy stops. Equating the expressions for h_λ and $\rho\dot{\epsilon} = \rho(U^3/L)$, we find that

$$\lambda_s \simeq \left(\frac{\eta}{\rho\dot{\epsilon}^{1/3}}\right)^{3/4} \simeq L\left(\frac{\eta}{\rho UL}\right)^{3/4} \simeq \frac{L}{R^{3/4}}, \tag{8.251}$$

where R is the Reynolds number evaluated at the largest scale. As long as the Reynolds number is large, we have $\lambda_s \ll L$.

If the velocity field is $\mathbf{v(x)}$, then the (scalar) correlation function of $\mathbf{v}$ can be defined as

$$\xi(\mathbf{x}) = \langle \mathbf{v(x + y)} \cdot \mathbf{v(y)} \rangle. \tag{8.252}$$

The power spectrum $S(\mathbf{k})$ is defined as the Fourier transform of $\xi(\mathbf{x})$. From relation (8.248), $v(x) \simeq \dot{\epsilon}^{1/3}x^{1/3}$, so that we expect $\xi(x)$ to scale as $\xi(x) \propto \dot{\epsilon}^{2/3}x^{2/3}$. The Fourier transform of $\xi(x)$ will vary with $k \simeq x^{-1}$ as $S(k) \propto x^3\xi(x) \propto \dot{\epsilon}^{2/3}x^{11/3} \propto \dot{\epsilon}^{2/3}k^{-11/3}$. The power per logarithmic band in k will be

$$S(k)k^2 \simeq \dot{\epsilon}^{2/3}k^{-5/3}, \tag{8.253}$$

which is known as Kolmogorov's spectrum. This result is used extensively in the modelling of turbulence.

9

Plasma Physics

9.1 Introduction

This chapter deals with the dynamics of electrically conducting fluids, usually called plasmas. The emphasis is on concepts that are of direct relevance to astrophysics. The basic ideas that are covered here will be used in several chapters of Vols. II and III.

9.2 The Mean Field and Collisions in Plasma

Several astrophysical systems are made of fully ionised gases, usually called plasmas, in which electromagnetic interactions between the constituents play a vital role. In this chapter, we treat fully ionised plasma as having two components: electrons of charge $-e$ and positive ions of charge Ze. We begin by discussing several assumptions and approximations that will be inherent in our description.

The effective use of statistical methods in the study of neutral gases relies on the fact that the interaction between constituent particles are of short range and random in nature. This assumes that the gas is sufficiently rarefied and can be treated as ideal. To treat a plasma as a fluid, it is necessary to impose a corresponding condition that, however, has some important conceptual differences from that of neutral gases.

The condition for a plasma to be treated as ideal requires that the random kinetic energy of the particles be large compared with the electrostatic potential energy between two particles, that is, $k_B T \gg e^2/\bar{r} \sim e^2 n^{1/3}$, where T is the temperature, n is the number density of particles, and $\bar{r} \approx n^{-1/3}$ is the mean interparticle distance. This condition can be expressed in terms of a quantity called the Debye length, which is defined as

$$\lambda_D \equiv \left(\frac{k_B T}{4\pi e^2 n} \right)^{1/2} = 6.9 \text{ cm} \left(\frac{T}{1\,\text{K}} \right)^{1/2} \left(\frac{n}{1\,\text{cm}^{-3}} \right)^{-1/2}$$

$$= 69 \text{ m} \left(\frac{T}{1\,\text{K}} \right)^{1/2} \left(\frac{n}{1\,\text{m}^{-3}} \right)^{-1/2}. \tag{9.1}$$

For the plasma to be nearly ideal, it is necessary that $e^2 n^{1/3}/k_B T \sim \bar{r}^2/4\pi \lambda_D^2 \ll 1$. Thus the mean distance between the particles must be small compared with the Debye length. Equivalently, the number of particles inside a Debye volume,

$$N_D \equiv (n\lambda_D^3) \simeq 360 \left(\frac{T}{1\,\mathrm{K}}\right)^{3/2} \left(\frac{n}{1\,\mathrm{cm}^{-3}}\right)^{-1/2}, \tag{9.2}$$

should be sufficiently large.

The Debye length has a simple and important physical meaning. Consider a fully ionised plasma at temperature T. If a positive charge Q is now introduced at the origin, then there will be a tendency for negative charges to gather preferentially around the charged particle and for positive charges to move preferentially away from it. This implies that, at sufficiently large distances, the effect of the positive charge Q will be shielded by the cloud of negative charge density. Let us determine the length scale over which the shielding operates. The Poisson equation for the plasma, with a charge Q added at the origin, will be

$$\nabla^2 \phi = 4\pi e(n_+ - n_-) + 4\pi Q\delta(\mathbf{r}), \tag{9.3}$$

where $n_\pm(\mathbf{x})$ is the number density of ions (electrons) in the plasma. In thermal equilibrium we can assume that $n_\pm \propto \exp(\pm e\phi/k_B T)$. Then,

$$n_+ - n_- = \bar{n}[\exp(e\phi/k_B T) - \exp(-e\phi/k_B T)], \tag{9.4}$$

where $\bar{n}$ is the mean density of particles. (The relative fluctuations in $n_\pm$ contained inside a sphere of radius r are small for sufficiently large r; hence the mean value of $n_\pm$, given by canonical distribution, provides an accurate description.) Now, far away from the charge, $(e\phi/k_B T) \ll 1$, giving

$$n_+ - n_- \simeq \left(\frac{2\bar{n}e}{k_B T}\right)\phi. \tag{9.5}$$

Hence Eq. (9.3) becomes

$$\nabla^2 \phi \cong \frac{8\pi e^2 \bar{n}}{k_B T}\phi + Q\delta(\mathbf{r}), \tag{9.6}$$

which has the solution

$$\phi = \frac{Q}{r}\exp\left(-\frac{\sqrt{2}r}{\lambda_D}\right), \tag{9.7}$$

where λ_D is the Debye length introduced in Eq. (9.1). It follows that, for the plasma to act collectively as a gas, the number of particles within the Debye length should be sufficiently large.

There is also a natural time scale that can be associated with fully ionised plasma. To arrive at an expression for this time scale, consider a thermal fluctuation that moves the electrons (relative to ions) by a small distance δx along the

x axis. This deposits a charge $Q \simeq e(nA\delta x)$ on a fictitious surface of area A perpendicular to x axis. This charge density leads to an electric field $E_x \simeq 4\pi(Q/A) \simeq 4\pi en\delta x$ that acts on the electrons in this small volume, pulling them back. Thus the electrons will oscillate in accordance with

$$m_e \frac{d^2\delta x}{dt^2} = eE_x = -4\pi e^2 n\delta x. \tag{9.8}$$

The characteristic frequency of oscillation (called the plasma frequency) is

$$\omega_p = \left(\frac{4\pi e^2 n}{m}\right)^{1/2} = 5.64 \times 10^4 \text{ Hz}\left(\frac{n}{1 \text{ cm}^{-3}}\right)^{1/2},$$

$$\nu_p = \frac{\omega_p}{2\pi} = 8.97 \text{ kHz}\left(\frac{n}{1 \text{ cm}^{-3}}\right)^{1/2}. \tag{9.9}$$

The connection between λ_D and ω_p can also be expressed in a different manner. The time scale t_D corresponding to the Debye length λ_D is $t_D = (\lambda_D/v)$, where v is the typical velocity of the electron. This is the time it takes for the electrons to move over the distance λ_D, neutralising any random fluctuation. Using $v^2 \simeq (k_B T/m)$ and the definition of λ_D, we get $t_D^2 = (k_B T/4\pi e^2 n)(m/k_B T) = (m/4\pi e^2 n) = \omega_p^{-2}$. These considerations show that λ_D and ω_p^{-1} are the fundamental length and the time scales of a plasma, respectively.

Having determined the condition for the validity of fluid description, we next consider the question of validity of the classical description, which requires that quantum-mechanical degeneracy effects be negligible; i.e., $k_B T \gg (\hbar^2 n^{2/3}/m)$, where m is the mass of the electron. Usually this condition is weak and is satisfied in gaseous astrophysical plasmas. In the case of stellar remnants we have to deal with a gas consisting of degenerate electrons and positively charged nuclei. Such a gas has the peculiar property that it becomes increasingly ideal as the density increases. The energy per electron of the Coulomb interaction between electrons and nuclei is $E_{\text{coul}} \approx (Ze^2/a)$, where $a \approx (ZV/N)^{1/3}$ is the mean distance between the electrons and the nuclei. The condition for ideal behaviour is that this energy should be small compared with the mean kinetic energy of the electrons, which is of the order of ϵ_F. The condition $E_{\text{coul}} \ll \epsilon_F$ translates into $(N/V) \gg (me^2/\hbar^2)^3 Z^2$ when expression (5.171) for ϵ_F is used. This clearly shows that the condition is more easily met at higher densities. It may be noted that the degeneracy temperature corresponding to the electron-gas density [with $(N/V) \gg (me^2/\hbar^2)^3 Z^2$] is $\sim$40 $Z^{4/3}$eV $\approx 0.5 \times 10^6 Z^{4/3}$ K.

The transport equation for each component of a classical plasma can be written in the generic form as

$$\frac{\partial f}{\partial t} + \mathbf{v} \cdot \frac{\partial f}{\partial \mathbf{r}} + \dot{\mathbf{p}} \cdot \frac{\partial f}{\partial \mathbf{p}} = C(f), \tag{9.10}$$

where the right-hand side denotes the collisional terms. Although the form of this equation is identical to that for neutral fluids, its interpretation for a plasma has some significant differences from that of neutral gases. For a neutral-gas in which self-gravity is ignored, the only forces that act on the particles are the collisional forces. These, in turn, arise because of short-range molecular interactions and are effective only when two particles are very close to each other. Hence we can set $\dot{\mathbf{p}} = 0$ on the left-hand side of transport equation (9.10) and treat the intermolecular forces as leading to random collisions by means of the $C(f)$ term on the right-hand side. Such an approach, however, is not possible (in general) for plasmas. Because the constituents are charged particles, they interact by means of long-range Coulomb forces and can influence each other even when the impact parameter is large. It is, of course, true that the Coulomb interaction is shielded for $r > \lambda_D$; but the basic assumption regarding the collective behaviour of plasma requires sufficiently large number of particles within a sphere of radius $r = \lambda_D$. These particles certainly feel the unshielded Coulomb force.

The Coulomb interaction between the particles cannot be incorporated into the collision term $C(f)$. By definition, the collision term represents a random stochastic process that causes the system to approach the state of equilibrium. The collisions with large impact parameters are usually a collective effect involving coherent action of many particles and hence cannot be considered random. It is therefore necessary to deal with the situation somewhat differently. The usual procedure adopted for the study of plasma is the following. The net electromagnetic field in the vicinity of a charged particle in a plasma is treated as a sum of mean macroscopic fields $\mathbf{E}$ and $\mathbf{B}$ and random stochastic fields $\mathbf{E}'$ and $\mathbf{B}'$. We obtain fields $\mathbf{E}$ and $\mathbf{B}$ by averaging over regions with dimensions that are large compared with $\bar{r}$ but small compared with λ_D. Fields $\mathbf{E}'$ and $\mathbf{B}'$ denote the deviation of the actual field from the average field and could be treated as causing random changes in the motion of the particles. In this description, fields $\mathbf{E}$ and $\mathbf{B}$ may be treated in the ordinary sense of a macroscopic electromagnetic field. The term $\dot{\mathbf{p}}$ on the left-hand side of transport equation (9.10) is now produced because of the Lorentz force that arises from fields $\mathbf{E}$ and $\mathbf{B}$. The effect of fields $\mathbf{E}'$ and $\mathbf{B}'$, on the other hand, can be treated as causing collisions and will be incorporated in the term $C(f)$. These collisions involve individual encounters with large impact parameters that cause small relative changes in the momenta of particles.

As an aside, it may be noted that a very similar situation arises in the study of self-gravitating systems such as, e.g., the collection of stars in a globular cluster. The long-range gravitational interaction between stars has to be separated into two parts, one of which is the mean gravitational field that is due to the smooth distribution of stars and another that represents the deviation from the average gravitational field and can be thought of as causing gravitational collisions. The techniques developed in this chapter also find application in Chap. 10 for dealing with such systems.

Just as in neutral gases, it is the collisional term on the right-hand side of transport equation (9.10) that drives the system towards thermodynamic equilibrium. The time scales for the plasma to reach an equilibrium will essentially be governed by $C(f)$. If the collisional frequency ν of the system is small compared with the frequency ω at which the macroscopic average fields **E** and **B** vary, then the plasma is said to be collisionless. On the other hand, if $\nu \gtrsim \omega$, then it is necessary to take into account collisions in plasmas. The collisions, in turn, can be classified as those between electrons, those between ions, and those between electrons and ions. The frequencies or time scales of these three kinds of collisions are different because of the large difference in the mass of the electron and the ion. It follows that the time taken by the electrons in the plasma to reach equilibrium among themselves could be quite different from the corresponding time taken by the ions or the time taken for electrons and ions to reach a common temperature.

We first consider the description of collision term $C(f)$ in the plasma and the approach to equilibrium by electrons and ions. Later on, in Section 9.4, we shall discuss collisionless evolution and other effects.

Exercise 9.1

Grain charge in a plasma: Consider a spherical dust grain of radius a located in a plasma at temperature T_{gas}, which is taken to be the same for electrons and ions. The electrons and ions that collide with the grain have a probability S for sticking to the grain. When the temperature is same for electrons and ions, electrons will be moving with greater speed and hence the grain will initially acquire an excess negative charge. Eventually this excess will cause a preferential gathering of ions rather than electrons around the grain and, in equilibrium, the rates of gain of positive charge and negative charge by the grain should be equal. Show that (1) the rate of gain of the positive charge by the grain is given by

$$\dot{Q}_+ = \frac{8}{\pi^{1/2}} \left(\frac{k_B T_{gas}}{2 m_i} \right)^{3/2} e n_i S_i \pi a^2 \left(1 + \frac{Z e^2}{a k_B T_{gas}} \right), \tag{9.11}$$

where n_i is the number density and S_i is the sticking probability for the ions. (2) Show that the rate of gain of the negative charge by the grain is

$$\dot{Q}_- = \left(\frac{8 k T_{gas}}{\pi m_e} \right)^{1/2} e n_e S_e \pi a^2 e^{-x_0}, \tag{9.12}$$

where n_e and S_e are the corresponding quantities for the electrons and $x_0 = (Z e^2 / a k_B T)$. In equilibrium, these two rates must be equal. Assuming that $S_e \approx S_i$, show that x_0 is determined by the transcendental equation

$$1 + x_0 = \left(\frac{m_i}{m_e} \right)^{1/2} \left(\frac{n_e}{n_i} \right) \exp(-x_0). \tag{9.13}$$

Solve this equation for a pure hydrogen plasma with $n_e = n_i$ and determine the charge on a dust grain $Z = x_0(k T_{gas} a / e^2)$.

9.3 Collisions in Plasmas

As noted in Section 9.2, the dominant collisions in plasma arise because of
the scattering of particles at large impact parameters, resulting in small relative changes in the momentum of the particle in each collision. Because of the
stochastic nature of such interaction, any given particle will perform a random
walk in the momentum space. Such a random walk can be thought of as a diffusion process in momentum space and can be studied by an equation of the form

$$\frac{df}{dt} = \frac{\partial f}{\partial t} + \mathbf{v} \cdot \frac{\partial f}{\partial \mathbf{x}} - \dot{\mathbf{v}} \cdot \frac{\partial f}{\partial \mathbf{v}} = -\frac{\partial j^{\alpha}}{\partial p^{\alpha}}. \tag{9.14}$$

The right-hand side is the divergence of a particle current j^{α} in momentum
space, which is characteristic of diffusive process. The problem now reduces to
that of determining the form of the current j_{α}.

Consider a collision in which two particles with momenta $\mathbf{p}_1$ and $\mathbf{p}_2$ scatter to
momenta $\mathbf{p}_1'$ and $\mathbf{p}_2'$. Let the momentum transfer in the collision be $\mathbf{q}$ so that the
scattering is from $(\mathbf{p}_1, \mathbf{p}_2)$ to $(\mathbf{p}_1 + \mathbf{q}, \mathbf{p}_2 - \mathbf{q})$. It is, however, more convenient
to characterise the scattering in terms of the mean momenta of the particles $\bar{\mathbf{p}}_1 =$
$\mathbf{p}_1 + (1/2)\mathbf{q}$ and $\bar{\mathbf{p}}_2 = \mathbf{p}_2 - (1/2)\mathbf{q}$. Let the quantity $w(\mathbf{p} + \mathbf{q}/2, \mathbf{p}' - \mathbf{q}/2; \mathbf{q})$
$f(\mathbf{p}) f(\mathbf{p}') d\mathbf{p}' d\mathbf{q}$ represent the number of collisions that take place in unit time
between some particle of momentum $\mathbf{p}$ and all particles with momentum in the
range $(\mathbf{p}', \mathbf{p}' + d\mathbf{p}')$ such that the momentum transfer is in the range $(\mathbf{q}, \mathbf{q} + d\mathbf{q})$.
[To simplify notation, we omit the $\mathbf{x}$ and t dependence of $f(\mathbf{x}, \mathbf{p}, t)$ and write
it as $f(\mathbf{p})$ if no confusion is likely to arise.] From time reversibility of the scat-
tering, we have the condition that $W(\bar{\mathbf{p}}_1, \bar{\mathbf{p}}_2; \mathbf{q}) = W(\bar{\mathbf{p}}_1, \bar{\mathbf{p}}_2; -\mathbf{q})$. Consider an
infinitesimal area element centered at $\mathbf{p}_1$ and perpendicular to the p_x direction
in momentum space. The flux of particles from left to right (i.e., the number of
particles per unit area that cross this surface from left to right in unit time) is
given by

$$I_L = \int_{q_x > 0} d\mathbf{q} \int_{(p_1)_x - q_x}^{(p_1)_x} dk_x \int_{\text{all}} d\mathbf{p}_2 W\left(\mathbf{k} + \frac{1}{2}\mathbf{q}, \mathbf{p}_2 - \frac{1}{2}\mathbf{q}\right) f(\mathbf{p}_2) f(\mathbf{k}). \tag{9.15}$$

The integrand describes the scattering from $(\mathbf{k}, \mathbf{p}_2)$ to $(\mathbf{k} + \mathbf{q}, \mathbf{p}_2 - \mathbf{q})$ in which the
particle with momentum $\mathbf{k}$ moves across the surface. To achieve this with a given
momentum transfer $\mathbf{q}$, the value of k_x must be in the range $[(p_1)_x - q_x, (p_1)_x]$.
We integrate over the momentum transfer $\mathbf{q}$, making sure it is a flow from left to
right (i.e., $q_x > 0$).

By similar reasoning, the flux of particles moving from right to left across this
surface is given by

$$I_R = \int_{q_x < 0} d\mathbf{q} \int_{(p_1)_x}^{(p_1)_x - q_x} dk_x \int_{\text{all}} d\mathbf{p}_2 W\left(\mathbf{k} + \frac{1}{2}\mathbf{q}, \mathbf{p}_2 - \frac{1}{2}\mathbf{q}; \mathbf{q}\right) f(\mathbf{p}_2) f(\mathbf{k}).$$
$$\tag{9.16}$$

The integrand in Eq. (9.16) is the same as that in Eq. (9.15) but the range is different. We are interested in the net flux $j_x \equiv I_L - I_R$. By a series of simple transformations, the range of integration in both I_R and I_L can be made the same. We have

$$
\begin{aligned}
I_R &= \int_{\text{all}} d^3 p_2 \int_{q_x > 0} d^3 q \int_{(p_1)_x}^{(p_1)_x + q_x} dk_x \, W\left(\mathbf{k} - \frac{1}{2}\mathbf{q}, \mathbf{p}_2 + \frac{1}{2}\mathbf{q}; -\mathbf{q}\right) f(\mathbf{k}) f(\mathbf{p}_2) \\
&= \int_{\text{all}} d^3 p_2 \int_{q_x > o} d^3 q \int_{(p_1)_x - q_x}^{(p_1)_x} dl_x \, W\left(\mathbf{l} + \frac{1}{2}\mathbf{q}, \mathbf{p}_2 + \frac{1}{2}\mathbf{q}; -\mathbf{q}\right) f(\mathbf{l} + \mathbf{q}) f(\mathbf{p}_2) \\
&= \int_{\text{all}} d^3 l' \int_{q_x > 0} d^3 q \int_{(p_1)_x - q_x}^{(p_1)_x} dl_x \, W\left(\mathbf{l} + \frac{1}{2}\mathbf{q}, \mathbf{l}' - \frac{1}{2}\mathbf{q}; -\mathbf{q}\right) f(\mathbf{l} + \mathbf{q}) f(\mathbf{l}' - \mathbf{q}).
\end{aligned}
\tag{9.17}
$$

In arriving at the first equality, we have changed $\mathbf{q}$ to $-\mathbf{q}$ in Eq. (9.16); in arriving at the second equality we have set $\mathbf{k} = \mathbf{l} + \mathbf{q}$; to arrive at the last equality, we have put $\mathbf{p}_2 = (\mathbf{l}' - \mathbf{q})$. Subtracting I_R from I_L, we can express the current j_x as

$$
\begin{aligned}
j_x = \int_{\text{all}} d\mathbf{l} \int_{q_x > 0} d\mathbf{q} \int_{(p_1)_x - q_x}^{(p_1)_x} dl_x \, W\left(\mathbf{l} + \frac{1}{2}\mathbf{q}, \mathbf{l}' - \frac{1}{2}\mathbf{q}, \mathbf{q}\right) \\
\times [f(\mathbf{l}) f(\mathbf{l}') - f(\mathbf{l} + \mathbf{q}) f(\mathbf{l} - \mathbf{q})].
\end{aligned}
\tag{9.18}
$$

To proceed further we make the assumption (which will be verified at the end of the calculation) that the collisions are soft so that most of the contribution to this integral comes from processes with small momentum transfer, i.e., from small $\mathbf{q}$. In this case, we can (1) replace W in Eq. (9.18) with $W(\mathbf{l}, \mathbf{l}', \mathbf{q})$; (2) expand $f(\mathbf{l} + \mathbf{q}) f(\mathbf{l}' - \mathbf{q})$ in a Taylor series in $\mathbf{q}$, retaining only up to linear terms in $\mathbf{q}$; and (3) replace the integral over l_x with the multiplication by the range of integration q_x. This brings $\mathbf{j}$ to the form

$$
\begin{aligned}
j_\alpha &= \frac{1}{2} \int d\mathbf{l}' \int d\mathbf{q} \, q_\alpha q_\beta W\left(f \frac{\partial f'}{\partial l'_\beta} - f' \frac{\partial f}{\partial l_\beta}\right) \\
&= \int d\mathbf{l} \, B_{\alpha\beta}(\mathbf{l}, \mathbf{l}')\left(f \frac{\partial f'}{\partial l'_\beta} - f' \frac{\partial f}{\partial l_\beta}\right),
\end{aligned}
\tag{9.19}
$$

where we have defined

$$
B_{\alpha\beta}(\mathbf{l}, \mathbf{l}') = \frac{1}{2} \int q_\alpha q_\beta W(\mathbf{l}, \mathbf{l}'; \mathbf{q}) \, d\mathbf{q}.
\tag{9.20}
$$

Note that $B_{\alpha\beta}$ can depend on only $\mathbf{k} \equiv \mathbf{l} - \mathbf{l}'$; further, for soft collisions, $B_{\alpha\beta}$ must be transverse to k^α as the momentum transfer is in the transverse direction. Hence we must have $B_{\alpha\beta} k^\alpha = 0$. The most general second-rank symmetric tensor, constructable from k_α and transverse to it, must have the form

$$B_{\alpha\beta} = \frac{1}{2}B\left\{\delta_{\alpha\beta} - \frac{k_\alpha k_\beta}{k^2}\right\}; \quad k = |\mathbf{k}|. \tag{9.21}$$

We therefore need to compute only

$$B = B_\alpha^\alpha = \frac{1}{2}\int q^2 W(\mathbf{l}, \mathbf{l}'; \mathbf{q})\,d\mathbf{q}. \tag{9.22}$$

If the angular-scattering cross section is given by $(d\sigma/d\Omega)d\Omega$, then from the definition of W it follows that $W\,d\mathbf{q} = |\mathbf{v} - \mathbf{v}'|(d\sigma/d\Omega)d\Omega$, so that

$$B = \frac{1}{2}\int q^2|\mathbf{v} - \mathbf{v}'|\frac{d\sigma}{d\Omega}\,d\Omega. \tag{9.23}$$

When the angle of deviation χ of relative velocity in the centre-of-mass system of the two particle is small, the momentum change is given by $q = \mu|\mathbf{v} - \mathbf{v}'|\chi$, where μ is the reduced mass of the particles. Then

$$B = \frac{1}{2}\mu^2|\mathbf{v} - \mathbf{v}'|^3\int\chi^2\left(\frac{d\sigma}{d\Omega}\right)d\Omega = \mu^2|\mathbf{v} - \mathbf{v}'|^3\sigma_t, \tag{9.24}$$

where

$$\sigma_t = \int(1 - \cos\chi)\left(\frac{d\sigma}{d\Omega}\right)d\Omega \approx \frac{1}{2}\int\chi^2\left(\frac{d\sigma}{d\Omega}\right)d\Omega \tag{9.25}$$

for $\chi \ll 1$. The differential cross section for small angles in a Coulomb field is given by (see Chap. 3, Subsection 3.8.1)

$$d\sigma \approx \frac{4(ee')^2\,d\Omega}{\mu^2(\mathbf{v} - \mathbf{v}')^4\chi^4} \approx \frac{8\pi(ee')^2}{\mu^2(\mathbf{v} - \mathbf{v}')^4}\frac{d\chi}{\chi^3}, \tag{9.26}$$

where e and e' are the charges of the colliding particles. This gives

$$\sigma_t = \frac{4\pi(ee')^2}{\mu^2(\mathbf{v} - \mathbf{v}')^4}L, \quad L \equiv \int\frac{d\chi}{\chi}. \tag{9.27}$$

Combining all these results, we finally get

$$B_{\alpha\beta} = \frac{2\pi(ee')^2}{|\mathbf{v} - \mathbf{v}'|}L\left[\delta_{\alpha\beta} - \frac{(v_\alpha - v'_\alpha)(v_\beta - v'_\beta)}{(\mathbf{v} - \mathbf{v}')^2}\right]. \tag{9.28}$$

The integral L appearing in these expressions is logarithmically divergent. The divergence at the lower limit is caused by scattering at large impact parameters and the slow decrease of the Coulomb field. This is an artifact because in real plasma the Coulomb field is shielded at impact parameters $b \gtrsim \lambda_D$. If the Coulomb scattering (in the r^{-1} potential) is considered applicable for $\chi > \chi_{min}$ then the lower limit of the integral for L should be taken as χ_{min}. The divergence at the upper limit arises because the formulas were written on the assumption that

χ is a small angle, which clearly breaks down when $\chi \gtrsim 1$. This suggests that we can take a upper limit to the integral to be approximately unity, thereby obtaining $L = \ln(1/\chi_{\min})$. The uncertainty in the value of L and hence the validity of all results dependent on it restrict the discussion to logarithmic accuracy, i.e., there could be relative errors of the order of $\mathcal{O}(1/L)$.

The precise value of $\chi_{\min}$ depends on whether the scattering should be treated classically or quantum mechanically. This problem has already been studied in Chap. 6, Section 6.9, in connection with bremsstrahlung. By a corresponding analysis, we can conclude that

$$L \cong \log\left[(\lambda_D \mu \bar{v}_{\rm rel}^2/|ee'|)\right] \quad \left(\text{for } (|ee'|/\hbar\bar{v}_{\rm rel}) \gg 1\right), \tag{9.29}$$

$$L \cong \log(\mu\lambda_D \bar{v}_{\rm rel}/\hbar) \quad \left(\text{for } (|ee'|/\hbar\bar{v}_{\rm rel}) \ll 1\right). \tag{9.30}$$

Numerically, L will be ~ 10–20 in most of the astrophysically relevant situations. The argument of the logarithm in relation (9.29) has a numerical value (in cgs units) that is approximately $1.3 \times 10^4 T^{3/2} n_e^{-1/2}$.

By using either of these forms in Eq. (9.28) and determining j_α from Eq. (9.19), we can write the evolution equation for f. Because the electron–electron collisions and electron–ion collisions are governed by different parameters, we should add up all the independent contributions to j^α. Then the final transport equation for plasma can be written in the form

$$\frac{\partial f}{\partial t} + \mathbf{v} \cdot \frac{\partial f}{\partial \mathbf{x}} + e\left(\mathbf{E} + \frac{\mathbf{v}}{c} \times \mathbf{B}\right) \cdot \frac{\partial f}{\partial \mathbf{p}} = -\mathrm{div}_p\, \mathbf{j}, \tag{9.31}$$

with

$$j_\alpha = \sum 2\pi (ee')^2 L \int \left(f\frac{\partial f'}{\partial p'_\beta} - f'\frac{\partial f}{\delta p_\beta}\right) \frac{(\mathbf{v}-\mathbf{v}')^2\delta_{\alpha\beta} - (v_\alpha - v'_\alpha)(v_\beta - v'_\beta)}{|\mathbf{v}-\mathbf{v}'|^3}\, d^3p'. \tag{9.32}$$

The summation is over electrons and ions and L is given by relations (9.29) and (9.30). This expression is valid as long as the characteristic length scale l over which the distribution function varies is large compared with λ_D and the characteristic time scale of variation of distribution function is large compared with the plasma frequency. Note that the expression within the large parentheses in Eq. (9.32) vanishes when $f(\mathbf{p})$ is a Maxwellian distribution.

Expression (9.32) also allows the determination of the relaxation time scales and mean free paths for electron–electron (e–e), electron–ion (e–i), and ion–ion (i–i) collisions. In scattering between a heavy and light particle, the energy of either particle changes very little. Hence the establishment of equilibrium among electrons alone or ions alone occurs much more rapidly than the establishment of equilibrium between electrons and the ions. This feature could give rise to an intermediate situation in which the electrons and the ions have a Maxwellian distribution of velocities with different temperatures T' and T, respectively. The

change in the energy of ions per unit time and unit volume will then be given by
the integral

$$\frac{dE}{dt} = \int \epsilon C(f) \, d^3 p = -\int \epsilon \, \mathrm{div}_p \, \mathbf{j} \, d^3 p = \int \mathbf{j} \cdot \left(\frac{\partial \epsilon}{\partial \mathbf{p}} \right) d^3 p = \int \mathbf{j} \cdot \mathbf{v} \, d^3 p,$$

(9.33)

where $\epsilon(\mathbf{p}) = (p^2/2m)$ is the kinetic energy of the particle and we have performed
an integration by parts and ignored the surface term. Substituting for j^α from
Eq. (9.32) and noting that the electron–electron and the ion–ion terms do not
contribute (as the distribution function for each of them is Maxwellian), we are
left with

$$j_\alpha = \int f f' \left(\frac{v_\beta}{k_B T} - \frac{v'_\beta}{k_B T'} \right) B_{\alpha\beta} \, d^3 p',$$

(9.34)

where we have used the Maxwellian distributions with temperatures T' and T
for electrons and ions, with the prime denoting quantities pertaining to electrons.
Using $B_{\alpha\beta} v'_\beta = B_{\alpha\beta} v_\beta$ in the expression for dE/dt, we get

$$\frac{dE}{dt} = \left(\frac{1}{T_i} - \frac{1}{T_e} \right) \int\int f f' v_\alpha v_\beta B_{\alpha\beta} \, d^3 p \, d^3 p',$$

(9.35)

where $T_e \equiv k_B T'$ and $T_i \equiv k_B T$ are the temperatures measured in energy units.
Further, because the mass of the electron is far less than that of an ion, we can
put $v'_\alpha - v_\alpha \approx v'_\alpha$ in $B_{\alpha\beta}$; the resulting expression is now independent of v_α and
the integrations can be carried out with

$$\int f v_\alpha v_\beta \, d^3 p = \frac{1}{3} \delta_{\alpha\beta} n \overline{v^2} = \delta_{\alpha\beta} \frac{n T_i}{M},$$

(9.36)

where M is the mass of the ion and n is the number density. Hence,

$$\frac{dE}{dt} = \frac{n T_i}{M} \left(\frac{1}{T_i} - \frac{1}{T_e} \right) \int f' B_\alpha^\alpha \, d^3 p'.$$

(9.37)

Substituting $B_\alpha^\alpha = (4\pi Z^2 e^4 L/v')$, which corresponds to the ionic charge of Ze,
in Eq. (9.37) and by using the relation

$$\int f' \frac{d^3 p'}{v'} = n' \sqrt{\frac{2m}{\pi T_e}},$$

(9.38)

we get

$$\frac{dE}{dt} = \frac{4 n n' Z^2 e^4 \sqrt{(2\pi m)} L_e}{M T_e^{3/2}} (T_e - T_i),$$

(9.39)

with L_e giving the electron Coulomb logarithm. The same expression with a
minus sign gives the decrease in the energy of the electron component in the

plasma. Writing $E' = (3/2)n'T_e$, we can express the rate of change of the electron temperature as

$$\frac{dT_e}{dt} = -\frac{T_e - T_i}{\tau_{ei}}, \qquad \tau_{ei} = \frac{3T_e^{3/2}M}{8n_i Z^2 e^4 L_e (2\pi m)^{1/2}}, \qquad (9.40)$$

where

$$L_e = \begin{cases} \log(\lambda_D T_e/Ze^2) & \text{(for } Ze^2/\hbar v_e \gg 1) \\ \log[\sqrt{(mT_e)}\lambda_D/\hbar] & \text{(for } Ze^2/\hbar v_e \ll 1). \end{cases} \qquad (9.41)$$

The quantity τ_{ei} is the relaxation time for the equilibrium between electron and ion components.

Let us next consider electron–electron collisions, for which the reduced mass μ is approximately the same as that of the electron mass and the speeds of electrons are much greater than those of ions, giving $\mu(\mathbf{v}_e - \mathbf{v}_i)^2 \sim mv_e^2 \sim T_e$. The scattering cross section is $\sigma_t \simeq 4\pi b_c^2$, with b_c determined by the condition $(e^2/b_c) \simeq T_e$. Hence the electron mean free path is given by

$$l_e \simeq \frac{1}{n\sigma_t} \simeq \frac{T_e^2}{4\pi e^4 n L_e} \approx 3.2 \times 10^6 \text{ cm } \frac{1}{L_e}\left(\frac{T}{1 \text{ K}}\right)^2 \left(\frac{n_e}{1 \text{ cm}^{-3}}\right)^{-1}. \qquad (9.42)$$

The mean free time for electron–electron collisions is

$$\tau_{ee} \equiv \frac{1}{\nu_{ee}} \equiv \frac{l_e}{v_e} \simeq \frac{T_e^{3/2}m^{1/2}}{4\pi e^4 n L_e}. \qquad (9.43)$$

Note that $(l_e/\lambda_D) \approx (1/L_e)(T_e/e^2 n^{1/3})^{3/2}$; the condition $l_e \gg \lambda_D$ implies that the collision frequency is small compared with the plasma frequency.

In a similar manner we can obtain the ion mean free path for the ion–ion collisions as

$$l_i \cong \frac{T_i^2}{4\pi Z^4 e^4 n L_i}, \qquad L_i = \log\left(\frac{\lambda_D T_i}{Z^2 e^2}\right). \qquad (9.44)$$

The corresponding mean free time is

$$\tau_{ii} = \frac{1}{\nu_{ii}} \simeq \frac{T_i^{3/2}M^{1/2}}{4\pi Z^4 e^4 n L_i}. \qquad (9.45)$$

It may noted that the relaxation time obtained above is in the approximate ratio $\tau_{ee}:\tau_{ii}:\tau_{ei} \simeq 1:(M/m)^{1/2}:(M/m)$ if $n' \simeq n$ and $T_e \simeq T_i$. The time scale τ_{ee}, which determines the relaxation time for the electron component, can also describe the rate of *momentum* transfer between electrons and ions. However, the establishment of thermal equilibrium between electrons and ions is governed by the rate of *energy* transfer that is given by τ_{ei} and not by τ_e.

Given the mean free path, we can also estimate the electrical conductivity σ, the thermal conductivity κ, and the coefficient of viscosity η along standard lines (see Chap. 8, Section 8.4). We get

$$\sigma \simeq \frac{T_e^{3/2}}{e^2 m^{1/2} L_e}, \quad \kappa \simeq \frac{T_e^{5/2}}{e^4 m^{1/2} L_e}, \quad \eta \simeq \frac{M^{1/2} T_i^{5/2}}{e^4 L_i}. \qquad (9.46)$$

All these quantities are independent of density. The numerical values for different equilibrium time scales are given by (with Boltzmann's constant reintroduced appropriately)

$$t_{eq,e \to e} = \frac{1}{4\pi} \frac{m_e^{1/2}(k_B T_e)^{3/2}}{e^4 n L_e} = \left(\frac{10^9 \text{ s}}{L_e/10}\right)\left[\frac{(k_B T_e/1 \text{ keV})^{3/2}}{n/1 \text{ cm}^{-3}}\right], \qquad (9.47)$$

$$t_{eq,i \to i} = \left(\frac{M}{m}\right)^{1/2}\left(\frac{T_i}{T_e}\right)^{3/2} t_{eq,e \to e} = \left(\frac{4.3 \times 10^{10} \text{ s}}{Z^4 L_i/10}\right)\left[\frac{(k_B T_i/1 \text{ keV})^{3/2}}{n/1 \text{ cm}^{-3}}\right], \qquad (9.48)$$

$$t_{eq,e \to i} = \left(\frac{9\pi}{8}\right)^{1/2} \frac{1}{Z^2}\left(\frac{m_p}{m_e}\right) t_{eq,e \to e} = \left(\frac{10^{12} \text{ s}}{Z^2 L_e/10}\right)\left[\frac{(k_B T_e/1 \text{ keV})^{3/2}}{n/1 \text{ cm}^{-3}}\right]. \qquad (9.49)$$

The electrical conductivity of a plasma is

$$\sigma \simeq \left(\frac{k_B T}{m}\right)^{3/2}\left(\frac{m}{e^2}\right) L_e^{-1} \simeq 2.5 \times 10^{17} \text{s}^{-1}\left(\frac{T}{10^6 \text{ K}}\right)^{3/2} L_e^{-1}, \qquad (9.50)$$

which, as noted above, is independent of density.

Exercise 9.2

Relaxation in astrophysical plasmas: A gaseous region of ionised hydrogen, usually called an HII region, has an electron number density $n_e = 100$ cm^{-3} and a kinetic temperature of $T = 10^4$ K. The age of such a region is typically 10^7 yr. Estimate the time scale for electrons and protons to achieve Maxwellian distribution individually and the time scale for the temperatures to be equalised. Take $L \approx 25$. Repeat the same analysis for a supernova remnant with $n_e = 1$ cm^{-3} and $T = 5 \times 10^7$ K and an age of ~ 400 yr. [Answers: For the HII region, $t_{eq,e-e} \approx 4 \times 10^{-6}$ yr; $t_{eq,i-i} \approx 1.6 \times 10^{-4}$ yr; and $t_{eq,i-e} \approx 7 \times 10^{-3}$ yr. Because these time scales are much smaller than the age of the system, complete kinetic equilibrium will exist in which electrons and ions have Maxwellian distributions and $T_e = T_i$. In the case of a supernova remnant, $t_{eq,e-e} \approx 10^2$ yr; $t_{eq,i-i} \approx 4 \times 10^3$ yr; $t_{eq,i-e} \approx 2 \times 10^5$ yr. Because $t_{eq,e-e}$ is of the order of the age of the remnant, electrons could have just achieved Maxwellian distribution; protons could not have achieved Maxwellian distribution and equipartition could not have taken place, leading to $t_i \neq t_e$.]

The transport coefficients obtained above use Coulomb scattering as the main collisional mechanism. Such a description may not be valid when the plasma hosts magnetic fields. In the presence of the magnetic field, we must compare the Coulomb collision frequency with the cyclotron frequency. If the collision frequency is large compared with the cyclotron frequency, then our analysis is valid and the effect of the magnetic field can be ignored. On the other hand, if the collision frequency is small compared with the cyclotron frequency, an electron will typically make many circuits along the magnetic-field lines before Coulomb scattering can affect its orbit. In such a case, the electrons will have great difficulty in travelling distances larger than the Larmor radius,

$$r_L \equiv \frac{v}{\omega_c} \simeq \frac{(3k_B T m_e c^2)^{1/2}}{eB} = 131 \text{ cm} \frac{\sqrt{k_B T / 1 \text{ keV}}}{B / 1 \text{ G}}, \tag{9.51}$$

in the direction perpendicular to the magnetic field. Then the electrical and thermal conductivities of the plasma will become anisotropic and will be suppressed significantly in the direction transverse to the magnetic field. The criterion for a significant effect to the magnetic field translates numerically into

$$\frac{B}{1 \text{ G}} \frac{(k_B T / 1 \text{ keV})^{3/2}}{(n / 1 \text{ cm}^{-3})} \gg 3 \times 10^{-17}. \tag{9.52}$$

Many astrophysical plasmas do satisfy this criterion and hence have anisotropic transport coefficients.

9.4 Collisionless Plasmas

The plasma may be treated as collisionless when the collision frequency is small compared with the time scale over which the mean fields on the left-hand side of the Vlasov equation varies. In this case, the electron and the ion distribution functions evolve according to

$$\frac{\partial f_e}{\partial t} + \mathbf{v} \cdot \frac{\partial f_e}{\partial \mathbf{x}} - e\left(\mathbf{E} + \frac{\mathbf{v} \times \mathbf{B}}{c}\right) \cdot \frac{\partial f_e}{\partial \mathbf{p}} = 0,$$

$$\frac{\partial f_i}{\partial t} + \mathbf{v} \cdot \frac{\partial f_i}{\partial \mathbf{x}} + Ze\left(\mathbf{E} + \frac{\mathbf{v} \times \mathbf{B}}{c}\right) \cdot \frac{\partial f_i}{\partial \mathbf{p}} = 0. \tag{9.53}$$

These equations have to be supplemented by Maxwell's equations,

$$\nabla \times \mathbf{E} = -\frac{1}{c}\frac{\partial \mathbf{B}}{\partial t}, \quad \nabla \cdot \mathbf{B} = 0, \quad \nabla \times \mathbf{B} = \frac{1}{c}\frac{\partial \mathbf{E}}{\partial t} + \frac{4\pi}{c}\mathbf{j}, \quad \nabla \cdot \mathbf{E} = 4\pi\rho, \tag{9.54}$$

and the definitions, for charge and current density,

$$\rho = e \int (Z f_i - f_e) d^3 p, \quad \mathbf{j} = e \int (Z f_i - f_e) \mathbf{v} d^3 p. \tag{9.55}$$

These equations have to be solved in a self-consistent manner.

The physical description of the self-consistent solutions to the above equations is greatly facilitated by the introduction of certain auxiliary fields and certain constants related to the material media discussed in Chap. 3, Subsection 3.11.4. We introduce electric polarisation $\mathbf{P}$ by the relations

$$\frac{\partial \mathbf{P}}{\partial t} = \mathbf{j}, \quad \nabla \cdot \mathbf{P} = -\rho, \tag{9.56}$$

and the electric induction by $\mathbf{D} = \mathbf{E} + 4\pi \mathbf{P}$. [The two relations in Eqs. (9.56) are connected by charge conservation $\dot{\rho} + \nabla \cdot \mathbf{j} = 0$.] The field $\mathbf{P}$ represents the dipole moment per unit volume of the medium that arises from the application of electric field $\mathbf{E}$; field $\mathbf{D}$ is the net field at a given location. If electric field $\mathbf{E}$ is sufficiently weak, then we would expect polarisation $\mathbf{P}$ to be linear in $\mathbf{E}$ in the following sense: if electric field $\mathbf{E}$ is enhanced by a factor α everywhere, polarisation $\mathbf{P}$ will also increase by the same factor α. Such a definition of linearity does not require that the relation be local in time or space. In general, polarisation $\mathbf{P}(t, \mathbf{r})$ will indeed depend on the values of $\mathbf{E}(t', \mathbf{r}')$, where $t' < t$ and $\mathbf{r}'$ can take any value. (This is a direct consequence of the fact that in a collisionless plasma, the movement of the particles is governed by the value of the field all along the trajectory.) Such a general linear relation implies that $\mathbf{D}$ and $\mathbf{E}$ can be related in the form

$$D_\alpha(t, \mathbf{r}) = E_\alpha(t, \mathbf{r}) + \int_0^\infty \int K_{\alpha\beta}(\tau, \mathbf{X}) E_\beta(t - \tau, \mathbf{r} - \mathbf{X}) \, d^3\mathbf{X} \, d\tau. \tag{9.57}$$

In writing such a relation, we assume that the plasma is (1) spatially homogeneous on the average and (2) in a steady state. When the fields are expanded in a set of plane waves varying as $\exp[i(\mathbf{k} \cdot \mathbf{r} - \omega t)]$, the linear relation assumes a simple, local form in Fourier space with

$$D_\alpha(\omega, \mathbf{k}) = \epsilon_{\alpha\beta}(\omega, \mathbf{k}) E_\beta(\omega, \mathbf{k}), \tag{9.58}$$

where the permittivity tensor $\epsilon_{\alpha\beta}$ is given by

$$\epsilon_{\alpha\beta}(\omega, \mathbf{k}) = \delta_{\alpha\beta} + \int_0^\infty \int K_{\alpha\beta}(\tau, \mathbf{X}) \exp[i(\omega\tau - \mathbf{k} \cdot \mathbf{X})] \, d^3\mathbf{X} \, d\tau. \tag{9.59}$$

Note that permittivity satisfies the condition

$$\epsilon_{\alpha\beta}(-\omega, -\mathbf{k}) = \epsilon_{\alpha\beta}^*(\omega, \mathbf{k}). \tag{9.60}$$

The utility of this tensor lies in the fact that some general conclusions can be obtained regarding its form. Because the only vector on which $\epsilon_{\alpha\beta}$ depends is $\mathbf{k}$, the most general form of the permittivity tensor is given by

$$\epsilon_{\alpha\beta}(\omega, \mathbf{k}) = \epsilon_t(\omega, k) \left(\delta_{\alpha\beta} - \frac{k_\alpha k_\beta}{k^2} \right) + \epsilon_l(\omega, k) \frac{k_\alpha k_\beta}{k^2}, \tag{9.61}$$

in terms of two functions $\epsilon_t(\omega, k)$ and $\epsilon_l(\omega, k)$. (A term proportional to $\epsilon_{\alpha\beta\gamma}k_\gamma$ will discriminate between **k** and $-$**k** and hence is not allowed.) The first term leads to an induction **D** that is transverse to the wave vector **k**, and the second term gives a contribution that is in the direction of **k**. Hence the scalar functions ϵ_t and ϵ_l are called transverse and longitudinal permittivities. From condition (9.60), it follows that

$$\epsilon_l(-\omega, k) = \epsilon_l^*(\omega, k), \quad \epsilon_t(-\omega, k) = \epsilon_t^*(\omega, k). \tag{9.62}$$

Further, in the limit of wave vector **k** going to zero, $\epsilon_{\alpha\beta}$ reduces to $\epsilon(\omega)\delta_{\alpha\beta}$ with the transverse and longitudinal permittivities becoming equal to the normal scalar permittivity of the material.

In arriving at the specific form for dielectric permittivity, it has been assumed that there are no distinctive directions in the plasma. This implies that (1) there are no external electromagnetic fields present in the plasma. (The word external refers to fields that exclude the fields produced by the charged particles in the plasma.) Usually, a plasma hosts large-scale magnetic fields that can provide a distinctive direction and hence invalidate the form of permittivity tensor obtained above. (2) The distribution function is isotropic in momentum space so that no distinctive flow direction is singled out.

In principle, the nonlocal dependence of **D** on **E** implies that the induction at one point is influenced by the electric field at all other points. In practice, however, we can introduce a correlation length r_{cor} beyond which this influence is negligible. The collisions, acting with frequency ν, perturb the free movement along the trajectory over a length scale $l \approx (\bar{v}/\nu)$, whereas the averaging of the oscillating field during the time of flight of the particle will operate at a length scale of $l \approx (\bar{v}/\omega)$. The correlation length will be given by a smaller of these two and – if $\nu \ll \omega$ – we have $r_{cor} \approx (\bar{v}/\omega)$. There will be a significant spatial dispersion when $kr_{cor} \gtrsim 1$ and negligible dispersion for $kr_{cor} \ll 1$. It should be stressed that r_{cor} in a plasma can be quite large compared with the interparticle distances. This allows us to provide a macroscopic description of spatial dispersion in terms of a well-defined permittivity even when the dispersion is *not* a small correction. This is to be contrasted with ordinary media (for example, solids) in which the correlation length is of the order of atomic dimensions; in such a case, spatial dispersion is only a small correction, unlike in the plasma.

The permittivity tensor determines the dispersion relation for the wave modes that plasma can support. Maxwell's equations for a dielectric medium, developed in Chap. 3, Subsection 3.11.4, take a simple form in Fourier space. For an electromagnetic field varying as $\exp[i(\mathbf{k}\cdot\mathbf{x} - \omega t)]$, Maxwell's equations become

$$\mathbf{k} \times \mathbf{E} = \frac{\omega\mathbf{B}}{c}, \quad \mathbf{k} \times \mathbf{B} = -\frac{\omega\mathbf{D}}{c}, \tag{9.63}$$

$$\mathbf{k} \cdot \mathbf{B} = 0, \quad \mathbf{k} \cdot \mathbf{D} = 0. \tag{9.64}$$

Substituting the first equation of Eqs. (9.63) into the second gives

$$\frac{\omega^2 \mathbf{D}}{c^2} = -\mathbf{k} \times (\mathbf{k} \times \mathbf{E}) = \mathbf{E}k^2 - \mathbf{k}(\mathbf{k} \cdot \mathbf{E}). \tag{9.65}$$

Writing $\mathbf{D}$ in terms of $\mathbf{E}$ and using the dielectric permittivity, we find that the component form of this equation becomes

$$E_\alpha k^2 - k_\alpha k_\beta E_\beta = \omega^2 D_\alpha / c^2 = \omega^2 \epsilon_{\alpha\beta} E_\beta / c^2 \tag{9.66}$$

or

$$\left\{ \left(\delta_{\alpha\beta} - \frac{k_\alpha k_\beta}{k^2} \right) - \frac{\omega^2}{k^2 c^2} \epsilon_{\alpha\beta} \right\} E_\beta = 0. \tag{9.67}$$

This equation has nontrivial solutions only if the determinant of the matrix in the braces vanishes:

$$\left| k^2 \delta_{\alpha\beta} - k_\alpha k_\beta - \left(\omega^2 \epsilon_{\alpha\beta} / c^2 \right) \right| = 0 \tag{9.68}$$

which gives the required dispersion relation. For a given real $\mathbf{k}$, this determines the frequency $\omega(\mathbf{k})$.

The meaning of this condition can be made clear if we consider separately transverse waves and longitudinal waves and use the form of $\epsilon_{\alpha\beta}$ in Eq. (9.61) Then Eq. (9.67) becomes

$$\left\{ \left(\delta_{\alpha\beta} - \frac{k_\alpha k_\beta}{k^2} \right) \left(1 - \frac{\epsilon_t \omega^2}{k^2 c^2} \right) - \frac{\omega^2}{c^2} \frac{k_\alpha k_\beta}{k^4} \epsilon_l \right\} E_\beta = 0. \tag{9.69}$$

If $k_\beta E_\beta = 0$ (transverse waves), we must have $\omega^2 = (k^2 c^2 / \epsilon_t)$. This determines the dispersion relation for transverse waves. On the other hand, if $\mathbf{E}$ is parallel to $\mathbf{k}$, then the term with ϵ_t does not contribute and this equation can be satisfied only with $\epsilon_l = 0$. In a plasma, we will see that $\epsilon_l = 0$ can indeed lead to a nontrivial dispersion relation, thereby allowing longitudinal waves. It is therefore important to determine the form of ϵ_t and ϵ_l for a plasma.

The explicit computation of the permittivity requires the solution of the transport equation for ions and electrons. In several contexts, however, we can ignore the motion of the ions (which is due to their larger mass) and study the dielectric polarisation of the plasma based on the movements of the electrons alone. For a weak field, the distribution function for electrons can be expressed in the form $f = f_0 + \delta f$, where f_0 is the stationary, isotropic, homogeneous distribution function and δf is the change that arises because of the self-consistent average field. Linearising the transport equation in δf, we get

$$\frac{\partial \delta f}{\partial t} + \mathbf{v} \cdot \frac{\partial \delta f}{\partial \mathbf{x}} = e \left(\mathbf{E} + \frac{\mathbf{v} \times \mathbf{B}}{c} \right) \cdot \frac{\partial f_0}{\partial \mathbf{p}} = e \mathbf{E} \cdot \frac{\partial f_0}{\partial \mathbf{p}}. \tag{9.70}$$

The second equality follows from the fact that, for an isotropic distribution function, $\partial f_0 / \partial \mathbf{p}$ has the same direction as $\mathbf{v}$ so that its scalar product $\mathbf{v} \times \mathbf{B}$ is zero. Taking δf and $\mathbf{E}$ to be proportional to $\exp[i(\mathbf{k} \cdot \mathbf{x} - \omega t)]$, we find that the amplitudes are related by

$$\delta f = \frac{e\mathbf{E}}{i(\mathbf{k} \cdot \mathbf{v} - \omega)} \cdot \frac{\partial f_0}{\partial \mathbf{p}}. \tag{9.71}$$

The charge and current densities in the plasma are contributed only by the perturbed part of the distribution function and are given by

$$\rho = -e \int \delta f \, d^3 p, \quad \mathbf{j} = -e \int \mathbf{v}\delta f \, d^3 p. \tag{9.72}$$

These quantities are also proportional to $\exp[i(\mathbf{k} \cdot \mathbf{x} - \omega t)]$, and the dielectric polarisation is related to charge and current densities by Eqs. (9.56):

$$i\mathbf{k} \cdot \mathbf{P} = -\rho, \quad -i\omega\mathbf{P} = \mathbf{j}. \tag{9.73}$$

Substituting the expression for δf into Eqs. (9.72), we get

$$i\mathbf{k} \cdot \mathbf{P} = -e^2\mathbf{E} \cdot \int \frac{\partial f_0}{\partial \mathbf{p}} \frac{d^3 p}{i(\mathbf{k} \cdot \mathbf{v} - \omega)}. \tag{9.74}$$

The integral has a pole at the point $\omega = \mathbf{k} \cdot \mathbf{v}$ and hence is ill defined. This arises because of the assumption that all quantities have been varying with time as $\exp(i\omega t)$ for all t. In any realistic situation, the perturbation must have been switched on slowly from $t = -\infty$ onwards. This is equivalent to assuming that the electric field, for example, has the time dependence $E \propto \exp(-i\omega t + t\delta)$ as $t \to -\infty$. The quantity δ is treated as a infinitesimal positive real number and the limit $\delta \to 0^+$ is taken at the end of the calculation. Such a modification ensures that perturbations vanish as $t \to -\infty$. (It is true that the modification makes the electric field and other quantities diverge as $t \to +\infty$; this, however, is of no consequence to the behaviour of the system at finite t, because, by the principle of causality, the physical variables at finite t cannot be influenced by the behaviour of variables at $t = +\infty$.) Mathematically, this modification corresponds to replacing ω with $\omega + i\delta$ with the understanding that $\delta \to +0$ at the end of the calculation. With this convention, the integral for the polarisation can be expressed in the form

$$i\mathbf{k} \cdot \mathbf{P} = -e^2\mathbf{E} \cdot \int \frac{\partial f_0}{\partial \mathbf{p}} \frac{d^3 p}{i(\mathbf{k} \cdot \mathbf{v} - \omega - i\delta)}. \tag{9.75}$$

This integral is now well defined because it has no poles along the real axis. In general, integrals of this form are evaluated with

$$\int_{-\infty}^{\infty} \frac{f(z)}{z - i\delta} \, dz = P \int_{-\infty}^{\infty} \frac{f(z)}{z} \, dz + i\pi f(0), \tag{9.76}$$

where P denotes the principle value of the integral.

To determine the longitudinal permittivity of the plasma, let us consider the case in which the field $\mathbf{E}$ and therefore $\mathbf{P}$ are parallel to $\mathbf{k}$, giving $4\pi\mathbf{P} = (\epsilon_l - 1)\mathbf{E}$.

Then, by using Eq. (9.75), we find that the longitudinal permittivity of the plasma with a stationary isotropic distribution function $f_0(p)$ is given by

$$\epsilon_l = 1 - \frac{4\pi e^2}{k^2} \int \mathbf{k} \cdot \frac{\partial f_0}{\partial \mathbf{p}} \frac{d^3 p}{(\mathbf{k} \cdot \mathbf{v} - \omega - i\delta)}. \tag{9.77}$$

With $\mathbf{k}$ taken along the x axis, the integral over $d^3 p$ can be split into an integral over dp_x and an integral over transverse directions. The dependence on p_y and p_z arises only from f_0. When

$$f(p_x) \equiv \int f_0(p) \, dp_y \, dp_z, \tag{9.78}$$

the longitudinal permittivity becomes

$$\epsilon_l = 1 - \frac{4\pi e^2}{k} \int_{-\infty}^{\infty} \frac{df(p_x)}{dp_x} \frac{dp_x}{kv_x - \omega - i\delta}. \tag{9.79}$$

The nature of this integral and relation (9.76) shows that the permittivity of a collisionless plasma is a complex quantity of the form $\epsilon_l = \epsilon_l^R + i\epsilon_l^I$, with the imaginary part given by

$$\epsilon_l^I = -4\pi^2 e^2 \int \frac{\partial f_0}{\partial \mathbf{p}} \cdot \frac{\mathbf{k}}{k^2} \delta(\omega - \mathbf{k} \cdot \mathbf{v}) \, d^3 p = -\frac{4\pi^2 e^2 m}{k^2} \left[\frac{df(p_x)}{dp_x} \right]_{v_x = \omega/k}. \tag{9.80}$$

The complex permittivity signifies that there is dissipation of electric energy in the medium. In general, for real $\mathbf{E}$ and $\mathbf{D}$, the mean amount of energy dissipated per unit time and unit volume is given by

$$Q = \frac{\partial}{\partial t} \left(\frac{E^2}{8\pi} \right) + \mathbf{E} \cdot \mathbf{j} = \frac{\mathbf{E} \cdot \dot{\mathbf{E}}}{4\pi} + \mathbf{E} \cdot \dot{\mathbf{P}} = \left(\frac{1}{4\pi} \right) \mathbf{E} \cdot (\dot{\mathbf{E}} + 4\pi \dot{\mathbf{P}}) = \frac{\mathbf{E} \cdot \dot{\mathbf{D}}}{4\pi}. \tag{9.81}$$

When complex numbers are used to describe the electric field, $\mathbf{E}$ has to be replaced with $(1/2)(\mathbf{E} + \mathbf{E}^*)$. The corresponding vector $\dot{\mathbf{D}}$ has the components

$$\frac{1}{2} i\omega \{ -\epsilon_{\alpha\beta}(\omega, \mathbf{k}) E_\beta + \epsilon_{\alpha\beta}(-\omega, -\mathbf{k}) E_\beta^* \}. \tag{9.82}$$

Using this result and averaging over one period, we can easily show that the dissipation is given by

$$Q = \frac{\omega}{8\pi} \epsilon_{\alpha\beta}^I(\omega, \mathbf{k}) E_\alpha E_\beta^* = \frac{\omega}{8\pi} \epsilon_l^I |\mathbf{E}|^2, \tag{9.83}$$

where the second relation is valid for the longitudinal field. In the present case, the dissipation is given by

$$Q = -|\mathbf{E}|^2 \frac{\pi m e^2 \omega}{2k^2} \left[\frac{df(p_x)}{dp_x} \right]_{v_x = \omega/k}. \tag{9.84}$$

This expression shows that dissipation can occur in a plasma in a manner that is fundamentally different from the normal dissipation of energy, which arises because of the collisions. The damping derived above (called Landau damping) is due to electrons whose speed in the direction of propagation of the electric wave is same as the phase velocity of the wave, $v_x = (\omega/k)$. (To be precise, we should also specify a small width in the velocity space in order to obtain finite number of electrons around this velocity. This can be taken care of by a more formal derivation like the one used in Chap. 7, Section 7.2, to study the rate of atomic transition. The work done will grow in proportion to time and will be limited eventually by the mean time between collisions; dividing by the total time will give the rate that is directly obtainable by the simpler derivation presented here.) Such electrons find the wave to be stationary in their rest frame; the wave can work on the electrons that will not average to zero when there is a gradient in the distribution function. For an isotropic plasma, it is straightforward to see that $(df(p_x)/dp_x)$ is negative: we have

$$\frac{df(p_x)}{dp_x} = \frac{d}{dp_x} \int_0^\infty f_0(p_x^2 + p_\perp^2)\pi\, d(p_\perp^2) = 2\pi p_x \int_0^\infty f_0'(p_x^2 + p_\perp^2)\, d(p_\perp^2).$$

$$(9.85)$$

Because $f_0(p^2) \to 0$ as $p^2 \to \infty$, it follows that $[df(p_x)/dp_x] = -2\pi p_x\, f_0(p_x^2)$, implying that $(df/dp_x) < 0$ when $p_x = (\omega/k) > 0$. Hence the process transfers energy from the wave to the electrons in an isotropic plasma. It should, however, be stressed that expression (9.84) is valid even for an anisotropic distribution for which it is in general possible for the energy to be transferred from the electrons to the waves.

To understand the origin of this damping mechanism, it is instructive to study the motion of the electrons in a weak electric field and the consequent work done by the field on the electron. Consider an electron moving along the x axis in a weak field of the form $E(t, x) = \mathrm{Re}\{E_0 \exp[i(kx - \omega t)]\exp(t\delta)\}$, which is also taken to be along the x axis. The unperturbed motion of the electron will be along the trajectory $x_0 = w_0 t$, where w_0 is the x component of the velocity of the electron. In the presence of the field, the position and the velocity will be perturbed by small quantities to the values $x = x_0 + \delta x$, and $w = w_0 + \delta w$. The linearised equation of motion,

$$m\frac{d\delta w}{dt} = -eE(t, x_0) = -e\,\mathrm{Re}\{E_0 e^{ikt(w_0 - \omega/k)}e^{(t\delta)}\}, \qquad (9.86)$$

has the solution

$$\delta w = -\frac{e}{m}\,\mathrm{Re}\frac{E(t, x_0)}{ik(w_0 - \omega/k) + \delta},$$

$$\delta x = -\frac{e}{m}\,\mathrm{Re}\frac{E(t, x_0)}{[ik(w_0 - \omega/k) + \delta]^2}.$$

$$(9.87)$$

The mean work q done by the field on a *single* electron per unit time is

$$q = -e\langle w E(t, x)\rangle = -e\langle (w_0 + \delta w)E(t, x_0 + \delta x)\rangle$$
$$\approx -e w_0 \left\langle \frac{\partial E}{\partial x_0}\delta x\right\rangle - e\langle \delta w \cdot E(t, x_0)\rangle, \tag{9.88}$$

or, in complex form,

$$q = -\frac{1}{2}e\,\mathrm{Re}\left\{ w_0\delta x \frac{\partial E^*}{\partial x_0} + \delta w \cdot E^*\right\}. \tag{9.89}$$

Substituting the expressions for δw and δx in (9.87) into Eq. (9.89) we get

$$q = \frac{e^2}{2m}|E^2|\frac{d}{dw_0}\frac{w_0\delta}{\delta^2 + k^2(w_0 - \omega/k)^2}. \tag{9.90}$$

To find the total work done, we have to integrate this expression over the distribution function of the particles, which gives

$$Q = \int_{-\infty}^{\infty} qf(p_x)\,dp_x = -\frac{1}{2}e^2|E^2|\int \frac{w_0\delta}{\delta^2 + k^2(w_0 - \omega/k)^2}\frac{df}{dp_x}\,dp_x, \tag{9.91}$$

where an integration by parts has been performed. It is now clear that the dissipation is related to the gradient of f, that is, it arises because of the existence of different amount of electrons on the two sides of the phase velocity ω/k. Taking the limit of $\delta \to 0$ by using

$$\lim_{\delta \to 0}\frac{\delta}{\delta^2 + z^2} = \pi\delta_D(z), \tag{9.92}$$

we get result (9.84). In this approach, we can say that particles with $v_x < \omega/k$ gain energy from the wave whereas particles with $v_x > \omega/k$ lose energy to the wave. The net damping arises because of the disparity in the number of electrons in these two regimes.

The original description of Landau damping, given above, was based on linear perturbation theory. When the wave amplitude becomes larger, it is possible for charged particles like the electrons to be trapped near the resonances produced by the above mechanism. Any particle, moving with a speed v close to the resonance with the wave, will have very little energy in the wave frame and hence can be trapped in the potential wells created by the electrostatic potential. Once such a particle is trapped, it is no longer capable of exchanging energy with the wave. Such electron trapping in a wave can cause them to be bunched together with important implications for radiation from the plasma. In general, charged particles emit radiation independently and the total power emitted by a system of N particles scales as N. However, charges are bunched together; they act as a single particle with total charge Nq and the total power will scale as N^2.

This feature also illustrates the limitations of studying linear perturbations in a plasma.

For a Maxwellian distribution, the explicit form of the longitudinal permittivity can be computed from Eq. (9.79) and is given by

$$\epsilon_l(\omega, k) = 1 + \frac{1}{(k\lambda_e)^2}\left[1 + F\left(\frac{\omega}{\sqrt{2}kv_e}\right)\right], \tag{9.93}$$

where

$$F(x) = \frac{x}{\sqrt{\pi}}\int_{-\infty}^{\infty}\frac{e^{-z^2}dz}{z - x - i\delta} = \frac{x}{\sqrt{\pi}}P\int_{-\infty}^{\infty}\frac{e^{-z^2}dz}{z - x} + i\sqrt{\pi}xe^{-x^2}, \tag{9.94}$$

with

$$v_e = \sqrt{(T_e/m)}, \quad \lambda_e = \sqrt{(T_e/4\pi n_e e^2)}. \tag{9.95}$$

The temperatures are measured in energy units; for conventional units, we should replace T_e with $k_B T_e$, etc. The asymptotic forms of $F(x)$ are easy to obtain. For $x \gg 1$, the Taylor expansion of $(z - x)^{-1}$ followed by term-by-term integration gives

$$F(x) \approx 1 - \frac{1}{2x^2} - \frac{3}{4x^4} + i\sqrt{\pi}xe^{-x^2}, \quad (x \gg 1). \tag{9.96}$$

For $x \ll 1$, a change of the integration variable to $u = z - x$ gives

$$\frac{x}{\sqrt{\pi}}P\int_{-\infty}^{\infty}\frac{e^{-z^2}dz}{z - x} = \frac{xe^{-x^2}}{\sqrt{\pi}}P\int_{-\infty}^{\infty}e^{-u^2 - 2ux}\frac{du}{u}$$

$$\approx \frac{x}{\sqrt{\pi}}P\int_{-\infty}^{\infty}e^{-u^2}\left(\frac{1}{u} - 2x\right)du, \tag{9.97}$$

so that

$$F(x) \approx -2x^2 + i\sqrt{\pi}x, \quad (x \ll 1). \tag{9.98}$$

The corresponding limiting form for the permittivity is

$$\epsilon_l = 1 - \frac{\Omega_e^2}{\omega^2}\left(1 + \frac{3k^2 v_e^2}{\omega^2}\right) + i\sqrt{\frac{\pi}{2}}\frac{\omega\Omega_e^2}{(kv_e)^3}\exp\left(-\frac{\omega^2}{2k^2 v_e^2}\right) \quad \left(\text{for } \frac{\omega}{kv_e} \gg 1\right), \tag{9.99}$$

where $\Omega_e = v/\lambda_e = (4\pi n_e e^2/m)^{1/2}$ is the plasma frequency of the electrons. In this limit the spatial dispersion is only a small correction to the permittivity and the imaginary part is exponentially small because the Maxwellian distribution cuts off the number of electrons that can participate in Landau damping in the high-frequency limit. The limiting form of permittivity for $\omega/k \gg v_e$ is independent of k and is given by

$$\epsilon(\omega) = 1 - \left(\frac{\Omega_e}{\omega}\right)^2. \tag{9.100}$$

In the limit of $k \to 0$, the distinction between ϵ_l and ϵ_t disappears and this form is clearly applicable to both transverse and longitudinal permittivities.

In the opposite limit of low frequencies, the permittivity is given by

$$\epsilon_l = 1 + \left(\frac{\Omega_e}{k v_e}\right)^2 \left[1 - \left(\frac{\omega}{k v_e}\right)^2 + i\sqrt{\frac{\pi}{2}}\frac{\omega}{k v_e}\right] \quad \left(\text{for } \frac{\omega}{k v_e} \ll 1\right). \tag{9.101}$$

The imaginary part is again small but not exponentially; this arises from the smallness of the phase volume occupied by the electrons satisfying $\omega = \mathbf{k} \cdot \mathbf{v}$.

All the above expressions were for the longitudinal part of the permittivity. The *general* expression for permittivity is given by

$$\epsilon_{\alpha\beta} = \delta_{\alpha\beta} - \frac{4\pi e^2}{\omega} \int \frac{v_\alpha}{\mathbf{k} \cdot \mathbf{v} - \omega - i\delta} \frac{\partial f_0}{\partial p_\beta} d^3 p. \tag{9.102}$$

When the transverse part is separated out as $\epsilon_t = (1/2)[\epsilon_{\alpha\alpha} - (\epsilon_{\alpha\beta} k_\alpha k_\beta / k^2)]$, the explicit expression turns out to be

$$\epsilon_t = 1 - \frac{2\pi e^2}{\omega} \int \mathbf{v}_\perp \cdot \frac{\partial f_0}{\partial \mathbf{p}_\perp} \frac{d^3 p}{\mathbf{k} \cdot \mathbf{v} - \omega - i0}. \tag{9.103}$$

For a Maxwellian distribution, direct integration again gives

$$\epsilon_t = 1 + \frac{\Omega_e^2}{\omega^2} F\left(\frac{\omega}{\sqrt{2} k v_e}\right) \tag{9.104}$$

with corresponding limiting forms. In particular, for $\omega \gg v_e k$, the transverse and the longitudinal permittivities are the same and are given by expression (9.100).

The contributions of ions to the permittivity are calculated in the same way and the two expressions can be added together. For example, the longitudinal part – taking into account both ions and electrons – will be

$$\epsilon_l - 1 = \frac{1}{(k\lambda_e)^2}\left[F\left(\frac{\omega}{\sqrt{2} k v_e}\right) + 1\right] + \frac{1}{(k\lambda_i)^2}\left[F\left(\frac{\omega}{\sqrt{2} k v_i}\right) + 1\right], \tag{9.105}$$

with

$$v_i = \left(\frac{T_i}{M}\right)^{1/2}, \quad \lambda_i = \frac{v_i}{\Omega_i} = \left[\frac{T_i}{4\pi n_i (Ze)^2}\right]^{1/2}, \quad \Omega_i^2 = \frac{4\pi n_i (Ze)^2}{M}. \tag{9.106}$$

These expressions are valid for a two-temperature plasma, in which ions and electrons are described by Maxwellian distributions with different temperatures. The limiting expressions can be easily found, as before.

Finally, let us consider the dispersion relations for the electromagnetic waves in the plasma described the permittivities derived above. For transverse waves, the dispersion relation is $\omega^2 = c^2 k^2 / \epsilon_t$ [see the discussion following Eq. (9.68)]. In the high-frequency limit, with $\omega \gg k v_e$, the oscillations are ordinary electromagnetic waves with ϵ_t given by Eq. (9.100) and the dispersion relation

$$\omega^2 = c^2 k^2 + \Omega_e^2. \tag{9.107}$$

The group velocity corresponding to this dispersion relation is $c_g \equiv (\partial \omega / \partial k) = c(1 - \Omega_e^2/\omega^2)^{1/2}$. If such a group of waves is travelling from a cosmic source towards Earth, through a plasma, the travel time to cover a distance r will be

$$t_\omega = \int_0^r \frac{ds}{c_g} = \int_0^r \frac{ds}{c} \left(1 - \frac{\Omega_e^2}{\omega^2} \right)^{-1/2}. \tag{9.108}$$

For $\omega \gg \Omega_e$, we can use the Taylor expansion of the square root and the definition of Ω_e^2 to get

$$t_\omega = \frac{r}{c} + \frac{2\pi e^2}{m_e \omega^2} \text{DM}, \tag{9.109}$$

where the quantity DM, called the dispersion measure of the plasma, is defined to be

$$\text{DM} \equiv \int_0^r n_e \, ds. \tag{9.110}$$

This quantity gives the column density of electrons along the line of sight. By using Eq. (9.109) we can relate the difference in the time of arrival, $\Delta \tau_D$, of a signal at two different frequencies to the dispersion measure. Numerically,

$$\Delta \tau_D = 1.34 \times 10^{-9} \, \mu s \left[\frac{\text{DM}}{\text{cm}^{-2}} \right] \left[\left(\frac{\nu_1}{1 \text{ MHz}} \right)^{-2} - \left(\frac{\nu_2}{1 \text{ MHz}} \right)^{-2} \right]. \tag{9.111}$$

More conventionally, n is measured in inverse cubic centimeters and l in parsecs, so that

$$\frac{\text{DM}}{\text{cm}^{-3} \, \text{pc}} = 2.410 \times 10^{-4} \left(\frac{\Delta \tau_D}{\mu s} \right) \left[\left(\frac{\nu_1}{1 \text{ MHz}} \right)^{-2} - \left(\frac{\nu_2}{1 \text{ MHz}} \right)^{-2} \right]^{-1}. \tag{9.112}$$

If the time delay at the two frequencies can be measured, then this equation allows for the dispersion measure to be determined along any given direction. We shall see in Vol. II that this process can serve as a useful tool in the diagnostics of the astrophysical plasmas.

For $\omega < \Omega_e$, the wave vector is imaginary, showing that the wave is exponentially attenuated with distance. Thus, for a sufficiently cold plasma, waves with $\omega < \Omega_e$ cannot propagate. For the ionosphere of Earth with $n_e \approx 10^6 \text{ cm}^{-3}$, waves with $\lambda \gtrsim 30$ m cannot propagate. Longer waves from space cannot reach ground; when transmitted from the ground they are reflected back by the ionosphere. For transverse waves with low frequencies ($\omega \ll k v_e$) and long wavelengths ($k\lambda_e \ll 1$), the primary contribution to the dispersion relations is

$$\omega = -i\sqrt{\frac{2}{\pi}}\frac{k^3 c^2 v_e}{\Omega_e^2}. \tag{9.113}$$

The purely imaginary value shows that the wave propagation cannot occur at low frequencies.

It was mentioned above that, because of spatial dispersion, a plasma can also sustain *longitudinal* waves, which we now analyse. The dispersion relation for these waves is given by the implicit equation $\epsilon_l(\omega, k) = 0$. When this condition is satisfied with a longitudinal electric field $\mathbf{E}$, we automatically have $\mathbf{D} = 0$. If we take $\mathbf{B} = 0$, then Maxwell's equations reduce to the condition $\nabla \times \mathbf{E} = 0$, which is automatically satisfied if the wave is longitudinal.

Because ϵ_l is complex, the roots of the equation $\epsilon_l(\omega, k) = 0$ can be taken to be the complex frequency $\omega = \omega_R + i\omega_I$. From expression (9.79) for ϵ_l, it can be seen that $\omega_I < 0$; because the amplitude is proportional to $\exp(-\omega_I t)$, the damping rate of the wave is given by $\gamma = -\omega_I$. When $\gamma \ll \omega_R$, propagating waves can exist. When $\omega \gg k v_e \gg k v_i$, the oscillations involve only the electron and the equation $\epsilon_l = 0$ can be solved by successive approximation. To the lowest order, we get $\omega = \Omega_e$, which represents longitudinal oscillations of electrons at plasma frequency. They are long waves that satisfy the condition $k\lambda_e \ll 1$. The next-order k−dependent corrections to the plasma oscillations are given by

$$\omega = \Omega_e\left(1 + \frac{3}{2}k^2\lambda_e^2\right). \tag{9.114}$$

The corresponding damping rate can be obtained by similar methods and we get

$$\gamma = \sqrt{\frac{\pi}{8}}\frac{\Omega_e}{(k\lambda_e)^3}\exp\left[-\frac{1}{2(k\lambda_e)^2} - \frac{3}{2}\right]. \tag{9.115}$$

The damping is exponentially small and can usually be ignored.

These expressions did not take into account the existence of ions. When both electrons and ions contribute, the longitudinal permittivity, to the necessary order of accuracy, can be taken as

$$\epsilon_l = 1 - \frac{\Omega_i^2}{\omega^2} + \frac{1}{(k\lambda_e)^2}\left(1 + i\sqrt{\frac{\pi}{2}}\frac{\omega}{k v_e}\right). \tag{9.116}$$

This is valid when $T_i \ll T_e$ and $v_i \ll (\omega/k) \ll v_e$. The dispersion relation obtained by setting $\epsilon_l = 0$ is

$$\omega^2 = \Omega_i^2\frac{k^2\lambda_e^2}{1 + k^2\lambda_e^2} = \frac{ZT_e}{M}\frac{k^2}{1 + k^2\lambda_e^2}, \tag{9.117}$$

where we have used the fact that $n_e = Zn_i$. For the longest waves, this dispersion relation reduces to

$$\omega = k\sqrt{\frac{ZT_e}{M}}, \quad (k\lambda_e \ll 1), \tag{9.118}$$

which are ordinary sound waves, called ion-acoustic waves in this context. The damping rate of these waves, which are essentially due to the electrons, is given by $\gamma = \omega\sqrt{(\pi Zm/8M)}$. The contribution of ions to γ is exponentially small. For shorter wavelengths in the range $(1/\lambda_e) \ll k \ll (1/\lambda_i)$ the dispersion relation reduces to $\omega \approx \Omega_I$. These waves are analogous to longitudinal oscillations of electrons, and they suffer only slight damping.

9.5 Waves in Magnetised Cold Plasmas

Most astrophysical plasmas host magnetic fields. Such an external magnetic field defines a special direction at any given location in the plasma, thereby making the properties of the plasma anisotropic. We consider several features of such a plasma in this section under the assumption that the plasma is cold, that is, the random motion that is due to finite temperature can be ignored. Under such circumstances, the properties of the plasma can be directly related to the motion of the charged particles.

Let us assume that the only anisotropy introduced in the plasma is due to a uniform, constant, external magnetic field $\mathbf{B}_0$. The equation of motion for an electron will then be

$$m\frac{d\mathbf{v}}{dt} = -e\mathbf{E} - e\left(\frac{\mathbf{v}}{c} \times \mathbf{B_0}\right), \tag{9.119}$$

where $\mathbf{E}$ is the internal, average, self-consistent field. Neglecting the spatial variation of the electric field in the region of motion of the particle, we can find a solution to this equation in which both the electric field and the velocity vary as $\exp(-i\omega t)$. The amplitudes will satisfy the relation

$$i\omega\mathbf{v} = e\mathbf{E}/m + e\mathbf{v} \times \mathbf{B}_0/c, \tag{9.120}$$

which can be solved for $\mathbf{v}$ to give

$$\mathbf{v} = -\frac{ie\omega}{m\left(\omega^2 - \omega_{Be}^2\right)}\left\{\mathbf{E} - \frac{\omega_{Be}^2}{\omega^2}\mathbf{b}(\mathbf{E} \cdot \mathbf{b}) - \frac{i\omega_{Be}}{\omega}(\mathbf{E} \times \mathbf{b})\right\}, \tag{9.121}$$

where $\mathbf{b} = \mathbf{B}/B$ is a unit vector in the direction of the magnetic field and $\omega_{Be} \equiv eB_0/mc$ is the Larmor frequency of the electrons. The corresponding polarisation is given by

$$-i\omega\mathbf{P} = -i\omega\frac{\mathbf{D} - \mathbf{E}}{4\pi} = \mathbf{j} = -en_e\mathbf{v}. \tag{9.122}$$

The contribution from the ions can be obtained in a corresponding manner.

The linear relation between **D** and **E** can now be written as

$$\mathbf{D} = \epsilon_\perp \mathbf{E} + (\epsilon_\| - \epsilon_\perp)\mathbf{b}(\mathbf{b}\cdot\mathbf{E}) + ig\mathbf{E}\times\mathbf{b}, \quad \mathbf{b} = \mathbf{B}/|\mathbf{B}| \quad (9.123)$$

in terms of the three scalars $\epsilon_\perp$, $\epsilon_\|$, and g, all of which are functions of ω and B_0. This form clearly takes account the anisotropy introduced by the magnetic field. The corresponding permittivity can be written in the form

$$\epsilon_{\alpha\beta} = \epsilon_\perp \delta_{\alpha\beta} + (\epsilon_\| - \epsilon_\perp)b_\alpha b_\beta + ig\epsilon_{\alpha\beta\gamma}b_\gamma \quad (9.124)$$

(where $\epsilon_{\alpha\beta\gamma}$ of the last term is the completely antisymmetric tensor in three dimensions and should not be confused with the permittivity $\epsilon_{\alpha\beta}$). Comparing Eqs. (9.121), (9.122), and (9.124), and including the contributions from the ions as well, we see that the permittivity is determined by the functions

$$\epsilon_\perp = 1 - \frac{\Omega_e^2}{\omega^2 - \omega_{Be}^2} - \frac{\Omega_i^2}{\omega^2 - \omega_{Bi}^2},$$

$$\epsilon_\| = 1 - \frac{\Omega_e^2 + \Omega_i^2}{\omega^2}, \quad (9.125)$$

$$g = \frac{\omega_{Be}\Omega_e^2}{\omega(\omega^2 - \omega_{Be}^2)} - \frac{\omega_{Bi}\Omega_i^2}{\omega(\omega^2 - \omega_{Bi}^2)},$$

where $\omega_{Bi} \equiv ZeB_0/Mc$ is the ion Larmor frequency. This completely solves the problem for a cold plasma with a constant magnetic field.

The ratio of Larmor frequencies of electrons and ions (or the plasma frequencies of electrons and ions),

$$\frac{\omega_{Bi}}{\omega_{Be}} = \frac{Zm}{M}, \quad \frac{\Omega_i}{\Omega_e} = \left(\frac{Zm}{M}\right)^{1/2}, \quad (9.126)$$

are small numbers. Nevertheless, the contribution of ions to the permittivity can be comparable with (or even larger than) that of electrons at low frequencies. As $\omega \to 0$ the two terms in g are of comparable order because $(\Omega_e^2/\omega_{Be}) = (\Omega_i^2/\omega_{Bi})$. In the transverse permittivity, the two terms are comparable in the range of $\omega \simeq \omega_{Bi}(M/m)^{1/2} \simeq (\omega_{Bi}\omega_{Be})^{1/2}$ and the ion contribution is negligible only if $\omega \gg (\omega_{Bi}\omega_{Be})^{1/2}$. In the longitudinal permittivity $\epsilon_\|$, the ionic part is always negligible.

In deriving the above results, we have assumed that the spatial variation of the electric field is negligible over the region in which the particles move. In the direction of the magnetic field, this region has a typical size of (v_T/ω), which is the typical distance traversed by a particle during one period of the wave. In the direction perpendicular to the magnetic field, the size of the region is decided by $r_B \simeq (v_T/\omega_B)$. Our approximation requires that the following conditions be satisfied: $(v_T|k_z|/\omega) \ll 1$ and $(v_T k_\perp/\omega_B) \ll 1$. It is also necessary that the frequency of the wave is not too close to the Larmor frequency of either ions or electrons so as to avoid a resonance situation.

Let us next consider the kind of waves that such a cold magnetoactive plasma can support. The dispersion relation for waves can be determined from condition (9.68). The solution to this equation in the presence of a magnetic field $\mathbf{B}_0$ gives rise to a wide variety of wave-propagation phenomena in the plasmas. The number of independent wave modes in a plasma can be determined as follows. In general, the electromagnetic field will introduce 2 degrees of freedom of oscillation whereas the electrons and ions will contribute 1 degree of freedom each. Thus there should be a total of four independent wave modes in a plasma. At very high frequencies all the modes decouple and the electromagnetic modes correspond to a propagating wave whereas the two matter modes corresponds to density variation (sound waves) of ions and electrons. In the case of a cold plasma ($T \approx 0$), the sound-wave modes disappear because of a lack of restoring force; in this case, there will be only two independent modes for the system. In general, the modes are mixed and cannot be separated out as electromagnetic or sound waves but the number of modes present in a system is conserved. To study them systematically, we first obtain the general solution to Eq. (9.68) and then treat different special cases.

Using the form of $\epsilon_{\alpha\beta}$ given in Eq. (9.124), we find that the dispersion relation becomes, after a simple calculation,

$$A(kc/\omega)^4 + B(kc/\omega)^2 + C = 0, \tag{9.127}$$

with

$$A = \epsilon_{\alpha\beta}k_\alpha k_\beta/k^2 = \epsilon_\perp \sin^2\theta + \epsilon_\parallel \cos^2\theta \equiv \epsilon_l, \tag{9.128}$$

$$B = -\epsilon_\perp\epsilon_\parallel(1 + \cos^2\theta) - (\epsilon_\perp^2 - g^2)\sin^2\theta, \tag{9.129}$$

$$C = \epsilon_\parallel(\epsilon_\perp^2 - g^2), \tag{9.130}$$

where θ is the angle between $\mathbf{k}$ and $\mathbf{B}_0$. For a given value of ω and θ, we obtain two values of k^2 that correspond to two types of waves that can propagate. We begin by discussing the propagation along the magnetic field ($\theta = 0$) and transverse to the magnetic field ($\theta = (\pi/2)$).

9.5.1 Propagation Along the Magnetic Field

In this case $\theta = 0$ and the roots of the dispersion relation are given by

$$\left(\frac{kc}{\omega}\right)^2 = \epsilon_\perp \pm g = 1 - \frac{\Omega_e^2}{\omega(\omega \pm \omega_{Be})} - \frac{\Omega_i^2}{\omega(\omega \mp \omega_{Be})}. \tag{9.131}$$

By using (9.66) we can easily see that these are transverse waves ($E_z = 0$) that are circularly polarised $[(E_y/E_x) = \mp i]$. The dispersion relation shows that the two circular polarisations have different characteristics of propagation. As a wave propagates along the direction of the magnetic fields, its electric vector rotates at a rate that depends on the magnetic field.

This leads to an important effect called Faraday rotation. To see this effect explicitly, let us consider the case with $\omega \gg \Omega_e$ and $\Omega_i \approx 0$. Then we can expand the dispersion relation in a Taylor series to obtain the two wave vectors as

$$k_\pm = \frac{\omega}{c}\left[1 - \frac{\Omega_e^2}{\omega^2}\left(1 \pm \frac{\omega_{Be}}{\omega}\right)^{-1}\right]^{1/2} \simeq \frac{\omega}{c}\left[1 - \frac{\Omega_e^2}{2\omega^2}\left(1 \mp \frac{\omega_{Be}}{\omega}\right)\right]$$

$$= \left(\frac{\omega}{c} - \frac{\Omega_e^2}{2\omega c}\right) \pm \frac{\Omega_e^2 \omega_{Be}}{2c\omega^2} \equiv k_0 \pm \Delta k, \qquad (9.132)$$

where the first term $k_0 = [(\omega/c) - (\Omega_e^2/2\omega c)]$ corresponds to the wave vector that has already been discussed in connection with the wave propagation in a nonmagnetoactive plasma. The extra term introduced by the magnetic field is denoted by $\Delta k = (2\pi e^3/m_e^2 c^2)(Bn_e/\omega^2)$. Consider now an electromagnetic wave that was originally linearly polarised in a direction transverse to the magnetic field. Any such linearly polarised wave can be thought of as a superposition of right and left circularly polarised waves. As the wave propagates along the magnetic field, these two components will introduce different amounts of phase change in the wave given by

$$\int_0^L k_\pm \, dl = \int_0^L k_0 \, dz \pm \int_0^L \Delta k \, dz. \qquad (9.133)$$

In other words, a linearly polarised wave will have its plane of polarisation rotated by an angle

$$\psi = \int_0^L \Delta k \, dz = \frac{2\pi e^3}{m_e^2 c^2 \omega^2}\int_0^L n_e B_0 \, dz \equiv \frac{2\pi e^3}{m_e^2 c^2 \omega^2}\langle RM \rangle \qquad (9.134)$$

while propagating a distance L along the magnetic field. The last equality defines a quantity RM, called rotation measure, which is an integral of the product of electron density and the magnetic field along the line of propagation. If the propagation is not strictly along the magnetic field, it is clear that the component of the magnetic field in the direction of propagation $B_\parallel$ is the relevant quantity in defining the rotation measure. Numerically, in conventional astronomical units,

$$\frac{\Delta\psi}{\mathrm{rad}} = 8.1 \times 10^5\left(\frac{\lambda}{10^2 \text{ cm}}\right)^2 \int_0^{L/\mathrm{pc}}\left(\frac{B_\parallel}{G}\right)\left(\frac{N_e}{\mathrm{cm}^{-3}}\right)d\left(\frac{z}{\mathrm{pc}}\right). \qquad (9.135)$$

As in the case of the dispersion measure, observations of Faraday rotation at two different frequencies can be used to determine the rotation measure by use of the above result in the form

$$\frac{RM}{rad\ m^{-2}} = 8.1 \times 10^5 \int_0^{L/pc} \left(\frac{B_\parallel}{G}\right)\left(\frac{N_e}{cm^{-3}}\right) d\left(\frac{z}{pc}\right)$$

$$= \left[\left(\frac{\Delta\psi_1}{rad}\right) - \left(\frac{\Delta\psi_2}{rad}\right)\right]\left[\left(\frac{\lambda_1}{m}\right)^2 - \left(\frac{\lambda_2}{m}\right)^2\right]^{-1}. \qquad (9.136)$$

We shall see in Vol. II that this serves as a diagnostic of astrophysical plasmas.

When $\theta = 0$, Eq. (9.127) is also satisfied by the condition $\epsilon_\parallel = 0$. This corresponds to the ordinary longitudinal plasma waves with $\omega \approx \Omega_e$, which have been already discussed in Section 9.4.

9.5.2 Propagation Perpendicular to the Magnetic Field

When $\theta = (\pi/2)$, the two roots of the dispersion relation are given by

$$\left(\frac{ck}{\omega}\right)^2 = \epsilon_\parallel, \quad \left(\frac{ck}{\omega}\right)^2 = \epsilon_\perp - \frac{g^2}{\epsilon_\perp}. \qquad (9.137)$$

The first of these corresponds to a wave in which the dispersion relation is independent of the magnetic field; this has been discussed earlier in Section 9.4. This is a transverse electromagnetic wave ($\mathbf{E}$ is perpendicular to $\mathbf{k}$) that is linearly polarised with the electric field parallel to $\mathbf{B}_0$. The second root corresponds to a wave in which the electric field is perpendicular to $\mathbf{B}_0$ but has both longitudinal and transverse components relative to $\mathbf{k}$. At high frequencies, with $\omega^2 \gg \omega_{Be}\omega_{Bi}$, the dispersion relation for this wave becomes

$$\left(\frac{ck}{\omega}\right)^2 = 1 - \frac{\Omega_e^2(\omega^2 - \Omega_e^2)}{\omega^2(\omega^2 - \omega_{Be}^2 - \Omega_e^2)}. \qquad (9.138)$$

9.5.3 Propagation Along a General Direction

For a general angle θ, we can always find frequencies for which the coefficient A vanishes. These correspond to the plasma resonance frequencies that satisfy

$$\epsilon_l \equiv \epsilon_\perp \sin^2\theta + \epsilon_\parallel \cos^2\theta$$

$$= 1 - \frac{\Omega_e^2 + \Omega_i^2}{\omega^2}\cos^2\theta - \left(\frac{\Omega_e^2}{\omega^2 - \omega_{Be}^2} + \frac{\Omega_i^2}{\omega^2 - \omega_{Bi}^2}\right)\sin^2\theta = 0. \qquad (9.139)$$

These are analogues of the longitudinal oscillations of the plasma discussed earlier in Section 9.4. In this limit of $A \to 0$, the roots of Eq. (9.127) go to $(-C/B)$ and $(-B/A)$, with one root diverging. Equation (9.139) is a cubic in

ω^2 and has three real roots. We can easily find two of them by ignoring the ionic contribution by using the fact that the ratio of ion frequency to electron frequency is a small number. This gives

$$\omega_{1,2}^2 \approx \frac{1}{2}(\Omega_e^2 + \omega_{Be}^2) \pm \frac{1}{2}[(\Omega_e^2 + \omega_{Be}^2)^2 - 4\Omega_e^2\omega_{Be}^2\cos^2\theta]^{1/2}. \qquad (9.140)$$

In the third root, ionic contribution cannot be ignored; however, by using $\Omega_e \gg \omega_{Bi}$, we can find this root as

$$\omega_3^2 \approx \omega_{Bi}^2\left[1 - \left(\frac{Zm}{M}\right)\tan^2\theta\right]. \qquad (9.141)$$

These expressions, however, are invalid when $\cos\theta \ll (m/M)$. In this case, ionic contribution is important in both ω_3 and ω_2 and the solutions are given by

$$\omega_2^2 \equiv \omega_{2h}^2 = \omega_{Be}^2\frac{\Omega_i^2 + \omega_{Bi}^2}{\Omega_e^2 + \omega_{Be}^2},$$

$$\omega_3^2 = \frac{\Omega_e^2\omega_{Bi}^2\cos^2\theta}{\Omega_e^2 + \omega_{Be}^2}. \qquad (9.142)$$

At $\theta = (\pi/2)$, they reach the values ω_{1h} and ω_{2h}, respectively, where $\omega_{1h}^2 = \Omega_e^2 + \omega_{Be}^2$. These are called the upper and the lower hybrid frequencies.

The general characteristics of the dispersion relation are largely decided by the position of ω_1, ω_2, and ω_3. It is easy to see that, in general, we will have five branches of the dispersion relation. Of the five branches, I and II reach the low-frequency region. In the limit of $\omega \to 0$, the phase velocities of these branches are

$$\left(\frac{\omega}{k}\right)_I = \frac{u_A|\cos\theta|}{(1 + u_A^2/c^2)^{1/2}},$$

$$\left(\frac{\omega}{k}\right)_{II} = \frac{u_A}{(1 + u_A^2/c^2)^{1/2}}, \qquad (9.143)$$

where

$$u_A = \frac{c\omega_{Bi}}{\Omega_i} = \frac{B_0}{(4\pi n_i M)^{1/2}} \qquad (9.144)$$

is called the Alfven speed. When $u_A \ll c$, these two phase velocities become u_A and $u_A|\cos\theta|$. This is discussed further in Section 9.7. The three branches I, II, and III are analogues of certain hydromagnetic waves that we shall discuss for a nonrelativistic plasma later on Section 9.7.

In the opposite case of high frequencies, the phase velocities of branches IV and V tend to c, corresponding to transverse high-frequency electromagnetic waves.

9.6 Magnetohydrodynamics

In the discussion so far, the plasma has been treated as being made of ions and electrons with separate distribution functions for the two components. In the fluid limit, the corresponding description would involve two different fluid velocities for the electronic and ionic components. In several astrophysical situations, it turns out that the electrons and ions of the fluid are "locked together" by the electrostatic forces so that the fluid can be described by a single velocity $\mathbf{v}(\mathbf{x}, t)$. In that case, the plasma can be treated as a single fluid and the analysis becomes considerably simpler. We now discuss the behaviour of charged fluids under such an approximation, which may be called magnetohydrodynamics (MHD).

9.6.1 The Assumptions and Equations of Magnetohydrodynamics

The fluid approximation necessarily involves the assumption that the mean free path of the particle is small compared with any other relevant length scales. Further, it is important to realise that there exists is an extremely tiny velocity difference $\mathbf{v}_{\mathrm{diff}} = \mathbf{v}_e - \mathbf{v}_i$ between the ionic velocity $\mathbf{v}_i$ and the electronic velocity $\mathbf{v}_e$. This difference produces an electric current $\mathbf{j}$ that, in turn, produces a magnetic field. In many astrophysical contexts, this difference $\mathbf{v}_{\mathrm{diff}}$ is very small. For example, in the outer zones of the Sun with $L \approx 2 \times 10^{10}$ cm, $n_e \approx 10^{23}$ cm^{-3}, and $B \approx 10^3$ G, the drift velocity is extremely tiny: $\mathbf{v}_{\mathrm{diff}} \simeq 10^{-12}$ cm s^{-1}. Thus we can ignore the $\mathbf{v}_{\mathrm{diff}}$ in discussing the macroscopic flow but must retain it while calculating the electromagnetic field.

To arrive at the approximate equations of MHD, it is necessary to make a careful analysis of the relative importance of different physical processes. The magnetohydrodynamic approximation is based on three key assumptions: (1) The electrons and ions are locked together so that, in the mean rest frame of the fluid, charge density ρ' is far less than the current density (divided by c), $|\mathbf{j}'/c|$. (2) In the local rest frame S' of the plasma, the current $\mathbf{j}'$ is described by Ohm's law, $\mathbf{j}' = \sigma \mathbf{E}'$; σ is assumed to be large so that even a tiny electric field can cause significant currents. (3) The fluid velocity is nonrelativistic so that we need to retain only terms that are lowest order in (v/c). Several conclusions follow from these assumptions.

Because the fluid velocity is nonrelativistic, the displacement current term $(1/c)(\partial \mathbf{E}/\partial t)$ can be neglected compared with $(4\pi \mathbf{j}/c)$. So the equations governing the magnetic field become

$$\nabla \cdot \mathbf{B} = 0, \qquad \nabla \times \mathbf{B} \cong \frac{4\pi}{c}\mathbf{j}. \tag{9.145}$$

The charge and current densities in the rest frame $(\rho', \mathbf{j}')$ are related to the corresponding quantities in the lab frame by Lorentz transformation. To the lowest order in (v/c), we have $\rho' \approx \rho - (\mathbf{j} \cdot \mathbf{v}/c^2)$ and $\mathbf{j}' \approx \mathbf{j} - \rho \mathbf{v}$. The condition

$\rho' \ll (j'/c)$ implies that $\rho \ll (j/c)$ in the lab frame. Comparing the equation $\nabla \cdot \mathbf{E} = 4\pi\rho$ with $\nabla \times \mathbf{B} = (4\pi \mathbf{j}/c)$, we conclude that $E \ll B$. In summary, the MHD approximation consists of the conditions

$$\rho \ll \left|\frac{\mathbf{j}}{c}\right|, \quad |\mathbf{E}| \ll |\mathbf{B}|, \quad \left|\frac{\partial \mathbf{E}}{\partial t}\right| \ll \left|\frac{\mathbf{j}}{c}\right|, \quad \left|\frac{\mathbf{v}}{c}\right| \ll 1. \tag{9.146}$$

Let us now derive the differential equations governing MHD under the above approximations. To find the equation satisfied by the magnetic field we start with the Maxwell's equation $(\partial \mathbf{B}/\partial t) = -c(\nabla \times \mathbf{E})$ and substitute for $\mathbf{E}$ in terms of $\mathbf{j}$ by using the relation

$$\mathbf{j} \simeq \mathbf{j}' \simeq \sigma \mathbf{E}' \cong \sigma \left[\mathbf{E} + \left(\frac{\mathbf{v}}{c} \times \mathbf{B}\right) \right]. \tag{9.147}$$

This gives

$$\frac{\partial \mathbf{B}}{\partial t} = -c \nabla \times \left[\frac{\mathbf{j}}{\sigma} - \left(\frac{\mathbf{v}}{c} \times \mathbf{B}\right) \right]. \tag{9.148}$$

We now substitute for $\mathbf{j}$ in terms of the magnetic field as $\mathbf{j} = (c/4\pi)(\nabla \times \mathbf{B})$. After some simple manipulation, this leads to

$$\frac{\partial \mathbf{B}}{\partial t} = \nabla \times (\mathbf{v} \times \mathbf{B}) - \nabla \times \left(\frac{c^2}{4\pi\sigma} \nabla \times \mathbf{B} \right) = \nabla \times (\mathbf{v} \times \mathbf{B}) + \frac{c^2}{4\pi\sigma} \nabla^2 \mathbf{B}. \tag{9.149}$$

In arriving at the last equation, we have used the fact that $\nabla \cdot \mathbf{B} = 0$. Note that if $\nabla \cdot \mathbf{B} = 0$ initially, this equation will preserve that condition, which is necessary for consistency. This equation determines the evolution of the magnetic field.

To determine the evolution of the system completely, we need to supplement these equations by the fluid equations for plasma. Among these, the equation of continuity, representing the conservation of mass, is unchanged in the presence of magnetic fields:

$$\frac{\partial \rho}{\partial t} + \nabla \cdot (\rho \mathbf{v}) = 0. \tag{9.150}$$

The Euler equation is modified because of the presence of electromagnetic force on matter. Because the electric field is negligible in the lab frame, the force on a charged particle is due to only the magnetic field and is given by $q(\mathbf{v} \times \mathbf{B})/c$. Hence the net force on a fluid element of unit volume can be written as $(\mathbf{j}_e \times \mathbf{B}/c)$. Substituting for $\mathbf{j}_e$ as $(c/4\pi)(\nabla \times \mathbf{B})$, we find the additional magnetic force to be $[(\nabla \times \mathbf{B}) \times \mathbf{B}]/4\pi$. Thus the Euler equation becomes

$$\rho \frac{d\mathbf{v}}{dt} = -\nabla p + \frac{1}{4\pi}(\nabla \times \mathbf{B}) \times \mathbf{B}. \tag{9.151}$$

The expression $(\nabla \times \mathbf{B}) \times \mathbf{B}$ has an elementary interpretation. Using a vector identity, we can write this force term as

$$\mathbf{f} = \frac{1}{4\pi}(\nabla \times \mathbf{B}) \times \mathbf{B} = \frac{1}{4\pi}(\mathbf{B} \cdot \nabla)\mathbf{B} - \frac{1}{8\pi}\nabla B^2. \tag{9.152}$$

The term $(\mathbf{B} \cdot \nabla)\mathbf{B}$ may be thought of as caused by magnetic tension $(B^2/4\pi)$ and tends to straighten a magnetic-field line that is curved. The second term can be thought of as originating because of an isotropic (magnetic) pressure $p_{\text{mag}} = B^2/8\pi$. More generally, when viscous forces are also present, the Euler equation can be written as

$$\frac{\partial(\rho v_i)}{\partial t} = -\frac{\partial \Pi_{ik}}{\partial x_k}, \tag{9.153}$$

where Π_{ik} is the stress tensor in the presence of magnetic field that includes the spatial part of the electromagnetic stress tensor for a magnetic field (see Chap. 3, Section 3.10):

$$\Pi_{ik} = \rho \, v_i v_k + p\delta_{ik} - \sigma_{ik} - \frac{1}{4\pi}\left(B_i B_k - \frac{1}{2}B^2\delta_{ik}\right) \tag{9.154}$$

where

$$\sigma_{ik} = \eta\left(\frac{\partial v_i}{\partial x_k} + \frac{\partial v_k}{\partial x_i} - \frac{2}{3}\delta_{ik}\frac{\partial v_l}{\partial x_l}\right) + \zeta\delta_{ik}\frac{\partial v_l}{\partial x_l} \tag{9.155}$$

is the standard stress tensor for a viscous fluid. Equations (9.149)–(9.151) and an equation of state $p = p(\rho)$ together constitute a set of eight equations for the eight unknowns ρ, p, $\mathbf{v}$, and $\mathbf{B}$. Once $\mathbf{B}$ is determined, we can calculate $\mathbf{j}$ from the relation $\mathbf{j} = (c/4\pi)(\nabla \times \mathbf{B})$ and $\mathbf{E}$ by using $\mathbf{E} = [\mathbf{j}/\rho - (\mathbf{v} \times \mathbf{B})/c]$. This completely solves the problem.

A certain peculiarity in the study of MHD needs to be emphasised. In conventional electrodynamics, we specify the charge and current densities and solve for the electric and the magnetic fields. In MHD, on the other hand, we solve for ρ, $\mathbf{v}$, and $\mathbf{B}$ by using the evolution equations. Once the magnetic field is known, $\mathbf{j}$ and $\mathbf{E}$ are calculated as secondary quantities. Hence, the magnetic field is of primary importance when plasma is treated in the MHD approximation.

The equation for heat transfer is also modified in the presence of electromagnetic fields. In the absence of any electromagnetic processes, we know that the rate of entropy increase is governed by [see Chap. 8, Eq. (8.61)]

$$\rho T \frac{ds}{dt} = \sigma_{ik}\frac{\partial v_i}{\partial x_k} + \nabla \cdot (\kappa \nabla T). \tag{9.156}$$

The left-hand side of this equation is the amount of heat generated per unit time and volume in a moving fluid element. On the right-hand side, the first term gives the amount of heat dissipated because of viscosity and the second one is due to thermal conduction. In the presence of electromagnetic fields, we have to add the work done on the conduction current $\mathbf{j}'$ by the electric field $\mathbf{E}'$ in the rest frame of the fluid. This contributes an amount $\mathbf{j}' \cdot \mathbf{E}' = (j'^2/\sigma) \simeq (j^2/\sigma)$. Adding this to the right-hand side we obtain the equation of energy conservation in MHD,

$$\rho T \frac{ds}{dt} = \sigma_{ik} \frac{\partial v_i}{\partial x_k} + \nabla \cdot (\kappa \nabla T) + \frac{j^2}{\sigma}, \qquad (9.157)$$

or, in terms of the magnetic field,

$$\rho T \left(\frac{\partial s}{\partial t} + \mathbf{v} \cdot \nabla s \right) = \sigma_{ik} \frac{v_i}{x_k} + \nabla \cdot (\kappa \nabla T) + \frac{c^2}{16\pi^2 \sigma} (\nabla \times \mathbf{B})^2. \qquad (9.158)$$

We can transform this equation for heat transfer, by using other equations, into an equation for conservation of energy. In ordinary fluid mechanics, we have

$$\frac{\partial}{\partial t} \left(\frac{1}{2} \rho v^2 + \rho \epsilon \right) = -\nabla \cdot \mathbf{q}, \quad q^a = \rho \left(\frac{1}{2} v^2 + w \right) v^a - v_b \sigma^{ab} - \kappa \frac{\partial T}{\partial x^a}, \qquad (9.159)$$

where ϵ and $w = (\epsilon + p/\rho)$ are the internal energy and the enthalpy per unit mass of the fluid, respectively. When the magnetic field is present, the energy density includes the magnetic energy $B^2/8\pi$ and the energy flux density includes the Poynting vector $(c/4\pi)\mathbf{E} \times \mathbf{B}$. Expressing the electric field in terms of the magnetic field, we obtain

$$\frac{\partial}{\partial t} \left(\frac{1}{2} \rho v^2 + \rho \epsilon + \frac{B^2}{8\pi} \right) = -\nabla \cdot \mathbf{q}, \qquad (9.160)$$

with

$$q^a = \rho \left(\frac{1}{2} v^2 + w \right) v^a + \frac{1}{4\pi} [\mathbf{B} \times (\mathbf{v} \times \mathbf{B})]^a$$

$$- \frac{c^2}{16\pi \sigma} [\mathbf{B} \times (\nabla \times \mathbf{B})]^a - v_b \sigma^{ab} - \kappa \frac{\partial T}{\partial x^a}. \qquad (9.161)$$

These equations become somewhat simplified if we assume that the fluid is incompressible. In this case, the complete set of equations reduces to

$$\nabla \cdot \mathbf{B} = 0, \quad \nabla \cdot \mathbf{v} = 0, \qquad (9.162)$$

$$\frac{\partial \mathbf{B}}{\partial t} + (\mathbf{v} \cdot \nabla) \mathbf{B} = (\mathbf{B} \cdot \nabla) \mathbf{v} + \frac{c^2}{4\pi \sigma} \nabla^2 \mathbf{B}, \qquad (9.163)$$

$$\frac{\partial \mathbf{v}}{\partial t} + (\mathbf{v} \cdot \nabla) \mathbf{v} = -\frac{1}{\rho} \nabla \left(p + \frac{B^2}{8\pi} \right) + \frac{1}{4\pi \rho} (\mathbf{B} \cdot \nabla) \mathbf{B} + \nu \nabla^2 \mathbf{v}, \qquad (9.164)$$

where $v = (\eta/\rho)$. Equation (9.158) is not needed for an incompressible flow unless we are specifically interested in the temperature distribution.

The equation for a magnetic field, (9.149) or (9.163), shows that the field $\mathbf{B}$ at any location changes because of two different processes. The term $\nabla \times (\mathbf{v} \times \mathbf{B})$, which is of the order of (vB/L), where L is the scale over which the magnetic field changes, makes the magnetic field flow along with the fluid. (This process is usually called advection.) The second term $(c^2/4\pi\sigma)\nabla^2 B$ has the order of magnitude $(c^2/4\pi\sigma)(B/L^2)$ and makes the magnetic field diffuse through the fluid. The ratio between these two terms defines the magnetic Reynold's number:

$$R_M = \frac{4\pi\sigma vL}{c^2}.$$ (9.165)

Clearly, when $R_M \gg 1$ the advection term dominates and when $R_M \ll 1$ the diffusion dominates. In the first case we can ignore the diffusion term and obtain

$$\frac{\partial \mathbf{B}}{\partial t} - \nabla \times (\mathbf{v} \times \mathbf{B}) = 0.$$ (9.166)

From the results of Chap. 8, Exercise 8.2, we can interpret this situation as the magnetic field being frozen in the field and carried away with it. On the other hand, when $R_M \ll 1$, we ignore the advection term and obtain the diffusion equation:

$$\frac{\partial \mathbf{B}}{\partial t} = \frac{c^2}{4\pi\sigma}\nabla^2 \mathbf{B} \simeq \frac{c^2}{4\pi\sigma L^2}\mathbf{B}.$$ (9.167)

This shows that the diffusion time scale is $t_{\text{diff}} \simeq 4\pi\sigma L^2/c^2$.

In many astrophysical contexts, the magnetic diffusion can be ignored and the magnetic field can be treated as frozen into the fluid and described by Eq. (9.166). For example, in the case of intergalactic plasma $\sigma \approx 3 \times 10^{14}\,\text{s}^{-1}$, $v \approx 200\,\text{km s}^{-1}$, and $L \approx 10$ kpc; this gives $R_m \approx 3 \times 10^{25} \gg 1$ and $t_{\text{diff}} \approx 4 \times 10^{39}$ s. In contrast, for a plasma in the laboratory, $\sigma \approx 10^{16}\,\text{s}^{-1}$, $v \approx 10\,\text{m s}^{-1}$, and $L \approx 3$ cm, giving $R_m \simeq 4$ and $t_{\text{diff}} \approx 10^{-3}$ s. This shows that diffusion can be important for laboratory plasma but not for astrophysical cases.

9.6.2 Batteries and Generation of the Magnetic Field

It should be noted that MHD Eq. (9.149) has no source term for the magnetic field; in other words, if $\mathbf{B} = 0$ initially, then $(\partial \mathbf{B}/\partial t) = 0$ initially and hence the magnetic field will never grow. Because of this fact, the generation of magnetic fields in different astrophysical objects has been a difficult problem. With future applications in mind, we derive the source term to the magnetic field that arises when the electron and the ion motions are treated as different to the lowest order of approximation.

In this case, the equations of motion for electron and proton fluids in a fully ionised hydrogen gas, say, are given by

$$\frac{d\mathbf{v}_e}{dt} = -\frac{\nabla p_e}{m_e n_e} - \frac{e}{m_e}\left(\mathbf{E} + \frac{\mathbf{v}_e \times \mathbf{B}}{c}\right) - \nabla\phi - \frac{(\mathbf{v}_e - \mathbf{v}_p)}{\tau}, \quad (9.168)$$

$$\frac{d\mathbf{v}_p}{dt} = -\frac{\nabla p_p}{m_p n_p} - \frac{e}{m_p}\left(\mathbf{E} + \frac{\mathbf{v}_p \times \mathbf{B}}{c}\right) - \nabla\phi + \left(\frac{m_e}{m_p}\right)\left(\frac{n_e}{n_p}\right)\frac{(\mathbf{v}_e - \mathbf{v}_p)}{\tau}.$$

$$(9.169)$$

The first terms on the right-hand sides arise from the pressure gradient of the respective fluids and the last terms are a phenomenological description of the collisions between the electrons and protons, which lead to finite conductivity. Subtracting the second equation from first and using the fact that $m_e \ll m_p$, we get

$$\frac{d}{dt}(\mathbf{v}_e - \mathbf{v}_p) = -\frac{\nabla p_e}{m_e n_e} - \frac{e}{m_e}\left(\mathbf{E} + \frac{\mathbf{v}_e \times \mathbf{B}}{c}\right) - \frac{(\mathbf{v}_e - \mathbf{v}_p)}{\tau}. \quad (9.170)$$

Introducing the current by the definition

$$\mathbf{J} = en_p\mathbf{v}_p - en_e\mathbf{v}_e = en(\mathbf{v}_p - \mathbf{v}_e), \quad n_e = n_p = n, \quad (9.171)$$

taking the curl of this equation, and rewriting the resulting term, we can easily show that

$$\frac{\partial \mathbf{B}}{\partial t} = c\nabla \times \left(\frac{\nabla p_e}{en_e}\right) + \nabla \times (\mathbf{v} \times \mathbf{B}) + \frac{c^2}{4\pi\sigma}\nabla^2 B, \quad \sigma = \frac{ne^2\tau}{m_e}. \quad (9.172)$$

This equation is identical to MHD Eq. (9.149) except for the first term on the right-hand side, which can act as the source term for the magnetic field. Such a production of a magnetic field is called a battery mechanism for the magnetic field.

9.6.3 Ambipolar Diffusion

The equation for the magnetic field also is modified if there is significant neutral component in the plasma. In such a case, the velocity of ions and the neutral particles will not be the same because the ions feel the Lorentz force from the magnetic field whereas the neutrals do not. However, because ions and neutrals are coupled together by normal molecular collisions in a fluid, there will arise a frictional drag force between these particles that can cause a diffusion of the magnetic-field lines, called ambipolar diffusion. To derive this effect, we begin by considering the rate of collisions of ions with neutrals, which is given by $n_i\langle v\sigma\rangle$, where σ is the elastic-scattering cross section for the collisions and v is the relative velocity. Such collisions lead to a rate of momentum transfer per unit volume given by

$$n_n \frac{m_n m_i}{(m_n + m_i)} (\mathbf{u}_i - \mathbf{u}_n) n_i \langle v\sigma \rangle, \tag{9.173}$$

where the subscripts i and n denote the variables pertaining to ions and neutral particles, respectively. This can be expressed as a frictional drag force per unit volume,

$$\mathbf{f}_d = \gamma \rho_n \rho_i (\mathbf{u}_i - \mathbf{u}_n), \tag{9.174}$$

where ρ denotes the mass density and

$$\gamma = \frac{\langle v\sigma \rangle}{(m_n + m_i)} \tag{9.175}$$

is a drag coefficient. In encounters with low relative velocities, an ion can induce a dipole moment in the neutral atom, thereby enhancing the effective cross section. The electric field $\mathbf{E} = (Zq/r^2)\hat{\mathbf{r}}$ that is due to an ion of charge Zq located at a distance r from the neutral atom will induce a dipole moment $\mathbf{p} = \alpha \mathbf{E}$ in the neutral atom. Clearly, the direction of $\mathbf{p}$ is along the line connecting the ion and the neutral atom. Using the general expression for the dipole potential given by conditions (3.146) and taking the gradient, we find the electric field of the polarised atom at the vicinity of the ion as

$$\mathbf{E}_{\text{pol}} = \frac{3\mathbf{n}(\mathbf{p} \cdot \mathbf{n}) - \mathbf{p}}{r^3} = -\frac{2\mathbf{p}}{r^3}, \qquad \mathbf{n} = -\frac{\mathbf{r}}{|r|}. \tag{9.176}$$

This causes a force $\mathbf{F} = Zq\mathbf{E}_{\text{pol}} = -2(Z^2 q^2 \alpha / r^5)\hat{\mathbf{r}}$. The corresponding potential energy of the interaction between the polarised atom and the ion is $U = -(Z^2 q^2 \alpha / 2r^4)$. The motion of the ion and the polarised atom under the action of such a potential can be determined in a straightforward manner by ordinary classical mechanics. In our case, we are interested in the ions scattering off the neutral atom that can be characterised by (1) a distance of closest approach r_p, (2) impact parameter b, and (3) initial velocity v. Conservation of energy and angular momentum gives the conditions

$$\frac{1}{2}\mu v^2 = \frac{1}{2}\mu v_p^2 - \frac{Z^2 q^2 \alpha}{2r_p^4}, \qquad \mu v b = \mu v_p r_p, \tag{9.177}$$

where $\mu = m_i m_n / (m_i + m_n)$ is the reduced mass and v_p is the speed at $r = r_p$. These equations can be easily solved to give the distance of closest approach as

$$r_p = \left\{ \frac{1}{2} \left[b^2 + \left(b^4 - \frac{4Z^2 q^2 \alpha}{\mu v^2} \right)^{1/2} \right] \right\}^{1/2}. \tag{9.178}$$

It is clear that $r_p \to 0$ when $b \to b_0 \equiv (4Z^2 q^2 \alpha / \mu v^2)^{1/4}$. Because most of the significant effects will occur at very close encounters we may take the effective scattering cross sections to be

$$\sigma = \pi b_0^2 = \frac{2\pi |Z| q}{v} \left(\frac{\alpha}{\mu} \right)^{1/2} \propto \frac{1}{v}. \tag{9.179}$$

Because $\sigma \propto (1/v)$, the rate of collisions $\langle \sigma v \rangle = 2\pi |Z^3| (\alpha/\mu)^{1/2}$ is independent of v. In the case of hydrogen atoms and protons, $\mu = (m_p/2)$ and $\alpha \approx (4\pi/3) a_0^3 \approx 6 \times 10^{-25}$ cm^3. This gives $\langle \sigma v \rangle \approx 3 \times 10^{-9}$ cm^3 s^{-1}, and the resulting value for γ turns out to be approximately $\gamma = 10^{14}$ cm^3 gm^{-1} s^{-1} = 10^{11} m^3 kg^{-1} s^{-1}.

In the opposite limit of fast encounters, σ is essentially determined by the geometric cross section and γ becomes

$$\gamma \approx \frac{4\pi (r_n + r_i)^2}{(m_n + m_i)} |\mathbf{u}_i - \mathbf{u}_n|. \tag{9.180}$$

In such a case, the drag force will vary quadratically with the relative velocity; but in most astrophysical systems, such high relative velocities (greater than ~ 10 km s^{-1}) seem unlikely.

The drag force by the neutrals can be significant in systems with very low fractional ionisation. In such systems, the Lorentz force $(4\pi)^{-1} [(\nabla \times \mathbf{B}) \times \mathbf{B}]$ and the drag force – being the dominant components – must balance each other. This condition gives the drift velocity of ions as

$$\mathbf{v}_d \equiv \mathbf{u}_i - \mathbf{u}_n = \frac{1}{4\pi \gamma \, \rho_n \, \rho_i} (\nabla \times \mathbf{B}) \times \mathbf{B}. \tag{9.181}$$

This is of the order of $v_d \approx v_A^2 / V$, where $v_A^2 \equiv B^2 / 4\pi \rho$ is called the Alfven speed (see Subsection 9.7.1) and $V = \gamma \rho_i L$ if the scale of variation of the magnetic field is L. In systems such as molecular clouds, $V \approx 6$ km s^{-1} and $v_A \approx 0.4$ km s^{-1}. If the magnetic field is frozen in the fluid of ions and electrons, it satisfies the equation

$$\frac{\partial \mathbf{B}}{\partial t} + \nabla \times (\mathbf{B} \times \mathbf{u}_i) = 0. \tag{9.182}$$

To determine the flow of a magnetic field with respect to the neutrals, we express $\mathbf{u}_i$ in terms of $\mathbf{u}_n$ by using Eq. (9.181), obtaining

$$\frac{\partial \mathbf{B}}{\partial t} + \nabla \times (\mathbf{B} \times \mathbf{u}_i) = \nabla \times \left\{ \frac{\mathbf{B}}{4\pi \rho_n \, \rho_i} \times [\mathbf{B} \times (\nabla \times \mathbf{B})] \right\}. \tag{9.183}$$

The right-hand side describes the slow diffusion of the magnetic field with respect to the neutrals with a diffusion coefficient $D \approx v_A^2 / (\gamma \rho_i)$.

The ambipolar diffusion also affects the flow of the neutrals. The Euler equation should now be modified by the addition of the drag force term $\mathbf{f_D}$ and the equation for heat balance should include the work done by this force. Because this force is essentially equal to the Lorentz force on the ions, the equations are the same as if the neutrals were also subjected to a Lorentz force. Of course, the origin of this force is due to the drag between charged components and neutrals by means of the molecular collisions.

9.6.4 Magnetic Virial Theorem

Finally, we mention the modifications to the virial theorem in the presence of a magnetic field. Equations (9.153) and (9.154) show that magnetic forces can be obtained from a stress tensor of the form

$$T_{ik} = \frac{B_i B_k}{4\pi} - \frac{|\mathbf{B}|^2}{8\pi} \delta_{ik}. \tag{9.184}$$

From the general form of the virial theorem obtained in Chap. 8, Section 8.2 [see Eqs. (8.25)] it is clear that magnetic field will make a contribution of

$$\mathcal{M} \equiv \int_V \frac{|\mathbf{B}|^2}{8\pi} d^3x \tag{9.185}$$

to the virial. In steady state, we now have the condition $2K + U + 3pV + \mathcal{M} = 0$.

To see the effect of the new term, let us consider the collapse of a cloud of mass M, radius R, and temperature T threaded through with a magnetic field B corresponding to a flux $\Phi = \pi R^2 B$. Assuming that the magnetic field is frozen to the fluid lines, the flux will be conserved during the collapse or expansion of the fluid. The magnetic contribution to the virial $\mathcal{M} \propto B^2 R^3 = \beta \Phi^2 / R$ where $\beta \approx (1/6\pi^2)$. Repeating the analysis of Chap. 8, Section 8.5, with magnetic contribution included, we find that the external pressure needed to confine the cloud is now given by

$$p = \frac{1}{4\pi} \left(-\alpha \frac{GM^2}{R^4} + \beta \frac{\Phi^2}{R^4} + 3 \frac{a^2 M}{R^3} \right). \tag{9.186}$$

We note that both the magnetic field and the gravity have the same R dependence but contribute with different signs. The net sign of the term proportional to R^{-4} depends on the ratio of the actual mass of the cloud M to a critical magnetic mass defined as $M_\Phi \equiv (\beta/\alpha)^{1/2} G^{-1/2} \Phi$. Numerical simulations show that $(\beta/\alpha)^{1/2} \approx 0.13$, which is close to the naive estimate of $(1/6\pi^2)^{1/2}$. Numerically

$$M_\Phi \approx 0.13 G^{-1/2} \Phi \simeq 10^3 M_\odot \left(\frac{B}{30 \, \mu G} \right) \left(\frac{R}{2 \, pc} \right)^2. \tag{9.187}$$

With this definition, the pressure needed to confine the cloud becomes

$$p = \frac{1}{4\pi} \left[\frac{\alpha G}{R^4} (M_\phi^2 - M^2) + 3 \frac{\alpha^2 M}{R^3} \right]. \tag{9.188}$$

From this expression it is clear that both terms are positive if the mass of the cloud is less than the magnetic critical mass M_Φ. In such a case, a collapse of the cloud is not possible. If $M > M_\Phi$, then collapse is possible but the magnetic field acts to dilute the effects of self-gravity. Therefore magnetic fields increase the resistance of the fluids to collapse if the field lines are frozen into the fluid.

9.7 Hydromagnetic Waves

Let us next consider the propagation of small disturbances in a fluid that is governed by MHD equations. We assume that the background solution is that of a fluid that has zero velocity, is homogeneous, and has a uniform constant magnetic field $\mathbf{B}_0$, and we ignore the effects of dissipative processes such as viscosity, thermal conductivity, and electrical resistivity of the medium. In such a case, the basic MHD equations become

$$\nabla \cdot \mathbf{B} = 0, \tag{9.189}$$

$$\frac{\partial \mathbf{B}}{\partial t} = \nabla \times (\mathbf{v} \times \mathbf{B}), \tag{9.190}$$

$$\frac{\partial \rho}{\partial t} + \nabla \cdot (\rho \mathbf{v}) = 0, \tag{9.191}$$

$$\frac{\partial \mathbf{v}}{\partial t} + (\mathbf{v} \cdot \nabla)\mathbf{v} = -\frac{1}{\rho}\nabla p + \frac{1}{4\pi\rho}(\nabla \times \mathbf{B}) \times \mathbf{B}. \tag{9.192}$$

The heat equation reduces to the conservation of energy in this particular case. Because the medium is homogeneous, we can take $s = $ constant. We write the perturbed quantities as

$$\mathbf{B} = \mathbf{B} + \mathbf{h}, \quad \rho = \rho + \rho', \quad p = p + p', \tag{9.193}$$

where the primed quantities denote the small perturbations around the unprimed background values. The velocity $\mathbf{v}$ (which is zero in the equilibrium configuration) will also be treated as a quantity of first-order smallness, although it is denoted without a prime. Because the flow is isentropic, $p' = (\partial p/\partial\rho)_s\rho' = u_0^2\rho'$, where u_0 denotes the speed of sound in the unperturbed medium. Linearising the equations to the lowest order in perturbed quantities, we obtain

$$\nabla \cdot \mathbf{h} = 0, \quad \frac{\partial \mathbf{h}}{\partial t} = \nabla \times (\mathbf{v} \times \mathbf{B}),$$

$$\frac{\partial \rho'}{\partial t} + \rho\nabla \cdot \mathbf{v} = 0, \tag{9.194}$$

$$\frac{\partial \mathbf{v}}{\partial t} = -\frac{u_0^2}{\rho}\nabla\rho' - \frac{1}{4\pi\rho}[\mathbf{B} \times (\nabla \times \mathbf{h})].$$

As usual, we look for solutions that are proportional to $\exp[i(\mathbf{k} \cdot \mathbf{x} - \omega t)]$. For such modes Eqs. (9.194) reduce to

$$-\omega \mathbf{h} = \mathbf{k} \times (\mathbf{v} \times \mathbf{B}), \quad \omega \rho' = \rho \mathbf{k} \cdot \mathbf{v}, \quad -\omega \mathbf{v} + \frac{u_0^2}{\rho} \rho' \mathbf{k} = -\frac{1}{4\pi\rho}[\mathbf{B} \times (\mathbf{k} \times \mathbf{h})].$$
$$(9.195)$$

Let the propagation direction specified by $\mathbf{k}$ be the x axis and let the plane defined by $\mathbf{k}$ and $\mathbf{B}$ be the x–y plane, with the angle between the two vectors being ψ. We also introduce the phase velocity of the wave $u = (\omega/k)$. Eliminating ρ' in the third equation of Eqs. (9.194) by using the second and writing the result in terms of the components, we get

$$u h_z = -v_z B_x, \quad u v_z = -\frac{B_x h_z}{4\pi\rho}, \tag{9.196}$$

$$u h_y = v_x B_y - v_y B_x, \quad u v_y = -\frac{B_x h_y}{4\pi\rho}, \quad v_x\left(u - \frac{u_0}{u}\right) = \frac{B_y h_y}{4\pi\rho}. \tag{9.197}$$

These equations clearly separate into two groups, with the first set involving only h_z and v_z and the second set involving only v_x, h_y, and v_y. It follows that these perturbations are independent and describe two different kinds of waves that can be sustained in MHD. Because the density perturbations satisfy $\rho' = (\rho v_x/u)$, it is the second set that generates density and pressure variations. We now discuss the properties of each of these waves separately.

9.7.1 Alfven Waves

Consider first the modes of the first set with $v_z \neq 0$ and $v_x = v_y = 0$. The compatibility of Eqs. (9.196) gives the dispersion relation

$$u \equiv u_1 = \frac{B_x}{\sqrt{4\pi\rho}}, \tag{9.198}$$

which can be written in vector notation as

$$\omega = \frac{\mathbf{B} \cdot \mathbf{k}}{\sqrt{4\pi\rho}}. \tag{9.199}$$

The quantity $v_A = (B^2/4\pi\rho)^{1/2}$ is called the Alfven speed. Note that $(B^2/4\pi\rho) = (B^2 V/4\pi)(\rho V)^{-1} = (T/M)$ is the ratio between the total magnetic tension $T = V(B^2/4\pi)$ in a volume V and the mass $M = \rho V$ contained in the same volume. This clearly shows that the restoring force arises from the magnetic tension of the field lines. The phase velocity of these waves (called Alfven waves) is given by $u = v_A \cos \psi$, and the group velocity $(\partial \omega / \partial k) = (\mathbf{B}/\sqrt{4\pi\rho})$ is independent of the direction of $\mathbf{k}$.

The phase velocity of an Alfven wave is along the wave vector $\mathbf{k}$ whereas the direction of propagation of the wave, in the sense of group velocity, is along the magnetic field. In this sense, all physical quantities such as the wave energy, etc., can be transported by an Alfven wave along the magnetic-field lines irrespectively of the direction of propagation specified by the wave vector $\mathbf{k}$. It is also clear that Alfven waves are transverse waves because the fluid displacement (which is along the z direction) is perpendicular to both $\mathbf{k}$ and $\mathbf{B}$.

9.7.2 Fast and Slow Megnetohydrodynamic Waves

Let us next consider modes in which $v_z = 0$. The compatibility of Eqs. (9.197) now requires that

$$\left(u^2 - u_0^2\right)\left(u^2 - \frac{B_x^2}{4\pi\rho}\right) = \frac{u^2 B_y^2}{4\pi\rho}. \tag{9.200}$$

Because the roots of a quartic equation $x^4 + px^2 + q = 0$ can be written as

$$x = \pm \frac{1}{2}\left\{\sqrt{(-p + 2\sqrt{q})} \pm \sqrt{(-p - 2\sqrt{q})}\right\}, \tag{9.201}$$

the roots of the above equation become

$$u_{2,3}^2 = \frac{\omega^2}{k^2} = \frac{1}{2}\left\{(v_A^2 + u_0^2) \pm [(v_A^2 + u_0^2)^2 - 4v_A^2 u_0^2 \cos^2 \psi]^{1/2}\right\}. \tag{9.202}$$

These represent two different waves; the upper sign, which gives a larger wave speed, yields the *fast* MHD wave and the lower sign gives the *slow* MHD wave. To understand their properties, it is convenient to consider two extreme cases with $\cos^2 \psi = 1$ and $\cos^2 \psi = 0$ separately.

When $\sin \psi = 0$ and $\cos \psi = \pm 1$, $\mathbf{k}$ is parallel to $\mathbf{B}$ and the dispersion relation becomes

$$\frac{\omega^2}{k^2} = \frac{1}{2}[(v_A^2 + u_0^2) \pm |v_A^2 - u_0^2|] = (v_A^2 \text{ or } u_0^2). \tag{9.203}$$

Here, the fast MHD wave is faster than an acoustic wave whereas the slow MHD wave is slower than the acoustic wave or an Alfven wave. In this particular case, the perturbations are characterised by the ratio

$$\rho' : v_x : v_y : h = 0 : 0 : \frac{\omega}{k} : \mp 1 \tag{9.204}$$

with the dispersion relation $\omega^2 = k^2 v_A^2$ for the Alfven-type wave. On the other hand, for the acoustic mode with $\omega^2 = k^2 u_0^2$, the corresponding ratios are

$$\rho':v_x:v_y:h = 1:\frac{\omega}{k}:0:0. \tag{9.205}$$

The Alfven-type wave is a transverse disturbance involving no change in density whereas the acoustic one is a longitudinal wave involving no compression of the magnetic field.

When $\cos\psi = 0$ and $\sin\psi = \pm 1$, $\mathbf{k}$ is perpendicular to $\mathbf{B}$ and the dispersion relation becomes

$$\frac{\omega^2}{k^2} = \left(v_A^2 + u_0^2 \quad \text{or } 0\right). \tag{9.206}$$

The slow MHD wave has now disappeared and the fast wave represents alternative compressions and rarefactions of both gas and the magnetic field. For the mode with the dispersion relation $\omega^2 = k^2(u_0^2 + v_A^2)$, the ratio of the amplitudes is given by

$$\rho':v_x:v_y:h = 1:\frac{\omega}{k}:0:\pm 1. \tag{9.207}$$

This is a longitudinal disturbance with compressions in both density and the magnetic field. For the zero-frequency mode, we still get the nontrivial ratios

$$\rho':v_x:v_y:h = 1:0:\frac{u_0}{v_A}\left(u_0^2 + v_A^2\right)^{1/2}:\mp\frac{u_0^2}{v_A^2}. \tag{9.208}$$

This mode still represents a transverse disturbance of the system that compresses the matter and displaces the field.

At intermediate angles, these waves are neither purely longitudinal nor purely transverse. Neither are they purely compressional or distortional. The general case is fairly complicated but does not contain any essentially new physics. In the limiting case of $B \ll (4\pi\rho u_0^2)$, we have $u_2 \approx u_0$ and $v_y \ll v_x$. These are ordinary sound waves propagated with velocity u_0. In this limit, $u_3 \approx u_1$ and the modes satisfy $v_x \approx 0$, $v_y \approx -(h_y/\sqrt{4\pi\rho})$. This is reminiscent of Alfven waves, but the vectors $\mathbf{v}$ and $\mathbf{h}$ are located in the plane defined by $\mathbf{k}$ and $\mathbf{B}$ instead of being perpendicular to the plane. In an incompressible fluid (which corresponds to the limit of $u_0 \to \infty$), only one type of wave occurs with two independent polarisations. The dispersion relation is that of Alfven waves and $\mathbf{v}$ and $\mathbf{h}$ are perpendicular to $\mathbf{k}$; they are related by $\mathbf{v} = -(\mathbf{h}/\sqrt{4\pi\rho})$.

It may be noted that the inequalities

$$u_3 \le u_1 \le u_2, \quad u_2 \ge u_0, \quad u_3 \le u_0, \tag{9.209}$$

always hold. Figure 9.1 shows the characteristics of the wave propagation in diagram form. If all the three kinds of waves are emitted from the origin at

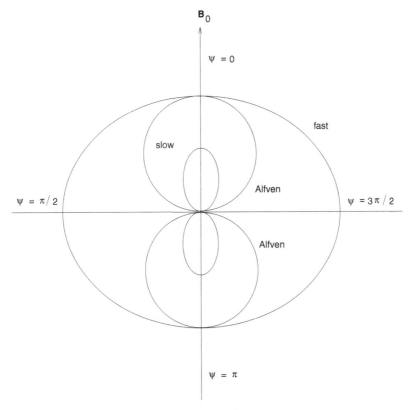

Fig. 9.1.   Polar diagram showing the propagation characteristics of fast, slow, and Alfven waves.

$t = 0$, then the Alfven, slow, and fast modes would have constant phases at the respective curves at a later time.

The wave solutions obtained above correspond to small perturbations and hence are not exact solutions. It is, however, possible to find *exact* solutions to the MHD equations that are in the form of propagating Alfven waves. These solutions exist in different polarisation states for the Alfven waves, and we consider a circularly polarised case, called torsional Alfven waves. To find these solutions, it is convenient to work in cylindrical coordinates (R, ϕ, z), where z is the direction of propagation. Such a wave solution will correspond to having all physical variables as functions of $z \pm v_A t$. Let us consider the following ansatz,

$$\rho = \text{constant}, \quad \mathbf{v} = R\Omega(z, t)\mathbf{e}_\phi, \quad \mathbf{B} = B_0\mathbf{e}_z + \beta\rho\mathbf{v}, \qquad (9.210)$$

as a possible solution to Eqs. (9.189)–(9.192). This solution is parameterised by three constants, ρ, B_0, and β. It can be directly verified that the above ansatz satisfies $\nabla \cdot \mathbf{v} = \nabla \cdot \mathbf{B} = 0$. Because dissipative effects are ignored, Eq. (9.166), which is applicable, has only one nontrivial component:

$$\beta \frac{\partial \Omega}{\partial t} = B_0 \frac{\partial \Omega}{\partial z}. \tag{9.211}$$

The three components of force Eq. (9.192) now becomes

$$-\rho R \Omega^2 = -\frac{\partial p}{\partial R} - \frac{\beta^2}{4\pi} 2 R \Omega^2, \tag{9.212}$$

$$\rho R \frac{\partial \Omega}{\partial t} = \frac{\beta B_0}{4\pi} R \frac{\partial \Omega}{\partial z}, \tag{9.213}$$

$$0 = -\frac{\partial p}{\partial z} - \frac{\beta^2}{4\pi} R^2 \Omega \frac{\partial \Omega}{\partial z}. \tag{9.214}$$

Equation (9.214) can be immediately integrated to give

$$p = -\frac{\beta^2}{8\pi} R^2 \, \Omega^2(z, t) + \mathcal{P}(R, t), \tag{9.215}$$

where $\mathcal{P}$ is an arbitrary function of t and R. However, substitution of this equation into Eq. (9.212) will require that (1) $\mathcal{P}$ be a constant and (2) $\beta = \pm (4\pi\rho)^{1/2}$ in order for the equation to be satisfied for all z. The magnetic field now has the form

$$\mathbf{B} = B_0 \left[\mathbf{e}_z \pm \frac{R\Omega(z, t)}{v_A} \mathbf{e}_\phi \right], \tag{9.216}$$

i.e., the ratio of the perturbed magnetic field to the original magnetic field is the same as that of the transverse fluid speed to the Alfven speed. We found in Section 9.6 that the same relation holds for linear Alfven solutions as well. The remaining equations now can be reduced to a single one that corresponds to

$$\frac{\partial \Omega}{\partial t} = \pm v_A \frac{\partial \Omega}{\partial z}, \tag{9.217}$$

which has the solution $\Omega(z, t) = F(z \pm v_A t)$, where F is an arbitrary function of its argument. It is clear that the perturbed magnetic field and other physical parameters have the same wavelike dependence. This solution represents a nonspreading wave form of arbitrary strength propagating along the z direction. The velocity field is clearly rotational and can carry angular momentum. This process can help a plasma threaded by a magnetic field to get rid of this angular momentum by means of Alfven waves.

In the discussion of nonmagnetic fluids, we noted that the steepening of sound waves can lead to the formation of shocks. In the presence of magnetic fields, more complicated structures of shocks can arise. Alfven wave solutions exist even for large amplitudes and hence can propagate without steepening in a medium, whereas slow and fast magnetosonic modes do steepen to form shocks. The jump condition at the shock front can, in principle, be derived by the usual continuity arguments. They are, however, now considerably complicated because of the existence of different magnetic fields on the two sides and by the fact that flow velocity may be at an angle to the magnetic field at the shock surface.

10

Gravitational Dynamics

10.1 Introduction

This chapter deals with the dynamics of systems dominated by gravitational force. Self-gravitating gaseous spheres and models of stellar and galactic systems are covered in this chapter. This will be needed in the study of compact stellar remnants and globular clusters in Vol. II and galactic dynamics in Vol. III.

10.2 Gravitational Interaction in Astrophysical Systems

Gravity plays a vital role in the nature and dynamics of several astrophysical systems. The actual description of such systems, however, varies considerably depending on the physical circumstance. We first discuss the different forms of description that are appropriate for various systems.

To begin with, it must be noted that the fundamentally correct description of gravitational interaction is fairly complex and requires a general theory of relativity, which will be developed in Chap. 11. General relativistic effects are important when the mass scale M and the length scale L involved in the problem are such that (GM/c^2L) is not negligible compared with unity. Such a situation arises in the final stages of stellar collapse and in cosmology. These two topics will *not* be covered in this chapter and we shall assume that the description based on Newtonian gravity is adequate.

In Newtonian theory, the gravitational force can be described as a gradient of a scalar potential. In that case, the evolution of a set of particles under the action of gravitational (and other) forces can be described by

$$\ddot{\mathbf{x}}_i = -\nabla\phi(\mathbf{x}_i, t) + m_i^{-1}\mathbf{F}, \quad \nabla^2\phi = 4\pi G \sum_i m_i \delta_D(\mathbf{x} - \mathbf{x}_i) \quad (10.1)$$

where $\mathbf{x}_i$ is the position of the ith particle, m_i is its mass, and $\mathbf{F}$ stands for all other (nongravitational) forces acting on the particle. For sufficiently large number of particles, it is useful to investigate whether some kind of statistical description of such a system is possible. The statistical mechanics of such a system is complicated by the following facts: (1) Gravitational force is always attractive and (2) it has a long range. While discussing physical processes in a plasma in Chap. 9, it was necessary to deal with electromagnetic forces that are also intrinsically long range. However, in a plasma, the long-range nature of Coulomb interaction undergoes Debye shielding because of the presence of charges of two different signs. This does not occur in a gravitating system because the masses of particles are always positive.

The long-range unscreened nature of gravity implies that the force acting on any given particle receives a contribution even from particles that are very far away. As a consequence of the attractive and long-range nature of gravity, it is not possible to apply the principles of statistical mechanics developed in Chap. 5 to such a system. If a self-gravitating system is divided into two parts, the total energy of the system cannot be expressed as the sum of the gravitational energy of the components. The formalism developed in Chap. 5 is based on the extensivity of the energy that is clearly not valid for gravitating systems. To make progress, we have to use different techniques that are appropriate for each situation.

The simplest context, in which gravitational effects can be taken into account fairly easily, corresponds to self-gravitating gaseous systems dominated by molecular collisions. For such systems, the mean free path – determined by molecular collisions – is much smaller than the scale over which other parameters vary. In that case, the gravitational field can be incorporated by the addition of the term $-\nabla \phi$ to the (Euler) force equation. This has already been done in Chap. 8 and poses no specific difficulty. The presence of molecular collisions leads to a small mean free path and a well-defined notion of collisional pressure for the system, and gravity plays the role of only a smooth force field in the evolution equation for the distribution function.

The situation becomes more complex when the system does not have molecular collisions (or some equivalent process). For example, in the study of a system of stars in a galaxy moving under their mutual gravitational field, physical collision between the stars is a very rare phenomenon. The mean free path for the physical collisions is much larger than the size of the system, and the only force that needs to be taken into account is that arising because of gravity. A very different description is required for such a system (compared with, for example, the case of a gaseous, self-gravitating star) and is constructed along the following lines.

To the lowest order of approximation, the gravitational potential at any given point can be approximated as that due to a smooth distribution of stars with a spatially varying density distribution. Each star may be assumed to move in a specific orbit in this mean, smooth potential. If the system is in steady state over the time scales of interest, then it is possible to populate the orbits in the potential in such a way as to provide a self-consistent solution. This is the approach that is usually taken in the description of systems such as galaxies, which are made up of a collection of stars moving under their mutual gravitational attraction.

The actual gravitational potential acting on a star at any given instant will, of course, be different from the smooth potential that is due to the granularity in the system. Hence the actual orbit of the star will deviate from the orbit calculated based on the smooth potential. Over a period of time, these deviations accumulate to make the description based on the smooth potential inadequate. As we shall see in Section 10.7, the time scale for this to happen is $t_c \approx (N/\ln N)t_{orbit}$, where t_{orbit} is the typical dynamical time scale of orbits in the smooth potential and N is the number of particles in the system. For $t \ll t_c$, gravitating systems can be described by a smooth potential. When $t \gtrsim t_c$, it is necessary to take the above effect into account and the evolution of gravitating systems becomes quite different.

This process can be thought of as an analogue of collisions in gaseous systems to the extent that this process can be described in statistical terms. The time scale t_c is usually called the gravitational collision time; the evolution of systems for $t \ll t_c$ is called collisionless and for $t \gtrsim t_c$ is called collisional. Although this terminology will be used in this chapter, it should be stressed that the word collisions used in this context should be interpreted with caution. There are significant differences between the process outlined above and the familiar molecular collisions.

We now take up the detailed study of gravitating systems in all the three different contexts described above.

10.3 Self-Gravitating Barotropic Fluids

We begin with a description of the simplest of the three astrophysical contexts described above, namely that of a self-gravitating fluid. The equations governing such a system have already been discussed in Chap. 8. General time-dependent solutions to these equations are impossible to obtain; however, some amount of progress can be achieved in the study of equilibrium configurations of such fluids, in which the gravitational force is balanced by the pressure. The relevant equation describing such a nonrotating fluid is [see Eq. (8.93)]

$$\frac{1}{r^2}\frac{d}{dr}\left(\frac{r^2}{\rho}\frac{dP}{dr}\right) = -4\pi G\rho, \tag{10.2}$$

where P and ρ are the pressure and the density of the fluid, respectively. In the astrophysical contexts within which this equation is useful, it is possible to treat the fluid as barotropic and provide an equation of state of the form $P = P(\rho)$. Given such an equation of state, Eq. (10.2) can be integrated to determine $P(r)$ and $\rho(r)$. We now discuss some important equations of state and their corresponding solutions.

10.3.1 Polytropic Equation of State

A particularly simple form of the barotropic fluid is given by $P = K\rho^{\gamma}$, where K and γ are constants. In this case, it is possible to use scaling arguments to obtain several important conclusions regarding the equilibrium configuration.

To do this, we transform the variables in Eq. (10.2) to an independent variable ξ and a dependent variable θ by the relations $\rho = \rho_c \theta^n$ and $r = \alpha \xi$, where $\rho_c = \rho(0)$ is the central density and the constants α and n are given by

$$\alpha = \left[\frac{n+1}{4\pi G} K \rho_c^{(1-n)/n} \right]^{1/2}, \quad \frac{1}{n} = \gamma - 1. \tag{10.3}$$

Straightforward algebra now converts Eq. (10.2) to a differential equation for $\theta(\xi)$:

$$\frac{1}{\xi^2} \frac{d}{d\xi} \left(\xi^2 \frac{d\theta}{d\xi} \right) = -\theta^n, \tag{10.4}$$

called the Lane–Emden equation. The boundary conditions for integrating this equation are as follows: From the definition, it is clear that $\theta = 1$ at $\xi = 0$ in order to produce the correct central density. Further, because the pressure gradient dP/dr has to vanish at the origin, it is necessary that $d\theta/d\xi = 0$ at $\xi = 0$. Given these two boundary conditions at the origin, the Lane–Emden equation can be integrated to find $\theta(\xi)$. The integration proceeds from the origin until a point $\xi = \xi_1$ at which θ vanishes for the first time. At this point, the density and the pressure vanish, showing that the surface of the gaseous sphere has been reached. All other physical variables can be expressed in terms of the behaviour of $\theta(\xi)$ near $\xi = \xi_1$.

Analytic solutions to Lane–Emden equations are known for the cases with $n = 0, 1, 5$ corresponding to $\gamma = \infty, 2, 6/5$. Denoting the solution for a particular value of n by $\theta_n(\xi)$, these analytic solutions are given by

$$n = 0, \quad \theta_0 = 1 - \xi^2/6;$$

$$n = 1, \quad \theta_1 = \frac{\sin \xi}{\xi}; \tag{10.5}$$

$$n = 5, \quad \theta_5 = (1 + \xi^3/3)^{-1/2}.$$

The solutions for $n = 0$ and $n = 1$ vanish at $\xi_1 = \sqrt{6}$ and $\xi_1 = \pi$, respectively. The case in which $n = 0$ corresponds to a sphere of constant density and the case in which $n = 1$ (corresponding to $\gamma = 2$) is of some relevance in modelling planetary interiors and brown dwarfs (see Vol. II). On the other hand, for $n = 5$, $\theta(\xi)$ does not vanish for any finite value of ξ and the matter distribution extends to infinity. Such a system is not of much relevance in describing gaseous spheres like stars but is used – in a different context – to model galaxies.

Using the transformation equation between the original variables and θ and ξ, we can express all other physical parameters in parametric form. The radius and the mass of the sphere are given by

$$R = \alpha \xi_1 = \left[\frac{(n+1)K}{4\pi G}\right]^{1/2} \rho_c^{(1-n)/2n} \xi_1, \tag{10.6}$$

$$M = 4\pi \alpha^3 \rho_c \left[-\xi^2 \frac{d\theta}{d\xi}\right]_{\xi=\xi_1} = 4\pi \left[\frac{(n+1)K}{4\pi G}\right]^{3/2} \rho_c^{(3-n)/2n} \left[-\xi^2 \frac{d\theta}{d\xi}\right]_{\xi=\xi_1}. \tag{10.7}$$

The mean density $\bar{\rho} = (3M/4\pi R^3)$ is related to the central density by

$$\bar{\rho} = \rho_c \left[-\frac{3}{\xi} \frac{d\theta}{d\xi}\right]_{\xi=\xi_i}. \tag{10.8}$$

The central pressure and the gravitational binding energy of this sphere are

$$P_c = \frac{GM^2}{R^4} \left[4\pi(n+1)\left(\frac{d\theta}{d\xi}\right)^2_{\xi=\xi_1}\right]^{-1}, \tag{10.9}$$

$$U_g \equiv \frac{1}{2} \int \rho\phi \, d^3\mathbf{x} = -\frac{3}{5-n} \frac{GM^2}{R}. \tag{10.10}$$

Eliminating ρ_c between Eqs. (10.6) and (10.7), we can find the mass–radius relationship for these systems:

$$M \propto R^{(3-n)/(1-n)}. \tag{10.11}$$

Scaling relations like expression (10.11) can be obtained more easily from dimensional arguments. In this particular case, by equating $\nabla P \simeq (P/R)$ to $(GM\rho/R^2)$ and using $P \propto \rho^{[1+(1/n)]}$, we get $\rho^{1/n} \propto (M/R)$. Now, by using $\rho \propto (M/R^3)$, we obtain the scaling $M \propto R^{(3-n)/(1-n)}$. This relation is independent of the detailed nature of the solution and reflects the fact that the Lane–Emden equation allows self-similar solutions. These results will be used extensively in Vol. II for the modelling of stars.

We will now provide a brief description of how models for gravitating systems, based on the polytropic equation, are constructed in practice. Let us assume that the equation of state $P = K\rho^\gamma = K\rho^{(n+1)/n}$ is given, thereby providing us with the numerical values of K and n. Knowing n, we can integrate the Lane–Emden equation from the origin to the first zero and obtain various properties related to $\theta(\xi)$. Some of the relevant combinations are listed in Table 10.1. To build a model, it is convenient to treat the central density ρ_c as a free variable. In that case, Eqs. (10.6)–(10.10) determine the radius, mass, central pressure, and the potential energy in terms of K, n, and ρ_c, and the model is determined in terms of one free variable ρ_c. If necessary, we can solve for ρ_c in terms of other parameters, say, M.

Table 10.1. Constants of Lane–Emden functions

n	ξ_1	$-\xi_1^2 \left(\frac{d\theta}{d\xi} \right)_{\xi = \xi_1}$	$\rho_c / \bar{\rho}$
0	2.4494	4.8988	1.0000
0.5	2.7528	3.7871	1.8361
1.0	3.14159	3.14159	3.28987
1.5	3.65375	2.71406	5.99071
2.0	4.35287	2.41105	11.40254
2.5	5.35528	2.18720	23.40646
3.0	6.89685	2.01824	54.1825
3.25	8.01894	1.94980	88.153
3.5	9.53581	1.89056	152.884
4.0	14.97155	1.79723	622.408
4.5	31.83646	1.73780	6189.47
4.9	169.47	1.7355	934800
5.0	∞	1.73205	∞

There are situations in which the value of K is not given a priori except that it is known to be a constant. Such a situation can arise, for example, if entropy has a constant value leading to $P \propto \rho^\gamma$; the proportionality constant is related to other parameters or boundary conditions of the problem and hence will not be known a priori. If this is the case, we can again follow the same procedure, but now treating ρ_c and K as *two* free parameters. If necessary, these two parameters can be solved for in terms of M and R, thereby expressing all other quantities in terms of mass and radius of the system.

Exercise 10.1

Jupiter and Saturn: The internal structure of planets like Jupiter or Saturn is thought to be described reasonably well by an equation of state $P = K\rho^2$, corresponding to $n = 1$, a polytrope that can be exactly solved. Use the known mass and radius of the planet to compute other physical parameters. Are the observed M and R values for Jupiter and Saturn consistent with this equation of state?

10.3.2 Degenerate Fermionic Systems

An important consequence of the scaling in expression (10.11) arises for degenerate fermionic systems that are self-gravitating. It was shown in Chap. 5 that the equation of state for such a system is well approximated by

$$P_{\text{NR}} = \frac{(3\pi^2)^{2/3}}{5} \frac{\hbar^2}{m_e} \left(\frac{Z}{A} \right)^{5/3} \left(\frac{\rho}{m_p} \right)^{5/3} \equiv \lambda_{\text{NR}} \rho^{5/3} \qquad (10.12)$$

in the nonrelativistic limit and by

$$P_R = \frac{1}{4}(3\pi^2)^{1/3}\hbar c \left(\frac{Z}{A}\right)^{4/3}\left(\frac{\rho}{m_p}\right)^{4/3} \equiv \lambda_R \rho^{4/3} \tag{10.13}$$

in the relativistic limit. These are clearly in polytropic form with the indices $\gamma = 5/3, 4/3$; the corresponding Lane–Emden indices are $n = (\gamma - 1)^{-1} = 3/2, 3$. When $n = 3/2$, expression (10.11) gives $M \propto R^{-3}$; in other words, the system has a smaller radius for a larger value of mass. The proportionality constant can be determined from the numerical results in Table 10.1 to give

$$MR^3 \cong \frac{92\hbar^6}{G^3 m_e^3 m_p^5}\left(\frac{Z}{A}\right)^5, \tag{10.14}$$

where m_e and m_p are the masses of electrons and protons, respectively, and Z and A are the atomic number and the atomic weight, respectively, of the material making up the sphere.

For $n = 3$, expression (10.11) shows that the mass M becomes independent of radius. The constant value of this mass, called the Chandrasekhar mass, is given by

$$M_0 \cong \frac{3.1}{m_p^2}\left(\frac{Z}{A}\right)^2\left(\frac{\hbar c}{G}\right)^{3/2} \simeq 5.8\left(\frac{Z}{A}\right)^2 M_\odot \tag{10.15}$$

if the material is entirely made of neutrons. We shall see in Vol. II that this mass acts as a limiting mass for compact remnants formed in stellar collapse. Systems with $M > M_0$ cannot be supported by degeneracy pressure against gravitational collapse and will possibly end up as black holes.

Equations of state (10.12) and (10.13) are the limiting forms of the more exact equation of state for a zero-temperature degenerate fermionic gas, which can be given in parametric form [see Eq. (5.156)]. Equation (5.156) can be used to integrate Eq. (10.2) numerically. A good fit for the resulting mass radius relation is given by

$$\frac{R}{R_0} = 2.02\left[1 - \left(\frac{M}{M_0}\right)^{4/3}\right]^{1/2}\left(\frac{M}{M_0}\right)^{-1/3}, \tag{10.16}$$

where M_0 is given by relation (10.15) and

$$R_0 = 0.01117\frac{Z}{A}R_\odot \tag{10.17}$$

10.3.3 Isothermal Sphere

When the Lane–Emden index n tends to infinity, the polytropic index γ goes to unity and the equation of state becomes $P \propto \rho$. Such a system is called an isothermal sphere. Because polytropic solutions become difficult to handle as $n \to \infty$, it is convenient to approach the isothermal sphere in a different manner.

Consider a self-gravitating ideal gas with constant temperature T, density $\rho(r)$, pressure $P(r) = \rho(r)k_B T$, and a gravitational potential $\phi(r)$ such that the density at any location is given by

$$\rho(r) = \rho_c \exp\{-\beta[\phi(r) - \phi(0)]\}, \tag{10.18}$$

with ρ_c denoting central density. This density distribution corresponds to a system in Boltzmann distribution at constant temperature. Combining the Poisson equation with Eq. (10.18) we get a differential equation for the gravitational potential:

$$\nabla^2 \phi = 4\pi G \rho_c e^{-\beta[\phi(\mathbf{x}) - \phi(0)]}. \tag{10.19}$$

Given the solution to this equation, all other quantities can be determined. As we shall see, this system has several peculiarities.

It is convenient to introduce the length, mass, and energy scale by the definitions

$$L_0 \equiv (4\pi G \rho_c \beta)^{1/2}, \quad M_0 = 4\pi \rho_c L_0^3, \quad \phi_0 \equiv \beta^{-1} = \frac{GM_0}{L_0}, \tag{10.20}$$

so that all other physical variables can be expressed in terms of the dimensionless quantities

$$x \equiv \frac{r}{L_0}, \quad n \equiv \frac{\rho}{\rho_c}, \quad m = \frac{M(r)}{M_0}, \quad y \equiv \beta[\phi - \phi(0)]. \tag{10.21}$$

It is easy to verify that the variables y, n, and m satisfy the equations

$$y' = \frac{m}{x^2}, \quad m' = nx^2, \quad n' = -\frac{mn}{x^2}. \tag{10.22}$$

In terms of $y(x)$ isothermal Eq. (10.19) becomes

$$\frac{1}{x^2}\frac{d}{dx}\left(x^2\frac{dy}{dx}\right) = e^{-y} \tag{10.23}$$

with the boundary condition $y(0) = y'(0) = 0$. Let us consider the nature of the solutions to this equation.

By direct substitution into Eq. (10.22), we see that $n = (2/x^2)$, $m = 2x$, and $y = 2\ln x$ satisfy these equations. This solution, however, is singular at the origin and hence is not physically admissible. The importance of this solution lies in the fact that other (physically admissible) solutions tend to this solution for large values of x. This result can be proved as follows: Consider a change of variable, from x to $q = \ln x$ in Eq. (10.23). This leads to

$$\frac{d^2y}{dq^2} + \frac{dy}{dq} = e^{2q-y}. \tag{10.24}$$

Changing further to the variable $X \equiv 2q - y$, we get

$$\frac{d^2 X}{dq^2} + \frac{dX}{dq} = -(e^X - 2) \equiv -\frac{\partial V}{\partial X}. \tag{10.25}$$

This equation describes the motion of a particle with position X at time q in a potential $V(X) = (e^X - 2X)$ with a damping term (dX/dq). Because of the damping, the kinetic energy $(dX/dq)^2$ will keep decreasing and the particle will approach the minimum of the potential at $X = \ln 2$. This minimum point corresponds to $y = 2q - X = 2 \ln x - \ln 2 = 2 \ln x + $ constant. Thus all solutions the approach the $y = 2 \ln x$ solution for large $q = \ln x$.

To obtain a more quantitative result, we can put $X = \ln 2 + \delta X$ and linearise Eq. (10.25) in δX. This gives

$$\left(\frac{d^2}{dq^2} + \frac{d}{dq} + 2 \right) \delta X = 0, \tag{10.26}$$

with the solution

$$\delta X = A e^{-q/2} \cos\left[\frac{\sqrt{7}}{2}(q + \alpha) \right] = \frac{A}{x^{1/2}} \cos\left[\frac{\sqrt{7}}{2}(\ln x + \alpha) \right], \tag{10.27}$$

which shows that δX decreases to zero in an oscillatory manner.

This asymptotic behaviour of all solutions shows that the density decreases as $1/r^2$ for large r, implying that the mass contained inside a sphere of radius r increases as $M(r) \propto r$ at large r. To find physically useful solutions, it is necessary to assume that the solution is cut off at some radius R. For example, we may assume that the system is enclosed in a spherical box of radius R. In any realistic system, such a confinement has to arise from physical considerations that invalidate the isothermality (assumed in the above discussion) for large r or that are due to the presence of external influences. In what follows, it will be assumed that the system has some cutoff radius R.

To find physically acceptable solutions and study their properties, it is convenient to proceed as follows: Equation (10.23) is invariant under the transformation $y \rightarrow y + a$; $x \rightarrow kx$ with $k^2 = e^a$. This invariance implies that, given a solution with some value of $y(0)$, we can obtain the solution with any other value of $y(0)$ by simple rescaling. Therefore only one of the two integration constants in Eq. (10.23) is really nontrivial. Hence it must be possible to reduce the degree of the equation from two to one by a judicious choice of variables. One such set of variables is

$$v \equiv \frac{m}{x}, \quad u \equiv \frac{nx^3}{m} = \frac{nx^2}{v}. \tag{10.28}$$

In terms of v and u, we have

$$v' = \frac{m'}{x} - \frac{m}{x^2} = \frac{v}{x}(u-1), \quad u' = \frac{3x^2 n}{m} - nx - \frac{n^2 x^5}{m^2} = \frac{u}{x}(3-v-u),$$

$$(10.29)$$

giving

$$\frac{u}{v}\frac{dv}{du} = -\frac{(u-1)}{(u+v-3)}. \qquad (10.30)$$

The boundary conditions $y(0) = y'(0) = 0$ translate into the following: v is zero at $u = 3$ and $(dv/du) = -5/3$ at $(3,0)$. The solution $v(u)$ has to be obtained numerically: It is plotted in Fig. 10.1 as the spiralling curve. The singular points of this differential equation are given by the intersection of the straight lines $u = 1$ and $u + v = 3$ on which the numerator and denominator of the right-hand side of Eq. (10.30) vanish; that is, the singular point is at $u_s = 1$, $v_s = 2$, corresponding to the solution $n = (2/x^2)$, $m = 2x$. It is obvious from the nature of the equations that the solutions will spiral around the singular point, corresponding to the oscillations in Eq. (10.27).

The nature of the solution shown in Fig. 10.1 allows us to put interesting bounds on physical quantities, including energy. To see this, we compute the

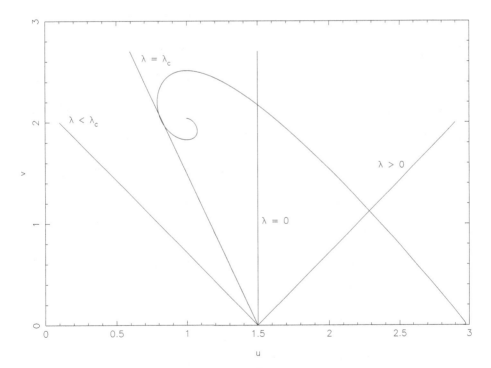

Fig. 10.1. Bound on RE/GM^2 for the isothermal sphere.

total energy E of the isothermal sphere. The potential and the kinetic energies are

$$U = - \int_0^R \frac{GM(r)\,dM}{r}\,\frac{dM}{dr}\,dr = - \frac{GM_0^2}{L_0} \int_0^{x_0} mnx\,dx,$$

$$K = \frac{3}{2}\frac{M}{\beta} = \frac{3}{2}\frac{GM_0^2}{L_0}m(x_0) = \frac{GM_0^2}{L_0}\frac{3}{2}\int_0^{x_0} nx^2\,dx, \qquad (10.31)$$

where $x_0 = R/L_0$. The total energy is therefore

$$E = K + U = \frac{GM_0^2}{2L_0} \int_0^{x_0} dx(3nx^2 - 2mnx)$$

$$= \frac{GM_0^2}{2L_0} \int_0^{x_0} dx\,\frac{d}{dx}\{2nx^3 - 3m\} = \frac{GM_0^2}{L_0}\left\{n_0 x_0^3 - \frac{3}{2}m_0\right\}, \qquad (10.32)$$

where $n_0 = n(x_0)$ and $m_0 = m(x_0)$. The dimensionless quantity (RE/GM^2) is given by

$$\lambda = \frac{RE}{GM^2} = \frac{L_0 x_0}{GM_0^2 m_0^2}\frac{GM_0^2}{L_0}\left\{n_0 x_0^3 - \frac{3}{2}m_0\right\} = \frac{1}{v_0}\left\{u_0 - \frac{3}{2}\right\}. \qquad (10.33)$$

Note that the combination (RE/GM^2) is a function of (u, v) alone. Let us now consider the constraints on λ. Suppose we specify some value for λ by specifying R, E, and M. Then such an isothermal sphere *must* lie on the curve

$$v = \frac{1}{\lambda}\left(u - \frac{3}{2}\right), \qquad \lambda \equiv \frac{RE}{GM^2}, \qquad (10.34)$$

which is a straight line through the point $(1.5, 0)$ with the slope λ^{-1}. On the other hand, because *all* isothermal spheres must lie on the u–v curve, an isothermal sphere can exist only if the line in Eq. (10.34) intersects the u–v curve.

For large positive λ (positive E) there is just one intersection (see Fig. 10.1). When $\lambda = 0$ (zero energy), we still have a unique isothermal sphere. [For $\lambda = 0$, Eq. (10.34) is a vertical line through $u = 3/2$.] When λ is negative (negative E), the line can cut the u–v curve at more than one point; thus more than one isothermal sphere can exist with a given value of λ. (Of course, specifying M, R, and E individually will remove this nonuniqueness.) However, as we decrease λ (more and more negative E) the line in Eq. (10.34) will slope more and more to the left; and when λ is smaller than a critical value λ_c, the intersection will cease to exist. *Thus no isothermal sphere can exist if (RE/GM^2) is below a critical value λ_c.* This fact follows immediately from the nature of the u–v curve and Eq. (10.34). The value of λ_c can be found from the numerical solution in Fig. 10.1. It turns out to be approximately -0.335. We shall see later in Section 10.8 that this result has implications for the evolution of gravitating systems.

10.3.4 Time-Dependent Isothermal Sphere Solutions

There is a simple generalisation of the isothermal sphere solution given above to a time-dependent case. Such a solution describes a scale-invariant collapse of a spherical isothermal cloud and is of some use in modelling the initial stages of star formation.

Because the only dimensional parameters of relevance to the problem are the gravitational constant G and the isothermal sound speed $a = (k_B T/m_0)^{-1/2}$, it is reasonable to expect solutions to time-dependent fluid equations that depend on only the parameter $x \equiv (r/at)$. The relevant equations can be written in the form

$$\frac{\partial M}{\partial t} + u\frac{\partial M}{\partial r} = 0, \quad \frac{\partial M}{\partial r} = 4\pi r^2 \rho, \tag{10.35}$$

$$\frac{\partial u}{\partial t} + u\frac{\partial u}{\partial r} = -\frac{a^2}{\rho}\frac{\partial \rho}{\partial r} - \frac{GM}{r^2}, \tag{10.36}$$

where $M(r, t)$ denotes the total amount of mass contained within a spherical shell of radius r. Using the parameters G and a, we can recast these equations into dimensionless form. We first define the dimensionless variables $\alpha(x)$, $m(x)$, and $v(x)$ by

$$\rho(r, t) = \frac{\alpha(x)}{4\pi G t^2}, \quad M(r, t) = \frac{a^3 t}{G}m(x), \quad u(r, t) = av(x). \tag{10.37}$$

In terms of these variables, the conservation of mass implies that

$$m + (v - x)\frac{dm}{dx} = 0, \quad \frac{dm}{dx} = x^2\alpha, \tag{10.38}$$

which can be combined to give

$$m = x^2\alpha(x - v). \tag{10.39}$$

Straightforward algebra now leads to two coupled nonlinear differential equations for $v(x)$ and $\alpha(x)$:

$$[(x - v)^2 - 1]\frac{dv}{dx} = \left[(x - v)\alpha - \frac{2}{x}\right](x - v), \tag{10.40}$$

$$[(x - v)^2 - 1]\frac{1}{\alpha}\frac{d\alpha}{dx} = \left[\alpha - \frac{2}{x}(x - v)\right](x - v). \tag{10.41}$$

These equations have to be integrated numerically, and the solutions are shown in Fig. 10.2. The static (singular) isothermal sphere discussed in Subsection 10.3.3 corresponds to the solution

$$v = 0, \quad \alpha = 2/x^2, \quad m = 2x, \tag{10.42}$$

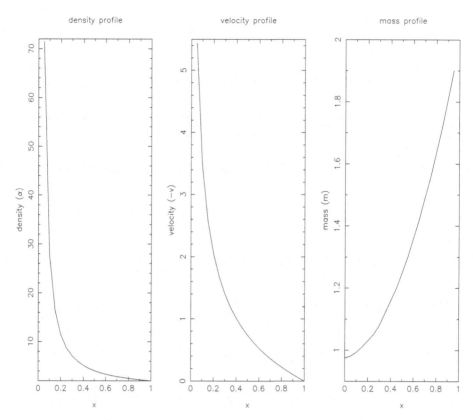

Fig. 10.2.   The density, velocity and mass profile of collapsing singular isothermal sphere.

which can be used as the initial conditions for solving these equations. At any time $t > 0$, we would have expected some amount of mass to have collapsed to the point singularity at the origin. In fact, the numerical integration gives $m(x) \rightarrow m_c = 0.975$ as $x \rightarrow 0$, giving $M(0, t) = m_c a^3 t / G$; that is the mass at the origin grows linearly with time at the accretion rate $\dot{M} = m_c a^3 / G$. Near the origin the velocity and the density fields approach the free-fall values given by $v(x) \rightarrow -(2m_c/x)^{1/2}$ and $\alpha(x) \rightarrow (m_c/2x^3)^{1/2}$. Far away from the origin, the information about the core collapse could not have reached. In fact, the numerical solution smoothly goes over to $v = 0$ at $x \geq 1$. For $x > 1$, corresponding to $r > at$, the solution is represented by the original singular isothermal sphere. In the in-between regime, away from the origin, the density profile is approximately r^{-2}. The form of the solutions is shown in Fig. 10.2.

10.3.5 Fluid Spheroids

The examples discussed so far were all nonrotating and spherically symmetric. It is also possible to obtain solutions to the equations describing self-gravitating fluids that are rotating. One simple class of such solutions is given below for future reference.

Because rotation of the fluid will distort the shape to an axisymmetric structure, we begin by considering the gravitational potential of a constant-density ellipsoid. Let the surface of an ellipsoid be given by

$$\frac{x^2}{a^2} + \frac{y^2}{b^2} + \frac{z^2}{c^2} = 1, \quad a > b > c. \tag{10.43}$$

Using the general formula for the gravitational potential that is due to a mass distribution,

$$\phi(\mathbf{x}) = -G \int \frac{\rho(\mathbf{y}) \, d^3 \mathbf{y}}{|\mathbf{x} - \mathbf{y}|},$$

we can write the potential outside a constant-density ellipsoid as

$$\phi = -\pi \rho abcG \int_\xi^\infty \left(1 - \frac{x^2}{a^2 + s} + \frac{y^2}{b^2 + s} - \frac{z^2}{c^2 + s}\right) \frac{ds}{R_s}, \tag{10.44}$$

where

$$R_s = \sqrt{(a^2 + s)(b^2 + s)(c^2 + s)}$$

and ξ is the positive root of the equation

$$\frac{x^2}{a^2 + \xi} + \frac{y^2}{b^2 + \xi} + \frac{z^2}{c^2 + \xi} = 1. \tag{10.45}$$

The potential in the interior of the ellipsoid is given by

$$\phi = -\pi \rho abcG \int_0^\infty \left(1 - \frac{x^2}{a^2 + s} - \frac{y^2}{b^2 + s} - \frac{z^2}{c^2 + s}\right) \frac{ds}{R_s}, \tag{10.46}$$

which differs from Eq. (10.44) in the lower limit of integration. Because of this difference, the coordinate dependence of the inside potential is through only the integrand and hence is quadratic. In other words, the gravitational potential of a constant-density ellipsoid is a quadratic function of the coordinates on the inside. The gravitational potential energy of this configuration is given by

$$U = \frac{1}{2} \int \rho \phi \, d^3 x = \frac{3Gm^2}{8} \cdot \int_0^\infty \left[\frac{1}{5}\left(\frac{a^2}{a^2 + s} + \frac{b^2}{b^2 + s} + \frac{c^2}{c^2 + s}\right) - 1\right] \frac{ds}{R_s}$$

$$= \frac{3Gm^2}{8} \int_0^\infty \left[\frac{2}{5} sd\left(\frac{1}{R}\right) - \frac{2}{5} \frac{ds}{R_s}\right] = \frac{3Gm^2}{10} \int_0^\infty \frac{ds}{R_s}. \tag{10.47}$$

(The integral over the volume of ellipsoid is elementary and m denotes the total mass of the body.) This expression reduces to simple closed forms for oblate

$(a = b > c)$ and prolate $(a > b = c)$ spheroids:

$$U = -\frac{3Gm^2}{5\sqrt{a^2 - c^2}} \cos^{-1}\frac{c}{a} \quad \text{(oblate)},$$

$$U = -\frac{3Gm^2}{5\sqrt{a^2 - c^2}} \cosh^{-1}\frac{c}{a} \quad \text{(prolate)}. \tag{10.48}$$

Let us now consider the equilibrium configuration of a homogeneous mass of fluid that is rotating as a whole. The condition for equilibrium is the constancy of the sum of the gravitational potential and the potential that is due to centrifugal forces on the surface of the body. This requires that

$$\phi - \frac{\Omega^2}{2}(x^2 + y^2) = \text{constant}, \tag{10.49}$$

where Ω is the angular velocity of the body about the z axis. This is the shape of an oblate ellipsoid of rotation with $a = b > c$. To determine its parameters, we substitute for the potential the expression obtained in Eq. (10.46) and eliminate z^2 by using Eq. (10.43), thereby obtaining the condition

$$(x^2 + y^2)\left[\int_0^\infty \frac{ds}{(a^2 + s)^2\sqrt{c^2 + s}} - \frac{\Omega^2}{2\pi\rho Ga^2c}\right.$$
$$\left. - \frac{c^2}{a^2}\int_0^\infty \frac{ds}{(a^2 + s)^2(c^2 + s)^{3/2}}\right] = \text{constant}. \tag{10.50}$$

Because the left-hand side has to be independent of x and y, it follows that the expression within the brackets must vanish. Performing the integrals, we can write this condition as

$$\frac{(a^2 + 2c^2)c}{(a^2 - c^2)^{3/2}}\cos^{-1}\frac{c}{a} - \frac{3c^2}{a^2 - c^2} = \frac{\Omega^2}{2\pi G\rho} = \frac{25}{6}\left(\frac{4\pi}{3}\right)^{1/3}\frac{J^2\rho^{1/3}}{m^{10/3}G}\left(\frac{c}{a}\right)^{4/3}, \tag{10.51}$$

where $J = (2/5)ma^2\Omega$ is the angular momentum of the body. This equation determines the ratio c/a for a given value of Ω or J. The dependence c/a on J is single valued and c/a increases monotonically with increasing J. The same relation can also be expressed in terms of the eccentricity e of the ellipsoid defined by $e^2 = 1 - (c/a)^2$. The result is

$$\frac{\Omega^2}{2\pi G\rho} = \frac{\sqrt{1 - e^2}}{e^3}[(3 - 2e^2)\sin^{-1}e - 3e\sqrt{1 - e^2}]. \tag{10.52}$$

For small eccentricities we have $\Omega^2 \approx (8\pi/15)G\rho e^2$, which, incidentally, can be used to relate the oblateness of the Earth to its rotation. The relationship between Ω and e is not monotonic; Ω attains a maximum value $\Omega_{\max} \approx 0.47\sqrt{2\pi G\rho}$ at $\epsilon = 0.93$.

An important measure that determines the overall stability and other features of a self-gravitating rotating ellipsoid is the ratio between the rotational kinetic energy T, given by

$$T = \frac{1}{2}\rho\Omega^2 \int (x^2 + y^2)\, d^3\mathbf{x} = \frac{4\pi}{15}\rho\Omega^2\sqrt{1 - e^2}\, a^5, \tag{10.53}$$

and the gravitational potential energy given by Eqs. (10.48). This ratio is

$$\frac{T}{|U|} = \frac{3}{2e^2} - 1 - \frac{3\sqrt{1 - e^2}}{2e \sin^{-1} e}, \tag{10.54}$$

which is a monotonically increasing function of the eccentricity.

Exercise 10.2

Potential of an ellipsoid: It was stated without proof that the inside potential of an ellipsoid is quadratic in the coordinates. Obtain this result along the following lines. Using the expansion of the Green's function for Poisson equation in terms of Legendre polynomials, show that the potential can be written in the form

$$\Phi = -2\pi G\rho \sum_{l=0}^{\infty} P_l(\cos\theta) \int_0^\pi \sin\theta' d\theta'\, P_l(\cos\theta') \left[\int_0^r \frac{(r')^{l+2}\, dr'}{r^{l+1}} + \int_r^R \frac{r'\, dr'}{(r')^{l-1}} \right], \tag{10.55}$$

where $R(\theta)$ is the surface defined by the equation

$$\frac{\sin^2\theta}{a^2} + \frac{\cos^2\theta}{c^2} = \frac{1}{R^2}. \tag{10.56}$$

Argue why only the terms with $l = 0$ and $l = 2$ should contribute in this expression. Evaluation of these two terms is elementary and leads to the result quoted in the text.

10.4 Collisionless Gravitating Systems in Steady State

We next consider systems in which the constituent particles evolve in a collisionless manner. A prototype of such a system will be the collection of $N(\gg 1)$ stars in a galaxy. If the stars are treated as a smooth fluid then we can introduce a distribution function $f(\mathbf{x}, \mathbf{v}, t)$ that gives the number of stars in a phase volume $d^3\mathbf{x}d^3\mathbf{v}$ at time t. The treatment of a collection of stars as a fluid necessarily implies that we are dealing with length scales that are large compared with the inter-star spacing in the system; further, the volume element $d^3\mathbf{x}d^3\mathbf{v}$ cannot be strictly infinitesimal if it has to contain a sufficient number of stars. These issues are of certain methodological significance, and one possible way of avoiding mathematical inexactness is to interpret $f(\mathbf{x}, \mathbf{v}, t)$ as a relative probability density. Such a sophistication, however, will not be adopted here as it is not of any significant physical consequence.

In general, the distribution function will evolve by the equation of the form $(df/dt) = C$, where C denotes the collision terms. For systems such as stars in a galaxy, physical collisions are negligible and the gravitational collisions are significant only for $t \gtrsim t_c \approx (N/\ln N)t_{\mathrm{dyn}}$. (This result will be derived in Section 10.7.) Except in the case of star clusters, t_c is much longer than the age of the universe. Hence systems such as galaxies can be assumed to evolve in a collisionless fashion, with $(df/dt) = 0$. Expanding out the time derivative and using $\dot{\mathbf{v}} = -\nabla\phi$, we can express the equations that describe a collisionless gravitating system as

$$\frac{df}{dt} = \frac{\partial f}{\partial \mathbf{x}} + \dot{\mathbf{v}} \cdot \frac{\partial f}{\partial \mathbf{v}} + \dot{\mathbf{x}} \cdot \frac{\partial f}{\partial \mathbf{x}} = \frac{\partial f}{\partial t} + \mathbf{v} \cdot \frac{\partial f}{\partial \mathbf{x}} - \nabla\phi \cdot \frac{\partial f}{\partial \mathbf{v}} = 0, \quad (10.57)$$

with the potential determined by

$$\nabla^2\phi = 4\pi G\rho, \quad \rho(\mathbf{x}, t) = m \int f(\mathbf{x}, \mathbf{v}, t)\, d^3\mathbf{v}. \quad (10.58)$$

Models for galaxies are based on the solution to coupled equations (10.57) and (10.58). Because these equations allow a wide variety of solutions, different classes of models are possible for the galaxies. Let us consider the solutions to these equations that are independent of time $f(t, \mathbf{x}, \mathbf{v}) = f(\mathbf{x}, \mathbf{v})$, postponing the discussion of time evolution to Section 10.7.

To produce such a steady-state solution, we can proceed as follows: Let $C_i = C_i(\mathbf{x}, \mathbf{v})$, $i = 1, 2, \ldots$, be a set of integrals of motion for the stars moving in the potential ϕ (which, right now, is not known). It is obvious that any function $f(C_i)$ of the C_i's will satisfy the steady-state Boltzmann equation; $(df/dt) = (\partial f/\partial C_i)\dot{C}_i = 0$, as $\dot{C}_i$ is identically zero. If we can now determine ϕ from f self-consistently and populate the orbits of ϕ with stars, we have solved the problem. Let us consider some specific examples to see how this idea works.

We begin with models in which the density and the potential of the system are spherically symmetric. In these calculations, it is convenient to shift the origin of ϕ by defining a new potential $\psi \equiv -\phi + \phi_0$, where ϕ_0 is a constant. (We will choose the value of ϕ_0 such that ψ vanishes at the boundary of the system.) The new potential satisfies the equation

$$\nabla^2\psi = -4\pi G\rho, \quad (10.59)$$

and the boundary condition $\psi \to \phi_0$ as $|\mathbf{x}| \to \infty$. We also define a "shifted" energy for the particles $\epsilon = -E + \phi_0$; because $\phi_0 = \psi + \phi$, $\epsilon = -E + \psi + \phi = -(1/2)v^2 + \psi$.

The simplest models are those in which $f(\mathbf{x}, \mathbf{v})$ depends on $\mathbf{x}$ and $\mathbf{v}$ through only the quantity ϵ so that $f = f(\epsilon) = f(\psi - \frac{1}{2}v^2)$. The density $\rho(\mathbf{x})$

corresponding to this distribution is

$$\rho(\mathbf{x}) = \int_0^{\sqrt{2\psi}} 4\pi v^2 \, dv f\left(\psi - \frac{1}{2}v^2\right) = \int_0^{\psi} 4\pi d\epsilon f(\epsilon)\sqrt{2(\psi - \epsilon)}. \quad (10.60)$$

The limits of integration are chosen in such a way as to pick only the particles bound in the system's potential. The right-hand side of Eq. (10.60) is a known function of ψ, once $f(\epsilon)$ is specified. The Poisson equation

$$\frac{1}{r^2}\frac{d}{dr}\left(r^2\frac{d\psi}{dr}\right) = -4\pi G\rho = -16\pi^2 G \int_0^{\psi} d\epsilon f(\epsilon)\sqrt{2(\psi - \epsilon)} \quad (10.61)$$

can now be solved – with some central value $\psi(0)$ and the boundary condition $\psi'(0) = 0$ – determining $\psi(r)$. Once $\psi(r)$ is known, all other variables can be computed.

Incidentally, note that a given $\rho(r)$ generates a unique $f(\epsilon)$ by the following procedure: Given $\rho(r)$, we can determine $\psi(r)$ and hence, by eliminating r, the function $\rho(\psi)$. Writing Eq. (10.60) as

$$\frac{1}{\sqrt{8\pi}}\rho(\psi) = 2\int_0^{\psi} f(\epsilon)\sqrt{\psi - \epsilon}\, d\epsilon \quad (10.62)$$

and differentiating both sides with respect to ψ, we get

$$\frac{1}{\sqrt{8\pi}}\frac{d\rho}{d\psi} = \int_0^{\psi}\frac{f(\epsilon)\, d\epsilon}{\sqrt{\psi - \epsilon}}. \quad (10.63)$$

This equation (called Abel's integral equation) has the solution

$$f(\epsilon) = \frac{1}{\sqrt{8}\pi^2}\frac{d}{d\epsilon}\int_0^{\epsilon}\left(\frac{d\rho}{d\psi}\right)\frac{d\psi}{\sqrt{\epsilon - \psi}}, \quad (10.64)$$

which determines $f(\epsilon)$. Although this procedure gives $f(\epsilon)$, there is no assurance that it will be positive definite. Three different choices for $f(\epsilon)$ have been extensively used in the literature to describe spherically symmetric systems; these models are explored in Exercise 10.3.

Exercise 10.3
Examples of spherical systems: (1) The simplest form of $f(\epsilon)$ is a power law with $f(\epsilon) = A\epsilon^{n - 3/2}$ for $\epsilon > 0$ and zero otherwise. Show that this corresponds to density distributions of the form $\rho = B\psi^n$ (for $\psi > 0$) with $B = (2\pi)^{3/2} A\Gamma(n - 1/2)\, [\Gamma(n + 1)]^{-1}$. (Clearly, we need $n > 1/2$ to obtain finite density. In this case, the resulting equation becomes the Lane–Emden equation discussed earlier in the case of gaseous spheres in Subsection (10.3.3).)

(2) Another favourite form of the distribution function is

$$f(\epsilon) = \frac{\rho_0}{(2\pi\sigma^2)^{3/2}}\exp\left(\frac{\epsilon}{\sigma^2}\right), \quad (10.65)$$

which is parameterised by two constants ρ_0 and σ. Verify that the mean-square velocity $\langle v^2 \rangle$ is $3\sigma^2$ and that the density distribution is $\rho(r) = \rho_0 \exp(\psi/\sigma^2)$. It is conventional to define a core radius and a set of dimensionless variables by

$$r_0 = \left(\frac{9\sigma^2}{4\pi G\rho_c} \right)^{1/2}, \quad l = \frac{r}{r_0}, \quad \xi = \frac{\rho}{\rho_c} \tag{10.66}$$

where ρ_c is the central density. Write the Poisson equation in terms of the dimensionless variables and show that it reduces to the equation for isothermal sphere.

(3) A modified version of isothermal sphere is called the King model and is based on the distribution function

$$f(\epsilon) = \frac{\rho_c}{(2\pi\sigma^2)^{3/2}} \left(e^{\epsilon/\sigma^2} - 1 \right), \quad \epsilon \geq 0, \tag{10.67}$$

and $f(\epsilon) = 0$ for $\epsilon < 0$. Because $f(\epsilon)$ vanishes for $\epsilon < 0$, this model may be thought of as a truncated isothermal sphere. Integrate this equation numerically from the origin by using suitable boundary conditions. The density will vanish at some radius r_t, called the tidal radius. If r_0 is the core radius as defined in the case of isothermal sphere, then the quantity $c = \log(r_t/r_0)$ is a measure of how concentrated the system is. Plot the density run for different values of c.

If we allow the distribution function to depend on the angular momentum J as well as on ϵ, then a wider class of models can be constructed to fit the same $\rho(r)$. To see this, assume that f depends on ϵ and J^2 only through the combination

$$Q = \epsilon - \frac{J^2}{2R^2}, \quad R = \text{constant}. \tag{10.68}$$

We can easily integrate out the angular variables in the velocity space and obtain the density distribution

$$\rho(r) = \frac{2\pi\sqrt{8}}{(1 + r^2/R^2)} \int_0^\psi f(Q)\sqrt{\psi - Q}\, dQ \equiv \frac{2\pi\sqrt{8}}{(1 + r^2/R^2)} \mu(\psi), \tag{10.69}$$

where the second equality defines $\mu(\psi)$. Comparing Eqs. (10.69) and (10.69), we see that

$$f(Q) = \frac{1}{\pi^2\sqrt{8}} \frac{d}{dQ} \int_0^Q \left(\frac{d\mu}{d\psi} \right) \frac{d\psi}{\sqrt{Q - \psi}}. \tag{10.70}$$

Thus, given any $\rho(r)$, we determine $\psi(r)$ and another function $\mu(r) \equiv \rho(r)$ $(1 + r^2/R^2)(2\pi\sqrt{8})^{-1}$. Eliminating r between $\mu(r)$ and $\psi(r)$, we obtain $\mu(\psi)$ and consequently $f(Q)$. This distribution function will lead to the density distribution $\rho(r)$ we started with. When $f = f(\epsilon)$, the velocity dispersion is isotropic with $\langle v_r^2 \rangle = \langle v_\theta^2 \rangle = \langle v_\phi^2 \rangle$. For a distribution function of the form $f(\epsilon, J^2)$, we have $\langle v_\theta^2 \rangle = \langle v_\phi^2 \rangle \neq \langle v_r^2 \rangle$; this is the key difference between the two cases. Similar modelling also works for a disklike system if the distribution function is taken to be of the form $f = f(\epsilon, J_z)$.

Exercise 10.4
Models for axisymmetric systems: (1) Consider an axisymmetric system with the distribution function

$$f(\epsilon, J_z) = \begin{cases} A J_z^n \exp(\epsilon/\sigma^2) & (\text{for } J_z \geq 0) \\ 0 & (\text{for } J_z \leq 0) \end{cases}. \tag{10.71}$$

Obtain the surface density of the disk and the circular velocity of the particles for this model. These are called Mestel disks. (2) A more complicated set of disk models (called Kalnajs disks) can be obtained from the distribution function that has the form

$$f(\epsilon, J_z) = A\left[(\Omega_0^2 - \Omega^2)a^2 + 2(\epsilon + \Omega J_z)\right]^{-1/2} \tag{10.72}$$

when the term in brackets is positive and zero otherwise. Show that in this case we are led to the following surface density and potential:

$$\Sigma(R) = \Sigma_0\left(1 - \frac{R^2}{a^2}\right)^{1/2}, \quad \phi(R) = \frac{\pi^2 G \Sigma_0}{4a}R^2 \equiv \frac{1}{2}\Omega_0^2 R^2, \tag{10.73}$$

with $\Sigma_0 = 2\pi A a \sqrt{\Omega_0^2 - \Omega^2}$. The parameter Ω is free and describes the mean (systematic) rotational velocity of the system, i.e., $\langle v_\phi \rangle = \Omega R$. (3) Another class of axisymmetric models can be constructed along the following lines: Consider a system with the density distribution $\rho(r, \theta) = \rho_0 S(\theta)(r_0/r)^2$, which is axisymmetric and falls as r^{-2}. Let $S(\theta)$ be normalised such that $\int_0^\pi S(\theta) \sin\theta \, d\theta = 2$. Show that the gravitational potential is given by

$$\Phi = v_0^2\left[\ln\left(\frac{r}{r_0}\right) + P(\theta)\right] \tag{10.74}$$

where $P(\theta)$ is related to $S(\theta)$ by

$$\frac{1}{\sin\theta}\frac{d}{d\theta}\left(\sin\theta\frac{dP}{d\theta}\right) = S(\theta) - 1. \tag{10.75}$$

As an example, consider a disk that is embedded in a spherical halo for which $S(\theta)$ can be taken to be $S(\theta) = \delta_D(\theta - \pi/2) + b$. The first term represents the matter density confined to the $\theta = \pi/2$ plane, and the second term, which is independent of θ, is due to a spherically symmetric halo. Obtain the gravitational potential in this case.

10.5 Moment Equations for Collisionless Systems

In many earlier chapters, it was shown that taking the moments of the Boltzmann equation could be of use in bringing out the underlying physics. This approach can also be used for a collisionless gravitating system. The first and the second moment equations for this nonrelativistic system will be identical in form to Eqs. (8.7) and (8.16), which were obtained in Chap. 8, Section 8.2. The first of these equations describes the conservation of mass, which can be written as

$$\frac{\partial n}{\partial t} + \nabla \cdot (n\bar{\mathbf{v}}) = 0, \tag{10.76}$$

where n and $\bar{\mathbf{v}}$ are defined as

$$n \equiv \int f \, d^3\mathbf{v}, \quad \bar{\mathbf{v}} = \frac{1}{n} \int f\mathbf{v} \, d^3\mathbf{v} \equiv \langle \mathbf{v} \rangle. \tag{10.77}$$

(The notation has been changed slightly from that of Chap. 8; the quantity n now denotes the *number* density of particles rather than the *mass* density and the quantity $\bar{\mathbf{v}}$ is the mean fluid velocity that was denoted by the symbol $\mathbf{V}$ in Chap. 8.) The momentum conservation equation becomes

$$\frac{\partial}{\partial t}(n\bar{v}^i) + \frac{\partial T^{ij}}{\partial x^j} + n\frac{\partial \phi}{\partial x_i} = 0, \tag{10.78}$$

where

$$T^{ij} = \int v^i v^j f \, d^3\mathbf{v} \equiv n\langle v^i v^j \rangle. \tag{10.79}$$

The three-dimensional stress tensor T^{ij} is now defined by taking a moment with respect to the number density of particles rather than the mass density. For practical applications, Eq. (10.78) is usually rewritten in an equivalent form as

$$n\left(\frac{\partial}{\partial t} + \bar{\mathbf{v}} \cdot \nabla\right)\bar{v}^j = -n\frac{\partial \phi}{\partial x^j} - \frac{\partial}{\partial x^i}(n\sigma_{ij}^2), \tag{10.80}$$

where the velocity dispersion tensor σ_{ij}^2 is defined as

$$\sigma_{ij}^2 = \langle (v_i - \bar{v}_i)(v_j - \bar{v}_j) \rangle = \langle v_i v_j \rangle - \bar{v}_i \bar{v}_j. \tag{10.81}$$

In a fluid system, this quantity will be related to the pressure of the fluid and an equation of state will provide a relationship between the pressure and density. In a collisionless system this is not possible and hence these equations do *not* form a closed set. In spite of this fact, the moment equations derived above are of substantial use in understanding the structure of our galaxy, and we will use them in Vol. III.

10.6 Approach of a Collisionless System to Steady State

We now consider the time evolution of a gravitating system, which, in general, is a more complicated problem compared with the steady-state situation discussed in Section 10.5.

Consider the evolution of a self-gravitating system at a time scale that is short compared with the collisional relaxation time scale. Let us assume that, at $t = 0$, the particles were located in some finite region $\mathcal{R}$ of the phase space.

As the particles move, the shape of $\mathcal{R}$ will become distorted and we expect the evolution to spread the phase density all over the phase space at finer and finer scales. This arises because of two kinds of processes: (1) Each particle is moving in the mean gravitational potential of the system. Because this potential is, in general, not harmonic, the orbital period will depend on the amplitude and the occupied region in the phase space will become "wound up" in the angular direction. (2) The gravitational potential is also changing with time because of which the energies of individual particles will not be constant. Hence the particles will also move along the radial direction in the phase space. We therefore conclude that, as time goes on, particles become mixed in phase space significantly.

To study the effects of such mixing, we divide the phase space into microcells of volume ω. These microcells are supposed to be so tiny that the distribution function f can be taken as a constant over the cell. Initially, we assume that the phase density is constant ($f = \eta$) in some region of the phase space and zero elsewhere. Therefore, initially, some microcells will contain $\eta\omega$ particles and others will have zero particles. The total number of particles is $N_{\text{tot}} = N\eta\omega$, where N is the number of *occupied* microcells.

We examine the phase density after averaging (also called coarse graining) it over cells of volume Ω (called macrocells). Let $\nu = (\Omega/\omega)$ be the number of microcells contained in any one macrocell. Consider any one particular macrocell, say, the ith one, that has n_i microcells occupied and $(\nu - n_i)$ cells empty. This can be done in

$$W_i = \frac{\nu!}{(\nu - n_i)!} \tag{10.82}$$

ways. The coarse-grained distribution is completely specified by the set of numbers $\{n_i\}$. To compute the number of ways of obtaining this configuration, we have to multiply the product of W_i's by the number of ways of splitting N microcells into $\{n_i\}$. This factor is $(N!/\prod n_i)$. Hence the number of ways of obtaining the particular coarse-grained configuration is

$$p = N! \prod_i \frac{\nu!}{n_i!(\nu - n_i)!}. \tag{10.83}$$

We now introduce a crucial assumption, viz., that particles are completely mixed up in phase space, giving equal a priori probability to all the configurations. In that case, the probability for any coarse-grained configuration is proportional to the number of ways of arriving at the configuration; we can therefore interpret p as proportional to this probability. We can obtain the most probable coarse-grained distribution function by maximising this probability with respect to n_i's, subject to the constraints that the total number of particles in the system and the energy of the system remain constant. We note that the expression for p is the same as that of a Fermi–Dirac gas [see Exercise 5.9] except for the (unimportant)

$N!$ term. Therefore the coarse-grained distribution corresponding to maximum probability will be the Fermi–Dirac distribution with

$$f_c(\mathbf{x}, \mathbf{v}) = \eta[\exp \beta(\epsilon - \mu) + 1]^{-1}, \tag{10.84}$$

where

$$\epsilon = \epsilon(\mathbf{x}, \mathbf{v}) = \frac{1}{2}v^2 + \phi(\mathbf{x}), \tag{10.85}$$

$$\nabla^2 \phi = 4\pi G \int f_c(\mathbf{x}, \mathbf{v}) \, d^3\mathbf{v}. \tag{10.86}$$

Equations (10.84)–(10.86) represent a self-gravitating gas of fermions with two parameters, μ and β. These parameters can be fixed in terms of the total energy and mass of the system.

In realistic situations the phase density will be significantly low compared with the maximum possible value, that is, we expect $f_c \ll \eta$ so that we can approximate the Fermi–Dirac distribution by an equivalent Maxwell–Boltzmann distribution and obtain

$$f_c(\mathbf{x}, \mathbf{v}) \simeq \eta \exp - \beta(\epsilon - \mu) \equiv A \exp - \beta\epsilon. \tag{10.87}$$

This equation, along with Eq. (10.86), implies that systems that have undergone collisionless relaxation are represented by isothermal spheres.

The mixing in phase space, described by the above process, can take place in a few orbital time scales. Hence this process operates at a far shorter time scale compared with the collisional time scale. Because of this reason, collisionless relaxation is also called violent relaxation. It must, however, be stressed that complete mixing in phase space does not occur for realistic systems and hence the argument leading to Eq. (10.84) is not strictly valid. Although violent relaxation does occur in realistic systems, the process may not reach completion and the distribution function could differ significantly from that in Eq. (10.84).

Such a collisionless evolution has to obey certain strong constraints. Let the phase-space density, i.e., the number of particles per unit phase-space volume, be $\mathcal{N}_0$ for the initial distribution and $\mathcal{N}$ for the final distribution. During a collisionless evolution, regions of high phase-space density get mixed with regions of lower phase-space density. Hence the value of the maximum phase-space density can only decrease in this process. Thus the phase-space density $\mathcal{N}$ must be less than the maximum value of $\mathcal{N}_0$.

In fact, this is a special case of a more general constraint that can be obtained as follows: Given a coarse-grained distribution function f_c, we define the volume of phase space with a phase density larger than q by the relation

$$V(q) = \int d\mathbf{x} \, d\mathbf{v} \theta[f_c(\mathbf{x}, \mathbf{v}) - q]. \tag{10.88}$$

where $\theta(z)$ is unity for $z > 0$ and is zero otherwise. Similarly, the mass contained in the region with a phase density larger than q is defined as

$$M(q) = \int d\mathbf{x} \, d\mathbf{v} \, f_c \theta(f_c - q). \tag{10.89}$$

Clearly

$$\frac{dM}{dq} = -\int d\mathbf{x} \, d\mathbf{v} \, f_c \delta_D(f_c - q) = -q \int d\mathbf{x} \, d\mathbf{v} \delta_D(f_c - q) = q \frac{dV}{dq}, \tag{10.90}$$

so that

$$M(V) = \int_0^V q(V') dV', \tag{10.91}$$

where $q(V)$ is the inverse function of $V(q)$. Given the distribution function f_c we can obtain $M(V)$ by eliminating q between relations (10.88) and (10.89).

Now suppose we are given two distribution functions f_1 and f_2 with corresponding $M_1(V)$ and $M_2(V)$. We can then show that f_1 could have evolved into f_2 if and only if $M_2(V) \le M_1(V)$ for all V. To do this, we need a preliminary result, which we derive first.

Consider any functional H of the coarse-grained distribution function f_c defined by the integral

$$H[f_c] = -\int C(f_c) \, d\mathbf{x} \, d\mathbf{v}, \tag{10.92}$$

where C is a *convex* function. Let the original (also called fine-grained) distribution function that, before averaging, corresponds to f_c be f; we also assume that at some initial moment $t = t_1$, $f_c(t_1) = f(t_1)$. Because the evolution is collisionless, $\dot{f} = 0$ and $\dot{C} = C'\dot{f} = 0$; clearly, C is conserved during evolution.

It can easily be shown that $H(t_2) \ge H(t_1)$ for all $t_2 > t_1$. We note that

$$H(t_2) - H(t_1) = \int d\mathbf{x} \, d\mathbf{v} \{C[f_c(t_1)] - C[f_c(t_2)]\}$$

$$= \int d\mathbf{x} \, d\mathbf{v} \{C[f(t_1)] - C[f_c(t_2)]\}$$

$$= \int d\mathbf{x} \, d\mathbf{v} \{C[f(t_2)] - C[f_c(t_2)]\}. \tag{10.93}$$

[The second equality follows from the fact that at $t = t_1$, $f_c(t_1) = f(t_1)$, the third from the conservation of $C(f)$ under time evolution.] However, the coarse-grained distribution function f_c is obtained by the averaging of f over the phase cells. For any convex function $C(f)$, the quantity $C(\langle f \rangle)$ will be less than $\langle C(f) \rangle$, where $\langle \cdots \rangle$ denotes an averaging process with positive semidefinite weights.

Therefore we may conclude that

$$\int d\mathbf{x}\, d\mathbf{v} C(f_c) \le \int d\mathbf{x}\, d\mathbf{v} C(f), \tag{10.94}$$

and hence

$$H(t_2) > H(t_1). \tag{10.95}$$

It should be stressed that $H(t)$, in general, is *not* a monotonically increasing function of t. The instant t_1 is special in the sense that we set $f_c = f$ at that instant. The above argument shows only that H functions will always have values higher than the value taken at this special instant. For two arbitrary instants $(t_2, t_3) > t_1$, nothing can be said about the relative values of $H(t_2)$ and $H(t_3)$.

The above condition shows that collisionless evolution must proceed in such a manner that *all* convex H functionals increase during the evolution. Suppose we are given two distribution functions $f_c(t_2)$ and $f_c(t_1) = f(t_1)$ with $t_2 > t_1$. Collisionless evolution could have evolved $f_c(t_1)$ to $f_c(t_2)$ only if *all* H functionals satisfy the condition $H[f_c(t_2)] > H[f_c(t_1)]$. We now show that if $M_2(V) \le M_1(V)$, then $H(f_2) > H(f_1)$ for all H functions. Starting with the relation

$$H(f_2) - H(f_1) = \int d\mathbf{x}\, d\mathbf{v}[C(f_1) - C(f_2)] = \int_0^\infty dV\{C(f_1) - C(f_2)\},$$

$$\tag{10.96}$$

and using the fact that, for convex functions, $C(f_1) - C(f_2) \ge (f_1 - f_2)C'(f_2)$, we get

$$H_2 - H_1 \ge \int_0^\infty dV(f_1 - f_2)C'(f_2) = \int_0^\infty dV \frac{d}{dV}(M_1 - M_2)C'(f_2).$$

$$\tag{10.97}$$

Integrating by parts, we find that

$$H_2 - H_1 \ge \{(M_1 - M_2)C'(f_2)\}_0^\infty - \int_0^\infty dV(M_1 - M_2)C''\left(\frac{df_2}{dV}\right)dV.$$

$$\tag{10.98}$$

The first term vanishes at both the limits; the second term is nonnegative because $(df_2/dV) \le 0$ and (for convex functions) $C'' > 0$. Thus $H_2 > H_1$.

We can also prove that if f_1 can evolve into f_2 then $M_2(V) < M_1(V)$. We can easily do this by using the convex function

$$C(f) = \begin{cases} 0 & f \le \phi \\ f - \phi & f > \phi \end{cases}, \tag{10.99}$$

where $\phi = f(V_0)$ with some arbitrary V_0. The fact that the H function constructed out of this particular $C(f)$ should be nondecreasing leads to the condition $M_2(V_0) \le M_1(V_0)$.

One useful corollary of the above constraint is the result obtained earlier: The maximum value of the phase density can only decrease during collision-less evolution. Because $f_{max} = f(V = 0)$, the Taylor expansion of $M(V)$ gives $M(V) \approx f_{max} V$; the condition $M_2 < M_1$ immediately leads to the conclusion that $f_{max}^{(2)} < f_{max}^{(1)}$, provided that $f_{max}^{(1)}$ is evaluated at the initial moment $t = t_i$ at which the coarse-grained and the fine-grained distributions are the same.

10.7 Collisional Evolution

It was mentioned earlier that the description of self-gravitating systems such as galaxies, globular clusters, etc., as collisionless is only an approximation, although a very good approximation. We now consider the corrections to these approximations arising from gravitational collisions. The basic formalism is identical to that used in Chap. 9, Section 9.3, to describe collisions in a plasma, and we shall merely highlight the differences.

The gravitational force acting on any particle in a self-gravitating system can be divided into two parts, $\mathbf{f}_{smooth}$ and $\mathbf{f}_{fluc}$. The $\mathbf{f}_{smooth}$ is due to the gravitational potential arising from the smooth distribution of matter. Because the matter is made up of individual particles, there will be a deviation from the smooth force $\mathbf{f}_{smooth}$, and this deviation is denoted by the fluctuating part of the force $\mathbf{f}_{fluc}$. Such a fluctuating force can be thought of as arising because of two different processes: close encounters and distant encounters. Let us consider the close encounters first.

Consider a gravitational encounter between two stars, each of mass m and relative velocity v. If the impact parameter is b, then the typical transverse velocity induced by the encounter is $\delta v_\perp \simeq (Gm/b^2)(2b/v) = (2Gm/bv)$. The deflection of the stars can be significant if $(\delta v_\perp/v) \gtrsim 1$, which occurs for collisions with impact parameter $b \lesssim b_c \simeq (Gm/v^2)$. Thus the effective cross section for collision with significant change of momentum is $\sigma \simeq b_c^2 \simeq (G^2m^2/v^4)$. The collisional relaxation time scale will be

$$t_{gc} \simeq \frac{1}{(n\sigma v)} \simeq \frac{R^3 v^3}{N(G^2 m^2)} \simeq \frac{NR^3 v^3}{G^2 M^2}. \tag{10.100}$$

If the system is virialised, then $(GM/R) \simeq v^2$, and we can write

$$t_{gc} \simeq N\left(\frac{R}{v}\right)\left(\frac{R^2 v^4}{G^2 M^2}\right) \simeq N\left(\frac{R}{v}\right). \tag{10.101}$$

Thus the collisional time scale is N times the crossing time R/v of the system.

There is, however, another effect that operates on a slightly shorter time scale. When the impact parameter b is far larger than b_c, individual collisions impart a small transverse velocity $(\delta v_\perp) \simeq (Gm/bv)$ to a given star. The effect of a large number of such collisions is to make the star perform a random walk in the velocity space. This will lead to a slow diffusion of particles in the momentum

space. As discussed in Chap. 9, Section 9.3 such a diffusion process can be
studied by an equation of the kind

$$\frac{df}{dt} = \frac{\partial f}{\partial t} + \mathbf{v} \cdot \frac{\partial f}{\partial \mathbf{x}} - \nabla\phi \cdot \frac{\partial f}{\partial \mathbf{v}} = -\frac{\partial J^\alpha}{\partial p^\alpha}. \tag{10.102}$$

The right-hand side is the divergence of a particle current J^α in momentum space,
which is characteristic of a diffusive process. The derivation of the form of J^α
proceeds exactly as in Chap. 9, Section 9.3. The final result can be expressed in
the form

$$J_\alpha = \frac{B_0}{2} \int d\mathbf{l'} \left\{ f \frac{\partial f'}{\partial l_\beta} - f' \frac{\partial f}{\partial l_\beta} \right\} \left\{ \frac{\delta_{\alpha\beta}}{k} - \frac{k_\alpha k_\beta}{k^3} \right\}, \tag{10.103}$$

where $\mathbf{k}$ is the momentum change $\mathbf{l} - \mathbf{l'}$ and $B_0 = (4\pi G^2 m^5 L)$, where L is the
usual logarithmic integral over the impact parameter:

$$L = \int_{b_1}^{b_2} \frac{db}{b} = \ln\left(\frac{b_2}{b_1}\right). \tag{10.104}$$

The values of b_2 and b_1 have to determined by physical considerations. For
gravitating systems, it is reasonable to take $b_2 \simeq R$, the size of the system.
As regards b_1, note that the effect of close encounters discussed above becomes important for $b < b_c$. Therefore we may take $b_1 \simeq b_c \simeq (Gm/v^2)$. Then
$(b_2/b_1) \simeq (Rv^2/Gm) = N(Rv^2/GM) \simeq N$ if the system is virialised. We now
consider several properties of the evolution described by this equation.

To begin with, it is clear that the collision term vanishes for the Maxwellian
distribution as it has the same form as in the case of plasma. This is obvious
from the form of Eq. (10.103), and, in fact, this is the key advantage of the form
in Eq. (10.103). But for self-gravitating systems, it is conventional to rewrite the
collisional term in a slightly different manner that can be achieved with a series
of algebraic manipulations.

It is clear that J_α will have one term proportional to f and one term proportional to $(\partial f/\partial l_\beta)$. Using

$$\frac{\delta_{\alpha\beta}}{k} - \frac{k_\alpha k_\beta}{k^3} = \frac{\partial^2 |\mathbf{k}|}{\partial k_\alpha \partial k_\beta} = \frac{\partial^2 k}{\partial k_\alpha \partial k_\beta}, \tag{10.105}$$

we can rewrite Eq. (10.103) as

$$J_\alpha = \frac{B_0}{2} f(\mathbf{l}) \int d\mathbf{l'} \frac{\partial f}{\partial l'_\beta} \cdot \frac{\partial^2 k}{\partial k_\alpha \partial k_\beta} - \frac{B_0}{2} \frac{\partial f}{\partial l_\beta} \cdot \int d\mathbf{l'} \frac{\partial^2 k}{\partial k_\alpha \partial k_\beta}. \tag{10.106}$$

Further, because

$$\frac{\partial}{\partial k_\alpha} = \frac{\partial}{\partial l_\alpha} = -\frac{\partial}{\partial l'_\alpha}, \tag{10.107}$$

J_α can be expressed in the form

$$J_\alpha = \frac{B_0}{2} f \int d\mathbf{l}' f_\beta' \frac{\partial^2 k}{\partial k_\alpha \partial k_\beta} - \frac{B_0}{2} f_\beta \cdot \left(\frac{\partial^2 \psi}{\partial k_\alpha \partial k_\beta} \right) \qquad (10.108)$$

where

$$\psi(\mathbf{l}) = \int d\mathbf{l}' f(\mathbf{l}')|\mathbf{l} - \mathbf{l}'|, \quad f_\beta' \equiv \partial f(\mathbf{l}')/\partial l_\beta'. \qquad (10.109)$$

We can transform the two terms in Eq. (10.108) into a more conventional form by integrating them by parts and using Eq. (10.107) repeatedly:

$$\int d\mathbf{l}' \frac{\partial f}{\partial l_\beta'} \cdot \frac{\partial^2 k}{\partial k_\alpha \partial k_\beta} = - \int d\mathbf{l}' f(\mathbf{l}') \frac{\partial}{\partial l_\beta'} \frac{\partial^2 k}{\partial k_\beta \partial k_\beta}$$

$$= + \int d\mathbf{l}' f(\mathbf{l}') \frac{\partial}{\partial k_\beta} \frac{\partial^2 k}{\partial k_\alpha \partial k_\beta} = \int d\mathbf{l}' f(\mathbf{l}') \frac{\partial}{\partial k_\alpha} \frac{\partial^2 k}{\partial k_\beta \partial k_\beta}$$

$$= \frac{\partial}{\partial l_\alpha} \frac{\partial}{\partial l^\beta \partial l_\beta} \int d\mathbf{l}' f(\mathbf{l}')|\mathbf{l} - \mathbf{l}'| = \frac{\partial}{\partial l_\alpha} \cdot \nabla_l^2 \psi(\mathbf{l}) \quad (10.110)$$

and

$$f_\beta \frac{\partial^2 \psi}{\partial k_\alpha \partial k_\beta} = \frac{\partial}{\partial k_\beta} \left(f \frac{\partial^2 \psi}{\partial k_\alpha \partial k_\beta} \right) - f \frac{\partial}{\partial k_\beta} \frac{\partial^2 \psi}{\partial k_\alpha \partial k_\beta}$$

$$= \frac{\partial}{\partial k_\beta} \left(f \frac{\partial^2 \psi}{\partial k_\alpha \partial k_\beta} \right) - f \frac{\partial}{\partial k_\alpha} (\nabla^2 \psi), \qquad (10.111)$$

where $\nabla^2 = (\partial^2 / \partial k^\alpha \partial k_\alpha)$ stands for the Laplacian in momentum space. Defining

$$\eta(\mathbf{l}) \equiv \nabla^2 \psi(\mathbf{l}) = \frac{\partial^2}{\partial l^\alpha \partial l_\alpha} \int d\mathbf{l} f(\mathbf{l}')|\mathbf{l} - \mathbf{l}'| = 2 \int d\mathbf{l}' \frac{f(\mathbf{l}')}{|\mathbf{l} - \mathbf{l}'|} \qquad (10.112)$$

and substituting Eqs. (10.110) and (10.111) into Eq. (10.108), we get

$$J_\alpha(l) = \frac{B_0}{2} f(\mathbf{l}) \frac{\partial \eta}{\partial l_\alpha} - \frac{B_0}{2} \left\{ \frac{\partial}{\partial l_\beta} \left(f \frac{\partial^2 \psi}{\partial l_\alpha \partial l_\beta} \right) - f \frac{\partial \eta}{\partial k_\alpha} \right\}$$

$$= B_0 f(\mathbf{l}) \frac{\partial \eta}{\partial l_\alpha} - \frac{B_0}{2} \frac{\partial}{\partial l_\beta} \left(f \frac{\partial^2 \psi}{\partial l_\alpha \partial l_\beta} \right)$$

$$\equiv a_\alpha(\mathbf{l}) f(\mathbf{l}) - \frac{1}{2} \frac{\partial}{\partial l_\beta} \{ \sigma_{\alpha\beta}^2 f \}, \qquad (10.113)$$

where $a_\alpha = (\partial \eta / \partial l_\alpha)$ and $\sigma_{\alpha\beta}^2 = (\partial^2 \psi / \partial l_\alpha \partial l_\beta)$, with

$$\nabla^2 \psi = \eta, \quad \nabla_l^2 \eta(\mathbf{l}) = \nabla_l^2 \left\{ 2 \int d\mathbf{l}' \frac{f(\mathbf{l}')}{|\mathbf{l} - \mathbf{l}'|} \right\} = -8\pi f(\mathbf{l}). \qquad (10.114)$$

To understand the physical meaning of a diffusion equation driven by a current J_α of the form given in Eq. (10.113), let us first consider a simpler equation of the same form:

$$\frac{\partial f(v, t)}{\partial t} = \frac{\partial}{\partial v}\left\{(\alpha v)f + \frac{\sigma^2}{2}\frac{\partial f}{\partial v}\right\} \equiv -\frac{\partial J}{\partial v}. \tag{10.115}$$

The J in this equation is similar in structure to J_α in Eq. (10.113) if we confine our attention to one dimension and set $a = \alpha v$. Let us consider the effect of the two terms inside the braces in Eq. (10.115) separately.

The second term $(\sigma^2/2)(\partial f/\partial v)$ has the standard form of a diffusion current proportional to the gradient in the velocity space. As time goes on, this term will cause the mean-square velocities of particles to increase in proportion to t, inducing the random walk in the velocity space. Under the effect of this term, *all* the particles in the system will have their $\langle v^2 \rangle$ increasing without bound. This unphysical situation is avoided by the presence of the first term $(\alpha v f)$ in J. This term acts as a friction term (called dynamical friction). The combined effect of the two terms is to drive f to a Maxwellian distribution with $\beta = (k_B T)^{-1} = (\alpha/\sigma^2)$. In such a Maxwellian distribution the gain made in Δv^2 that is due to diffusion is exactly balanced by the losses that are due to dynamical friction. When two particles scatter, one gains the energy lost by the other; on the average, we may say that the one that has lost the energy has undergone dynamical friction whereas the one that has gained energy has achieved diffusion to higher v^2. The cumulative effect of such phenomena is described by the two terms in $J(v)$.

The above points can be easily illustrated by the explicit solution of Eq. (10.115). Suppose we take a initial distribution $f(v, 0) = \delta(v - v_0)$ peaked at a velocity v_0. The solution of Eq. (10.115) with this initial condition is

$$f(v, t) = \left[\frac{\alpha}{\pi\sigma^2(1 - e^{-2\alpha t})}\right]^{1/2} \exp\left[-\frac{\alpha(v - v_0 e^{-\alpha t})^2}{\sigma^2(1 - e^{-2\alpha t})}\right], \tag{10.116}$$

which is a Gaussian with the mean

$$\langle v \rangle = v_0 e^{-\alpha t} \tag{10.117}$$

and dispersion

$$\langle v^2 \rangle - \langle v \rangle^2 = \frac{\sigma^2}{\alpha}(1 - e^{-2\alpha t}). \tag{10.118}$$

At late times ($t \to \infty$), the mean velocity $\langle v \rangle$ goes to zero while the velocity dispersion becomes (σ^2/α). Thus the equilibrium configuration is a Maxwellian distribution of velocities with this particular dispersion, for which $J = 0$. To see the effect of the two terms individually on initial distribution $f(v, 0) = \delta(v - v_0)$, we can set α or σ to zero. When $\alpha = 0$, we get pure diffusion:

$$f_{\alpha=0}(v, t) = \left(\frac{1}{2\pi\sigma^2 t}\right)^{1/2} \exp\left\{-\frac{(v-v_0)^2}{2\sigma^2 t}\right\}. \tag{10.119}$$

Nothing happens to the steady velocity v_0; but the velocity dispersion increases in proportion to t, representing a random walk in the velocity space. On the other hand, if we set $\sigma = 0$, then we get

$$f_{\sigma=0}(v, t) = \delta(v - v_0 e^{-\alpha t}). \tag{10.120}$$

Now there is no spreading in velocity space (no diffusion); instead the friction steadily decreases $\langle v \rangle$. Note that the time scale for the operation of the dynamical friction is $\alpha^{-1} = (a/v)^{-1}$.

Equations (10.114) allow the computation of the coefficient of dynamical friction α if f is known. For a Maxwellian distribution of velocities,

$$f(v) = \frac{n_0}{(2\pi q^2)^{3/2}} \exp\left(-\frac{v^2}{2q^2}\right), \tag{10.121}$$

we can integrate the second equation in Eqs. (10.114) to find η and then compute $a_\mu = (\partial\eta/\partial l_\mu)$. This gives

$$a(v) = -\left(\frac{32\sqrt{\pi}}{3}L\right)(G^2 m^2 n_0)\left(\frac{v}{q^3}\right)$$

$$\times \left[\frac{3\sqrt{\pi}}{4}\left(\frac{q}{v}\right)^3 \mathrm{erf}\left(\frac{v}{\sqrt{2}q}\right) - \frac{3}{2\sqrt{2}}\left(\frac{q}{v}\right)^2 e^{-v^2/2q^2}\right]. \tag{10.122}$$

For small (v/q), the expression in brackets has the Taylor expansion $[1 - (3/5)(v/q)^2 + \mathcal{O}(v^4/q^4)]$. Therefore, in this limit, we have

$$a(v) \cong -\frac{32\sqrt{\pi}}{3} n_0(Gm)^2 L \frac{v}{q^3}\left(1 - \frac{3}{5}\frac{v^2}{q^2} + \cdots\right). \tag{10.123}$$

The relaxation time scale t_c is determined by

$$t_c^{-1} \equiv \alpha(v) \equiv -\frac{a(v)}{v} \cong \frac{16}{3\sqrt{\pi}}\left(\frac{\ln N}{N}\right)\left(\frac{q}{R}\right). \tag{10.124}$$

This is the time scale specified by $C(f)$. This result shows that the time scale for gravitational collisions to change the structure of the system significantly is $(N/\ln N)$ times the typical orbital time scale, except for numerical factors of the order of unity.

For future applications, we shall rewrite the dynamical friction term in a different form. It is clear from Eq. (10.117) that the mean velocity changes according to the law $\langle \dot{v} \rangle = -\alpha \langle v \rangle$. The expression for $a(v) = \alpha v$ obtained in relation (10.123) is applicable for a system made of particles of equal mass. A similar effect will occur even when a particle of mass M moves through a system of particles (of mass m) described by a Maxwellian distribution function of the form in Eq. (10.121). In this case, the velocity of the particle with mass M changes according to the law

$$\frac{d\mathbf{v}_M}{dt} = -\frac{4\pi L G^2(M+m)n_0 m}{v_M^3}\left[\text{erf}(X) - \frac{2X}{\sqrt{\pi}}e^{-X^2}\right]\mathbf{v}_M, \quad (10.125)$$

where $X = v_M/(\sqrt{2}q)$. [We have only replaced m^2 with $(M+m)m$ in Eq. (10.122).] This formula is usually used in the limit of $M \gg m$. In that case, the dynamical friction depends on only the mass density $\rho_0 = mn_0$ and is independent of the mass of the individual particles that form the Maxwellian background distribution.

Finally we comment on one curious property of dynamical friction. We note that the second equation in Eqs. (10.114) is essentially Coulomb's law in the velocity space, with $f(\mathbf{l})$ as the source. Suppose that the velocity distribution is spherically symmetrical, i.e., $f(\mathbf{l}) = f(|\mathbf{l}|)$. In that case, the force $(\partial\eta/\partial l_\alpha)$ at any given $\mathbf{l}$ is due to only the particles inside the (velocity) sphere of radius $|\mathbf{l}|$. In other words, dynamical friction force on a particle with momentum l is contributed *only* by the particles with lower momentum if $f(\mathbf{l}) = f(|\mathbf{l}|)$.

10.8 Dynamical Evolution of Gravitating Systems

Because gravitating systems do not possess extensivity in energy, the standard formalism of statistical mechanics developed in Chap. 5 cannot be applied to such systems. The question arises as to how we can view the dynamical evolution of self-gravitating systems of N-point particles.

In the previous sections this question was addressed in several parts. During the initial stages of evolution such a system undergoes violent relaxation and approaches a steady-state. Assuming that such general considerations as the maximisation of phase volume is applicable, we may ask what kind of steady-state configuration is preferred. The following caveat, however, should be kept in mind: The steady-state distribution obeys the collisionless Boltzmann equation for $t \ll t_c$ and several different kinds of solutions are possible to this equation. The particular configuration attained by any given system could, in principle, depend on the initial configuration. In the statistical mechanics of the systems interacting by short range, initial conditions do not play a significant role in determining the steady-state configuration. However, it is not clear whether such a result holds for systems dominated by long-range interactions.

Consider a system of N particles interacting with each other through the two-body potential $U(\mathbf{x}, \mathbf{y})$. The entropy S of this system, in the microcanonical description, is defined through the relation

$$e^S = g(E) = \frac{1}{N!} \int d^{3N}x \, d^{3N}p \, \delta(E - H)$$

$$= \frac{A}{N!} \int d^{3N}x \left[E - \frac{1}{2} \sum_{i \neq j} U(\mathbf{x}_i, \mathbf{x}_j) \right]^{\frac{3N}{2}}, \qquad (10.126)$$

where we have performed the momentum integrations and replaced $(3N/2 - 1)$ with $(3N/2)$. We approximate expression (10.126) in the following manner.

Let the spatial volume V be divided into M (with $M \ll N$) cells of equal size, large enough to contain many particles but small enough for the potential to be treated as a constant inside each cell. (It is assumed that such an intermediate scale exists.) Instead of integrating over the particle coordinates $(\mathbf{x}_1, \mathbf{x}_2, \ldots, \mathbf{x}_N)$, we sum over the number of particles n_a in the cell centred at $\mathbf{x}_a$ (where $a = 1, 2, \ldots, M$). Using the standard result that the integration over $(N!)^{-1} d^{3N}x$ can be replaced with

$$\sum_{n_1=1}^{\infty} \left(\frac{1}{n_1!} \right) \sum_{n_2=1}^{\infty} \left(\frac{1}{n_2!} \right) \cdots \sum_{n_M=1}^{\infty} \left(\frac{1}{n_M!} \right) \delta \left(N - \sum_a n_a \right) \left(\frac{V}{M} \right)^N \quad (10.127)$$

and ignoring the unimportant constant A, we can rewrite Eq. (10.126) as

$$e^S = \sum_{n_1=1}^{\infty} \left(\frac{1}{n_1!} \right) \sum_{n_2=1}^{\infty} \left(\frac{1}{n_2!} \right) \cdots \sum_{n_M=1}^{\infty} \left(\frac{1}{n_M!} \right) \delta \left(N - \sum_a n_a \right)$$

$$\times \left(\frac{V}{M} \right)^N \left[E - \frac{1}{2} \sum_{a \neq b}^{M} n_a U_{ab} n_b \right]^{\frac{3N}{2}}$$

$$\approx \sum_{n_1=1}^{\infty} \sum_{n_2=1}^{\infty} \cdots \sum_{n_M=1}^{\infty} \delta \left(N - \sum_a n_a \right) \exp S(\{n_a\}), \qquad (10.128)$$

where

$$S(\{n_a\}) = \frac{3N}{2} \ln \left[E - \frac{1}{2} \sum_{a \neq b}^{M} n_a U(\mathbf{x}_a, \mathbf{x}_b) n_b \right] - \sum_{a=1}^{M} n_a \ln \left(\frac{n_a M}{eV} \right). \qquad (10.129)$$

In arriving at the last expression we have used Sterling's approximation for the factorials. We obtain the mean-field limit now by retaining in the sum of Eq. (10.128) only the term for which the summand reaches the maximum value, subject to the constraint on the total number, that is, we claim that

$$\sum_{\{n_a\}} e^{S[n_a]} \approx e^{S[n_{a,\max}]}, \qquad (10.130)$$

where $n_{a,\max}$ is the solution to the variational problem

$$\left(\frac{\delta S}{\delta n_a}\right)_{n_a=n_{a,\max}} = 0 \quad \text{with} \quad \sum_{a=1}^{M} n_a = N. \tag{10.131}$$

Imposing this constraint by a Lagrange multiplier and using expression (10.129) for S, we obtain the equation satisfied by $n_{a,\max}$:

$$\frac{1}{T}\sum_{b=1}^{M} U(\mathbf{x}_a, \mathbf{x}_b)n_{b,\max} + \ln\left(\frac{n_{a,\max}M}{V}\right) = \text{constant}, \tag{10.132}$$

where we have defined the temperature T as

$$\frac{1}{T} = \frac{3N}{2}\left[E - \frac{1}{2}\sum_{a\neq b}^{M} n_a U(\mathbf{x}_a, \mathbf{x}_b)n_b\right]^{-1} = \beta. \tag{10.133}$$

We see from Eq. (10.129) that this expression is also equal to $\partial S/\partial E$; therefore T is indeed the correct thermodynamic temperature. We can now return back to the continuum limit by the replacements

$$n_{a,\max}\frac{M}{V} = \rho(\mathbf{x}_a), \quad \sum_{a=1}^{M} \to \frac{M}{V}\int. \tag{10.134}$$

In this limit, extremum solution (10.132) is given by

$$\rho(\mathbf{x}) = A\exp[-\beta\phi(\mathbf{x})]; \quad \text{where} \quad \phi(\mathbf{x}) = \int d^3\mathbf{y}\,U(\mathbf{x}, \mathbf{y})\rho(\mathbf{y}), \tag{10.135}$$

which, in the case of gravitational interactions, become

$$\rho(\mathbf{x}) = A\exp[-\beta\phi(\mathbf{x})], \quad \phi(\mathbf{x}) = -G\int\frac{\rho(\mathbf{y})\,d^3\mathbf{y}}{|\mathbf{x} - \mathbf{y}|}. \tag{10.136}$$

Equations (10.136) represent the equilibrium configuration for a gravitating system in the mean-field limit. The constant β is already determined through Eq. (10.133) in terms of the total energy of the system. The constant A has to be fixed in terms of the total number (or mass) of the particles in the system.

An important point needs to be noted about the mean-field result we have obtained: The various manipulations in Eqs. (10.126)–(10.132) tacitly assume that the expressions we are dealing with are finite. Unfortunately, for gravitational interactions *without* a short-distance cutoff, the quantity e^S and hence all the terms we have been handling are divergent. We should therefore remember that a short-distance cutoff is needed to justify the entire procedure and that Eqs. (10.136) – which is based on a strict r^{-1} potential and does not incorporate any such cutoff – can be only approximately correct.

This result also shows that the isothermal sphere discussed in Subsection 10.3.3 has a special status as a solution to the mean-field equations. It was, however, shown in Subsection 10.3.3 that isothermal spheres cannot exist if $(RE/GM^2) < -0.335$. Even when $(RE/GM^2) > -0.335$, the isothermal solution need not be stable. We can investigate the stability of this solution by studying the second variation of the entropy.

Such a detailed analysis shows that the following results are true. (1) Systems with $(RE/GM^2) < -0.335$ cannot evolve into isothermal spheres. Entropy has no extremum for such systems. (2) Systems with $(RE/GM^2) > -0.335$ and $\rho(0) > 709\,\rho(R)$ can exist in a metastable (saddle-point state) isothermal sphere configuration. Here $\rho(0)$ and $\rho(R)$ denote the densities at the centre and the edge, respectively. The entropy extrema exist but they are not local maxima. (3) Systems with $(RE/GM^2) > -0.335$ and $\rho(0) < 709\,\rho(R)$ can form isothermal spheres that are local maxima of entropy. These results are not of direct significance to astrophysical systems. However, in a modified form, these considerations seem to play a role in the late time evolution of globular clusters. (See Vol. II.)

11

General Theory of Relativity

11.1 Introduction

This chapter develops the general theory of relativity and discusses some of its key applications that will be of relevance to astrophysics. It uses concepts developed in Chaps. 3 and 4 and will be needed in the discussion of physics of compact objects (Vol. II) and cosmology (Vol. III). In this chapter, Latin indices such as $a, b, \ldots, i, k, \ldots$, vary over 0, 1, 2, 3, and the Greek indices such as $\alpha, \beta, \ldots$, vary over 1, 2, 3.

11.2 Inescapable Connection between Gravity and Geometry

Combining special relativity with Newtonian gravity turns out to be far more difficult than we would have imagined a priori. Any attempt to do so inevitably suggests a geometrical description for gravity. In this section we examine some simple attempts to combine Newtonian gravity and special relativity and see how a geometrical description arises in the process.

The action for a particle of mass m, in an external gravitational field characterised by the Newtonian potential $\phi(t, \mathbf{x})$, can be taken to be

$$A = \int dt \, \frac{1}{2} m v^2 - \int dt \, m\phi. \tag{11.1}$$

The simplest generalisation of this action, in the context of special relativity, will be

$$A = -mc^2 \int dt \sqrt{1 - \frac{v^2}{c^2}} - \int dt \sqrt{1 - \frac{v^2}{c^2}} \, m\phi, \tag{11.2}$$

in which dt is replaced by the proper time ds. Because the same parameter m multiplies the kinetic- and potential-energy terms of the action, it has no influence on the equations of motion. Therefore the trajectories $\mathbf{f}(t)$ of material particles (with the same initial conditions) will be independent of the properties of the particle and will depend on only the gravitational potential $\phi(x^i)$. In the limit of $c \to \infty$, the particles at an event $\mathcal{P}$ will experience the *same* acceleration $g^\alpha \equiv -(\partial\phi/\partial x^\alpha)|_\mathcal{P}$, where $\alpha = 1, 2, 3$. If we now choose a new set of spatial coordinates

$$\xi^\alpha = x^\alpha - f^\alpha(t) \cong x^\alpha - \frac{1}{2}g^\alpha t^2 \quad (\alpha = 1, 2, 3) \tag{11.3}$$

near $\mathcal{P}$, then in the new frame the particles will experience no acceleration. It follows that the trajectories of particles in a given gravitational field are indistinguishable *locally* from the trajectories of free particles viewed from an accelerated frame. Hence the gravitational field is locally indistinguishable from a suitably chosen noninertial frame as far as the laws of mechanics are concerned. Note that we could arrive at this result only because the trajectories $\mathbf{f}(t)$ are independent of any property of the particle. This is *not* the case, for example, in electromagnetism in which the trajectories depend on the q/m ratio for the particles.

Because accelerated frames seem to mimic the effects of a gravitational field locally, understanding the features of accelerated frames will help in arriving at the correct description of gravity. We begin by asking how we should construct coordinate systems that are appropriate for observers who are not inertial; that is, we want to generalise our concept of Lorentz transformations so as to include observers who are moving arbitrarily with respect to each other.

We can do this by using the procedure developed in Chap. 3, Section 3.3. Let (T, X, Y, Z) be an inertial coordinate system. Consider an observer travelling along the X axis in a trajectory $X = f(\tau)$, $T = h(\tau)$, where f and h are specified functions and τ is the proper time in the clock carried by the observer. Let $\mathcal{P}$ be some event with Minkowski coordinates (T, X) to which the observer assigns the coordinates (t, x). Then, it was shown in Chap. 3 that

$$X - cT = X_A - cT_A = f(t_A) - ch(t_A) = f(t - x/c) - ch(t - x/c), \tag{11.4}$$
$$X + cT = X_B + cT_B = f(t_B) + ch(t_B) = f(t + x/c) + ch(t + x/c). \tag{11.5}$$

Given f and h, these equations can be solved to find (X, T) in terms of (x, t).

Let us now apply this to an observer travelling along the X axis with a uniform acceleration g. The equation of motion

$$\frac{d}{dT}\left(\frac{v}{\sqrt{1 - v^2/c^2}}\right) = g \tag{11.6}$$

leads to the velocity

$$v = gT\left(1 + \frac{g^2 T^2}{c^2}\right)^{-1/2} = \frac{dX}{dT} \tag{11.7}$$

if we take the initial condition to be $v = 0$ at $T = 0$. Integrating this equation again and setting $X = 0$ at $T = 0$, we find that

$$X = \frac{c^2}{g}\left[\sqrt{1 + (g^2 T^2/c^2)} - 1\right].$$ (11.8)

The proper time τ in the clock carried by the observer is

$$\tau(T) = \int_0^T dT\sqrt{1 - v^2/c^2} = \frac{c}{g}\sinh^{-1}\left(\frac{gT}{c}\right).$$ (11.9)

Using this, we can express the trajectory (11.8) as

$$T = \frac{c}{g}\sinh\left(\frac{g\tau}{c}\right), \quad X = \frac{c^2}{g}\left[\cosh\left(\frac{g\tau}{c}\right) - 1\right].$$ (11.10)

By shifting the spatial origin by $c^2 g^{-1}$ (i.e., by using $X + c^2 g^{-1}$ rather than X), we can write the trajectory in a more symmetric form as

$$gT = \sinh g\tau \equiv g\,h(\tau), \quad gX = \cosh g\tau \equiv g\,f(\tau)$$ (11.11)

in units with $c = 1$. Equations (11.4) and (11.5) now become

$$X - T = g^{-1}\exp[-g(t - x)], \quad X + T = g^{-1}\exp[g(t + x)],$$ (11.12)

or,

$$X = g^{-1}e^{gx}\cosh gt, \quad T = g^{-1}e^{gx}\sinh gt.$$ (11.13)

This provides the transformation between the inertial coordinate system and that of a uniformly accelerated observer.

These transformations are clearly nonlinear and hence do not preserve the form of the line element ds^2. Using

$$dT^2 - dX^2 = d(T - X)d(T + X) = e^{2gx}(dt^2 - dx^2),$$ (11.14)

we get the line element in the accelerated frame:

$$ds^2 = dT^2 - dX^2 - dY^2 - dZ^2 = e^{2gx}(dt^2 - dx^2) - dy^2 - dz^2.$$ (11.15)

If we change to a new space coordinate $\bar{x}$ with $(1 + g\bar{x}) = e^{gx}$ and $e^{gx}dx = d\bar{x}$, we get

$$ds^2 = (1 + g\bar{x})^2 dt^2 - d\bar{x}^2 - dy^2 - dz^2.$$ (11.16)

This result shows that noninertial frames are described by intervals of the form $ds^2 = g_{ik}(x)\,dx^i\,dx^k$, where $g_{ik}(x)$, called the metric tensor, will, in general, depend on t and $\mathbf{x}$.

The two results we have obtained – namely, that (1) gravitational fields are locally indistinguishable from accelerated frames and that (2) accelerated frames are described by a line element of the form in Eq. (11.16) – combine to provide an important conclusion: The gravitational field affects the rate of clocks in such a way that the clocks slow down in strong gravitational fields. To the lowest order in (ϕ/c^2), we now show that

$$\Delta t = \Delta T \left(1 - \frac{\phi}{c^2}\right), \tag{11.17}$$

where ΔT is the time interval measured by a clock in the absence of the gravitational field and Δt is the corresponding interval measured by a clock located in the gravitational potential.

To see this, let us consider two frames of reference S and S', with S' moving along the positive x axis with acceleration g. An observer in S' will feel all particles in his or her vicinity to be moving in the $-\hat{x}$ direction with an acceleration g. Hence he or she will attribute this motion as arising because of a gravitational potential $\phi = gx$ in the vicinity. Let there be a series of clocks along the x axis in S and let the observer in S' compare the readings of his or her clock with these stationary clocks as he or she passes each one of them. From Eq. (11.9) we have $t = g^{-1} \sinh^{-1}(gT/c^2)$, giving

$$\Delta t = \frac{\Delta T}{\sqrt{1 + g^2 T^2/c^2}} \cong \Delta T \left(1 - \frac{1}{2}\frac{g^2 T^2}{c^2}\right). \tag{11.18}$$

However, $x = (1/2)gT^2$ is the location of the observer at this instant. So we can also write

$$\Delta t \cong \Delta T \left(1 - \frac{gx}{c^2}\right) = \Delta T \left(1 - \frac{\phi}{c^2}\right). \tag{11.19}$$

In the frame S', the observer will attribute the difference between ΔT and Δt to the presence of a gravitational potential ϕ and conclude that rate of the flow of clocks is affected by the gravitational field.

Considering the importance of this result, we provide a more physical argument that leads to the same conclusion. We know that the annihilation of electrons and positrons can lead to γ rays and that under suitable conditions we can produce $e^+ e^-$ pairs from the radiation. This fact can be used to devise a suitable thought experiment that will prove that the gravitational field must necessarily affect the rate of flow of clocks.

For the purpose of the argument given below, we may assume that the gravitational potential near the surface of the Earth varies linearly with height; then the potential difference between two points A and B, separated by height L, will be gL, where g is the acceleration that is due to gravity. Let us assume that a pair of sufficiently high-energy photons, each of energy $\hbar\omega$, was converted into an $e^+ e^-$ pair at point A. These particles move from point A to point B, losing the gravitational potential energy mgL per particle. (We have assumed that B is above A.) At B we annihilate the two particles and produce two photons that are sent back to A. The sequence of events described above has restored the original configuration. Because the amount of energy available in the form of particles was lower at B than at A (because of the variation in gravitational potential), it follows that the energy of the photons must change in going from B to A. If this is not the case, we could use the above sequence of events to increase the energy of the system repeatedly. Because the energy of the photon is proportional to the frequency, the frequency of the photons at A and B must be related by $\omega_A = \omega_B(1 + gL/c^2)$.

Because the frequency shift derived above is independent of $\hbar$ (even though the original argument was quantum mechanical), it follows that the result should be true even in the classical limit. However, in the classical limit, we can consider photons as making up an electromagnetic wave with some frequency. We can determine this frequency by counting the number of crests N of a wave train that crosses an observer in a time interval Δt. In a static gravitational field, the head of the wave train and the tail of the wave train will propagate in an identical manner. Hence two observers located at A and B will attribute different frequencies to the wave *only if* the reference clocks that they have at A and B run at different rates. To reproduce the frequency shift derived above, the time intervals of the clocks have to obey the relation obtained in relation (11.19).

This conclusion has far-reaching implications. Consider the line interval ds between two infinitesimally separated events in space–time. When the spatial interval $d\mathbf{x}$ between the two points vanishes, ds measures the lapse of proper time at a given spatial location. If a global Lorentz time exists, then $ds = dt$, which is the same for all observers at rest. If the gravitational field affects the rate of flow of clocks, then ds should vary from point to point even when $d\mathbf{x} = 0$. This is possible only if $ds^2 = g_{00}(\mathbf{x})\,dt^2$, where g_{00} is a nontrivial function of the spatial coordinates. From our previous argument, we can even fix the form of g_{00}; to arrive at the correct flow of time, we need $g_{00}(x) = (1 + 2\phi/c^2)$. It follows that the line interval between two events in the space–time cannot have the form $ds^2 = (dt^2 - d\mathbf{x}^2)$ but should *at least* be modified to a form

$$ds^2 = \left(1 + \frac{2\phi}{c^2}\right)c^2 dt^2 - d\mathbf{x}^2 \qquad (11.20)$$

in the presence of a gravitational field.

The same conclusion can be obtained from another argument, as follows. The action for a particle in a gravitational field,

$$A = -mc^2 \int dt \sqrt{1 - \frac{v^2}{c^2}} - \int dt \sqrt{1 - \frac{v^2}{c^2}}\, m\phi, \qquad (11.21)$$

can be rewritten as

$$A = -(mc) \int (cdt)\left[\sqrt{1 - \frac{v^2}{c^2}}\left(1 + \frac{\phi}{c^2}\right)\right]$$

$$\cong -(mc) \int \left[\left(1 + \frac{2\phi}{c^2}\right)c^2 dt^2 - d\mathbf{x}^2\right]^{1/2}, \qquad (11.22)$$

provided that $\phi \ll c^2$. The last form suggests that, if we introduce a line interval with

$$ds^2 = g_{ik}dx^i dx^k = \left(1 + \frac{2\phi}{c^2}\right)c^2 dt^2 - dx^2, \qquad (11.23)$$

then we can express the action as

$$A \simeq -mc \int \sqrt{g_{ik}dx^i dx^k} = -mc \int ds. \qquad (11.24)$$

In the case of a uniformly accelerated observer (corresponding to a constant gravitational field) we find that

$$ds^2 = [1 + (g\bar{x}/c^2)]^2 dt^2 - dx^2 \approx [1 + 2(g\bar{x}/c^2)] dt^2 - dx^2, \quad (11.25)$$

which matches the result derived above.

All the above features show that the space–time interval cannot retain its usual special relativistic form in the presence of a gravitational field. Such a modification of the metric tensor is absolutely essential if we have to obtain physically reasonable conclusions. For this reason, we have to treat a gravitational field as geometrical in origin.

11.3 Metric Tensor and Gravity

The above arguments show that a weak gravitational field cannot be distinguished from a modified space–time interval as far as mechanical phenomena are concerned. Einstein made a bold generalisation of this result by postulating that all aspects of gravitational physics allow a geometrical description. We thus generalise the tentative conclusion of the previous section to include arbitrarily strong gravitational fields and *all* possible physical phenomena. We now explore some elementary consequences of this postulate.

To begin with, we accept the conclusion obtained in Section 11.2 and postulate that the gravitational field will modify the space–time interval to the form

$$ds^2 = g_{ik}(x^a) dx^i dx^k, \qquad (11.26)$$

where the metric tensor $g_{ik}(x^a)$ characterises a specific gravitational field. We must first understand the geometrical features of the space–time described by such a metric.

A metric tensor such as $g_{ab}(x^i)$ – which depends on the coordinates – can arise in two separate contexts. If a flat space–time is described in noninertial coordinates, then $g_{ab}(x^i)$ will not be constant. (This was shown in Section 11.2 for a uniformly accelerated frame.) On the other hand, a nonconstant $g_{ab}(x^i)$ can also describe a curved space–time. To illustrate these two results in a familiar context, consider a two-dimensional space with coordinates (α, β) and the line interval $ds^2 = (d\alpha)^2 + (\alpha)^2(d\beta)^2$; the metric clearly depends on α. However, this space is really the flat two-dimensional space in polar coordinates. If we transform to $X = \alpha \cos(\beta)$, $Y = \alpha \sin(\beta)$ then the metric will take the familiar form $ds^2 = dX^2 + dY^2$. On the other hand, consider a line element $ds^2 = (d\alpha)^2 + \sin^2(\alpha)(d\beta)^2$ that represents the interval on the surface of a two sphere (with $\alpha = \theta$; $\beta = \phi$) and describes a genuinely curved surface. This line element cannot be reduced to the flat form by any coordinate transformation.

Coming back to four dimensions, it is *not* possible, in general, to change the coordinates from x^i to some other x'^i such that ds^2 in Eq. (11.26) reduces to the Minkowski line-element form. Hence the form in Eq. (11.26) represents, in general, a *curved* space–time. To prove this claim, let us consider a flat space–time with inertial coordinates X^i and metric $\eta_{ik} = \text{dia}\,(1, -1, -1, -1)$. We make a coordinate transformation from X^i to a new set of coordinates x^k that are arbitrary functions of the original coordinates. The line interval in the new coordinates now becomes

$$ds^2 = \eta_{ik}dX^i dX^k = \eta_{ik}\frac{\partial X^i}{\partial x^a}\frac{\partial X^k}{\partial x^b}dx^a dx^b \equiv g_{ab}(x)dx^a dx^b. \quad (11.27)$$

Note that the transformation from X^i to x^k involves only four arbitrary functions. Therefore g_{ab} in the line interval can contain only four independent functions. In general, $g_{ab}(x)$ has 10 independent functions. [This can be seen as follows. When a and b range through 0, 1, 2, 3, and the pair (a, b) can take $4 \times 4 = 16$ different values. Because g_{ab} is symmetric, only the diagonal values, 4 in number, and half of the nondiagonal values, $(16 - 4)/2 = 6$ in number, are independent.] It is therefore impossible to obtain an arbitrary set of $g_{ab}(x)$ by transforming the coordinates from the inertial frame to the noninertial frame. Conversely it is not possible to reduce an arbitrary metric $g_{ab}(x)$ to η_{ab} by a coordinate transformation.

It is, however, possible to choose coordinate systems such that, *near* any event $\mathcal{P}$, the following conditions are satisfied. (1) $g'_{ik}(\mathcal{P}) = \eta_{ik}$, (2) $(\partial g'_{ik}/\partial x^{k'})_{\mathcal{P}} = 0$. Consider a coordinate transformation from the coordinates x^i to $x^{i'}$ around an event $\mathcal{P}$. We expand $x^{i'}$ in terms of x^i in a Taylor series as

$$x^{i'} = A^i + B^i_k x^k + C^i_{jk}x^j x^k + D^i_{jkl}x^j x^k x^l + \cdots. \quad (11.28)$$

We now attempt the following: (1) We choose the values of B^i_k such that the transformed metric tensor g'_{ik} at the event $\mathcal{P}$ is the same as η_{ik}, and (2) We choose C^i_{jk} such that the first derivatives of g'_{ik} at $\mathcal{P}$ vanish. The first requirement amounts to 10 independent conditions, and the second requirement has $4 \times 10 = 40$ conditions. In B^i_k we have 16 independent parameters and in C^i_{jk} we have $4 \times 10 = 40$ independent parameters (because $C^i_{jk} = C^i_{kj}$ by definition). Using these 16 parameters of B^i_k, we can certainly satisfy the 10 conditions $g'_{ik} = \eta_{ik}$. The six parameters that are left free correspond to the Lorentz transformations and rotations (three components of the velocity vector and three Euler angles) that are still allowed. The 40 parameters in C^i_{jk} can be used to impose the 40 conditions $(\partial g'_{ik}/\partial x^{a'}) = 0$. Hence we can always choose the coordinates in such a way that the metric reduces to the inertial form in an infinitesimal *region* around an event $\mathcal{P}$.

It is not possible, however, to make the second derivatives vanish around the chosen event. To see this, let us first count the number of independent components in D^i_{jkl}, which is completely symmetric in j, k, and l. For a given value of i, the number of possible independent combinations of jkl is the same as the number of ways of choosing three objects out of four, disregarding the order. This is given by $[4!/(3!1!)] = 20$. Because i can take 4 values, we have a total of $4 \times 20 = 80$ independent components in D^i_{jkl}. There are $10 \times 10 = 100$ independent components in the second derivatives $\partial^2 g'_{ik}/\partial x^{a'}\partial x^{b'}$. However, because D^i_{jkl} has only 80 independent components, we cannot make all the second derivatives of g'_{ik} vanish at $\mathcal{P}$. We can set 80 of them to preassigned values, leaving 20 independent components. It follows that 20 independent pieces of information about the curvature of the space–time are contained in the second derivatives of the metric. (We make this idea more formal in Section 11.6.)

Because g_{ik} cannot be reduced to any preassigned form, no coordinate system has a preferred status in the presence of a gravitational field. Hence all laws of physics must be expressed in terms of physical quantities that are coordinate independent. When we developed special relativity in Chap. 3, we saw that four vectors, tensors, etc., satisfy this requirement as far as Lorentz transformations are concerned. We have to generalise the definitions of these quantities to incorporate *arbitrary* coordinate transformations.

Let us start with the definition for a four vector. The simplest four vector is a tangent vector v^i to any curve $x^i(\lambda)$ defined by $v^i = (dx^i/d\lambda)$. All other four vectors can be defined as quantities that transform in the same manner as the tangent vector v^i. When the coordinate system is changed from x^i to $x^{\prime i}$, the components of the tangent vector v^i to a curve $x^i(\lambda)$ changes to

$$v^{\prime i} = \frac{dx^{\prime i}}{d\lambda} = \frac{\partial x^{\prime i}}{\partial x^a}\frac{dx^a}{d\lambda} = \frac{\partial x^{\prime i}}{\partial x^a}v^a. \tag{11.29}$$

This defines the transformation law for any vector v^a under the coordinate transformation; that is, the four vector $v^a(x)$ is defined as a set of four functions that change according to the law

$$v^{\prime a}(x^{\prime i}) = \left(\frac{\partial x^{\prime a}}{\partial x^b}\right)v^b(x^i) \tag{11.30}$$

under a coordinate transformation $x^i \rightarrow x^{\prime i}$. Note that, in general, the coefficients $(\partial x^{\prime a}/\partial x^b)$ depend on the event x^i. Hence vectors located at different points in the space–time transform differently under a general coordinate transformation.

Consider next a quantity $T^{ab} = v^a v^b$. From the known transformation law for v^a we can determine the transformation law for T^{ab}. We find that

$$T'^{ab} \equiv v'^a v'^b = \frac{\partial x'^a}{\partial x^c} \frac{\partial x'^b}{\partial x^d} v^c v^d = \frac{\partial x'^a}{\partial x^c} \frac{\partial x'^b}{\partial x^d} T^{cd}. \tag{11.31}$$

The quantity T^{ab} is clearly a second-rank tensor, although it is a special kind of tensor. The transformation law arrived at above, however, is stated entirely in terms of the components T^{ab} and does not depend on the vector v^a. It is therefore natural to define a transformation law for a tensor T^{ab} based on Eq. (11.31).

The definition can be generalised easily for higher-rank tensors. To every index we introduce a factor $(\partial x'^a / \partial x^b)$ in such a way as to relate the component a in the new coordinates to a component b in the old coordinates.

Let us next consider the metric tensor and the process of raising and lowering the indices. Treating the metric tensor g_{ik} as a matrix we can define its inverse g^{ik} such that $g^{ik} g_{kl} = \delta^i_l$. We can obtain the transformation law for g^{ik} by noting that the relation $g^{ik} g_{kl} = \delta^i_l$ should hold in all coordinate frames. Because

$$g'_{ik} = \frac{\partial x^a}{\partial x'^i} \frac{\partial x^b}{\partial x'^k} g_{ab} \tag{11.32}$$

it follows that we must have

$$g'^{ik} = \frac{\partial x'^i}{\partial x^a} \frac{\partial x'^k}{\partial x^b} g^{ab}. \tag{11.33}$$

We can define the operation of raising and lowering an index of a tensor by using an appropriate form of metric tensor, that is, given T^{ik}, we define two new tensors $T^i_{\ k}$ and T_{ik} by the relations

$$T^i_{\ k} \equiv g_{ka} T^{ia}, \quad T_{ik} \equiv g_{ia} g_{kb} T^{ab}. \tag{11.34}$$

When we transform from x^i to x'^i, the quantity $T^i_{\ k}$ transforms as

$$T'^i_{\ k} = g'_{ka} T'^{ia} = \frac{\partial x^b}{\partial x'^k} \frac{\partial x^c}{\partial x'^a} g_{bc} \cdot \frac{\partial x'^i}{\partial x^m} \frac{\partial x'^a}{\partial x^n} T^{mn}$$

$$= g_{bc} T^{mn} \frac{\partial x^b}{\partial x'^k} \frac{\partial x'^i}{\partial x^m} \delta^c_n = g_{bn} T^{mn} \frac{\partial x^b}{\partial x'^k} \frac{\partial x'^i}{\partial x^m}$$

$$= T^m_{\ b} \frac{\partial x^b}{\partial x'^k} \frac{\partial x'^i}{\partial x^m}. \tag{11.35}$$

This relation shows that $T^i_{\ k}$ transforms correctly when the coordinates are changed. Similar result can be obtained for T_{ik}. Thus the raising and lowering of indices by use of g_{ik} is a legitimate tensorial operation.

Finally, we derive some results related to the determinant g of the metric tensor that will prove useful. We can obtain the differential dg of the determinant g by taking the differential of each component of the tensor g_{ik} and multiplying it by its cofactor in the determinant, i.e., by the corresponding minor. On the other hand, we obtain the tensor g^{ik} by taking the minors of g_{ik} and dividing by determinant g. Hence the minor of g_{ik} is gg^{ik}, giving $dg = gg^{ik}dg_{ik}$. It follows that

$$\frac{\partial g}{\partial x^a} = gg^{ik}\frac{\partial g_{ik}}{\partial x^a} = -gg_{ik}\frac{\partial g^{ik}}{\partial x^a}. \tag{11.36}$$

The second equality arises from the fact that $\partial(g_{ik}g^{ik})/\partial x^a = 0$.

11.4 Particle Trajectories in a Gravitational Field

In special relativity, we can obtain the equation of motion for a free particle by varying the action

$$A = -m\int ds = -m\int \left(\eta_{ab}\frac{dX^a}{d\lambda}\frac{dX^b}{d\lambda}\right)^{1/2} d\lambda. \tag{11.37}$$

The variation of A gives $(d^2X^a/d\lambda^2) = (du^a/d\lambda) = 0$, where $u^a = (dX^a/d\lambda)$ is the four velocity that is the tangent vector to the trajectory $X^a(\lambda)$. This equation can be equivalently written as

$$\frac{du^a}{d\lambda} = \left(\frac{dX^b}{d\lambda}\right)\left(\frac{\partial u^a}{\partial X^b}\right) = u^b\left(\frac{\partial u^a}{\partial X^b}\right) \equiv u^b\left(u^a_{,b}\right) = 0. \tag{11.38}$$

Because u^b transforms as a vector under coordinate transformation, this equation will retain its form under arbitrary coordinate transformations, provided that the derivative $u^a_{,b} \equiv (\partial u^a/\partial x^b)$ transforms as a tensor. The transformation law for this quantity can be obtained directly. We have

$$u'^a_{,b} = \frac{\partial u'^a}{\partial x'^b} = \frac{\partial x^c}{\partial x'^b}\frac{\partial}{\partial x^c}\left\{\frac{\partial x'^a}{\partial x^d}u^d\right\}$$

$$= \left(\frac{\partial x^c}{\partial x'^b}\frac{\partial x'^a}{\partial x^d}\right)u^d_{,c} + \left[\left(\frac{\partial x^c}{\partial x'^b}\right)\frac{\partial^2 x'^a}{\partial x^c\partial x^d}\right]u^d. \tag{11.39}$$

If $u^a_{,b}$ has to transform as a tensor, the second term has to vanish. This term vanishes for Lorentz transformations (for which x'^i are linear functions of x^i) but not for arbitrary coordinate transformations. Hence $u^a_{,b}$ is not a tensor under general coordinate transformation and Eq. (11.38) is not a valid tensor equation (in the sense that its form will not remain the same in an arbitrary coordinate system).

Another way of understanding this result is as follows: The derivative $u^a_{,b}$ involves subtracting $u^a(x^b)$ from $u^a(x^b + \Delta x^b)$. Because the transformation law for vectors at x^b and $x^b + \Delta x^b$ are different, the subtraction of vectors located at two different events is not a tensorial operation.

To obtain the correct equation of motion, which is generally covariant, we proceed as follows. In any small region around an event, we can always choose an inertial coordinate system. In such a small region, the form of the action in Eq. (11.37) remains valid. However, in this small region $\eta_{ab} = g_{ab}$ and so we can also write the action in the form

$$A = -m \int ds = -m \int \sqrt{g_{ab}dx^a dx^b} = -m \int \left(g_{ab}\frac{dx^a}{d\lambda}\frac{dx^b}{d\lambda} \right)^{1/2} d\lambda.$$
(11.40)

Because *this* action is made of proper tensorial quantities it will remain valid in any coordinate system. Therefore the equations derived from this action will provide us with the correct generalisation of Eq. (11.38).

Because we can invoke special relativity at a local region around any event, it follows that u^i is timelike for the massive particles and zero for massless particles. Also, because the Lagrangian multiplying $d\lambda$ in Eq. (11.40) is independent of λ, the energy $g_{ik}u^i u^k$ is a constant along the trajectory. For massive particles we choose this constant to be unity, which is equivalent to choosing $\lambda = s$, the proper time. The Euler–Lagrange equations then give

$$\frac{d}{ds}\left[\frac{1}{2}(g_{ik}u^i u^k)^{-1/2}2g_{ab}u^b \right] = \frac{1}{2}(g_{ik}u^i u^k)^{-1/2}\frac{\partial g_{ik}}{\partial x^a}u^i u^k.$$
(11.41)

Using $g_{ik}u^i u^k = 1$ and writing $(dg_{ab}/ds) = (\partial g_{ab}/\partial x^i)u^i$, we find that

$$g_{ab}\frac{du^b}{ds} = \frac{1}{2}\frac{\partial g_{ik}}{\partial x^a}u^i u^k - \frac{\partial g_{ab}}{\partial x^i}u^i u^b = \frac{1}{2}\left(\frac{\partial g_{ik}}{\partial x^a} - \frac{\partial g_{ak}}{\partial x^i} - \frac{\partial g_{ai}}{\partial x^k} \right)u^i u^k.$$
(11.42)

In arriving at the last equality, we have used the symmetry of $u^i u^k$ to write the second term in a symmetric manner. This equation may be written as

$$\frac{du^i}{ds} = -\frac{1}{2}g^{ik}\left(-\frac{\partial g_{mn}}{\partial x^k} + \frac{\partial g_{mk}}{\partial x^n} + \frac{\partial g_{kn}}{\partial x^m} \right)u^m u^n \equiv -\Gamma^i_{mn}u^m u^n,$$
(11.43)

where the second equality defines the quantities Γ^i_{km} (called the affine connection or Christoffel symbols) in terms of the derivatives of the metric. Note that Γ^i_{km} is symmetric in the lower indices; $\Gamma^i_{km} = \Gamma^i_{mk}$. Equation (11.43) (called the geodesic equation) governs the motion of material particles in a given gravitational field and can be thought of as the generalisation of the force equation in Newtonian gravity.

As an example of the geodesic equation, let us consider the motion of a particle in a constant gravitational field. Such a field is described by a metric that is independent of the time coordinate x^0. It is convenient, in this case, to introduce the notation $h \equiv g_{00}$, $g_\alpha \equiv -(g_{0\alpha}/g_{00})$, and $\gamma_{\alpha\beta} \equiv -(g_{\alpha\beta} + hg_\alpha g_\beta)$. It can be easily verified that the line interval on the three space $x^0 = \text{constant}$ is given by $dl^2 = \gamma_{\alpha\beta}dx^\alpha dx^\beta$. Further, the inverse $\gamma^{\alpha\beta}$ of the three metric $\gamma_{\alpha\beta}$ is given by $\gamma^{\alpha\beta} = -g^{\alpha\beta}$. With these notations we can compute directly the right-hand side of Eq. (11.43) and obtain the force f_α that is acting on a particle. In vector notation, this is given by

$$\mathbf{f} = \frac{mc^2}{\sqrt{1 - v^2/c^2}} \left\{ -\nabla \ln \sqrt{h} + \sqrt{h} \frac{\mathbf{v}}{c} \times (\nabla \times \mathbf{g}) \right\}. \tag{11.44}$$

The curl in this expression is calculated with the formula

$$(\nabla \times \mathbf{a})^\alpha = \frac{1}{2\sqrt{\gamma}} e^{\alpha\beta\gamma} \left(\frac{\partial a_\gamma}{\partial x^\beta} - \frac{\partial a_\beta}{\partial x^\gamma} \right), \tag{11.45}$$

where $e^{\alpha\beta\gamma}$ is the completely antisymmetric tensor in three dimensions. The first term reduces to $-\nabla\phi$ when $v \ll c$ and $h \approx 1 + (2\phi/c^2)$, and we retain only the leading-order term. The second term is analogous to the Coriolis force corresponding to an angular velocity $\Omega = (c/2)\sqrt{h}\nabla \times \mathbf{g}$. This is a purely relativistic effect.

Exercise 11.1
Filling in the details: Prove Eq. (11.44).

The geodesic equation suggests an interesting (and useful) generalisation of the concept of a derivative. We can rewrite Eq. (11.43) as

$$\frac{du^i}{ds} = \left(\frac{dx^a}{ds} \right) \left(\frac{\partial u^i}{\partial x^a} \right) = u^a \left(\frac{\partial u^i}{\partial x^a} \right) = -\Gamma^i_{ja} u^j u^a \tag{11.46}$$

or

$$u^a \left[\frac{\partial u^i}{\partial x^a} + \Gamma^i_{ja} u^j \right] \equiv u^a u^i_{;a} = 0, \tag{11.47}$$

where the first equality defines the symbol with the semicolon $u^i_{;a}$. Because the action principle is covariant, this equation must be valid in any frame of reference. Hence the quantity

$$u^a_{;b} \equiv u^a_{,b} + \Gamma^a_{ib} u^i \tag{11.48}$$

must transform as a tensor. From the transformation law for g_{ik}, it can be directly verified that Γ^a_{bc} transforms as

$$\Gamma'^a_{bc} = \frac{\partial x'^a}{\partial x^i} \frac{\partial x^k}{\partial x'^b} \frac{\partial x^m}{\partial x'^c} \Gamma^i_{km} + \frac{\partial x'^a}{\partial x^i} \frac{\partial^2 x^i}{\partial x'^b \partial x'^c}. \tag{11.49}$$

Comparing the second term in Eq. (11.49) with the second term in Eq. (11.39), we see that $u^a_{,b}$ does transform as a tensor even though neither Γ^a_{bc} nor $u^a_{,b}$ is a tensor. Relation (11.48) allows us to define a covariant derivative, which is the generalisation of an ordinary derivative to curvilinear coordinates. We use the definition of the covariant derivative in Eq. (11.48) for *any* vector field $v^i(x)$.

We can generalise the notion of a covariant derivative to other tensorial quantities by assuming that the chain rule of differentiation should remain valid. To begin with, because the ordinary derivative of a scalar $\phi_{,a} = (\partial \phi / \partial x^a)$ does transform like a covariant vector, we define $\phi_{;a} = \phi_{,a}$ for scalars. Next, because $(u^i u_i)$ is scalar, $(u^i u_i)_{,b} = (u^i u_i)_{;b}$. Using the chain rule on both sides and the known form of $u^i_{;b}$, we find that

$$u_{i;b} = u_{i,b} - \Gamma^k_{ib} u_k. \tag{11.50}$$

This defines the covariant derivative of a covariant vector.

The covariant derivatives of higher-rank tensors are defined by these results and the chain rule. Consider, for example, the tensor $T^a_b = u^a u_b$. Using $T^a_{b;i} = (u^a u_b)_{;i} = (u^a_{;i}) u_b + u^a (u_{b;i})$ and the known forms for the covariant derivatives of vectors, we get

$$T^a_{b;i} = T^a_{b,i} + \Gamma^a_{ki} T^k_b - \Gamma^k_{bi} T^a_k. \tag{11.51}$$

This will be the definition for $T^a_{b;i}$ for an *arbitrary* T^a_b. A similar procedure is adopted for higher-rank tensors.

It can be verified by direct computation that $g_{ik;b} = 0$. This fact allows us to interchange the order of covariant differentiation and the process of raising and lowering the indices. For example,

$$u_{a;b} = (g_{ai} u^i)_{;b} = (g_{ai;b}) u^i + g_{ai} \left(u^i_{;b} \right) = g_{ai} \left(u^i_{;b} \right). \tag{11.52}$$

The calculation of covariant derivatives can be simplified by two identities that we now derive. From the definition of Γ^i_{km} it follows that

$$\Gamma^a_{ba} = \frac{1}{2} g^{ad} \frac{\partial g_{ad}}{\partial x^b} = \frac{1}{2g} \frac{\partial g}{\partial x^b} = \frac{\partial}{\partial x^b} (\ln \sqrt{-g}), \tag{11.53}$$

where we have used Eq. (11.36). Similarly,

$$g^{bc} \Gamma^a_{bc} = g^{bc} g^{ad} \left(\frac{\partial g_{db}}{\partial x^c} - \frac{1}{2} \frac{\partial g_{bc}}{\partial x^d} \right) = -\frac{1}{\sqrt{-g}} \frac{\partial (\sqrt{-g} g^{ab})}{\partial x^b}. \tag{11.54}$$

Consider now the covariant *divergence* of a vector $A^a_{;a}$. From the definition of the covariant derivative we get

$$A^a_{;a} = \frac{\partial A^a}{\partial x^a} + \Gamma^a_{ca} A^c = \frac{\partial A^a}{\partial x^a} + A^c \frac{\partial (\ln \sqrt{-g})}{\partial x^c} = \frac{1}{\sqrt{-g}} \frac{\partial (\sqrt{-g} A^a)}{\partial x^a}, \tag{11.55}$$

where we have used Eq. (11.53). Further, if A^a is a gradient of some function ϕ, $A^a = \partial^a \phi$, then $A^a_{;a}$ will represent the covariant Laplacian $\phi^{;a}_{;a}$ of the scalar ϕ. We see that

$$\phi^{;i}_{;i} = \frac{1}{\sqrt{-g}} \frac{\partial}{\partial x^i} \left(\sqrt{-g} g^{ik} \frac{\partial \phi}{\partial x^k} \right). \tag{11.56}$$

In particular, we note that $\phi^{;i}_{;i}$ is *not* given by $g^{ik}\partial_i\partial_k\phi$. The above considerations remain valid in any dimensions.

A similar result can be derived for the covariant derivative of an antisymmetric tensor A^{ab}. We have

$$A^{ab}{}_{;b} = \frac{\partial A^{ab}}{\partial x^b} + \Gamma^a_{db}A^{db} + \Gamma^b_{db}A^{ad} = \frac{\partial A^{ab}}{\partial x^b} + \Gamma^b_{db}A^{ad}, \qquad (11.57)$$

as $\Gamma^a_{db}A^{db} = 0$ for an antisymmetric A^{db}. Using Eq. (11.53), we get

$$A^{ab}{}_{;b} = \frac{1}{\sqrt{-g}}\frac{\partial(\sqrt{-g}A^{ab})}{\partial x^b}. \qquad (11.58)$$

We also note that

$$V_{a;b} - V_{b;a} = \frac{\partial V_a}{\partial x^b} - \frac{\partial V_b}{\partial x^a} \qquad (11.59)$$

for any vector field V_a because the term involving Γ's cancels out in this expression.

From our formula for covariant derivative, we also get, for the stress tensor T^{ab},

$$T^k{}_{i;k} = \frac{\partial T^k_i}{\partial x^k} + \Gamma^k_{lk}T^l_i - \Gamma^l_{ik}T^k_l = \frac{1}{\sqrt{-g}}\frac{\partial\left(\sqrt{-g}\,T^k_i\right)}{\partial x^k} - \Gamma^l_{ki}T^k_l. \qquad (11.60)$$

Expanding out Γ^l_{ki}, we find that the last term is equal to

$$-\frac{1}{2}\left(-\frac{\partial g_{ki}}{\partial x^l} + \frac{\partial g_{kl}}{\partial x^i} + \frac{\partial g_{il}}{\partial x^k}\right)T^{kl}. \qquad (11.61)$$

Because of the symmetry of T^{kl}, the first and the third terms in the parentheses cancel, giving

$$T^k{}_{i;k} = \frac{1}{\sqrt{-g}}\frac{\partial\left(\sqrt{-g}\,T^k_i\right)}{\partial x^k} - \frac{1}{2}\frac{\partial g_{kl}}{\partial x^i}T^{kl}. \qquad (11.62)$$

The covariant derivative can be used to define a useful construct called parallel transport. Consider a curve $x^i(\lambda)$ passing through an event $\mathcal{P}$ in the space–time. Let the coordinates of $\mathcal{P}$ be $x^i(0)$. If $v^j(x)$ is a vector field, then we can interpret the quantity

$$v^a{}_{;b}\left(\frac{dx^b}{d\lambda}\right) = v^a{}_{,b}\left(\frac{dx^b}{d\lambda}\right) + \Gamma^a_{ib}v^i\left(\frac{dx^b}{d\lambda}\right) \qquad (11.63)$$

as the generalisation of the directional derivative of the vector field along a particular direction specified by the tangent vector $(dx^b/d\lambda)$. Suppose we are given a vector k^a at one event $\mathcal{P}$ and some curve $x^a(\lambda)$ passing through $\mathcal{P} = x^a(0)$.

We can then solve the differential equation

$$\frac{dv^a}{d\lambda} + \Gamma^a_{ik}(\lambda)\frac{dx^i}{d\lambda}v^k = 0, \tag{11.64}$$

{where $\Gamma^a_{ik}(\lambda) = \Gamma^a_{ik}[x(\lambda)]$ etc.,} with the boundary condition $v^a(\lambda = 0) = k^a$. The solution $v^a(\lambda)$ allows us to define a vector field along the curve $x^a(\lambda)$, given the vector k^a at only one point. If the space–time is actually flat, then this construction is equivalent to moving the vector from event to event, maintaining the same Cartesian components; i.e., the vector is moved "parallel" to itself. The above construction generalises the notion of parallel transport to curved space–time.

The idea of parallel transport gives an alternative interpretation to geodesic equation (11.43) or (11.47). In flat space a "straight" line can be defined in two ways: (1) It is the shortest distance between two points, or (2) it is that curve for which the tangent vector always points in the same direction, i.e, the tangent vector moves parallel to itself as we move along the curve. The generalisation of (1) to curved space was obtained in Eqs. (11.40)–(11.43), in which we extremised the interval between the events; we found that the "straight line" in curved space satisfies Eq. (11.43). To generalise (2) to curved space, we should demand that the tangent vector be parallel transported along the curve, i.e., $u^i_{;a}u^a = 0$. Using the definition of the covariant derivative, we see that this is indeed the same as Eq. (11.43), showing that both definitions (1) and (2) coincide in this formalism.

Exercise 11.2
Index gymnastics: (1) Prove that $g_{ab;c} = 0$. (2) Prove transformation law (11.49). (3) Calculate Γ^a_{bc} for the two-dimensional space (corresponding to the surface of sphere) with line interval $dl^2 = (d\theta^2 + \sin^2\theta d\phi^2)$.

Exercise 11.3
Parallel transport of a vector: The vector tangent to the curve $\phi = 0$, located at the north pole of a two sphere, is parallel transported along the following curves: (1) along $\phi = 0$ from the north pole to the point $\theta = \theta_0$; (2) along the curve $\theta = \theta_0$ from $\phi = 0$ to $\phi = \phi_0$; (3) along the curve $\phi = \phi_0$ from $\theta = \theta_0$ to the north pole. Calculate the angle through which the original vector has rotated.

Exercise 11.4
Geodesic on a two sphere: Find the geodesics on the two sphere with line interval $dl^2 = d\theta^2 + \sin^2\theta\, d\phi^2$ by solving Eq. (11.43).

11.5 Physics in Curved Space–Time

The description of a gravitational field in terms of g_{ik} assumes that all the laws of physics can be suitably generalised from flat space–time to curved space–time. The procedure for doing this is as follows: Around any event $\mathcal{P}$ we choose a locally inertial frame in which laws of special relativity are valid. By writing these laws in a covariant manner using suitable tensors, we will have a description that is valid in any arbitrary coordinate system around $\mathcal{P}$. Because the form of the equations is expected to be same (locally) in an accelerated frame and in the presence of gravity, the above procedure allows us to describe physics in curved space–time. This was the procedure we followed in Section 11.4 to describe the motion of a particle in a curved space–time.

We now carry out this procedure for the case of a free electromagnetic field and obtain the form of Maxwell's equations in an arbitrary space–time. Consider the action for the electromagnetic field in flat space–time that is Lorentz invariant:

$$\mathcal{A}_{em} = -\frac{1}{16\pi} \int F_{ab} F^{ab} d^4 X, \quad F_{ab} = A_{b,a} - A_{a,b}. \quad (11.65)$$

This action can be rewritten in a form that is generally covariant by the following procedure: We replace the volume element $d^4 X$ with the covariant volume element in arbitrary coordinates is $\sqrt{-g} d^4 x$; we also change the ordinary derivative $A_{b,a}$ to the covariant derivative $A_{b;a}$, etc. This leads to the action

$$\mathcal{A}_{em} = -\frac{1}{16\pi} \int F_{ab} F^{ab} \sqrt{-g} d^4 x, \quad F_{ab} = A_{b;a} - A_{a;b}. \quad (11.66)$$

We can obtain the electromagnetic-field equation in a curved space–time by varying this action, which gives $F^{ab}{}_{;b} = 0$. Using identities (11.58) and (11.59) derived in Section 11.4, we can write these equations can be written as

$$\frac{1}{\sqrt{-g}} \partial_a (\sqrt{-g} F^{ab}) = 0, \quad F_{ab} = \frac{\partial A_b}{\partial x^a} - \frac{\partial A_a}{\partial x^b}. \quad (11.67)$$

These equations describe the influence of gravitational field on the electromagnetic phenomenon. The second of these equations can also be rewritten in terms of F_{ik} as

$$F_{ik;l} + F_{li;k} + F_{kl;i} = \frac{\partial F_{ik}}{\partial x^l} + \frac{\partial F_{li}}{\partial x^k} + \frac{\partial F_{kl}}{\partial x^i} = 0. \quad (11.68)$$

(The first equality can be easily verified for the antisymmetric tensor F_{ik}.) Thus the first pair of Maxwell's equations has the same form in curved space–time as in flat space–time. We can also introduce the electric and the magnetic fields by the definitions $E_\alpha \equiv F_{0\alpha}$ and $B^\alpha = -(1/2\sqrt{\gamma})\epsilon^{\alpha\beta\gamma} F_{\beta\gamma}$, where γ is the determinant of the metric of three-dimensional space.

As a specific example of the effects of curved space–time, consider a metric separated out in the form

$$ds^2 = g_{ik} dx^i dx^k = g_{00} dt^2 + 2g_{0\alpha} dt dx^\alpha + g_{\alpha\beta} dx^\alpha dx^\beta. \quad (11.69)$$

For the sake of convenience, we again introduce the notation $h \equiv g_{00}$ and define a vector **g** with components $g_\alpha \equiv -(g_{0\alpha}/g_{00})$. The three-dimensional metric $\gamma_{\alpha\beta}$

is defined as $[-g_{\alpha\beta} + h g_\alpha g_\beta]$. Given these definitions, it is straightforward to work out various components of the Maxwell's equations in the $(3 + 1)$ form. The equation

$$\frac{\partial F_{ik}}{\partial x^l} + \frac{\partial F_{li}}{\partial x^k} + \frac{\partial F_{kl}}{\partial x^i} = 0 \tag{11.70}$$

becomes

$$\nabla \cdot \mathbf{B} = 0, \quad \nabla \times \mathbf{E} = -\frac{1}{\sqrt{\gamma}} \frac{\partial}{\partial t} (\sqrt{\gamma} \mathbf{B}). \tag{11.71}$$

The equation

$$\frac{1}{\sqrt{-g}} \partial_a (\sqrt{-g} F^{ab}) = 0 \tag{11.72}$$

can be written as

$$\nabla \cdot \mathbf{D} = 0, \quad \nabla \times \mathbf{H} = \frac{1}{\sqrt{\gamma}} \frac{\partial}{\partial t} (\sqrt{\gamma} \mathbf{D}) \tag{11.73}$$

where

$$\mathbf{D} = \frac{\mathbf{E}}{\sqrt{h}} + \mathbf{H} \times \mathbf{g}, \quad \mathbf{B} = \frac{\mathbf{H}}{\sqrt{h}} + \mathbf{g} \times \mathbf{E}. \tag{11.74}$$

In static gravitational fields with $\mathbf{g} = 0$ and other metric components independent of time, these equations become

$$\nabla \cdot \mathbf{B} = 0, \quad \nabla \times \mathbf{E} = -\frac{\partial \mathbf{B}}{\partial t}, \quad \nabla \cdot \left(\frac{\mathbf{E}}{\sqrt{h}}\right) = 0, \quad \nabla \times (\sqrt{h}\mathbf{B}) = \frac{1}{\sqrt{h}} \frac{\partial \mathbf{E}}{\partial t}. \tag{11.75}$$

These equations are identical to Maxwell's equations in material media with electric and magnetic permeability $\epsilon = \mu = (1/\sqrt{h})$.

Exercise 11.5
Filling in the details: Prove Eqs. (11.71) and (11.73).

11.6 Curvature

The discussion so far has used only the fact that the metric tensor depended on the coordinates. We saw earlier in Section (11.3) that this could arise either because of the use of curvilinear coordinates in flat space–time or because of a genuine curvature of the space–time. Because the gravitational field should really be connected with the latter, we have to arrive at a suitable characterisation of curvature that is coordinate independent.

One possible way of doing this is as follows: We begin by noting that there is a vital difference between the covariant derivative and the ordinary derivative: Unlike ordinary derivatives, covariant derivatives do not commute, i.e., $A^a{}_{;b;c} \neq A^a{}_{;c;b}$. The covariant derivative $A^a{}_{;b}$ of a vector field is

$$A^a{}_{;b} = A^a{}_{,b} + \Gamma^a_{mb} A^m. \tag{11.76}$$

Differentiating once again by treating $A^a{}_{;b}$ as a second-rank tensor, we get

$$A^a{}_{;b;c} = \left(A^a{}_{;b}\right)_{,c} + \Gamma^a_{nc} A^n{}_{;b} - \Gamma^m_{bc} A^a{}_{;m}, \tag{11.77}$$

which reduces to

$$A^a{}_{;b;c} = A^a{}_{,b,c} + \left(\Gamma^a_{bm} A^m\right)_{,c} + \Gamma^a_{cn}\left(A^n{}_{,b} + \Gamma^n_{mb} A^m\right) - \Gamma^m_{bc}\left(A^a{}_{,m} + \Gamma^a_{nm} A^n\right). \tag{11.78}$$

Computing the difference $(A^a{}_{;b;c} - A^a{}_{;c;b})$, we find that the terms with first derivatives of A^a vanish, giving

$$A^a{}_{;b;c} - A^a{}_{;c;b} = -\left(\Gamma^a_{ic,b} - \Gamma^a_{ib,c} + \Gamma^a_{kb}\Gamma^k_{ic} - \Gamma^a_{kc}\Gamma^k_{ib}\right)A^i \equiv -R^a_{ibc} A^i. \tag{11.79}$$

Because the left-hand side is a tensor and A^i is a vector, it follows that R^a_{ibc} is a tensor.

Now suppose that the space–time is actually flat. In that case, we can always choose an inertial coordinate system in which the Γ's vanish at *all* events. Then the derivatives of the Γ's will also vanish, making $R^a_{ibc} = 0$. However, because this is a tensor equation, it will be valid in any other coordinate system. Thus we find that, in flat space–time, we must have $R^a_{ibc} = 0$. If $R^a_{ibc} \neq 0$ the space–time will be curved. This condition provides the invariant characterisation of curvature we were seeking; the tensor R^a_{ibc} is called the curvature tensor.

The curvature tensor can also be given another physical meaning: Consider two curves $x^i_A(\lambda)$ and $x^i_B(\lambda)$ that intersect at the events $\mathcal{P}$ and $\mathcal{Q}$. Given any vector at the event $\mathcal{P}$ we can parallel transport it to $\mathcal{Q}$ along either of these two curves. The resultant vector will not, in general, be the same. This feature again distinguishes curved space–time from flat space–time and is related to the nonvanishing of the curvature tensor. To prove this, we can proceed as follows.

The parallel transport of a vector along a curve is equivalent to solving the partial differential equation

$$\frac{\partial v^a}{\partial x^c} = -\Gamma^a_{bc}(x) v^b \tag{11.80}$$

[which is same as Eq. (11.64)]. In general, such an equation cannot be integrated to give a unique vector field $v^a(x)$. This equation will have an acceptable, unique solution only if the integrability conditions

$$\frac{\partial^2 v^a}{\partial x^b \partial x^c} = \frac{\partial^2 v^a}{\partial x^c \partial x^b} \tag{11.81}$$

are satisfied. [Note that we are essentially demanding that $dv^a = -\Gamma^a_{bc}(x)v^b$ $(x)\,dx^c$ be an exact differential.] Differentiating Eq. (11.80) with respect to x^d and using Eq. (11.80) again, we find that

$$\frac{\partial^2 v^a}{\partial x^d \partial x^c} = -\left[\frac{\partial \Gamma^a_{lc}}{\partial x^d} - \Gamma^a_{bc}\Gamma^b_{dl}\right]v^l. \tag{11.82}$$

Hence the integrability condition is equivalent to $R^a_{ldc}v^l = 0$, where

$$R^a_{ldc} = \frac{\partial \Gamma^a_{lc}}{\partial x^d} - \frac{\partial \Gamma^a_{ld}}{\partial x^c} + \Gamma^a_{bd}\Gamma^b_{cl} - \Gamma^a_{bc}\Gamma^b_{dl}. \tag{11.83}$$

If this result should hold for an arbitrary v^a, then R^a_{ldc} has to vanish. In general, this quantity will *not* vanish and hence parallel transport will not give a unique vector except in flat space–time. In a flat space–time we can always choose a system of coordinates such that $\Gamma^i_{kl} = 0$. In such a coordinate system R^i_{klm} will vanish. Because R^i_{klm} is a tensor, it follows that it will vanish in any coordinate system in flat space–time and parallel transport will be unique.

The converse is also true. Suppose we are given that R^i_{klm} vanishes identically in a given space–time. We then choose some event in the space–time and construct a locally inertial Cartesian coordinate system around it. Because parallel transport leads to a unique vector field when $R^i_{klm} = 0$, we can parallel transport this Cartesian coordinate system to any other event in the space–time in a unique manner. This permits us to construct a global inertial coordinate system in the space–time, proving that the space–time is flat. Thus the vanishing of R^i_{klm} is both necessary and sufficient for the space–time to be flat.

The curvature also has an effect on the behaviour of geodesics. Consider a flat two-dimensional space in which two geodesics (which are straight lines) start from some point A with an angle θ between them. We know that this angle does not change as we move along the geodesics, and obviously the geodesics will not intersect again. Consider next a similar situation on the surface of a sphere: Two geodesics (which are great circles) start from some point A of the surface and make an angle θ at A. These geodesics, however, will *not* maintain the same angle between them. In fact, they will eventually intersect again. This is yet another manifestation of curvature and we will expect the curvature tensor to determine the rate at which the separation between the nearby geodesics (called the geodesic deviation) changes. Let us work out the expression for this.

Consider a bunch of particles moving along geodesics in a space–time. Let $x^i = x^i(s, v)$ denote the family of geodesics, where s represents the proper time along the geodesics and v labels a particular geodesic. The vector

$n^i = (\partial x^i/\partial v)\,\delta v \equiv v^i\,\delta v$ denotes the deviation between two neighbouring geodesics parameterised by the values v and $v + \delta v$. From the definition of the covariant derivative and the relation $(\partial u^i/\partial v) = \partial v^i/\partial s$ (where u^i is the tangent vector to the geodesic), it follows that

$$u^i{}_{;k}v^k = v^i{}_{;k}u^k. \qquad (11.84)$$

Consider now the variation of the separation vector v^i between two neighbouring geodesics. We have

$$\frac{D^2 v^i}{Ds^2} \equiv \left(v^i{}_{;k}u^k\right)_{;l}u^l = \left(u^i{}_{;k}v^k\right)_{;l}u^l = u^i{}_{;k;l}v^k u^l + u^i{}_{;k}v^k{}_{;l}u^l. \qquad (11.85)$$

Using Eq. (11.84) in the second term of Eq. (11.85) and changing the order of covariant derivatives in the first term we get

$$\frac{D^2 v^i}{Ds^2} = \left(u^i{}_{;l}u^l\right)_{;k}v^k + u^m R^i_{mkl}u^k v^l. \qquad (11.86)$$

The first term vanishes because $u^i{}_{;l}u^l = 0$ along the geodesics and we are left with

$$\frac{D^2 v^i}{Ds^2} = R^i_{klm}u^k u^l v^m. \qquad (11.87)$$

This result shows how the curvature generates the geodesic deviation.

From the definition of R^i_{klm} we can construct the tensors $R_{iklm} \equiv g_{ia}R^a_{klm}$ and $R_{ik} \equiv R^j_{ijk}$ and the scalar $R \equiv g^{ik}R_{ik}$. The tensor R_{ik}, called the Ricci tensor, satisfies an important identity (called the Bianchi identity):

$$\left(R^i_k - \frac{1}{2}\delta^i_k R\right)_{;i} = 0. \qquad (11.88)$$

This identity is most easily proved in a locally inertial frame around some event. In such a frame, $\Gamma^i_{kl} = 0$, and we have

$$R^n_{ikl;m} = \frac{\partial R^n_{ikl}}{\partial x^m} = \frac{\partial^2 \Gamma^n_{il}}{\partial x^m \partial x^k} - \frac{\partial^2 \Gamma^n_{ik}}{\partial x^m \partial x^l}. \qquad (11.89)$$

For this expression it is straightforward to verify that

$$R^n_{ikl;m} + R^n_{imk;l} + R^n_{ilm;k} = 0. \qquad (11.90)$$

Because this is a tensor equation, it will remain valid in any coordinate system if it is valid in the locally inertial frame. Contracting this equation on ik and ln (ie., by setting $i = k$ and $l = n$) we obtain the relation $R^l_{m;l} = (1/2)(\partial R/\partial x^m)$. This can be rewritten as Eq. (11.88).

We saw in Section 11.3 that when we choose the locally inertial coordinate system around some event, we are, in general, left with 20 nonzero functions involving the second derivatives of the metric tensor. Because the curvature tensor can be expressed in terms of the second derivatives of the metric in the locally inertial frame, it follows that R^i_{klm} will have only 20 independent components. This can be easily verified: We begin by noting that R_{iklm} is antisymmetric in ik and lm but symmetric under pair exchange:

$$R_{iklm} = -R_{kilm} = -R_{ikml} = R_{lmik}. \tag{11.91}$$

It is also easy to see that the cyclic sum of any three indices vanishes, i.e.,

$$R_{iklm} + R_{imkl} + R_{ilmk} = 0. \tag{11.92}$$

Thus the pairs of indices ik and lm can take six different sets of values that are independent. Hence there are 6 components of R_{iklm} with identical values for ik and lm and $(6 \times 5)/2 = 15$ components of R_{iklm} with different values for ik and lm giving a total of 21 independent components that respect the symmetries of Eq. (11.91). The components in which all the indices are different are, however, not independent of one another because of the identity in Eq. (11.92). There is one independent constraint that these components have to satisfy, namely,

$$R_{0123} + R_{0312} + R_{0231} = 0. \tag{11.93}$$

Hence the number of independent components of R_{iklm} is $(21 - 1) = 20$, as expected.

11.7 Dynamics of Gravitational Field

In Newtonian gravity, the gravitational field is described by one scalar function $\phi(t, \mathbf{x})$ that can be determined if the density distribution $\rho(t, \mathbf{x})$ is given. The action for the Newtonian gravitational field is

$$A = \frac{1}{8\pi G} \int (\nabla \phi)^2 d^3\mathbf{x}\, dt + \int \rho \phi d^3\mathbf{x}\, dt, \tag{11.94}$$

which, when varied with respect to ϕ, gives the field equation $\nabla^2 \phi = 4\pi G\rho$. In general relativity, the gravitational field is described by g_{ik} and the field equations are expected to be differential equations that determine this quantity in terms of the matter variables. We now determine the action principle and the field equations in general relativity.

11.7.1 Action for the Gravitational Field

We can take the total action A_{tot} as a sum of two terms, A_g, which represents the gravitational part, and A_m, which depends on the metric as well as the matter variables. We have already seen that we can obtain the matter part of the action in curved space–time by generalising the corresponding special relativistic action. We therefore have to determine only A_g.

The action principle must be such that the equations of motion obtained from it involve only up to second derivatives of g_{ik}. Further, the action should reduce to the Newtonian action for the metric given in Eq. (11.23). This would suggest that A_g should contain up to quadratic terms in the first derivatives of g_{ik}.

However, no nontrivial scalar quantity can be constructed from the metric and its first derivatives alone. This can be proved as follows: Consider any scalar quantity $f(g_{ik}, \partial g_{ik}/\partial x^l)$ that depends on only the metric tensor and its first derivative. Around any event in space–time we can construct a locally inertial coordinate system in which $g_{ik} = \eta_{ik}$ and $(\partial g_{ik}/\partial x^l) = 0$. Hence the function f will become a constant at any event in the locally inertial frame. If the function is a scalar, then its value will remain the same in any other coordinate system. In other words, any covariant scalar function built out of the metric tensor and its derivatives will be a trivial constant.

Thus we conclude that we must look for scalar quantity that contains the metric, its first derivatives, and second derivatives. If the function is linear in the second derivative then it is possible to express the integral for the action as one arising from a Lagrangian involving only the metric and its first derivatives and another term that contributes only at the boundaries. By considering suitable variations of the metric tensor, we can ignore the contribution from the boundary term and thus obtain equations of motion that involve only up to second derivatives in the metric tensor. Such an action can be taken to be proportional to

$$A_g = \int R\sqrt{-g}\,d^4x, \qquad (11.95)$$

that is, we show that the integrand can be written in the form

$$R\sqrt{-g} = \sqrt{-g}\,g^{ik}\left(\Gamma^m_{il}\Gamma^l_{km} - \Gamma^l_{ik}\Gamma^m_{lm}\right) + \frac{\partial\sqrt{-g}\,w^i}{\partial x^i} = \sqrt{-g}\,\mathcal{G} + \frac{\partial\sqrt{-g}\,w^i}{\partial x^i}, \qquad (11.96)$$

where the first term $\sqrt{-g}\,\mathcal{G}$ is quadratic in the derivatives of g_{ik} and the second term is a four divergence of some four-vector w^i. To prove this, we start with the expression for $\sqrt{-g}\,R$,

$$\sqrt{-g}\,R = \sqrt{-g}\,g^{ik}R_{ik} = \sqrt{-g}\left\{g^{ik}\frac{\partial\Gamma^l_{ik}}{\partial x^l} - g^{ik}\frac{\partial\Gamma^l_{il}}{\partial x^k} + g^{ik}\Gamma^l_{ik}\Gamma^m_{lm} - g^{ik}\Gamma^m_{il}\Gamma^l_{km}\right\}, \qquad (11.97)$$

and rewrite the first two terms on the right-hand side by separating out a total derivative using

$$\sqrt{-g}\,g^{ik}\frac{\partial\Gamma^l_{ik}}{\partial x^l} = \frac{\partial}{\partial x^l}\left(\sqrt{-g}\,g^{ik}\Gamma^l_{ik}\right) - \Gamma^l_{ik}\frac{\partial}{\partial x^l}\left(\sqrt{-g}\,g^{ik}\right), \qquad (11.98)$$

$$\sqrt{-g}\,g^{ik}\frac{\partial\Gamma^l_{il}}{\partial x^k} = \frac{\partial}{\partial x^k}\left(\sqrt{-g}\,g^{ik}\Gamma^l_{il}\right) - \Gamma^l_{il}\frac{\partial}{\partial x^k}\left(\sqrt{-g}\,g^{ik}\right). \qquad (11.99)$$

Then Eq. (11.97) becomes

$$R\sqrt{-g} = \sqrt{-g}\mathcal{G} + \frac{\partial\sqrt{-g}\,w^i}{\partial x^i}, \tag{11.100}$$

where

$$\sqrt{-g}\mathcal{G} = \Gamma_{im}^m \frac{\partial}{\partial x^k}(\sqrt{-g}\,g^{ik}) - \Gamma_{ik}^l \frac{\partial}{\partial x^l}(\sqrt{-g}\,g^{ik}) - \left(\Gamma_{il}^m \Gamma_{km}^l - \Gamma_{ik}^l \Gamma_{lm}^m\right) g^{ik}\sqrt{-g}, \tag{11.101}$$

$$w^i = g^{ab}\Gamma_{ab}^i - g^{ik}\Gamma_{kl}^l. \tag{11.102}$$

We can simplify the expression for $\sqrt{-g}\mathcal{G}$ by using the identity derived in Section 11.4 and the relation

$$\frac{\partial g^{ik}}{\partial x^l} = -\Gamma_{ml}^i g^{mk} - \Gamma_{ml}^k g^{im}. \tag{11.103}$$

The first two terms of $\sqrt{-g}\mathcal{G}$ reduce to

$$\sqrt{-g}\left(2\Gamma_{ik}^l \Gamma_{lm}^i g^{mk} - \Gamma_{im}^m \Gamma_{kl}^i g^{kl} - \Gamma_{ik}^l \Gamma_{lm}^m g^{ik}\right)$$

$$= \sqrt{-g}\,g^{ik}\left(2\Gamma_{mk}^l \Gamma_{li}^m - \Gamma_{lm}^m \Gamma_{ik}^l - \Gamma_{ik}^l \Gamma_{lm}^m\right)$$

$$= 2\sqrt{-g}\,g^{ik}\left(\Gamma_{il}^m \Gamma_{km}^l - \Gamma_{ik}^l \Gamma_{lm}^m\right). \tag{11.104}$$

Using this expression in Eqs. (11.101) and (11.96), we can now write the full Lagrangian $R\sqrt{-g}$ as

$$R\sqrt{-g} = \mathcal{L} + \frac{\partial}{\partial x^c}\left[\sqrt{-g}\left(g^{ik}\Gamma_{ik}^c - g^{kc}\Gamma_{km}^m\right)\right], \tag{11.105}$$

where

$$\mathcal{L} \equiv \sqrt{-g}\mathcal{G} = \sqrt{-g}\,g^{ik}\left(\Gamma_{i\ell}^m \Gamma_{km}^\ell - \Gamma_{ik}^\ell \Gamma_{\ell m}^m\right). \tag{11.106}$$

The full action for the system containing gravity and matter is

$$A = \frac{1}{c_1}\int R\sqrt{-g}\,d^4x + \int \mathcal{L}_m(q^A, g_{ik})\sqrt{-g}\,d^4x, \tag{11.107}$$

where q^A denotes the matter variables. To determine the constant $(1/c_1)$ that multiplies the first term, we have to evaluate the action for a weak gravitational field and identify it with the action in Eq. (11.94). We saw in Section 11.2 that the metric describing a weak gravitational field is given by

$$ds^2 = \left(1 + \frac{2\phi}{c^2}\right)c^2 dt^2 - dx^2 - dy^2 - dz^2, \tag{11.108}$$

with $\phi = \phi(t, \mathbf{x}) \ll c^2$. We have to evaluate $R\sqrt{-g}$ to the lowest nontrivial order in ϕ/c^2 for this metric.

In this line element, the spatial components of the metric tensor are constants and hence their derivatives vanish. The Christoffel symbols Γ^i_{kl} are made entirely of the derivatives of g_{00}. Among these derivatives, the time derivatives are of one order lower compared with spatial derivatives because time derivatives involve a factor ct. Therefore, to the lowest order, the nontrivial contribution to the Γ's arises from only the spatial derivatives g_{00}. Therefore the only contribution, to the lowest order, is

$$\Gamma^\alpha_{00} \simeq -\frac{1}{2} g^{\alpha\beta} \frac{\partial g_{00}}{\partial x^\beta} = \frac{1}{c^2} \frac{\partial\phi}{\partial x^\alpha}. \tag{11.109}$$

By the same argument, the only nonzero component (to leading order) in R_{ik} is

$$R_{00} = R^0_0 = \frac{\partial \Gamma^\alpha_{00}}{\partial x^\alpha} = \frac{1}{c^2} \nabla^2 \phi. \tag{11.110}$$

Thus we have

$$\sqrt{-g} R \cong g^{00} R_{00} \cong \left(1 - \frac{2\phi}{c^2}\right) \frac{\nabla^2\phi}{c^2} = \frac{\nabla^2\phi}{c^2} - \frac{2\phi\nabla^2\phi}{c^4}$$

$$= \frac{\nabla^2\phi}{c^2} - \frac{2\nabla \cdot (\phi\nabla\phi)}{c^4} + \frac{2(\nabla\phi)^2}{c^4}. \tag{11.111}$$

The first two terms are total divergences that may be ignored, leaving us with

$$\sqrt{-g} R \cong \frac{2(\nabla\phi)^2}{c^4}. \tag{11.112}$$

The action in Newtonian gravity is $(8\pi G)^{-1}(\nabla\phi)^2$; to reproduce the correct Newtonian limit, we need to take the Lagrangian in general relativity as $c^4(16\pi G)^{-1} \sqrt{-g} R$. This completely determines the action for our system.

11.7.2 Field Equations for the Gravitational Field

To obtain the dynamical equations for the gravitational field, we have to vary the action with respect to the metric tensor g^{ik}. The gravitational part of the action gives the contribution

$$\delta \int R\sqrt{-g} d^4x = \delta \int g^{ik} R_{ik} \sqrt{-g} d^4x$$

$$= \int \{R_{ik}\sqrt{-g}\delta g^{ik} + R_{ik} g^{ik}\delta\sqrt{-g} + g^{ik} \sqrt{-g}\delta R_{ik}\} d^4x. \tag{11.113}$$

Using the fact that

$$\delta\sqrt{-g} = -\frac{1}{2\sqrt{-g}}\delta g = -\frac{1}{2}\sqrt{-g} g_{ik}\delta g^{ik}, \tag{11.114}$$

we get

$$\delta \int R\sqrt{-g}d^4x = \int \left(R_{ik} - \frac{1}{2}g_{ik}R\right)\delta g^{ik}\sqrt{-g}d^4x + \int g^{ik}\delta R_{ik}\sqrt{-g}d^4x.$$

(11.115)

To proceed further, we need to evaluate the quantity $g^{ik}\delta R_{ik}$. This is most easily done in a locally inertial frame in which $\Gamma^i_{kl} = 0$. Using the definition of R_{ik}, we find that

$$g^{ik}\delta R_{ik} = g^{ik}\left\{\frac{\partial}{\partial x^l}\delta\Gamma^l_{ik} - \frac{\partial}{\partial x^k}\delta\Gamma^l_{il}\right\} = g^{ik}\frac{\partial}{\partial x^l}\delta\Gamma^l_{ik} - g^{il}\frac{\partial}{\partial x^l}\delta\Gamma^k_{ik} = \frac{\partial w^l}{\partial x^l}$$

(11.116)

with

$$w^l = g^{ik}\delta\Gamma^l_{ik} - g^{il}\delta\Gamma^k_{ik}.$$

(11.117)

We now use the fact that even though Γ^i_{kl} is not a tensor $\delta\Gamma^i_{kl}$ is a tensor. To see this, we have to note only that $\Gamma^i_{kl}A^k dx^l$ is the change in a vector under parallel displacement between two infinitesimally separated points; hence $\delta\Gamma^i_{kl}A^k dx^l$ is the difference between two vectors obtained by two parallel displacements [one with Γ^i_{kl} and the other with $(\Gamma^i_{kl} + \delta\Gamma^i_{kl})$] between two infinitesimal points. The difference between two vectors at the same point is a vector and hence $\delta\Gamma^i_{kl}$ is a tensor. Because it is a tensor we may write the above expression in any coordinate system by replacing $(\partial w^l/\partial x^l)$ with $w^l_{;l}$. This gives

$$g^{ik}\delta R_{ik} = \frac{1}{\sqrt{-g}}\frac{\partial}{\partial x^l}(\sqrt{-g}w^l),$$

(11.118)

and consequently

$$\int g^{ik}\delta R_{ik}\sqrt{-g}d^4x = \int \frac{\partial(\sqrt{-g}w^l)}{\partial x^l}d^4x.$$

(11.119)

This integral can be transformed into a surface integral. We now assume that we consider only the variations δg^{ik} for which this surface term vanishes. In that case, this term does not contribute and the variation of the gravitational action gives

$$\delta \int R\sqrt{-g}d^4x = \int \left(R_{ik} - \frac{1}{2}g_{ik}R\right)\delta g^{ik}\sqrt{-g}d^4x \equiv \int G_{ik}\delta g^{ik}\sqrt{-g}d^4x.$$

(11.120)

The variation of the matter part of the action can be expressed in the form

$$\delta A_m = \frac{1}{2}\int T_{ik}\delta g^{ik}\sqrt{-g}d^4x.$$

(11.121)

This equation, in fact, defines the second-rank symmetric tensor T_{ik}. Setting the total variation to zero, we obtain the equations for the gravitational field:

$$G_{ik} = R_{ik} - \frac{1}{2} g_{ik} R = \frac{8\pi G}{c^4} T_{ik}. \qquad (11.122)$$

The above equation shows that T^{ik} is the source of the gravitational field in Einstein's theory. It also follows from the definition that it is a symmetric tensor. In the Newtonian limit the component T^{00} should reduce to the mass density. This suggests that we interpret T^{ik} as the energy-momentum tensor of the matter field. In fact, Eq. (11.121) may be considered as the definition of the energy momentum in a fully relativistic situation. Before studying the details of the equations of motion, we first discuss some general properties of T_{ik}.

11.7.3 Properties of the Energy-Momentum Tensor

We saw in section 11.6 that the quantity on the left-hand side of Eq. (11.122) has zero covariant divergence [see Eq. (11.88)]. For the validity of Eq. (11.122), we must have $T^i_{k;i} = 0$, irrespective of the detailed form of the matter Lagrangian. This result can be proved as follows: Consider an infinitesimal coordinate transformation from x^i to $x'^i = x^i + \xi^i(x)$, where $\xi^i(x)$ are considered to be infinitesimal quantities. Under this transformation the metric tensor changes to

$$g'^{ik}(x'^l) = g^{lm}(x^l) \frac{\partial x'^i}{\partial x^l} \frac{\partial x'^k}{\partial x^m} = g^{lm} \left(\delta_l^i + \frac{\partial \xi^i}{\partial x^l} \right) \left(\delta_m^k + \frac{\partial \xi^k}{\partial x^m} \right)$$

$$\approx g^{ik}(x^l) + g^{im} \frac{\partial \xi^k}{\partial x^m} + g^{kl} \frac{\partial \xi^i}{\partial x^l}. \qquad (11.123)$$

In this expression g'^{ik} is a function of x'^l and g^{ik} is a function of x^l. We can express g'^{ik} as a function of the original coordinates x^l by expanding $g'^{ik}(x^l + \xi^l)$ in a Taylor series in powers of ξ^l and retaining up to linear order. This will give

$$g'^{ik}(x^l) = g^{ik}(x^l) - \xi^l \frac{\partial g^{ik}}{\partial x^l} + g^{il} \frac{\partial \xi^k}{\partial x^l} + g^{kl} \frac{\partial \xi^i}{\partial x^l}. \qquad (11.124)$$

This expression provides the net change in the component g^{ik} treated as a function of the coordinates x^l. It is easy to verify that the last three terms can be combined to give $\xi^{i;k} + \xi^{k;i}$. Therefore the change in the metric tensor under an infinitesimal coordinate transformation can be expressed in the form

$$g'^{ik} = g^{ik} + \delta g^{ik}, \qquad \delta g^{ik} = \xi^{i;k} + \xi^{k;i}. \qquad (11.125)$$

Similarly

$$g'_{ik} = g_{ik} + \delta g_{ik}, \qquad \delta g_{ik} = -(\xi_{i;k} + \xi_{k;i}). \qquad (11.126)$$

Consider now the variation in the matter action A_m when the coordinates are changed by an infinitesimal amount. The change of coordinates will induce a certain variation δq in the matter variables and a variation $\delta g^{ik} = \xi^{i;k} + \xi^{k;i}$ in the metric. The net change in the matter action is

$$\delta A_m = \left(\frac{\delta A_m}{\delta q}\right)_g \delta q + \left(\frac{\delta A_m}{\delta g^{ik}}\right)_q \delta g^{ik}. \tag{11.127}$$

However, when the equations of motion for the matter variables are satisfied, we know that $(\delta A / \delta q)_g = 0$. Further, because action is a scalar, we know that the total variation δA_m should vanish. Hence we must have

$$0 = \delta A_m = \frac{1}{2} \int T_{ik} \delta g^{ik} \sqrt{-g}\, d^4 x \tag{11.128}$$

when $\delta g^{ik} = \xi^{i;k} + \xi^{k;i}$. Using Eq. (11.125) and the symmetry of T_{ik}, we can write

$$0 = \frac{1}{2} \int T_{ik}(\xi^{i;k} + \xi^{k;i})\sqrt{-g}\, d^4 x = \int T_{ik}\xi^{i;k} \sqrt{-g}\, d^4 x$$

$$= \int (T_i^k \xi^i)_{;k} \sqrt{-g}\, d^4 x - \int T_{i;k}^i \xi^i \sqrt{-g}\, d^4 x. \tag{11.129}$$

The first term can be expressed in the form

$$\int \frac{\partial}{\partial x^k}\left(\sqrt{-g} T_i^k \xi^i\right) d^4 x, \tag{11.130}$$

which could be converted into an integral over a hypersurface. Because ξ^i may be taken to vanish on this hypersurface, this integral drops out, allowing us to conclude that

$$T_{i;k}^k = 0. \tag{11.131}$$

Let us work out the explicit form of T_{ik} for the electromagnetic field. In this case, the matter action is

$$A = -\frac{1}{16\pi} \int F_{ab}F^{ab} \sqrt{-g}\, d^4 x = -\frac{1}{16\pi} \int F_{ab}F_{dc}g^{ad}g^{cb} \sqrt{-g}\, d^4 x. \tag{11.132}$$

On varying g_{ab}, we get

$$\delta A = -\frac{1}{16\pi} \int d^4 x [2F_{ab}F_{dc}g^{ad}\delta g^{bc}\sqrt{-g} + F_{mn}F^{mn}\delta(\sqrt{-g})]$$

$$= -\frac{1}{16\pi} \int d^4 x \left[2F_{ab}F^a{}_c\delta g^{bc}\sqrt{-g} + F_{mn}F^{mn}\frac{1}{2}\frac{(-1)}{\sqrt{-g}}(-g)g_{bc}\delta g^{bc}\right]$$

$$= \frac{1}{16\pi} \int d^4 x \sqrt{-g}\delta g^{bc}\left[-F_{ab}F^a{}_c + \frac{1}{4}F_{mn}F^{mn}g_{bc}\right], \tag{11.133}$$

so that

$$4\pi T_{bc} = -F_{ab}F^a{}_c + \frac{1}{4}F_{mn}F^{mn}g_{bc}, \qquad (11.134)$$

which is the expression for the energy-momentum tensor of the electromagnetic field used in Chap. 3; T_0^0 denotes the energy density and T_0^m denotes the energy flux along the m direction, etc. The conditions $T^a_{b;a} = 0$ can be easily shown to give the equations $F^a_{b;a} = 0$, which are the Maxwell's equations in curved space–time derived in Section (11.5). In this sense, Einstein's equations imply the equations of motion for the source term.

Exercise 11.6

Energy-momentum tensor for free particles: Consider a system of relativistic particles with the action $(c = 1)$

$$A = -\sum_A m_A \int ds_A, \qquad (11.135)$$

where m_A and s_A are the mass and the proper time of the Ath particle, respectively. Vary this action with respect to g_{ab} and obtain $T^{ab} = \rho u^a u^b$, where ρ is the mass density of the particles.

11.7.4 General Properties of the Gravitational-Field Equations

We have shown that for any matter action, $T^i_{k;i} = 0$. Hence the right-hand side of Einstein's equation has zero covariant divergence. We have seen above, [see Eq. (11.88)] that the left-hand side also has vanishing divergence. Thus the 10 equations in Eq. (11.122) are constrained by the 4 identities of Eq. (11.88), leaving 6 independent equations. Also note that, among the 10 variables g^{ik}, 4 can be assigned specific values by a suitable choice of coordinates. Thus there are only six independent functions we need to solve for, which is the same as the number of independent equations available.

Structurally, Einstein's equations are 10 second-order partial differential equations with the independent variables x^i. Thus, a priori, we would have thought that the values of the metric tensor g^{ik} and its first time derivatives $(\partial g^{ik}/\partial t)$ need to be specified as initial conditions. This is, however, not true because of certain peculiar features in Einstein's equations.

It is clear from the definition of R_{iklm} that the second derivatives with respect to time are contained in only the components $R_{0\alpha0\beta}$ in which they enter through the term $(-1/2)\ddot{g}_{\alpha\beta}$. This shows that the second derivatives of the metric components $g_{0\alpha}$ and g_{00} do not appear in R_{iklm} and hence in Einstein's equations. Further, even the second derivatives of $g_{\alpha\beta}$ appear only in the space–space part of Einstien's equations. The time–time part and the space–space part contain time derivatives only up to first order. To prove this claim, note that Eq. (11.88) can be written in the form

$$\left(R_i^0 - \frac{1}{2}\delta_i^0 R \right)_{;0} = -\left(R_i^\alpha - \frac{1}{2}\delta_i^\alpha R \right)_{;\alpha} \tag{11.136}$$

where $i = 0, 1, 2, 3$ and $\alpha = 1, 2, 3$. The highest time derivatives appearing on the right-hand side of these equations are second derivatives; hence it follows that the quantities within the parentheses on the left-hand side can contain only first-order time derivatives. Hence the time–time and time–space components G_0^0 and G_α^0 contain only the first time derivatives of the metric tensor.

Furthermore, the space–time and the time–time equations do not contain the first derivatives $\dot{g}_{0\alpha}$ and $\dot{g}_{00}$ but only $\dot{g}_{\alpha\beta}$. This is because, among all Christoffel symbols, only $\Gamma_{\alpha,00}$ and $\Gamma_{0,00}$ contain these quantities; but these appear only in the components $R_{0\alpha0\beta}$, which drop out from the time–time and the time–space parts of Einstein's equations.

The above considerations show that (1) the space–time and the time–time parts of Einstein's equations involve only $\dot{g}_{\alpha\beta}$ as the highest-order time derivatives, (2) the space–space part contains $\ddot{g}_{\alpha\beta}$, and (3) the time derivatives of g_{00} and $g_{0\alpha}$ do not appear in Einstien's equations.

It is therefore possible to assign as initial conditions the functions $g_{\alpha\beta}$ and $\dot{g}_{\alpha\beta}$ at some time $t = t_0$. The space–time and the time–time parts of Einstein's equation will then determine the initial values of $g_{0\alpha}$ and g_{00}. (They are not freely specifiable and are determined by the constraint equations.) The initial values of $\dot{g}_{0\alpha}$ remain arbitrary. These are the valid initial data for integrating Einstein's equations.

The above discussion used an arbitrary coordinate system. If we use liberty in the choice of coordinate system, it is possible to show that the gravitational field has only 2 genuine degrees of freedom per event. This can be seen as follows: We first use the four coordinate transformations $x^i \to x'^i$ to arrange $g_{00} = 1$ and $g_{0\alpha} = 0$ in the neighbourhood of any event. This reduces the metric to the form

$$ds^2 = dt^2 + g_{\alpha\beta}dx^\alpha dx^\beta, \tag{11.137}$$

leaving six components of $g_{\alpha\beta}$ nonzero. Consider now, the infinitesimal coordinate transformation $t \to t' = t + f$, $x^\alpha \to x'^\alpha = x^\alpha + \lambda^\alpha$. This will change the metric coefficients to

$$g'_{00} = 1 + 2\dot{f}, \quad g'_{0\alpha} = f_{,\alpha} + g_{\alpha\beta}\dot{\lambda}^\beta, \quad g'_{\alpha\beta} = g_{\alpha\beta} + \lambda^\gamma_{,\alpha}g_{\gamma\beta} + \lambda^\gamma_{,\beta}g_{\alpha\gamma}. \tag{11.138}$$

Demanding that the coordinate transformation should not change the conditions

$g_{00} = 1$ and $g_{0\alpha} = 0$ will lead to the constraints

$$\dot{f} = 0, \qquad f_{,\alpha} = -g_{\alpha\beta}\dot{\lambda}^{\beta}. \tag{11.139}$$

The first equation implies that f is a function of space alone; $f = f(\mathbf{x})$. Using this, we can integrate the second equation to give

$$\lambda^{\alpha} = p^{\alpha}(\mathbf{x}) - f_{,\beta} \int^{t} dt\, g^{\alpha\beta}, \tag{11.140}$$

where $p^{\alpha}(\mathbf{x})$ are arbitrary functions. Thus on any given spatial hypersurface, we have the freedom to choose the four functions $f(\mathbf{x})$ and $p^{\alpha}(\mathbf{x})$ in order to bring four components of $g^{\alpha\beta}(\mathbf{x}, t)$ to preassigned values. Thus only two components of $g_{\alpha\beta}$ remain arbitrary to be propagated forward in time.

Finally, we can obtain an interesting result by combining field equations with the geodesic deviation equation (11.87). Let us choose a coordinate system that is locally inertial along the trajectory of the first observer. In such a frame, the observer's four velocity will be $u^{i} = \delta^{i}_{0}$ and the spatial component of the geodesic deviation equation (11.87) becomes

$$g^{\alpha} \equiv \frac{D^{2}v^{\alpha}}{Ds^{2}} = R^{\alpha}_{abi}u^{a}u^{b}v^{i} = R^{\alpha}_{00\beta}v^{\beta}. \tag{11.141}$$

(Note that $R^{a}_{b00} = 0$ because of antisymmetry in the last two indices and hence only the spatial part of v^{β} contributes on the right-hand side.) Taking the divergence of this relative acceleration and using Einstein's equations, we get (with $c = 1$)

$$\frac{\partial g^{\alpha}}{\partial v^{\alpha}} \equiv \nabla_{\mathbf{v}} \cdot \mathbf{g} = R^{\alpha}_{00\alpha} = -R_{00} = -8\pi G\left(T_{00} - \frac{1}{2}T\right). \tag{11.142}$$

In the locally inertial coordinates we are using $(T_{00} - (1/2)T) = (1/2)(\rho + T^{\alpha}_{\alpha})$, where α is summed over the spatial indices. In the case of an ideal fluid, with $T_{ab} = (\rho + p)u_{a}u_{b} - g_{ab}p$, this equation becomes

$$\nabla \cdot \mathbf{g} = -4\pi G(\rho + 3p). \tag{11.143}$$

This equation does not require any approximation regarding the gravitational field and is always valid in the frame of a freely falling observer. It shows that the effective gravitational mass in the frame of a freely falling observer is $\rho + 3p$.

Finally, we recast Einstein's equations in a form that is useful for understanding the effects of general relativistic corrections to Newtonian gravity. For this purpose, it is convenient to describe the gravitational field in a manner similar to that of the electromagnetic field. The analogues of scalar and vector potentials in electromagnetism are provided by the functions $\phi = (g_{00} - 1)/2$ and $A_{\alpha} \equiv g_{0\alpha}$. [These quantities are related to h and $\mathbf{g}$ used earlier in Section (11.5), by $\phi = (h - 1)/2$, $\mathbf{A} = \mathbf{g}$; we have changed only the notation.] Defining a gravitoelectric field by $\mathbf{E}_{gr} \equiv -\nabla\phi$ and a gravitomagnetic field by $\mathbf{B}_{gr} \equiv \nabla \times \mathbf{A}$, we can express Einstein's equations in the form

$$\nabla \cdot \mathbf{E}_{gr} \cong -4\pi G\rho, \quad \nabla \cdot \mathbf{B}_{gr} \cong 0, \quad \nabla \times \mathbf{E}_{gr} \cong 0,$$

$$\nabla \times \mathbf{B}_{gr} \cong -4\left(4\pi G\rho\, \mathbf{v} - \frac{\partial \mathbf{E}_{gr}}{\partial t}\right) \tag{11.144}$$

if we retain only the lowest-order contributions from the gravitoelectric and magnetic fields. [Units have been chosen with $c = 1$ and it is assumed that $B_{gr} = \mathcal{O}(v E_{gr})$.] To the same order of approximation, the geodesic equation becomes

$$\frac{d\mathbf{v}}{dt} \cong \mathbf{E}_{gr} + \mathbf{v} \times \mathbf{B}_{gr}, \tag{11.145}$$

which we can obtain from Eq. (11.44) by retaining leading-order terms. The numerical coefficient and signs are different in relations (11.144) compared with electromagnetism; the Faraday induction term is absent whereas the displacement current is present to the order of approximation we are working. To avoid misunderstanding, it is stressed that electromagnetic and gravitational fields are totally different in structure and the above description should be treated as only a simple mnemonic device to understand the effects of lowest-order relativistic corrections.

Exercise 11.7
Analogy with electromagnetic fields: (1) Show that Einstein's equations reduce to the forms of relations (11.144) when leading-order approximation is taken. [Hint: Approximate the action for gravitational field to the necessary order and then vary it to get the equations.] (2) Obtain relation (11.145) from Eq. (11.44).

11.8 Schwarzschild Metric

The gravitational-field equations are extremely complicated and hence physically relevant exact solutions are available in only very few cases. Among these solutions, the most important one is that corresponding to a spherically symmetric distribution of matter. We now discuss this solution, called the Schwarzschild metric. We use units with $c = 1$.

Let the source T_{ik} be zero outside a spherical region of radius L and be spherically symmetric inside. This suggests that the metric will also be spherically symmetric both inside and outside. Further, for $r > L$, the space–time is empty. We are interested in determining the metric in the region $r > L$ that is due to the source at $r < L$.

The most general spherically symmetric metric can be expressed in the form

$$ds^2 = A(r, \bar{t}) d\bar{t}^2 + B(r, \bar{t}) dr^2 + C(r, \bar{t}) dr d\bar{t} - r^2(d\theta^2 + \sin^2\theta d\phi^2).$$

(11.146)

This line element is constructed so that surfaces of $t =$ constant, and $r =$ constant have an area $4\pi r^2$. In fact, this criterion uniquely fixes the coordinate r. We can now make a transformation from $\bar{t}$ to t such that the quadratic form in $dr d\bar{t}$ is diagonalised. This will lead to a metric that can be written in the form

$$ds^2 = e^{\nu} dt^2 - e^{\lambda} dr^2 - r^2(d\theta^2 + \sin^2\theta d\phi^2),$$

(11.147)

where ν and λ are functions of r and t. For this metric, the nonvanishing Christoffel symbols are

$$\Gamma^1_{11} = \frac{\lambda'}{2}, \quad \Gamma^0_{10} = \frac{\nu'}{2}, \quad \Gamma^2_{33} = -\sin\theta\cos\theta,$$

$$\Gamma^0_{11} = \frac{\lambda}{2}e^{\lambda-\nu}, \quad \Gamma^1_{22} = -re^{-\lambda}, \quad \Gamma^1_{00} = \frac{\nu'}{2}e^{\nu-\lambda},$$

$$\Gamma^2_{12} = \Gamma^3_{13} = \frac{1}{r}, \quad \Gamma^3_{23} = \cot\theta, \quad \Gamma^0_{00} = \frac{\dot{\nu}}{2},$$

$$\Gamma^1_{10} = \frac{\dot{\lambda}}{2}, \quad \Gamma^1_{33} = -r\sin^2\theta\,e^{-\lambda}.$$

(11.148)

Here the prime denotes a derivative with respect to r and an overdot denotes differentiation with respect to t. The components of G^i_k can be computed in a straightforward manner, and we find

$$G^1_1 = -e^{-\lambda}\left(\frac{\nu'}{r} + \frac{1}{r^2}\right) + \frac{1}{r^2},$$

$$G^2_2 = G^3_3 = -\frac{1}{2}e^{-\lambda}\left(\nu'' + \frac{\nu'^2}{2} + \frac{\nu'-\lambda'}{r} - \frac{\nu'\lambda'}{2}\right) + \frac{1}{2}e^{-\nu}\left(\ddot{\lambda} + \frac{\dot{\lambda}^2}{2} - \frac{\dot{\lambda}\dot{\nu}}{2}\right),$$

$$G^0_0 = -e^{-\lambda}\left(\frac{1}{r^2} - \frac{\lambda'}{r}\right) + \frac{1}{r^2},$$

$$G^1_0 = -e^{-\lambda}\frac{\dot{\lambda}}{r}.$$

(11.149)

All other components vanish identically. So Einstein's equations in empty space, outside a spherically symmetric source, become

$$e^{-\lambda}\left(\frac{\nu'}{r} + \frac{1}{r^2}\right) - \frac{1}{r^2} = 0, \quad e^{-\lambda}\left(\frac{\lambda'}{r} - \frac{1}{r^2}\right) + \frac{1}{r^2}, \quad \dot{\lambda} = 0.$$

(11.150)

The last equation shows that λ is independent of time; further, adding the first two equations, we see that $\lambda' + \nu' = 0$, implying that $\lambda + \nu = f(t)$, where $f(t)$ is a function of time alone. In metric (11.147), we still have the freedom of changing the time coordinate from t to any other arbitrary function of t. By using this freedom, we can set $f(t) = 0$. Hence we find that $\nu = -\lambda$, with both ν and λ depending on only r.

This result shows that all spherically symmetric solutions to vacuum Einstein's equations are time independent. For example, consider a star that is collapsing (or oscillating) radially in a time-dependent manner. Outside the star, the metric has to be independent of time. This result is known as Birkoff's theorem.

Integrating the first two equations in Eqs. (11.150), we find that

$$e^{\nu} = e^{-\lambda} = 1 + \frac{\text{constant}}{r}. \tag{11.151}$$

We expect the gravitational field, at large distances from the spherical body, to be weak and describable by Newtonian theory. In this limit, the metric should take the form $g_{00} \cong 1 + 2\phi = 1 - 2GM/r$. Therefore the constant in Eq. (11.151) should have the value $2GM$, where M is the mass of the central object. [In normal units, $2GM$ will become $(2GM/c^2)$.] Thus the metric can be expressed in the form

$$ds^2 = \left(1 - \frac{2GM}{r}\right) dt^2 - \left(1 - \frac{2GM}{r}\right)^{-1} dr^2 - r^2(d\theta^2 + \sin^2\theta d\phi^2). \tag{11.152}$$

This is called the Schwarzschild metric.

The analysis, incidentally, allows us to derive another result. Consider the metric that is due to a thin spherical shell of mass M_{shell}. Because space–time inside and outside the shell are spherically symmetric and empty, the metric must have the form of Eq. (11.152) with (possibly) different constants M_{in} and M_{out} both inside and outside the shell. From the conditions at $r \to \infty$, it is clear that $M_{\text{out}} = M_{\text{shell}}$. The regularity at $r = 0$ will require that $M_{\text{in}} = 0$. Hence the space–time inside a spherical shell must be flat.

If the source is more complicated than a spherical shell, Einstein's equation has to be explicitly integrated for $r < L$ to find the metric in this region. The constants of integration have to be chosen in such a way that the metric matches with the Schwarzschild metric at the surface of the spherical body. As a specific example, let us consider a source that is a barotropic ideal fluid described by an energy density ρ and pressure p and an equation of state $p = p(\rho)$. The time–time component of Einstein's equations now gives

$$\frac{1}{r^2}\frac{d}{dr}[r(1 - e^{-\lambda})] = 8\pi G\rho. \tag{11.153}$$

By introducing a function $m(r)$ by a relation

$$e^{\lambda(r)} \equiv \left[1 - \frac{2Gm(r)}{r}\right]^{-1}, \tag{11.154}$$

We can integrate Eq. (11.153) to give

$$m(r) = \int_0^\infty dr\, 4\pi r^2 \rho(r).$$ (11.155)

The remaining Einstein's equations now reduce to

$$\frac{dv}{dr} = \frac{2Gm(r)}{r^2}\left[1 + \frac{4\pi r^3 p(r)}{m(r)}\right]\left[1 - \frac{2Gm(r)}{r}\right]^{-1},$$ (11.156)

$$\frac{dp}{dr} = -\frac{\rho(r)m(r)}{r^2}\left[1 + \frac{p}{\rho}\right]\left[1 + \frac{4\pi r^3 p}{m}\right]\left[1 - \frac{2Gm}{r}\right]^{-1}.$$ (11.157)

Equations (11.155)–(11.157) together with the equation of state form a set of four equations for the four functions $p(r)$, $\rho(r)$, $m(r)$, and $v(r)$. The metric is given by

$$ds^2 = e^{v(r)}\,dt^2 - \frac{dr^2}{[1 - 2Gm(r)/r]} - r^2(d\theta^2 + \sin^2\theta d\phi^2).$$ (11.158)

These equations describe static, spherical distributions of highly relativistic matter. They reduce to the Newtonian equations when $p \ll \rho$, $r \gg (2Gm/c^2)$, and $v \approx [1 + (2\phi/c^2)]$, where ϕ is the Newtonian gravitational potential. If the matter distribution is confined to a radius $r < L$, then matching the metric coefficient g_{rr} at $r = L$ shows that

$$M = \int_0^L 4\pi r^2 \rho(r)\,dr.$$ (11.159)

This relation allows us to determine the value of the parameter M in the outside Schwarzschild geometry in terms of the source. The integration in this expression is with respect to the volume element $4\pi r^2 dr$ whereas the proper spatial volume element in the space–time is $4\pi r^2 e^{\lambda/2} dr$. The difference between these two integrals represents the gravitational binding energy of the body.

For a region outside a spherically symmetric distribution of matter, the Schwarzschild metric provides an exact solution to Einstein's equation. When the matter distribution is not strictly spherically symmetric, the metric will differ from the Schwarzschild form. However, at sufficiently large distances from the body, we can think of the Schwarzschild metric as providing the lowest-order correction to the flat space–time metric. More generally, we can write

$$g_{ik} = g_{ik}^{(0)} + h_{ik}^{(1)} + h_{ik}^{(2)} + \cdots,$$ (11.160)

where $g_{ik}^{(0)} \equiv \eta_{ik}$ denotes the flat Minkowski metric, $h_{ik}^{(1)}$ denotes the first-order corrections that fall as $(1/r)$, and $h_{ik}^{(2)}$ denotes the second-order corrections that fall as $(1/r^2)$, etc. We can determine these corrections systematically by substituting these ansätze into Einstein's equations and solving order by order. The lowest-order corrections to flat space–time is obviously given by the Schwarzschild metric, expanded in Taylor series to the required order. Hence

$$h_{00}^{(1)} = -\frac{r_g}{r}, \quad h_{\alpha\beta}^{(1)} = -\frac{r_g}{r} n_\alpha n_\beta, \quad h_{0\alpha}^{(1)} = 0, \tag{11.161}$$

where $\mathbf{n}$ is a unit vector in the direction of $\mathbf{r}$. In the second order, we get corrections that arise from the nonlinear combinations of first-order terms as well as genuine second-order terms that correspond to the solutions of the linearised equations to the necessary order. We can again determine the first kind of nonlinearities directly by expanding the Schwarzschild metric to the required order; we get

$$h_{00}^{(2)} = 0, \quad h_{\alpha\beta}^{(2)} = -\left(\frac{r_g}{r}\right)^2 n_\alpha n_\beta. \tag{11.162}$$

The remaining nonlinear term can be shown to contribute to $h_{0\alpha}^{(2)}$ and can be written in terms of the parameterisation introduced in Section 11.5. The correction term is

$$\mathbf{g} = \frac{2G}{c^3 r^2} \mathbf{n} \times \mathbf{M}, \tag{11.163}$$

where $\mathbf{M}$ is the total angular momentum of the body. [The derivation of Eq. (11.163) requires linearisation of Einstein's equations, which will be discussed in Section 11.12.] This term can exert a Coriolis force on a distant particle with a strength that corresponds to the one in a frame with the angular velocity [see Eq. (11.44)]

$$\boldsymbol{\Omega} \approx \frac{c}{2} \nabla \times \mathbf{g} = \frac{G}{c^2 r^3} [\mathbf{M} - 3\mathbf{n}(\mathbf{M} \cdot \mathbf{n})]. \tag{11.164}$$

This expression also gives the rate of precession of an inertial frame anchored to fixed stars because of the rotating source, which will be $\Omega_{\text{prec}} = -\boldsymbol{\Omega}$. This precession of inertial frame is called the Lens-Thirring effect.

Exercise 11.8
Variational principle for pressure support: Show that we can derive Eq. (11.157) by extremising

$$M(r) = 4\pi \int_0^r \rho\, r'^2\, dr' \tag{11.165}$$

with respect to an adiabatic, Eulerian variation in which the number of baryons,

$$N = \int_0^R 4\pi r^2 n(r) \left(1 - \frac{2GM_r}{c^2 r}\right)^{-1/2} dr, \tag{11.166}$$

remains constant. [Hint: (1) Vary $M + \lambda N$, where λ is a Lagrangian multiplier, and use the adiabatic condition $(\delta n/\delta \rho) = nc^2/(P + \rho c^2)$. (2) Solve for λ in terms of other variables in the resulting equation. Because λ is a constant, set its derivative to zero.]

11.9 Orbits in the Schwarzschild Metric

The metric in Eq. (11.152) describes the gravitational field around any massive body, which may be approximated as spherically symmetric. The behaviour of particles and light rays in this metric will show characteristic features of general relativity that are absent in Newtonian gravity. We now study the motion of particles and light rays in this metric.

The motion of a particle of mass m in a metric g^{ik} can be most easily tackled by use of the relativistic Hamilton–Jacobi equation. We know that the momentum and the energy of a particle can be expressed in terms of the derivatives of the action $A(x^i)$ of a particle treated as a function of the end points. In a relativistic situation, this relation has the form $p_i = (\partial A/\partial x^i)$. Because $p_i p^i = m^2$, the relativistic Hamilton–Jacobi equation becomes

$$g^{ik} \frac{\partial A}{\partial x^i} \frac{\partial A}{\partial x^k} - m^2 = 0. \tag{11.167}$$

In the case of Schwarzschild metric, this equation is

$$\left(1 - \frac{2GM}{r}\right)^{-1} \left(\frac{\partial A}{\partial t}\right)^2 - \left(1 - \frac{2GM}{r}\right)\left(\frac{\partial A}{\partial r}\right)^2 - \frac{1}{r^2}\left(\frac{\partial A}{\partial \phi}\right)^2 = m^2. \tag{11.168}$$

We have assumed that the motion takes place in the plane $\theta = \pi/2$ and set $(\partial A/\partial \theta) = 0$. This Hamilton-Jacobi equation can be solved by the ansatz

$$A = -\mathcal{E}t + L\phi + A_r(r), \tag{11.169}$$

where L and $\mathcal{E}$ denote the angular momentum and the energy of the particle. Substituting Eq. (11.169) into the Hamilton–Jacobi equation and solving, we find that

$$A_r = \int \sqrt{\mathcal{E}^2\left(1 - \frac{2GM}{r}\right)^{-2} - \left(m^2 + \frac{L^2}{r^2}\right)\left(1 - \frac{2GM}{r}\right)^{-1}}\, dr. \tag{11.170}$$

The trajectory of the particle is determined by the equations $(\partial A/\partial \mathcal{E}) = $ constant and $(\partial A / \partial L) = $ constant. This gives $r(t)$ and $\phi(r)$ as

$$
t = \frac{\mathcal{E}}{m} \int dr \left(1 - \frac{2GM}{r}\right)^{-1} \left[\left(\frac{\mathcal{E}}{m}\right)^2 - \left(1 + \frac{L^2}{m^2 r^2}\right)\left(1 - \frac{2GM}{r}\right)\right]^{-1/2},
$$

(11.171)

$$
\phi = \int dr \left(\frac{L}{r^2}\right) \left[\mathcal{E}^2 - \left(m^2 + \frac{L^2}{r^2}\right)\left(1 - \frac{2GM}{r}\right)\right]^{-1/2}. \quad (11.172)
$$

From Eq. (11.171) we get

$$
\left(1 - \frac{2GM}{r}\right)^{-1} \frac{dr}{dt} = \frac{1}{\mathcal{E}}\left[\mathcal{E}^2 - V_{\text{eff}}^2(r)\right]^{1/2}, \quad (11.173)
$$

with

$$
V_{\text{eff}}^2(r) = m^2 \left(1 - \frac{2GM}{r}\right)\left(1 + \frac{L^2}{m^2 r^2}\right). \quad (11.174)
$$

Similarly, from Eq. (11.172) we can determine $(d\phi/dr)$ that, when combined with the expression for (dr/dt), gives

$$
r^2 \dot{\phi} = \left(\frac{L}{\mathcal{E}}\right)\left(1 - \frac{2GM}{r}\right). \quad (11.175)
$$

Figure 11.1 gives a plot of V_{eff}/m against $(rc^2/GM) = r/M$ (in units with $G = c = 1$) for different values of L. Several important aspects of the motion can be deduced from this figure:

(1) For a given value of L and $\mathcal{E}$ the nature of the orbit will, in general, be governed by the turning points in r, determined by the equation $V_{\text{eff}}^2(r) = \mathcal{E}^2$. For a given $L > 4GMm$, the function $V_{\text{eff}}^2(r)$ has one maximum and one minimum. Further, $V_{\text{eff}} \to m$ as $r \to \infty$ for all values of L. If the energy $\mathcal{E}$ of the particle is lower than m then there will be two turning points. The particle will orbit the central body with a perihelion and an aphelion. This is similar to elliptic orbits in Newtonian gravity.
(2) If $m < \mathcal{E} < V_{\text{max}}(L)$ there will be only one turning point. The particle will approach the central mass from infinity, reach a radius of closest approach, and travel back to infinity. This is similar to the hyperbolic orbits in Newtonian gravity.
(3) If $\mathcal{E} = V_{\text{max}}(L)$, then the orbit will be circular at some radius r_0 determined by the condition $V'(r_0) = 0$, $V(r_0) = \mathcal{E}$. This gives the radius of the circular

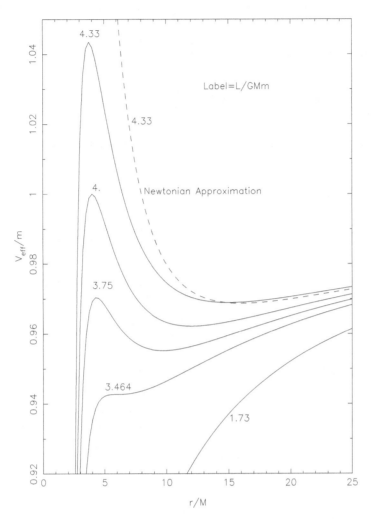

Fig. 11.1.   The effective potential for particle motion in a Schwarzschild metric.

orbit and the energy as

$$\frac{r_0}{2GM} = \frac{L^2}{4G^2M^2m^2}\left(1 \pm \sqrt{1 - \frac{12G^2M^2m^2}{L^2}}\right),$$

$$\mathcal{E}^2 = \frac{L^2}{GMr_0}\left(1 - \frac{2GM}{r_0}\right). \tag{11.176}$$

The upper sign refers to the stable orbit and the lower to the unstable orbit. The stable orbit closest to the centre has the parameters $r_0 = 3GM$, $L = 2\sqrt{3}GMm$, and $\mathcal{E} = [\sqrt{(8/9)}]m$.

(4) If $\mathcal{E} > V_{\text{max}}(L)$, the particle falls to the centre. This behaviour is in sharp contrast to Newtonian gravity in which a particle with nonzero angular momentum can never reach $r = 0$.

(5) As the angular momentum is lowered, V_{max} decreases and for $L = 4GMm$ the maximum value of the potential is at $V_{\text{max}} = m$. In this case all particles from infinity will fall to the origin. Particles with $\mathcal{E} < m$ will still have two turning points and will form bound orbits.

(6) When L is reduced still further, the maxima and the minima of $V(r)$ approach each other. For $L \leq 2\sqrt{3}GMm$, there are no turning points in the $V(r)$ curve. Particles with lower angular momentum will fall to the origin irrespective of the energy.

The above discussion shows that the key effects of general relativity will be felt for orbits close to $r \approx 2GM$. For the Sun, this radius is ~ 3 km. Because most of the planetary orbits are at much larger distances, the general relativistic effects are quite small in the solar system and can be determined as a small correction to the Newtonian result. Two crucial, verifiable predictions emerge from this analysis, which we now discuss.

Exercise 11.9

Capture of photons by a Schwarzschild black hole: Consider a photon travelling at an angle Ψ to the radial direction in a Schwarzschild metric. (1) Show that (i) if the initial location of the photon is at $r < 3GM$ and the motion is inwardly directed, it will be captured by the black hole; (ii) if the initial location is at $r > 3GM$ and is outwardly directed, the photon will escape to infinity. (2) If the photon is at $r < 3GM$ and is moving outward, then the condition for escape is

$$\sin \Psi < \frac{3\sqrt{3}\,GM}{r}\sqrt{1 - \frac{2GM}{r}}. \qquad (11.177)$$

(3) If the photon is at $r > 3GM$ and moving inwards, the condition for escape is

$$\sin \Psi > \frac{3\sqrt{3}\,GM}{r}\sqrt{1 - \frac{2GM}{r}}. \qquad (11.178)$$

11.9.1 Precession of the Perihelion

Let us first consider the corrections to the orbit of a planet arising from general relativity. To do this it is best to start from the equation for the orbit, Eq. (11.172), expressed in the form

$$\left(\frac{du}{d\phi}\right)^2 = \frac{1}{L^2}\left[\frac{\mathcal{E}^2}{c^2} - \left(m^2c^2 + \frac{L^2}{r^2}\right)\left(1 - \frac{2GM}{c^2 r}\right)\right], \qquad (11.179)$$

where $u = (1/r)$ and the c-factors are reintroduced. Differentiating this equation with respect to ϕ, we get

$$\frac{d^2u}{d\phi^2} + u = \frac{GMm^2}{L^2} + \frac{3GM}{c^2}u^2. \qquad (11.180)$$

The first term on the right-hand side is purely Newtonian and the second term is the correction from general relativity. This correction term changes the nature of the orbits in several ways. To begin with, it changes the relationship between the parameters of the orbit and the energy and the angular momentum of the particle. More importantly, it causes the elliptical orbit of Newtonian gravity to precess slowly. If the original orbit has low eccentricity (i.e, nearly circular, as most planetary orbits are), then precession can be determined as follows. Let the radius of the circular orbit be r_0, for which $u = (1/r_0) \equiv k_0$. For the actual orbit, $u = k_0 + u_1$, where we expect the second term to be a small correction. Changing the variables from u to u_1, where $u_1 = u - k_0$, we can write Eq. (11.180) as

$$u_1'' + u_1 + k_0 = \frac{GMm^2}{L^2} + \frac{3GM}{c^2}\left(u_1^2 + k_0^2 + 2u_1k_0\right). \qquad (11.181)$$

We now choose k_0 to satisfy the condition

$$k_0 = \frac{GMm^2}{L^2} + k_0^2\frac{3GM}{c^2}, \qquad (11.182)$$

which determines the radius of the original circular orbit in terms of the other parameters. Now the equation for u_1 becomes

$$u_1'' + \left(1 - \frac{6k_0GM}{c^2}\right)u_1 = \frac{3GM}{c^2}u_1^2. \qquad (11.183)$$

This equation is exact. We now use the fact that the deviation from circular orbit, characterised by u_1, is small, and we ignore the right-hand side of Eq. (11.183). Solving Eq. (11.183), with the right-hand side set to zero, we get

$$u_1 \cong A \cos\left[\left(1 - \frac{6GM}{c^2r_0}\right)^{1/2}\phi\right]. \qquad (11.184)$$

The term multiplying the angle ϕ causes the precession of the orbit. The angle by which the orbit rotates between two successive moments of same value of u_1 is given by

$$\delta\phi \approx 2\pi\left[1 - (6GM/c^2r_0)\right]^{-1/2}.$$

Note that this result does not require relativistic effects to be small and relies on only the assumption that the original orbit was nearly circular. If we further assume that $r_0 \gg (GM/c^2)$, then $k_0 \approx (GMm^2/L^2)$, and the precession is given by

$$\Delta\phi = 2\pi\left(1 - \frac{6G^2 M^2 m^2}{c^2 L^2}\right)^{-1/2} \cong 2\pi + 6\pi\left(\frac{GMm}{Lc}\right)^2. \quad (11.185)$$

The second term on the right-hand side gives the angle by which the major axis rotates per orbit. For Mercury and Earth, the precession is ~43″ per century and 3.8″ per century, respectively. The result for Mercury has been one of the crucial tests of the general theory of relativity.

Exercise 11.10 *Perihelion shift for an oblate Sun*: If the Sun has a quadrapole moment, then a perihelion shift to the orbit of Mercury will arise even in Newtonian theory. In such a case, the gravitational potential that is due to the Sun can be expressed in the form

$$U = \frac{M_\odot}{r}\left(1 - J_2 \frac{R_\odot^2}{r^2}\frac{3\cos^2\theta - 1}{2}\right), \quad (11.186)$$

in units with $G = c = 1$ and J_2 is a quadrapole moment parameter. Show that, in this case, the total precession per orbit is given by

$$\delta\phi = \frac{6\pi M_\odot}{a(1-e^2)} + J_2 \frac{3\pi R_\odot^2}{a^2(1-e^2)^2}, \quad (11.187)$$

where a and e are the semimajor axes and the eccentricity of the original elliptic orbit, respectively. Show that the first term is the same as the result derived in the text and that the second term arises because of the oblateness of the Sun.

11.9.2 Deflection of an Ultrarelativistic Particle

The corrections discussed above are for a bound elliptical orbit in Newtonian theory. A different effect in general relativity arises when an ultrarelativistic particle passes near a central mass, as in the case of a light ray travelling near the Sun. In this context, we are interested in the deflection of particle from a straight-line trajectory that is due to the gravitational attraction of the central mass. When this deflection is small, it can be computed by perturbing the straight-line trajectory.

The starting point is again the equation for the orbit given by Eq. (11.180). The undeflected straight-line solution to this equation is given by $u_0 = b^{-1}\cos\phi$, which is obtained when both terms on the right-hand side are neglected. (Here b is the impact parameter of the orbit.) To find the deflection that is due to the gravitational field, we try a solution to Eq. (11.180) of the form $u = b^{-1}\cos\phi + v(\phi)$, with $bv \ll 1$. Substituting this ansatz into Eq. (11.180) and retaining only the leading-order terms, we get

$$v'' + v \cong \frac{GMm^2}{L^2} + \frac{3GM}{2c^2b^2}(\cos 2\phi + 1). \tag{11.188}$$

The first term on the right-hand side describes the deflection that is due to purely Newtonian attraction and the second term gives the general relativistic correction. This equation is identical to that of a harmonic oscillator perturbed by an external force and can be easily solved. Substituting the solution in our ansatz for u, we get

$$u = \frac{1}{b}\cos\phi - \frac{GM}{c^2b^2}\cos^2\phi + \frac{GMm^2}{L^2} + \frac{2GM}{c^2b^2}. \tag{11.189}$$

The first term is the undeflected trajectory and the other three terms give the deflection that is due to the gravitational field of the central mass. To find the angle of deflection when the particle moves past the central mass, we set $u = 0$ (which corresponds to $r = \infty$) and solve for ϕ. To the leading order, we can ignore the $\cos^2\phi$ term and obtain

$$\cos\phi \cong \frac{GMm^2b}{L^2} + \frac{2GM}{c^2b} \equiv q. \tag{11.190}$$

To the same order of accuracy, the solution to this equation is given by $\phi = (\pi/2) \pm q$ and the net deflection will be $\delta\phi \approx 2q$. To relate q to parameters of the particle at large distances, we note that the angular momentum L is given by bp_∞, where $p_\infty = \gamma m v_\infty$ is the momentum of the particle at large distances. Using these relations in the definition of q, we find that the deflection is given by

$$\delta\phi = 2q = \frac{2GM}{bv_\infty^2}\left(1 + \frac{v_\infty^2}{c^2}\right). \tag{11.191}$$

When $c \to \infty$, this reduces to the Newtonian deflection of $(2GM/bv_\infty^2)$. The second term in the parentheses gives the correction that is due to general relativity. For photons, with $v_\infty = c$, this leads to a deflection of $\delta\phi \approx (4GM/c^2b)$, which is twice the Newtonian value. This was another prediction from general relativity that has been successfully tested.

Note that the deflection in Eq. (11.191) has a very different velocity dependence compared with the deflection of a *relativistic* charged particle moving in the Coulomb field (see Chap. 3, Subsection 3.8.1), even though both have the same velocity dependence in the nonrelativistic limit (that is, when $c \to \infty$). However, as $v \to c$, the electromagnetic deflection vanishes while the gravitational deflection becomes twice the nonrelativistic value. This difference is due to the fact that the electromagnetism is based on a vector field whereas gravity is based on a second-rank tensor field.

The deflection of light by the gravitational field leads to an important phenomenon called gravitational lensing, which will be discussed in Vol. III.

Exercise 11.11

Angular shift of the direction of stars: What is actually observed in the observations involving the deflection of light by the Sun is the shift in the angular position of a star. Show that this shift is given by

$$\delta\alpha = \frac{2GM_\odot}{c^2 r_{\text{ES}}} \left(\frac{1 + \cos\alpha}{1 - \cos\alpha}\right)^{1/2}, \tag{11.192}$$

where α is the unperturbed direction of the star when viewed from Earth and r_{ES} is the distance between the centres of Earth and Sun. Estimate this quantity $\delta\alpha$ when the direction of the ray of light varies from $\alpha = \pi$ (opposite to the Sun's direction) through $\alpha = (\pi/2)$ (perpendicular to the Earth-Sun line) to the grazing incidence touching the rim of the Sun.

11.9.3 Post-Newtonian Precession

The spin angular-momentum vector of a body will be affected by the curvature of the space–time, just as any other vector will. In particular, a spinning gyroscope orbiting a massive body will undergo precession of the spin vector that can serve as a sensitive test of general relativity. We now obtain the general formula for the precession of the spin for a gyroscope orbiting a massive body.

The equation of motion for the spin four vector S^i has a simple form relative to an instantaneous local Lorentz frame. During an infinitesimal increment $d\tau$ of the proper time, any spatial component S^α will vary by the amount $dS^\alpha = \Gamma^\alpha_{ij} S^j u^i d\tau = \Gamma^\alpha_{\beta 0} S^\beta d\tau$ because in the local Lorentz frame $u^i = (1, 0)$ and $S^j = (0, \mathbf{S})$. Hence the precession of the spatial component of the spin vector can be obtained from the expression for $\Gamma^\alpha_{\beta 0}$ in the local Lorentz frame. This can be computed in a straightforward manner to the lowest nontrivial order of accuracy for a metric in the form

$$ds^2 = (1 + 2\phi) dt^2 - (1 - 2\phi)\delta_{\alpha\beta} dx^\alpha dx^\beta - 2g_\alpha dx^\alpha dt. \tag{11.193}$$

The result is

$$\Gamma_{\alpha,\beta 0} = \frac{3}{2}(\phi_{,\alpha} v_\beta - \phi_{,\beta} v_\alpha) + \frac{1}{2}(g_{\beta,\alpha} - g_{\alpha,\beta}) + \frac{1}{2}(a_\alpha v_\beta - a_\beta v_\alpha), \tag{11.194}$$

where $\mathbf{v}$ is the three velocity and $\mathbf{a}$ is the three acceleration of the spinning object. (We stress the fact that the components are evaluated in the instantaneous Lorentz frame.) Using this expression, we find that the equation of motion $(dS^\alpha/d\tau) = \Gamma^\alpha_{\beta 0} S^\beta$ reduces to the form $(d\mathbf{S}/d\tau) = \mathbf{\Omega} \times \mathbf{S}$, where

$$\mathbf{\Omega} = -\frac{1}{2}(\nabla \times \mathbf{g}) + \frac{1}{2}(\mathbf{a} \times \mathbf{v}) + \frac{3}{2}(\nabla\phi \times \mathbf{v}). \tag{11.195}$$

Each of the terms has a simple significance. The first term, $-(1/2)(\nabla \times \mathbf{g})$, arises because of the rotation of the central body, which causes the inertial frames to be dragged along with respect to fixed stars. This has already been worked out in Section 11.8 [see relation (11.164)]. The second term is a purely special relativistic effect called Thomas precession, which was derived in Chap. 3, Exercise 3.4. The last term arises for two separate reasons. Part of this term, $(1/2)(\nabla\phi \times \mathbf{v})$, arises because of precession in the gravomagnetic field whereas another part, $(\nabla\phi \times \mathbf{v})$, arises because of the curvature of space, which is described below.

These three terms lead to different amounts of total precession for different objects. To begin with, the first term is of purely geometrical origin and is independent of the orbital elements of the gyroscope. The second term will contribute whenever there is a nongravitational source of acceleration but will vanish for a gyroscope in free-fall orbit around a massive body. Such an orbiting gyroscope will respond to only the first and third terms. For a gyroscope at rest on the surface of the Earth, say, there is partial cancellation between the second and the third terms (because $\mathbf{a} = -\nabla\phi$), leading to a net precession of $\nabla\phi \times \mathbf{v}$.

The precession obtained above has a fairly simple physical meaning that is highlighted by the following alternative derivation. Consider a gyroscope that is in a circular orbit around a massive body of mass M. To the lowest order, the orbital speed of the gyroscope is given by $v = (GM/r)^{1/2}$, where r is the radius of the orbit. In the instantaneous inertial frame of the gyroscope, there will be a gravitomagnetic field of magnitude $\mathbf{B}_{gr} = (\mathbf{v} \times \nabla\phi)/c^3 = -(v^2/c^3)\Omega$, where Ω is the angular velocity. From equations of motion (11.145), it is clear that a spinning body located in a gravitomagnetic field behaves as a magnetic dipole in a magnetic field. The classical gyromagnetic ratio $(q/2m) = (1/2)$ for gravity because $q = m$. The spin-orbit precession will therefore be at an angular frequency of

$$\Omega_{\text{sp}-\text{orb}} = -\frac{1}{2}c\mathbf{B}_{gr} = \frac{1}{2}\frac{v^2}{c^3}\Omega = \frac{1}{2}\left(\frac{G^3 M^3}{c^4 r^5}\right)^{1/2}\hat{\Omega}. \qquad (11.196)$$

In addition to this, curvature in space will also induce a precession when the angular-momentum vector of the gyroscope is transported around in an orbit. A simple way of deriving this result is as follows. The two-dimensional orbital plane has the metric

$$ds^2 = \frac{dr^2}{1 + 2(GM/c^2r)} + r^2 d\phi^2 \simeq \left[\left(1 + \frac{GM}{c^2 r}\right)dr\right]^2 + r^2 d\phi^2$$

$$\cong dl^2 + l^2\left(\frac{d\phi}{1 + (GM/c^2 r_0)}\right)^2 \qquad (11.197)$$

near the orbit $r = r_0$, with $l = (1 + GM/c^2 r_0)r$. Embedding this space in the three-dimensional Euclidean space, we can work out the trajectory of the gyroscope in the embedding space. Because of the curvature of the orbital plane, a cone that is tangent to this space at the location of the orbit will have a deficit in the angle compared with the Euclidean embedding space, as is clear from the last part of Eq. (11.197). This deficit angle of the cone will appear as a precession angle per orbit when we identify the beginning and the end of the orbit. We thus see that the precession will be by an angle $(2\pi GM/c^2 r)$ per orbit. Equivalently, the precession rate will be

$$\Omega_{\text{curv}} = \frac{2\pi GM}{c^2 r} \frac{\Omega}{2\pi} = \frac{v^2}{c^2} \Omega. \qquad (11.198)$$

Adding the two contributions, we find that the net precession becomes

$$\Omega_{\text{tot}} = \frac{3v^2}{2c^2} \Omega = \frac{3}{2} \left(\frac{G^3 M^3}{c^4 r^5} \right)^{1/2}, \qquad (11.199)$$

which is the third term in Eq. (11.195).

Exercise 11.12
Relativistic precession: Fill in the details and derive Eqs. (11.194) and (11.195).

11.10 Gravitational Collapse and Black Holes

The Schwarzschild metric that is due to a mass M has a characteristic length scale given by $r_g \equiv (2GM/c^2)$. The g_{00} component vanishes, and g_{rr} diverges at $r = r_g$, showing that this radius must play a key role in the description of physics in Schwarzschild metric. It must, however, be remembered that the Schwarzschild metric describes the *empty* region outside a spherical source; in the nonempty region occupied by the source, the metric will be quite different. [If the source is static then the inside metric is given by Eq. (11.158).] If the size of the source L is larger than r_g, then the Schwarzschild metric is not valid around $r = r_g$ and the singular behaviour of the metric coefficients at $r = r_g$ is not of any concern. However, if the size of the source is such that $L < r_g$, then several peculiar features come into play, which need to be discussed.

Such a situation can arise, for example, in the gravitational collapse of matter. Consider, for simplicity, a sphere of dust with $p = 0$, $\rho \neq 0$, that was occupying a spherical region of radius $r = L_1$ at time $t = t_1$. Let the total mass of the dust sphere be such that $L_1 > r_g$. At $t = t_1$, the metric at $r > L_1$ is given by the Schwarzschild metric, which is everywhere well behaved because of our assumption that $L_1 > r_g$. This initial configuration, however, cannot be static because there is no pressure gradient to balance the gravitational attraction. We expect the dust sphere to collapse under its own gravitational force, thereby decreasing its size. When its surface contracts to a radius of less than r_g, the metric outside becomes ill behaved at $r = r_g$ and the question arises as to what the new features are that such a collapse introduces. This issue, related to the formation and the dynamics of a black hole, will be discussed in Vol. II.

11.11 The Energy-Momentum Pseudotensor for Gravity

In Einstein's general theory of relativity, gravitational effects are manifested as being due to the curvature of space–time. This curvature is produced by the distribution of matter in space–time through Einstein's equations. In this sense, the gravitational field is conceptually very different from all other fields, like for example, the electromagnetic field.

The curvature of the space–time affects other matter such as particles, electromagnetic waves, etc. Hence, the material bodies, electromagnetic fields, etc., will not form a closed system in curved space–time. As a consequence, the energy and the momentum of such matter will not, in general, remain constant. It was shown in Subsection 11.7.3 that the energy-momentum tensor for matter satisfies the equation $T^k_{i;k} = 0$. Using the identity of Eq. (11.62), derived in Section 11.4, we can write this equation as

$$T^k_{i;k} = \frac{1}{\sqrt{-g}} \frac{\partial \left(T^k_i \sqrt{-g} \right)}{\partial x^k} - \frac{1}{2} \frac{\partial g_{kl}}{\partial x^i} T^{kl} = 0. \qquad (11.200)$$

This equation does not represent a conservation law because of the presence of the second term involving the derivatives of the metric tensor. It is not possible to convert this equation – say, by integrating over a four volume – into a form that represents the conservation of some physical quantity. This is, however, to be expected because the presence of curvature will affect the matter.

The question arises as to whether we can construct an energy-momentum tensor t_{ik} for the gravitational field such that the total energy-momentum tensor of matter plus gravitational field will be conserved. There are several difficulties in realising such a goal. To begin with, it is conceptually difficult to understand how energy and momentum can be attributed to the curvature of space–time. Operationally we would have expected the energy-momentum tensor t_{ik} to be made of the metric tensor and its first derivatives. In a locally inertial frame around any event, the metric tensor will reduce to its flat space–time value and the first derivatives will vanish; therefore we can always choose a coordinate system around any event such that $t_{ik} = 0$. It immediately follows that t_{ik}, if it were to be a true tensor, will vanish identically.

It is, however, possible to find expressions for t_{ik} such that it may be interpreted as the energy-momentum tensor of the gravitational field in a specific coordinate system. Such expressions are not unique, nor is t_{ik} a tensorial quantity. Nevertheless, it may be of use in some specific and limited context – especially if the space–time geometry becomes flat at large distances from a collection of particles, say. One such expression is given by

$$t^{ik} = \frac{c^4}{16\pi G}\Big\{\big(2\Gamma^n_{lm}\Gamma^p_{np} - \Gamma^n_{lp}\Gamma^p_{mn} - \Gamma^n_{ln}\Gamma^p_{mp}\big)(g^{il}g^{km} - g^{ik}g^{lm})$$

$$+ g^{il}g^{mn}\big(\Gamma^k_{lp}\Gamma^p_{mn} + \Gamma^k_{mn}\Gamma^p_{lp} - \Gamma^k_{np}\Gamma^p_{lm} - \Gamma^k_{lm}\Gamma^p_{np}\big)$$

$$+ g^{kl}g^{mn}\big(\Gamma^i_{lp}\Gamma^p_{mn} + \Gamma^i_{mn}\Gamma^p_{lp} - \Gamma^i_{np}\Gamma^p_{lm} - \Gamma^i_{lm}\Gamma^p_{np}\big)$$

$$+ g^{lm}g^{np}\big(\Gamma^i_{ln}\Gamma^k_{mp} - \Gamma^i_{lm}\Gamma^k_{np}\big)\Big\}. \tag{11.201}$$

This quantity satisfies the equation

$$\frac{\partial}{\partial x^k}(-g)(T^{ik} + t^{ik}) = 0, \tag{11.202}$$

which can be verified by straightforward but tedious computation. This relation shows that there exists a conservation law for the quantities

$$P^i = \frac{1}{c}\int(-g)(T^{ik} + t^{ik})\,dS_k = \frac{1}{c}\int(-g)(T^{i0} + t^{i0})\,dV \tag{11.203}$$

in the given coordinate system. In the second expression, we have chosen the $x^0 = $ constant surface to do the integral. Let us assume that the matter is confined to some finite region around the origin of a coordinate system. Then we can choose our coordinates such that the metric reduces to the flat space–time values at a large distance from the origin. (This, of course, does not yet uniquely specify the coordinates as we can make different choices near the origin.) Let us now consider the value of P^i we obtain by integrating over a large volume of space whose surface is at a distance far away from the origin and located in a region in which the space–time is nearly flat. In that case, the values of P^i are independent of the choice of coordinates near the origin and can be thought of as the total four momentum inside this region. Thus, whenever the matter is confined to finite region of space and we use coordinates that are asymptotically flat, it is possible to define the concept of total four momentum of matter plus gravity. This quantity P^i is also *not* a true four vector but it behaves like a four vector under Lorentz transformations in the asymptotically flat region of space time.

11.12 Gravitational Waves

Because Einstein's theory of gravity incorporates the principle of relativity, there exists a finite velocity of propagation for the gravitational interaction. Changes in the configuration of the source cannot influence regions of space–time that are not causally connected.

The gravitational field in the absence of matter is described by the equation $R_{ik} = 0$. Because R_{ik} is a complicated nonlinear function involving the derivatives of g_{ik}, this equation possesses highly nontrivial solutions. In fact, the most general solution to this equation is not known. Among the class of solutions, there exists a subset that may be thought of as describing the propagation of gravitational influence with the speed of light. These solutions are called gravitational waves and are similar to the solutions of free Maxwell's equations that describe the electromagnetic waves. However, because Maxwell's equations are linear, it is easy to find the most general solution that describes the propagation of an electromagnetic wave. This, however, is not possible in the case of gravity. In general, because Einstein's theory is nonlinear, the gravitational wave will interact with itself in a complicated manner as it propagates through space–time.

Although the exact discussion of gravitational waves is fairly complicated, it turns out that this is not really necessary in astrophysical situations. In most of the realistic situations, the gravitational waves emitted by an astrophysical system will be quite weak and the part of the metric tensor describing the gravitational waves will have components that are small compared with unity. In such a case, the gravitational waves may be treated as small perturbations in the background space–time and the relevant equations can be solved. Even the background space–time can be taken to be flat while the propagation of gravitational waves is studied at large distances from the source. Let us begin by studying this case of gravitational waves propagating in a flat space–time.

The metric corresponding to such a situation may be taken as $g_{ik} = g^0_{ik} + h_{ik}$, where the first term on the right describes the Minkowski space–time and h_{ik} is the metric that is due to the gravitational wave. In what follows, all operations of raising and lowering of indices will be done with the background metric g^0_{ik}. It is assumed that all components of h_{ik} are small compared with unity, and we can use perturbation theory, retaining terms that are of lowest order in h_{ik}.

This decomposition, of course, does not completely specify h_{ik}. It was shown in Subsection 11.7.3 that under the coordinate transformation $x^i \to x^{i'} = x^i + \xi^i$ with small ξ, the metric tensor changes to

$$h'_{ik} = h_{ik} - \frac{\partial \xi_i}{\partial x^k} - \frac{\partial \xi_k}{\partial x^i}. \tag{11.204}$$

Using this arbitrariness in the choice of coordinates, we can impose the following conditions on h_{ik}:

$$\frac{\partial \psi^k_i}{\partial x^k} = 0, \quad \psi^k_i = h^k_i - \frac{1}{2}\delta^k_i h, \tag{11.205}$$

where $h = h^i_i$ is the trace of the tensor h_{ik}. (This condition is similar to the gauge condition $\partial_i A^i = 0$, which is used in electromagnetism.) In fact, these conditions still do not fix a unique choice of the reference frame. If some set of h_{ik} satisfies these conditions, then so will another set h'_{ik}, if we take ξ_i to be solutions of the equation $\Box \xi_i = 0$, where, in Cartesian coordinates, with $g^{lm(0)} = \eta^{lm}$,

$$\Box = -g^{lm(0)} \frac{\partial^2}{\partial x^l \partial x^m} = \nabla^2 - \frac{1}{c^2} \frac{\partial^2}{\partial t^2}. \tag{11.206}$$

We will have occasion to use this freedom later in this section.

To determine the form of h_{ik}, we have to substitute our expansion for g_{ik} into Einstein's equations and linearise it in h_{ik}. To linear order, the curvature tensor and the contracted Ricci tensor are given by

$$R_{iklm} = \frac{1}{2}\left(\frac{\partial^2 h_{im}}{\partial x^k \partial x^l} + \frac{\partial^2 h_{kl}}{\partial x^i \partial x^m} - \frac{\partial^2 h_{km}}{\partial x^i \partial x^l} - \frac{\partial^2 h_{il}}{\partial x^k \partial x^m} \right), \tag{11.207}$$

$$R_{ik} = \frac{1}{2}\left(-g^{lm(0)} \frac{\partial^2 h_{ik}}{\partial x^l \partial x^m} + \frac{\partial^2 h'_i}{\partial x^k \partial x^l} - \frac{\partial^2 h'_k}{\partial x^i \partial x^l} - \frac{\partial^2 h}{\partial x^i \partial x^k} \right), \tag{11.208}$$

where no additional conditions have been imposed on h_{ik}. On using gauge conditions (11.205), the Ricci tensor becomes simpler:

$$R_{ik} = -\frac{1}{2} g^{lm(0)} \frac{\partial^2 h_{ik}}{\partial x^l \partial x^m} = \frac{1}{2}\Box h_{ik}. \tag{11.209}$$

Equating this expression to zero, we find that the perturbations $\Box h_{ik}$ satisfy the wave equation $\Box h_i^k = 0$. This gravitational field, like an electromagnetic field, propagates in vacuum with the speed of light.

To study the properties of the gravitational waves, let us consider solutions to this equation that depend on only x and t. The wave equation then reduces to

$$\left(\frac{\partial^2}{\partial x^2} - \frac{1}{c^2} \frac{\partial^2}{\partial t^2} \right) h_i^k = 0, \tag{11.210}$$

and the solution will be any arbitrary function of $t \pm (x/c)$. If we choose the positive x direction as the propagation of plane waves, then h_{ik} are functions of $t - (x/c)$. Auxiliary conditions (11.205) give $\dot{\psi}_i^1 - \dot{\psi}_i^0 = 0$. Integrating this, we can write the conditions as

$$\psi_1^1 = \psi_1^0, \quad \psi_2^1 = \psi_2^0, \quad \psi_3^1 = \psi_3^0, \quad \psi_0^1 = \psi_0^0. \tag{11.211}$$

As mentioned above, conditions (11.205) do not completely specify the system; it is possible to make a coordinate transformation of the form $x^{i'} = x^i + \xi^i[t - (x/c)]$. These transformations can now be used to make $\psi_1^0, \psi_2^0, \psi_3^0$, and $\psi_2^2 + \psi_3^3$ vanish. From Eqs. (11.211), it follows that the components $\psi_1^1, \psi_2^1, \psi_3^1$, and ψ_0^0 also vanish. The remaining quantities $\psi_2^3, \psi_2^2 - \psi_3^3$ cannot be made to vanish by any choice of reference frame and represent the irreducible degrees of freedom of the gravitational wave. With these conditions, the trace $\psi \equiv \psi_i^i$ also vanishes, implying that $\psi_i^k = h_i^k$ and $\psi_2^2 = \psi_3^3$.

Because the plane gravitational wave is determined by the two quantities h_{23} and $h_{22} = -h_{33}$, (which have components only in the y–z plane) it follows that the waves are transverse. The polarisation of the wave is determined by a symmetric, traceless second-rank tensor in the y–z plane. The two independent polarisations may be chosen to be the cases in which one of the two quantities h_{23}, $(1/2)(h_{22} - h_{33})$, differs from zero. These two polarisations differ from each other by a rotation through an angle $\pi/4$ in the y–z plane.

Because the gravitational waves are propagating in a flat background space–time, we managed to choose a preferred Lorentzian coordinate system to describe them. The auxiliary conditions on the coordinate system remove all additional degrees of freedom. In this particular coordinate system, we can compute the expression for the gravitational energy-momentum pseudotensor that may be interpreted as describing the energy and the momentum of the gravitational wave. The computation is straightforward and is based on our noting that, for a plane wave satisfying the auxiliary condition, all the nonzero terms in t^{ik} are contributed by the term

$$\frac{1}{2} g^{il} g^{km} g_{np} g_{qr} g^{nr}_{,l} g^{pq}_{,m} = \frac{1}{2} h^{n,i}_q h^{q,k}_n. \tag{11.212}$$

The final result is

$$t^{ik} = \frac{c^4}{32\pi G} h^{n,i}_q h^{q,k}_n. \tag{11.213}$$

The energy flux along the x axis is given by ct^{01}:

$$ct^{01} = \frac{c^3}{16\pi G}\left[\dot{h}^2_{23} + \frac{1}{4}(\dot{h}_{22} - \dot{h}_{33})^2\right]. \tag{11.214}$$

The above discussion was confined to free gravitational waves that are propagating in the space–time, which is analogous to free electromagnetic waves. In the case of electromagnetism, we can relate the radiation of waves to the accelerated motion of the charged particles. To achieve the corresponding results in gravity, we have to solve the perturbed Einstein's equations in the presence of matter. In the presence of matter, the evolution of perturbations will be governed by an equation of the form

$$\frac{1}{2}\Box \psi^k_i = \frac{8\pi G}{c^4} \tau^k_i, \tag{11.215}$$

with

$$\psi^k_i = h^k_i - \frac{1}{2}\delta^k_i h. \tag{11.216}$$

The τ^k_i denotes auxiliary quantities that are obtained from the stress tensor T^k_i in the linear limit. The τ^0_0 and τ^0_α are obtained from the corresponding components of T^i_k by direct linearisation; the τ^α_β will contain, along with terms obtained from T^α_β, other terms transposed from the left-hand side of Einstein's equation. Auxiliary conditions (11.211) imply that $(\partial \tau^k_i/\partial x^k) = 0$, which, in expanded form, is given as

$$\frac{\partial \tau_{\alpha\gamma}}{\partial x^\gamma} - \frac{\partial \tau_{\alpha 0}}{\partial x^0} = 0, \qquad \frac{\partial \tau_{0\gamma}}{\partial x^\gamma} - \frac{\partial \tau_{00}}{\partial x^0} = 0. \qquad (11.217)$$

Because basic equation (11.215) is identical to the equation satisfied by the vector potential in electromagnetism, we can write the general solution in the form

$$\psi^k_i = -\frac{4G}{c^4} \int \left(\tau^k_i\right)_{t-(R/c)} \frac{dV}{R} \approx -\frac{4G}{c^4 R_0} \int \left(\tau^k_i\right)_{t-(R_0/c)} dV, \qquad (11.218)$$

where the second form is valid for the field at sufficiently large distances from a system of bodies that are moving nonrelativistically. Hereafter, we omit the subscript $t - (R_0/c)$ with the understanding that the integrands have to be evaluated at the retarded time.

It is possible to evaluate the integrals on the right-hand side and express the result in a simpler form. To do this, we begin by multiplying the first equation in Eqs. (11.217) by x^β and integrating over all space:

$$\frac{\partial}{\partial x^0} \int \tau_{\alpha 0} x^\beta dV = \int \frac{\partial \tau_{\alpha\gamma}}{\partial x^\gamma} x^\beta dV = \int \frac{\partial (\tau_{\alpha\gamma} x^\beta)}{\partial x^\gamma} dV - \int \tau_{\alpha\beta} dV.$$

$$(11.219)$$

The first term on the right-hand side vanishes on use of the divergence theorem. Taking half the sum of the resulting equation with the one obtained by transposing the indices, we get

$$\int \tau_{\alpha\beta} dV = -\frac{1}{2} \frac{\partial}{\partial x^0} \int \left(\tau_{\alpha 0} x^\beta + \tau_{\beta 0} x^\alpha\right) dV. \qquad (11.220)$$

A similar operation on the second equation of Eqs. (11.217), after it is multiplied by $x^\alpha x^\beta$, leads to

$$\frac{\partial}{\partial x^0} \int \tau_{00} x^\alpha x^\beta dV = -\int \left(\tau_{\alpha 0} x^\beta + \tau_{\beta 0} x^\alpha\right) dV. \qquad (11.221)$$

Comparing the two results, we get an expression for the required integral in terms of the derivatives of τ_{00}:

$$\int \tau_{\alpha\beta} dV = \frac{1}{2} \frac{\partial^2}{\partial x_0^2} \int \tau_{00} x^\alpha x^\beta dV. \qquad (11.222)$$

Substituting back into Eq. (11.218) and using $\tau_{00} = \rho c^2$ and $t = (x^0/c)$, we get

$$\psi^{\alpha\beta} = -\frac{2G}{c^4 R_0} \frac{\partial^2}{\partial t^2} \int \rho x^\alpha x^\beta dV. \qquad (11.223)$$

This equation gives the metric perturbations in terms of the characteristics of the source. To find the amount of energy radiated by the source, we first evaluate the components $h_{23} = \psi_{23}$ and $h_{22} - h_{33} = \psi_{22} - \psi_{33}$, which are given by

$$h_{23} = -\frac{2G}{3c^4 R_0}\ddot{D}_{23}, \quad h_{22} - h_{33} = -\frac{2G}{3c^4 R_0}(\ddot{D}_{22} - \ddot{D}_{33}), \quad (11.224)$$

where the mass quadrupole tensor $D_{\alpha\beta}$ is defined by

$$D_{\alpha\beta} = \int \rho(3x_\alpha x_\beta - \delta_{\alpha\beta}r^2)\,dV. \quad (11.225)$$

(The definition of a quadrupole moment here is completely analogous to that in electromagnetism.) The energy flux along the x axis becomes

$$ct^{10} = \frac{G}{36\pi c^5 R_0^2}\left[\left(\frac{\dddot{D}_{22} - \dddot{D}_{33}}{2}\right)^2 + \dddot{D}_{23}^2\right]. \quad (11.226)$$

The expression falls as $1/R_0^2$, which is to be expected for a radiation field. We find the energy emitted into a solid angle $d\Omega$ by multiplying this expression by $R_0^2 d\Omega$; it is independent of R_0. Also note that the lowest-order gravitational radiation is quadrupole, unlike electromagnetism in which the lowest order is dipole. This is because the time derivative of the mass dipole moment $\sum_i m_i \mathbf{x}_i$ vanishes for a closed system due to momentum conservation. (In fact, the dipole radiation for a system of charges with the same q/m ratio will also vanish for the same reason; see Chap. 6, Subsection 6.9.3.)

The two terms in the above expressions clearly represent the two independent polarisations. To write this result in an invariant form, we introduce the unit polarisation tensor $e_{\alpha\beta}$ that satisfies the conditions

$$e_{\alpha\alpha} = 0, \quad e_{\alpha\beta}n^\beta = 0, \quad e^{\alpha\beta}e_{\alpha\beta} = 1, \quad (11.227)$$

where $\mathbf{n}$ is a unit vector in the direction of propagation of the wave. In this case, the intensity of radiation of a given polarisation into a solid angle $d\Omega$ is given by

$$\frac{dI}{d\Omega} = \frac{G}{72\pi c^5}(\dddot{D}_{\alpha\beta}e^{\alpha\beta})^2. \quad (11.228)$$

To find the net angular distribution, we must average over all the polarisation states and multiply by 2. The averaging is straightforward but tedious and is done with the formula

$$\overline{e_{\alpha\beta}e_{\gamma\delta}} = \frac{1}{4}\{n_\alpha n_\beta n_\gamma n_\delta + (n_\alpha n_\beta \delta_{\gamma\delta} + n_\gamma n_\delta \delta_{\alpha\beta})$$
$$- (n_\alpha n_\gamma \delta_{\beta\delta} + n_\beta n_\gamma \delta_{\alpha\delta} + n_\alpha n_\delta \delta_{\beta\gamma} + n_\beta n_\delta \delta_{\alpha\gamma})$$
$$- \delta_{\alpha\beta}\delta_{\gamma\delta} + \delta_{\alpha\gamma}\delta_{\beta\delta} + \delta_{\beta\gamma}\delta_{\alpha\delta}\}. \quad (11.229)$$

The result is

$$\frac{dI}{d\Omega} = \frac{G}{36\pi c^5}\left[\frac{1}{4}(\dddot{D}_{\alpha\beta}n^\alpha n^\beta)^2 + \frac{1}{2}\dddot{D}_{\alpha\beta}^2 - \dddot{D}^{\alpha\beta}\dddot{D}_{\alpha\gamma}n_\beta n^\gamma\right], \quad (11.230)$$

with $\dddot{D}_{\alpha\beta}^2 \equiv \dddot{D}_{\alpha\beta}\dddot{D}^{\alpha\beta}$. Finally, we find the total radiation emitted in all directions by integrating this expression over all directions, and we can do this by using the formulas

$$\overline{n_\alpha n_\beta} = \frac{1}{3}\delta_{\alpha\beta}; \quad \overline{n_\alpha n_\beta n_\gamma n_\delta} = \frac{1}{15}(\delta_{\alpha\beta}\delta_{\gamma\delta} + \delta_{\alpha\gamma}\delta_{\beta\delta} + \delta_{\alpha\delta}\delta_{\beta\gamma}). \quad (11.231)$$

This gives the net energy loss of the system per unit time as

$$\frac{d\mathcal{E}}{dt} = -\frac{G}{45c^5}\dddot{D}_{\alpha\beta}^2. \quad (11.232)$$

As an example of the use of this formula, let us consider the radiation emitted by two bodies that are moving in circular orbits around their common centre of inertia. Choosing the coordinate origin at the centre of mass, we find that the radius vectors of the two bodies are given by

$$\mathbf{r}_1 = \frac{m_2}{m_1 + m_2}\mathbf{r}, \quad \mathbf{r}_2 = \frac{m_1}{m_1 + m_2}\mathbf{r}, \quad \mathbf{r} = \mathbf{r}_1 - \mathbf{r}_2. \quad (11.233)$$

Taking the plane of motion to be the x–y plane, we find that the components of the quadrupole tensor are

$$D_{xx} = \mu r^2(3\cos^2\psi - 1), \quad D_{yy} = \mu r^2(3\sin^2\psi - 1), \quad (11.234)$$

$$D_{xy} = 3\mu r^2\cos\psi\sin\psi, \quad D_{zz} = -\mu r^2, \quad (11.235)$$

where $\mu = [m_1 m_2/(m_1 + m_2)]$ and (r, ψ) are the polar coordinates in the x–y plane. For circular motion, r is a constant and $\dot\psi = (G(m_1 + m_2)/r^3)^{1/2} \equiv \omega$. Let $\mathbf{n}$ be a unit vector defined by the polar and azimuthal angles (θ, ϕ). We take the two polarisation states as: (1) $e_{\theta\phi} = 1/\sqrt{2}$ and (2) $e_{\theta\theta} = -e_{\phi\phi} = 1/\sqrt{2}$. Knowing the components of $e_{\alpha\beta}$ and $D_{\alpha\beta}$, we can evaluate the expression (11.228) as a function of time. Averaging this expression over one orbital period, we find the results for the two polarisation states as

$$\frac{dI_1}{d\Omega} = \frac{G\mu^2\omega^6 r^4}{2\pi c^5}4\cos^2\theta, \quad \frac{dI_2}{d\Omega} = \frac{G\mu^2\omega^6 r^4}{2\pi c^5}(1 + \cos^2\theta)^2. \quad (11.236)$$

The net radiation is the sum of the two results:

$$\frac{dI}{d\Omega} = \frac{G\mu^2\omega^6 r^4}{2\pi c^5}(1 + 6\cos^2\theta + \cos^4\theta). \quad (11.237)$$

Integrating over all directions, we find the total energy loss as

$$-\frac{d\mathcal{E}}{dt} = I = \frac{32G\mu^2\omega^6 r^4}{5c^5} = \frac{32G^4 m_1^2 m_2^2(m_1 + m_2)}{5c^5 r^5}. \quad (11.238)$$

The loss of energy through gravitational radiation will make the two bodies slowly approach each other in a secular fashion. Because the energy of a circular orbit is given by $\mathcal{E} = -(Gm_1m_2/2r)$, the rate of approach is

$$\dot{r} = \frac{2r^2}{Gm_1m_2}\frac{d\mathcal{E}}{dt} = -\frac{64G^3m_1m_2(m_1+m_2)}{5c^5r^3}. \tag{11.239}$$

A similar computation can be done for the more complicated case of an elliptical orbit. If e is the eccentricity and a is the semimajor axis of the orbit, then the total energy radiated by the system when averaged over an orbital period is given by

$$-\frac{\overline{d\mathcal{E}}}{dt} = \frac{32G^4m_1^2m_2^2(m_1+m_2)}{5c^5a^5}\frac{1}{(1-e^2)^{7/2}}\left(1 + \frac{73}{24}e^2 + \frac{37}{96}e^4\right). \tag{11.240}$$

We note that the rate of energy loss increases rapidly with the eccentricity of the orbit.

When a system of bodies – in approximately stationary motion – is emitting gravitational waves, there is also a steady loss of angular momentum from the system. In general, this rate can be written in the form

$$\frac{\overline{dM_\alpha}}{dt} = \sum\overline{(\mathbf{r}\times\mathbf{f})_\alpha} = \sum\overline{e_{\alpha\beta\gamma}x^\beta f^\gamma}, \tag{11.241}$$

where $\mathbf{f}$ is the frictional force acting on the particles because of the emission of gravitational radiation defined through the relation

$$\frac{\overline{d\mathcal{E}}}{dt} = \sum\overline{\mathbf{f}\cdot\mathbf{v}}. \tag{11.242}$$

In all these expressions, the summation is over all the particles in the system. To determine $\mathbf{f}$, we write

$$\frac{\overline{d\mathcal{E}}}{dt} = -\frac{G}{45c^5}\overline{\dddot{D}_{\alpha\beta}\dddot{D}^{\alpha\beta}} = -\frac{G}{45c^5}\overline{\dddot{D}^{\alpha\beta}D_{\alpha\beta}^{(V)}}, \tag{11.243}$$

where the superscript (V) stands for the fifth derivative of the quantity. Substituting the expression

$$\dot{D}_{\alpha\beta} = \sum m(3x_\alpha v_\beta + 3x_\beta v_\alpha - 2\delta_{\alpha\beta}\mathbf{r}\cdot\mathbf{v}) \tag{11.244}$$

into Eq. (11.243) we find that

$$f_\alpha = -\frac{2Gm}{15c^5}D_{\alpha\beta}^{(V)}x^\beta. \tag{11.245}$$

Comparison now gives

$$\frac{\overline{dM^\alpha}}{dt} = \frac{2G}{45c^5}e^{\alpha\beta}{}_\gamma\overline{D_{\beta\delta}^{(V)}D^{\gamma\delta}} = \frac{2G}{45c^5}e^{\alpha\beta}{}_\gamma\overline{\dddot{D}_{\beta\delta}\dddot{D}^{\delta\gamma}}. \tag{11.246}$$

As an example, we can use the above formula to determine the rate of loss of angular momentum for two bodies moving in elliptical orbits. In this case, we obtain

$$-\frac{\overline{d\,M_\alpha}}{dt} = \frac{32G^{7/2}m_1^2m_2^2\sqrt{m_1+m_2}}{5c^5a^{7/2}}\frac{1}{(1-e^2)^2}\left(1+\frac{7}{4}e^2\right). \quad (11.247)$$

It can be easily verified that in the case of circular motion with $\epsilon = 0$, $\dot{\mathcal{E}} = \dot{M}\omega$, as it should. We shall see in Vol. II that the secular changes in the orbital parameters of a binary pulsar can lead to an indirect detection of gravitational waves.

12

Basics of Nuclear Physics

12.1 Introduction

This chapter introduces several concepts from nuclear physics that are used in different areas of astrophysics, especially in the study of stellar evolution. The ideas developed here will be used in the theory of stellar structure and evolution (Vol. II) and in cosmology (Vol. III).

12.2 Nuclear Structure

The nuclei of atoms contain protons and neutrons that differ – most importantly – in their electrical charge; the proton is positively charged whereas the neutron is neutral. They are both fermions with spin half and their masses are almost equal ($m_p c^2 = 938.3$ MeV for a proton and $m_n c^2 = 939.6$ MeV for a neutron). Protons, being positively charged, will repel each other and a collection of protons will fly apart because of this repulsion. They are kept together along with neutrons in a nucleus because of the existence of a stronger, attractive, nuclear force. As far as nuclear force is concerned, protons and neutrons behave in very similar manner. We may say that the nuclear force provides the necessary binding energy to keep the atomic nucleus together.

A fundamental many-body theory describing the nuclear force that acts between the nucleons is not yet fully developed. Hence it is not possible to study aspects involving nuclear forces from first principles. In what follows we rely extensively on phenomenological facts and experimental results.

Different atomic nuclei are characterised by different values of the binding energy per nucleon, $\bar{E}_b$. This quantity $\bar{E}_b$ denotes the mean amount of energy that has to be supplied to a nucleon for it to be ejected out of the nucleus. More precisely, it can be computed as the difference between the sum of the masses of the constituent nucleons and the mass of the bound nuclei. If $M(Z, A)$ is the mass of a nucleus with atomic weight $A = N + Z$ and atomic number Z, then the binding energy is defined through the relation

$$M(Z, A) = [(A - Z)\, m_n c^2 + Z(m_p + m_e)\, c^2 - A\bar{E}_b]. \tag{12.1}$$

This quantity $M(Z, A)$ is given empirically by the fitting formula

$$M(Z, A) = m_u c^2 \left[b_1 A + b_2 A^{2/3} - b_3 Z + b_4 A \left(\frac{1}{2} - \frac{Z}{A} \right)^2 + \frac{b_5 Z^2}{A^{1/3}} \right], \tag{12.2}$$

with $m_u = 1.66057 \times 10^{-24}$ gm, denoting $1/12$ of the mass of the C^{12} nucleus, and the different coefficients are given by $b_1 = 0.991749$, $b_2 = 0.01911$, $b_3 = 0.000840$, $b_4 = 0.10175$, and $b_5 = 0.000763$. (It is convenient to use energy units, such as mega-electron-volts, to measure the masses of nucleons and the nuclei so that the difference in the masses can directly give the energies that are involved.) Each of the terms in this expression has a simple, physical explanation.

The dominant contribution to $\bar{E}_b$ is proportional to A and hence to the volume of the nucleus. The difference between b_1 and unity is largely due to the volume binding energy. The term involving b_2 corresponds to a surface tension that corrects for the fact that nucleons near the surface have fewer neighbours compared with those inside. The origin of the term containing b_4 can be understood along the following lines. If we assume that the nucleus is made of two ideal, nonrelativistic, Fermi gases, one with Z protons and the other with $N = (A - Z)$ neutrons, then it is easy to show that the energy (excluding the rest energy) of the system will scale as

$$E = \frac{3}{5} E'_F \left[\left(\frac{2N}{A} \right)^{2/3} N + \left(\frac{2Z}{A} \right)^{2/3} Z \right]. \tag{12.3}$$

Here E'_F is the Fermi energy (excluding the rest energy), which we define by putting A nucleons inside a volume V with four nucleons per momentum state. This expression E is minimised for $N = Z$ (that is, for nuclei with an equal number of neutrons and protons) and the lowest-order corrections around the minimum will be proportional to $[(1/2) - (Z/A)]^2$. This gives a term with coefficient b_4 in our expression. The term b_3 is quite small and arises from the correction $(m_n - m_p - m_e)/m_u$. Finally, the term with b_5 gives the Coulomb self-energy of the system.

Expressions (12.2) and (12.1) allow us to obtain an empirical relation for $\bar{E}_b$ that can be compared with the observed binding energy per nucleon for different elements. The latter quantity is denoted by $f(A)$, and Fig. 12.1 shows the variation of f with the atomic weight of the nucleus. It is clear that except for very light nuclei and those with some specific values for A, and Z, the mean binding energy is $f \approx 8$ MeV. The curve $f(A)$ reaches a maximum around $A = 56$, corresponding to the iron nucleus. The increase of the curve $f(A)$ for $A < 56$ is essentially a surface effect: In tightly packed nuclei, particles in the

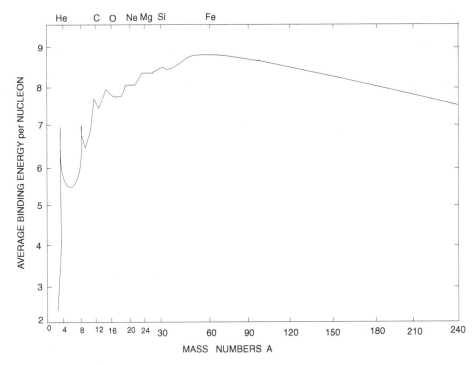

Fig. 12.1.   Binding energy per nucleon versus atomic weight.

interior experience more force compared with the particles near the surface. As the atomic weight increases, the number of particles in the interior grows faster than the number of particles at the surface, thereby providing a net increase in the binding energy. The decrease in $f(A)$ for $A > 56$ arises because of the Coulomb repulsion between the protons. This effect dominates over the nuclear forces for sufficiently large nuclei. The structure of $f(A)$ shows that the fusion of light nuclei (with $A < 56$) or the fission of heavy nuclei (with $A > 56$) will lead to the release of excess binding energy.

Most properties of a nucleus with Z protons and N neutrons can be understood through use of a model called the shell model in which it is assumed that the interaction of any given nucleon with the rest of the nucleus can be approximated by a spherical potential $V(r)$. The simplest choice for such a potential is that of a spherical well of depth W and radius R. The energy levels of a particle in such a spherical well can be found by standard procedures of quantum mechanics. The levels are characterised by two integers n and l and are given (approximately) by the expression

$$E(n, l) \approx -W + \frac{\hbar^2}{2mR^2}\left[\pi^2\left(n + \frac{l}{2}\right)^2 - l(l + 1)\right]. \qquad (12.4)$$

We can estimate the radius R of a nucleus with an atomic weight A can be estimated, to a very good approximation, by assuming that the nucleus is made of A spheres, each of radius r_0, in a tightly packed configuration; the value of r_0 is well approximated by the Compton wavelength of the nucleon, (h/mc). This corresponds to $R \equiv r_0 A^{1/3} \approx 1.4 \times 10^{-13} A^{1/3}$ cm, leading to a density of $\rho_{nuc} \approx 1.4 \times 10^{14}$ gm cm^{-3}. To estimate the energy scale W, we can proceed as follows. The nucleons in the nuclear well behave as independent fermions of spin half at zero temperature. From Chap. 5, Subsection 5.9.2, we know that such a system is characterised by a Fermi momentum p_F. The volume of phase space available to the fermions is $(4\pi/3)p_F^3 V$, where $V = (4\pi/3)R^3$ is the volume of the nucleus. Dividing by $(2\pi\hbar)^3$, we get the number of cells in each of which two protons and two neutrons can be accommodated. Assuming that the number of protons is equal to the number of neutrons, we get the relation $4(4\pi/3)(p_F/2\pi\hbar)^3 V = A$. Using the expression for the radius R, we find

$$p_F = \left(\frac{3\pi^2 A}{2V}\right)^{1/3} \hbar = \frac{(9\pi)^{1/3}\hbar}{2r_0} = 1.4 \times 10^{-14} \text{ g cm s}^{-1}. \qquad (12.5)$$

This corresponds to a Fermi energy of ~ 25 MeV. If ~ 8 MeV per nucleon is required for pulling out a nucleon from the high-lying states, it follows that the total depth of the potential well is $W \approx (25 + 8)$ MeV $= 33$ MeV. Thus the nuclear interactions can be thought of as providing a spherical potential well in the nucleus with a depth of $W \approx 30$ MeV and a radius of $R = r_0 A^{1/2}$, with $r_0 \approx 10^{-13}$ cm.

Each set of numbers (n, l) in Eq. (12.4) denotes a shell in which two fermions can be accommodated. Of particular importance to nuclear physics are the values of Z at which some of the shells are completely filled and the energy gap to the next shell is large. These occur for $Z = 2, 8, 20, 28, 50, 82$, and 126 and are called magic numbers of nuclear structure. The binding energies of these nuclei are significantly higher (i.e., they are more tightly bound; see Fig. 12.1) than their neighbours. Some of the examples of such nuclei are $\text{He}_2^4 = (1s_{1/2})^2$, $\text{O}_8^{16} = (1s_{1/2})^2 (1p_{3/2})^4 (1p_{1/2})^2$, $\text{Ca}_{20}^{40} = (1s_{1/2})^2 (1p_{3/2})^4$ $(1p_{1/2})^2 (1d_{5/2})^6 (2s_{1/2})^2 (1d_{3/2})^4$, etc., where the shell structure is denoted by standard notation (see Chap. 7). The subscript and the superscript to the symbol indicating the nucleus give Z and $A \equiv N + Z$, respectively.

The discussion so far did not take into account the electromagnetic interaction of the particles. For a charged particle (with charge $Z_1 e$) that is attempting to enter or leave the nucleus (with charge $Z_2 e$), we also have to consider the Coulomb interaction energy $E_{\text{Coul}} \approx (Z_1 Z_2 e^2 / r)$. At $r = r_0$, this energy is $E_{\text{Coul}}(r_0) \approx Z_1 Z_2$ MeV. Figure 12.2 shows the form of the interaction potential felt by a charged nuclear particle interacting with a nucleus. For $r < r_0$, the potential is attractive because of the strong nuclear forces. For $r > r_0$, the potential is repulsive because of Coulombic interaction.

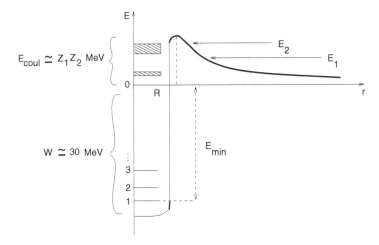

Fig. 12.2. The potential felt by two nuclei interacting by means of Coulomb and strong interactions.

12.3 Thermonuclear Reactions

Nuclear reactions between particles can occur under different physical circumstances, leading to different consequences. The description in Section 12.2 shows that, in an interaction between two nuclei of charges Z_1 and Z_2, the height of the repulsive part of the potential is of the order of $Z_1 Z_2$ MeV. If the typical kinetic energies of the nuclei are greater than this value, then the interaction can proceed without any difficulty and will be governed essentially by the nuclear forces. This can happen in some astrophysical cases in which very high-energy particles (i.e., particles with energies greater than a few mega-electron-volts) are involved. The equivalent temperature, if the kinetic energy arises because of thermal motion, is higher than 10^{10} K.

A different case arises in the interaction of nuclei at somewhat lower temperatures, say, $\sim 10^7$ K. At this temperature, the typical energies of the particles are significantly lower than the peak of the interaction potential in Fig. 12.2. Classically, it will not be possible for two nuclei with kinetic energies in the kilo-electron-volt range to come sufficiently close to each other, overcoming the Coulombic repulsion. Hence, classically they cannot interact by means of the strong nuclear forces and, for example, undergo nuclear fusion. The situation, however, is quite different in quantum mechanics. Quantum mechanics allows for a particle to tunnel through a potential barrier with a low but nonzero probability. Hence even particles with energies lower than the peak energy in Fig. 12.2 will be able to interact with each other and undergo nuclear reaction. This process plays a vital role in energy generation of stellar interiors, and we shall discuss it in some detail.

Consider a reaction of a nucleus X with a particle a leading to a nucleus Y and a particle b. Such a reaction, $a + X \rightarrow Y + b$, can be denoted in a condensed form as $X(a, b)Y$. (Even though we have called a and b particles and X and Y nuclei, similar considerations apply even when a and b are also nuclei.) For such a nuclear reaction to occur, it is necessary that X and Y overcome the Coulomb barrier, say, by quantum tunneling. The nuclear reaction of this kind can be thought of as proceeding through an intermediate compound nucleus, which we denote by the symbol Z^*. The above reaction can now take place in two steps: The original reactants a and X combine to form the nucleus Z^*, which then decays to the final products Y and b. In principle, the compound nucleus Z^* could decay into a wide variety of final products such as $(Y, b), (X, a), (Y_1, b_1), (Y_2, b_2), \ldots$, of which (Y, b) is just one possibility. (Note that the second possibility corresponds to the scattering of X and a without any transformation.) Each of these decay modes is called a channel and each has its own decay rate Γ_i; the total decay rate of the compound nucleus Γ will be the sum of decay rates through all the channels. The reactions $a + X \rightarrow Z^*$ and $Z^* \rightarrow Y + b$ occur independently and the probabilities for either of them can be computed with the laws of quantum mechanics.

It is clear from the above description that the properties of the intermediate nuclei Z^*, especially the energy levels, will play an important role in governing both the reactions (a, X) and (Y, b). From general quantum-mechanical considerations, we would expect the energy levels to be of the kind indicated by the horizontal lines in Fig. 12.2. The levels for $E < 0$ are relatively stable; a nucleus in level 2 can, however, decay to the ground state by the emission of a gamma-ray photon. The energy levels at $E > 0$ are, however, metastable. A particle in any of these levels has a finite probability of tunnelling through the Coulomb barrier and emerging outside the nucleus. For this reason, these energy levels will be broader than the energy levels with $E < 0$. The width of the energy level Γ is related to the lifetime τ of the level by $\Gamma = (\hbar/\tau)$. Above a certain energy $E_{\max}$, the width Γ becomes comparable with the energy-level separation and we are led to a continuum of energy states.

The existence of metastable energy levels in a compound nucleus Z^* leads to the following phenomena. Consider the interaction $X + a$ as the the energy E of the relative motion of X and a is increased gradually. The reaction probability will increase with E because it will be easier for the particles to penetrate the Coulomb barrier for higher values of E. This increase will be a relatively smooth function of E as long as the value of E is not near any of the metastable energy levels. If, however, the energy E is close to one of the metastable levels of the compound nucleus (like E_2 in Fig. 12.2), then the colliding particles will be in resonance with a metastable state and can form the compound nucleus much more easily. At such resonance energies E_r, the reaction rate and the reaction cross section are significantly enhanced.

Let us now consider the factors that determine the reaction rate $r_{\alpha\beta}$ for the reaction $X(\alpha, \beta)Y$. Such a reaction rate (in number of events per unit time per unit volume) can be written in the form

$$r_{\alpha\beta}(v) = n_\alpha n_X \sigma_{\alpha\beta}(v)v, \tag{12.6}$$

where $n_\alpha v$ is the flux of incident particles, n_X is the number density of target particles, and $\sigma_{\alpha\beta}(v)$ is the reaction cross section. (For the sake of definiteness, we are using the frame of reference in which X is at rest. It is also assumed that α and X are different particles. In the case in which they are identical, the reaction rate should be divided by a factor of 2 to avoid double counting.) When the particles do not have definite velocities but are described by a distribution function in energy, then it is necessary to average this rate over the energy distribution. If the particles are described by a Maxwell–Boltzmann distribution, then the distribution of the centre-of-mass energy of the $\alpha-X$ pair is given by (see Chap. 5, Exercise 5.3)

$$\mathcal{P}(E) = \frac{2}{\pi^{1/2}} \frac{E^{1/2}}{(k_B T)^{3/2}} e^{-E/k_B T}. \tag{12.7}$$

In this case, the averaged reaction rate can be written as

$$r_{\alpha\beta} = n_\alpha n_X \langle \sigma v \rangle_{\alpha\beta}, \tag{12.8}$$

where

$$\langle \sigma v \rangle_{\alpha\beta} = \frac{\int_0^\infty \mathcal{P}(E)\sigma_{\alpha\beta}(E)v(E)\, dE}{\int_0^\infty \mathcal{P}(E)\, dE}. \tag{12.9}$$

The velocity v is now in the centre-of-mass frame. Using the form of $\mathcal{P}(E)$ and substituting $v = (2E/m)^{1/2}$, we get

$$\langle \sigma v \rangle_{\alpha\beta} = \left(\frac{8}{\pi m}\right)^{1/2} (kT)^{-3/2} \int_0^\infty \sigma_{\alpha\beta}(E)\, e^{-E/k_B T} E\, dE, \tag{12.10}$$

where m is the reduced mass of the reacting particles.

The problem now reduces to computing $\sigma_{\alpha\beta}(E)$, which can be expressed in a convenient form as

$$\sigma_{\alpha\beta}(E) = \pi \, \bar{\lambda}^2 g \left(\frac{\Gamma_\alpha}{\Gamma}\right)\left(\frac{\Gamma_\beta}{\Gamma}\right) f(E), \tag{12.11}$$

where the various terms have the following interpretation: The quantity Γ_α denotes the decay rate of the compound nucleus by means of the channel $Z* \to X + \alpha$; Γ_β denotes the decay rate of the compound nucleus by means of the channel $Z* \to Y + \beta$; and Γ denotes the total decay rate of $Z*$ through *all* channels. The factor g is of the order of unity and contains information about the spin states of the reactants and the compound nucleus. The quantity

$$\pi \, \bar{\lambda}^2 = \frac{\pi \hbar^2}{2Em} = \frac{0.657}{E(\text{MeV})} \frac{1}{\mu} \text{ b}, \tag{12.12}$$

represents the geometric cross section of a particle with de Broglie wavelength
$\bar{\lambda}$ and $\mu = A_\alpha A_X/(A_\alpha + A_X)$. The unit used here is a barn (denoted by b),
which is defined as 1 barn $= 10^{-24}$ cm^2. The origin of the geometrical cross
section and the relative probabilities Γ_α/Γ and Γ_β/Γ for the formation and
the decay of compound nuclei are easy to understand. All other physical effects
are parameterised by the function $f(E)$, which is called the shape factor. If the
reaction proceeds through a resonance, then the shape factor is given by

$$f(E) = \frac{\Gamma^2}{(E - E_r)^2 + (\Gamma/2)^2}, \tag{12.13}$$

which is characteristic of a resonance phenomenon. If no resonant interaction is
involved, the $f(E)$ is a constant or a slowly varying function of energy.

We first consider the nonresonant interactions in Subsection 12.3.1 and then
comment on the resonant ones in Subsection 12.3.2.

12.3.1 Nonresonant Reaction Rates

In applications to stellar physics, we are interested in temperatures of approximately 10^7–$10^9 K$, which correspond to thermal energies in the range $1 < k_B T <$
100 keV. When two particles in this energy range undergo nuclear fusion, they release the excess binding energy. Because binding energies are in mega-electron-
volts, it follows that the nuclear-energy release in fusion is considerably higher
than the typical energies of the entrance channel. Again, because the Coulomb
barrier heights are of the order of mega-electron-volts, particles X and α can
combine only through quantum-mechanical tunneling. The end products, however, have energies in mega-electron-volts and hence are not affected by the
Coulomb barrier to the same extent.

It follows from these considerations that the probability of X and α combining will depend on the tunneling rate through the barrier. A simple quantum-
mechanical calculation shows that this rate should be proportional to

$$\Gamma_\alpha(E) \propto e^{-2\pi\eta}, \tag{12.14}$$

where

$$\eta = \frac{Z_\alpha Z_X e^2}{\hbar v} = 0.1574 Z_\alpha Z_X \left(\frac{\mu}{E}\right)^{1/2}. \tag{12.15}$$

The energy E and reduced mass μ are measured in this expression in mega-
electron-volts; for $E \approx k_B T \approx 100$ keV, $\mu \approx 1$ and $Z_\alpha Z_X \approx 2$, we find that $2\pi\eta \gtrsim$
12. At this range, the exponential factor provides the strongest dependence on
the energy. As far as the products are concerned, the Coulomb barrier is not of
importance and we can take both Γ_β and Γ to be independent of E. The only other
energy-dependent contribution is from $\bar{\lambda}^2 \propto (1/E)$. Putting all these together, we

can write the nonresonant cross section in the form

$$\sigma_{\alpha\beta}(E) = \frac{S(E)}{E} e^{-2\pi\eta} \tag{12.16}$$

where $S(E)$ is a slowly varying function of energy. For practical computations, it is necessary to determine $S(E)$ experimentally (although it is not an easy task). Substituting Eq. (12.16) into Eq. (12.10), we find that

$$\langle\sigma v\rangle_{\alpha\beta} = \left(\frac{8}{\pi m}\right)^{1/2} (kT)^{-3/2} \int_0^\infty S(E) \exp\left[-\left(\frac{E}{k_B T} + \frac{b}{E^{1/2}}\right)\right] dE, \tag{12.17}$$

with $b = 0.99 Z_\alpha Z_X \mu^{1/2}$ (MeV)$^{1/2}$. If $S(E)$ is a slowly varying function then it can be pulled out of the integral, and the integral reduces to one of the form

$$I = \int_0^\infty \exp\left(-\frac{E}{k_B T} - \frac{b}{E^{1/2}}\right) dE \equiv \int_0^\infty e^{q(E)} dE. \tag{12.18}$$

The integrand is the product of two exponentials, one of which decreases sharply with E while the other one rises with E. It is clear that the integrand has a very sharp maximum (called the Gamow peak) at some value $E = E_0$. Simple analysis shows that

$$E_0 = \left(\frac{1}{2} b k_B T\right)^{2/3} \cong 1.22 \left(Z_\alpha^2 Z_X^2 \mu\right)^{1/3} \left(\frac{T}{10^6 \text{ K}}\right)^{2/3} \text{ keV}. \tag{12.19}$$

To evaluate the integral in Eq. (12.18) we Taylor expand the function $q(E)$ around the maximum $E = E_0$, retaining up to quadratic order

$$q(E) \cong -\tau - \frac{1}{4}\tau\left(\frac{E}{E_0} - 1\right)^2, \quad \tau \equiv \frac{3E_0}{k_B T}. \tag{12.20}$$

The integration now gives

$$I \cong \left(\frac{2}{3}k_B T\right) \tau^{1/2} e^{-\tau} \int_{-\sqrt{\tau}/2}^\infty e^{-\xi^2} d\xi \cong \frac{2\sqrt{\pi}}{3}\left(\tau^{1/2}e^{-\tau}\right) k_B T. \tag{12.21}$$

In arriving at the last equality, we have extended the lower limit of integration to $-\infty$ as most of the contribution to the integral comes from near $\xi = 0$. In this case, the final answer becomes

$$\langle\sigma v\rangle_{\alpha\beta} = \frac{0.72 \times 10^{-18} S}{\mu Z_\alpha Z_X} e^{-\tau} \tau^2 \text{ cm}^3 \text{ s}^{-1}$$

$$= \frac{0.72 \times 10^{-18} S a^2}{\mu Z_\alpha Z_X} T_6^{-2/3} e^{-(a/T_6^{1/3})} \text{ cm}^3 \text{ s}^{-1}, \tag{12.22}$$

where $a = 42.49(Z_\alpha^2 Z_X^2 \mu)^{1/3}$, $T_6 = (T/10^6 \text{ K})$, and S is measured in kilo-electron-volts times barns. In what follows, we repeatedly use the notation T_n for $(T/10^n \text{ K})$.

This expression provides a reasonable estimate for the average cross section. The next order corrections multiply the above result by $[1 + (5/12\tau) + \cdots]$.

As a realistic example of a nonresonant nuclear reaction, consider a collision between hydrogen and carbon-12, (i.e., $p + {}^{12}C$) at a temperature of 2.2×10^7 K. At this temperature, $T_6 = 22$, $k_B T = 1.896$ keV, and $E_0 = 30.8$ keV. The full width of the integrand is $\Delta \approx 2.3(E_0 k_B T)^{1/2} \approx 17.6$ keV. This means that roughly half of the reactions arise from reactants that have the energy in the range $22 \lesssim E \lesssim 40$ keV.

Once the velocity-averaged cross section is known, the reaction rate can be found with Eq. (12.8). It is conventional to write this relation in terms of the density of material and the mass fractions X_i of the particles α and X in the form

$$r_{\alpha\beta} = \rho^2 N_A^2 \frac{X_\alpha X_X}{A_\alpha A_X} \langle \sigma v \rangle_{\alpha\beta}. \tag{12.23}$$

If the amount of energy released in the reaction is Q, then the amount of energy released per second per unit mass of matter is given by

$$\epsilon_{\alpha\beta} = \frac{r_{\alpha\beta} Q}{\rho}. \tag{12.24}$$

The temperature dependence of ϵ is the same as that of reaction rate and is given by $T^{-2/3} \exp(-aT^{-1/2})$. By taking the logarithmic derivative of these expressions, we can express the energy-generation rate in power-law form in a restricted range:

$$\epsilon = \epsilon_0 \rho^\lambda T^\nu, \tag{12.25}$$

where

$$\lambda = \left(\frac{\partial \ln \epsilon}{\partial \ln \rho}\right)_T = 1, \quad \nu = \left(\frac{\partial \ln \epsilon}{\partial \ln T}\right)_\rho = \frac{a}{3T_6^{1/3}} - \frac{2}{3}. \tag{12.26}$$

For the reaction ${}^{12}C(p, \gamma){}^{13}N$ at $T_6 = 20$ the index ν is ~ 16. Such a strong dependence on temperature is characteristic of the Coulomb barrier.

12.3.2 Resonant Reaction Rates

In the case of resonant reaction, $S(E)$ contains the resonance factor $[(E - E_r)^2 + (\Gamma/2)^2]^{-1}$, which contributes essentially around $E = E_r$. Treating this factor as a Dirac delta function, we can approximate the average cross section as

$$\langle \sigma v \rangle_{\alpha\beta} = \frac{\hbar^2}{2m} \left(\frac{8\pi}{m}\right)^{1/2} (k_B T)^{-3/2} e^{-E_r/k_B T} g \Gamma_\alpha(E_r) \Gamma_\beta(E_r)$$

$$\times \int_0^\infty \frac{dE}{(E - E_r)^2 + (\Gamma/2)^2}. \tag{12.27}$$

Extending the lower limit of integration to $-\infty$, we get the result for resonant reactions as

$$\langle\sigma v\rangle_{\alpha\beta} = \hbar^2 \left(\frac{2\pi}{mk_BT}\right)^{3/2} \left(g\frac{\Gamma_\alpha\Gamma_\beta}{\Gamma}\right) e^{-E_r/k_BT}. \tag{12.28}$$

The quantity $\Omega = (g\Gamma_\alpha\Gamma_\beta/\Gamma)$ has been tabulated from experimental results. Numerically, the averaged cross section is given by

$$\langle\sigma v\rangle_{\alpha\beta} = \left(\frac{2\pi\hbar^2}{mk_BT}\right)^{3/2} \frac{\Omega}{\hbar} e^{-E_r/k_BT}$$

$$= 2.56 \times 10^{-13} \frac{\Omega}{(\mu T_9)^{3/2}} e^{-11.605\, E_r/k_BT} \text{cm}^3\,\text{s}^{-1}, \tag{12.29}$$

where Ω and E are in mega-electron-volts. The rest of the analysis proceeds exactly as before. The temperature exponent for a resonant reaction has the form

$$v = \frac{11.61\, E_r}{T_9} - \frac{3}{2}. \tag{12.30}$$

In the above discussion the effect of Coulomb barrier in the nuclear reactions has been computed with the electrostatic potential, which varies as $1/r$. This result, however, is not exact as the Coulomb interactions between ions will be shielded by the free electrons that are present in the plasma. It was shown in Chap. 9, Section 9.2, that the shielding changes the potential into the form

$$\phi(r) = \frac{Ze}{r} e^{-\kappa_d r}, \tag{12.31}$$

where

$$\kappa_d = \left[\frac{4\pi e^2}{k_BT}(Z^2n_i + n_e)\right]^{1/2} \tag{12.32}$$

is the inverse of the Debye length. For a particle with energy E, the characteristic impact parameter is approximately $r_t = (Z^2e^2/E)$; if $E \approx 10$ keV and $Z \approx 1$, then $r_t \approx 10^{-11}$ cm and $\kappa_d r_t \ll 1$ in most stellar interiors. In this case, we can approximate the potential in the form

$$\phi(r) \approx \frac{Ze}{r}(1 - \kappa_d r) = \frac{Ze}{r} - Ze\,\kappa_d. \tag{12.33}$$

Thus the net effect of the Debye screening is to reduce the electrostatic potential energy by the factor $U_0 \approx Z^2e^2\kappa_d$. Equivalently, we may say that the effective energy of the interaction has been increased from E to $E + U_0$ as far as the scattering cross section is concerned. Hence, to incorporate the effects of Debye shielding, we only have to replace $\sigma_{\alpha\beta}(E)$ with $\sigma_{\alpha\beta}(E + U_0)$ in Eq. (12.10).

Shifting the variable of integration from E to $(E - U_0)$ gives the result

$$\langle \sigma v \rangle_{\alpha\beta} \propto \int_{U_0}^{\infty} (E - U_0)\sigma_{\alpha\beta}(E)\,e^{-E/k_B T}\,e^{U_0/k_B T}\,dE. \qquad (12.34)$$

The dominant contribution to this integral occurs from around $E \approx E_0$ [of Eq. (12.19)] in the case of a nonresonant reaction and from around $E \approx E_r$ [of Eq. (12.27)] in the case of a resonant reaction. Both E_0 and E_r are larger than U_0 when these approximations are valid. Hence we conclude that the most important effect of the shielding is to change the result by an extra factor $\exp(U_0/k_B T)$, that is,

$$\langle \sigma v \rangle_{\alpha\beta} \text{ (with screening)} = \langle \sigma v \rangle_{\alpha\beta} \text{ (unscreened)} \times e^{U_0/k_B T}. \qquad (12.35)$$

This correction is important and in some of the reactions this could increase the rate by $\sim 25\%$. Numerically,

$$\frac{U_0}{k_B T} \cong 5.92 \times 10^{-3} Z_1 Z_2 \left(\frac{\rho}{T_7^3}\right)^{1/2}. \qquad (12.36)$$

When the screening effects are strong, i.e., when $(U_0/k_B T) \gtrsim 1$, the treatment is much more difficult. In the limit of strong screening, an approximate formula for $(U_0/k_B T)$ is given by

$$\frac{U_0}{k_B T} \cong 0.0205\big[(Z_1 + Z_2)^{5/3} - Z_1^{5/3} - Z_2^{5/3}\big]\frac{(\rho/\mu_e)^{1/3}}{T_7}, \qquad (12.37)$$

where $\mu_e^{-1} \equiv \sum(X_i Z_i/A_i)$. These expressions show that the screening factor $\exp(U_0/k_B T)$ increases substantially for increasing ρ and decreasing T. For the shielded reaction rates, the density and the temperature exponents are given by

$$\lambda = 1 + \frac{1}{3}\frac{U_0}{k_B T}, \qquad \nu = \frac{a}{3T_6^{1/2}} - \frac{2}{3} - \frac{U_0}{k_B T} \qquad (12.38)$$

Thus, for high densities and moderate to low temperatures (for example, $\rho > 10^6$ g cm^3 and $T < 10^7$ K), the temperature sensitivity decreases and the density sensitivity increases. In such a situation the reaction rates depend mainly on density rather than on temperature; such reactions are called pycnonuclear reactions.

The discussion of nuclear reactions so far was based on the assumption that both the reactants are charged particles. This is clearly not true if, for example, we consider one of the reactants as a neutron. Nuclear reactions triggered by the capture of a neutron by a nucleus is believed to be responsible for the generation of several elements with $A > 60$. For such reactions the Coulomb barrier is absent and $\langle \sigma v \rangle$ remains nearly constant.

In the above discussion we concentrated on nuclear reactions involving two reactants in the initial state. Another class of nuclear reactions that is also of importance in astrophysics is the decay of a nucleus through electromagnetic or weak interaction processes. Some brief comments about these decays are given below.

The electromagnetic decay of a nucleus proceeds fundamentally along the same line as the decay of an excited state of an atom and hence an order-of-magnitude estimate can be obtained by methods similar to the one used in Chap. 1, Subsection 1.4.2. Rewriting relation (1.29) for dipole radiation as

$$Q_{ED} \approx \frac{\omega^3 d^2}{\hbar c^3} = \left(\frac{q^2}{\hbar c}\right)\left(\frac{\omega}{c}R\right)^2 \omega = \alpha(kR)^2\omega \qquad (12.39)$$

(where $d = qR$ is the nuclear dipole moment and $k = \omega/c$ is the wave number) and noting that, for nuclear reactions involving energy E, the wave number is $k = (E/\hbar c) \approx 0.5 \times 10^{11}(E/0.1 \text{ MeV})$ cm^{-1} and $R \approx 10^{-13}$ cm, we get $(kR)^2 \approx 2.5 \times 10^{-5}(E/1 \text{ MeV})^2$. This gives an estimate of $Q_{ED} \approx 0.5 \times 10^{15}$ $(E/1 \text{ MeV})^3$ s^{-1}.

In a more rigorous calculation, the dipole moment has to be estimated as the matrix element between the initial and the final states of the dipole moment operator and the result can be expressed as

$$Q_{ED} \approx 0.5 \times 10^{15} \text{ s}^{-1} \left(\frac{E}{1 \text{ MeV}}\right)^3 \left(\frac{|\langle F|\mathbf{d}|I\rangle|^2}{q^2 R^2}\right). \qquad (12.40)$$

As in the case of atomic transitions, the existence of electric dipole decay will be governed by the selection rules that control the existence of a non zero matrix element $\langle F|\mathbf{d}|I\rangle$ between the two relevant states. If a dipole transition is forbidden, the reaction can still proceed through quadrupole or higher orders; in that case the result will be multiplied (approximately) by a factor $(kR)^{n-2}$, where $n = 2$ for dipole, etc. Another possible decay scheme would be through magnetic dipole transitions that are governed by the nuclear magnetic moment $\mu_N \approx (q\hbar/2m_p c)$. This differs from the electric dipole moment by the replacement of $R \approx 10^{-13}$ cm $= 1$ F with $(\hbar/m_p c) \approx 0.1$ F (F is the abbreviation for the unit Fermi). Hence, the rate for magnetic dipole will be approximately $(0.1)^2$ times lower than the electric dipole rate:

$$Q_{MD} \approx 0.5 \times 10^{13} \text{ s}^{-1} \left(\frac{E}{1 \text{ MeV}}\right)^3 \left(\frac{|\langle F|\mathbf{m}|I\rangle|^2}{\mu_N^2}\right). \qquad (12.41)$$

Let us next consider briefly the decays through weak interactions. The weak interactions are mediated by three kinds of particles – called $W^\pm$ and Z bosons – just as electromagnetic interactions are mediated by the photon. In general, an interaction will operate over a length scale $l \approx (\hbar/Mc)$, where M is the mass of the particle mediating the interaction. Because a photon has zero mass, electomagnetic interactions have an infinite range; the W and Z bosons have masses in the range of 50 GeV and hence the weak interactions have an extremely small range. For all practical purposes, such an interaction can be treated as a point interaction at nuclear energies and the relevant coupling constant governing the interaction is called the Fermi coupling constant G_F with a numerical value $G_F/(\hbar c)^3 \approx 1.166 \times 10^{-11}$ MeV^{-2}.

Three of the most important nuclear reactions proceeding by means of weak interactions are the (1) β decay $[A(Z, N) \rightarrow A(Z+1, N-1) + e^- + \bar{\nu}_e]$, (2) β^+ decay $[A(Z, N) \rightarrow A(Z-1, N+1) + e^+ + \nu_e]$, and (3) an electron capture in which the nucleus makes a transition $A(Z, N) \rightarrow A(Z-1, N+1)$, capturing an inner-shell electron. In the β decay, a neutron in the nucleus converts into a proton emitting an electron and an antineutrino; an amount of energy $E_{\beta^-} = [M(Z, N) - M(Z+1, N-1)]c^2$ will be carried away as the kinetic energy of the electron and the neutrino. In β^+ decay, an $e^- e^+$ pair is created from the vacuum and the e^- is absorbed by the nucleus and the resulting inverse beta decay emits a neutrino as well as the e^+ of the original pair. This can release an amount of energy $E_{\beta^+} = [M(Z, N) - M(Z-1, N+1)]c^2 - 2m_e c^2$, where the second term arises from the energy required for creating an electron–positron pair. Obviously this reaction can proceed only if E_{β^+} is positive. Finally, the electron capture releases the energy $E_{EC} = [M(Z, N) - M(Z-1, N+1)]c^2$.

The decay rates for all these processes can be computed with the weak interaction Hamiltionian just as atomic decay rates are computed from the electromagnetic Hamiltonian. For example, the expression for the β decay rate will be

$$Q = \sum \left(\frac{2\pi}{\hbar}\right) |\langle \text{final nuclear, emitted particles}| H_{\text{weak}} |\text{initial nuclear}\rangle|^2$$

$$= \left(\frac{2\pi}{\hbar}\right) |\langle \text{final nuclear}| H |\text{initial nuclear}\rangle|^2 \times (\text{density of states}), \quad (12.42)$$

where the first factor comes from the relevant nuclear matrix element and the second facor takes into account the density of states for the neutrino and the electron. The latter can be written as

$$\int \frac{d^3 p_\nu}{(2\pi\hbar)^3} \frac{d^3 p_e}{(2\pi\hbar)^3} \delta_D(E_\nu + E_e - E_0) F(Z, E_e)$$

$$= \frac{1}{4\pi^4 \hbar^6} \int dp_\nu dp_e \, p_\nu^2 p_e^2 F(Z, E_e) \delta_D(E_\nu + E_e - E_0). \quad (12.43)$$

The phase-volume factors are easy to understand with the Dirac delta function ensuring the conservation of energy (E_0 denotes the energy difference between the initial and the final nuclear levels); the extra factor $F(Z, E_e)$ arises from the Coulomb interaction of the emitted electron with the nucleus and is, in general, a complicated expression. Writing

$$p_\nu^2 dp_\nu = \left(\frac{E_0 - E_e}{c}\right)^2 \frac{dE_\nu}{c} = \left(\frac{E_0 - E_e}{c}\right)^2 \frac{dE_e}{c}, \quad (12.44)$$

we can express the rate of decay

$$Q = \frac{m_e^5 c^4}{2\pi^3 \hbar^7} \int_0^{\sqrt{E_0^2 - m_e^2 c^4}} |\langle F|H|I\rangle|^2 F(Z, E_e) \left(\frac{E_0 - E_e}{m_e c^2}\right)^2 \frac{p_e^2 dp_e}{(m_e c)^3}$$

$$\equiv \frac{m_e^5 c^4}{2\pi^3 \hbar^7} |\langle F|H|I\rangle|^2 f(Z, E_0), \tag{12.45}$$

where $f(Z, E_0)$ is called the Fermi integral. For strong transitions, the matrix element can be estimated as $|\langle F|H|I\rangle|^2 \approx G_F^2$ and f is of the order of unity. The half-life $\tau_{1/2}$ of the decay is usually defined by $\tau_{1/2} \equiv [\ln 2/Q]$. From our expression for Q we find that

$$f(Z, E_0)\tau_{1/2} \approx (2\pi^3)(\ln 2) \frac{\hbar}{(m_e c^2)^5} \left(\frac{\hbar^3 c^3}{G_F}\right)^2 \simeq 4 \times 10^3 \text{ s}, \tag{12.46}$$

which is the characteristic time scale for weak decays. A similar analysis can be performed for any other weak interaction process.

12.4 Specific Thermonuclear Reactions

There are certain nuclear reactions that play a special role in stellar physics. These reactions are essentially fusion reactions involving light elements that lead to a slow buildup of heavier elements in stellar interiors. We now discuss several features of these specific reactions.

12.4.1 Hydrogen Burning: PP Chain

A nuclear reaction in which an element Z is converted into other elements is usually called Z burning when the reaction is thermonuclear. In this terminology, the hydrogen can be burned by means of two main sequences of reactions, called the PP chain and the CNO cycle. The PP chain is discussed in this subsection and the CNO cycle will be covered in Subsection 12.4.2.

The PP chain starts with two protons combining together to form a deuterium nucleus, which in turn combines with another proton to form a He-3 nucleus:

$$^1H + {}^1H \longrightarrow {}^2H + e^+ + \nu, \quad {}^2H + {}^1H \longrightarrow {}^3He + \gamma. \tag{12.47}$$

The first of these reactions requires that one of the protons undergo beta decay, thereby forming a neutron, while the two protons are strongly interacting. The beta decay is a process governed by weak interactions and hence is extremely slow.

The He-3 nucleus, formed in the above process, can end up as a He-4 nucleus through (essentially) three different channels, conventionally called PP1, PP2, or PP3. These reactions are shown in Table 12.1. The PP1 channel uses two He-3 nuclei and hence requires the initial reaction in expressions (12.47) to occur

Table 12.1. *The proton–proton chains*

$$
\text{PP--I} \quad \left\{
\begin{array}{l}
{}^1\text{H} + {}^1\text{H} \longrightarrow {}^2\text{H} + e^+ + \nu_e \\
{}^2\text{H} + {}^1\text{H} \longrightarrow {}^3\text{He} + \gamma \\
{}^3\text{He} + {}^3\text{He} \longrightarrow {}^4\text{He} + {}^1\text{H} + {}^1\text{H}
\end{array}
\right.
$$

$$
\text{PP--II} \quad \left\{
\begin{array}{l}
{}^3\text{He} + {}^4\text{He} \longrightarrow {}^7\text{Be} + \gamma \\
{}^7\text{Be} + e^- \longrightarrow {}^7\text{Li} + \nu_e(+\gamma) \\
{}^7\text{Li} + {}^1\text{H} \longrightarrow {}^4\text{He} + {}^4\text{He}
\end{array}
\right.
$$

$$
\text{PP--III} \quad \left\{
\begin{array}{l}
{}^7\text{Be} + {}^1\text{H} \longrightarrow {}^8\text{B} + \gamma \\
{}^8\text{B} \longrightarrow {}^8\text{Be} + e^+ + \nu_e \\
{}^8\text{Be} \longrightarrow {}^4\text{He} + {}^4\text{He}
\end{array}
\right.
$$

twice. The starting point for PP2 and PP3 requires the He-4 nucleus, which should either be present primordially or could have been produced by an earlier PP1 reaction. The first reaction of PP1 or PP2 produces a Be-7 nucleus; the branching between PP2 or PP3 occurs depending on the reaction of this Be-7 nucleus.

The different reactions of the PP chain occur at wildly different rates, depending on the temperature. The intrinsically slowest one is the first reaction in the chain. The next one, ${}^2\text{H}(p, \gamma){}^3\text{He}$, on the other hand, is extremely fast so that it reaches thermal equilibrium at most of the relevant temperatures in stellar interiors. (Hence most of the deuterium is destroyed in the stars and the abundance is kept at a very low level, consistent with thermal equilibrium.) The last reaction of PP1 is faster than the first but not as fast as the second. In most circumstances, the abundance of ${}^3\text{He}$ can be calculated with thermal equilibrium. When the temperature is increased, the equilibrium abundance of ${}^3\text{He}$ decreases until the first reaction of PP2 starts dominating. The second reaction of PP2 involves an electron capture by ${}^7\text{Be}$, which is less sensitive to temperature compared with ion capture reaction. The alternative route for ${}^7\text{Be}$ is by means of proton capture, which is the first reaction of PP3. The rates for ${}^7\text{Be}(e^-, \nu){}^7\text{Li}$ and ${}^7\text{Be}(p, \gamma){}^8\text{B}$ are comparable around $T \simeq 2.4 \times 10^7$ K. At higher temperatures, PP3 dominates over PP2. ${}^8\text{B}$ is unstable to positron decay with a half-life of 0.8 s. The final reaction of PP3 involves the decay of ${}^8\text{Be}$ to two helium nuclei; ${}^8\text{Be}$ is an extremely unstable element with a lifetime of only $\sim 10^{-16}$ s.

To find the rate of energy production in the PP chain, we need to multiply the energy released in each of the channels by the corresponding rate, taking into account the branching probabilities for each of the channels. It may be noted that some of these reactions involve the emission of neutrinos that can carry part of the energy. The scattering cross section of neutrinos of energy E by normal matter is given by $\sigma \approx 10^{-44}(E/m_e c^2)^2$ cm^2. This is extremely tiny for the neutrinos in the mega-electron-volt range. The corresponding mean free path for neutrinos is given by

$$l_\nu = \frac{1}{n\sigma_\nu} = \frac{\mu m_u}{\rho \sigma_\nu} \approx 2 \times 10^{20} \text{cm} \left(\frac{\rho}{1\,\text{g}\,\text{cm}^{-3}} \right)^{-1}. \tag{12.48}$$

For normal stellar matter with $\rho \approx (1 - 10^6)$ g cm^3, $l_\nu \approx 3000 R_\odot - 100$ pc. It is therefore clear that the neutrinos escape easily from a star, carrying the energy with them. (It turns out that the situation is very different during a stellar collapse in final stages when the densities can reach nuclear densities of $\rho \approx 10^{14}$ g cm^3. The high-energy neutrinos in such situations will have a short mean free path and can play a vital role in redistributing the energy.) Assuming that the neutrinos escape from the star, we must compute the net yield of a reaction by subtracting the energy carried away by the neutrinos. Because different reactions involve different amount of energies carried away by the neutrinos, the net energy release (excluding the energies carried away by the neutrinos) is different for the three chains. The effective energy release, taking into account the above complications and the different branches, is described fairly accurately by

$$Q_{\text{eff}}(p-p \text{ chains}) = 13.116\left[1 + 1.412 \times 10^8 (1/X - 1)\,e^{-4.998/T_9^{1/3}}\right] \text{MeV}, \tag{12.49}$$

where X is the hydrogen mass fraction. As for the rate of reaction, it is dominated by the slowest in the chain, namely, the first one $^1\text{H}(^1\text{H}, e^+\nu_e)^2\text{H}$. This reaction is nonresonant and the temperature dependence arises essentially because of the Coulomb barrier between the two protons. The result also depends on the weak interaction strength and could be expressed as

$$r_{p-p \text{ reaction}} = \frac{1.15 \times 10^9}{T_9^{2/3}} X^2 \rho^2 e^{-3.380/T_9^{1/3}} \text{ cm}^{-3}\,\text{s}^{-1}. \tag{12.50}$$

Combining these two results, we find that the effective energy-generation rate goes as

$$\epsilon_{\text{eff}}(p-p \text{ chains}) = r_{pp} Q_{\text{eff}}/\rho \approx \frac{2.4 \times 10^4 \rho X^2}{T_9^{2/3}} e^{-3.380/T_9^{1/3}} \text{ erg g}^{-1}\,\text{s}^{-1}. \tag{12.51}$$

(It is conventional to use c.g.s units in astrophysical contexts and hence Eq. (12.51) is given in c.g.s units. To convert this result and similar ones to SI units, note that 1 erg g^{-1} = 10^{-4}J kg^{-1}.) The temperature index for the energy-generation rate is

$$\nu_{p-p \text{ reaction}} = \frac{11.3}{T_6^{1/3}} - \frac{2}{3}, \tag{12.52}$$

which gives $\nu \approx 4$ for $T_6 \approx 15$. The PP chain has only a relatively weak dependence on the temperature.

Table 12.2. *The carbon–nitrogen–oxygen cycles*

$$^{12}\text{C} + {}^{1}\text{H} \longrightarrow {}^{13}\text{N} + \gamma$$
$$^{13}\text{N} \longrightarrow {}^{13}\text{C} + e^{+} + \nu_{e}$$
$$^{13}\text{C} + {}^{1}\text{H} \longrightarrow {}^{14}\text{N} + \gamma$$
$$^{14}\text{N} + {}^{1}\text{H} \longrightarrow {}^{15}\text{O} + \gamma$$
$$^{15}\text{O} \longrightarrow {}^{15}\text{N} + e^{+} + \nu_{e}$$
$$^{15}\text{N} + {}^{1}\text{H} \longrightarrow {}^{12}\text{C} + {}^{4}\text{He}$$

–or–

$$^{15}\text{N} + {}^{1}\text{H} \longrightarrow {}^{16}\text{O} + \gamma$$
$$^{16}\text{O} + {}^{1}\text{H} \longrightarrow {}^{17}\text{F} + \gamma$$
$$^{17}\text{F} \longrightarrow {}^{17}\text{O} + e^{+} + \nu_{e}$$
$$^{17}\text{O} + {}^{1}\text{H} \longrightarrow {}^{14}\text{N} + {}^{4}\text{He}$$

12.4.2 Hydrogen Burning: CNO Cycle

The PP chain described above did not require the presence of elements heavier than hydrogen to start with. In contrast, it is possible to have a sequence of reactions converting hydrogen into helium in the presence of heavier elements in which the heavier elements act only as catalyst, that is, the heavier elements remain unchanged at the end of the cycle. One prominent cycle of this kind involves carbon, nitrogen, and oxygen and proceeds along the reaction shown in Table 12.2.

The first set of six reactions in this cycle involves the interaction of hydrogen with carbon and nitrogen. At the end of the cycle ^{12}C comes out unaffected and four hydrogen nuclei have effectively combined to form a helium nucleus. The second set of four reactions in Table 12.2 arises when ^{16}O enters the fray. This could happen either because of the presence of ^{16}O atoms present primordially or because of the reaction $^{15}\text{N}(p, \gamma)^{16}\text{O}$. At the end of the 10 reactions, helium is produced at the expense of hydrogen, with C, N, and O acting as the catalyst. The first set of six reactions is called the CN cycle and the full set of 10 reactions is called the CNO cycle.

The calculation of the energy production rate for the CNO cycle is fairly complicated and the results have not yet been established very accurately. The details of the calculation depends, among other things, on whether CN and CNO cycles have reached equilibrium as well as the abundance of different isotopes of C, N, and O. It is generally believed that the rate is essentially determined by the reaction $^{14}\text{N}(p, \gamma)^{15}\text{O}$. In that case, a useful estimate for energy-generation rate in these cycles is given by

$$\epsilon_{\text{CNO}} \approx \frac{4.4 \times 10^{25} \rho X Z}{T_9^{2/3}} \, e^{-15.228/T_9^{1/3}} \quad \text{erg g}^{-1} \text{ s}^{-1} \tag{12.53}$$

where Z is the mass fraction of elements heavier than helium. This cycle has a very high temperature sensitivity and the index is given by

$$\nu(\text{CNO}) = \frac{50.8}{T_6^{1/3}} - \frac{2}{3}, \qquad (12.54)$$

which is ~ 18 for $T_6 \approx 20$. For a solar mass star, with $T_6 = 15$, $X = 0.7$, and $Z = 0.02$, the energy generation by the PP chain is nearly 10 times larger than the energy generation by the CNO cycle. However, as the temperature increases, the CNO cycle starts dominating over the PP chain. The crossover occurs around $T \simeq 1.4 \times 10^7$ K.

12.4.3 Helium Burning

We next consider reactions in which helium is burnt to form higher elements. These reactions are important in stellar evolution studies in order to calculate the abundance of heavy elements produced in the stars. The first reaction in this chain is $^4\text{He} + {}^4\text{He} \rightarrow {}^8\text{Be}$. This is the inverse of the last reaction $^8\text{Be} \rightarrow {}^4\text{He} + {}^4\text{He}$, which terminated the PP chain. Because ^8Be is unstable with a lifetime of 10^{-16} s, the question arises as to how efficiently the production of ^8Be can occur.

To begin with, it must be noted that the ground state of ^8Be is ~ 92 keV higher than the mass of the reactants. In other words, the α particles must have sufficient kinetic energy to form ^8Be in the first place. Further, to increase the rate of forward reaction, it would help if the reaction can proceed through a resonance. To check the resonance condition, we should estimate the temperature at which the Gamow peak [see Eq. (12.19) with $Z_\alpha = 2$, $\mu = 2$], located at $E_0 \approx 3.9 \times T_6^{2/3}$ keV, corresponds to the ground-state energy of 92 keV of ^8Be. This happens at $T \approx 1.2 \times 10^8$ K. For temperatures higher than this value, the formation rate of ^8Be will begin to match the decay rate, leading the abundance of ^8Be to the equilibrium value. To estimate this equilibrium abundance, we can proceed as described in Chap. 5, Subsection 5.12.3, dealing with nuclear statistical equilibrium. Because the reaction being considered is $\alpha + \alpha \leftrightarrow {}^8\text{Be}$ we need to make the following modifications in Saha's equation: (1) n_p and n_e should be replaced with n_α and n_H should be replaced with the number density of ^8Be, which may be denoted by $n(^8\text{Be})$. (2) The statistical factors are unity for both α particles and ^8Be because they all have zero spin. (3) The ionisation potential has to be replaced with the energy difference in the reaction, which is (-91.78) keV, and finally, (4) the mass of the electron m_e should be replaced with the reduced mass $m_\alpha^2/m(^8\text{Be}) \approx m_\alpha/2$. With these replacements, the equilibrium concentration is given by

$$\frac{n_\alpha^2}{n(^8\text{Be})} = \left(\frac{\pi m_\alpha kT}{h^2}\right)^{3/2} e^{-Q/k_B T} = 1.69 \times 10^{34} T_9^{3/2} e^{-1.065/T_9}. \qquad (12.55)$$

For $T_9 \approx 0.1$, $n(^8\text{Be})/n_\alpha$ is only $\sim 7 \times 10^{-9}$.

The next crucial stage in helium burning involves converting ^{8}Be to ^{12}C through the reaction ^{8}Be$(\alpha, \gamma)^{12}$C. This is an exothermic reaction with an energy release of 7.367 MeV and proceeds through an excited state of ^{12}C* at an energy of 7.654 MeV. Once again, the decay of ^{12}C* by the emission of a gamma-ray photon is subdominant to the inverse reaction of ^{12}C* to ^{8}Be and an α particle. However, at high temperatures, a small set of ^{12}C* does build up and decays through the gamma-ray emission. The rate of this reaction is therefore essentially the same as the formation rate of the ground state of ^{12}C.

The combination of the two reactions, namely $\alpha + \alpha \rightarrow {}^8$Be and ^{8}Be$(\alpha, \gamma)^{12}$C, results in the conversion of three α particles into a carbon nucleus. The net reaction releases the energy, as the second reaction releases more energy than is used up in the first. The net energy release is $Q = 7.367 - 0.0918 = 7.275$ MeV. We can compute the rate for this energy production by combining the rates for the individual processes described above, and we get

$$\epsilon_{\alpha\alpha} = \epsilon_{3\alpha} = \frac{5.1 \times 10^8 \rho^2 Y^3}{T_9^3} e^{-4.4027/T_9} \text{ erg g}^{-1} \text{ s}^{-1}. \qquad (12.56)$$

This reaction is highly temperature sensitive, with the temperature and density indices being

$$\lambda_{3\alpha} = 2, \quad \nu_{3\alpha} = \frac{4.4}{T_9} - 3. \qquad (12.57)$$

For $T_8 = 1$, the temperature index is ~ 40, which is considerably larger than the exponent for hydrogen burning. This implies that, as a fuel, helium is much more explosive in a nuclear reaction than hydrogen. (It must be emphasised that the above discussion is somewhat oversimplified for several reasons. We have not taken into account the effects of Coulomb screening or the detailed structure of the sampling of the resonance.)

Once ^{12}C is formed, a further reaction of the kind ^{12}C$(\alpha, \gamma)^{16}$O and ^{16}O$(\alpha\gamma)$^{20}Ne can proceed, forming the oxygen and neon nuclei. These reactions rates are not very precisely known and approximate fitting formulas are given by

$$\epsilon(\alpha, {}^{12}\text{C}) = \frac{1.5 \times 10^{25} Y X_{12} \rho}{T_9^2} (1 + 0.0489 T_9^{-2/3})^{-2}$$

$$\times \exp\left[-32.12 T_9^{-1/3} - (0.286 T_9)^2\right] \text{ erg g}^{-1} \text{ s}^{-1}, \qquad (12.58)$$

$$\epsilon(\alpha, {}^{16}\text{O}) = \frac{6.69 \times 10^{26} Y X_{16} \rho}{T_9^{2/3}}$$

$$\times \exp\left[-39.757 T_9^{-1/3} - (0.631 T_9)^2\right] \text{ erg g}^{-1} \text{ s}^{-1}. \qquad (12.59)$$

All these reactions deplete the helium and produce carbon, oxygen, and neon.

12.4.4 Burning of Heavier Elements

The processes similar to those described above can lead to the burning of heavier elements such as carbon, neon, oxygen, silicon, etc. These reactions are important at temperatures of $T_9 \gtrsim 0.5$. Most of these reactions are fairly complex and can proceed through very many channels. The energy production rates are often computed by different approximations and may not be quite reliable. As an example of complexity, consider the reaction of two carbon atoms that can proceed by means of many different channels such as

$$
\begin{aligned}
{}^{12}\text{C} + {}^{12}\text{C} &\longrightarrow {}^{24}\text{Mg} + \gamma, & (13.931) \\
&\longrightarrow {}^{23}\text{Mg} + n, & (-2.605) \\
&\longrightarrow {}^{23}\text{Na} + p, & (2.238) \\
&\longrightarrow {}^{20}\text{Ne} + \alpha, & (4.616) \\
&\longrightarrow {}^{16}\text{O} + 2\alpha, & (-0.114) \qquad (12.60)
\end{aligned}
$$

where the numbers in parentheses denote the Q value of the reaction in mega-electron-volts. The relative probabilities for these channels are very different and depend on the temperature. The γ decay that leads to ^{24}Mg and the endothermic decays are improbable compared with the third and fourth channels which lead to the production of ^{23}Na and ^{20}Ne. These reactions can be followed by ^{23}Na$(p, \alpha)^{20}$Ne and ^{23}Na$(p, \gamma)^{24}$Mg, both of which are exothermic with an energy release of 2.379 and 11.691 MeV, respectively. A similar set of reactions can also take place with ^{16}O. Depending on the temperature and the density, the net result of these reactions is to form ^{20}Ne and somewhat lesser amounts of ^{23}Na and ^{24}Mg. The energy production rate is

$$
\epsilon({}^{12}\text{C} + {}^{12}\text{C}) = \frac{1.43 \times 10^{42} Q \eta \rho X_{12}^2}{T_9^{3/2}} e^{-84.165 T_9^{-1/3}} \quad \text{erg g}^{-1} \text{ s}^{-1}, \qquad (12.61)
$$

where Q is the yield of the reaction in mega-electron-volts and η is the fractional probability for each of the channels. Assuming that the third and fourth reactions in expresions (12.60) are the most dominant ones and using the values $\eta = 0.56,\ 0.44$ for the third and fourth reactions, we can find the temperature and density exponents for this reaction:

$$
\nu = \frac{28}{T_9^{1/3}} - \frac{3}{2}, \quad \lambda = 1. \qquad (12.62)
$$

In a similar manner, we can have different reactions involving oxygen nuclei for which some of the dominant channels are

$$
\begin{aligned}
{}^{16}\text{O} + {}^{16}\text{O} &\longrightarrow {}^{32}\text{S} + \gamma, & (16.541) \\
&\longrightarrow {}^{31}\text{P} + p, & (7.677) \\
&\longrightarrow {}^{31}\text{S} + n, & (1.453) \\
&\longrightarrow {}^{28}\text{Si} + \alpha, & (9.593) \\
&\longrightarrow {}^{24}\text{Mg} + 2\alpha, & (-0.393). \qquad (12.63)
\end{aligned}
$$

The reactions ending up with ^{28}Si, ^{31}S, and ^{31}P are the most dominant ones, with branching fractions of 0.21, 0.18, and 0.61, respectively. The energy production rate for this set of reactions is given by

$$\epsilon(^{16}\text{O} + ^{16}\text{O}) = \frac{1.3 \times 10^{52} Q \eta \rho X_{16}^2}{T_9^{2/3}} e^{-135.93 T_9^{-1/3}}$$

$$\times e^{(-0.629 T_9^{2/3} - 0.445 T_9^{4/3} + 0.0103 T_9^2)}. \tag{12.64}$$

It is clear that these reactions are important only when $T_9 \gtrsim 1$, which corresponds to energy greater than ~ 0.1 MeV. At these temperatures, the photons at the high-energy tail of the Planck distribution have energies comparable with those of low-lying quantum states of some of the nuclei. The interaction between these high-energy photons and the nuclei can excite the latter to unstable states that subsequently break through a decay. Such a process is called the photodisintegration of the nucleus and is the analogue of photoionisation in atoms. The existence of such reactions complicates matters at these high temperatures. For example, the reaction $^{20}\text{Ne}(\alpha, \gamma)^{16}\text{O}$ can be a source of α particles by means of the photodisintegration of neon. Such an α particle will again combine in a sequence of reactions $^{20}\text{Ne}(\alpha, \gamma)^{24}\text{Mg}(\alpha, \gamma)^{28}\text{Si}$. All the sets of reactions eventually lead to a pool of ^{16}O, ^{24}Mg, and ^{28}Si.

When the temperatures are higher than 3×10^9 K, several complex sequences of nuclear reactions and photodisintegration can occur. These processes gradually build up heavier nuclei such as ^{27}Al and ^{24}Mg, working up to ^{56}Fe. Because the binding energy per nucleon reaches a maximum at ^{56}Fe, it can survive further transformations to a certain extent. The net effect can be thought of as the conversion of two ^{28}Si to a single ^{56}Fe. For $T_9 \gtrsim 5$, photodisintegration can break up even the ^{56}Fe.

12.4.5 Neutron Capture Reactions

The nuclear reactions discussed so far are capable of progressively synthesising several elements by starting from light elements. This process, however, cannot produce elements significantly heavier than iron, as the binding energy per nucleon decreases for such elements. There are other reactions that are capable of generating heavier elements, starting from elements that are to the left of the iron peak (i.e., with $Z < 56$). One set of such reactions relies on the capture of neutrons by nuclei – a process that is not affected by the Coulomb repulsion.

Consider a collection of nuclei that are hit by a flux of neutrons. Absorption of a neutron by an element (Z, A) leads to the reaction

$$(Z, A) + n \rightarrow (Z, A + 1) + \gamma, \tag{12.65}$$

thereby forming a nucleus $(Z, A + 1)$. If this nucleus is stable, then the process can continue with the absorption of another neutron, leading to $(Z, A + 2)$, $(Z,$

$A + 3$), ..., etc. At some stage in this sequence, possibly even at the first step, we may reach an element that is unstable because of beta decay:

$$(Z, A + 1) \rightarrow (Z + 1, A + 1) + \bar{\nu} + e^-. \tag{12.66}$$

If this happens then the further evolution will depend on stability or otherwise of the elements $(Z + 1, A + 1)$. The crucial parameter that determines the actual channel through which such reactions proceed is the neutron flux.

The mean time between the capture of two neutrons is of the order of $\tau_{\text{cap}} \approx (\sigma_n \Phi_n)^{-1} = (n_n \langle \sigma_n v \rangle)^{-1}$, where Φ_n is the neutron flux, n_n is the number density of the neutrons, v the speed of neutrons, and σ_n is the neutron absorption cross section. For $\sigma \approx 10^{-25}$ cm^2 and $v \approx 3 \times 10^8$ cm s^{-1}, we have $\tau_{\text{cap}} \approx 10^9$ yr n_n^{-1}. If τ_{cap} is smaller than the lifetime τ_β of the nucleus against beta decay, then the further addition of neutrons is very likely. On the other hand, if $\tau_{\text{cap}} > \tau_\beta$, the nucleus is likely to undergo beta decay before forming the next isotope. The τ_β generally ranges from a fraction of a second to a few years and is obviously independent of Φ_n; τ_{cap}, on the other hand, varies inversely as the neutron flux Φ_n and can be larger than τ_β for low fluxes. In such a situation, beta decay occurs before the second neutron is captured and the nuclei formed by this process lie along the continuous path of stable nuclei in a nuclear chart. Such elements are called s–process elements, with s signifying a slow neutron capture. On the other hand, if the neutron flux is large, heavier and heavier isotopes can be synthesised by the addition of neutrons before eventually leading to an unstable nucleus in which a (γ, n) reaction occurs before the (n, γ) reaction occurs. This leads to a sequence of r–process elements (with r standing for rapid) that lie along the neutron-rich side of stable elements. For $n_n \lesssim 10^5$ cm^{-3}, $\tau_{\text{cap}} \gtrsim 10^4$ yr is larger than any beta decay lifetime; such low neutron fluxes correspond to the s process. When $n_n \gtrsim 10^{22}$ cm^{-3}, $\tau_n \lesssim 10^{-6}$ s; such a neutron-rich environment will be dominated by the r process. We shall see in Vol. II that these processes play an important role in the late stages of stellar evolution.

Notes and References

Some general references relevent to all chapters are the following.

1. F.H. Shu (1991), *Physics of Astrophysics* (University Science, Mill Valley, CA), Vols. I and II.
2. M. Harwitt (1988), *Astrophysical Concepts* (Springer-Verlag, New York).
3. J.I. Katz (1987), *High Energy Astrophysics* (Addison-Wesley, Palo Alto, CA).
4. L.D. Landau and E.M. Lifshitz (1975), *Course of Theoretical Physics* (Pergamon, New York), Vols. 1–10. In these Notes, the nth volume of this series is denoted by LLn.
5. G.B. Rybicki and A.P. Lightman (1979), *Radiative Processes in Astrophysics* (Wiley, New York).
6. M.S. Longair (1992), *High Energy Astrophysics*, Vol. 1 (1994), Vol. II (Cambridge University Press, New York)

Chapter 1

(1.1–1.4) Many of the introductory textbooks in astrophysics provide simple derivations of physical processes similar to the ones given here. In particular, see F.H. Shu (1982). *The Physical Universe* (University Science, Mill Valley, CA), B.W. Caroll and D.A. Oslie, *Modern Astrophysics* (Addison-Wesley, New York), and Ref. 2, cited above.

(1.5) The derivation of masses and sizes of planets, stars, galaxies, etc., has been usually attempted with in the context of anthropic principle. See for example, B.J. Carr and M.J. Rees (1979), *Nature (London)* **278**, 605; J.D. Barrow and F.J. Tipler (1982), *The Anthropic Principle* (Oxford University Press, Oxford, UK); also see L.M. Celnikier (1989), *Basics of Cosmic Structures* (Editions Frontiers, Paris).

Chapter 2

(2.1) The best introduction to classical mechanics is LL1. A more "modern" approach can be found in V.I. Arnold (1989), *Mathematical Methods of Classical Mechanics* (Springer-Verlag, New York).

(2.3.1) The reference to Feynman's plate is given in R.P. Feynman (1985), *Surely You're Joking, Mr. Feynman* (Norton, New York).

(2.3.3) A detailed discussion of the reduced three-body problem can be found in M. Henon (1983), Les Houches School on *Chaotic Behaviour of Deterministic Systems* (North-Holland, Amsterdam), p. 55, and in C. Marchal (1990), *The Three-Body Problem* (Elsevier, Amsterdam). Several fitting formulas giving distances to Lagrange points exist in literature. See for example, B.W. Carroll and D.A. Ostlie (1996), ibid (1.1), p. 687.

(2.4) The separability of the Hamilton–Jacobi equation is discussed in P.M. Morse and H. Feshbach (1953), *Methods of theoretical physics* (McGraw-Hill, New York), Vol. 1.

(2.5–2.7) A lucid introduction to issues related to the KAM theorem can be found in the article by M. Berry in R.S. Mackay and J.D. Meiss (1987), *Hamiltonian Dynamical Systems: A Reprint Selection* (Adam-Hilger, Bristol, UK).

(2.8) See the article by Henon cited in (2.3.3) above.

Chapter 3

(3.1) A very concise but adequate introduction to special relativity and electrodynamics is available in LL2.

(3.12.4) A comprehensive discussion of electrodynamics of material media can be found in LL8.

(3.16) See, for example, V. I. Tatarski (1961), *Wave Propagation in a Turbulent Medium* (McGraw-Hill, New York); some of these aspects are also discussed in P. Lena (1988), *Observational Astrophysics* (Springer-Verlag, New York).

Chapter 4

(4.1) A general reference to this chapter is LL2.

(4.2.1) This derivation for the radiation field was first given by J.J. Thomson (1907), in *Electricity and Matter* (Archibald Constable, London), Chap. 3. It is repeated in F.S. Crawford (1968), *Waves*, Vol. III of Berkeley Physics Course (McGraw-Hill, New York).

(4.2.2) The derivation given here is based on A. Gupta and T. Padmanabhan (1998), *Phys. Rev. D* **57**, 7241; also see F. Rohrleich (1965), *Classical Charged Particles* (Addison-Wesley, Reading, MA).

(4.5) A clear and concise introduction to the behaviour of photons, bosons, and fermions can be found in the first three chapters of LL4.

Chapter 5

(5.1) The operational part of statistical mechanics is available in LL5 although there is very little discussion of the conceptual issues. A detailed discussion of ergodicity and related issues can be found in R.S. Mackay and J.D. Meiss, ibid (2.5) above.

(5.6) Most of the material discussed here related to thermodynamics indices, Γ's, etc. is relevant in the study of stellar structure and is often discussed in books dealing with stellar structure. See, for example, C.J. Hansen and S.D. Kawaler (1994), *Stellar Interiors: Physical Principles, Structure and Evolution* (Springer Verlag, New York).

(5.8–5.10)   R.K. Pathria (1995) *Statistical Mechanics* (Pergamon, New York), is a good source of useful formulas related to fermionic and bosonic integrals.

Chapter 6

(6.1)   The radiative process is covered in many books dealing with astrophysics. See, for example, Ref. 1.

(6.4)   The Klein–Nishina formula, as well as processes dealing with pair production, is discussed in LL4. The electromagnetic properties of dust grain are summarised in Evans (1993), *The Dusty Universe* (Ellis Harwood, New York).

(6.5)   A detailed description of inverse Compton scattering and related issues can be found in G.R. Blumenthal and R.J. Gould (1970), *Rev. Mod. Phys.* **42**, 237.

(6.9)   The classical theory of bremsstrahlung is described in LL2 and the quantum version in LL4. Thermal bremsstrahlung and a description of Gaunt factors are given in several books on radiative processes: see, for example, Ref. 5.

(6.10)   A detailed discussion of several radiative processes as well as convenient fitting formulas are given in K. Rohlfs and T.L.Wilson (1996), *Tools of Radio Astronomy* (Springer-Verlag, Berlin).

(6.11)   The synchrotron integrals for a power-law spectrum of electrons are given, for example, in Ref. 6.

(6.12)   A rigorous derivation of photoionisation is given in LL4. The approximate formula for the recombination rate is from Ref. 5.

(6.13)   Collisional ionisation and related issues are covered in D. Emerson (1996), *Interpreting Astronomical Spectra* (Wiley, New York).

Chapter 7

(7.1)   A general discussion of astronomical spectroscopy can be found in D. Emerson, ibid (6.13) above. Some of the Exercises and Fig. 7.2 are adapted from the Web resources provided by C.L. Sarazin.

(7.2)   The derivation of the natural widths of spectral lines is given in LL4.

(7.4)   Information about different energy levels can be found in D. Emerson, ibid (6.13) above. The fine structure of hydrogen atom energy levels arising from relativistic effects is given in LL4.

(7.5)   The selection rules for different atomic transitions are described in LL4.

Chapter 8

(8.1)   An excellent reference for fluid mechanics is LL6.

(8.4)   A rigorous derivation of the transport coefficients for a neutral gas can be found in LL10.

(8.14)   The Rayleigh criterion for convection is covered in LL6.

Chapter 9

(9.1) The general reference for plasma physics is LL10. The magnetohydrodynamics is described in LL8.

Chapter 10

(10.1) An excellent discussion of gravitational dynamics can be found in J. Binney and S. Tremaine (1987), *Galactic Dynamics* (Princeton University Press, Princeton, NJ). A detailed discussion of statistical mechanics applied to gravitating systems can be found in T. Padmanabhan (1990), "Statistical mechanics of gravitating systems," *Phys. Rep.* **285**, 188.

(10.3.2) The fitting formula for $R(M)$ is from P. Eggleton (1982), given as a private communication in J.T. Truran and M. Livio (1986), *Ap.J.* **308**, 721.

(10.3.3) The limit on $(GM^2/R|E|)$ was first derived by V.A. Antonov (1962), *Vestn. Leningrad Gos. Univ.* **7**, 135 [An English translation is available in J. Goodman and P. Hut, eds. (1985), *Dynamics of Globular Clusters*, IAU Symposium, Vol. 113 (Reidel, Dordrecht, The Netherlands.)] The simple derivation given here is based on T. Padmanabhan, (1990), ibid. (10.1) above.

(10.3.4) A discussion of time-dependent isothermal spheres can be found in Ref. 1.

(10.3.5) The equilibrium configuration for self-gravitating fluids is discussed in many places. See, for example, J. Binney and S. Tremaine, cited above; LL2; also see S. Chandrasekhar (1969), *Ellipsoidal Figures of Equilibrium* (Yale University Press, New Haven, CT) for a mathematically thorough presentation.

(10.6) The process of violent relaxation was discovered by D. Lynden-Bell. This discussion is based on the papers by D. Lynden-Bell (1967), *MNRAS* **136**, 101; S. Tremaine, M. Henon and D. Lynden-Bell (1986), *MNRAS* **219**, 285.

(10.7) For a general review, see T. Padmanabhan (1990), ibid. (10.1) above.

(10.8) Much of this discussion is based on the work by Antonov, ibid. (10.3.3) above; T. Padmanabhan (1990), ibid. (10.1) above, and D. Lynden-Bell and R. Wood (1968), *MNRAS* **138**, 495.

Chapter 11

(11.1) The best place to learn about general relativity is from the last four chapters of LL2. For a more detailed working practise, the following book is excellent: A.P. Lightman, W.H. Press, R.H. Price, and S.A. Teukolsky (1975), *Problem Book in General Relativity and Gravitation* (Princeton University Press, Princeton, NJ). Also see Ya.B. Zel'dovich and I. Novikov (1983), *Relativistic Astrophysics* (University of Chicago Press, Chicago), Vol. II.

(11.8) The details of the perturbation theory for the metric tensor can be found in LL2.

(11.11) The expression for the energy-momentum pseudotensor for gravity is from LL2.

(11.12) The loss of energy that is due to the radiation of gravitational waves for a system of points moving in an elliptical orbit is from P.C. Peters (1964), *Phys. Rev.* **B 136**, 1224. Also see LL2.

Chapter 12

(12.1) Most books on stellar astrophysics describe some details of nucleosynthesis. A particularly good reference is D.D. Clayton (1983), *Principles of Stellar Evolution and Nucleosynthesis* (University of Chicago Press, Chicago).

(12.2–12.4) Much of the details on the nuclear reactions are given in C.J. Hansen and S.D. Kawaler (1994), ibid. (5.6) above.

Index

Printed in the United States
By Bookmasters